B &
onship

BU5

Josiah = Sarah
dgwood I   Wedgwood
730-95   1734-1815

usannah
1765-1817

John Bartlett Allen = Elizabeth Hensleigh
1733-1803   1738-90

Josiah II = Elizabeth
1769-1843   (Bessy)
            1764-1846

Catherine (Kitty) = Sir James
1765-1830   Mackintosh
                    1765-1832

Caroline = Edward
1768-1835   Drewe
            1756-1810

John Hensleigh

Thomas
771-1805
Catherine
(Kitty)
1774-1823
Sarah
lizabeth (Sarah)
1778-1856

...ndi
842

Caroline
— Sarah = Josiah D
1800-88   1795-188

Charles = Charlot
Langton 1797-18
1801-86

Elizabeth
(Bessy)
1799-1823

Frances = Fran
Mosley   (Fra
d.1874   1800-

— Charles = Emm
Robert   1808-
1809-82

Robert
1806-64

THE CORRESPONDENCE OF
CHARLES DARWIN

Editors
FREDERICK BURKHARDT   SYDNEY SMITH

Associate Editors
DAVID KOHN   WILLIAM MONTGOMERY

Managing Editor
STEPHEN V. POCOCK

Assistant Editors
CHARLOTTE BOWMAN   JANET BROWNE
ANNE SECORD

Research Associates
ANNE SCHLABACH BURKHARDT
NORA CARROLL STEVENSON

This edition of the Correspondence of Charles Darwin is sponsored by the American Council of Learned Societies. Its preparation is made possible by the co-operation of the Cambridge University Library and the American Philosophical Society.

Advisory Committees for the edition, appointed by the Council, have the following members:

The principal sources of funds for editing this volume have been the National Endowment for the Humanities, the National Science Foundation, and the Andrew W. Mellon Foundation. The National Endowment's grants (Nos. RE-23166-75-513, RE-27067-77-1359, RE-00082-80-1628, and RE-20166-82) were from its Program for Editions; the National Science Foundation's support of the work was under grants Nos. SOC-75-15840 and SES-7912492. Any opinions, findings, conclusions or recommendations expressed in this publication are those of the authors and do not necessarily reflect the views of the grantors.

Charles Darwin, 1840.

THE CORRESPONDENCE OF
CHARLES DARWIN

VOLUME 1   1821–1836

The right of the
University of Cambridge
to print and sell
all manner of books
was granted by
Henry VIII in 1534.
The University has printed
and published continuously
since 1584.

CAMBRIDGE UNIVERSITY PRESS

CAMBRIDGE

LONDON   NEW YORK   NEW ROCHELLE

MELBOURNE   SYDNEY

Published by the Press Syndicate of the University of Cambridge
The Pitt Building, Trumpington Street, Cambridge CB2 1RP
32 East 57th Street, New York, NY 10022, USA
10 Stamford Road, Oakleigh, Melbourne 3166, Australia

First published 1985

Printed in Great Britain at
the University Press, Cambridge

Library of Congress catalogue card number: 84-45347

British Library Cataloguing in Publication Data
Darwin, Charles
The correspondence of Charles Darwin.
Vol. 1: 1821–1836
1. Darwin, Charles 2. Naturalists—
England—Biography
I. Title II. Burkhardt, Frederick
III. Smith, Sydney, *19--*
575'.0092'4 QH31.D2
ISBN 0 521 25587 2

Library of Congress Cataloguing in Publication Data
Darwin, Charles, 1809–1882.
The correspondence of Charles Darwin.

1. Darwin, Charles, 1809–1882. 2. Naturalists—
England—Correspondence. I. *Burkhardt*, Frederick,
1912– . II. Smith, Sydney, 1911– . III. Title.
QH31.D2A33 1984 575'.0092'4 [B] 84-45347
ISBN 0 521 25587 2

CONTENTS

ILLUSTRATIONS

LISTING OF LETTERS INCLUDED IN THIS VOLUME

The following list is in the order of the entries in the *Calendar of the Correspondence of Charles Darwin*. The *Calendar* numbers listed cover those entries that are dated in the *Calendar* within the range of dates covered by this volume. The dates given are the dates ascribed to the letters in the *Correspondence*, so the list enables a user of the *Calendar* to locate a particular letter in the *Correspondence* regardless of any redating carried out since publication of the *Calendar*. A date given in italic type indicates that that item is *not* transcribed in full in this volume.

FOREWORD

Charles Darwin's life and work are of widespread and ever-increasing interest to the general public as well as to scholars in many fields of enquiry, and his letters are a basic source of information about that life and work. Although selections of the letters have been published by various editors, an authoritative and comprehensive edition of the correspondence, such as a figure of Darwin's stature clearly warrants, has never previously been attempted. A considerable number of letters remain unpublished, and some, even of the more important collections, are out of print. Of those that are readily available, few have been published in accordance with modern standards of editing, and many contain transcription errors or have been wrongly dated or identified.

The objective of *The Correspondence of Charles Darwin* is to provide definitive transcripts of all the available Darwin letters. The principles and practices employed in editing the letters are described in the Note on Editorial Policy, following the Introduction. To the extent that the editors are successful, this policy results in a text that reproduces, as closely as print will permit, what Darwin actually wrote. His spelling and punctuation, which are especially idiosyncratic in the early letters, are preserved, and the alterations made while writing a letter are recorded in the apparatus. The printed transcript represents Darwin's final version of the text, but by using the record of alterations the full complexities of the original, with its deletions, insertions, and other emendations, can be reconstructed for specialised study.

The *Correspondence* includes letters written to Darwin, by his family, friends, and scientific correspondents. These letters place him in his social and intellectual contexts and clarify the extensive scientific network of which he was a part. The nature of his work and his poor health left him more than usually dependent upon correspondence in carrying out his investigations. Presenting both sides of the exchanges shows the extent and importance of the contribution made to his work by a diverse group of correspondents. Insights into Darwin's assessment of the value of facts and ideas communicated to him are provided by recording any annotations made by him on letters he received.

At the beginning of each volume of the *Correspondence* a list of the letters in the volume will be printed. This will include letter dates listed against the numbers that were assigned to the letters in the *Calendar of the correspondence of Charles Darwin*. The *Calendar* was compiled by the editors to provide a concise guide to the correspondence as a whole. It details and summarises the 13,925 letters located prior to its publication. Users of the *Calendar* wishing to locate a particular letter in the *Correspondence* can do so by reference to the list of *Calendar* numbers against letter date.

Each volume of the *Correspondence* will include an introduction, editors' footnotes, appendixes, a bibliography, a register giving brief biographical details of persons referred to in the letters and listing the letters to and from correspondents, and an index. The appendixes may include a variety of material pertinent to the letters, but will always include translations of non-English letters and a chronology of the period of Darwin's life covered by the volume.

INTRODUCTION

The first volume of the *Correspondence* contains the letters of the years 1821 – 36. Beginning with one written to Darwin at the age of twelve, they continue through his school-days at Shrewsbury, his two years as a medical student at Edinburgh University, the undergraduate years at Cambridge, and the five years of the voyage of H.M.S. *Beagle*. The volume ends with the twenty-seven-year-old traveller's return to England and his enthusiastic reception into the scientific community there.

Although some of Darwin's letters from these years may be familiar from *Life and letters*, *More letters* and Nora Barlow's volumes, *Charles Darwin and the voyage of the Beagle* and *Darwin and Henslow*, all now gain a new interest from their juxtaposition with a substantial number of previously unpublished letters. Of the letters from Darwin, sixty-six appear in print for the first time, so far as is known, among them thirty-nine letters to his cousin William Darwin Fox, eleven letters to members of his family, and three to Robert FitzRoy.

Almost all of the letters to Darwin in this volume are previously unpublished. Besides those from his brother Erasmus and his sisters, there are letters from Shropshire friends, contemporaries at Edinburgh and Cambridge, and from acquaintances made during the *Beagle* voyage. Darwin himself does not appear as a letter-writer until October 1825, but long before then the letters from Mary Congreve, Erasmus, and his sisters have provided the reader with illuminating details of the character and interests of the Shrewsbury schoolboy. Later, the letters from his sisters reveal much about life at The Mount, the family home in Shrewsbury, and of the role his family played in Darwin's early life; those from Sarah and Fanny Owen, full of lively humour and pert social observation, illuminate the close friendship between their family and the Darwins and the informality that was the consequence of such closeness. Together these letters give a vivid picture of the social life of the Shropshire gentry of the 1820s and 1830s.

The letters from William Darwin Fox, Frederick William Hope, and Thomas Campbell Eyton, among others, illustrate the scientific pursuits of the clergy and the gentry, display the differences between the scientific worlds of the metropolis and the provinces, and herald the emergence of new scientific organisations and publications. The letters written to Darwin during the voyage of H.M.S. *Beagle* kept him informed of such events as the founding of the Entomological Society and the early meetings of the British Association for the Advancement of Science.

In the earliest letters Darwin was already keenly interested in natural history and an avid collector. At home these interests were recognised and, at the least, well accepted. The signatures in a little botanical lexicon, the *Clavis Anglica linguæ botanicæ* (Berkenhout 1764), now in the Rare Book Collection in the Cambridge University Library, indicate that Robert Waring Darwin gave his own copy to his son in 1820, and in the early 1820s Darwin's brother Erasmus made him his assistant and collaborator in the chemical laboratory they installed in the garden of The Mount. From Cambridge Erasmus wrote to his younger brother detailing desirable experiments and equipment and keeping him informed of interesting specimens seen or purchased from mineral sellers.

When himself a medical student at Edinburgh University, Darwin devoted much of his time to natural history and particularly to collecting marine specimens from the Firth of Forth with Robert Edmond Grant, his zoological mentor, and his friends John Coldstream and Frederick Bream Glasspoole. Later, at Cambridge, led by the example of his cousin, Fox, he soon became an enthusiastic and respected collector of beetles. Fox also introduced him to John Stevens Henslow and Darwin was a regular presence at the botanical lectures, excursions, and undergraduate parties organised by the professor of botany. And it was Henslow who, encouraging Darwin to broaden his scientific horizons, persuaded him to commence the study of geology. It was at this time, too, that Darwin first began to display the remarkable ability he had of persuading others to help him in his scientific pursuits. In September 1828 he had John Maurice Herbert and Thomas Butler following his detailed instructions in search of particular beetles at Barmouth, requests that prompted Fox to comment: 'what deuced goodnatured fellows your friends at Barmouth must be' (see letter to J. M. Herbert, [13 September 1828]). The willing responses of those asked for help here, as later in Darwin's life, are indicative of the warm and lasting friendships that he was always capable of forming.

The opportunity to become a naturalist on a voyage of exploration arose because of Henslow's recognition of the abilities Darwin had displayed during his years at Cambridge and his consequent recommendation of the young graduate. Although Robert Waring Darwin initially doubted the value of the enterprise, he offered his son every assistance as soon as his objections had been reasonably answered. During the voyage of H.M.S. *Beagle* Darwin's letters convey the excitement and enthusiasm of a keen and careful collector let loose in a new and challenging land; an ardent naturalist who, as the voyage progressed, became wholly committed to a life of scientific enquiry. Coupled with this commitment was Darwin's growing recognition of his ability to contribute significantly to the study of natural history. He was careful to collect among those groups of organisms least known in Europe, and supplemented all his collecting with extensive records and observations. In palaeontology his finds were spectacular, a triumph he shared with his sisters and which added to his confidence in the overall success of the undertaking. In geology he was soon led to speculate around his observations and tried out his theories on Henslow in the privacy of their letters. Well before the end of the voyage he was aware of the value of at least some of his work, and envisaged the future publication of the results. In one memorandum he had already worked out the main outlines of his theory of the formation of coral reefs, a theory which, had he done nothing more, would have placed him in the front ranks of nineteenth century geologists. His *Journal of researches*, which became a classic travel book, was substantially finished by the end of the voyage, though its publication was delayed until FitzRoy completed his volume of the *Narrative* in 1839.

When Darwin returned to England in October 1836 it was with the firm intention of writing up his geological observations and theories, the importance of which were clear to him from his communication with Henslow and his study of Lyell's uniformitarian views set out in the *Principles of geology* (1830–3), and of finding able naturalists to tackle the task of sorting and identifying his zoological, mineralogical, and botanical collections. The *Beagle* voyage gave rise to a series of geological and zoological books and papers over the following years that testify to the wealth and quality of Darwin's collections and observations. But more than this, the *Beagle* material was to provide Darwin with the basis for his life's work, for it was as a result of his observations and the judgements passed by systematists on some of his specimens that Darwin became a committed transmutationist a few months after his return to England.

During the voyage itself, although generalisations and theories are plentiful in his records, there are no indications that he had given up a belief in immutable, created species. The first stirrings of doubt about the fixity of species occur in the well-known passage in the ornithological notes written during the summer of 1836, when, homeward bound, he was concentrating on organising his notes on the birds he had collected. Then, Darwin wrote of the Galápagos mocking-bird specimens: 'When I see these Islands in sight of each other, & possessed of but a scanty stock of animals, tenanted by these birds, but slightly differing in structure & filling the same place in Nature, I must suspect they are only varieties . . . If there is the slightest foundation for these remarks the zoology of Archipelagoes—will be well worth examining; for such facts would undermine the stability of Species' (DAR 29.2: 73–4; 'Ornithological notes' p. 262).

In the winter of 1836 the question of the stability of species awaited further consideration. As Darwin was able properly to consider the results of his and others' examinations of the *Beagle* specimens the suggestion of instability developed into the conviction that species were mutable. By the spring of 1837 Darwin was a transmutationist and had embarked on the long task of collecting facts and constructing theories to explain this view of the origin of species.

ACKNOWLEDGMENTS

The editors are grateful to Mr George Pember Darwin for permission to publish the Darwin letters and manuscripts. They also thank the Syndics of the Cambridge University Library and other owners of the manuscript letters who have generously made them available

The research and editorial work for this edition as a whole has been supported by grants from the National Endowment for the Humanities and the National Science Foundation. The Andrew W. Mellon Foundation has provided grants to match NEH funding, and in 1981 awarded a grant to Cambridge University that has made it possible to put the entire Darwin correspondence into machine-readable form. Timely research assistance has been furnished by grants from the Royal Society of London and the British Academy. Bern Dibner and the late Mary S. Hopkins provided financial assistance for the purchase of books and other necessary project expenses.

The Cambridge University Library and the American Philosophical Society have generously made working space and many services available to the editorial staff.

Since the project began in 1975 we have been fortunate in benefiting from the interest, experience, and practical help of many people in many places, and the editors hope that they have adequately expressed their thanks to them individually as the work proceeded. There are some, however, who have helped over so long a period that it would be ungracious not to thank them personally in the work which has profited so much from their co-operation.

Over the years, George F. Farr, Jr, Assistant Director of the Division of Research Grants of NEH, and Ronald J. Overmann, Program Director for History and Philosophy of Science of NSF, have given far more time and attention to the project than their formal duties required. The editors also appreciate the interest and encouragement they received from John E. Sawyer, President of the Andrew W. Mellon Foundation and his colleague, James Morris.

Without the expert help of John L. Dawson of the Literary and Linguistic Computing Centre of Cambridge University the computerisation of the correspondence would not have been possible, for the work on both the *Calendar* and the *Correspondence* required the solution of many novel technical problems. Thanks are also due to the members of the LLCC staff, Cinzia Caballero, Monique Johnson, and Yeshe Zangmo.

The computer of the Cambridge University Library, as well as that of the University, was used in the preparation of this volume. W. D. S. Motherwell, head of automation at the Library arranged for the installation of terminals in the project's office in the Manuscripts Department and solved the operational problems that arose.

This book was typeset at the Oxford University Computing Service on a Monotype Lasercomp. We are grateful to Susan Hockey for enabling the typesetting of the volume to be carried out, and to the Oxford staff involved.

Sir Hedley Atkins and Philip Titheradge of the Darwin Museum at Down House, Downe, Kent have welcomed the editors on numerous visits and responded most generously to frequent requests for information.

Libraries all over the world have given help that was literally indispensable by making available photocopies of Darwin correspondence and other manuscripts in their collections. The institutions that furnished copies of letters for this volume are listed on page xxiii. To all of them the editors are grateful.

Among the librarians whose co-operation has been of especial importance, the foremost is Peter J. Gautrey who has charge of the Darwin Archive at the Cambridge University Library. His great knowledge of Darwiniana has always been readily shared with the editors, as it has with countless other Darwin scholars. Other members of the staff of Cambridge University Library we wish to thank for their willing response to the editors' needs are Arthur Owen, Elizabeth Leedham-Green, Margaret Pamplin, Jayne Ringrose, Gerry Bye, Alan Purvis, Godfrey Waller, Janice Fairholm, Roger Fairclough, Cynthia Webster, Daphne O'Dell, Noni Hyde-Smith, Hazel Bell, and Kim Dickson.

At the American Philosophical Society Library a splendid collection of Darwiniana and works in the history of science has been continuously available to the editorial staff since the project was started. Whitfield J. Bell Jr, Secretary of the Society until 1983, has served on the U.S. Advisory Committee for the project and has done his utmost to further its work. The editors have also benefited on a daily basis from the co-operation of Edward Carter II, Murphy D. Smith, Roy C. Goodman, Stephen Catlett, Willman Spawn, Carl F. Miller, and Bertram Dodelin, all of the APS Library.

Douglas W. Bryant, formerly of the Harvard University Library, Rodney Dennis of the Houghton Library, Constance Carter of the Science Division of the Library of Congress, James Henderson, Walter Zervas, and Joseph Mask of the New York Public Library have all been exceptionally helpful in providing material from their great collections.

The editors often had recourse to assistance from Maldwyn J. Rowlands, Rex E. R. Banks, Mrs Valerie C. Phillips and Dorothy Norman of the British Museum (Natural History) Library; Gavin Bridson, Librarian of the Linnean Society of London; John Thackray and Rosemary Evans of the Geological Society of London Library; Victor T. H. Parry, Sylvia FitzGerald and Irene Smith of the Library of the Royal Botanic Gardens, Kew; Anthony M. Carr, Miss Williams and Mrs Ion of the Local Studies Library, Shrewsbury; Anthony P. Shearman, City Librarian, of Edinburgh; and Ian H. Frazer and Martin Phillips of the Keele University Library.

The editors owe thanks also to Kate Smith for her help in transcribing early letters, and to Eleanor Moore for her assistance during the search for letters. Among the scholars who have been consulted in editing this volume are Paul H. Barrett, who in addition to his advice and counsel as a member of the U.S. Advisory Committee, gave us the benefit of his own research on Darwin's early letters, P. Thomas Carroll, Ralph Colp Jr, Richard B. Freeman, Mario Di Gregorio, Jane Oppenheimer, Martin Rudwick, Silvan S. Schweber, David Stanbury, Frank J. Sulloway, and Garry J. Tee. Judith Butcher, Fredson Bowers, and G. Thomas Tanselle were frequently called upon for advice on editorial method.

During the decade since work began on the Darwin Correspondence Project a number of able assistants have helped with research, checking, filing, and other essential tasks. We are grateful to: Doris E. Andrews, Pamela J. Brant, Deborah Fitzgerald, Dorothy Hoffman, Christine M. Joyner, Joan W. Kimball, Barbara A. Kimmelman, Nancy Mautner, John A. Reesman, and Edith Stewart.

Finally, special thanks are due to Hedy Franks and Alison Soanes for their skill and patience in deciphering and keyboarding manuscript material of considerable complexity.

LIST OF PROVENANCES

The following list gives the locations of the original versions of the letters printed in this volume. The editors are most grateful to all the institutions and individuals listed for allowing access to the letters.

American Philosophical Society, Philadelphia, Pennsylvania
Bathurst District Historical Society, Bathurst, NSW, Australia
Bodleian Library, Oxford, England
Bower, T. H. W., private collection
British Library, Manuscripts Division, London, England
Christ's College Library, Cambridge, England
DAR, Darwin Archive, University Library, Cambridge, England
Fitzwilliam Museum, Cambridge, England
Hope Entomological Collections, University Museum, Oxford, England
Hydrographer of the Navy, Ministry of Defence, Taunton, Somerset, England
Keele University Library, Keele, Staffs., England (Wedgwood/Mosley collection
  by courtesy of the Trustees of the Wedgwood Museum, Barlaston, Stoke-on-
  Trent.)
Menendez, Armando Braun, private collection
National Library of Australia, Canberra, Australia
Royal Botanic Gardens, Kew, Richmond, Surrey, England
Royal College of Surgeons of England, London, England
Ximenes Rare Books (dealers)

in the navy. ushered in by right
honourable Lords + gallant Captains. —
This letter is a most unconnected tissue
of facts; but whilst I recollect. I must
put you in mind that you have my
snuff box. + I have yours; doubtless yours
is the most valuable: but mine. was
the gift of M.r Owen. + he is the
Father of Fanny. + Fanny, as all the
world knows, is the prettiest. plumpest.
Charming personage. that Shropshire possesses,
ay + Burraghton too. always excepting
the blooming Bessy. + now that I know.
a most pleasant train of ideas are
excited in your. mind. I will not
interrupt them by writing any more
But believe me my dear Fox
Yours most sincerely
Chas: Darwin;

Page of the letter from CD to W. D. Fox, [28 December 1828].

(Courtesy of the Master and Fellows of Christ's College, Cambridge.)

A NOTE ON EDITORIAL POLICY

The first and chief objective of this edition is to provide complete and authoritative texts of Darwin's correspondence. Insofar as it is possible the letters have been dated, arranged in chronological order, and the recipients or senders identified. Darwin seldom wrote the full date on his letters and, unless the addressee was well known to him, usually wrote only 'Dear Sir' or 'Dear Madam'. In the 1840s, after the adoption of adhesive postage stamps, the separate covers that came into use with them were usually not preserved and thus the dates and the names of many recipients of Darwin's letters have had to be derived from other evidence. The notes made by Francis Darwin on letters sent to him by the recipients have been helpful, as have matching letters in the correspondence, but many dates and recipients have had to be deduced from the subject-matter or references in the letters themselves. These tasks, together with the deciphering of Darwin's handwriting, have been the most troublesome problems.

The editors seek to provide a sound basis for future scholarly work on Darwin and his correspondents. Accordingly, the footnotes are intended to make the content of a letter clear, to relate it to other letters on the same subject or to Darwin's writings, and to provide the reader who wishes to pursue the subject of a letter with references to sources and other amplifying information.

The method of transcription employed in this edition is adapted from that described by Fredson Bowers in 'Transcription of manuscripts: the record of variants', *Studies in Bibliography* 29 (1976): 212–64. This system is based on accepted principles of modern textual editing and has been widely adopted in literary editions.

The case for using the principles and techniques of this form of textual editing for historical and non-literary documents, both in manuscript and print, has been forcefully argued by G. Thomas Tanselle in 'The editing of historical documents' in *Studies in Bibliography* 31 (1978): 1–56. The editors of the *Correspondence* have followed Dr Tanselle in his conclusion that 'a scholarly edition of letters or journals should not contain a text which had editorially been corrected, made consistent, or otherwise smoothed out' (p. 48), but they have not wholly subscribed to the statement made earlier in the article in which he says, 'In the case of notebooks, diaries, letters and the like, whatever state they are in constitutes their finished form, and the question of whether the writer "intended" something else is irrelevant' (p. 47). The editors have preserved the spelling, punctuation, and grammar of the original, but they have found it impossible to set aside entirely the question of authorial intent. One obvious reason is that in reading Darwin's writing, there must necessarily be reliance upon both context and intent. Even when Darwin's general intent is clear, there are cases in which alternative readings are, or may be, possible and therefore the transcription decided upon must to some extent be conjectural. In this work, when the editors were uncertain of their transcription they have enclosed the doubtful text in italic square brackets.

It was also necessary to consider Darwin's intent in coping with the transcription problem posed by his practice in the early letters of sprinkling his sentences with points. These would normally be interpreted as full stops, but their number and position makes it clear that they could not possibly have been intended to function as such. The reproduction on p. xxiv, of part of letter to W. D. Fox, [24 Dec 1828], illustrates this problem. At an early stage in their work, the editors interpreted these points to be 'pen rests' marking places where Darwin paused to take thought. This may indeed be the explanation for some of them. Where the points may have been intended to act as commas they have been silently changed; those for which no grammatical function could be conjectured have been omitted, with the omissions recorded in the Manuscript Alterations and Comments section of the apparatus.

This treatment of a particular and rather exceptional problem is consistent with the major editorial decision to adopt the so-called 'clear-text' method of transcription, which so far as possible keeps the text free of brackets recording deletions, insertions, and other alterations in the places in which they occur. These changes are recorded in the Manuscript Alterations together with any comment the editors have to make regarding the original text. The notes are keyed to the printed text by paragraph and line number. All lines above the first paragraph of the letter (i.e., date, address, or salutation) are referred as paragraph 'o'. Separate paragraph numbers are used for subscriptions and postcripts. The textual notes enable the reader who wishes to do so to reconstruct the manuscript versions of Darwin's letters while furnishing printed versions that are uninterrupted by editorial interpolations. They record all alterations made by Darwin in his letters and any editorial amendments made in transcription. For copies and drafts of Darwin letters included in the *Correspondence* no attempt has been made to record systematically all alterations to the text, but major revisions of drafts and ambiguous passages in copies are noted. The editors believe it would be impracticable to attempt to go further without reliable information about the text of the original or the final versions of the letters involved. The letters to Darwin have been transcribed without recording the writer's alterations unless they reflect significant changes in substance; in such cases footnotes will bring them to the reader's attention.

Misspellings have been preserved, even when it is clear that they were unintentional as, for instance, 'lawer' for 'lawyer'. Such errors often indicate excitement or haste, or, over a series of letters, they may exhibit in the aggregate a habit of carelessness in writing to a particular correspondent or about a particular subject.

Capital letters have also been transcribed as they occur except in certain cases, such as 'm' and 'c', which are frequently written somewhat larger than others as initial letters of words. In these cases the normal practice of the writers has been followed. If there is doubt about Darwin's intention, the capital has been retained.

In some instances, that are not misspellings in a strict sense, editorial corrections have been made. In his early manuscripts and letters Darwin consistently wrote 'bl' so that it looks like 'lb' as in 'albe' for 'able', 'talbe' for 'table'. Because the form of the letters is so consistent in different words, the editors consider that this is most unlikely to be a misspelling but must be explained simply as a peculiarity of Darwin's handwriting. Consequently, the affected words have been transcribed as normally spelled and no record of any alteration is given in the apparatus. Elsewhere, though, there are misformed letters that the editors have noted because they do, or could, affect the meaning of the word in which they appear. The main example is the occasional, inadvertent crossing of 'l'. When the editors are satisfied that the intended letter was 'l' and not 't', as, for example, in 'stippers' or 'istand', then 'l' has been transcribed, but the actual form of the word in the manuscript has been given in an alteration note.

Editorial interpolations are in square brackets. Italic square brackets enclose conjectured readings and descriptions of illegible passages. To avoid confusion, in the few instances in which Darwin himself used square brackets, they have been altered to parentheses, with the change noted.

Material that is irrecoverable because the manuscript has been torn or damaged is indicated by angle brackets 〈 〉; any text supplied within them is obviously the responsibility of the editors.

Finally, two kinds of change made silently by the editors should be mentioned. In the early letters, when postage was expensive, and especially in the letters written during the *Beagle* voyage, space is seldom wasted, and paragraphs are often not indicated. Darwin sometimes marked a change of subject by leaving a somewhat larger space than usual between sentences; sometimes he employed a longer dash. In these cases and in very long stretches of text, when the subject is clearly changed, a new paragraph is started by the editors.

The other unrecorded type of change concerns the position of insertions made in the margin, or at the head or foot of a letter. Such text is transcribed and placed where the editors believe it was intended to be read. If the material was added after the main body of the text was written, this fact is recorded as a manuscript alteration, but the location of the added material in the manuscript is not given unless the editors consider its placing to be significant or Darwin's intention to be unclear.

Some Darwin letters and an occasional letter to Darwin are known only from entries in the catalogues of book and manuscript dealers or mentions in other published sources. Whatever information these sources provide about the content of such letters has been reproduced.

For every Darwin letter, the text that is available to the editors is always given in full. Some other items, however, are not printed in their entirety. Most memoranda and other documents that are not letters have been summarised, as have those few letters to Darwin that the editors consider can be presented adequately in shortened form.

The format in which the full transcriptions are printed in the *Correspondence* is as follows:

1. *Order of letters*. The letters are arranged in chronological sequence.

2. *Headline*. This gives the name of the sender or recipient of the letter and its date. The date is given in standard form, but those elements not taken directly from the manuscript are supplied in square brackets.

3. *The letter text*. Whenever possible transcriptions have been made from manuscript. If the manuscript was inaccessible but some photocopy or other facsimile version was available, that version has been used as the source. Other copies and published texts have been transcribed when they have provided the only known version of the letter.

The transcribed text follows as closely as possible the layout of the source, although no attempt is made to produce a type facsimile of the manuscript; that is, word-spacing and line-division in the running text are not adhered to. Dates and addresses given by authors are transcribed as they appear. If no address is given anywhere in the letter the editors have supplied one, when able to do so, in square brackets at the head of the letter. Addresses on printed stationery are transcribed in italics, often in a simplified form. Addresses and valedictions have been run into single lines to save space, but the positions of line-breaks in the original are marked by vertical bars.

4. *Physical description*. All letters are signed autograph letters (ALS) unless otherwise described. A letter that is signed by Darwin but written in the hand of an amanuensis is noted by the abbreviation 'LS'. If possible the writer is identified. If Darwin has made corrections or additions in his own hand, the abbreviation used is LS(A). If the source is other than the manuscript, details are given. Postmarks are recorded only when they are the evidence for the date or address of the letter.

5. *Source*. The final item provides the provenance of the text. Some sources are here given in abbreviated form (e.g., DAR 140: 18), but all are also listed in full in the Provenance List on p. xxiii, unless the source is a published work. Publication references, as generally, are in author—date form, with full titles supplied in the bibliography.

6. *Darwin's annotations*. In the early letters to Darwin only a few carry any comments by Darwin, but after the voyage, when his correspondence was more

closely related to his work, he frequently made marginal notes or he scored passages that he thought were significant. Most of these notes are short, but they occasionally run from a paragraph to several pages, written on separate paper and attached to the letter. For the more extended annotations the interested reader will be directed to the Manuscript Alterations for the specified letter; otherwise Darwin's annotations are recorded immediately following the letter on which they were made.

Darwin's notes occasionally take the form of a draft reply. In such cases, if the final form exists in a preserved letter, the draft is not reproduced, but its existence and significant differences from the final form are noted. If the final form has not been found, the draft reply is reproduced in lieu of the letter.

Darwin's annotations will be transcribed in 'diplomatic' style. In this method the alterations in the text are recorded in brackets at the places in which they occur. For example, an entry that reads:

6.1   See Daubeny ['vol.1' *del*] for *description of
        volcanoes in [*interl*] S. A.   6.3 *ink*

means that Darwin noted, in ink, in the margin of paragraph 6 of the letter, between lines 1 and 3, 'See Daubeny vol.1 for S. A.' and then deleted 'vol.1' and inserted 'description of volcanoes in' after 'for'. The asterisk before 'description' marks the beginning of the interlined phrase, which ends at the bracket. The asterisk is used when more than one word is interlined. The final text can be read simply by skipping the bracketed material. The notes are keyed to transcriptions by paragraph and line number.

7. *Abbreviations.* 'CD' is used for references to Charles Darwin throughout.

A list of all abbreviations used in editing this volume is given on p. xxxii.

For Darwin's books, articles, and collections of his letters, short titles are used (e.g., *Descent*, 'Formation of mould', *LL*, *Darwin and Henslow*). Short titles (e.g., *Alum. Cantab.*, *Wellesley index*, *DNB*) are also used for some standard reference works; for all other works author—date references are used. The full titles of all the books referred to are given in the bibliography.

The full name and brief biographical notes on persons mentioned in the text can be found in the Biographical Register and Index of Correspondents. If the name in the text is incomplete or incorrectly spelled, the full correct form is generally given in a footnote.

Future volumes of the *Correspondence* will be published in chronological order. Each volume will be self-contained, having its own index and biographical notes. References will be supplied for persons and subjects mentioned, even though some repetition of material in earlier volumes is involved. A comprehensive index is planned for the final volume.

THE CORRESPONDENCE OF
CHARLES DARWIN
1821 — 1836

ABBREVIATIONS

A	autograph, i.e., in the hand of the sender
L	letter
mem	memorandum
S	signed by the sender
AL	autograph letter
ALS	autograph letter signed
LS	letter in hand of amanuensis, signed by sender
LS(A)	letter in hand of amanuensis with additions by author
CD	Charles Darwin
CUL	University Library, Cambridge
del	deleted
illeg	illegible
underl	underlined

TRANSCRIPTION CONVENTIONS

[some text]	'some text' is an editorial insertion
⌈some text⌉	'some text' is the conjectured reading of an ambiguous word or passage
[some text]	'some text' is a description of a word or passage that cannot be transcribed, e.g, *3 words illeg*
⟨ ⟩	word(s) destroyed
⟨some text⟩	'some text' is a suggested reading for a destroyed word or passage
⟨*some text*⟩	'some text' is a description of a destroyed word or passage, e.g., *3 lines excised*

From Mary Congreve[1] 27 October [1821][2]

My dear M.[r] Charles

I find I have only just time to thank you for your entertaining letter, as if I take time to write what I intended I shall not be able to get it franked[3] & I'm sure it will not be worth the postage, I should have liked to have seen the good Gentleman *Grin* that you mention there is no doubt but those that were out of the Scrape were much amused, I assure you I wish'd much you had been of our party on thursday night at the play, I think you would have been highly entertained both with the Coronation,[4] and the entertainment of Monsieur Tonson,[5] I never laugh'd so much at a play I think, I dare say you have been much amused with M.[r] Alexander[6] & I hope I shall hear some specimenes of his art from you when I return, as I dare say it is practiced in School Lane, so god bless you as I am obliged to conclude this ever believe me | Yours truly M Congreve

Saturday 27th Oct

I think you will not be able with all your Greek knowledge to read this precious Scrawl

DAR 204.16

[1] The writer is referred to in an obituary notice in the *Salopian Journal*, 5 February 1823, as 'eldest sister to the late General Sir William Congreve, Bart and aunt to the present.' See also the letter from E. A. Darwin, 5 [March 1823].
[2] The letter is dated from the reference to Mr Alexander (see n. 6, below).
[3] Peers, Members of Parliament, and Officers of State had the privilege of free mail. 'Franks' (covers addressed, dated, and signed by the holder of the privilege) were distributed freely among constituents and friends. In 1821 Miss Congreve's nephew, Sir William Congreve, was M.P. for Plymouth.
[4] A pageant representation of the coronation of George IV, produced by Robert William Elliston at the Drury Lane Theatre (Genest 1832, 9: 96).
[5] A farce by Moncrieff first performed on 20 September 1821 at Drury Lane (*ibid.*, 9: 96 − 7).
[6] The *Salopian Journal* of 17 October 1821 records the performance of Monsieur Alexandre, a ventriloquist, at the Royal Free Grammar School of Shrewsbury on 12 October. School Lane, now called School Gardens, was immediately adjacent to the school.

From Erasmus Alvey Darwin 25 [October 1822]

[Cambridge][1]
25th

Dear Bobby.[2]
I feel sorry we sent to y[e] Glass House, for there is a shop here with every sort of thing, it quite made my mouth water to see all the jars & stopcocks & all sorts of things, graduated tubes, blow pipes, cubic inch measures, test tubes, & y[e] Lord knows what besides. There is one thing I shall certainly get, a thermometer graduated from 50 below Zero to y[e] boiling point of Mercury, & the

bottom part of ye scale has a hinge so as to turn back & leave ye bulb exposed to be put in any corrosive mixture, & what is best of all it only costs £1 ″4. I have also found out another very nice little man. He was 14 years assistant to ye late Dr Clarke,^{3} ye great mineralogist. I have bought 2 or 3 litle stones from him; 2 specimens of uranite which is a very scarce stone, & some leaf copper, & a very odd looking thing like a petrefaction called a brain stone from its similarity to ye brain. He sells things very cheap & so if you will mention any stones I can probably get them.^{4} I saw some of ye fossil shells in chalk (1. shilling), which are ye latest formation I beleive. very pretty garnets ye same size as we saw at Mr Cottons^5 at 1.s. a piece—

I forgot to tell you that I got a very nice little blow pipe made of glass for 1 ″6, which gives a great deal of heat, & will be very useful if we do not get a better one.—

Hary^6 came down to Cam last night quite unexpectedly & made his bed upon a sofa in Henesly's^7 Rooms.—

If Blunt^8 gets ye things before I come write me word what they are, every individual thing, that I may know wha⟨t⟩ to get, & whether they are nice ones.—

I am getting on very comfortably here Settlled in fruiterers, where ye jellies & puffs & cakes, & buns &c would tempt the most obdurate sinner, quite as much as ye Lumberland pigs with knives & forks stuck in their backs.—^{9}

I know you dont like long letters & I have nothing more to say so good bye be a good boy & you shall have a sugar plum I remain yours | affect E. Darwin

Postmarks: C 26 OC 26 1822; Cambridge 52 1822^{10}
DAR 204.1

^{1} Erasmus, CD's elder brother, was admitted to Christ's College, Cambridge, in February 1822 (*Alum. Cantab.*).

^{2} Members of the family called CD 'Bobby', as well as 'Charley', during his boyhood. The letter is addressed to 'R. Darwin Esq. Revd Dr Butlers, Shrewsbury'. CD was a boarder at Shrewsbury School, of which Samuel Butler was headmaster from 1798 to 1836. Under Butler's direction Shrewsbury had become one of the leading schools in England.

^{3} Edward Daniel Clarke, Professor of Mineralogy at Cambridge, 1808—22.

^{4} In the *Autobiography*, p. 45, CD says that during his school years he 'continued collecting minerals with much zeal, but quite unscientifically—all that I cared for was a new *named* mineral, and I hardly attempted to classify them.'

^{5} Identified in Judd 1909, p. 340, as Richard Cotton, a Shrewsbury naturalist. See *Autobiography*, p. 52, for CD's account of the deep impression made by Mr Cotton's solemn assurance 'that the world would come to an end' before anyone would be able to explain how the bell-stone, a well-known boulder, came to be at Shrewsbury, the nearest rock like it being no nearer than Cumberland or Scotland.

^{6} Henry Allen Wedgwood, cousin of the Darwin brothers. See genealogical chart.

^{7} Hensleigh Wedgwood, brother of the above.

^{8} Thomas Blunt, Shrewsbury chemist, believed by the Darwins to be the 'best chemist in the world' (*Emma Darwin* 2: 118).

^{9} A reference to Pieter Bruegel the Elder's 'Das Schlaraffenland' or 'Land of Cockayne' (1567). The painting depicts, among other gluttons' fantasies, pigs as described by Erasmus. 'Lumberland' is a slip either for 'Slumberland' or for 'Lubberland', a literal translation of the title.

[10] Until the 1830s the post office stamps of many towns included the mileage to London— that being the most frequent destination. The recipient, who normally paid the postage, would be charged according to the distance on a seven-zone mileage scale.

From E. A. Darwin 14 November 1822

[Cambridge]
Nov. 14th 1822.

Dear Bobby.

I am going to honour you again by writing, & first of all, I think it would be an improvement in the Lab:[1] to have some more shelves fixed up. The places I have thought of are, 1st a shelf to go over the window

I have drawn a picture of what I mean, & have put some bottles on the shelf to ornament it.

It is to extend from wall to wall, so that the shelf of y[e] lamp & also of y[e] cork bottles must come down NB. the little shelf for y[e] test glasses may stay up, & if it is possible, let there be one over the door, so as to be a continuation of y[e] one over the fire place, & if that takes place, the little shelf on which the pounded marble &c is, must also come down, but the little ones above may stay up. Perhaps the shelf over the window had better not be more than a foot wide, otherwise I shall be knocking my head against it.— Get it done as soon as you conveniently can, & pay for it, & I will then have the remainder of y[e] cow[2]

 sent up to me, & I will spend it in chemical instruments.— I have

ordered a small goniometer (an instrument to measure angles) so that we shall be able to seperate the different crystalls in your cab: I have not yet procured any of the minerals you mentioned. Hensleigh has promised me a bit of asphaltum which he obtained at Teignmouth. I dont think you have any, it is a kind of bitumen very light, will burn, & is of a snuff colour.—

I have bought a book which will be very useful. there are directions for finding out the names of minerals &c. &c. & the rules are not very difficult.—[3] I dare say I shall be able obtain some specimens of rocks for you, for Professor Sedgwick[4] said that at the Gog Magog hills (about 4 miles distant) there were a vast number of specimens, which I shall certainly some day go & explore.— I am attending Professor Cumming's[5] Lectures on Chemistry which are very entertaining. I have written all his experiments down as far as we have proceded, which we shall be able to try over again   This is y[e] shape of y[e] Lecture Room

aa the seats
c.c. 2 large tables
b. ye Lecturer
dddd, a long place like a blacksmith chimney, under wh: are furnaces &c
G. a private room.—

He has 3 men to assist him so that we get over a good deal of ground in an hour. he lectures every day, all this term. I think my *vacation* & your *holidays* will about fit, as my term divided, a few days ago i.e. half the term is over.—

I have just been to chapel, in wh: respect we are better off than you, as we are only obliged to attend 7 times a week. Some evenings preceding St^s days we go to Chapel in surplices, wh: look for all the world like sheets, & indeed one man of St Johns went, for a wager, into chapel dressed in a sheet, & sat before the master without being discovered. I was yesterday matriculated, in wh: you swear that you will not do this & that, wh: one does every day, you swear you wont wear a hat, wont ride, wont play at cricket &c before 12 o clock & half a hundred other things.[6] They have affixed a notice, that if you submit yourself humbly after any offence they wont consider you guilty of perjury. I suppose you have heard of black Frank Lightons[7] being rusticated. When he came up his Tutor asked him, why he had not read the things he was desired; he answered "why to be frank with you, one gentleman should be frank to another, in the first part of the vacation I was at some very pleasant places and it was impossible quite impossible, you know to read; & in the last, what with shooting & hunting it was impossible"— "Then perhaps Mr Leighton it wont be impossible for you to go down to the country" replied the Tutor, & so down he went.

Give my love to Massie, Pearson, J Lawson,[8] & all others. I remain your affectionate brother | E. Darwin.

Thursday. 14th—

Postmark: Cambridge 52
DAR 204.1

[1] In the *Autobiography*, pp. 45−6, CD wrote: 'Towards the close of my school life, my brother worked hard at chemistry and made a fair laboratory with proper apparatus in the tool-house in the garden, and I was allowed to aid him as a servant in most of his experiments. He made all the gases and many compounds, and I read with care several books on chemistry, such as Henry [Henry 1810] and Parkes's *Chemical Catechism* [Parkes 1806]. The subject interested me greatly, and we often used to go on working till rather late at night. This was the best part of my education at school'. CD's tutelage in chemistry lasted from 1822 to October 1825, when the brothers entered Edinburgh University.

[2] Apparently a reference to a fund set aside for laboratory expenses. On the cover of the letter, perhaps in CD's hand, is written: 5. 2. 6
                                                      44 17 6
                                                         2 6.

[3] Possibly Phillips 1816. The third edition (1823) is preserved in CD's library at Down House.
[4] Adam Sedgwick, Woodwardian Professor of Geology.
[5] James Cumming, Professor of Chemistry.
[6] By signing the matriculation book undergraduates swore to observe the laws, statutes, and privileges of the University. It is the statutes, many of which were obsolete, to which Erasmus refers.
[7] Probably Francis Leighton of Shrewsbury, admitted pensioner at Trinity College, Cambridge in 1821 (*Alum. Cantab.*).
[8] Edward Massie, Henry Pearson, and John Lawson were schoolfellows of Erasmus and CD at Shrewsbury School (*Shrewsbury School Register*).

From E. A. Darwin 8 December 1822

[Cambridge]

Dear Bobby.

I should not have written till I had received an answer about my going to fetch the girls, only I received a message, to acknowledge immediately the receipt of the drafts.—

You give me a very good account of the Lab:, it is quite wonderful how perverse the glass house people are, for I particularly told them to make the test tubes round at the bottom & they must need's grind them flat. If the cow is not utterly consumed the next milking, it would be a very good thing to buy as many of the large green, stoppered, bottles, as possible, & have them all filled with distilled water, & then I shall not be obliged to *persuade* you quite so often to run down to Blunt for some water.

Professor Cummings shewed us some of the most beautiful crystall's you can imagine, in fact they were artificial diamonds. The way he made them was, by putting a few drops of Sulphuret of Carbon at the bottom of a well stopped bottle, & the liquid, being the most volatile of any, crystallized against the sides of the vessel. They were far more brilliant than any diamonds I ever saw, but if air was admitted they melted away.—

I shall not bring you any minerals down this time as they would stand a very good chance of being knocked about, rather more than their constitution can bear.— I will bring you some wavellite,[1] & uranite & likewise a petrifaction from a well near here, as I have bought them already.—

What I shall buy in London will be principally stop cocks, & jars with stop cocks, for that is what we are principally deficient in, & we shall then be able to get over the ground rather quicker in our gas experiments.— The lab will look very insignificant after all the grand things I see here.— I hope you practise making gases, and have learnt to manage the argand Lamp,[2] quite beautifully.— In lecture the other day, 3 or 4 men got perfectly tipsy with the nitrous oxide or laughing gas, (that one made from nitrate of ammonia) and made themselves utterly ridiculous. they all recovered in about a minute—one man said he felt just as if he could fly, & when he was drinking the gas, he began to jump, & twiddle his fingers, & make a kind of half laughing & screaming noise, & y[e]

Professor gave a good tug, before he would let the bladder be taken out of his mouth.—

Tell Susan[3] I will bring the book down. I have nothing more to say of news or nonsense. | I remain yours affectionately | E. Darwin.

Dec. 8th 1822.

DAR 204.1

[1] Hydrous phosphate of aluminium, named after Dr William Wavel, its discoverer.
[2] A lamp invented *c.* 1782 by Aimé Argand.
[3] Susan Elizabeth Darwin, CD's sister.

From E. A. Darwin 5 [March 1823]

[Cambridge]
Wednesday 5th.

My dear Charles.

I beg many apologies for not writing sooner, but I have ye same excuse as you had. I am very glad to hear that all the glass & earthenware apparatus has arrived safe, & return my thanks for ye payment of ye Extra Money. You say you are very full. I should recommend ye first two or three shillings there are to spare in ye Lab, to have a shelf put up either in ye place over ye retort shelf or else under it; or both. The 10£ which Poor Miss Congreve has left & of which we are to be the trustees, will come in very nicely for an Air Pump or some such thing. As far as I can make out you are not carrying on any experiments in ye Lab. I will recommend you a few which will employ you some time, & will not be expensive. To obtain pure alumina from Alum, & pure Silex from Rock Crystal, & likewise to obtain pure metals from ye Metal of Commerce, such as Copper, Lead, Tin, &c &c, or any which you find easy. Have you my "Ure's" Dict[1] in ye Lab?— If it is there (& if not it is lost), that & Brande[2] will be ye best books for particular directions. You could try also to obtain ye silver in a sixpence free from ye Copper. While I remember, & if it is not too late, get a red Cabbage[3] dried by Blunt, cut into strips, & then put into green bottles, (such as ye lime was in) & well sealed up. I hope there are good many Evaporating Dishes come, for I forget how many we ordered. The shape of ye Crucibles I like very much, & more especially their having covers is a very great improvement upon ye old ones. How does ye old fireplace get on. I have formed a plan when that is quite done for, it is to have a little stove (ye same shape as that in ye Servant hall) with a moveable top so as to have a sand bath or a kettle or any thing else to fit on ye top. They have oil gas, & they condense into vessels to carry about. A curious looking liquid seperates from ye condensed gas. It resembles Naptha in appearance, & is rather volatile, it burns with a great light & smoke. It smells

The young Erasmus Alvey Darwin.

Robert Waring Darwin.

worse than ye most rancid oil you can conceve. Professor Henslow (on Mineralogy) has twice shewn us ye experiment of ye test of Arsenic by burning it with a blow pipe but I was so far off, that I neither time smelled ye garlic odour which they describe He has 16 guineas worth in one set of wooden crystals, besides many others. The Lectures are very entertaining, & this is his first course so that he will have improved by ye time you come up.[4]

Give my Love to Massie, & say I hope his hand is better. Ask W Downes[5] to write a few arguments, if he has time— | I remain yours affectionately | E Darwin

P.S. The first page of your letter was very well written so I give you a √

Postmark: C 6 ⟨MR⟩ 6 1823
DAR 204.1

[1] Ure 1821. A copy of the second edition of this work (1823) is preserved in the Manuscripts Department, Cambridge University Library, where a large proportion of the books in CD's library, particularly those containing his annotations, are now located. Hereafter this collection will be referred to as Darwin Library—CUL. Other works are located at Down House (Darwin Library—Down), and a few are at the Cambridge University Botany School (Darwin Library—Botany School). The copy of Ure 1823 is unannotated except for a note, dated March 1880, on an endpaper, concerned with precipitating the lime in the well-water at Down.
[2] Brande 1819.
[3] 'Brassica Rubra. The red cabbage affords a very excellent test both for acids and alkalis' (Ure 1821).
[4] John Stevens Henslow, later to become CD's mentor, was elected Professor of Mineralogy in 1822. He began his lectures after the division of Lent Term, 15 February 1823 (*Cambridge University calendar*, 1823, p. 38). See Winstanley 1940, pp. 32−41, for a discussion of the jurisdictional and legal issues raised by his election.
[5] William James Downes.

From E. A. Darwin 18 May 1823

[Cambridge]
May 18th. A.D. 1823

Dear Bobby.

If you come at ye time you propose To Wit ye 17th., you will not be able to see anything, for every body will be gone down except a few reading men. The best plan I think will be for me to come down (30th) & then we will come up for a couple of days or so by ye 1st of July when the Commencement will take place, that is to say when ye Noblemen & such like take their degrees & every thing is very gay. The difference of expense in staying up till ye 17th. & going down & up, will be rather in my favour, not to mention ye ride. If this plan is not approved of, you must bring up a good supply of money, for by that time I shall not have so much as a brass farthing to buy some broth with when you come starving off ye coach. If it is approved of, you must in your answer to this letter in close a ten pound note to frank me down.[1] N.B. both case are ye same to me so suit your own fancy. I have got a few Devils toe nail's[2] for you which are common enough here abouts. P.S. If you come up in July you will fall in with ye Newmarket meeting which inter nos is not half such good amusement as ye Shrewsbury Races. Hensleigh & I went over ye other day in a tilbury & we found it rather stupid, but I have been since on a good horse & found four miles of beautiful turf no mean galloping. It was also some amusement listening to the legs (i.e. black)[3] who are upon ye course in swarms, willing to ease ye Cantabs of as much money as they please.—

I have Witherings Botany[4] in 4 vol. apropos, I wish you would ask my Father if I shall bring him a copy of "Donns Catalogue of Plants,"[5] as there is a new Edition by Linly[6] & I have some idea he told me to get him one, th⟨ou⟩gh I am not at all sure.

Vale⟨as⟩. E. Darwin

P.S. Write soon or I will smak your b——m.

2nd PS. Dont come up in ye boots you had last Christmas.

Tell me what day ye speeches are.[7]

Postmark: Cambridge ⟨5⟩2
DAR 204.1

[1] 'To facilitate the coming and going of (a person)' (*OED*).
[2] A local name for a Belemnite.
[3] Swindlers.
[4] Withering 1796. William Withering, physician and botanist, was well known for his introduction of the foxglove into the pharmacopoeia.
[5] Donn 1823. A copy of this edition is preserved in Darwin Library — CUL. It has few annotations by CD; a large number are in another hand, probably that of his father, for whom botany was a hobby.
[6] John Lindley. CD's copies of several of Lindley's works are preserved at Darwin Library — Down.
[7] A reference to Speech Day at Shrewsbury School.

From Emily Catherine Darwin[1] [*c*. June 1823][2]

[Maer][3]

My dear Bobby,

I am afraid you were very angry at my not writing to you by my Sisters but upon my word I did not open my box and find your very entertaining letter till after they were gone.— I am very much obliged to you for it. The ink is so thick and pen so bad that you must excuse this bad hand—

I am very glad that you will go this tour to Cambridge—it will be so very pleasant for you—

We spent a pleasant day at Parkfields[4] the other day—and came back the next day to breakfast— You have no idea how I long to see you again my dear Charles and Laboratory also— I have won 3^s = 6^d at Cards at Parkfields but I lost 2^s = 6^d at Maer— How snug the Laboratory will be in Winter!! How does Mineralogy, Botany, Chemistry and Entmology,[5] go on?

Charlotte[6] is coming home today though I dislike her in general yet the party is today so dull without Marianne[7] and Susan that I shall be very glad to see her again—

I hope you will be able to read this— I have the good news to tell you that Aunt Kitty has very goodnaturedly brought some very pretty Specime⟨ns⟩ of di⟨ffer⟩ent Fossils and Spars.[8] My ⟨s⟩pirits are pretty good in spite of the thoughts of the Incantations & Whippings of Mrs Mayer,—[9] This Morning I heard that when they were naughty she locked them up with their hands tied behind them. *Pleasing Prospects*—[10]

Pray write to me dearest Bobby I shall so like to hear from you.— Of Course you will burn this letter the moment you have read it—and if you show this letter to any body I will never write to you any more— I am very sorry to write you such a short stupid one— We set off at six th⟨is⟩ evening— yr ever affectionate | E. C. Darwin—

DAR 204.16

[1] CD's younger sister, usually called 'Catherine' or 'Catty' by the family.
[2] Dated from the references to the Cambridge tour, which was planned for early July 1823 (see letter from E. A. Darwin, 18 May 1823), and to Kitty Wedgwood, who died in 1823.
[3] Maer Hall, near Newcastle-under-Lyme, Staffordshire, home of Josiah Wedgwood II.
[4] The home of Sarah, Mrs Josiah Wedgwood I, (CD's grandmother) and her daughters, Catherine (Kitty) and Sarah Elizabeth (Sarah), near Barlaston, Staffordshire. Mrs Wedgwood had lived there until her death in 1815; Kitty died there in 1823; Sarah Elizabeth moved in 1827 to Camp Hill, a house built for her near Maer Hall.
[5] In the *Autobiography*, p. 45, CD remembers observing insects 'with some little care' as early as 1819, but gives his second cousin, William Darwin Fox, credit for introducing him to entomology at Cambridge in 1828 (*ibid.*, p. 63).
[6] Charlotte Wedgwood, second daughter of Josiah Wedgwood II.
[7] Marianne Darwin, CD's eldest sister.
[8] A general term for a number of crystalline minerals.
[9] Mrs Mayer's school was at Greville House, on Paddington Green, London. CD's cousins, Frances and Emma Wedgwood, had been placed there in January 1822. Henrietta Litchfield, commenting on the teaching, says, 'In French history they never got beyond Charlemagne, as with every new girl the class began again at the beginning with Clovis.' (*Emma Darwin* 1: 141).
[10] The passage 'This Morning . . . *Prospects*—' has been heavily crossed through in pencil and across it in another hand is written 'A mistake it was another school'.

From John Wood Warter[1] 23 December 1824

Dear Charley/—

As I had not an opportunity of seeing you on the Saturday Night previous to your breaking up, I must now like "a friend misericord" just warn you against the Encouragement of idleness, to which I know you are ever fully inclined. I s^d— advise you to read some of Xenophon's Philosophical Treatises which you will find in the Dalzel[2] p. 297 in my edition, immediately following the Excerpta

Rhetorica. I have great reason to believe that Iliff[3] will set you some part from that selection: and if he does not, the words that you will meet with in Xenophon, are continually used by all other authors, and consequently will in the end prove of advantage. I can form no idea with regard to Wakefield[4] as a master— he'll never fill the place of his good predecessor Sheepshanks,[5] our mutual friend, tho' when at School he shewed me very much attention. Read Hor: Od: 1 — 24[6] and think of Sheepshanks, bearing in mind that "*desiderio*" in the first line means "*regret*" and answers to the Greek πόθος. If you will ride over here, I shall at all times feel much gratification in helping you over any difficulties that occur, if *I can*, for you well know my abilities:— "My *loom*" as Shakespeare has somewhere observed "is of *a mingled yarn*",[7]— to speak in more plain English I am very stupid. Do you remember ever quibbling with me as to the right pronuntiation of *sloth*? You, if such a trifle has not fled your mind, w^d always call it short—thus slŏth.— I will transcribe two lines from our sweet poët Goldsmith which will indubitably confi⟨rm my⟩ opinion
"The robe that wraps his limbs ⟨in⟩ silken *sloth*
"Has robb'd the neighb'ring fields of half their *growth*"
These lines you will find in the Deserted Village, and I beg you will henceforward call it "*slōth*"—*mei memor*. When you write remember me very kindly to Old Strol—[8]

Hastily, Your very Sincere Friend | John Warter

Cruck-Meole[9]
Dec: 23rd 1824 | Thursday Morning—

DAR 204.16

[1] A Shrewsbury schoolfellow, 1820 — 4 (*Shrewsbury School Register*).
[2] Andrew Dalzel's selections from Greek writers (Dalzel 1789 — 97) were standard textbooks in the classical curriculum of the schools of the period.
[3] Frederick Iliff.
[4] John Mort Wakefield.
[5] Thomas Sheepshanks.
[6] Horace, Ad Virgilium, de Morte Quinctilii Vari, *Odes*, Bk 1, Ode 24: 1. 1, 'Quis desiderio sit pudor aut modus tam cari capitis' ('What restraint or limit should there be to grief for one so dear?', translated by C. E. Bennett. Loeb Classical Library. 1968).
[7] Probably a misquotation of 'The web of our life is of a mingled yarn, good and ill together', Shakespeare, *All's well that ends well*, 4. 3. 68 — 9 (Arden edition).
[8] A nickname given to E. A. Darwin at school. John Price, in a letter of reminiscences to Francis Darwin, says of Erasmus: 'Why he was called *John* Darwin or Strŏl must go down the stream of time with other school mysteries.' (DAR 112 (ser. 2): 104).
[9] A village 4½ miles south-west of Shrewsbury. 'Cruckmeole Hall is a handsome brick mansion, the property and residence of Henry Diggory Warter, Esq.' (Bagshaw 1851, p. 680).

From E. A. Darwin [10 January 1825]

[Cambridge]

My Dear Bobby.

Comment vous portez vous, Monsieur? Pretty well. Je suis ravi. Ah. Ah. What a charming shot you must be to have killed that young tyger, that you mentioned; it must have been very horrid, only suppose you had missed him & that he had run at you. Bye the Bye. How came this tiger to be loose? You dont mention that. Is there any there any menagerie at Styche?[1] But I begin to suspect this is like some other wonderful birds that you see with your pocket telescopes, or eyes whichever you please to call them.

There was a travelling mineral-seller here some time ago who made my mouth water with such a magnificent piece of amethystine quartz. The crystal was 4 or 5 inches long & so beautiful, only the wretch asked a pound for it. He had an immense number, tho' not variety of stones, & I did see anything which you had not got.

I cant say *my cabinet* increases very much I have got a piece of petrified sponge & shall perhaps have a curious piece of coal which J. Price[2] pulled out of his fire

I have given you a sketch of it, the figs (a) are little circular hollows and those (β) are little compressed tubercles. They are curiously mixed together in some places. He is going to shew it to the Prof. of Geology.

You cant think what a pretty bird the spotted Woodpecker is, if you never saw it There was one being stuffed here some time ago. There is a very beautiful work now publishing which you would like. Griffiths Animal Kingdom[3] in Quarterly parts. 24 shillings each the best copies $\frac{1}{2}$ a guinea the worst.

Price & I are going to anatomize cows h⟨ ⟩ &c very vigorously as soon as he he ⟨ha⟩s time which I dont think you woul⟨d⟩ fanc⟨y.⟩

I this morning saw a curiously diseased liver &c of a child who had died with a jaundice— I send you the sixpence & pray dont make a beast of yourself with it

Let me hear from you again. | Love ad omnes | Greet Spark[4] | yours, | E. D.

P.S. I can not make out whether you mean your P.S. for a quiz.[5] If there was a list it was an old list of my own.

2nd P.S. In the last No of ye An. of. Phil. The Editors so quizzed "self adjusting candles" &c for the Blowpipe that I hardly know if I shall venture to send an account of my admirable candlestick.[6]

3rd P.S. If you know of any person coming to to Cambridge get him to carry up my "Samouelles Entomolog. Companion"[7]. N.B it is a red book—

Postmarks: C 10 JA 10 1825; Cambridge 52
DAR 204.1

[1] Styche Hall was the seat of the Clive family (Bagshaw 1851, p. 297).
[2] John Price.
[3] Cuvier 1827—35.
[4] Darwin family dog.
[5] 'A practical joke'; 'a jest or witticism' (*OED*).
[6] In a book review mention is made of 'a blowpipe with a self-adjusting candlestick (a self-adjusting fiddlestick to Mr François Cramer's violin would be about as necessary or useful)' (*Annals of Philosophy* n.s. 9 (1825): 58).
[7] Samouelle 1819. The copy in Darwin Library—CUL has annotations in another hand designed to make reference to the work more convenient. On the end-paper CD has written 'p. 338 Explanation of terms'.

From E. A. Darwin [17 January 1825]

[Cambridge]

My dear Bobby.

I have been indulging myself in the pleasure of scheming, & I here send you the fruits of it. The Lab:, as we know to our cost, wants a dry place to keep our apparatus in, and the above drawing is a plan for that purpose, but from not re-membering the dimensions I have been obliged to draw it by my eye, & not by any scale. A.A you must understand are the chimneys as they now stand, & I propose under the arch to open a window looking into the drying yard, this is represented at C. Next, to inclose a small space sufficient to sit or turn round it with lath & plaster. I should put the laths double with a space between of one or two inches as figured at B,B. I have drawn the door at D, which I make as narrow as possible, & perhaps double like a Rumford ash-door (wh. you may see in ye scullery). I should floor it with wood raised up an inch from the bricks, & having a chimney on both sides, if any out-of-doors place can be dry I should think this would. There might be a shelf before the window, which would be an excellent place for weighing &c. and being South would do very well for any experiments with the Sun's rays, which is also a desideratum in our Lab.

I have drawn this with dotted lines, those on each side may be either shelves or drawers.

Go to the Lab. with a rule in your hand & consider about it, & let me know what you think.

Price in drinking tea observed a black heavy sediment constantly form, & by degrees he collected a small pill box full. He gave me a few grains to examine, & I found it be entirely attracted by the magnet. Is not this curious? We are going to try how much per cent is in the tea, & then send it to Children[1] (Editor of An. of Phil.) to examine it & what else he may please to do.

I wish when you write you would send me a piece of Plat. Wire. Put it in the fold of the first page & gum it fast & I dare say it will pass muster with the post office—

Good Bye. Let me hear soon | yours, E. D.

N.B. Keep the plan at all events.

Postmarks: C 17 JA 17 1825; Cambridge 52
DAR 204.1

[1] John George Children.

From E. A. Darwin [24 January 1825][1]

[Cambridge]

Dear Bobby.

I want you to try an experiment for me, & you remember you promised to do any I asked you & moreover to do them *immediately*. In a french book on physiology that I am reading it says that the affinity of saliva is so great for oxygen, that if gold leaf be triturated with it, it becomes oxidated. Now this I doubt, & I want you to try the fact wether it is so or no, & save you trouble I will write the experiment down

A. Triturate a small bit of gold leaf in pure water, & if that has no effect add a little gum.

B. Triturate it with saliva (give this a fair trial)

C If in either case it turns brown or yellow, in order to determine whether it is an oxide or simply in a state of fine division, the simplest way will I think be to digest it in Nitric Acid wh. will dissolve the oxide, & not the metal.

D. In order to be sure about the solubility of ox. of Gold in Nitric A^d, precipitate the N— Mur. of Gold with Lime water, & after washing the precip on a filter digest it in Nitric A^d

Just try also whether the precip: after being heated red-hot in Sp^t Lamp in a glass tube for a few minutes is then soluble in Nitric Acid.—

I here send you a very correct picture of a curious stone I got out of the fire. There is another piece wh. exactly corresponds. In taking it out of the fire it dropped on the hearth, & divided so as to shew the impressions wh. I have represented, most probably vegetable remains. It appears to be quartz, & towards A is most beautifully coloured apparently with iron for towards B in the interi⟨or⟩ is a dingy red. Between A & B it is striated The ⟨da⟩rk blotch lower down appears to be carbona⟨ ⟩ matter, & this together with the vegetable remain⟨s⟩ I thought never were found in quartz. It is not quartz I now see, but a kind of micaceous slate. The glistening deceived me.

Good Bye. yours truly. E. D.

P.S. If any body at home wants a copy of Bewicks Quadrupeds,[2] there is a very decent copy at Deightons 2nd Hand for 17^{s} Demy 8vo.

P.S. 2nd | I want you also to try if the per-oxide of iron is attracted by ye magnet, as also the sulphate.

I told you about the iron found in the tea by Price, I to-day tried the proportion & found it to be 13 per cent. Is not this an enormous imposition?

P.S. 3rd | I kept this open to wait for your letter which has just arrived, & for wh. many thanks.

It takes an additional night to go thro' Cambridge.

I should conceive ye genera to be nil nisi fumus

There is no mineral seller here now

If you will send me the size of bottle I'll get it, but if you go to London you can do it better at Newman's.

Let me hear more about your school-plans. I should vote for your staying on till at least 18 years

Postmarks: 24 ⟨ ⟩; Cambridge 52
DAR 204.1

[1] The letter is dated from the reference to Price's finding the iron in tea, mentioned in the letter from E. A. Darwin, [17 January 1825], and from the reference to school plans, since it had not yet been decided that CD should leave school (see letter from E. A. Darwin, [24 February 1825]).
[2] Bewick 1790.

From E. A. Darwin [1 February 1825]

[Cambridge]

Dear Bobby.

Many thanks for your letter. The bill is all right, & I shall be much obliged to the D.ʳ if he will pay it. It is for those things wh. I told you he kept me so long waiting for.

There is another bill I have in town, which I should wish paid in the same manner if it would not be too much trouble, to wit—

> W & S. Ewart & Co
> Wine Merchants
> 19. Swallow Street
> London

£13″12.0.

You may ask this latter gentleman to send an acknowledgment, for I have known him send a bill twice over.

If this last commission should be troublesome, never mind it.

Good Bye | yours, | E. D.

P.S. | Let's hear from you. Obs. The P.S. is as short as the letter & for the same reason, which I shall leave you to guess.

This for D.ʳ D—

I lent M.ʳ Tighe 13£ in your name & wh. I believe he wrote about, & is I hope all right | I remain | your lendpenny | & spendpenny | son. | E. D—

Postmarks: C 1 FE 1 1825; Cambridge 52
DAR 204.1

From E. A. Darwin [24 February 1825]

[Cambridge]

My Dear Bobby.

Many thanks for your letter The examination[1] turned out as I expected in one respect, but how in the name of heaven T.B[2] came to be where he is I cant conceive.

You have I suppose by this time felt the good effects of Peiles schorlarship:[3] has it turned the D.ʳˢ head?

I dont know whether to be sorry or glad about your Edinborough plans.[4] I think it is ten thousand pities if you do not come to Cambridge (wh. seems to be the case), & I shall venture to add that it is a pity you leave school so soon, but to this latter doctrine you will hardly give credit. It will be very pleasant our being together, we shall be as cozy as possible, & I almost think that when you have arrived at the dignity of a 'Varsity' man, that I shall leave of licking you. We shall have some good amusement in scheeming out our plans next summer, & I shall vote setting off as soon as may be after your half-year finishes, for I suppose you wont return for the Quarter, & there will be no earthly use staying at Shrewsbury, for it is out of the question reading there, the Lab on one hand, &

no room on the other, whereas if we get domiciled at E—— we can both read like horses that is to say if we like it, & I must apropos, when & where are you to graduate? but this I suppose you dont know.

I am getting a little case-hardening in anatomy; for yesterday seeing a body ex⟨ ⟩ & being Junior they gave me a ⟨ ⟩ deal of the dirty work, & I was not the least annoyed while an old Physician also present kept leaving the room perpetually. I dont fancy it w^ld. have suited your stomack especially before breakfast—

Write to me again whether you have anything or nothing or both to say In the meantime good Bye | yours, E. D—

Postmarks: Cambridge C 24 FE 24 1825; Cambridge 52
DAR 204.1

[1] The sixth form examinations at Shrewsbury School took place on 6 February 1825 (Shrewsbury School archives).
[2] Thomas Butler, son of the headmaster, was placed fourth in the February examinations. In August 1825 he went on to achieve the highest standing of the school (Examination Records, Shrewsbury School archives).
[3] Thomas Williamson Peile, who had attended Shrewsbury School, won the Davies scholarship in classics at Cambridge in 1825.
[4] 'As I was doing no good at school, my father wisely took me away at a rather earlier age than usual, and sent me (October 1825) to Edinburgh University with my brother' (*Autobiography*, p. 46).

From E. A. Darwin[1] [8 March 1825]

[Cambridge]

My Dear Bobby,

Much obliged for y^r letter, & lest I should forget as I did before, you may take Hunter[2] or any other book you choose. If you wont be very angry I will send you once more into the plate-closet to hunt out some books, which you may send to me per Van carefully wrapt up. 1st. any or all of Cullen's Works, if they do not come to many vols, but at all events his Nosology & Practice of Physick.[3] 2nd ye 1st vol 4to of Haller's Physiologia[4] wh. you will find in ye top shelf, 3^rdly any small work you can find of 1 or 2 vol 8vo on Physiology,[5] & 4^thly ye Mss catalogue of Books, in order that I may avoid procuring duplicates of any works, & 5^thly if ye packet does not look very large send me Linnæi Systema Naturæ[6] wh. is among ye books in ye Morning Room & unbound.

Apropos to Linnæus I wish you would look in ye English Systema Vegetab:[7] & copy me out the specific description of Pinus sylvestris.

If the D^r should be at leizure I wish you would skim over the Mss Catalogue & see if there are any other works it would be worth while sending up.

I hope you dont object to schedule of jobs very grievously.

From whom did you hear about the rabbits foot, as your account was the first I heard about it?

Next week I am going to dine with Prof. Haviland[8] when I shall probably pick up something or other, about Edinborough, wh. I will let you know.

The next time you go home please to let them know that I am alive & well, y[e] very same morning I received just such a letter from Susan, & I verily believe that I wrote last in both cases. I shall at all events be giving y[e] D[r] a practical proof shortly that I am alive by sending for some money. Touching wh. The glass Bill has not been paid, try if you can find the bill in my chest of drawers & have it done. The man's name is Johnson or Bery or both, I forget wh.

Good Bye give my Love to Doctrss P.[9] & tell her she must not have Spark | E Darwin

Postmarks: C 8 MR 8 1825; Cambridge 52
DAR 204.1

[1] The letter was originally addressed to 'C. Darwin Esq., Dr. Butlers, Shrewsbury'. This was changed in a different hand and ink to read 'Gas Darwin Esq.'. 'Gas' was the nickname given to CD by his schoolmates because of his chemistry work with Erasmus (see *Autobiography*, p. 46).
[2] Probably a reference to a work by either William Hunter, anatomist, or his brother, John Hunter, surgeon and anatomist. A copy of J. Hunter 1778 is in Darwin Library—Down.
[3] Cullen 1814, Cullen 1776—84.
[4] Haller 1757—66.
[5] Above 'Physiology' is written 'Golias' in another hand, possibly CD's. 'Golias' has not been identified.
[6] Linnaeus 1789—96 (Darwin Library—Down).
[7] Linnaeus 1783a.
[8] John Haviland, Regius Professor of Medicine.
[9] Their eldest sister, Marianne, had married Henry Parker, M.D. in November 1824.

From E. A. Darwin [June 1825][1]

[Cambridge]

Dear Bobby,

You really are a beautiful correspondent. As to the tackle you are quite welcome to have it all except the line whose beauties you don't appear to appreciate properly.

Are you making any plans with regard to Edinbororough? any nice little stony excursions? We'll have rare fun in planning if we dont have any in performing them. I've made five hundred already for my share. How shall you fancy going there by sea? it is 14 hour's passage from Liverpool to Glascow. We can stay a day or two at y[e] Ile of Man if either of us feel inclined to give up the ghost, which wont be very unlikely I think. I shall be down in about 3 weeks & then Gracious Heavens! how *you* will chatter but.

Tell Susan that I dont know whether she corresponds with Edward Holland,[2] but that he is the only channel thro' which I hear of her. How came Susan to come away before min' aunt?[3] Did she quarrel with Jessy[4] or was it because Tom[5]

had the rheumatism in his jaws. The Hollands are all here with whom I am now going to dine & must therefore conclude.

My Love ad omnes, & I hope the D^r is better | Yours truly, E. D.

Postmark: Cambridge 52
DAR 204.1

[1] The letter is dated from the reference to Erasmus's coming down in about three weeks. The 1825 Easter Term ended 6 July.
[2] A second cousin, Edward Holland, was at the time an undergraduate at Trinity College.
[3] Louisa Jane (Jane) Wedgwood.
[4] Jessie Wedgwood.
[5] Thomas Josiah Wedgwood, Jessie's brother.

To Robert Waring Darwin   [23 October 1825][1]

[Edinburgh]
Sunday morning.

My dear Father

As I suppose Erasmus has given all the particulars of the journey I will say no more about it, except that alltogether it has cost me 7 pounds— We got into our lodgings yesterday evening, which are very comfortable & near the College— Our Landlady, by name Mrs Mackay, is a nice clean old body, and exceedingly civil & attentive— She lives in "11 Lothian Street Edinburg"[2] & only four flights of steps from the ground floor which is very moderate to some other lodgings that we were nearly taking— The terms are 1£—6^s for two very nice & *light* bedrooms & a nice sitting room; by the way, light bedrooms are very scarce articles in Edinburg, since most of them are little holes in which there is neither air or light. We called on Dr Hawley[3] the first morning, whom I think we never should have found had it not been a good natured D^r of Divinity who took us into his Library & showed us a map, & gave us how find him: Indeed all the Scotchmen are so civil and attentive, that it is enough to make an Englishman ashamed of himself—

I should think Dr Butler or any other fat English divine would take two utter strangers into his library and show them the way! When at last we found the Doctor & having made all the proper speeches on both sides we all three set out and walked all about the town; which we admire excessively; indeed Bridge Street is the most extraordinary thing I ever saw, and when we first looked over the sides we could hardly believe our eyes, when, instead of a fine river we saw a stream of people—

We spend all our mornings in promenading about the town, which we know pretty well, and in the Evenings we go to the play to hear Miss Stephens,[4] which is quite delightful. She is very popular here, being encored to such a degree that she can hardly get on with the play— On Monday we are going to Der F.[5] (I do not know how to spell the rest of the word)— Before we got into our lodgings we were staying at the Star Hotel in Princes St. where to my surprise I met with an old school fellow whom I like very much; he is just come back from a walking tour in Switzerland, and is now going to study for [his degree?].

The introductory lectures begin next Wednesday, and we were matriculated for them on Saturday: we pay 10.[s] & write our names in a book, & the ceremony is finished; but the Library is not free to us till we get a ticket from a Professor—[6]

We just have been to church and heard a sermon of only 20 minutes. I expected from Sir Walter Scott's account, a soul-cutting discourse of 2 hours & a half—

I remain Y.[r] affectionate son | C. Darwin.

Copy[7]
DAR 154

[1] Dated from the reference to lectures beginning 'next Wednesday'. Lectures for the 1825—6 session at the University began on Wednesday, 26 October (Ashworth 1935, p. 97).

[2] The site is now occupied by the Lecture Theatre of the Royal Scottish Museum (information supplied by Mr Antony P. Shearman, City Librarian of Edinburgh).

[3] Probably Richard Maddock Hawley. The copy and Francis Darwin's version both read 'Hanley', but no doctor by that name has been located. In her letter of [26 October 1825], Catherine Darwin clearly wrote 'Hawley'.

[4] Catherine Stephens, a leading soprano of the time. During October 1825 she appeared at the Theatre-Royal, Edinburgh.

[5] *Der Freischütz* by Carl Maria Friedrich Ernst von Weber was first performed at the Theatre-Royal, Edinburgh on 29 December 1824 (Harvard College Library Theatre Collection).

[6] The anonymous author of two articles in the *St. James's Gazette*, 'Darwin in Edinburgh.—I.', 16 February 1888, and 'Darwin in Edinburgh.—II.', 17 February 1888, states that the University Library record books for 1825—6 (since lost) showed that CD and Erasmus borrowed more books than other students. The *St. James's Gazette* passage on CD's borrowing reads: 'They included Good's "Study" [Good 1822] (with which his list opened), Pemberton on Viscera [Pemberton 1806] (which Erasmus also took out), Young's Philosophy [Young 1807], Fleming's Zoology [Fleming 1822], Kerr [?Kerr 1792] and Foster [not identified], two volumes on entomology, Wood on Insects [Wood 1821], Brook's Conchology [Brookes 1815], and Newton's Optics [Newton 1704]. . . What may be called the only lay work he took out was Boswell's Johnson [Boswell 1791], which he seems to have read all through, carrying off a volume at a time.' (16 February 1888, p. 5). In the following session

(1826 – 7), however, CD did not pay the library deposit and borrowed no books at all (17 February 1888, p. 7). He paid his fees for two classes 'Practice of Physic' and 'Midwifery' and attended Robert Jameson's lectures in Geology and Zoology, which he remembered as having been 'incredibly dull' (*Autobiography*, p. 52).

7 The original of the letter has not been found. A copy made for Francis Darwin (by CD's sister, Caroline) has been compared with the text published in *ML* 1: 5 – 7. The copy has been followed rather than the printed text wherever Francis diverged from it to conform to the editorial style of *ML*. Where he differs from the copyist's transcription of words, his version has been adopted as more probable, unless clear evidence for an alternative rendering exists, as in the case of 'Edinburg' and 'alltogether' (see letter from Catherine Darwin, [26 October 1825]). In *ML* 1: 7 'his [degree?]' was supplied in a space left blank in the copy. See also the 'Textual Notes' for this letter.

From Catherine Darwin   [26 October 1825]

[Shrewsbury]
Thursday Evening—

My dear Charles,

Papa and Susan are gone to dinner at the Salts';[1] and I, being alone, will write this page to you, and give you an account of all our gay proceedings of late. This letter bids fair to be very complex, I think; Papa has written a message to Dr Hawley on the bottom flap, which you must tear off, and give to him. We were very glad to hear of your safe arrival; and that you are so much pleased with Edinburgh— Papa sends his best thanks for the very nice letter he received from you this morning. I shall hope soon to hear from you again, my dear Bobby, and, remember that every little particular will be interesting— I must just mention that Edin*burg***h** is spelt with an *h* at the end; and, *altogether* has only **one** *l*, not *all*together, as you spelt it in your letter to Papa; do not be very cross at this, dear Charley.— I assure you, I miss you very much, both to walk with me, and also for *you know what*; and as for poor Shelah[2] and Spark, they look the pictures of melancholy, and are really grateful for any notice, from any body.— I have not yet told you how gay we have been lately; we have been as constant play goers as you can have been; for Macready[3] has been here for the three last nights, and all our hearts, that were not quite gone before, are lost now— On Monday he acted William Tell, a beautiful play, & written by the same Liverpool gentleman as your favourite Virginius.[4] He takes a little boy about with him, who acts the part of William Tell's son very well indeed. On Tuesday, it was Othello, which is a disagreeable Play, I think to see; and besides we had such a horrid Desdemona; we were in great want of your *lovee* Mrs Brudenell,[5] who, it is said here, is gone to the Edinburgh Theatre. I hope you will see her there. On Wednesday (Macready's Benefit) we had Virginius again, which he acted more beautifully than any thing I ever saw in my life, and which I quite agree with you, in thinking by far his best character. He had promised to act in the Farce that night;[6] but was so much offended at the empty houses & slight applause on Monday & Tuesday, that he would not.— He looks ill, & is said to be much out of health; he is going to his house in North Wales for a short time, to recruit his strength.— Fanny Wedgwood came over here yesterday, on purpose to see him, and Caroline[7] returned with her and Emma[8] to Maer this morning, so we are quite a small party again—

What was the name of the schoolfellow you met with at the Hotel? how surprised you must have been—

Good bye, dear Charles, pray write soon— My love to Erasmus; ever yr[s] | E.— C.— D—

Postmarks: Shrewsbury OC 26; [another stamp partially illegible but probably Edinburgh's] W 30 E 1825
DAR 204.2

[1] Probably the Thomas Salts. Salt was a partner of the Shrewsbury law firm of Dukes and Salt.
[2] Darwin family dog.
[3] William Charles Macready, who was at the height of his fame at this time.
[4] *Virginius* (1820) and *William Tell* (1825) by James Sheridan Knowles. His plays were 'a genuine and frequently popular attempt to blend heroic grandeur with a domesticity familiar to his audiences ("the scene is the Forum, but the sentiments those of the 'Bedford Arms' ")'. (*McGraw-Hill encyclopaedia of world drama* 2 (1972): 484).
[5] London playbills of September and October 1824 have Mrs Brudenell acting, mainly in comedies, at the Covent Garden and the Haymarket Theatre, but not in Macready's company (Harvard College Library Theatre Collection).
[6] At benefit performances, the proceeds of which went to the acting company, it was Macready's custom to add a light comedy to the programme (Archer 1890, p. 61).
[7] Caroline Sarah Darwin, CD's elder sister.
[8] Emma Wedgwood.

From Catherine & Susan Darwin 4 December [1825]

Shrewsbury—
Sunday Morning, December 4th

My dear Charles,

I am quite ashamed of myself for not having written to you before now; but I have been putting off doing so from day to day for a long time, and I am afraid I have not now much to tell you— The principal piece of news I have to tell you is that last week Papa received the melancholy account of poor M[rs] Maling's[1] death in Peru; she died in August at Valparaiso of a complaint peculiar to the climate; of which she had been ill since January, but had concealed it carefully from her friends in England only saying "that she no longer had her vulgar, robust state of health"; so that her death was a great shock to poor M[rs]Darwin, her mother, who had been looking forward to her return about this time— Papa thinks that you need not go into mourning, on account of the distance.—

I have been spending a fortnight at Overton,[2] from whence I returned on Friday; it was horrid, rainy weather the whole time, as is always the case there I think; but notwithstanding which I had a very pleasant visit there, and like my brother, the D[r], more than I did before— Baby[3] is very well, and fat, and has the prettiest colour in his little cheeks— I do not think that even *you* could *pretend* to be disgusted with your little nephew, if you were to see him smile and laugh so prettily now— By the time you return, next summer he will be toddling about— You see I have quite caught the *mania* of Overton, as *Papa* and *Mama* are both of course doating; but I must now give you a piece of news about *your* favourite *child*, which I am afraid will prove a blow to you; i.e. that Spark is gone to Overton; at least till your return next summer, as they were in want of a watch dog, and Czar is finally going, having bit another person— I am afraid this intelligence will be a shock to all your nerves, and will spoil a good many breakfasts; but all I can tell you for your comfort is that D[r] Parker is very fond of her, and means to take the feeding of her entirely in his own care, and there are a profusion of rats and mice about for her to kill; and if you wish, she shall come back here to meet you on your return, next summer dear Bobby.—

Caroline and Susan are going to Woodhouse[4] tomorrow to spend the week— Fanny Owen is copying two beautiful pictures, a Coreggio and a Murillo, and Caroline is going to attempt them in miniature— We are reading aloud a very entertaining novel lately published "Matilda, a Tale of the Day"; it is written by Lord Normanby, nephew to Capt Maling; if it should fall in your way, I advise you to read it, as it is very clever and witty in many parts, and only one octavo volume— The new Society[5] is opened, and we have had several books from it; among others "the Story of a Life" by the au⟨thor⟩ of Rec⟨ollec⟩tions of the Peninsula[6] which you liked so much; ⟨bu⟩t you never read such affected stuff, quite unreadable.— Three little boys dined here the other day, Allan, Browne, and H. Hutchings, and Susan played the pretty to them all evening; are you not sorry you missed such a charming evening?— I met Watkins[7] in the street the other day; I never saw such a dandy as he is grown, his hair all frizzed out in the most absurd, coxcombish style and his hat on one side— I did not feel at all inclined to call him "M[r] Will.-Willough-Watkins" The only bit of gossip I can tell you, (and I am afraid you will not care for that) is that Miss Derby, the pretty flirting heiress, is really going to be married at once to M[r] Pryce Owen.—[8]

I shall hope soon to hear from you, my dear Bobby; any little particular will be interesting— What capital luck you are in, just to fall in with all the good London performers at Edinburgh, Liston,[9] Miss Stephens, and Macready; give me some account of the latter, if you see him in any particularly good or new characters; you know I am always interested in him— and now Good Bye, my dear Charley— Love to Erasmus. Believe me, | Ever y[rs]— | E. Catherine Darwin

My dear Charley. I am very glad to hear you are such a good boy about your French, and I hope now that you are reading something more interesting than the *Baroness & Countess'* silly letters, you like French better than you did— you must tell me how you spend y[r] days as I shall like to hear.— Catherine has written you such a long letter that she must have told you all the news. | best love to Eras. ever y^rs. Susan—

Papa says Erasmus may wear Flannel next his skin in cold weather by all means & that he may *sleep* in it also, tho he does not think that very adviseable—but in warm weather he very much objects to it.[10]

Postmarks: Shrewsbury DE 5 ⟨ ⟩ 153; DEC ⟨ ⟩ 7 E 1825
DAR 204.2

[1] Harriot Maling, seventh child of CD's grandfather, Erasmus, by his second wife, Elizabeth.
[2] Residence of Marianne and Dr Henry Parker.
[3] Robert Parker.
[4] 'Woodhouse is a beautiful mansion of white freestone, the seat of William Mostyn Owen, [Sr] . . . delightfully situated on a gentle eminence, commanding fine views, and surrounded by park-like grounds beautifully wooded. The mansion is approached by a noble portico, supported by four circular columns; and there is a fine avenue of beech and other trees on the south side of the park.' (Bagshaw 1851, p. 204). Woodhouse is situated in Rednal, Shropshire, 13 miles north-west of Shrewsbury.
[5] There were several private reading societies in Shrewsbury, besides a well-stocked Public Subscription Library (Howell 1816, p. 87).
[6] Sherer 1823 and 1825.
[7] Frederick Watkins, CD's Shrewsbury schoolmate and later fellow undergraduate at Christ's College.
[8] Mary Darby and Edward Pryce Owen were married 6 December 1825 (*Alum. Cantab.*).
[9] John Liston.
[10] The final paragraph is written on the flap in Caroline Darwin's hand.

From Caroline & Susan Darwin 2 [January 1826]

[Shrewsbury]
Monday. 2^d.

My dear Charles

Catherine desires her love and many thanks for your letter—and Papa says you are a very good lad for remembering to mention Johnson,[1] who he wanted to hear something about. You are in good luck to have so many good actors— Shrewsbury is going to make an attempt at gayety this week. The new Hunt Ball, & 2 concerts with no performers worth hearing, & if you were at home you would not even have the fun you subtracted from them last year in seeing *chair men killed* or drawn to pieces as the concerts are to be held at the Theatre & not at the Circus— M^r & Mrs Mathew come to us & three Mr Clives[2]—so Susan & Emma Wedgwood who is staying with us will have plenty of parteners.— the bells are ringing & making a great noise, so we suppose M^r and Mrs Price Owen are arrived at the Archdeacons[3] I think Miss Derby was no more a flame of y^rs than Mr P. Owen an object of my admiration, so we will wish them happy to-gether.— I did see a few days ago a young gentleman who I thought I own, quite exquisite, M^r. Gibbon, so handsome, & so conscious of it that he could not speak or turn his head with out thinking he was a study for a painter & model to a Sculptor. he told me a story of a young lady saved from being drowned by a newfoundland dog—who dragged her to land—and added, "I always call that dog a gallant fellow." I give this as a specimen of the good taste of his conversation. Papa turned this delightful dandyfied young man into me in the Morning room—my long tête à tête was at length interrupted by Susan who came in huddled up in an old woman's grey cloak— She started on seeing him—but sat down, & presently observed—"Caroline there is an old man wandering up & down the walks. I suppose he has lost himself, I dare say he is gone". this old man was Mr Gibbon's father—so I do not think Miss Susan will get him for a partener after speaking so disrespectfully of his Papa.— this is sad dull stuff to write about dear Charles—but you know the difficulty of a letter with no particular fact to communicate— Mrs Shelah condscends to pay me much more attention than when you were at home— She does not get much exercise beyond her daily walk into town & a little romping with any odd apple which she entreats me to throw down the bank for her to pick up.— Spark we hear from Overton is in high preservation, & so is your little nephew who of course you have a much stronger affection for. I am going to Woodhouse next Monday to copy a picture & intend to pay Marianne a visit in my way home— I was very near being seized as a poacher the last time I was at Woodhouse, walking in the dusk in the little wood near the house, & looking at some pheasants which I had startled, but it was soon my turn to be startled, for a man jumped over a wall & ran to me, calling "stop, stand," which I did, he did so likewise, then we heard Mr Owen's voice in the distance, "tell y^r name, who are you"— I gave my name, he did not hear me but again roared out "who are you", the man mean while stood quite still, He, He, He, Ha Ha Ha, Ho, Ho, Ho, then touched his hat and told me 'Master set him on'— We have had no more new books, but have just ordered Coleridgs—"Aids to reflection,"[4] which I hear very highly spoken of by several people.— Mores Sherridan[5] is very interesting, & well written I think. Susan likes it particularly, but unfortunately any particular part that is mentioned in conversation she can not quite call to mind, his marriage for instance, "Dear me! Married was he? Well, I suppose I missed that chapter", and many other incidents "Dear me! well, I suppose I missed that page"— Do you read your magnificent work on Zoology in that manner? How do you like the lectures? are you busy and industrious— I suppose poor French has no chance now—

Susan has begged the flaps, so goodbye dear Bobby wishing you & Eras a happy new year & many returns of it. Ev^r Y^r affec Sister | C. Darwin

Papa desires his love to the Boy Tactus and the man Daggy

My dear Charley— Your message to me in your letter to Catty promising to write to me in such a good hand, is so tempting that I shall certainly send you a full account of my Ball on Thursday & all the particulars of my flirtation with my partners particularly yr friend *Snortamalus* who I saw to day in great force.— Caroline's amusing little anecdote at my expense on the other page I beg to contradict as it is *quite false.*

There is a little Book an account of "The loss of the Kent"[6] which is very interesting & I advise you to read it if you can meet with it.— Louisa & Clare Leighton dined here the other day we wanted you and Eras very much to help to romp & flirt with them— I assure you Clare and I talked very sentimentally about you & pitied yr hard fates very much at being so far away from us—

Good byee my dear old Charley | Susan Darwin

Postmark: Shrewsbury Jan 3 ⟨182⟩6
DAR 204.2

[1] Henry Johnson, who attended Shrewsbury School (*Shrewsbury School Register*), studied medicine at Edinburgh and, after receiving his M.D. in 1829, returned to Shrewsbury to practise.
[2] Probably Robert Herbert, Henry, and Edward Clive, of Styche, nephews of Robert Clive of India.
[3] Hugh Owen, Archdeacon of Shropshire, Edward Pryce Owen's father.
[4] Coleridge 1825.
[5] Thomas Moore, *Memoirs of the life of the Right Honourable Richard Brinsley Sheridan* (Moore 1825).
[6] *The loss of the Kent Indiaman, by fire in the Bay of Biscay* (1826).

To Caroline Darwin 6 January 1826

Edinburgh.
Jan. 6th | 1826—

My dear Caroline,

Many thanks for your very entertaining letter, which was a great relief after hearing a long stupid lecture from Duncan on Materia Medica— But as you know nothing either of the Lectures or Lecturers, I will give you a short account of them.— Dr. Duncan is so very learned that his wisdom has left no room for his sense, & he lectures, as I have already said, on the Materia Medica, which cannot be translated into any word expressive enough of its stupidity. These few last mornings, however, he has shown signs of improvement & I hope he will "go on as well as can be expected."[1] His lectures begin at eight in the morning.— Dr. Hope[2] begins at ten o'clock, & I like both him & his lectures *very* much. (After which Erasmus goes to Mr. Lizars[3] on Anatomy", who is a charming Lecturer) At 12, the Hospital, after which *I* attend Munro on Anatomy— I dislike him & his Lectures so much that I cannot speak with decency about them.[4] He is so dirty in person & actions.— Thrice a week we have what is called Clinical Lectures, which means lectures on the sick people in the Hospitals—these I like *very* much.—[5] I said this account should be short, but I am afraid it has been too long like the Lectures themselves.—

I will be a good *boy*, and tell something about Johnson again (not but what I am very much surprised that Papa should so forget himself as to call me, a Collegian in the University of Edinburgh, a boy.) he has changed his lodgings for the third time, he has got very cheap ones, but I am afraid it will not answer, for they must make up by cheating.— I hope you like Erasmus' official news, he means to begin every letter so.— You mentioned in your letter that Emma was staying with you, if she is not gone ask her to tell Jos.[6] that I have not succeeded in getting any [titanium],[7] but that I will try again. Tell Katty and Susan I shall be very grateful if they will write to me, it is so pleasant receiving letters; and I hope, although our correspondence has begun late, you will send me many more nice affecting letters about dear little black nose. Erasmus thinks I shall have more pleasure in seeing it than all the rest of the families put together. You seem to hold the same opinion with regard to my dear little nephew.— I want to know how old I shall be next Birthday. I believe 17, & if so I shall be forced to go abroad for one year since it is necessary that I shall have completed my 21st year before I take my degree. Now you have no business to be frowning & puzzling over this letter for I did not promise to write a good hand to you.

I remain your af— dear Caroline, | C. Darwin.

Love to Papa & tell him I am going to write to him in a few days—

Copy
DAR 154

[1] CD refers to Andrew Duncan, the younger. A more favourable estimate of him is cited in Ashworth 1935. For a summary of the sources and background of all that is known of CD's student days at Edinburgh, see Shepperson 1961.

[2] Thomas Charles Hope.

[3] The copy reads 'Lezars', which Francis Darwin changed to 'Sizars' (*ML* 1: 7). However, CD almost certainly referred to the surgeon John Lizars, later Professor of Surgery at the Royal College of Surgeons, Edinburgh, author of Lizars 1822.

[4] In the *Autobiography*, p. 47, CD says of Alexander Monro, the third: '[He] made his lectures on human anatomy as dull, as he was himself, and the subject disgusted me. It has proved one of the greatest evils in my life that I was not urged to practise dissection, for I should soon have got over my disgust; and the practice would have been invaluable for all my future work. This has been an irremediable evil, as well as my incapacity to draw.'

[5] Little manuscript material survives from CD's student days at Edinburgh. Some of his lecture notes of 1825 − 6, on medicine and chemistry, are preserved in DAR 5, which also contains an account of a zoological walk to Portobello. DAR 129 contains a diary of zoological and botanical observations made in 1826. A notebook in DAR 118 of zoological observations made in March 1827 records his observations on *Flustra* and *Pontobdella muricata* which he reported to the Plinian Natural History Society of the University of Edinburgh on 27 March 1827 (not 1826, as CD states in the *Autobiography*, p. 50). See *Collected papers* 2: 285 − 91 for a complete transcription by P. H. Barrett of the notebook, and Ashworth 1935, pp. 102 − 5, for an account of the Plinian Society meetings and CD's membership, drawn from the minute-books of the Society.

[6] Josiah Wedgwood III.

[7] A blank space was left in the copy after the word 'any'; 'titanium' was inserted by Francis Darwin.

From Catherine Darwin 15 January [1826]

Overton—
January 15th. | Sunday Morning.

My dear Charles,

Since you are so very flattering as to say you wish to hear from me, I am sure I can do no less than send you an account of our proceedings lately, tho' I am afraid there is not much to be told— Caroline sends you her love and thanks for your very nice and agreeable letter.— Your description of the Lecturers is not very promising, but I hope Dr Duncan will continue to improve— Caroline and I have been spending a very pleasant week at Woodhouse, (where C— was painting) and yesterday we came on here to pay Marianne a *pop* visit, (which is her *pet* delight) till tomorrow, when we return home— The first object I saw here was Spark, grown considerably fatter, and most comfortably established quite at home; but what will you say for your *child*, when I tell you that she growled and snarled at us, and had even the impudence to snap Caroline's finger! I think when you hear this, you will be almost staggered in your belief that she will recollect you; I must say, I thought her indifference to me particularly unkind, considering how much *I* always preferred her to *your* favourite, *the old Duchess*.— The name of "Shelah, Shelah" produced a very visible effect upon her, as she pricked up her ears, and looked excessively puzzled— Dr Parker is very fond of her, and she is privileged to go into any room in the house, that she pleases.— Now that I have given you this long and minute account of Spark, perhaps I may tell you what a little fat, merry, rosy thing your little nephew is become—

What bitterly cold weather we have had lately; it is almost impossible to keep warm an inch from the fire; and you must be still worse off at Edinburgh— I had very good fun on the ice at Woodhouse; Francis and Arthur are at home now, as it is the holidays, and they drew the girls and myself about on sledges; and with a helping hand, sliding is very pleasant.— *I* was *particularly* unfortunate in getting innumerable tumbles, which I assure you caused no few peals of laughing, which was not a little provoking to *me* as *I* lay *sprawling* on the ice— Have you been skaiting at all? The frost you were wishing for, to dry the walks, has come at last, in good earnest, and it is prophesied here, will last for ever— William Owen is at Woodhouse now; he had been over to Ireland to get his hunters, and returned, just as we left it on Saturday— Do you know a Miss Monro, in Edinburgh? The Owens are going to play William a practical joke; i.e. he pretended to them that he could not endure Edinburgh or its amusements, when he was there, with his regiment; but they have discovered from Edward Clive, (who danced with Sarah at the Ball, & who is a great friend of William's) that he was passionately in love with a Miss Monro, and that he was the only officer who went out in Edinburgh; but that he was alway⟨s⟩ out, with Miss M.— The Owens mean to forge an invitation to William from a neighbouring place, mentioning that some Scotch friends are staying with them, and amongst the rest, a Miss Monro; and the Owens hope that William will ride over to call upon Miss M., and when he arrives there, will find how he has been *bamboozled*.— I am sure, dear Charley, you need never apologize to *me*, for writing a bad hand, for I shall give you great credit, if you are able either to read or to understand this long story.—

I never saw such merry, agreeable girls as Fanny and Sarah are; talking so easily and naturally, and so full of fun and nonsense— They are very much admired, and get plenty of partners at the Balls; but they are not at all more reconciled to England, and are longing to return to France.—

I forgot to tell you that I saw your friend Major Bayley, not long ago, who desired to be *most particularly* remembered to you; he is quite well again.—

We see a great deal of the Leightons; Susan has been paying them daily visits in her solitude; Clare is gone off in pleasantness I think; she is grown more affected, and full of *manner*, and *very* pert to Louisa— Frank[1] is at home now; I like him very much, he is so merry and goodnatured and has lost the finery he used to have.—

I *know* you will be 17, the 12th of next month.—

I dare say some of us will be writing very soon to the Wedgwoods, and then we will send your message to Jos.—

Goodbye, my dear old Boy; write to me soon; and believe me to be | Ever yr affectionate | E. Catherine Darwin.

My best love to Ras.—

Postmark: JAN W 18 E 1826
DAR 204.2

[1] Francis Knyvett Leighton, not to be confused with the Francis Leighton in the letter from E. A. Darwin, 14 November 1822, also of Shrewsbury.

To Susan Darwin 29 January [1826]

Edinburgh.
Jan 29.

My dear Susan

The whole family have been so very good in writing to me so often that I do not know whom to begin to thank first, so to save trouble I return my humble thanks to you all, from my Father down to little Katty.— The Gaieties of Edinburgh are now just beginning, last week there was an Assembly, & shortly there will be another. Erasmus & ⟨ ⟩ intended to have gone to the first, but mean to make it up by going to the next.— We also have been very dissipated.— We dined at Dr Hawley's on Saturday, & had a very pleasant party, after which we went to the Theatre, with a Mr Greville I believe a relation of the great Botanist, Dr Greville.[1] Dr Hawley has procured some information about my Fathers questions, & will write it shortly to him. Next Friday we are going to the old

D.ʳ Duncan,[2] & I hope it will be a pleasanter party than the last; which a very specimen of stupidity. What an extraordinary old man he is, now being past 80, & continuing to lecture. D.ʳ Hawley hints that he is rapidly failing. I have been most shockingly idle, actually reading two novels at once. a good scolding would do me a vast deal of good, & I hope you will send one of your most severe one's.— What an entertaining book Granby is;[3] do you remember Lady Harriet talking about inhaling ⟨Ni⟩tric Oxide? Johnson has actually done it, & describes the effects as the most intense pleasure he ever felt. We both mean to get tipsey in the Vacation.—. The old M.ʳ Wedgwood, I see in Ure's Chem. Dic.,[4] did nothing else but hold his nose & kick. It occasionally brings on fainting. Erasmus knows a man in Cambridge, who when in that state had the faculty of hearing, but not of motion or speech & to his horror, heard them consulting whether they should open the Temporal Artery.— Poor Eyre[5] is most danger-ously ill of a Fever, a friend of his has written to Erasmus stating that Eyre has a great wish to see him.— Erasmus is in a state of great perplexity what what to do, since he could not set out directly on account of not having sufficient money & by the time he could get there the fever, in all probability would have terminated ⟨one w⟩ay or another. So he has written to his ⟨ ⟩*[ther]* to find out some more particulars. ⟨We⟩ met when we dined at D.ʳ Hope's, a M.ʳ Stuart, whose father was twin brother to Lord M*u*rray or M*o*ray I suppose this can be no relation to the Lady Jane Stuart My Father mentioned.— Would you also tell my Father, that I should be much obliged if would pay Wilding's bill, & am quite ashamed that I forgot it myself. I am going to learn to stuff birds, from a blackamoor[6] I believe an old servant of D.ʳ Duncan: it has the recommendation of cheapness, if it has nothing else, as he only charges one guinea, for an hour every day for two months.

I rem. n— | Charles Darwin

Catherine mentions that D.ʳ Parker is very fond of Spark I hope he will not forget, that she is Erasmus', & that he returns at the end of April.—

P.S. This letter was written & sealed & I have reopened it, to acknowledge & thank for the monies. Good Bye—

Postmark: JAN H 30 A 1826
DAR 92: 3–4

[1] Robert Kaye Greville.
[2] Dr Andrew Duncan, the elder.
[3] Lister 1826.
[4] Ure 1823, p. 598. The article on 'Nitrogen' quotes from Davy 1800, but the reference is to Thomas Wedgwood, not, as CD supposes, to Josiah Wedgwood I of Etruria (see B. and H. Wedgwood 1980, p. 112).
[5] Samuel Parr Eyre died 1 February 1826 (*Shrewsbury School Register*).
[6] Identified by R. B. Freeman as John Edmonston, a freed slave who had been taught taxidermy by the naturalist — traveller Charles Waterton (Freeman 1978–9; Waterton 1825, p. 153).

From Marianne Parker [23 February 1826]

[Overton]

My dear Charles

I think this is the first correspondence we have ever had together & it will sound odd to begin my letter with telling you that I never in my life was more sorry to write a letter than I am to write this one—but it is the case— I believe you know that we begged to have poor little Spark for a short time 'till we found another, but I dont think you were told that the day after she came to us, she ran away, & though we made every possible enquiry & search we heard nothing of her for a fortnight when we found her established at a gentlemans' house in this neighbourhood & the grand pet of all the family— We got her back again, but we soon perceived that she had during her absence from us become with pup— We were very sorry for it, as we knew you did not wish it. Last Monday the poor little thing was taken ill, & after the birth of one puppy she died— You cannot think how sorry we have all been about it— Every body in the house had got so fond of her, & she was such a nice little Dog. I hope you will write to me my dear Charles—for though I have been very sorry for poor Spark's death on her own account, I have been still more so on yours, & altogether it has vexed me more than any thing that has happened for a long time— Shelah is going to have a family, but I am afraid they will not replace poor little black nose—

I should have written to you at once, but I thought they would have told you from home⟨.⟩ Dr. Parker desires me to say that he has now got a little black terrier puppy of a particularly good breed about the size of Spark, & that you are very welcome to her if you would like to have her.

I must just tell you that your little nephew & namesake Bobby is very well & the nicest little man that ever was seen. You must remember that you have promised to be fond of him when he is two years old & upon those conditions I will excuse you from caring about him now—

I was so sorry to hear of poor Eyre's death— I had a letter from Miss Congreeve[1] whose governess is one of Eyre's sisters & makes enquiries about his illness &c.—supposing that Erasmus was at Cambridge— Are there any brothers left, or was poor Eyre the last. Did Erasmus regret very much that he was not with him during his illness—

I want to know how you both like Edinburgh— I suppose you like it better than the Drs[2] Eras not so well as his dear C. C— C—[3]

Good bye My dear Charles— My best love to Eras: & hoping to hear from you soon. Yrs. aff: | M P—

Postmarks: Overton 175; Shrewsbury FE 23 1826
DAR 204.2

[1] Sarah Adelaide Congreve.
[2] Dr Butler's, i.e., Shrewsbury School.
[3] Christ's College, Cambridge.

From Caroline Darwin [27 February 1826]

[Shrewsbury]
Monday

My dear Charles

It is some time since we have had a letter from you, & I want to hear whether you have been blown away in any of the stormy nights we have had latterly—for if we feel the storms of wind, I should think you would in a house with an uncountable number of stories & in Scotland moreover.— Mr Young[1] has been acting here in his way to Edinburgh I should very much liked to have accompanied him & paid you a visit, for I want very much to see you & Eras again. it seems particularly odd being so long without your running up once from School— Mr Young has not at all rivalled Macready in my affections though I admired his acting very much, there was a better house for his benefit than I have seen very long in Shrewsbury. A great many of Doctor Butlers young gentlemen patronized him & M^r and Mrs Iliffe—the latter is a prettyish, *pinkish*, primish, looking young lady, & had a proper conscious bride like look all evening—& I dare say flattered herself that she and Mr Young divided the attention of the house— There was likewise a new actress who sang a duet with her lover the burden of which was "sweet little Barbara, sweet, sweet, sweet, little Barbara" I quite longed to tell her it would have been much more modest if *she* had not joined in that part of the song but left her lover to sing it alone— I heard that 7 boys have run away from the schools already this half year, I do not think Dr Butler will much like that.—

You will have heard from Marianne of poor little Sparks melancholy death— I am rather frightened for Shelah. I find she is going to have puppies & is so very large, that I shall be very glad when she has them & is well.— I am sure poor Sparks death will be a sad grief to you, independant of the sorrow for the little dog herself— I do not think you will understand me but I do not know how to express myself clearly—

Susan has been staying at Maer some time she will come home I believe tomorrow week— We are surprised that your friend Archdeacon Corbet[2] has not called an antislavery meeting in this town—he has always done so before whenever there have been meetings held on the subject & it is strange he should not now—[3] The Assizes[4] will be the week after next. The Owens of Woodhouse come to us for it— Harry[5] also will make the circuit for the first time— I wish he may get a brief but I do not think there is much chance this time— We have been very busy in the flower garden, planting sweet peas &c. I flatter myself it will look much gayer this year than it did last—that I know **you** will think it may easily do, I have remembered your admiration of the Holyhocks at Maer & have been buying some, so that at least we will not be outdone in that flower— We are going to have pipes laid to have a supply of water in the flower garden, so next summer your goodnature will not be so often taxed with, "Charles it is very hot." ("very hot indeed") you unthinkingly answer) "Dear Bobby, the

ground is so dry that the pans of water you brought half an hour ago did hardly
any good, would you bring one more?"

Papa pays frequent visits to the garden to see a *Leucojum vernum* which is now
in blow & rather a rare plant. Have you ever seen Battys engravings of Germany
and France,[6] the buildings must be very fine in the former place—

My best love to Erasmus & Believe me Y^rs affec | Caroline Darwin

Papa sends his love to you both—

Postmark: Shrewsbury FE 27 ⟨182⟩6
DAR 204.2

[1] Charles Mayne Young, a popular comic actor who also played tragic roles with great success; he
appeared in *Much ado about nothing* at the Theatre-Royal, Edinburgh on 4 March 1826.
[2] Joseph Corbett.
[3] An Anti-Slavery Society had been established in 1823 by William Wilberforce, Thomas Fowell
Buxton, and others, to work towards the abolition of slavery in the British colonies.
[4] Assize week brought families to Shrewsbury from all over the county and was the occasion for many
social events.
[5] Henry Allen Wedgwood was called to the Bar 10 February 1826.
[6] Lieutenant-Colonel Robert Batty, amateur artist, published *French scenery* (1822), *German scenery*
(1823), *Scenes of the Rhine, Belgium and Holland* (1826) 'all of which have been much esteemed'
(Williamson 1903 – 5).

From E. A. Darwin [9 March 1826]

[Glasgow]

Dear C

You asked me if I had time in plenty to write from Glasgow, & time I have
in plenty enough to write more than ever you could read. The Steam Vessel does
not sail till the middle of Friday so that I have two days & a half complete to
spend here & nothing in the world to do with myself. I shall in consequence give
you a most detailed account of the journey. I got into the Tun a little before nine
o'clock. There is little or nothing to be seen the whole way at least I saw nothing
being confined to the cabbin from the rain which more or less was almost
incessant. They have the providence to put books in the Cabin & in the course of
the day I read a volume of Guy Mannering of Mid Lothian,[1] L'Amie Inconnue
The Lounger & several No's of the Monthly Review[2] & also divers Newspapers.
They give a very decent breakfast & at one 'oclock I got some biscuits &
'decoction of what do you call it to wit porter which served for the first boat.
The canal goes thro' a mountain by means of a tunnel ½ mile long & which has
quite a beautiful effect. It then by means of a vast number of locks goes down
a Hill, but the passengers get out & walk down to the second boat. I was then
the only Cabin passenger, the others having dropt away one by one.

About four dinner. Cold veal, ham & beef ham & roast potatoes & some most
excellent bottled Ale which I recommend to your attention. Another passenger

now joined us & when it grew dark we played at backgammon & drank some toddy till we arrived. I got a porter, took his number & left him to bring my luggage up at his leisure to the Eagle, in Maxwell St. which cost 1/6. The Steam Vessel is ye Henry Bell, from Greenock. The Sovreign Steam Boat goes from the Broomielaw at Glasgow to Greenock at 12 noon. You take your passage at Laird & Co 25 York St Glasgow.— This is all copied from one of the papers. Another vessel I believe sails on Tuesday but I do not know any thing about it. I found my Backgammon friend very useful as he told me all about the vessels & shewed me the way to the inn.

I went to the college in High St which is a very handsome old building & heard Thomson[3] from 10 to 11. His room is not bigger than Alisons[4] & has not as many pupils I think & altogether a most vile turn out. He Lectures in the most singular manner about two words & then a pause sitting quite quietly in his chair. Most of the students wear red gowns & are a step worse than the Edinenses. I shall not try for Ure[5] but the Andersonian Institution[6] is in John St. I believe I saw the D^r himself but am not the least sure.

Valeas. Read & Burn.

I see a concert advertised here the performers Miss Paton Mr Thorne! Miss Noel!!— Miss Dyer came from Glasgow originally.

Postmark: Glasgow 405 9 MAR 1826
AL
DAR 204.1

[1] *Guy Mannering* was published anonymously in 1815. Sir Walter Scott's authorship remained a closely guarded secret until 1827.

[2] *L'amie inconnue* has not been identified. *The Lounger: a periodical paper, published at Edinburgh in the years 1785 and 1786* (by H. Mackenzie and others); Erasmus probably read one of the editions in two or three volumes that were issued in later years. *The Monthly Review (or Literary Journal)* (1749–1825); new series, London (1826–8).

[3] Thomas Thomson, Regius Professor of Chemistry at Glasgow University.

[4] William Pulteney Alison, Professor of 'Institutes of Medicine' at Edinburgh University.

[5] Andrew Ure, editor of *A dictionary of chemistry*, was Professor of Chemistry at Anderson's Institution, Glasgow.

[6] Anderson's Institution, founded in 1796, later affiliated with the University of Glasgow. In 1964 it was given a charter as the University of Strathclyde.

From E. A. Darwin [9 March 1826]

[Glasgow]

Dear C.

Of course you will be *delighted* to pay the postage of another note, being a second edition of my stupidity.

I have given up my intention of going by sea, & I was meditating to send you such an account of my not being sick but for want of something just at present to tell you I am obliged to tell you the truth, which is this.

I went to the office to get my place when I learnt that the vessel would not sail till Saturday & not then for certain because of the uncertainty of the weather: the man assured me it would not signify for if the vessel set off on Saturday we should be in Liverpool by Monday morning, which is rather too much of a good thing. I have sent my box to go by sea & it does not require to be directed to any person in Liverpool: If you also should alter your mind I dare say you can send it from Edinburgh directed in the following kind of fashion.

C. D.
Laird & Co 25 York St. Glasgow
By the Henry Bell Steam Boat
Liverpool
To be sent on to
Shrewsbury

I went this morning & dawdled an hour or so over the Hunterian Museum[1] which is well worth going to being only a 1/. & nobody to bother you by lionising. There are most splending *foreign* birds Birds of Paradise Humming Birds which quite dazzle the eyes. English Birds Lions &c, 4 ornithrync⟨hi⟩ numbers of snakes, min⟨erals⟩ shells, anatomical prep⟨ ⟩ pictures, &c. &c. &c. &c.

The Students here actually play at foot ball within the precincts of the college: you never did see such a set set since you was born & please God never again.

I intend travelling the same route as we came bye indeed there is little choice without a considerable circuit.

Good Bye once again.[2]

Postmark: Glasg⟨ow⟩ 9 MAR 1826 405
AL
DAR 204.1

---

[1] The bequest of William Hunter to the University of Glasgow. The collection included geological, zoological, and ethnographical specimens, medical preparations, a library, manuscripts, coins, and works of art.
[2] Instead of a signature, there is a scrawl resembling a leaf or feather.

From Marianne Parker [*c.* 13 March 1826]

[Overton]

My dear Charles

I have a great mind to begin my letter to you with as severe a scolding as you gave me— how could you for a sensible young man who has left School & settled at Edinburgh, think of finding fault with a person for not directing a letter properly, & then neglect to give them the right direction— I could not have answered your letter 'till I had written to Shrewsbury to enquire how I ought to direct if Caroline luckily had not come over on Monday to stay a day with me— She desired me to put in M^rs. Macky's name but did not tell me what number in Lothian St—but my last letter found you so I have not many fears for this—

I will tell you every particular I can think of relating to poor little Spark. She ran away from us very early in the morning the day after she came here She had been shut up as Mark said he did not think she would bear tying up, & sombody in going out of the kitchen where she was, had left the door open. She made her escape before we were up but as soon as we knew of it, we sent a great many people after her, & offered a guinea reward to whoever could find her, the consequence of which was we had many little black & white dogs brought us but no Spark—& for a whole fortnight we could learn nothing of her—when one day D^r. Parker met D^r. Wynne the Rector of Overton & Bangor who said that he had heard we had lost a dog, & that, his servant had found a very pretty little black & white one on the road about a fortnight ago, which he had no doubt was ours as it proved— I am sorry to say none of her puppies were born alive, there was but one born & that was dead. She did not seem to suffer much pain but was restless & uneasy for two days before she died The puppies were too large for her was the cause of her death— I am glad you did not wish to have D^r. P.'s little puppy as it also is dead— We are so unlucky in puppies, that I hope we never shall have any more— I am very busy with my flower garden which I am altering & improving very much, but I am grived to find all the labels Erasmus wrote for me are washed out, & I am afraid I shall not get him to christen any more.

Baby's Mine & D^r. P's love to you & Eras | Y^rs aff. | M P.

I hope to hear from you again my dear Charles as I sh^ld be very sorry if our correspondence stops here

Postmarks: Overton ⟨   ⟩ 17⟨5⟩; MAR H 15 ⟨E⟩ 18⟨2⟩6
DAR 204.2

From Caroline Darwin [22 March 1826]

[Shrewsbury]

My dear Charles

I hope you are a little reconciled to Erasmus's absence by this time & I am very glad to to think you will not now have much more solitude. it is a great pity that Eras thought he should dislike his having acquaintances in Edinbourgh & so refused to take introductions. Experience, Eras tells me, he thinks will make you wiser than he was, & that you will not object to letters of introduction when you return to Ednbourgh. We have had a very merry gay week which I will give you an account of— On Wednesday Harry arr'd & on Thursday Hensleigh & the Owens of Woodhouse. Hensleigh has had his head shaved & we were puzzled now that his hair has begun to grow again & his wig not very new, whether he would look best as a new born babe, with a bald head, or with an old greenish wig too small for his head, the latter was determined upon contrary to my opinion, the consequence was, one eving we were all playing at cards, Hensleigh very eager, put forth his head & hand, so the wig fell back, Hensleigh twitched it & thought he had replaced it properly, but the wig was obstinate & sat on the point of his head the rest of the night, much to Fanny Owen's amusement. On Thursday the Leightons & W. Owen dined & the tag rag had a very merry eving till 12 oclock, cards, dancing &^{tc} the next day the account came of Mrs Lethbridge's (neice to Mr Owen) death & the girls were not allowed to go into court which was a great grief to them, but they rambled about the town all day. We had a large party to dinner—16. the Smythe Owens Mr Pelham & 2 lawers, a friend of Harrys & a friend of the Owens, we all played at cards again in the evening. the next day, Saturday—the girls again went rambling about the town longing to go into court. they met Mr Owen & after some entreaty he consented to take them in, but they must not tell Mrs Owen, so in they went— mean while Mrs Owen who was also about the town met first one gentleman then another who said "how long have you come out of court." "I have not been in at all". "Nay, is that possible I have just seen you or yr daughters" Mrs Owen when this had been said several times suspected some one had taken in the Miss Owens, & meeting Mr Owen told him what she thought, he pretended to go into a passion & said, "What my daughters in court, impossible, nonsense"— at last she found out all & was not very indignant.— they all went on Saturday. On Sunday after church Harry went by the Mail on his circuit. On Monday we had a party of Erasmus's friends—Hildyard, Wakefield, Wingfield & Vaughan,[1] I liked them all except the latter—whose manners I thought as bad as many of his opinions—he evidently rather laught at the suffrings of the poor slaves & admires some book which John Bull quotes which is I suppose taking the slave holder's part—[2] Moreover, he seemed ashamed of doing his duty as a clergyman, & said *packs of cards* he should be more likely to purchase than bibles & prayer books, & rather implied all *very* religious people were hypocrits, &c. I sat between him & Hildyard who I like very much & who when he saw I would not answer Vaughan when he laughed on religious subjects immediately joined in the conversation & so pleasantly that he quite confirmed my liking to him.—

dear Charles I hope you read the bible & not only because you think it wrong not to read it, but with the wish of learning there what is necessary to feel & do to go to heaven after you die. I am sure I gain more by praying over a few verses than by reading simply— many chapters— I suppose you do not feel prepared yet to take the sacrament—

it made me feel quite melancholy the other day looking at your old garden, & the flowers, just coming up which you used to be so happy watching I think the time when you & Catherine were little children & I was always with you or thinking about you was the happiest part of my life & I dare say always will be— Eras. is gone today with Hensleigh to pay Marianne a visit at Overton— Marianne will be very glad to see him

I am very sorry for your disappointing about the puppy it is a frightful fat little thing—

Good bye dear Charley | Y^r affec sister | Caroline Darwin

do not when you write allude to the last part of my letter about Mr Vaughan &^{tc} it would seem so spitefully gossiping to write what he said so many miles—

Postmark: Shrewsbury MR 22 ⟨182⟩6 153
DAR 204.2

[1] Frederick Hildyard, John Mort Wakefield, Edward John Wingfield, and (probably) William Vaughan. All attended Shrewsbury School with Erasmus and CD. In 1826 Hildyard and Wakefield were Assistant Masters at Shrewsbury.

[2] *John Bull*, a popular Tory weekly, notorious for its disregard of the laws of libel, was founded in 1820 by Theodore Hook (Andrews 1859). It eventually became more restrained in its journalistic practices and, by 1859, was more representative of the Conservative press.

From Susan Darwin [27 March 1826][1]

[Shrewsbury]
Monday Evening

My dear Charley—

I was very well pleased to receive your last letter as I hope you are getting in better humour with Edinburgh now that Spring is come.— Do go some day to see Roslyn Castle, which I believe is within a few miles from Edinburgh, & it is very well worth seeing— At present they have made a Diorama of it, in London.— My reason for writing so soon is, that I have a message from Papa to give you, which I am afraid you won't like; he desires me to say that he thinks your plan of picking & chusing what lectures you like to attend, not at all a good one; and as you cannot have enough information to know what may be of use to you, it is quite necessary for you to bear with a good deal of stupid & dry work: but if you do not discontinue your present indulgent way, your course of study will be utterly useless.— Papa was sorry to hear that you thought of coming home before the course of Lectures were finished, but hopes you will not do so.—

Erasmus has seen a great deal of Hildyard since his return. the latter dined here yesterday. he seems a very merry sort of person— Price is come to take his place of master— I fear I can give you very little School gossip. I can only inform you that I never go into town without meeting your two School Beauties (Watkins & Turner)[2] arm in arm parading the streets of Shrewsbury— I shall be very happy to renew my acquaintance & *flirtation* with Massie this summer. I wish it was in my power to do you a good turn by inviting the Miss Massies.

We have been making Erasmus very useful, taking him to Doctor all our sick poor people. he goes pretty often to visit poor Mrs O. Jones who has been quite confined to her bed for some time. the other day when he was there, she gave him an old Guinea wrapped up, saying "she wished very much to give him his first fee, & hoped he would keep it as a love token from her"— This little mark of friendship & kind manner of doing it pleased Eras very much.

I am very glad to hear you are becoming a Politician[3] & I shall be still better pleased when I hear you are studying "The Morning Chronicle" in place of *odious* John Bull: what is the use or pleasure of reading a paper which everybody abuses & allows to be so unfair? I cannot tell you much about Mr Hume[4] except what I daresay you know, that he is a Scotch Member & a great person for reforming all money abuses & taxes in Government &c.— If you can get the Edinburgh Review for last November, do read "Lord Suffield on the Game Laws" which is one of the articles in it, and which we all like extremely.—[5]

We all get up early & are good & industrious. Eras is very pleasant & affectionate. He has been staying three days at Overton, & came back very full of his little nephew's charms, allowing that he had no idea babies ever were such nice little creatures.

We have written to invite a great many Galtons[6] to come to see ⟨u⟩s next Month—& we are expecting Eliza & Jessie Wedgwood. I don't know how the house will hold them all.—

Puppy gets so fat, he can niether *stand or go*. we had an alarm that he was poisoned last week, by some paint, as the House & premises are infested with a troop of little painters with Mr Pierce at the head of them, who are painting the windows &c &c. Caroline scolded all the painters round & afterwards found her alarm was false.—

I am so sleepy I must wish you Good night | ever y^r affect | Susan

Remember to write *slow* and then you will form each letter distinctly.—

We have had a nasty East wind for the last ten days.—

Will you in your next letter answer the message about a Book & Constable[7] which Erasmus gave you in his letter—

Don't waste your time by going to the Play as it must prevent your getting up early, or attending to y^r studies.—

For this next month devote yrself to wisdom & you will be much happier.

Postmarks: Shrewsbury MR ⟨182⟩6; MAR H 30 E 1826
DAR 204.2

[1] Dated 27 March from reference to Erasmus's having returned from Overton (a trip on which he started 22 March) and from postmarks.

[2] James Farley Turner, Shrewsbury schoolmate. He and CD renewed their friendship at Christ's College, Cambridge, in 1827. In the *Autobiography*, p. 67, CD refers to him as 'one of my sporting friends'.

[3] No record of any political activity by CD has been found.

[4] Joseph Hume.

[5] Edward Harbord, 3d Baron Suffield. His book (Harbord 1825) advocated changing the game laws to permit owners to sell their game. Poaching would then become theft and public opinion would support enforcing the law. The review appeared in November 1825, pp. 242 – 62. *Wellesley Index* 1: 467 lists its author as Henry Brougham or perhaps Sydney Smith.

[6] The Galtons were close relatives and friends of the Darwin family. Violetta Darwin, a daughter of Dr Erasmus Darwin's second marriage, had married Samuel Tertius Galton. They had four daughters and three sons, one of whom was Francis Galton, later to become a close associate and correspondent of CD.

[7] Probably Archibald Constable, the Edinburgh publisher.

To Caroline Darwin 8 April [1826][1]

[Edinburgh]
April 8th

My dear Caroline

I dare say I shall not be able to finish this letter, but I cannot help writing to thank you for your very nice and kind letter. It makes me feel how very ungrateful I have been to you for all the kindness and trouble you took for me when I was a child. Indeed I often cannot help wondering at my own blind Ungratefulness. I have tried to follow your advice about the Bible, what part of the Bible do you like best? I like the Gospels. Do you know which of them is generally reckoned the best? Do write to me again soon, for you do not know how I like receiving such letters as yours. The weather has been very pleasant for these few last days, but nevertheless I long to return very much. D.ʳ Hope has been giving some very good Lectures on Electricity &c. and I am very glad I stayed for them. The Classes are beginning to thin. I think I shall stay about nine days or a fortnight longer. But as my money will not last out for the latter period I should be much obliged if my Father would send up a £5 or £10 pound bill. I should also be much obliged if it could be sent up **directly**: as you will not receive this letter for three days, and I get the answer in another three.

I hope Eras. has got his books safely, let me know in the next letter, as I mean to go as far as Glasgow by the canals and from thence on "terra firma" to Shrewsbury, but sending my books per sea. How shockingly untidy this letter is, but I think you have a sympathy for untidiness, not that I mean to say that yours are half as bad as this.

I will send my next and last John Bull to Shrewsbury for your edification. I hope you received the last and studied it well. I said in the beginning of this very tidy epistle that I never should be able to finish it, which is now verified.

Love to all. Do not show this. | Your affect., Dear Caroline | Charles Darwin

Nota Benè. I do not so very much care about the money, as what I have will last for nine days: not long after which I hope to be in Shrewsbury. Anyhow let

somebody write soon that there may be no mistake, as I will if I receive the
money. Direct as formerly to *Mrs. Mackays, 11 Lothian St.*

Copy
DAR 154

1 The copy is dated 1828, but the letter is clearly an answer to the letter from Caroline Darwin, [22
  March 1826].

From Catherine & Caroline Darwin   11 April [1826]

[Shrewsbury]
Tuesday Evening— | April 11th.—

My dear Charles,

For fear you should entirely forget that there is such a person as myself in
the land of the living, I will begin an epistle to you, though our correspondence
really has come to such a dead stop, that it requires some courage so to do.— I am
very glad that we shall be able to *renew our acquaintance* in so short a time. I long
to see you again, my very dear Bobby, and I hope we shall have some very nice
walks and talks together again.— You will have heard from Erasmus that
Eliza and Jessie Wedgwood are staying with us now, and we expect Uncle John
and Aunt Jane on Monday— I should like you to see Jessie again; I am sure you
would like her; she is so very merry and pleasant, and I think very pretty—

We have all been taking to gardening very vigorously, and shall expect some
very elegant compliments from you on its beauty; and I assure you it is very
gay, and much admired.—

Erasmus' Friend, Mr Pryce (who is come to be a Master for three years at the
Schools) and 2 other unfortunate Masters Mr Wakefield, and a Mr *Hubble-
Bubble*[1] or some such name, dined here yesterday— We all liked Pryce; he seems
very gentlemanly and agreeable.— it is a pity he should *stupify* for so long a time
"teaching the young ideas how to shoot."— I am reading your favourite
Southeys Peninsular War, and like it exceedingly— Do you remember the account
of the seige of Saragoza?[2] one cannot conceive any thing more dreadful— The
2d Volume is come out, and we are going to order it in the Society; it really is the
greatest labour of Hercules I ever undertook.— I am glad you are become a
politician, and hope you will teach me a little. Susan is hotter than ever about
Slavery— John Bull pretending that the Slaves lead a life of *Comfort* and *Happiness* really seems as if he was quizzing the subject.— Little Harriet Perkins
is dead, which is a very happy release from all her sufferings in my opinion.—
We make Erasmus very useful by taking him about among the Poor; he has
acquired a very high reputation for doctoring them, and has got a great many
Patients.— I suppose you have heard what a gay party we were at the Assizes,
and how capitally Harry flirted with the Miss Owens— Fanny Owen has quite

the preference to Sarah among all the gentlemen, as she must have every where; there is something so very engaging and delightful about her.—

Shelah's puppy is really quite a sight to be seen; it is as broad as it is long, and Shelah also has been sadly corrupted, and nearly rivals her child in fat— That savage Turk is gone from Overton to my great joy; I think he was as much your abomination as mine.— Archdeacon Owen has got St Mary's in Mr Blakeway's place;[3] it is not yet known who will have St Julian's.— Caroline wants the rest, so I must stop— Good Bye my dear Bobby. | Ever yrs affectionately | E. Catherine Darwin

Thank you dear Charles for your very kind affection⟨ate⟩ letter, you can not tell how much I value your love— I am very glad we shall have you at home again soon. & Papa is very glad that you have remained to attend all the Lectures, as he is sure they will be useful to you. he wants to know if you are thinking of any excursion in your way home to see any new country that you did not pass through as you went to Edinburgh— Papa is rather surprised at your going to Glasgow as there is nothing worth seeing in your way there, & he should have thought it a better plan to leave as much of your heavy luggage as you can in Edinburgh as you will return there & so not have the plague of the carriage of it backwards & forwards— shall you travel down with Johnson, or alone? Mr Galton[4] has sent you a letter of introduction to Mr Horner[5], a friend of his, & a very clever agreeable man, Erasmus tells me you will very gladly accept any letters of introduction—particularly to pleasant people like the Horners, & if you will call with it I dare say they will be very civil to you.

Papa's love & he has sent you £20—which he supposes will answer your purpose— Eliza & Jessie Wedgwood are staying here & desire their love to you— Uncle John & Aunt Jane come on Monday; yesterday 3 of Eras's friends dined here. Price, Wakefield & a man whose name we never could remember or pronounce better than Hubblebuble— There were at dinner some things which were called Mushrooms—but we suspected them to be *toadstools*—they were black, & tough & Jessie after eating some the day before has complained of *Vertigo* & stomach ache Eras allows to be signs of poison— These unfortunate young men Messrs Wakefield & Hubblebuble eat of the said black dish—in the evening they became silent, at ten oclock tottered out of the room, & no more has been heard of them.

But to change from nonsense to grave matters. I must say dear Charles how glad I am you have been studying the bible— I agree with in liking St Johns the best of the Gospels I am very fond of that short Epistle of St James, as well as St Johns— I often regret myself that when I was younger & fuller of pursuits & high spirits I was not more religious—but it is very difficult to be so habitually—

I think you met the Tollets[6] once at Maer. poor Gorgiana Tollet has had an abscess in her arm, & to save her life has been obliged to have it cut off, she bore the operation with out a scream or groan— Good bye dear Charles, I did ⟨no⟩t shew you[r] letter | Ever your affection⟨ate⟩ | Caroline Darwin

The plant is Ranunculus Ficaria?

Postmark: Shrewsbury AP 14 1826 153
DAR 204.2

¹ Nathan Hubbersty.
² Southey 1823, ch. 9, 'The siege of Zaragoza', pp. 398–424.
³ John Brickdale Blakeway.
⁴ Probably Samuel Tertius Galton.
⁵ Leonard Horner, geologist and organiser of Whig meetings in Edinburgh at this period.
⁶ The Tollets of Betley Hall near Maer were close friends of the Wedgwoods (see *Emma Darwin* 1: 52).

From John Price [July 1826]¹

Thursday

Dear Darwin

Many thanks for your friendly letter of yesterday w^h. I am, thank God able to answer favorably, with my own hand. I have no trace of my disorder but extreme weakness from depletion, w^h is also rapidly going—no one could be more speedy & I am satisfied the attack has cleared my constitution & made a new man of me. I have not felt for a day since Xmas, so well & clear as I do at this moment —farewell blue devils & indigestion & aching shoulders. I shall now take *extreme* care of my health. I wish you would tell your Father that I forgot to mention that I drank a pint of ale a day during those 3 months, when my health was

so *particularly good*, & w^h. he told me ought therefore to be a *pattern* for my manner of living: at the same time I do not in the least wish to *rebel* if he thinks malt liquor bad for me, but I have a prejudice in favor of *cwrw*² w^h. I have seen made over & over again whereas wine may be the devil's dam's milk, for ought I know: I am going to commence the *dinner pills*, having disbanded all other physic. Please to call at *Bisco's*³ for the prescription, & ask the cuckoldy knave what business he had to keep it, & send it me soon, or I shall be run out. I left my bed on Saturday, having suffered after you left far worse than ever, Stranguary⁴ being added to my other ailments from excessive blistering, & a very fair imitation of the clap it is, to tell the truth. I hope you hear good acc^ts of your sister & Stroll to whom give my best regards when you write. I hope you will be in Shrewsbury when I return— I never thought to see you again, being not far from death, as my poor Mother also thought, who tended me like a guardian angel night and day to the injury of her own health w^h. is however recovering fast with my own. We go to the sea side as soon as I can move. Buxton is too far. I am glad you derived some pleasure from your tour w^h I much wish I could have joined & should have been glad to assist you by speaking the British language: "*chattering*" I do not understand— I am in your debt for the

transplanter w^h pleases my Mother in theory vastly. She begs to be kindly remembered & is sorry you had such a comfortless visit, but you knew the cause & I hope will not be deterred from a second trial under better auspices. Present my best comp^{ts} to your Father & the s⟨is⟩ters & tell the former that I have disinherited my *undutiful child* & never mean to countenance him again— And do not forget to ask about John Barleycorn. I have shaved my skull & started a devilish smart crop wig, w^h makes my upper works very imposing, but my legs, my dear fellow, are deplorably reduced, & would qualify me to enact Trinculo[5] to a shaving. I talk of planting them in a box & watering them, but perchance roast beef, butter milk &c will make 'em grow quite as fast: not forgetting a modicum of cwrw if allowed— We expect my Father from his living tonight. I walk out every day & the view of these glorious mountains & smell of their air is life to me. I am tired & have in deed filled my sheet—

Believe me dear D. Yours v truly J. Price

DAR 204.3

[1] The references to CD's tour and Price's expected return to Shrewsbury make the summer of 1826 a probable date. According to his 'Journal' CD toured North Wales with Nathan Hubbersty in mid-June of 1826 (see Appendix I). Price was then an Assistant Master at Shrewsbury School.
[2] Welsh for 'beer'.
[3] Richard Briscoe, Chemist, Castle Street, Shrewsbury (Pigot's Salop directory 1825).
[4] Strangury: a disease of the urinary organs (*OED*).
[5] 'I'll pull thee by the lesser legs: if any be Trinculo's legs, these are they.' Shakespeare, *The Tempest*, 2. 2. 104−6 (Arden edition).

From James Farley Turner [28 July 1826]

Park Hall.— [Kidderminster]

Dear Charley

Having an idle hour or two on my hands before we sit down to dinner, I cant do anything that gives me more pleasure than in fulfilling my promise of writing a few lines to you before my departure for Shrewsbury— The Holydays are really come to a very narrow span, & the time will soon be here when I shall come back for the last time but one, w^h I am happy to say is finally fixed; so that one more speeches over & I shall vanish from thence. This hot weather I suppose you like most other people study to keep as quiet as possible. As far as regards myself however, I am obliged now to keep myself so, as last Sunday I fainted in Church, obliged to be carried out directly &tc, w^h soon brought me to my senses, an unpleasant thing I must confess, & by no means wish to experience the same thing again. Watkins wrote to me the other day & told me a good deal of news w^h would be uninteresting to you & therefore shall not relate it: by the way he has seen Hubberstie in Matlock, who is very well after his fatigues in Wales. A Brother of mine & myself do nothing but shoot Rabbits, w^h is no despicable

sport in the cool of the Evening: there are abundance, so that we manage to swell
out our pockets pretty well. What you do with yourself I cant conceive, unless it
be walk up & down that refreshing, cool (& a dozen other epithets) walk with
Shee. Have you been at the Woodhouse lately, I think you talked of visiting it
soon when I saw you last. I hope you will excuse so short & uninteresting an
epistle; but there being nothing in this part of the country that would be
interesting to you, in case I was to pen it, you must excuse it. I am not certain yet
whether I shall come on Saturday or Sunday, but most probably the former, so
that I may have the pleasure of spending some part of the ensuing day with you.
The Loxdales did not half like my coming up both evenings when I was staying, at
least this is what I have learnt since, much I care about their likings
& dislikings!!!—

I conclude with hoping this will reach you & yours safe—& believe me Your
sincere Friend | J. F. Turner.

Postmark: Kidderminster JY 28 1826
DAR 204.16

From E. A. Darwin　[29 September 1826]

4 Chapel St West. Curzon St.

My Dear Bobby,

From shere want of employment I write to tell you my plans about lectures[1]
which I have fixed much to my own satisfaction. Luckily I have none *very* early,
Dr Chambers[2] on Practice of Physick from nine to ten alternate days: as much
time as I like in the dissecting rooms till one. From 1 to 2 demonstrations by
Cæsar Hawkins[3] & from 2 to 3 Mr Bell[4] on Anatomy &c. In the evening on
alternate days Mr Brodie[5] on Surgery This gives me two hours one day & four
the next which I think I shall be able to go through with.

I am ashamed to say that I have already given up my thoughts of graduating
next spring, but it is by Dr Hollands[6] opinion who is just at present my oracle.

I find that John Bull & the Age[7] are published the same day & hold precisely
the same opinions only that ye Age is ten times the most scurrilous of the two &
I have therefore taken in that alone. Do you approve?

As soon as you get to Edinburgh if it does not prove too troublesome I'll match
the London climate against yours and see which has most drizzle. To make a fair
estimate we'll consider a London fog equivalent to a Scotch mist.

Liston[8] has of course been very popular amongst you rustics. I was at the Haymarket on Tuesday & heard several exclamations at the thinness of the house
in consequence of his departure. I was reminded of you also by a s⟨nu⟩ffy
old gentleman who in the most interesting part of the play unfortunately
dropt all his snuff and was obliged to leave instantly.

Tell Hubbersty I delivered his parcel and also ask Price if he knows Fred. H——'s address as I should be glad to have it and tell him to write to me if he is a Christian I have sold my insects for two thirds & am to have some of Smiths Botany[9] in exchange— I yesterday saw a cryptogamic work but instead of engravings the actual dried plants. They were as dear and very much inferiour to engravings in my humble opinion

Good Bye. | E. D.

Postmark: H SE 29 1826
DAR 204.1

[1] Erasmus had enrolled in the Great Windmill Street School of Anatomy, London.
[2] William Frederick Chambers.
[3] Caesar Henry Hawkins.
[4] Charles Bell taught anatomy at the Middlesex Hospital and the Great Windmill Street School of Anatomy. His *Essays on the anatomy of expression in painting* (Bell 1806) first aroused CD's interest in the subject (*Descent*, Introduction, p. 5). A copy of the third edition (Bell 1844), is in Darwin Library — CUL.
[5] Benjamin Collins Brodie.
[6] Henry Holland, a successful, fashionable London physician. A distant relative and friend of the Darwin family.
[7] *The Age* was founded in May 1825. It called itself 'the leading weekly Conservative journal' and rivalled *John Bull* in scandal-mongering and popularity. Publication ceased in October 1843.
[8] John Liston played regularly at the Haymarket Theatre from 1805 until 1830. His most famous part was that of 'Paul Pry' at the Haymarket, 13 September 1825.
[9] James Edward Smith's *English botany* (Smith 1790 — 1814) dealt with all known British plants except the fungi.

From E. A. Darwin 10 October [1826]

4 Chapel St West | Curzon St.
Oct 10th—

Dear Bobby

You must excuse this dirty outside sheet of paper which has been polishing a table for the last week as it is the only one that I have, & I write having nothing to say as you profess to like receiving any thing that pretends to be a letter.

At ye demonstrations this morning who should I see walk in but Smith the pleasant looking clerk of Graham[1] in the hospital, & at the Anatomical Lecture he came & sat by me and entered into conversation. He said that he was come to London & knew nobody & that knowing my face so well he felt quite an old acquaintance I said I felt just the same & so we struck up a friendship & went and dined together at a french house and afterwards went for a lounge to Armstrong[2] who gave I think the very best lecture I ever heard here he met with to his great joy an Edinburgh friend who also seems a very decent sort of person. Long ye friend of Hodson is I find Surgeon to the Lock Hospital & also that he was a pupil of Laenec[3] which may give him some right to give himself airs.

Smith & I did not quite agree about Dr Graham he called him a *charming* person, of all epithets in the world, but I made him agree with me in the detail that a great deal of his practice was bad & his manners harsh.

There certainly must be some radical difference in the London & Edinburgh lectures, for I can attend four hours in the day without the least weariness: the students are no comparison better behaved than on the other side of the Tweed, the other day Mayo[4] got animated about some old bones & kept on for an hour & a half without the slightest symptom of impatience being shown. I have not heard a single scrape & excepting at the close of ye opening lecture no applause. I went & lionisized St Bartholomew with Moor which is really worth seeing it appears to be so well conducted. He seems quite at home & very well acquainted with the patients, one poor dropsical wretch in particular to my astonishment who seemed more dead than alive & had been tapped that day got quite amused by listening to him & complained that he made her wish to laugh.

The dissection is going on languidly there is but one subject come in yet & there are six engaged before the one I have put my name down to: they are cheap compared with Edinburgh being £8"8 which however when it comes to be multiplied three or four times is a heavy draw back.

Let me hear your Edinburgh plans when you have fixed them.

Good Bye.—

P.S. Lataika[5] Snuff is all the go: shall I send you a packet per coach?

AL
Postmark: O OC 12 1826
DAR 204.1

[1] CD's class cards, preserved in Edinburgh University Library, include one for Clinical Lectures by Dr [Robert] Graham and Dr [William Pulteney] Alison.

[2] Probably John Armstrong, the elder.

[3] Théophile-René-Hyacinthe Laennec. Leading French physician and anatomist.

[4] Herbert Mayo. He and Caesar Henry Hawkins bought the Great Windmill Street School of Anatomy from Charles Bell in 1826.

[5] Latakia, a Syrian tobacco from Latakia.

From Sarah Harriet Owen 31 [December 1827]

6. Marine Square. [Brighton]
Monday 31st.—

My dear Charles.

I was very glad to receive your letter, though it gave an additional pang to my guilty conscience for not having sooner fulfilled my promise of writing to you, but I hope you do not think that *forgetfulness* has been the cause of my silence. I have said to myself every day for the last fortnight, "I *will* write to Charles

Darwin." & every day some awfully sudden event has prevented my doing so, but as to my forgetting *you*, M.ʳ Charles *that* is quite out of the question—

I am afraid I have nothing very interesting to tell you, though I have been very gay since came here, & like Brighton *very* much, we are out almost every night, which *I* think very wonderful, considering we knew so few people here when we first came. Last week we were at *three* Balls, besides Parties, & are engaged to four or five more, you make such particular enquries after *Red coats & shootables*[1] that I cannot resist telling you (like M.ʳˢ Mathew) that they are most of them "remarkably frightful," & the *scarlet shootables* & Scorpions[2] are also *remarkably useless* as very few of them go out & dance. I had the supreme bliss of *chauncing* to meet my very favourite shootable yesterday, viz M.ʳ *Charles Jones*, we met out walking & were equally surprised to see each other, he fancied I was at Eaton[3], & I thought him safe in London, he turned back & walked with me for an hour by *Shrewsbury Clocks*[4] & of course was very agreeable &c &c &c &c *as usual* I am afraid he is only here for a day or two, he called on us yesterday but we were out, & did not see him— I have only had one *ramrodding* excrescence[5] since I came here, & that was with the Foxhounds, & excellent sport we had, I wish we could have the Spaniard here, instead of at the *Forest*, where he is now quite useless, there are very good horses to be hired here, & we intend to have another gallop soon—

Brighton is so *horridly* windy that there is very little pleasure in walking here, & besides we are in sad want of a *walking* Chaperon, & I *often* wish you were here to take pity on two *desolate Flams*. Caddy[6] wrote me word you were to go to the Forest the first hard frost, but I have not yet heard your name amongst the fashionable arrivals, remember you will owe two or three long visits by the time we return, I do not exactly know when that will be, I should think not before the beginning of February, I hope you will not be departed to your *"brilliant establishment"* at Cambridge—[7] We dined yesterday at Sir George Anson's, & had a very pleasant evening, Francis Anson is very goodnatured & agreeable & a great favourite with us all, he *chaunces* to drop in every Morning to see what we are about, & enquire if he can be useful in any way, I do not know what we should do without him. Then there is Cap.ᵗ Fremantle who is very goodnatured & civil, but I do not like him half so well as Francis— As for ⟨   ⟩[8]

⟨   ⟩ Darwin Hall & believe me, my dear Charles. | ever Yours sincerely | S. H Owen

The *quantity* of this will I hope atone for the *quality*, if you ever have patience to read it all—

Incomplete
DAR 204.4

---

[1] Presumably officers and other targets for unmarried young ladies. As later letters make clear, 'shootable' is also used by the Owen girls generally for 'suitable'.

[2] In Lister's popular novel *Granby* (Lister 1826) the term 'scorpions' is used for 'That detrimental class . . . younger brothers' (*OED*).

[3] Eaton Mascott, near Shrewsbury, home of the Williams family. Sarah Owen married Edward Hosier Williams in 1831.

[4] 'by Shrewsbury clock', a colloquial phrase lessening or even cancelling the period of time (Partridge 1973, p. 840). Shakespeare used the term in *Henry IV* Part I, 5. 4. 147 (Arden edition) in Falstaff's fabricated account of his fighting with Percy for 'a long hour by Shrewsbury clock' at the battle of Shrewsbury.

[5] Excursion or outing. Other examples of the Owen–Darwin private language that occur in the correspondence are: broadcloth = man; black broadcloth = clergyman; budget = letter; clandecently = clandestinely; foiblesse = weakness, foible; halter = altar; hare = suitor; insinivating = insinuating; insinivation = invitation; leetle = little; muslin = woman; mystery = rumour, gossip; punchon = penchant; statue = statute (of lunacy).

[6] Probably a nickname for Caroline Owen.

[7] CD had been admitted pensioner to Christ's College, 15 October 1827 but did not come into residence at Cambridge until January 1828 (*Biographical Register of Christ's College*, vol. 2).

[8] At least one page is missing. The final lines are written vertically across the first page.

From Fanny Owen [January 1828]

[Brighton]

My dear Charles,

I never was so horror struck as to receive your leetle note the other day My conscience upbraided me so much as really to prevent my eating any *breakfast* altho' a plate of *hot toast* was smoking before me. I reproach myself bitterly with ingratitude for your entertaining budgets, in having delayed so long sending you an effusion. pray forgive the *penitent* Housemaid. She is truly sorry for her crime— I hope you had good sport at the Forest I hear you are become an *undeniable* shot. I was very much surprised to hear from Sarah that you have decided to become a *DD* instead of an *MD*.[1] you never let *me* into the secret— We are very dissipated here, at a Ball or Party almost every night, which as you may suppose I find not *bad sport* in *its way*, but all must come to an end and I *fear* we shall be dragg'd away to the shades of the Forest and leave Brighton in the higths of its gaiety—no time is yet fix'd however. Poor Caddy tells me she is very triste at home having to *fight it out* with all the Boys who are very impudent— Pray tell me some Shrewsbury scandal You well know if the Ho⟨use⟩maid has *a foiblesse* it is for a *mystery*— The W family[2] I believe are certainly coming here *bag & baggage* & very soon have you seen any of them the *Hero Ted* is at Eaton & I believe too the *odious* Richard— I have not been riding for some time, the last time I did was with the Hound's which answer's my idea of "*Bliss on Earth*," it is such exquisite fun gallopping on the Downs as hard as one can go—and there are generally about *two hundred people* out— I wish we had the Spaniard here, he is *wasting his sweetness* in the *desert air* of the Forest— I hope you will not be gone from Home when I return I shall expect *cart loads* of *black mysteries* after so long an absence so pray collect all you can— How is Maam Burton,[3] & who could have invented the report that her dear little Eliza was going to be *dragg'd* to the *Halter* by a Splendid *Shootable*— Is Mʳ Upper still as brilliant &

elegant as when I left him— I must conclude this horride scrawl but I have a horride pen and am in a *horride* hurry—but will send you another effusion soon. I am too stupid & sleepy tonight to do any thing but sleep in the arm chair— I am sure you will say what a *stupid prosy* creature the Housemaid is become, but remember it is *1828* & she is a year *older*—

adieu my d^r Postillion believe me ever y^rs sin^ly Fanny O

Burn this—

DAR 204.4

[1] 'After having spent two sessions in Edinburgh, my father perceived or he heard from my sisters, that I did not like the thought of being a physician, so he proposed that I should become a clergyman' (see *Autobiography*, pp. 56—7, where CD describes his scruples about following his father's advice).
[2] The Williams family of Eaton Mascott. The '*Hero Ted*' is probably Edward Hosier Williams.
[3] Probably Mary Burton, wife of Henry Burton, Vicar of Atcham, Shropshire.

From Fanny Owen [late January 1828]

[Brighton]

My dear Postillion,

You must have thought the *Housemaid* most *awfully* changed by the Brighton Air to suppose for a moment that she had dashed with *indignation* your agreable letter into the flames and determined to treat its author with *merited disdain*. The real reason why I did not long ago send you a *leetle* effusion was that you told me you were going to *Cambridge* the end of the month, now that word **Cambridge** sounded most formidable to my ears; I had visions of Mrs *Burton constantly* starting up before me, which seemed to say, "Dear me Ma'am would you believe it Miss *Fanny Owen corresponds with a young man Ma'am* at *the University*" this horrid phantom appear'd to me so often and worked upon my *heated fancy* to such a degree that after *mature deliberation* I thought it best to *remain silent* and if the Postillion should be in a *fury* at *my ingratitude* to explain it all away when we shall meet once more at the Forest—& now you see M^rs. Burton has been the cause of all this I hope you will forgive the poor Housemaid— Of course you have heard of the wonderful event going to take place viz the Woodhouse family *actually* going to set forth on their travels— Heaven grant the *old Van* may not fall to peices before they reach *London*: but I think it will be a miracle if it ever reaches its destination—

I suppose Shrewsbury is full of *gossip* and *scandal* as there has just been a meeting of all the surrounding families, to *stare at* and talk *scandal* of their neighbours— I suppose all the Shrewsbury young Ladies worked their *fingers ends* off for the Bazaar, Caddy Pemby of course amongst the number & I am very curious to hear what *they did* and how the *valuable* articles sold—it must have been good fun seeing *Miss Clarissa* behind the Counter—

I was at a very gay Ball on Wednesday night plenty of *red Coats* & *Moustachio's* which *you say must* be pleasant I like Blue & silver better—

I have been riding a great deal too lately which is very pleasant. When M.ʳ Hill was here he got me a very nice Horse belonging to a *useful friend* and since I have been riding with the Anson's—

I must conclude this for I must write to my much injured Catherine— I only hope I may *appease you* both & excuse this scrawl but I have *such a Pen* and besides never could write like any thing but what *I am*, a *Housemaid* so d.ʳ Postillion ever y.ʳ | F Owen

Burn this as soon as read—or tremble at my fury and revenge—

DAR 204.4

From Sarah Owen 18 February [1828]

The Forest.
Monday Feb.ʸ 18.ᵗʰ—

My dear Charles,

If you have thought of me at all since your arrival at *Gay* Cambridge, I fear it has only been to reproach me with *breach of promise* &c &c in not fulfilling the engagement I so solemnly made to intrude upon your time, & interrupt more agreeable studies by "a letter from Sarah Owen, bless me! what *can* she write to *me* about." Indeed M.ʳ Charles, had I followed my own inclination, you would have been troubled in this way *many* days, nay, weeks ago but since I received your *very* agreeable effusion at Brighton, I have been so busy in various ways, that I have hardly had any *spare time* on my hands— Perhaps you know that we staid at Brighton a week longer than *Papa* intended, on account of a very gay Fancy ball given by **five & twenty** Batchelors, indeed it would have been too provoking to have missed it, for it was an *ad*mirable rigadoon, every body in Fancy dresses or Uniforms, Fanny and I went as *She Turks*, that is in the dress of *Fatima* if you ever *chaunced* to see it, this Festivity took place on the first of February, & lasted till 7 o'clock A.M. on the 2.ᵈ of Feb.ʸ We had *3* balls, three successive nights before & even *my constitution* began to fail whatever M.ʳˢ Burton may say or think to the contrary—

We left Brighton on the 4.ᵗʰ & *very sorry* I was to go away so soon, I have however no *right* or cause to complain, for we went there for three weeks & staid ten— We had three *very* pleasant days in London from Monday till Friday Morning We were to have left it on Thursday but the Williams begged so hard for another day, that Papa was obliged to *give in*. We dined with them *every* day, & went one evening to the French Play, which I enjoyed very much, (as you will easily believe), & the last night we went, a party of *twelve*, to Covent Garden, & the great *Ricciardo*[1] gave us a *farewell Supper* in his rooms, which are very magnificent— The party there consisted of M.ʳ Williams, Emma, M.ʳˢ Bruce *two* eldest Mr Williams, M^r & Mrs Hopkinson, Mr Powell[2], Lord Kinnaird, a *very* nice Shootable, & last, though not least Mr Charles Jones!! I had the *pleasure* of meeting him each day at dinner, & he was *quite* as agreeable as usual— Mr Hopkinson sang two or three excellent songs, & "the evening was passed in the utmost *armony*" &c &c &c. We left London very early the following Morning, & arrived here on Saturday night, found Owen, Aunt Laura Caddy & Emma waiting to receive us, & here my *Personal Narrative*[3] ends, for since that moment, I have not been outside the Barriers of the Forest, or seen anybody except Mr Lloyd Kenyon, Mr Tom *Ditto*, & the Rector of the Parish, & you know that neither of these three pieces of *broad* Cloth can afford me the least assistance in filling up a *vacant* sheet of Paper like this now before me—

The Tranquillity here is indeed truly awful, & there is no saying how long it may last, I wrote last week to implore Caroline Susan or Catherine to come to our assistance, but I am sorry to say they returned a *flat* denial, as M^r & Mrs Wedgwood &c &c. are now staying with th⟨em⟩— And now, Charles, I must say a few ⟨con⟩*fidential* words, on the *two* great mysteries in *your* Family, viz, the *Crack* in your foundations & Walls, & the *black broad Cloth* affair—to begin with the *Crack*, which from the best authority, seems to *widen* & become more serious & alarming every day, Heaven knows what effect this Frost & Snow may have on the sloping banks of the Meandering Severn!! I hear that Susan's sensitive feelings suffered a severe shock not *many days* ago, on hearing two *unfeeling* Men conversing (near the walk overhanging the *precipice*) on the melancholy fact, they concluded by remarking "it is a very pretty house, *more's the pity* we shall so soon see it in the river"— for Heaven's sake, Charles, breathe not a syllable of this fatal information to any mortal ear, you must be aware of the *tenderness of this subject*, particularly to the Parties concerned. Secondly, have you heard of the *Vicar of Pool's*[4] sudden accession of Fortune; 11,000£ *I* heard, you may perhaps persist in thinking this is no business of *yours*, but *I* also persist in my original opinion on the *mysterious* subject. "Time will shew" the truth of my suspicions. pray tell me if **your** *sentiments* continue the same— I heard of your paying two visits to the Forest during our absence I am glad *some people were missed*, as that circumstance may perhaps induce you to stay longer, when *some people* are to be found in Paradise— A sad Catastrophe occurred here last night, one of the black Carriage horses which has been indisposed for some days, *expired*, deeply lamented by all whom he had *drawn* or *followed* after. I hope the Andalusian may not catch the disorder, I have not yet rode him since we returned,—but as the Frost seems going, I shall have that pleasure soon. I hear he is grown very frisky, &c &c, which is so much the better—

I really must conclude this scrawl; you have wasted too much time already in reading it, pray put it into the Fire which is now blazing in your room, & sit down to your Folio's again & as soon as you have an idle half hour, remember that *I* hope to hear from you— Of course I need not remind you of your promise not to shew my stupid effusions to *any-body* on any account *whatsomever*, I will therefore conclude. I remain *always* | Yours most truly | S. H. Owen.

You must have heard all the Shropshire Marriages—Miss I. Forester & Mr Biddulph, Miss Mary Burton & Mr Richards, "little Lingen" & Miss Edwardes &c &c if it were not trifling with *your* feelings, Mr. Charles, I would add Miss Clare Leighton "that *rising* star of Ton—" & Lord Hill—which is an undoubted fact, but *don't* mention it at present. I shall expect an effusion very soon.—

Owen returned today from a ⟨   ⟩ *bliss* at Chillington.

Postmark: Shrewsbury FE 19 1828
DAR 204.4

[1] Possibly Richard Williams, the '*odious Richard*' of Fanny Owen's letter [January 1828], who was living in Albemarle Street, London, at this time (*Boyle's Court and Country Guide* London 1828).

[2] John Allan Powell, solicitor, with whom Edward Hosier Williams was in partnership (*Boyle's Court and Country Guide* London 1828).

[3] Apparently a reference to Humboldt 1814–29, one of CD's favourite books, which he may have told Sarah he was reading.

[4] Possibly William Clive, Vicar of Welshpool from 1819, who married Marianne Tollet in 1829. The Clives of Styche were friends of the Darwin family.

From Fanny Owen [9 March 1828]

The Forest
Sunday—

Dear Charles—

I tremble at having the boldness to begin a letter to you after having allowd your last long effusion to remain such a time unanswer'd, but as I believe I have told you before I *cannot* & *will not* play **second** fiddle & as Sarah wrote to you very soon after we came home I thought I had better delay doing so *till* **now**— not that I have collected a bit of news or gossip to tell you don't expect such an article for you will be disappointed. I am *more stupid* than *any body* ever was *before* or I should hope ever will be again—not a *whisper*, not the *ghost* of a *mystery* have I worth retailing— The only time we have been out since our return was last week to *Tommy* Hunts[1] quite *brilliant festivities*. Sarah Owen & yr. humble servant were there for two days, *tolerably* good fun considering it was at the *Prêtre's*, and Lloyd Kenyon the Hero of the fête, it being a sort of *farewell Rigadoon* before that *interesting* youth join'd his regiment which he did a few days ago to our **unspeakable dispair**—for Oh! how he used to come and *sit* and *eat* and *sit* till we thought he must have *grown to his chair*—

We have tried in vain to get your sisters here—for a little *larking* & *scandal* but as I suppose you know the Wedgewoods have been there for a long time thank heaven's they go next week and Caroline & Catherine have promis'd to ride over to pay us a visit— I wish you were to be of the party but *"brighter hours will come"* and I look forward to y^r paying us a long visit in y^r next vacation— I am schocked to inform you that I have not yet learn't to play at *Billiards*, I hear you exclaim what a lazy creature, but it is not *my fault* I assure you, for I can get nobody to give me some lesson's, Owen votes it a *fat bore*—so I suppose I must *trust to fate* for *a master* or wait till *you come*— I have not been riding *near so much* as I *wish*, for we have only the *Andalusian* for the *whole* Family, which is really *too bad*, and puts me in a *d——l of a temper*, the Governor is going to buy another but cannot find one *shootable* about here, I hope he *will* soon or I shall *get furious*—

William Hill is the only visitor we have had, he has been here almost ever since we came home, but departed this morning per Coach to London on his road to **Paris** and *Italy happy fellow*!!— The Assizes are on the 20th and we expect a *visitation* from that *insinivating Counsellor* Mr Alexander.[2] There is going to be a grand feast at Tedsmore[3] on Thursday where we are going, the *old Piebald* (as Mrs *B. O* is now *christen'd*) is I think grown more *hideous* than ever.— one evening at Tom Hunts we danc'd, & Owen *waltzed* with *her* which was so ridiculous we could not *help laughing outright*, & Mr Kynaston behaved so ill that he really did all he could to *knock* down the *poor old Piebald* by *running* against her *as if by accident*, till at last her *little Husband* found it out and was *very indignant*—

I hope you like Cambridge pray tell me all about it & when you expect to come home—

I must conclude for I find myself getting *awfully dull* and *prosy*, but what can you expect from an *unfortunate exile* of the Forest *pity* and *forgive* is all I ask—& believe me my d^r Postillion ever y^r faithfull Housemaid | Fanny O—

For *Heaven's* sake burn this, or if it falls into the hands of any of the *young men*, what would they think, of a *Housemaid* writing to Mr Charles Darwin— Ma'm Burton *would die of it* I *think*—

Postmark: Shrewsbury MA 11 1828 153
DAR 204.4

[1] Possibly Thomas Hunt, Rector of Wentnor and West Felton, Shropshire.
[2] Possibly William John Alexander.
[3] Tedsmore, about 2 miles south of Woodhouse, the home of Thomas Bulkeley Bulkeley Owen (Bagshaw 1851, p. 205).

From Sarah Owen 30 March [1828]

Woodhouse.
Sunday March 30th—

My dear Charles.

I have been very idle & ungrateful in not thanking you sooner for your *very* interesting *leetle* effusion. I should certainly have written to you last week, if I had not witnessed the departure of a *very considerable* dispatch composed by Catherine & directed to M.ʳ Charles Darwin, & as I knew she had therein deposited all the news &c. &c. of the Forest that she could *scrape* together, I thought I had better defer troubling you for a few days longer & *now* Heaven only knows how I shall fill this sheet of paper, for we have been in such a state of awful tranquillity since I wrote to you, that I despair of sending you any thing worth Postage—

Of course you know that Caroline & Catherine were here for a few days last week, and during those few days, I spared neither pains or trouble to sift the report I told you of, to the bottom, I grieve to tell you, that my efforts were unsuccessful, they *both* solemnly & positively denied the fact, as far as *Caroline* was concerned, & since that time I have been haunted & tormented with another & a *newer* idea. Can it be *Susan* do you think? I ponder deeply, but can come to no result, pray give me your opinion of the matter. I am happy to say Susan comes here tomorrow, I have not seen her since the memorable old Hunt & Bazaar, *which impoverished* so *many misguided shootables* M.ʳ Charles—

Two of *the* M.ʳ Clives, George & the little Orange Nabob[1] came here last Monday for a day or two, they seem very goodnatured *&c &c.*— On Tuesday the whole Congreve Family stormed the Fortifications, & I had to mount the *strictest* guard till yesterday when they took their departure to my great joy, I had them almost entirely to myself for Fanny was ill, & did not leave *Paradise Row* till they left the lower Regions, her's has certainly been a *well timed* sickness— The party was increased on Friday by the arrival of M.ʳ Henry Burton, & M.ʳ Dowdeswell a nephew of M.ʳ Deans who is staying at Atcham— "Sure such a pair were never seen" for M.ʳ D. is, if possible more "remarkably frightful" than M.ʳ B. though I am sure you will not easily believe this, & certainly without *ocular* proof it must be difficult to do so— I believe "little Lingen" still continues single though people seem determined to choose him a Miss Shootable since he took possession of the *large* room at the Abbey, as they declare it is quite *un*possible he can fill it *alone*—

I suppose you have been officially informed of the Marriage which is shortly to take place in *your* Family!! so I need hardly mention names, I only hope it may be the means of my devouring ½ a pound at least of *P.C.*— *I* have suspected it **long ago**!!! Miss Butler was married last week, the *yong* gentlemen presented her with a piece of plate, on the appy event, & in return, received a dinner & 2 days additional holidays at Easter, they must each lament that D.ʳ Butler has *no more* Single daughters—[2]

I am very sorry you do not come down *this vacation*, though perhaps if you had, you would not think it *worth while* to visit the Forest, *out of the shooting Season*, now that all the *Muslin* is collected there. I wish you could *chaunce to come by* on Tuesday, as we are meditating an excrescence to Wrexham Fair, to buy Bargains, & see all the Beasts & Beastesses &c—

I have been riding a good deal lately, Fanny has not been out for the last fortnight & I have had the Andalusian all to myself, he is in great beauty now & very frisky & restive— Papa has bought a *wonderful* pony, dark chesnut, with a white mane & tail, it is only two years old, so it will be of little use for a year to come.

As for Aunt *Pedigree*,[3] if it was not for the credit of the Family, we should certainly take out a *Statue* of Lunacy against her, I think I told you she had bought a *bit of blood* for her own riding, she keeps it at Mr Haycock the Butcher's, though every thing shootable has been provided for its accommodation here, she has had two *horse breakers* in attendance feeds the horse herself 4 times a day, & carries the key of the Stable in her pocket, she is feeding & physicking the poor thing to death, & will not ride it herself or allow any body else to do so, I am quite sure she will never summon courage to mount it herself, & it *never* will be of the least use to her—

I hear your Favorite, Miss Clare is astonishing the weak minds of all the Barristers this Circuit, if I were Mr Charles Darwin, I should be quite jealous.!!! I think Caroline & Catherine will have a very pleasant junket to the Isle of Wight. have you heard that Susan positively declines leaving "sweet ome" though Erasmus offered to take her to Paris, or any part of the Continent. *I* cannot but think this *most mysterious*. By the bye, Charles, what *can* you mean by saying you have not *set your eyes* on Muslin since you left Shropshire. Have the Doctors, Proctors, Deans &c &c &c neither Wives or Daughters!! or are the Cambridge shootables thought too *dangereux* to be seen by them? pray satisfy my curiosity on this point the next time you write— I almost forgot to apologize for my great *denseness* in directing to Trinity College instead of Christs but somehow or other I had a *running in my head* that it was Trinity, however if it only delayed my scrawl a few days longer, it could make little difference to any body & as Mr Cotton says, "it will all be the same 50 years hence—

Have you heard that Tom Parker had a dreadful *Cross* some time ago, he has had a brain fever in consequence, & is now quite deranged & at Worcester, it is a shocking thing for them all. Mr Mathew has also been very ill, & I believe is still in danger, Dr Darwin & Dr Parker attend him—

I do think *I* have *scraped* together all the news I have heard for a Month past, & *my little all* will be very dull to you I am afraid, but as you begged me to write, it is partly your own fault, for bringing this trouble on yourself. If you wish the *grievance* to continue I *hope* & beg you will write to me *very* soon, you need not make the old excuse of having nothing to say, for you see before your eyes a whole sheet of hot press filled with *nothing at all*, so try to imitate the good example I set you, & if you *remember the promise you made*, you may believe me my dear Charles ever your's sincerely | S. H O—

I forgot to "*mention*" that the Flower-Garden, notwithstanding ancient prophecies, is remarkably flourishing & promising & will be *brilliant* by the time you see it. Two or even *3* flowers have actually made their appearance, & the

Fortifications are proof against *Hares* of *every* description— I forget whether I informed you of the Marriage w.ʰ took place in this Family a little while ago, your Friend M.ʳ Starkie *clandecently* led to the Halter Miss Anna Bates, daughter & Coheiress of M.ʳ Bate the Cow man, & M.ʳˢ Bate, the Family Washerwoman— M.ʳ Starkie narrowly escaped being dismissed the service, but he is now forgiven, & kills Cats & Rabbits as usual—

Now that you have arrived at the end of this scrawl, go back to your Quarto's & folios my dear Charles, & think no more of me or my *Foolscap*.. Goodbye—

Postmark: Shrewsbury AP 1 1828 153
DAR 204.4

[1] Probably George Arthur Clive, brother of Robert, Henry, and Edward (see letter from Caroline & Susan Darwin, 2 [January 1826], n. 2). The 'little Orange Nabob' perhaps refers to Robert or to another brother, Richard, both of whom spent some time in the Indian Civil Service (*Eton School Lists*).

[2] Samuel Butler had two daughters, Mary and Harriet. Mary Butler married Edward Bather (*DNB*) in 1828.

[3] An allusion to Oliver Goldsmith's *She stoops to conquer* (1773).

To William Darwin Fox[1] 12 [June 1828]

Friday 12.ᵗʰ—[2]

My dear Fox

I am dying by inches, from not having any body to talk to about insects:—my only reason for writing, is to remove a heavy weight from my mind, so now you must understand, what you will perceive before you come to the end of this; that I am writing merely for my own pleasure & not your's.— I have been very idle since I left Cambridge in every possible way & amongst the rest in Entomology. I have however captured a few insects, about which I am much interested: My sister has made rough drawings of three of them: I. fig: is I am nearly sure, the same insect as H⟨oa⟩r,[3] of Queens took in a Willow tree, & which Garland[4] did not know. I took [it] under the bark of rail, was very active, striking looking insect, took 3 specimens I think this is an admirable prize

II. fig: is an *extremely* common insect; of the family of scarabidæ. Do you know it's name?—

III. fig: A *most* beautiful Leptura (?) very like the Quadrifasciata, only the body is of the same size throughout.— I tell you all these particulars, as I am anxious to know something about these little g⟨ ⟩s.— I was not *fully* aware of your extreme value before I left Cambridge. I am constantly saying "I do wish Fox was here".— And I again say, I hope you will come & pay me visit before the summer is over— My Father desired me to say, that he should be at anytime most happy to see you.—

I have taken 3 species of Coccinellæ, one, the same as Hoar took in the Fens, which you said was rare, & another with 7 *white*! marks on each elytron.— I will mention, as I believe you are interested about it, that I have seen the Cocc: bipunctata (or dispar) 4 or 5 in actu coitus with a black one with *4* red marks (I believe most of the black ones you have got have ⟨ ⟩ marks, & hence I suppose a different species) also, which is very singular, I have frequently seen two of the bipunctata's in actu.— I Have taken Clivina Collaris, fig ⟨3⟩ Plate III of Stephens;[5] also a beautiful copper-coloured Elater (with

Antennæ pectin⟨ate⟩ like this. Do you want any of the

Byrrhus Pillula? I can get any number.—

My dear Fox I must again beg your pardon for sending such a very selfish stupid letter: but remember I am your pupil, so you must forgive me.— I hope you will write to me soon, & tell me every thing you have been doing, & more particularly how you are in health, as to your eyes, & body.— How was poor little Fan: how was No 16!? what do you intend doing this summer? in short write me a good long letter about yourself & all other insects: My plans remain the same as formerly. I am going to Barmouth[6] for two months.— If you have not written to my brother, write to him before the 22[d] & direct, Poste Restante, Munich I hope you will be able to send me a better account of Miss Fox, when you write; tell me whether you intend going to Tenby?

I should not send this very shamefully stupid letter, only I am very anxious to get some *crumbs* of information about yourself & the insects.

Believe me my dear Fox | Yours most sincerely | Charles Darwin

Shrewsbury

of a fine bluish black colour, but is not so broad as made in this drawing

rather lighter coloured & more metallic the legs are left out.—

this is a very good representation

5 tarsi

4 Tarsi

I. fig is more like a Pyrochroa or a very narrow Blaps than any thing I can compare it to.—

I.I. fig: be sure to give me the name of this insect

Postmark: Shrewsbury JU 13 1828 153
Christ's College Library, Cambridge (Fox 1)

[1] CD's second cousin and his 'entomological tutor' during CD's first year at Cambridge. Fox's mother, Ann, was the daughter of Dr Erasmus Darwin's brother, William Alvey Darwin.

[2] Friday was the 13th.

[3] The middle letters of the name are obliterated by a blot. It is likely that CD wrote 'Hoar' (see below in letter), but he probably refers to William Strong Hore, who was an undergraduate at Queens' College 1826—30. Letters to W. D. Fox, [7 January 1829] and [25—9 January 1829] give support to this possibility. In both, CD asks to be remembered to 'Hore'.

[4] Possibly Lewes (or Lewis) Garland.

[5] Stephens 1827—46. *Clivina collaris* is listed in *Mandibulata* 1: 40, plate iii, fig. 3.

[6] On Cardigan Bay, Gwynedd, North Wales. CD spent the summer on a reading tour with some undergraduate friends and George Ash Butterton, of St John's College, a private tutor in mathematics (see *LL* 1: 166, which quotes a letter of reminiscences from John Maurice Herbert to Francis Darwin, now in DAR 112).

To W. D. Fox [30 June 1828]

[Shrewsbury]

My dear Fox,[']

I do not know whether to begin by giving you a good scolding, or by thanking you over & over again for your most acceptable parcel. Your letter was perfect & informed me of every thing I wanted to hear about & you really were most goodnatured to think of forwarding Way's[1] letter.—

And now for the scolding; but I must confess I am not very certain about the grounds of it. Did I pay you the 10"6^d for the Poor Irishman?[2] if I did not; you must have been aware of it,—& therefore you acted like a shabby fellow, in allowing me to be so.— I will pay you the 10"6^d & the 1£ for the pins, (which were just what I wanted & I am extremely obliged for you thinking of me.) when we meet at Cambridge.—

I am very glad you have had so pleasant a time in London. it almost makes me wish to be there: but nevertheless I had much rather be wandering with you in the wilds of Derbyshire; than be idling in that horrid smoky wilderness London.— I envy you very much in going to the Opera, for I have at last got a very decided taste for Music; When Roper[3] was here we had a concert from morning to night: his visit here was very pleasant as it is quite delightful to hear anyone sing with such spirit, & excellent good taste.—

And now for Insectology, Way need not doubt my ardour as it is I think redoubled; but my success does not equal what it did in Cambridge.

Fig: (I.) is this a Cychrus? I make it so by Lamarck.[4] (N.B. when you recognize an insect by my description, always say whether you possess it). fig II is a most splendid Cerambre—what is it? fig III, do you know this insect? I have also take another very large grey looking Curculio? Also a Cerambyx twice as big as that common one, with horn coloured elytra & the stomach marked with bands something like a Cock chafer. About 19 new Carabi— several Coccinellæ, & other insects which I cannot describe.—

Taken in plenty, the little beetle that destroy Scotch firs. What is its name.—

The Crysom: (fulgida(?)[5] the same as your one specimen, is the *greatest* profusion & am *quite* convinced it is a different insect from the little one you took in plenty in the Fens, as it never lives but on one plant, some sort of dead nettle.— of course I procure duplicate specimens for you of all insects that I can.

I have adopted Way's plan of having a store box, & only having a few specimens of each insect in my regular case:—

We heard from Erasmus the other day just when he was on the point of leaving Vienna he is now on his road to Munich or Milan all letters must be sent to the latter place. I do not think he much admires Vienna, for it has been literally a perfect state of solitude for him, latterly he has got to know one single family Vivenot, a German Doctor, very pleasant people but who the devil would not be pleasant, when one has been silent for nearly a month?— *If* you have written to him either at Vienna or Munich, you need not be alarmed about their arrival, as he has left orders that all letters may be forwarded.— On Wednesday I set out on my Entomo-Mathematical expedition to Barmouth; by the blessings of Providence I hope *the science* will not drive out of my poor noddle the Mathematics.— Talking of *the science* I must tell you, that since beginning this letter, I think, sir, upon my soul sir, I will take my oath sir, (as Way would say), that I have discovered, that I posses a valuable insect, viz Melasis Flabellicornis, as the description & habits in Samouelle[6] & Lamarck *perfectly* agree.— look and see, what Samouelle says, "once taken (continued from the cover)[7] in Norfolk, by J. C. Curtis"[8] &c &c; also I have taken 2 new Elaters; one with reddish brown elytra, the other with yellow elytra, red thorax & black head.— Also, have taken Tritoma Bipustulatum, vide Samouell. Pl. II.;[9] also a small black Byrrhus, with brownish band across the back; also a small Silpha, with a red spot on each elytron; also have twice !**seen**! the Bombylius Minor, but curses on my clumsiness missed catching them.—

You will hardly believe how *bumptious* I am grown, in the height of my presumption, I put down the Leptura Arietes, as the Saperda Lineato Colls. because it had a yellow ring round its neck! & as for the Melandrya I thought it at least a new species if not Genus.— I must confess I felt a pang of jealousy when I read over the splendid list of yours & Way's Captures.—

What a pity it is that you do not collect letters, like insects, birds & beasts. What a glorious specimen this would be of an interesting, neatly written & elegantly composed epistle: but my dear Fox remember this, although my prosing unscientific details about insects may be very tiresome to you, do not for one instant suppose, that your letters are so to me, for you cannot conceive, with what great pleasure I look out for an Entom: Letter now that I have nobody to talk to.

I return Way's letter & must again thank you for it, when you write remember me most kindly & tell him, if he has a fit of idleness, he knows how glad I should be to hear from him.—

When you write, & I hope it will be soon, direct to Shrewsbury, as usual Believe me my dear Fox | Yours ever sincr | C. D.

Shall I or will you write to 〈   〉 M.A., if you say nothing about 〈   〉 I shall, as soon as I hear from y〈   〉 alls well Wed.[10]

| jet black taken under stones in a boggy situation One of the carabidæ Thorax drawn too large | the insect is more beautiful than the drawing | larger than Curc. niger, but black, marked with yellow dots. Taken on a Pine tree |

I   II   III

Postmark: Shrewsbury JU 30 1828 153
Christ's College Library, Cambridge (Fox 2)

[1] Albert Way, undergraduate at Trinity College. He and Fox introduced CD to insect collecting.

[2] Not identified. Probably the same as 'Our old Irish friend' referred to in the letter to W. D. Fox, [10 April 1829].

[3] Probably Henry W. Roper.

[4] Lamarck 1815−22. *Cychrus* is described in vol. 4, pp. 516−17. Volumes two and four, with CD's annotations, are preserved in Darwin Library−CUL, as are the eleven volumes of the second edition (1835−45).

[5] *Chrysomela fulgida* (Samouelle 1819, p. 416). CD's copy is preserved in Darwin Library−CUL.

[6] Samouelle 1819, pp. 160, 415.

[7] CD had left the mid-section of page 4 blank for the address but, after adding a sheet to his letter, he returned to the unused portion of page 4.

[8] Samouelle 1819, p. 160: 'In England it has been once taken by Mr. J. Curtis, of Norwich, an excellent artist and an industrious entomologist'.

Robert Edmund Grant, *c.* 1838—41.

(Courtesy of University College Archives, London.)

Sketches by Albert Way, a Cambridge contemporary of CD.

(DAR 204.3. Courtesy of the Syndics of the Cambridge University Library.)

The Mount, the Darwin family home.

Maer Hall, home of Josiah Wedgwood II.

Woodhouse, home of the Mostyn Owen family.

(Photographs courtesy of Hensleigh Wedgwood and William Mostyn-Owen.)

[9] Samouelle 1819, Plate II, fig. 9, and 'Explanation of plates', p. 445.
[10] 'Shall . . . Wed.' written on the outside fold of the cover.

From Fanny Owen [July? 1828]

My dear Postillion,

Of course you are well initiated in all the *black mysteries* which I am sure are going on at the Forest, I think *you know* that *if* I *have* a *punchon* it is for a *thorough good mystery*, so I implore you to write me a full account of all that has been going on *lately* between the noble Houses of the *Forest, Bliss Castle* & *Darwin-Hall*.[1] I am dying *by inches* to know *all*, & your *well known* and *laudable thirst* for useful *knowledge* has I am sure made you acquainted with *every particular* before this so pray sit down and relieve me from my dreadful *state* of *haggitation*— I have heard of your honoring the Forest with your presence, & regretted very much I was not there to win some of your *shigs* at écarté, or to make a *beast of myself* in the *strawberry beds*. I am afraid your *holidays* will be over before I *grace* Shropshire with my *presence*, for I don't know at all how soon that will be, my *bereaved Parents* want me to *return*, and the Dean's want to keep me—how it will be settled I don't know—

Have you been at the Sale I hear that *brilliant Hare* M.ʳ George Scott intended to buy a great deal of furniture if it *went cheap*, he means to *furnish Cronkhill* and "*see Batchellors to dinner*". Shall *you* be included do you think, among the select *batchellors*— Pray tell me all the gossip—how goes on Miss *Clarissa*, Abyssinia & *Donadavid Dallas*,[2] as for Caddy Pemby and M.ʳ Gibby I am tired of their *names*, for they never *get on any further*— but tell me *any* thing or *every* thing you know, the *smallest* intelligence will be *thankfully* received— I am sure if *Ma'm* Burton could hear of my *corresponding* with M.ʳ Charles Darwin, she would say *dear me* Ma'm *what a sense* of *propriety* Miss Fanny Owen must have, indeed I think it is quite incorrect, but it *is to be hoped* it may not *reach her ears*. I must conclude now, for my paper is full and my pen *split* up *half* an inch which puts me in *horride* passion—

so I remain d.ʳ Postillion | y.ʳ dutiful Housemaid | Fanny Owen—

I must put this in an envellope *well sealed* sisters I know *peep so*, and pray make a mystery of it, tell them there is something *very* particular—

DAR 204.4

[1] Woodhouse, Maer Hall, and The Mount. The Mount, overlooking the Severn in Shrewsbury, was the residence of the Darwins. The house is now occupied by the Inland Revenue District Valuer and Valuation Officer, Salop. The grounds have been taken over by a suburban development of private residences.
[2] Davidona Eleanor Dallas.

To W. D. Fox [29 July 1828]

Barmouth
Tuesday

Dear Fox

What excuse have you to offer for not having answered my letter long before this? I hope it is nothing worse than idleness; or what would be still better, I hope it arises from your being 10 fathoms deep in *the Mathematics*, & if you are God help you, for so am I, only with this difference I stick fast in the mud at the bottom & there I shall remain in statu quo.—[1]

But you see *I* can afford time to write a single letter.— but I will not mention, how long I have been hoping in vain to receive a letter from my old master; more especially at this present time, when, if you remember; we talked of meeting on the top of some Welch mountain between Barmouth & Tenby. I will not flatter you, by saying how much I wish to hear what you yourself are doing & how entomology & your 101 other pursuits are going on? After this aweful Philippic I hope I shall soon receive a letter.—

I have been at Barmouth ever since the first of July, & like it very much; the scenery & therefore the walks are quite delightful. I only wish you would make a trip here, & I would Cicerone you up & down the mountains, untill you had not a particle of wind left in your lungs.— I go on very badly with Mathematics; neither have I succeeded so well as I expected in entomology: but I must mention a few of the most conspicuous insects or rather those which I can at all describe. Amongst the Lepidoptera, the Pterophorus Pentadactylus (Lamarck), a most extraordinary moth,—several large moths, one of which is figured in Samouelle; brown with a white band & spots; a beautiful small copper coloured butterfly, with a little tail like the Machæon: allso a sphinx Zigena (Lam:).

Inter Coleoptera—. A carabus, larger than the Violaceus, jet black & smooth.— Of the Carabidæ, one very odd one; black polished; very long; thorax, orbicular, & sep⟨arate⟩ from the abdomen, of this length . caught near sea side.— A Donacia, with calves as big as mine; several large Cimices, one most beautiful, scarlet & black;— a red & black Ptinus, with a thorax like a night-cap; et multa alia.— How go on the Chrysalises *If* you answer this letter, tell me of your success in *the science*, (of *names*, I am afraid).—

We have heard several times from Erasmus, he has changed his plans so often; that to follow him in his course would be to pursue a Machæon on a windy day. I have nothing more to write about, but I hope soon to write again.— Heaven knows this effusion will not tempt you to write again: but I must take my chance.—

My dear old Fox I long to see you again, but I suppose it will not be before Term begins | Believe me your most sincere & loving $\dfrac{\text{cousin}}{n^2}$ | Charles Darwin

(n, being an unknown quantity, this formula will exactly express our relationship)

Endorsement: Charles Darwin Aug^st 4 1828 WDF
Christ's College Library, Cambridge (Fox 3)

[1] In the *Autobiography*, p. 58, CD comments on his weakness in mathematics: 'I attempted mathematics, and even went during the summer of 1828 with a private tutor (a very dull man) to Barmouth, but I got on very slowly. The work was repugnant to me, chiefly from my not being able to see any meaning in the early steps in algebra. This impatience was very foolish, and in after years I have deeply regretted that I did not proceed far enough at least to understand something of the great leading principles of mathematics; for men thus endowed seem to have an extra sense.'

To W. D. Fox [19 August 1828]

[Barmouth]
Tuesday Evening

My dear old Fox

How very unfortunate it was that our letters crossed on the road, more especially as mine contained such very severe remarks. I should indeed be most ungrateful, if after reading your letter, I did not repent of them. To give the devil his due, I must say you are, (excepting always punctuality) a perfect pattern for a correspondent. You answer so distinctly & satisfactorily all one's question's. You see I am already beginning to harp on Entomology.— But before that I must thank you for your most kind invitation to Osmaston[1] for the Music Meeting. I shall be extremely happy to come, my only difficulty at present is how to get there as on the 7th I shall be in remote part of Staffordshire. Talking of that, upon my soul it is only about a fortnight to the first. And then if there is bliss on earth that is it.—[2]

I suppose the Music Meeting will be very glorious as this will be my first, but I must say I expect much greater pleasure in seeing you, & all your beasts, & last, but not least all the insects.— I suppose I had better bring my gun as I hope we shall have one or two shots together.—

Looking over some insects the other day I found one of my *quondam*-Nebris, & eheu eheu I believe it turns out to be a (Nosodendrum of Lam:) as what I thought were the Palpi, I now believe to be very short thick Antennæ. Your description however agrees pretty well.— I have taken both varieties of the Pine destroyer.— Since I wrote, have taken a dirty-purple coloured Cicindela, with squarish white markings, is it Sylvatica? Also, a *most splendid* **Elater** not Buprestis, by Marsham[3] I make it out clearly to be the Cyaneus, but he gives no references, from which I infer it v⟨ery r⟩are. Elytra & thorax metallic purple blu⟨e,⟩ abdomen greenish blue, legs yellow, taken at a great altitude.—

I hope, if you can, that you will answer this letter directly, as on the 27th. I leave for Shrewsbury, direct Post Office Barmouth: Look over my former letters, & answer my Entomolo: questions. It is quite absurd how interested I am getting about the science.— The great black Carabus, considerably bigger than Violaceus which I mentioned in former letter was taken at great altitude.—

I am going to mention a few insects, which are very doubtful.— 2 most beautiful insects, Cryptocephalus sericeus, & Lebia Cyanocephalus, a Cara-bidæ.— allso, a Bembidion, (Littorale). 3 sorts of dung beetles like the Vaccæ.— a beautiful, square built Donacia, &c &c &c &c. I mean to take all my insects to Cambridge, & then you will see all these *wonders*.—

Give me some instructions about keeping Crysalises, as I possess some of ditto.— I have got so much to say & so much to see at Osmaston that it will be God's Mercy if I go away alive. nevertheless I much hope for the experiment to be tried A part of your letter has given me a Panic, you say you do not know when you shall return to Cambridge I *most sincerely* hope it only means at what time in *October* I should be quite lost without my good old Entomological Tutor.—

I have been abusing you for unpunctuality when I forget you might apply the argumentum ad hominem. The reason I delayed answering is that I have been on an expedition for a few days. For you must know that I am become a "Brother of the Angle" under the superintendence of M[r] Slaney[4] (MP. for our town of Shrewsbury), who pronounces me a very flourishing Pupil.—

Do write soon, & believe me, dear Fox | Yours most sincerely | Chas Darwin

I need not tell you to mention to your Father how very much obliged I am for his kind invitation, & how happy I shall be to accept it.—

Perhaps you had better direct to Shrewsbury as there is hardly time to receive a letter before the 27th.

Postmark: Barmouth 223
Endorsement: C. Darwin Aug 23, 1828 WDF
Christ's College Library, Cambridge (Fox 4)

[1] Osmaston Hall, near Derby, was the Fox family home.
[2] A reference to the start of the shooting season (see *LL* 1: 167).
[3] Marsham 1802. Volume one is in Darwin Library—CUL.
[4] Robert Aglionby Slaney.

To John Maurice Herbert [13 September 1828]

[Osmaston, near Derby]
Saturday Evening

My dear old *Cherbury*[1]

I am about to fulfill my promise of writing to you, but I am sorry to add there is a very selfish motive at the bottom. I am going to ask you a great favor, & you cannot imagine how much you will oblige me by procuring some more spe-cimens of some insects which I dare say I can describe. In the first place I must inform you I have taken some of the rarest of the British Insects, & their being found near Barmouth is quite unknown to the Entomological world: I think I shall write & inform some of the crack Entomologists.

But now for Business: **several** more specimens if you can procure them without much trouble, of the following insects. The violet black coloured beetle found on Craig Storm[2] under stones, allso a large smooth black one, very like it: a bluish, metallic coloured, globular, dung beetle, which is *very* common on the hill sides: Also, if you *would* be so very kind as to cross the ferry, & you will find a great number under the stones on the waste land of a long smooth jet black beettle (a great many of these): also in same situation a very small pinkish insect, with black spots, with a curved thorax projecting beyond the head: also upon the marshy over the ferry land near the sea under old sea weed, stones, &c &c, you will find a small yellowish transparent beetle, with 2 or 4 blackish marks on back, under these stones there are two sorts, one much darker than the other, the li⟨ght⟩ coloured is that which I want: These 2 last insects are **exessively rare**: & you really will *extremely* oblige me by taking all this trouble pretty soon: Remember me most kindly to Butler,[3] tell him of my success, & I daresay both of you will easily recognize these insects: I hope his caterpillars go on well. I think many of the Crysalises are well worth keeping: I really am quite ashamed so long letter all about my own concerns: but do return good for evil & & send me a long account of all your proceedings:

In the first week I killed 75 head of game; a very contemptable number, but there are very few birds, I killed however, a brace of Black Game: Since that time I have been staying at the Foxes near Derby: it is a very pleasant house & the Music Meeting went off very well: I want to hear how Yate[4] likes his Gun, & what use he has made of it?

If the bottle is not large you can buy another for me, & when you pass through Shrewsbury you can leave these treasures & I hope if you possibly can will stay a day or two with me, as I hope I need not say how glad I shall be to see you again: Fox remarked what deuced goodnatured fellows your friends at Barmouth must be; & if I did not know that you & Butler were so I would not think of giving you so much trouble.

believe me my dear Herbert | Your most sincere | Charles Darwin

Remember me to all friends

How is Buz & Bossy. I am afraid you yourself must be grown terribly bumptious:

Direct to Shrewsbury: if there is any thing you want I can send it for you to Barmouth: such as gloves &c &c &c | Good Bye

Postmark: Derby SE 14 1828
American Philosophical Society 1

[1] CD's nickname for Herbert, an allusion to Edward Herbert, 1st Baron Cherbury.
[2] A hilltop near Barmouth.
[3] Thomas Butler (see letter from E. A. Darwin, [24 February 1825], n. 2) had entered St John's College, Cambridge, after leaving Shrewsbury School.
[4] Charles Yate, Herbert's private tutor in the reading party at Barmouth (see *LL* 1: 166).

To W. D. Fox [October 1828]

[Shrewsbury]
Tuesday

My dear Fox

I think I mentioned that I had a few stuffed birds, & as they would be of much more use to you than to me, I have taken the liberty of sending them to the "*Osmaston Museum*" & hope they will arrive safe.— I have also sent a few insects, a Carabus with 6 punctures—taken at Maer, & another Leistus of a light brown colour.— Tell me what you think of these insects, also of a common black (but new to me) carabus.— N.B. The Terne was shot on Maer pool last September.—

So much for Natu History, & excepting that I have been doing little else, & therefore my letter must be as stupid as I myself am. I staid two days at Maer, where I left orders about your birds, & on Monday returned to sweet home. Home is doubtless very sweet, but like all good things one is apt to cloy on it; accordingly I have resolved to go to Woodhouse for a week. This is to me a paradise, about which, like any good Mussulman I am always thinking; the black-eyed Houris however, do not merely exist in Mahomets noddle, but are real substantial flesh & blood. Formerly I used to have two places, Maer & Woodhouse, about which, like a wheel on a pivot I used to revolve. Now I am luckier in having a third, & I hope I need not say that third is Osmaston: I must say, although for the 10th time, & although you doubtless would elegantly term it humbug, I do not know when I have spent 3 pleas⟨anter⟩ weeks. Would you be so kind as to present to your sister Emma, a few franks, which I have rummaged out, & I hope some few of them will be new. Mr Joseph splendid example was before me, he indeed gave original sonnets, I alas must be contented with my minor contribution

I hope, when you write, you will give me a most minute account of every thing alive & dead about Osmaston Remember me to your sister Julia & tell her how much amusement I received from her permission of peeping, & far from being able to forestall her in the article of news her letter afforded me as much pleasure as it did Catherine

Remember me most kindly to M^r & Mrs Fox & most tenderly (it is your own term) to the rest of your family & believe me my dear old Fox | yours sincerely | Chas. Darwin

I want to know the name of a butterfly, which you have got, its wings are *most wonderfully jagged*, & of a reddish colour, after an immense chase with all the servants in the house I at last captured it—

Look over your butterflies & you will soon perceive what I mean—

Endorsement: Charles Darwin Oct. 1828 WDF
Christ's College Library, Cambridge (Fox 5)

To J. M. Herbert [3 October 1828]

[Shrewsbury]
Friday

My dear Herbert

Very many thanks for your letter. owing to a mistake of not forwarding it, I only received a few days ago: I am very much obliged for all your labours in *the science*, & I daresay when I receive the treasure, it will be just as it should be. I saw Butler the day before yesterday, & he gave me an account of all your Welsh proceedings. Your last fatal trip to Drus ar (duzdy)[1] appears rather to have cooled your courage, for, from what I hear you appear to have been very studious & quiet, after I left Barmouth. When I arrived home, I received such a letter from old Whitley.[2] "weakness", "folly", "irresolution", "idleness" & many other long & terrible words filled up the measure of his epistle: I sent a most submissive & humble answer: but in return, merely received an acknowledgement of my *extreme candour*, another charge of the same description as before: I suppose you are at present enjoying all the delights, which, by the many true & *faithful* accounts, we have all heard Montgomeryshire must posses:[3] I suppose you are, in your accustomed manner, delighting all the little dear female hearts & receiving in return, those gentle insinivating smiles, which Squinny, Miss Jones & ("as hot as love") used to cast on you: I hear many a parting tear was shed at your sorrowful exit; poor dear squinny *could* only ⟨I⟩ suppose, shed *one*, but He⟨aven⟩ knows, that is not her fault: I have been enjoying myself very much in Derbyshire, the Music meeting went off very well. I shot pretty well: my Entomological pursuits succeded, & the Miss Foxes are very pleasant girls, & so altogether I have spent a very pleasant month

But now for business. I hear from Butler that you have all the beettles & butterflies When you pass through Shrewsbury, you can give them me. of course you will pay me a visit I am *almost* sure of being at home. I shall go up early, but not quite so soon as the 10th— Butler goes on Tuesday next

Believe me my dear Herbert | Yours, Charles Darwin

Exert your Zeal & try to add a few more beetles to the bottle. turn over a few old stones & roots of trees. Get the lady with a strong imagination to procure the beettle with a face so very dreadful

I cannot recollect your Christian name, so I shall Christen you G[4]

Postmark: Shrewsbury OC 4 1828 153
American Philosophical Society 2

[1] [Bwlch] Drws Ardudwy, a pass in Gwynedd, Wales.
[2] Charles Thomas Whitley, Herbert's first cousin and a close friend of CD at Cambridge. See *LL* 1: 49, 166.
[3] Herbert's home was in Montgomeryshire (now Powys).
[4] The letter is addressed to 'G. Herbert Esqr, Court Calmore, Welch Pool'.

To the Darwin Family[1] 3 October 1828

[The Mount, Shrewsbury]

To the

Charitable & Humane
The Case of a Distressed Sportsman—
1828. Oct 3—

Charles Darwin gent—humbly petitions all benevolently disposed persons to pay attention to his case—

Whereas he the aforesaid formerly gained a respectable livelihood by destroying hares, pheasants partridges & woodcocks, with the aid of a double barrelled gun, & the said gun becoming dangerous & liable to destroy the aforesaid Charles Darwins legs arms, body & brains & consequently unfit for use—he is reduced to lay his deplorable case before the charitable & humane being utterly unable to raise the sum requisite for the purchase of new Double barrd Gun— Value £20—

	£.	s.	d.
R W Darwin	5 =	0 =	0
Miss Darwin	5 .	0 —	0
Miss Susan Darwin	5 "	0 .	0
Miss Cath Darwin	5 "	0 "	0.

L
DAR 204.3

[1] The 'petition' is in Caroline Darwin's hand. The subscribers' names are in their own hands.

From Fanny Owen [26 October 1828]

[Woodhouse]

Dear Charles—

I assure you I should not have allow'd your *long* and *insinivating* dispatch to remain so long unanswer'd, but really have lately had such a great **press** of *business*, and my *keeper* been so active in *stirring me up* to my painting & *keeping* me *there* that I have hardly had a moment to myself— I wish you had been here last night for we went to the Play at Oswestry & had very good fun, it was quite a *miracle* that we persuaded Papa to let us go knowing his *antipathy* to that *kind* of *hinnocent* amusement, but the wonder ceased when we found that the *gay, dashing handsome, dissipated General Despard*[1] was the *only shootable in* the *Play House*— We went six inside the family Van, your two sisters, Mrs Parker who is very *corpulent* and *we three* only fancy what a *suffocation*! I really thought some valuable lives would have been lost for the *Black hole* at *Calcutta* must have been a *fool*[2] to it—but happily all escaped uninjured, and very good fun and plenty of laughing we had—but it is *brutal* to tell you all this— I hear you have been paying a *domestic* visit at Overton. Why did you not *drop in* here on your way? Owen I find has written to try to *bribe* you here on Monday. I'm sorry to say I shan't see you if you come for we are all going tomorrow to Erthig for the Birthday at Wynnstay—[3] The *purpose* of this note I find I have completely forgotten, but better *late* than *never*, it was to tell you how very much oblig'd to you I was for the Books, and also the *Swallow tail* which has absolutely *astounded* my weak mind, there is *something* about it so *werry pecoolier* My poor screens have been quite put by this week. Papa keeps me hard at work at that *odious picture* but I'll have my *revenge* for I'll be *paid well for it*. I wont pass the *best part* of *my life* in a *dirty cheese room* for nothing. *No! No!*

I have not rode since I saw you, but have *nobody* to *ride with* so *no wonder*!! A *quiet life* I have *now* and *that* is *something pleasant* and what I have not been used to for *some time past* The Williams (I believe about a *dozen* of them) are coming here next week, and the week *after* we are going to Eaton so I think we shall be *beWilliams'd*— I wish I had any thing to tell you either in the shape of *gossip* or mystery to fill this sheet, but I am as *stupid* as L.K.[4] ever *could be* & have not one *bright* idea to tell you— Not one game of Billiards have I had since I play'd with you. I can get nobody to play with & am a⟨fr⟩aid for *want* of *practice* shall forget all my *fine strokes*— *Burn this* as soon as you have *made out* the nonsense, remember it is a *four* instead of a *three corner'd note* so *must serve* for a *long time*—

Bilieve me in haste yrs truly an *"Unbelieving Jew"*

Paradise Row
½ past 12— Saturday night

AL
Postmark: Shrewsbury OC 26 1828 153
DAR 204.4

[1] John Despard, who was 83 years old in 1828.
[2] 'to be as nothing compared to' (*OED*).
[3] Wynnstay was the estate of Sir Watkin Williams Wynn, near Llangollen, Clwyd. Erthig was the large estate of the Yorke family in nearby Wrexham. Both are now National Trust properties.
[4] Possibly Lloyd Kenyon, 1804–36 (see letters from Sarah Owen, 18 February [1828] and from Fanny Owen, [9 March 1828]).

To W. D. Fox [29 October 1828]

[Shrewsbury]
Wednesday

My dear Fox

I should have answered your letter before this, but when you know the reason, you will say it is a good one.— But first, I must grieve over the loss both of your hawk & stuffed birds the first is the most serious one, but I cannot help grudging the fate of the others, as one gets a sort of affection for anything procured by one own self: I go to Cambridge on Thursday, & shall arrive there on Friday evening, & I hope it will not be long before you are there, & now for the glorious news. I have been introduced, & if I may presume to say so, struck up a friendship with M.ʳ Hope:[1] I met him at dinner, & I find he knows all my Scotch friends, & we had so much entomological talk, that he asked me bring over all my insects to Netley.[2] When we meet I will tell you the result, but I must mention, what he said, that for 4 or 5 years back, he had not seen such a rich case collected in one year ∴ yours must be still richer: I hope you had all the duplicates of carabi taken at Barmouth, as most are very rare, & some new he believes he is a perfect specimen of an Entomologist, so generous, straitforward. I could not prevail upon him to have any new ones, except 2 flies, he believes both new to England. My head is quite full of Entomology. I **long** to empty some information out of it into yours.— He thinks he can give me 3 or 400 speci⟨es⟩ at Christmas. the other day he sent 700 to D.ʳ Fleming.[3] In the spring he wants me to come an expedition, all over the Welch mountains & he insures me to find many *new* insects. I could write all day about him; but I long to see you! he has given me a great many water beettles.

Yours affectionat, | Ch Darwin

Catherine sends something more than her love to Julia, & entreats she will write soon.

I suppose you will soon be up at X.ˢᵗ Coll: as the *delightful! mornings!* at the Priory[4] are come to end.—

Postmark: Shrewsbury OC 3⟨0⟩ 1828 153
Christ's College Library, Cambridge (Fox 6)

[1] Frederick William Hope.
[2] Netley Hall, the Hope family seat, about 5 miles south of Shrewsbury.
[3] Probably John Fleming.
[4] Breadsall Priory, near Derby, home of Elizabeth Collier Darwin, widow of CD's grandfather Erasmus.

To E. A. Darwin   21 December 1828

[Shrewsbury]

(Preceded by a letter signed S.E.D.)[1]

My dear Erasmus

Susan has given me this third sheet to write to you upon; but why I should do so I know not. Apologies are come much too late for such shamefully ungrateful conduct as mine has been and excuses I have none. You are become like a stranger to me. I do not know from what point to begin with After you left Cambridge I got into very nice rooms in College, far more comfortable than lodgings,[2] as you will find when you come next to Cambridge. I imbibed your tastes[3] about prints, and put it into practice, and have bought some very good prints, which I long for you to see. I came down from Cambridge yesterday which I left very empty. I think I shall go and see old Edinburg this X'tmas, before all my friends entirely leave it. I left Fox up in Cam. in great awe and tribulation about his degree, which is to be made very much more strict, so that they give out that at least 50 will be plucked:[4] I live almost entirely with Fox and Entomology goes on most surprisingly. Price is in Cambridge and has got some pupils, he intends writing to you very soon. Kennedy[5] is a tutor at St. Johns: Whitley will be either first or second next year: I see a good deal of him and like him very much. This will be but a poor finishing to Susan's letter, and must be awfully stupid. When is there any prospect of your coming back again? I long very much to see the Bachelor of Medicine again.[6] Fanny Owen is as charming as ever and deserves your string of German epithets more than ever.

My dear old Erasmus | Yours most affectionately | C. Darwin

Copy
DAR 154

[1] The dates and note were written at the top of the copy by the copyist. Susan's letter to Erasmus was apparently not preserved.
[2] CD had taken lodgings above the shop of W. Bacon, tobacconist, in Sydney Street (*LL* 1: 163). The site, marked by a plaque, is now occupied by a branch of Boots the Chemists. His rooms at Christ's were on the south side of the first court (see *Christ's College Magazine*, Easter Term, 1909, pp. 184–208).
[3] The copyist wrote 'tastes' and, in the margin, 'obscured by seal'.
[4] The term 'plucked' originally meant to be refused a degree for any of several reasons, but it gradually came to be used only for failure in the examination.
[5] Benjamin Hall Kennedy.
[6] Erasmus passed the M.B. examination at Cambridge in 1828.

To W. D. Fox [24 December 1828]

[Shrewsbury]
Wednesday

My dear Fox

I am sorry I did not write to you, as I promised, on Monday, but we have had people in the house so could not.— I & M.ʳ Dash arrived quite safe here on Saturday morning He rises in my opinion hourly, & I would not sell him for a 5£ Pound note.— it would have excited your envy & spleen to have seen him on the scent of a covey of Birds, & the style in which he went down when I held up my hand.— I have begun to entomologize & have taken thirty specimens of Platynus Angusticollis: the best silpha: Dromius agilis, & a few other insects.—

You cannot imagine how pleased my Father was with the Death-Head's;[1] to use his own words "if he himself had thought for a week he could not picked out a present so acceptable". He came in very tired from Nottinghamshire, & the sight of them acted like a charm— I hope it is not the Entomological Mr. Boothby, (who married Lord Vernons daughter), who is so very ill: My Father in his way back again, called at the Priory: where Mrs Darwin gave a magnifiqùe account of you & your character to my Father.—

Catherine had a letter the other day from Bessy Galton,[2] which contained a flaming account of Erasmus[3] entry in the Navy, ushered in by right honourable Lords & gallant Captains.—

This letter is a most unconnected tissue of facts; but whilst I recollect, I must put you in mind that you have my snuff box, & I have yours; doubtless yours is the most valuable: but mine was the gift of Mr Owen, & he is the Father of Fanny, & Fanny, as all the world knows, is the prettiest, plumpest, Charming personage that Shropshire posseses, ay & Birmingham too, always excepting the blooming Bessy, & now that I know a most pleasant train of ideas are excited in your mind, I will not interrupt them by writing any more

But believe me my dear Fox | Yours most sincerely | Chas: Darwin.

Do write soon & tell me how Entomology goes on.— Catherine is ruminating some message to you, which will express a proper proportion of kindliness & decorum mingled in due measure. I cannot wait, so frame one according to your own liking.— I most cordially hope your degree goes on well. write a most minute & particular account of all your doings.

Postmark: Shrewsbury ⟨DE⟩ 25 1828 153
Endorsement: Decr 27, 1828.— C. Darwin
Christ's College Library, Cambridge (Fox 7)

[1] Death's Head moth (*Acherontia stropos*).
[2] Elizabeth Ann Galton.
[3] Erasmus Galton, Bessy Galton's brother.

To W. D. Fox [7 January 1829]

[Shrewsbury]
Wednesday

My dear Fox

You must have thought me a most ungrateful wretch to have taken not the slightest notice of your offer of getting up the morning of my departure from Cambridge; the reason was, (& I told Impey[1] particularly to mention it to you) that I was called so late that in the hurry I forgot all about it, & in despair I rushed to your rooms, but found the staircase so dark, that I had not time to crawl up them.—

I am much obliged for your giving me so minute an account of your Entomological proceedings, although they have excited no small degree of envy & spleen: I literally can take nothing, excepting one small insect, something like the Trufle beettle, which Henslow gave us.—

The Chrysalis goes on very well, it is much more lively, so that if touched, it will roll itself about.— & it appears to me the parts about the head are very much more distinct than they formerly were.— The process is,—a flower pot-full of dampish mold, & over that a stratum of not very dry sand, on which the Chrysalis is placed; an inverted glass vessel, which I suppose prevents too much evaperation, is placed over it,—the whole kept in a warm room.—

I suppose you have heard poor Mr Boothby is dead he expired whilst my Father was in the house.— Lord Vernon also is there he is quite paralytic on one side.— What an excessively nice old gentleman he appears to be, it makes one pity him so much more.—

From the bottom of my heart I pity your dismal state in Cambridge: I suppose there is scarcely any body up there, & wit⟨h⟩ this odious weather, & still worse degree, how it must make you long to be at dear old Osmaston.—

How very odd it is you do not hear from Chapman—² You can scarcely wish, half so much as I do, for it to be settled favourably, as I should think a curacy in Derbyshire would be pleasanter to you, but to me it will make all the difference whilst I remain at Cambridge

My Father has given me 20£ towards going to Edinburgh, but I still remain in doubt. it is so very long a journey to take in winter.— Remember me kindly to Hore & Jem. Holden—³ I long to hear of your being AB. as I have no doubt I soon shall.—

I remain dear Fox yours sincerely | Chas. Darwin

Postmark: Shrewsbury JA 8 1829 153
Christ's College Library, Cambridge (Fox 8a)

¹ Impey was CD's 'gyp' (college servant). See *Emma Darwin* 2: 166.
² Possibly Benjamin Chapman, a fellow undergraduate at Christ's College.
³ James Richard Holden.

To W. D. Fox [25 – 9 January 1829]

[Shrewsbury]

My dear Fox

I waited till to day for the chance of a letter, but I will wait no longer: I must most sincerely & cordially congratulate you on having finished all your labours. I think your place a *very good* one,¹ considering by how much you have beaten many men, who had the start of you in reading— I do so wish I were now in Cambridge (a very selfish wish however, as I was not with you in all your troubles & misery) to join in all the glory & happiness, which dangers gone by can give.— how we would talk walk & entomologize— Sappho should be the best of bitches & Dash of dogs: there should be "peace on earth—good will to men" (which by the way, I always think the most perfect description of happiness that words can give) I was very sorry to see Holdens name amongst what I suppose to be the plucked men,[2] & amazed to see Pulleins name no where.—[3]

It would be superfluous to thank you for the Newspaper, as I am sure you must know how very anxious I was hear the issue. I believe (?) we have some bets pending about M.[r] Philpot[4] (who it seems has astonished the knowing ones) & about yourself, which must remain in doubt till I go to Cambridge & look at my betting book.— I received the swan about a week ago, & should have written to acknowledge & thank for it before; only that I thought it better to wait till the examination was over. We have not yet eat it & it is probable it will keep sometime longer.—

My Father has had a bad fit of the gout together with a good deal of fever. he has been confined to his bed for a week.— he was very much pleased with the Hooper & as he said himself, if we had thought for a month we could not have made two more acceptable presents than the Deaths head & swan.—

I think it is probable we shall get for you both the common & Pine Marten, the latter alive, which I should think you would like, as th⟨ey⟩ are very interesting animals to keep tame, so an Irish gentleman tells me.—

Now for Entomology.— a beautiful Agonum with dark blood red elytra.— a small Elaphrus; an insect like the Pederus ripalis,[5] only with a white mark on each elytron.—cum paucis aliis.— in your next letter tell me how you get on in the science.— My life is very quiet & uniform, & what makes it more so, my lips have lately taken to be bad,[6] which will prevent my going to Edinburgh.— my Studies consist of Adam Smith & Locke, in the latter of which I suppose you are an adept, & I hope you properly admire it— About the little Go[7] I am in doubt & tribulation. I have had very little shooting. I went to Woodhouse for a week, & on the first day & first shot one of the young Owens cut his eye so badly with a Copper Cap; that he has been in bed for a week.— I think I never in my life time was half so much frightened: I am sure I have written enough about myself & my own concerns: make a handsome return tell me every thing about yourself. I am exceedingly anxious to know what steps you have taken about the Curacy? where you are going? how long you intend staying at Osmaston & the Larches?[8] in short do give me an outline for the 2 or 3 next months. I hope you will write to me directly Erasmus is in Rome & likes it very much. he wants me to join him in Geneva. he intends wintering next year in Paris

Remember me kindly to Whitley if you see him, Pullein & Hore, & believe me | My dear Fox | Yours most sin | C. Darwin.

if you do not get the Curacy I do not know when we shall meet again Qære. How much of this is legible?

Postmark: Shrewsbury JA 2⟨   ⟩ 1829 153
Christ's College Library, Cambridge (Fox 8)

[1] Fox ranked 88th on the list of 160 candidates (*Cambridge Chronicle and Journal* . . . , Friday, 23 January 1829, 2d edition; repeated, Friday, 30 January 1829).

[2] James Richard Holden's name appears fifth in a group of fourteen names, without numbers, following the 160 that are ranked. In the list printed 30 January 1829 his name along with others appears with an asterisk, which, a footnote explains, designates men who 'have one or more terms to keep previous to being *admitted* to their degrees, although they passed their examination in the above order of arrangement' (*ibid.*, 30 January 1829). No explanation is given of the separate unnumbered listing of 23 January.

[3] Robert Pulleine sat for his examination the following March (see letter to W. D. Fox, [26 February 1829]).

[4] Henry Philpott. CD's allusion is to his having achieved Senior Wrangler, that is, the student taking first place in the examinations for mathematical honours.

[5] No such species of *Paederus* exists. CD probably intended to write 'riparius'.

[6] A skin eruption about CD's mouth and on his hands was a recurrent malady. John Maurice Herbert, in a letter to Francis Darwin, stated that in his youth CD took small doses of arsenic for it. 'He told me that he had ment[d] this treatment to his Father—& that his Father had warned him that the cure might be attended with worse consequences—I forget what he said the risk was, but I think it was of partial paralysis.' (Letter from J. M. Herbert to Francis Darwin, 12 June 1882, DAR 112: 61v.). For a comprehensive study of CD's illnesses throughout his life, see Colp 1977.

[7] Undergraduates' name for the University 'Previous Examination', taken in the second year. 'The subjects of examination are one of the four Gospels or the Acts of the Apostles in the original Greek, Paley's Evidences of Christianity, one of the Greek and one of the Latin Classics' (*Cambridge University calendar*, 1829, p. 169). Of the works CD was studying, Locke's *An essay concerning human understanding* appeared in the B.A. examination for those who were not candidates for honours. No work by Adam Smith appeared on the list of examination subjects for either the 'Little Go' or B.A.

[8] The Larches, Sparkbrook, Birmingham, was the home of Samuel Tertius and Violetta Galton until 1832. (See Francis Galton *Memories of my life* (London: 1908), pp. 2–3, 18).

To W. D. Fox [26 February 1829]

Cambridge
Thursday

My dear Fox

When I arrived here on Tuesday, I found to my great grief & surprise, a letter on my table which I had written to you about a fortnight ago; the stupid porter never took the trouble of getting the letter forwarded: I suppose you have been abusing me for a most ungrateful wretch: but I am sure you will pity me now, as nothing is so vexatious as having written a letter in vain.—

Last Thursday I left Shrewsbury for London & staid there till Tuesday, on which day I came down here by the Times:[1] The two first day I spent entirely with M[r] Hope.—& did little else but talk about & look at insects: his collection is most magnificent & he himself is the most generous of Entomologists he has given me about 160 new species, & actually often wanted to give me the rarest insects of which he had only two specimens: He made many civil speeches, & hoped you will call on him some time with me, whenever we should happen to be in London. He greatly compliments our exertions in Entomology & says we have taken a wonderfully great number of good insects

On Sunday I spent the day with Holland,[2] who lent me a horse to ride in the park with.—

On Monday evening I drank tea with Stephens:[3] his cabinet is more magnificent than the most zealous Entomologist could dream of: He appears to be a very goodhumoured pleasant little man.— I am glad to find that neither he nor Mr Hope have much opinion of Mr Jennynys;[4] they seem to think him very selfish & illiberal.— Whilst in town, I went to the Royal Institution Linnean society, & Zoological gardens, & many other places where Naturalists are gregarious.— If you had but been with me, I think London would be a very delightful place; as things were it was much pleasanter, than I could have supposed such a dreary wilderness of houses to be.—

I shot whilst in Shrewsbury a Dundiver, (female Goosander as I suppose you know). Shaw has stuffed it, & when I have an opportunity I will send it to Osmaston: There have been shot also 5 Waxen Chatterers, 3 of which Shaw has for sale, would you like to purchase a specimen? I have not yet thanked you for your last very long & agreable letter: It would have been still more agreable had it contained the joyful intelligence, that you were coming up here; my two solitary breakfastes have already made me aware how very very much I shall miss you— How goes on Entomology? You ought to take a vast number of Mycetophagos insects in the Ash Wood

Your new Dyticus is a most extraordinary one: I have ordered a 15£ Cabinet, & when we get the insects all arrayed together; we shall be able to make them much better out. I have taken this Vacation the Diaporis Anea, an insect which only Stephens & Hope posses.—

By Grahams[5] decided advice, I do not go in for my little Go.— Hoar left Cambridge for Devonshire this morning. Pulleine goes in for his degree today, in appearance he is in very good spirits, but his manners are quite different, & I think he has not at all recovered his most unfortunate accident.— Wedgwood[6] is elected a fellow.— & this is all the news I can scrape together; Cambridge is in statu quo I left it, rain, sleet & cold winds, alternating.

Believe me | My dear old Fox | Most sincerely Yours | C. Darwin

Write soon, for I long to hear what the Bishop says:

Postmark: Cambridge FE 26 182⟨9⟩
Christ's College Library, Cambridge (Fox 9)

[1] The 'Times' mail coach left London for Cambridge at 3 p.m. (*Cambridge University calendar*, 1829).
[2] Edward Holland.
[3] James Francis Stephens.
[4] In the *Autobiography*, pp. 66 – 7, CD writes of Leonard Jenyns, 'At first I disliked him from his somewhat grim and sarcastic expression . . . but I was completely mistaken and found him very kindhearted, pleasant and with a good stock of humour.'
[5] John Graham, Fellow, later Master, of Christ's College and CD's tutor.

[6] Hensleigh Wedgwood had taken his B.A. (8th Wrangler) in 1824. Because he was last in the Classical Tripos, that place became known as the 'wooden wedge', corresponding to 'wooden spoon', the traditional designation for the last place in the Mathematical Tripos (*Alum. Cantab.*). He was elected Fellow of Christ's College in 1829.

From John Coldstream[1] 28 February 1829

My dear Darwin

I confess I was rather illpleased to receive your letter in place of yourself, as I had been looking for your arrival in Edinburgh every day for some weeks before; Mr Barker having apprised me of your intention to visit us about Christmas. However, as you may guess, even the letter received a hearty welcome. I hope this will find you restored to health, and engaged in planning a descent upon Scotland in the summer holidays. You ask me what I have been doing since I left Paris. I remained in France till June, when I set out for Prussia with the hope of ascending the Rhine, and travelling thro' Switzerland & the North of Italy. But I took ill in Westphalia and was obliged to hurry home thro' Holland. I returned to Leith about the end of July, in a state of health which prevented me from deriving pleasure even from Natural History. I recovered very slowly, but now am quite well.

During the winter, I have been almost entirely engaged in practice, and have persevered in my abjuration of Zoology, with a few reservations in favor of walks along the Seashore & occasional visits to the Black rocks.[2] I found these localities, abounding in so many interesting associations, and offering an ever varying succession of animals, to be more powerfully alluring even than the Jardin des Plantes. Perhaps there is no real occasion for the *greif* which you express, in imagining that I have entirely forsworn Natural History: I am still ready to agree with you in the sentiment 'that no pursuit is more becoming for a physician than Nat: Hist:' and I feel myself as much, if not more, inclined than ever, to look into the World of Nature around me, but I feel, that as I must soon enter upon the public & the responsible discharge of the duties of my profession, I owe it as a duty of the highest importance, both to others & to myself, to make up as perfectly as possible, the deficiencies in *useful knowledge* which, I have too good reason to fear, my devotion to science (unprofitable as it has been) has occasioned. You may depend upon this, *at least*, that "the good old Cause of Zoology" is not slighted by me; and of course, still less so, the friends in whose company I formerly prosecuted it.—

Glasspoole[3] left Paris in May with the intention of going through Switzerland & Germany: I heard of his being in London in September; when I believe Dr Grant[4] saw him: and I have since heard that he has settled in *Brighton*: or, at least, that he is living there. I was happy to receive intelligence regarding Mr Hope: I am greatly indebted to him for a valuable collection of types of the British genera of Coleoptera, which he sent me, probably with the view of stimulating me in the study of Entomology; but I have done very little in it. I am not unmindful of Mr Hope's kindness, and although my long silence may appear to him a mark of ingratitude, and unpardonable carelessness, I am not without some expectation of being able at some future time to redeem my lost character in his eyes. You will probably have seen Dr Grant by the time you receive this. If you should be in London, be so good as present my kindest regards to him, and say, that I received his letter, and intend answering it shortly. You ask me for Edinburgh news, but truly I know of none worth giving you; excepting, indeed, that the *Plinian*[5] is flourishing most vigorously.— There are several very industrious young naturalists in it at present; and Brown[6] continues to lend his matured experience in the presidency. David Ritchie has succeeded his father as minister of the Parish of Tarbolton in Ayr.[7] Your request about the Carabidae I cannot answer. The only entomological information I can give you, is, that I had an excellent opportunity in Westphalia of examining the singular habits of the Bombyx processionaria (larva) described by Reamur.[8] I verified most fully all his interesting description. The caterpillars were particularly abundant in the woods of Westphalia last year, and destroyed the leaves of almost all the oaks. You are, of course, aware of the publication of Dr Fleming's "British Animals",[9] and Mr Stark's "elements of Natural History."[10] D^r F's is a very valuable work but Mr Stark's, being more general and more complete, will probably enjoy the larger share of popularity. Dr Grant is engaged in the publication of "*Outlines* of Zoology & Comparative Anatomy."—[11]

Be so good as write me again soon, and tell me something of the present state of Natural History in Cambridge. Have you had any opportunity of studying marine Zoology since you left this?—

With every affectionate wish I remain, | My Dear Darwin, | Yours most truly | John Coldstream

Leith.
28th. Feb. 1829.

DAR 204.3

[1] CD's fellow medical student and friend at the University of Edinburgh. In the *Autobiography*, p. 48, CD describes him as 'prim, formal, highly religious and most kind-hearted: he afterwards published some good zoological articles [for Todd's *Cyclopaedia of anatomy and physiology*].' He proposed CD for membership of the Plinian Society (see n. 5 below).

[2] In a notebook in which CD recorded observations during the spring of 1827 there is the following entry: 'Observed with Mr Coldstream at the black rocks at Leith an Asterias rubens' (DAR 118, p. 12).

[3] Frederick Bream Glasspoole.

[4] Robert Edmond Grant, CD's close associate and mentor in zoology at Edinburgh (see Jespersen 1948−9 and *Autobiography*, pp. 49, 51). In an article published in July 1827, on 'the ova of the *Pontobdella muricata*' Grant gave CD credit for being the first to identify them: 'The merit of having first ascertained them [the ova] to belong to that animal is due to my zealous young friend Mr Charles Darwin of Shrewsbury, who kindly presented me with specimens of the ova exhibiting the animal in different stages of maturity' (Grant 1827; Darwin Library − CUL: 'Philosophical tracts'). This was the first scientific discovery by CD to be reported in print, though CD was not the first to make it (see Ashworth 1935: 105−6).

[5] The Plinian Natural History Society was founded in 1823 by Robert Jameson. Ashworth 1935 has a detailed account of the Society. For mention of CD's reports to it, see the letter to Caroline Darwin, 6 January 1826, n. 5.

[6] William Alexander Francis Browne.

[7] In December 1826 Ritchie successfully opposed the election of Georges Cuvier and Johann Friedrich Blumenbach to honorary membership of the Plinian Society (Gruber and Barrett 1974, p. 39).

[8] René-Antoine Ferchault de Réaumur. The *Bombyx* larva is described in Réaumur 1734−42, 1: 172, pl. V.

[9] Fleming 1828 (Darwin Library − CUL).

[10] Stark 1828.

[11] No publication with this title has been located.

To W. D. Fox [15 March 1829]

[Cambridge]

My dear Fox

I am afraid if you judge from the time I take in answering your letters you will think I am not very impatient to receive them. Your last letter however was at no time more acceptable to me, as our correspondence, through the Porters stupidity had nearly come to an untimely end;—

I am much grieved to find how likely it is my prophecy will be verified, & am still more so to hear the cause. I hope however by this time that M[rs] Bristoe[1] is quite recovered.— I am glad to hear that you are reading divinity: I should like to know what books you are reading, & your opinions about them; you need not be afraid of preaching to me prematurely.—

J. Price is reading very hard on the same subject, & seems to find it no ordinary labour.— I suppose you have heard by this time both from Holden & Pulleine: the latter, as you will perceive is half mad with joy: I saw him the next morning & he literally could not sit still in any posture for five minutes together.— I am anxious to hear how you like Clifton, & what you are doing there? I suppose your unsettled life will prevent you doing much of anything, & amongst the rest of Entomology: I had such a dose of "the Science" in London; that I have hardly recovered from it even to this time: this sort of scientific seediness after a nights debauch joined with M[ss] Polo & Habours[2] idleness, has, *at present*, entirely put a stop to my Entomological proceedings: I am leading a quiet every-day sort of a life; a little of Gibbons history in the morning & a good deal of *Van. John*,[3] in the evening this with an occasional ride with Simcox[4] & constitutional with Whitley, makes up the regular routine of my days.—

I see good deal both of Herbert & Whitley, & the more I see of them, increases every day the respect I have for their excellent understandings & dispositions. They have been giving some very gay parties—nearly sixty men there both evenings.— You ask me about the money? I know no more than the man in the moon what the sum is, indeed, I had entirely forgotten all about it till I received your letter. I thought I had *occasionally* heard you mention the regularity with which you keep your accounts. I must confess, their *utility is most striking*.—

I have paid Clayton[5] for the memorable Swan. After all it was hardly kept long enough: it was pretty good, but tasted like neither flesh nor fowl, but something half way, like Venison with Wild Duck— Have you heard anything more of the old sinner the incumbent, not that I exactly understand how his not paying for dilapidation influences you, as if I understood right, you were to live in a farm house situated some miles from any other human habitation. I hope you will write soon, & give me an account of all your "outgoings" & "incomings".—

Give my most kind respects to M[rs] Fox & the rest of your family & believe me my dear old Fox, yours most sincerely | C. Darwin

Recollect: that Deo Volente whether your Parsonage boast of a roof or not I shall pay you a visit this summer.

Sunday Evening

Postmark: Cambridge 1829; C 17 MR 17 1829
Christ's College Library, Cambridge (Fox 10)

[1] Mary Ann Bristowe, Fox's older sister.
[2] Messrs Polo and Harbours sold specimens of insects to undergraduate entomologists (see letters to W. D. Fox, 1 April [1829] and [10 April 1829]). Polo was a apparently a nickname. In an anecdote about his father, George Darwin says: 'Amongst his Cambridge expeditions I remember his speaking of going down to the fens . . . with a sporting sort of guide who went by the name of Marco Polo, because he carried a leaping pole with a flat board fastened at the bottom for leaping the ditches' (DAR 112 (ser. 2): 17).
[3] A corruption of 'vingt-et-un', the popular card game.
[4] No Simcox is listed in *Alum. Cantab.* or in the *Cambridge University calendar*. Possibly a familiar name for George Simpson (see letters to W. D. Fox, [18 May 1829], 'Simpson', [3 January 1830], 'old Simpcox').
[5] Robert Clayton, Cambridge fishmonger and wild-fowl dealer.

To W. D. Fox 1 April [1829]

X[st] College [Cambridge]
April 1st

My dear Fox

In your letter to Holden you are pleased to observe, "that of all the blackguards you ever met with I am the greatest": Upon this observation I shall make no remark, excepting that I must give you all due credit for acting on it most rigidly: And now I should like know, in what one particular are you less of Blackguard than I am? You idle old wretch why have you not answered my last letter which I am sure I forwarded to Clifton nearly 3 weeks ago? If I was not really very anxious to hear what you are doing: I should have allowed you to remain till you thought it worth while to treat me like a gentleman.—

And now having vented my spleen in scolding you, & having told you, what you must know, how very much & how anxiously I want to hear how you & your family are getting on at Clifton, the purport of this letter is finished. If you did but know how often I think of you & how often I regret your absence, I am sure I should have heard from you long enough ago: I find Cambridge rather stupid, & as I know scarcely any one that walks, & this joined with my lips not being quite so well, has reduced me to sort of Hybernation, which almost equals "poor little Whitmores" melancholy case.—[1] Old Whitley has begun to take your place, & we have just commenced a regular series of constitutionals.—

Entomology goes on but poorly: a few Dromius & Agonum's, together with the Pæcilus (with red thighs) make the g⟨reat⟩ part of what I have collected this ter⟨m⟩. I have caught M^r. Harbour letting Babington[2] have the first pick of the beettles; accordingly we have made our final adieus, my part in the affecting scene consisted in telling him he was a d——d rascal, & signifying I should kick him down the stairs if ever he appeared in my rooms again: it seemed altogether mightily to surprise the young gentleman.—

I have no news to tell you, indeed, when a correspondence has been broken off like ours has been, it is difficult to make the first start again.— Last night there was a terrible fire at Linton eleven miles from Cambridge; seeing the reflection so plainly in sky, Hall,[3] Woodyeare[4] Turner[5] & myself thought we would ride & see it we set out at ½ after 9, & rode like incarnate devils there, & did not return till 2 in the morning. altogether it was a most awful sight.—

I cannot conclude, without telling you, "that of all the blackguards I ever met with, you are ye greatest & the best | C Darwin.—

1 o'clock—going on very well

Postmark: Cambridge AP 2 1829
Christ's College Library, Cambridge (Fox 11)

[1] Ainslie Henry Whitmore. The reference may be to his having been 'rusticated'.
[2] Charles Cardale Babington.
[3] Jeffry Brock Hall.
[4] John Fountain Woodyeare.
[5] James Farley Turner.

To W. D. Fox [10 April 1829]

[Cambridge]
Friday

My dear Fox

The very day I received your letter, I had given up all hopes of receiving one, & had written to you again at Clifton, in order to receive some account of all your proceedings: I suppose, as you are now at Bath, you will not have the pleasure of receiving it: that will not however be a very serious loss to you: as it contained nothing that might not as well have been left unsaid, excepting the old burthen of all my letters, that I wished to hear what you are doing.—

I have been in such a perfect & absolute state of idleness, that it is enough to paralyze all ones faculties; riding & walking in the morning, gambling at Van John to a most disgusting extent in the evenings, compose the elegant & instructive routine of my life. Lord help me, with this basis, how is it likely that I should be able to write a letter: for six weeks past have I been intending daily to write to my injured brother, & I cannot screw my courage to the sticking point, & so you ought to be much flattered at my writing at all to you. Therefore you must take the intention for the deed, & excuse my abominably stupid scrawls: Markham has not sold Sappho accordingly I have given her to him: There are no letters for you: Pullein wrote to you at Clifton & wants an answer, direct Rev. D^r. Burrows, no 30 Fitzroy Sq^r. London where he is going to reside for 6 months in order to read divinity: The only thing that is talked about in Cambridge, is little Go, which has been unusually strict. most men are in the first class, except Woodyeare, but a wonderfull number were called in a second time:

Your friend M^r. Wale, the proctor, was most gloriously hissed, & pelted with mud:[1] he was driven so furious that even his own gyp dared not go near him for an hour.—

I shall go up to Town for a few days with old Whitley next week, & I wish I could afford to put your very pleasant scheme into execution. But all my allowance is spent, & 2 Tutors bills may be expected.—in due time.— I think it is probable that you may see Wilmer[2] there sometime this Easter Vacation.— Polo has not brought me many insects lately, but the Pieman has brought me a great many, inter alia Chlænius holoriseus.—[3] I am sadly in want of somebody to entomologize with. I wish A. Way was here. Holland saw him in Switzerland from whence he was going into Italy. Who do you think walked into my room the other day? Our old Irish friend, dressed most respectably, he said M^r. Wynne's father had not helped him but that he had got into a very good place at Oxford, & was on his way to Ireland. he said he had written to you, he appeared very grateful & begged me most particularly to express his thanks to you: I do not think we have often spent a small sum of money more usefully:

Chapman tells me that he intends writing to you very soon: I cannot comprehend your plans. Whose Curacy is it you wish to get?

I am very sorry to hear that Mrs Fox has been unwell, but I most sincerely hope she will soon be recovered

Yours most affectionately | C. Darwin

Postmarks: Cambridge ⟨A⟩P 11 ⟨1829⟩; C 13 AP 13 1829
Christ's College Library, Cambridge (Fox 12)

[1] Alexander Malcolm Wale. 'In Apr. 1829 he was mobbed by undergraduates who pursued him from the Senate House to the gates of St John's, hissing and groaning, and though some of the offenders were summoned before the Vice-Chancellor's Court, the Proctors were not satisfied and sent in their resignations' (*Alum. Cantab.*).
[2] Bradford Wilmer.
[3] *Chlaenius holosericeus.*

To W. D. Fox [12 April 1829]

Sunday

My dear Fox

I received your letter yesterday morning, & it was with the most sincere grief I read its contents: I am well aware how illtimed all consolation must at present appear. But yet knowing how very unhappy you must be at your sisters most alarming illness; I cannot help troubling you with these few lines, if it were merely to thank you for writing to me: I should indeed be sorry, if anybody, for whom I have such a strong friendship & respect as I must always feel, my dear Fox, for you, should be suffering under the most poignant grief, & I not aware of it.—

I most fervently hope, by the time this letter reaches Clifton, that your sister will be better. When you have a *spare* moment do write to me & tell me how she is going on.—

I wrote to you at Bath only the day before I received your last letter, but little did I think, when I wrote it, that it would find you in such a melancholy situation

Believe me my dear Fox | Yours most affectionately | Charles Darwin

Cambridge

Postmark: Cambridge AP 14 1829
Christ's College Library, Cambridge (Fox 13)

To W. D. Fox [23 April 1829]

Thursday

My dear Fox

I have delayed answering your last letter for these few days, as I thought that under such melancholy circumstances my writing to you would be probably only giving you trouble.—

This morning I received a letter from Catherine informing me of that event, which indeed from your letter I had hardly dared to hope would have happened otherwise.—[1] I feel most sincerely & deeply for you & all your family: But at the same time, as far as anyone can, by his own good principles & religion be supported under such a misfortune, you, I am assured, well know where to look for such support. And after so pure & holy a comfort as the Bible affords, I am equally assured how useless the sympathy of all friends must appear, although it be as heartfelt & sincere, as I hope you believe me capable of feeling.—

At such a time of deep distress, I will say nothing more, excepting that I trust your father & M^rs. Fox bear this blow as well as under such circumstances can be hoped for.—

I am afraid it will be a long time, my dear Fox, before we meet. till then believe me, at all times | Yours most affectionately | Charles Darwin

Cambridge

Postmark: Cambridge AP 23 1829
Christ's College Library, Cambridge (Fox 14)

[1] Fox's sister, Mary Ann Bristowe, died 19 April 1829.

To W. D. Fox [18 May 1829]

[Cambridge]
Monday Morning

My dear Fox

I was exceedingly glad to receive your last letter, as I had been for some time past very anxious to hear how you & the rest of your family were after having undergone such a painful affliction.— I should have written before only I waited to get the Bills you mentioned.— Claytons amounts to 2£″7″10, but Bakers' has not arrived yet.—

I have procured for you from Polo a beautiful specimen of the Ring Owzel, & gave 5 shillings, also a young male ring tail, with half the plumage blue & half like the female ring tail price 5^s—also 2 fine specimens of the black tern, price 3^s—also a Royston Crow.—

Now being upon Nat: history, I will go on: All the Crysalis you left in the deal Box, I never discovered till too late, & found all the moths utterly destroyed. The Cossus has eat its way out & is lost.—

Aiken[1] has got all the skins of your birds: he also had my Crysalis, which go on very well, & to which I have added a great many more.— I have taken, or rather had collected for me vast numbers of insects: a good many new ones to

Cambridge, both large & small.— have since Christmas filled nearly three Standish Boxes.—

But Hoare's success is *most* brilliant: Carabus intricatus, an insect fit for Paradise; a beautiful Lamprias: Badister cephalotes, & multa alia: My success also has been very good amongst the water beettles. I think I beat Jenyns in Colymbetes.—

I want much to hear what you have done in Entomology? I heard from old Pulleine the other day. he wrote to enquire about you. I told him he had better direct to Osmaston so I daresay before long you will hear from him.

Before I tell you my plans for the next 2 or 3 months, I must most cordially thank you for your very kind invitation to Osmaston. You used accuse me of *making speeches* about Osmaston. This accusation although proved to be most palpably absurd by the length of my former visit, shall be made doubly so, by the pleasure with which I accept your invitation. On the 5th of next month I go to Shrewsbury, either through London, in order to chaperon home my sisters, or else d⟨ ⟩ through Birmingham. A *few* days of home, & then I go a three week trip with Mr Hope either in S or W Wales: after that home again (& Woodhouse & Fanny), then to Maer, & then I hope to have the pleasure, my dear Fox of seeing you again, after our long absence: this till August will fill my time up pretty well.—

As to what I have been & am now doing, the less that is said about it, the better: my time is solely occupied in riding & Entomologizing. Simpson or Hall are both industrious when compared with me. I cannot help thinking what a great deal of good advice I should receive were you but here.— I hope you will write soon & tell me all about every thing: I am sorry, although I think you are very wise in not going into orders at present Your mind at such a time must be very unfit for so serious an undertaking: Make any use of me & my rooms in future arrangements that may please you:

Give my most kind remembrances to your father, & I hope that he & Mrs Fox are going on pretty well: | Believe me my dear Fox | Yours most sincerely | Chas. Darwin.—

Postmark: Cambridge MY 18 1829
Christ's College Library, Cambridge (Fox 15)

[1] Not identified. He was perhaps a dealer in natural history specimens.

From Francis Knyvett Leighton [3 June 1829]

My dear Charles,

I do not know whether you take much interest in the trials of strength, which are about to take place between the sister Universities,—but, if you are inclined to take this opportunity of exploring our "classic shades" I shall be most happy to be your *chaperon* to Henly for the boat race,[1] & to the Magdalen cricket ground for the cricket match on the following day—& afterwards I will be your cicerone thro' the lions of Oxford & its neighbourhood. I cannot promise you much fun, as I am living here in a very unsettled manner, having for some time broken up my *College establishment*, but I can give you board & lodging, if you will be content with Bachelor's fare.

I hear the Cantabs are very confident of success, & there has been so much dissension among our crew & mismanagement among the directors, that I think there is every probability of your being triumphant. However I have sufficient exprit de corps to support my colours, &, if any of your betting characters are inclined to bet 3 to 1 on the Cambridge Heroes, I shall be obliged to you if you would take the bet for me as far as 5£ will go—

I trust I shall receive a favorable answer— Write immediately & tell me the day & hour you mean to come here & I will have all things in readiness for you— Y^{rs} very truly | F. K. Leighton.

Magd:[alen] Coll: [Oxford]
Wednesday.

Postmark: Oxford JU 3 1829
DAR 204.3

[1] The first Oxford—Cambridge Boat Race took place at Henley on 10 June 1829. Oxford won easily.

To W. D. Fox   7 June [1829]

[Cambridge]
Sunday. June 7th.

My dear Fox

You did not give me time in your last letter, in order to see about your commissions, then write to you & wait for an answer: accordingly I have been obliged to act to the best in my power, & the imperfections must be supplied next October: I booked & sent of per waggon, a portmanteau & deal box yesterday, containing I hope & trust every thing except your skins of birds & prints; Pulleine had left such imperfect directions with Aiken that we did not know yours from his: & we agreed that they had better be kept in my rooms till either Pulleine or you could see them: they are all safely packed up together with the ring Owzel: the other skins I have sent with the boxes (but one of the terns was too far gone). About the prints I thought they would go better some other time than with the boxes & tea chests.—

You will find all the Keys in the deal box, which I had well nailed up & corded: & now for money matters;[1]

	£	s	d.
	£	s	d.
Clayton	2 "	7 "	8[2]
Baker	2 "	3	
Brett		10 "	6
Dos for plate		" 16 "	
	"	5 " 17 "	6

Paid for you for skins		s	
& &c &c	"	" 15 "	
Debt.	2 "		
	£	2 " 15	
	£	5 " 17 "	6
	"	8 " 12 "	6

Accordingly I have paid for you altogether, 8£ " 12 " 6 & as you sent me 10£, I owe you, 1£ " 7 " 6, which I will pay you when I see you in the Summer I have got all the receipts safe in Cambridge:—

The man from whom you bought poor little Fan told me he saw the other day in Bedfordshire, in very good keep & a very great favourite under such circumstances, I thought it better to leave her where she was.—

So much for your affairs. I am afraid I have executed them imper⟨fect⟩ly but it is to the best of my power.—

And now for myself. I leave Cambridge tomorrow morning, for London—& on Tuesday I go to Shrewsbury, from thence I go into Wales with Mr Hope to Entomologize. I will write to you again, whilst on our expedition: my success has been very splendid in the science, as you will say, when you next see my insects. I will write soon again. Yours affectionately | Chas. Darwin

Christ's College Library, Cambridge (Fox 16)

[1] The accounts are in a boxed-off section at the lower right of the sheet.
[2] A mistake for £2 7s. 10d. (see letter to W. D. Fox, [18 May 1829]).

To W. D. Fox [3 July 1829]

Shrewsbury
Friday

My dear Fox

I should have written to you before, only that whilst our expedition lasted I was too much engaged, & the conclusion was so unfortunate, that I was too unhappy to write to you. this weeks quiet at home, & the thoughts of Woodhouse next week has at last given me courage to relate my unfortunate case.—

I started from this place about a fortnight ago to take an Entomological trip with Mr Hope through all North Wales: & Barmouth was our first destination. the two first days I went on pretty well, taking several good insects, but for the rest of that week, my lips became suddenly so bad, & I myself not very well, that I was unable to leave the room, & on the Monday I retreated with grief & sorrow back again to Shrewsbury. The two first days I took some good insects, Cillenum latterale, (Stephens,[1] 16th No) Heterocerus 2. species. Calathus latus, (bad specimen of), & crocopus: Opatrum tibiale: Cerambyx nebulosus: & many others: But the days that I unable to go out, M^r. Hope did wonders several rare species, of the genera Helobia, Carabus, Omaseus, Elater Nothiophilus, & to day I have received another parcel of insects from him, such Colymbetes, such Carabi, & such magnificent Elaters, (2 species of the bright scarlet sort). I am sure you will properly sympathise with my unfortunate situation; I am determined I will go over the same ground that he does before Autumn comes, & if working hard will procure insects I will bring home a glorious stock.—

It is so long a time since I have heard from you, that I am very anxious to receive a letter I want to know what you have been doing lately, & what your plans are for the rest of the summer. Catherine wants to know whether M^rs. Fox & Julia are returned to Osmaston? I am most dreadfully stupid here, so you must excuse this equally stupid letter. I hope you will write soon, & then I will promise to write a longer one.—

Catherine sends her kind ⟨remem⟩brances to you & the rest of your family, in which I beg leave to join & believe me My dear Fox, Yours most sincerely, | Chas. Darwin

I hope the boxes have arrived safe: keep an account of how much money I am indebted to you, as I have already forgotten

I forgot to tell you, that my Father has got two Martens, I believe both species, ready mounted at M^r. Shaws, & I have got a dundiver, which can be sent anytime you like

Postmark: Shrewsbury JY 4 1829
Christ's College Library, Cambridge (Fox 17)

[1] Stephens 1827—46, *Mandibulata. Cillenium laterale* is described in the issue of 1 July 1828 (vol. 2: 4); in the bound volumes at the Cambridge University Library the issues are not numbered. Of the eleven volumes, only volumes one and two of *Haustellata* are preserved in one bound volume at Darwin Library—CUL. CD's annotations are not numerous; they consist mainly of adding common names or colour identifications to the descriptions.

To W. D. Fox [15 July 1829]

My dear Fox

Many thanks for your letter, it was very good of you to write so soon, the reason of my writing now is not to fulfill my promise of a long letter, for that is, I am afraid impossible, but to tell you, that, yours & Pulleins birds skins are in my room. M.ʳ Aiken seemed to be rather troubled with the size of the Boxes, for indeed they quite filled his little room, so I believe he has one box, & I have two, (in one of which is the ring-Owzel), but I am sorry to say the Key is locked up in one of my drawers. But if you were to write to Impey, he could send the boxes (first getting them broken open & packed up by Mʳ Brett) to Osmaston— I am very sorry I have given so much trouble. But I did not think it fair to leave such Leviathan-like boxes in M.ʳ Aikens house.— I am also sorry to hear that some of your insects got loose. I thought I had packed them pretty tight:—

It is terrible news to me to hear that you have been idle in the good cause, but I hope soon that your zeal will be renewed. You will see my name in Stephens' last number.[1] I am glad of it if it is merely to spite M.ʳ Jenyns.— I shall like very much looking over all your insects again, & I think I shall be able to name a great number of the *genus* Colymbetes: I have a great number of insects at Cambridge for you, & you shall have the duplicates of every thing I take this summer. whenever I take any thing good, you may rely upon it, that you always are uppermost in my mind: I shall not soon forget my first entomological walks with you, & poor little Way: I cannot help often wishing to hear something about him.—

I have some thoughts of going for a week into Wales again this summer: would it be impossible for you to come here & then go on with me. We should do wonders amongst the Carabi.— I told Polo to collect, but I am afraid he will never do much more as an Entomologist he is grown far too idle. You had better try your hand at him, for I despair.— You shall receive before very long, the two Martens, and my Dundiver.—

Recollect what the sum of money is that I am indebted to you, as I do not keep account, but you pretend to do, so put it to the proof, and do not forget.—

I am going to Maer next week in order to Entomolog & shall stay there a week: for the rest of this summer I intend to lead a perfectly idle & wandering life, always taking care to have as little of home & as much of Woodhouse as possible. You see I am much in the same state that you are, with this difference you make good resolutions & never keep them. I never make them, so cannot keep them: It is all very well writing in this manner, but I must read for my little Go. Graham smiled & bowed so very civilly, when he told me that he was one of the six appointed to make the examination stricter, & that they were determined they would make it a very different thing from any previous examination that

from all this, I am sure, it will be the very devil to pay amongst all idle men &
Entomologists. Erasmus we expect home in a few weeks time; he intends passing
next winter in Paris:—

Be sure you order the 2 lists of insects, published by Stephens, one printed on
both sides, & the other, only on one. You will find them very useful in many
points of view.

Dear old Fox, yours | C. Darwin

Postmark: Shrewsbury JY 16 1829
Christ's College Library, Cambridge (Fox 18)

[1] It is not clear to which number of Stephens 1827−46 CD refers. The 1 June 1829 number on
Haustellata, vol. 2, has an entry (p. 200) '*Graphiphora plecta* "Cambridge".— C. Darwin, Esq.' The
number dated 15 June 1829 of *Mandibulata*, vol. 2, Appendix (p. 194) lists 13 species of beetles
reported by CD. The report of the moth, if it appeared early in June, would be the first time CD's
name appeared in print, but in the *Autobiography*, p. 63, where he writes of his delight at first seeing
the magic words, "captured by C. Darwin, Esq.", his memory is of beetle collecting. Unfortunately
the 'magic words' do not help in making a clear decision about the date or the publication; they do
not occur in either list, nor have they been found elsewhere in *Illustrations of British entomology*. A total
of 35 species of insects, 34 of them beetles, are listed as taken by CD. (See also Freeman 1977, pp.
19−20.)

To W. D. Fox   29 [July 1829]

[Shrewsbury]
Wednesday 29th.—

My dear Fox

I take this opportunity of writing a few lines to you. It was by a mistake of
Mr Shaw that the beasts & birds were not sent to you before this time.— I hope
you will approve of their stuffing & that they will arrive safe & sound. The
one with the smaller tail is from Ireland, & the other from Wales. Which is the
Pine & which is the common Marten appears to me hard to ascertain.—

I returned from Maer on Saturday evening, & was most exceedingly grieved to
hear from Catherine that there is no hope of seeing you & your family this
summer at Shrewsbury I want very much to hear from you as I do not quite
understand your plans. You are going into Yorkshire, sometime in the end of
August, & I want to know how soon you will return? for I am very sorry to say
that I do not see how it is possible for me to come to Osmaston, between this
time & your intended trip to Yorkshire. The reason of this is that my sister
Mrs Parker has the misfortune of having two unchristened children, & I am
going after the first week in August to stand Godfather to them: I was thinking
(& I hope you will excuse the coolness of my offering myself) that if you were
at home: I could pay you a visit on my road to Cambridge.— When you write tell
me what you think of this scheme.— I hope it will answer:

I do hope that when you are in Yorkshire, you will redeem your character in

Entomology, it is a long time since you have mentioned taking any good insects.—
I think you would be surprized if you were to see all my insects.—

Pray mention in your next letter how Uncle Edward[1] is, we were all very sorry
to hear he had been so unwell: Catherine begs that you will give her love Julia &
many thanks for her kind wish of seeing her at Osmaston, but is very sorry that
it is impossible.

Believe me dear Fox | Yours most sincerely | Chas. Darwin

Endorsement: C. Darwin July 29, 1829
Christ's College Library, Cambridge (Fox 19)

[1] Edward Darwin of Mackworth, Derbyshire.

To W. D. Fox   26 [August 1829]

[Shrewsbury]
Wednesday 26th.—

My dear Fox

I returned from Black Game shooting at Maer on Monday, & was most
exceedingly glad to find your letter, when I came home: I am very much obliged
for your most kind & welcome invitation to Osmaston. And, as by good luck
your visit into Yorkshire has been delayed, if it will suit you, I will come there
early in September.— I intend going to Maer for the first, & shall stay there
about a week, after which I will come through Newcastle to Derby.— I will
write & let you know when I can find out about coaches &c &c

But first, to know whether it is perfectly convenient I must have a letter
directed to me at J. Wedgwood Esq. Maer Hall *Newcastle under Line.* You must
write nearly as soon as you receive this.—

I am afraid it is almost too late for Entomologizing, but I shall like a quiet
week at Osmaston better than anything else: I had so very pleasant a visit the
last time I was there; that every thing has a charm to me from Pig Park to
your' Sanctum Sanctorum.—

The Wedgwoods[1] whom you saw at Kingscote are staying at Maer. they have
been for the last year living with Mrs Sismondi[2] near Geneva & there they saw
the Ways who have a house there.— They did not see little Albert. his direction
is Les Dèlices,[3] Près de Genève.— If you have any virtue & are not *too much*
engaged, you will of course immediately write to him.—

You will be surprised to hear that I have been again to Barmouth. my sisters
took a sudden freak[4] to go to the sea-side so I chaperoned them there. The weather
was dreadfully bad so I did not do much amongst the beetles.— We will work
gloriously, when I am at Osma⟨ston.⟩ I shall extremely like going regularly

through your collection, & I think I shall be able to name a good many.— I will bring you the duplicates of this summer work, such as they are.—

My Father of course meant to send the Martens as they were as I also did the Dundiver, and we are glad they arrived safe.— My Father is glad to hear that Mrs Darwin is tolerably well: it must have been a terrible blow to her.—

The Owens of Woodhouse, the idols of my adoration, have been staying at Buxton, & are now returned for the 1st of September. They give a most amusing account of the state decrepitude to which the Beau Garcons are there reduced.—

But if I begin to talk about Woodhouse & la belle Fanny, I never shall conclude this letter, so believe me my dear Fox, Yours sincerely | Chas. Darwin

My Father sends his best regard to you.— Be sure write soon, & direct to Maer.

Postmark: Shrewsbury AU 27 1829
Christ's College Library, Cambridge (Fox 20)

[1] The John Wedgwoods (see *Emma Darwin* 1: 215).

[2] Jessie Sismondi, wife of the historian, Jean Charles Léonard Simonde de Sismondi.

[3] Les Délices was one of Voltaire's houses, near Geneva. He bought and named the property in 1755. 'It is the palace of a philosopher with the gardens of Epicurus: it is a delicious retreat' (Besterman 1969, p. 338).

[4] 'A sudden causeless change or turn of the mind . . .' (*OED*).

To W. D. Fox   4 September [1829]

[Maer]
September 4th.

My dear Fox

I was very glad to receive your letter & to find that you can receive me next week at Osmaston.— I write these few lines, to tell you that I shall be in Derby on Tuesday Evening, but I am afraid my visit must be very short. Erasmus came home on Tuesday, & intends only staying a fortnight or 3 weeks at Shrewsbury & then goes for the winter to Paris; so that I shall scarce be able to see anything of him.—

I long to Entomologize with you, & have brought my duplicates with me.—

There are literally no Partridges in this part of the country. I have been out two days & have only killed 3 birds a day.—

Yours affectionately | Chas. Darwin

Postmark: Newcastle SE 5 1829
Christ's College Library, Cambridge (Fox 21)

To W. D. Fox [22 September 1829]

[Shrewsbury]
Tuesday

My dear Fox

I had hoped to have been able, before writing to you, to have seen M.ʳ Hope, and then I should have had something to write about, as it is I am in my usual predicament of having nothing to say.— When at Osmaston, Erasmus plans were, you are aware, to go to Paris for the winter. accordingly I hurried home like a dutiful brother; I might have spared myself that trouble; as it was not likely his plans would continue the same for a whole month: He now intends to give up Doctoring for the present. & take rooms & live in London

He is come back very agreeable & industrious, & foreign parts have not all spoilt him.— he means to go the Birmingham Music meeting, from thence to London, making a detour to Kenilworth, Warwick & Oxford.—& before xˢᵗmas he is coming to me in Cambridge.— All this is excessively pleasant, & he means to have an air-cushion in his rooms for me to sleep on, which will make London a delightful place.—

I myself am beginning to doubt about going to Birmingham on account of the expence, but if you will make up your mind to go, so will I.—

M.ʳ Hope is gone up to London, so that I shall not be able to get anymore information about the insects we took together.—but I trust you have been to the Moor (take your net & sweep), & that you pretty often look over & shake the Fungi.—

I had a letter from Turner yesterday & he says he has actually collected a case full of insects. Do not let yourself be beaten by such a novice as he is.—

I intend going to Cambridge about the 15.ᵗʰ of October, & I do hope you will contrive to be there before very long.—

You must write *soon* & tell me all your plans.—

Give my love to the little Puss & remember me most kindly to all your family: & Believe me dear Fox | Yours most sincerely, | Charles Darwin

Write to me directly.

Postmark: Shrewsbury SE 23 1829
Christ's College Library, Cambridge (Fox 22)

To W. D. Fox [15 October 1829]

Christ College [Cambridge]
Thursday

My dear Fox

I am afraid you will be very angry with me for not having written during the Music Meeting, but really I was worked so hard that I had no time: I arrived here on Monday, & found my rooms in dreadful confusion, as they have been taking up the floor, & you may suppose that I have had plenty to do for these two days.—

The Music Meeting was the most glorious thing I ever experienced: & as for Malibrand,[1] words cannot praise he[r] enough, she is quite the most charming person I ever saw.— We had extracts out of several of the best Operas, acted in character, & you cannot imagine, how very superior it made the concerts to any I ever heard before.—

J. de Begnis[2] acted Il Fanatico[3] in Character, being dressed up, an extraordinary figure gives a much greater effect to his acting. He kept the whole theatre in roars of laughter.— I liked Madam Blasis[4] very much, but nothing will do after Malibrand; who sung some comic songs, & persons heart must must have been made of stone not to have lost it to her.—

I lodged very near the Wedgwoods & lived entirely with them, which was very pleasant, & had you been there it would have been quite perfect.— It knocked me up most dreadfully, & I never will attempt again to go to two things the same day.—

I passed through London on Sunday, but Mr Hope was not there: Weaver in Birmingham has a very good collection, but does not know much about them, & I bought 15^s worth of insects of him: one insect, if it is British, quite surpasses my knowledge.— Would you believe it, that I have the Endomychus coccineus, a bad specimen however, given me by Mr Hope?— I hope you have found something in the Fungi: any how tell me when you, what you have done in Entomology.—

Cambridge is most wonderfully empty. Wilmer & the Rev— Robinson[5] are the only men up that you know: Poor Robinson has got to go in again for his degree as he had not entirely kept one of his terms.—

Your conjecture was perfectly right about the letter, when I came home I found it in my great big top-coat, to my sisters no small disgust & indignation.— I suppose you have sent Stephens Catalogues to Shrewsbury, from whence I have ordered them to be sent here: it is a pity, as they have come so late, that they were sent for at all.— Do write soon & tell me your plans. I earnestly hope that you mean to come up early & stay for some time: it will be so very snug being together once more.—

Yours most sincerely | Charles Darwin

Postmark: Cambridge ⟨O⟩C 16 1829
Christ's College Library, Cambridge (Fox 23)

[1] Maria Felicia Garcia Malibran.
[2] Giuseppe de Begnis.
[3] *Il fanatico per la musica*, opera by Antonio Sacchini. De Begnis sang the aria 'I violini'. (Birmingham Musical Festival programme, 8 October 1829, Birmingham Central Libraries.)
[4] Virginia de Blasis.
[5] George Alington Robinson who was ordained deacon in 1829 (*Alum. Cantab.*).

To W. D. Fox [3 November 1829]

[Cambridge]
Tuesday

My dear Fox

I am ashamed at not having written to you before, & more especially as Aiken gave me some very important messages to you, about Cocks & Hens,[1] no less than ten days ago: Imprimis, there are two hatches: no mother: price of larger 3ˢ each & of the lesser chickens 2"6. He is afraid you will think them very dear, but could not get them cheaper.—

I am sure you will be very sorry to hear, that my Father has been very ill, but is now I hope steadily but slowly getting better: The cause of his illness is a deep seated Erysipelus in his neck, attended with great deal of Fever. I am afraid he has undergone a great deal of suffering: & that it will be some time before he quite recovers the effect of it. I should have gone down to Shrewsbury, only that his Medical attendants particularly insisted that he should be kept as quiet as possible: My sisters appear to have been terribly alarmed, as I suppose on Wednesday he was in imminent danger.—

Erasmus has been staying here for a few days, & we passed the time very pleasantly chiefly in the Fitzwilliam[2] & he would have stayed some time longer, only that he went to London, in order to be in readiness to go down to Shrewsbury if my Father should be worse: None of the Cam: Entomologists have done anything to speak of, but I am going to Bottisham to see Mr Jenyns cabinet, & I believe he is coming to see mine: Professor Henslows[3] parties go on very well. the last was the pleasantest I was ever at:

I have been very idle since I have come up, but have had some good hunting: Rob⟨in⟩son has been riding to th⟨e⟩ ad⟨mir⟩at⟨ion⟩ ⟨of⟩ every body. I think he is the best rid⟨er⟩ I ever saw.—

I am very sorry that you are so undecided as to when you will come up to Cambridge. I do hope it will eventually take place. I shall go down very early in order to see my Father.—& return early in the next Term to read for my little Go.

Yours very sincerely | C. Darwin

Postmark: Cambridge NO 3 1829
Christ's College Library, Cambridge (Fox 24)

[1] Fox was apparently starting to work on animal breeding, which later became an important subject of CD's correspondence with him.

[2] Richard, 7th Viscount Fitzwilliam, had bequeathed his collection of pictures, books, and illuminated manuscripts to the University, together with £100,000 for a building. The collection, to which other donations were added, was temporarily housed in the old Free School in Free School Lane. The present Fitzwilliam Museum building was begun in 1837.

[3] John Stevens Henslow had been appointed to the Chair of Botany in 1825 (see Winstanley 1940, pp. 39−40), and was by this time one of the most influential professors in the University. As CD's friend

and mentor in natural history, Henslow was the most important influence in CD's early scientific development.

To W. D. Fox [3 January 1830]

Christ Coll: [Cambridge]
Sunday Evening

My dear Fox

I shall think myself very lucky if this letter is not returned unopened, for my shameful conduct well deserves it. I should have written long enough ago, only that when in Cambridge, I thought I should have had more to say in London, & when there I thought it better to put it off till I came here. So owing to these two weighty reason, it is nearly six weeks since I received your last, & it would serve me right, if it were six weeks till I received another, but I hope better things.—

Cambridge is very empty, but for a change, I daresay it will be very pleasant: any how the bachelors dinners, I have no doubt will be very good.— I came here on Friday evening, having been staying with my brother for the last three weeks in Town. I never was there for so long a time before & enjoyed it much more than I expected. There is literally nothing at all going on, & I cannot mention a single thing that I did. The time however glided away, much in the same way as it does at Osmaston: I performed one Herculean task, having nearly finished Clarissa Harlowe, the most glorious novel ever written, & I advise you begin it as soon as you can.—[1] M.ʳ Hope was in Town for the last week, & I worked away very hard at entomology. I went two evenings to M.ʳ Stephens, & I begin to like him very much.— a M.ʳ Waterhouse[2] gave me some most glorious insects, inter alia, 2 species of Trichius: Your Tarus is basalis.—

I hope you have not thrown away the Fungi: but I believe we collected it rather too late. You can have no conception what magnificent insects some of the Fungiverous are.— M.ʳ Jenynys has been looking at my insects, & he was very grateful for a good many insects I gave him.— I wrote to Baker about the Shovellers & tracheæ: M.ʳˢ Field bill is: 5£"7·6 & Bretts for packing up a case, 12ˢ"6. & I paid 2"6 for a grebe that he has got to skin: so you must keep an account of how we stand together, when I have paid Brett.—

The men are all reading at a most wonderful pace. Old Simpcox counts every minute in the 24 hours.— I am afraid Whitley has a very poor chance of being Senior W. but I trust he will be second.— Herbert I believe will be in the first six.—[3] I hope you will write soon & tell me what your future plans are.—for from the weekly variations, I can generally make a guess what they really will be. This indeed with my Brother is my only way of finding out:—

I shall if possible withstand temptation & not ride this term. I had one or two most glorious days hunting at the end of last. But the last day I had two such aweful rolls as nearly knocked my lungs out. This puts me in mind to give you a good scolding for imagining me such a blackguard as to forget my old friends & do nothing but poke about the stable & look knowing. Forgive my negligence as I do you for making so unjust a supposition & believe me dear Fox | Yours very sincerely | Chas Darwin

I saw old Pulleine, he looked very fat & rosy. I forgot to mention, I dined with Sir J. Mackintosh[4] & had some talk with him about Phrenology,[5] & he has entirely battered down the very little belief of it that I picked up at Osmaston. He says, as long as Education is supposed to have any effect in decreasing the power of any organ of the brain, he cannot see how it ever can be proved true[6]

Postmark: Cambridge JA 3 1830
Christ's College Library, Cambridge (Fox 25)

[1] Richardson 1747–8.
[2] George Robert Waterhouse.
[3] Charles Thomas Whitley achieved Senior Wrangler and 2d Class in Classics. John Maurice Herbert was 7th Wrangler.
[4] Sir James Mackintosh, 'the best converser I ever listened to' (*Autobiography*, p. 55).
[5] According to the phrenological doctrine, as elaborated by Franz Joseph Gall, the shape of the skull reflects the 'organs' or faculties of the brain. Phrenology attained considerable popularity in England—by 1832 there were 29 phrenological societies and an influential journal edited by George Combe. See *EB* 21: 534–40.
[6] Mackintosh's argument appears to be that if education can reduce the powers of the organs of the brain—for instance, those of acquisitiveness or secretiveness—then these powers are not innate and no causal relation exists between the organs and the shape of the skull.

To W. D. Fox [13 January 1830]

[Cambridge]
Wednesday

My dear old Fox

I observe that it has happened once or twice before, that at the same time that my guilty conscience has goaded me into writing, your stock of patience has been worn quite thread bare.— Owing to this exact proportion our letters unfortunately clashed on the road, & to show my real penitence I make this fresh start.—

I opened your letter with trepidation & awe, & as all good books say, that a forgiving spirit is the truest revenge, so I found it; for your very kind letter made me more ashamed of myself than any well deserved & severe speeches could have done.— I wish it had so happened that I had gone to Shrew: this Vacation, if it had merely been to have paid you a visit, for idle as I am, I would not have let you come all the way to Birmingham merely to see me, (always with the supposition that there was no, "metal more attractive"[1] to be found in that metalliferous district) I hope you had a pleasant visit at the Larches. You do not say much about it. I hope you will mention in what sort of state you brought your heart back again to Osmaston? in a sad raggy condition I am afraid.—[2]

I forget whether I mentioned that I have ordered a Cabinet. I long to begin about arranging & naming my insects; amongst the Carabidæ, I think I have a third in number.— I have lately taken Demetrias: imperialis!!! & Endomychus coccineus.— What do you mean by saying Miller has sent you Saperda Carcharias; why you have already got that, it is the insect you bought of Finch, large & mottled with brown.—

There is a much better entom: than Miller in Bristol, of the name of Millard, but I hear very stingy.— N.B. Mr Miller is suspected of having got a good many foreign insects in his collection from Germany, where he formerly collected.—

My Father has been for some time quite well, but has since had a little gout.— I thought I had mentioned, that both the Deaths Heads died in the Summer.—

I rode over the other day to pay Mr Jenynys a visit. he has a nice snug little house, & lives very comfortably, & was altogether very civil, but not particularly liberal. I gave him an awesome lot of insects, & he gave me ⟨2⟩ good ones, & 2 or 3 more common ones, refused me a specimen of the Necroph. sepultor, (the Fen one of which he can get plenty) although he has 7 or 8 specimens by him at that time.— There is a perfect specimen of liberality for you: a true disciple of the Curtis school:[3] My Dytici & Colym: astonished his weak mind: I do most sincerely hope you will at once fix a time for coming up here, & do let it be soon. I long to see you: & to go through my insects will be no small fun.— I shall call you irresolute & weak, if in your next letter you do not fix a definite time.—

The men are all in a dreadful plight, from fear & anxiety, but nevertheless Cam: is remarkably pleasant, & if you were here, to have our old breakfasts together it would be delightful.— I never saw Whitmore in such good spirits as he is at present: he goes walking about quite independently, & drinks brandy to the nth—

It is quite curious, when thrown into contact with any set of men, how much they continue improving in ones good opinion, as one gets ackquainted with them. This was an argument used, in a religious point of view, by a very clever Clergyman in Shrews. to encourage sociability (he himself being very fond of society), for he said that the good always preponderates over the bad in every persons character, & he thought, the most social men were generally the most benevolent, & had the best opinion of human nature. I have heard my father mention this as a remarkably good observation, & I quite agree with him.—

Chapman & many other men often ask after you:

My dear Fox | Yours sincerely | C. D

Postmark: Cambridge J⟨   ⟩30
Endorsement: Charles Darwin January 16, 1829. [Corrected to 1830 in purple crayon.]
Christ's College Library, Cambridge (Fox 26)

[1] Shakespeare, *Hamlet*, 3. 2. 116 (Arden edition).

[2] CD alludes to Fox's interest in Bessy Galton.

[3] Probably William Curtis. See Curtis 1771 and 1772. Darwin Library — CUL contains Curtis's *The Botanical Magazine or Flower-Garden Displayed*, vols. 1 and 2 (in one volume), London, 1793.

From Fanny Owen   27 January [1830]

[Woodhouse]

My dear Charles—

I received your *leetle* message from Catherine when I was staying at the *Darwin Hotel* for the Hunt week I have delay'd so long complying with your *modest* (but I cant help thinking *unaccountable* request) that I fear you are thinking me an *ill tempered creature: sulky, idle &c &c* but I have been so busy lately *or so cold & frost bitten* that I have never had time to scribble you a few lines— I am *quite convinced* that there does not exist so *dull* & stupid a letter *composer* as me, & how any body in *their senses* can wish for one of my productions is more than my *weak mind* can comprehend. I can only say I feel extremely *flatter'd* by such a compliment, and if you are put to sleep in reading the effusion you have only *yourself* to *blame* for it—

I have very little *news* or *scandal* I am afraid to tell you— Shropshire as usual does not abound with either— I never did *enjoy any thing* more than my my week at *Darwin Hotel* We were so *perfectly* at our *ease*, and *comfortable* two Balls there were which I liked *very much*, the intermediate evening we went to a little sort of *Fandango*—(or as Owen calls it a "**Blow out**," at the Leightons, which was pleasant for Clare with her *usual goodnature sang* to the *admiring* and *awe struck multitude* almost all night. What more could we wish for? She pitched her voice in its *shrillest* key and oh ye powers!! Heavens & Earth. I wish you could have heard her—but why did you not come home this Xmas? I fully expected to have seen you—but I suppose some *dear little Beetles*, in Cambridge or London kept you away— I know when a *Beetle is in the case* every other *paltry* object *gives way*—if I could have sent to tell you I had found a **Scrofulum morturorum** perhaps you might have been induced to come down!— how does the *mania* go on, are you as constant *as ever*? don't think me impertinent, but tell me—*pity* and *forgive* my ignorance— I am to day left solo at home with the Boys, the Governor Mama S. & Car being gone for two days to Erthig (the Yorkes). I thought it would be a *dull affair*, so got off going and am very *comfortable mistress* of the Forest— I think this horrid frost never will end. I never was so tired of any thing one can do nothing but sit over the fire and read Novels. I have been *improving my mind* very much in this way, but at Oswestry nothing much newer than "The *Mysteries of Udolpho*"[1] is to be had— Your sister Caroline has been here for 3 or 4 days. She went away this morning— Catherine is at Maer but I am telling you ⟨wh⟩at you know perhaps— Oh that I had a mystery to unfold, but indeed n⟨ot⟩ the ghost of one has appear'd to me for an age— that *eternal* never *to be ended* Williams affair goes on as usual,[2] I never was so sick of any thing, and see no chance of it coming to an end—any thing wd be better than this horrid state of uncertainty, and always hearing the same thing—the *name* now quite puts me in a fury— *Human patience* can bear it no longer, & my *little stock* has been *long exhausted* on *that subject*! Owen is now on leave of absence, and as you may suppose is in a gᵗ rage at the Frost— he has no less than *five horses* and *two men* here living on *"the Family"*—did you ever know such impudence—but the Governor bears it all with Christian *fortitude* & *resignation*. I think it is a shame he should drain all the substance of the Family— Hard times these! *upon my word* Charles—Xmas *may be* a *joyful* time but I'm sure does not *seem so to me*—my finances are in a most pitiable state & the *Mortgages* on *my Estate incalculable*—the *Duns* begin to ride—and the *King's Bench*[3] stares me in *the face*—what a situation—? but I can't say my spirits *as yet* suffer *much*. I drive *"dull care"* away wonderfully—and hope Providence will carry me through my difficulties—it *often has*, so I must trust to it— I dare say you are in the same *sitivation* so can have a fellow feeling for *me*— What a **horrid disgusting thing** money is—I hate the name of it—*dont* you—it is fit for *vulgar souls*—not **Beetle** *Hunters*—and *Paint brush Drivers*!!!— Arthur went back to Hertford[4] last week—he shot a good deal whilst at home but I believe not much sport they say there is very little Game—hardly any Pheasants—

I think I have given you prose enough for once so will now put you out of yʳ misery.— you must excuse my Dulness but it is yᵗ own fault—a letter you *would have* so a *letter* you *have got*—and a pretty production it is— I never *could* and never *shall* write *"like a Lady."* *Miss Casteau* has often rung in my ears—so good bye | believe me dᵗ Charles | yʳˢ ever truly | F. Owen

Jan 27ᵗʰ—

Burn this I *do* desire *Sir*!!!!!!!

Postmark: Shrewsbury JA 27 1830
DAR 204.4

---

[1] Radcliffe 1794.

[2] Probably a reference to the protracted engagement of Sarah Owen (see letter from Fanny Owen, [26 September 1831]).

[3] A prison for debtors as well as a court of law.

[4] Arthur Owen was attending the East India Company College at Haileybury, a few miles from Hertford.

To W. D. Fox [25 March 1830]

[Cambridge]
Thursday

My dear Fox

I am through my little Go,!!!![1] I am too much exalted to humble myself by apologising for not having written before.— But, I assure you before I went in & when my nerves were in a shattered & weak condition, your injured person often rose before my eyes & taunted me with my idleness. But I am through through through. I could write the whole sheet full, with this delightful word.— I went in yesterday, & have just heard the joyful news.— I shall not know for a week, which class I am in.— The whole examination is carried on in a different system. It has one grand advantage, being over in one day. They are rather strict; & ask a wonderful number of questions:

And now I want to know something about your plans: *of course* you intend coming up here: what fun we will have together, what beetles we will catch, it will do my heart good to go once more together to some of our old haunts: I have two very promising pupils in Entomology, & we will make regular campaigns into the Fens; Heaven protect the beetles & Mr Jenyns, for we wont leave him a pair in the whole country. My new cabinet is come down & a gay little affair it is.—

And now for the time, I think I shall go for a few days to Town, to hear an Opera & see Mr Hope; not to mention my brother also whom I should have no objection to see.— If I go pretty soon, you can come afterwards but if you will settle your plans definitely, I will arrange mine. So send me a letter by return of post:— And I charge you let it be favourable, that is to say come directly.—

Holden has been ordained, & drove the Coach out on the Monday, I do not think he is looking very well.— Chapman wants you & myself to pay him a visit, when you come up, & begs to be remembered to you. You must excuse this short letter, as I have no end more to send off by this days post.—

I long to see you again, & till then | My dear good old Fox | I am yours most sincerely | C. Darwin

Postmark: Cambridge MA 25 1830
Christ's College Library, Cambridge (Fox 27)

[1] The results were published in two lists: those who passed with credit, and those to whom the Examiners had *only not refused* their certification of approval. CD's name appeared in the first list (Cambridge University Archives Pass Lists, Previous Examinations 1824—83, Exam. L. 9).

To W. D. Fox [1 April 1830]

Cambridge
Thursday

My dear Fox

I should have answered your letter sooner than this, only that for the few last days I have been very busy in taking sundry long walks: The information your letter contained about Erasmus, settled my plans & I immediately made up my mind to remain here during the whole vacation. And I have since heard from M.ʳ Hope, who is staying at Southend, that he should not return to London for some weeks, so that there is not a single inducement for me to leave Cambridge: Here then I shall patiently vegetate living in the hope of soon seeing you:

If it had but been possible for you to have come up here during this Vacation, how very snug & quiet we should have been, but in another respect, vid: Entomology, it is better as it is, as this cold & rainy weather has driven all the beetles back to their homes.—

Do continue steadily to try to start for Cambridge, before the end of this month, for if you have many more delays, all the time will slip by. Let me hear from you some little time before you come; And I most fervently hope that time is near.— but there is no use in hoping, as I am sure, if you find it possible, we shall meet before the end of the month.—

I suppose you do not know much more of your plans for the summer, than you do of those for the spring. M.ʳ Bristow told Erasmus that you intended going to the Sea side with your sisters: Is this quite fixed? & where is the place?—

M.ʳˢ Field does not appear to be any great hurry for her money. she says she never received any from Graham.— I think I told you the amount of the bill.—

I want to know did you receive with your birds the ring Owzel?— M.ʳ Stephens has been very ill with a fever, which is the reason that he has not been publishing lately. I have sent up all my Nitidulæ &c for him to examine.— I find I get on very slowly with my cabinet, & shall be very glad of your assistance. I have only yet got to the Amaræ.[1]—

Barker,[2] whom we met at Sir Francis[3] is come up as Fellow Commoner[4] to Downing. is his Mother alive? for he talked one day as if Derbyshire was no longer a home to him.— I have been seeing a good deal lately of Prof: Henslow; I took a long walk with him the other day: I like him most exceedingly, he is so very goodnatured & agreeable.—

This letter is strange rag-tag & bob tail affair, & a poor return for your agreeable ones, but with it go my best wishes for you my dear Fox & believe me | Y.ʳˢ most sincerely | C. Darwin

Postmark: Cambridge AP 5 1830
Christ's College Library, Cambridge (Fox 28)

[1] CD was using Stephens 1829, *Mandibulata*, vol. 1, in cataloguing his specimens (see letter to W. D. Fox, [9 May 1830], n. 4). In his copy, which is preserved at the Cambridge University Library, there are notes opposite each species in his collection, with the locations where they were taken. Occasionally names appear, e.g., 'Hope', 'Waterhouse', 'Bought of Weaver 1829'.

[2] Thomas Alfred Barker.

[3] Francis Sacheverel Darwin. Francis Galton mentions him as 'originally a physician, but for many years living in a then secluded part of Derbyshire, surrounded by animal oddities; half-wild pigs ran about the woods, tamed snakes frequented the house, and the like' (Galton 1874, p. 47).

[4] Fellow Commoners (aristocrats and other 'men of family') were the first of three ranks in which students were matriculated, the others being Pensioners (ordinary paying students) and Sizars, who received aid for which they had to perform services, often of a menial nature (*Cambridge University calendar*, 1831).

To W. D. Fox [9 May 1830]

[Cambridge]
Sunday

My dear Fox

I am very sorry to find that all our plans are likely to vanish into air.— It is most unfortunate your being obliged to go with your Father & Mother to Cheltenham, for the weather is so fine, the beetles so numerous, our zeal so ardent that the Science would have received a benefit never to be forgotten.— But it is a shame to talk about would have taken place: let us think of the future.—

I am very much obliged for your invitation to Osmaston, but I do not see how it is possible for me to accept it My brother has returned from Shrewsbury to London, & I have agreeed to pay him a visit, which if I do not do, I shall not see him again for a good while.— And after that I really must proceed to " 'sweet home' ".— for my unfortunate relations have been bereaved of my presence, for eight long Months.— But I have a plan which will remove a good many difficulties.— Why, in the name of providence, not pay me a visit in Salop??? You know how glad we all shall be to see you.— You can start early from Derby, so through Lichfield to Birmingham, & then per the Wonder[1] to Shrewsbury, where at 11 oclock the same night you will arrive safe & sound, & make me glad by your sight.— What plan can be more easy or natural? It is too unreasonable even to Hope, that you will come to Cambridge for so short a time as four days, so that I shall not see till we meet at Shrewsbury. If you are half as eager as I am for that occurrence, you will not make any sort of excuse.— As soon as you return to Derby go to Mr Hey,[2] & find the habitat Odontomyia & Dromius & promised specimens.—

My plans for the summer are totally undecided.— write soon be sure do not say no to Shrewsbury scheme else I will never forgive you.— till then, My dear old Fox | Yours very sincerely | C. D.

Wilmer has left Cambridge, & begged to be most kindly remembered to you.— he has sent to Baker for you a Falcon out of Norfolk.— I have got your prints.—

Chapman send Ditto.—he is going abroad.— I took the other day Elater sanguineus. What do think of that? I have 19 species in genus Amara.—

I have seen a good deal of Henslow lately & the more I see of him the more I like him I have some thoughts of reading divinity with him the summer after next.[3] Adieu

As far as to the Bembidídæ, this in toto 339 species,[4] out of which I have 208 of them— What do you think of that?

Postmark: C ⟨1⟩o MY 10 1830
Christ's College Library, Cambridge (Fox 29)

[1] The London, Birmingham, Shrewsbury mail coach.
[2] The Rev. Samuel Hey of Derbyshire, mentioned in Stephens 1827 – 46, *Mandibulata* 3: 373 as having rediscovered the rare *Chrysomela cerealis* and as having directed 'his friend and neighbour Mr. Fox to the locality'.
[3] If John Maurice Herbert's memory was accurate, CD's plan to take orders represented a change in his religious convictions. Earlier, at about the time of the reading tour at Barmouth (letter to W. D. Fox, 12 [June 1828]), according to Herbert, CD had serious doubts about this career. Herbert remembered 'an earnest conversation' he had with CD 'about going into Holy Orders; & I remember his asking me with reference to the question put by the Bishop in the Ordination service: "Do you trust that you are inwardly moved by the Holy Spirit &c" whether I could answer in the affirmative; & on my saying "I could not", he said, "neither can I, & therefore I can not take orders." ' (Letter of reminiscences to Francis Darwin, 2 June 1882, DAR 112 (ser. 2): 63 – 4).
[4] The number 339 corresponds to the number listed in Stephens 1829, 1: 36, after which the descriptions of the family Bembidiidae begin.

To W. D. Fox [31 May 1830]

[Cambridge]
Monday

My dear Fox,

I write these few lines to give you an outline of my plans, to apologise for not having written long enough ago, & lastly to say how very glad I am to hear that my plan of your paying me a visit in Shrewsbury, meets with your approbation.— I shall go up to Town on Friday, or Saturday, stay a few days there, & then on Shrewsbury: from whence I will again write to you.—

M.ʳ Hope has paid me a visit but went back disgusted at the weather. he wants me to make an expedition to the New Forest.—but I do not think I shall: after I once I get to Shrewsbury the sooner you come the better, the only thing at all likely to interfere is Mʳ Hope & Eyton[1] want me to come an Entomological trip into N. Wales. Cannot you perhaps join us? Think of it.—

Of Course you have heard of the new species of wild Swan, discovered in England, by M.ʳ Yarrell.[2] I have bad stuffed specimen of it, for 10.ˢ— bad as it is, you may think yourself, lucky in getting. Yarrell himself, has pronounced it to be the new sort, so there can be no doubt.—

I have had a letter from old Simpson, & he begs to kindly remembered to you: I am packing up Pulleins birds: I have got for you a dusky Grebe.

I will write again when I have more time. I only send this scrawl that you may know how very happy I am at the thoughts of seeing you at Shrewsbury.— | Dear old Fox | Yours | C. Darwin

Mr Hope has taken P. Podalirius in Shropshire, & seen it several times

Postmark: Cambridge MY 31 1830
Christ's College Library, Cambridge (Fox 30)

[1] Thomas Campbell Eyton.
[2] William Yarrell.

To W. D. Fox   25 [August 1830]

Barmouth.
Wednesday 25th.—

My dear Fox

I have been intending to write every hour for the last fortnight, but *really* have had no time: I left Shrewsbury this day fortnight ago, & have since that time been working from morning to night in catching fish or beetles. This is literally the first idle day I have had to myself: for on the rainy days I go fishing, on the good ones Entomologizing. You may recollect that for the fortnight previous to all this, you told me not to write, so that I hope I have made out some sort of defence for not having sooner answered your two long & very agreeable letters: I am delighted that we agree so well about all the Wedgwoods, & as you say it is so much pleasanter talking about any place or persons that both parties have seen.— I trust that Charlotte[1] has driven out of the field your musical enchanter at Ashby: Cannot you imagine how very pleasant I must find Maer, not one drawback. I have not even the least dread of the *Parish* a sensation, which it appears, you were not quite void of. I should enjoy most exceedingly to spend a week there with you: it is more probable now than that originally you should have been at Maer at all.—

I suppose you are now living in Cambridge with Prof: Henslowe. I shall enjoy hearing from you some gossip about him & his wife[2] & I hope you will return good for evil & write soon to me.— I am afraid you must find Cambridge rather stupid. it must be such a contrast from your former life there.— of course every thing my rooms contain is most entirely at your service: you do not know how it would please me if there is any chance of your being there in the first week in October, by which time I would be up, if it were merely to eat our breakfasts together as we did in good old times.—

I shall be at Maer for the first week or 10 days in September, & write to me there, (direction is Maer, Newcastle, *Staffordshire*) & tell me your plans.—

And now I give you some account of our Welch trip. Hope 2 Eytons (father & son) & myself arrived at Capel Curig on a Thursday. On Monday we went up Snowdon, & were disgusted with it we took 3 Cerculis a few glabrati & nothing else. Hope & Eyton went off next day to Barmouth, Old E & myself staid a few days longer & had some pretty good trout fishing. I joined Hope & young E. here on Friday, but they returned to Shrewsbury on Monday for a cricket match, & here I am enjoying my solitude extremely: I am quite disgusted with Hopes egotism & stupidity: how I wish you were here, the very thought of a days entomologizing with you, is quite refreshing.— Try & prolong your stay in Cam. to October. I long to spend a little time with you quietly. I do not call your visit to Shrewsbury anything, there were so many people making a row & confusion.— The Entomologists did not do a very great deal here, took some good Colymbetes, Cicin: hybrida, & Eyton took a most beautiful Crysomela. I have taken some good Leiodes by the way ask Prof: Henslow, what is the time to get Leiodes cinnamomia. I want some very much.— You ought also to take some good water beetles this time of the year.—

When you write tell me what you read. Remember me most kindly to Prof: Henslowe— Excuse this patch-work letter & believe me dear Fox. | Yours, sinc, | C. D.—

Postmark: Barmouth; C 27 AU 27 1830
Christ's College Library, Cambridge (Fox 31)

[1] Charlotte Wedgwood.
[2] Harriet Henslow, sister of Leonard Jenyns.

To W. D. Fox [8 September 1830]

Maer.
Wednesday

My dear Fox

I cannot help writing a few lines to say how very sorry I am to hear such bad news about M[rs] Fox, & to beg of you to write to me very soon to me again to say how she is going on.— And I most sincerely hope to receive a favourable account.— I return to Shrewsbury on Friday: soon after which I shall expect a letter.—

I have had very bad shooting this season, the first day I killed 10 brace, since which I have not on any day bagged three.— But I find Maer just as delightful as ever & Charlotte just as agreeable as ever: id est, as agreeable as possible.— I have done a good deal in Entomology, & to day have worked so hard, that I can hardly see to write.—

I am very much obliged for the tin Box. The Enicoceri are invaluable.— But beetles, partridges & every thing else, are as nothing to me, now that I have got a horse.— I am positively in love with him, stands 16 hands & one inch, and I think will make a very good hunter.—

But now I have got on the subject of horses I had better wish you good night.— I shall be very anxious for a letter so I hope you will not disappoint me I sent your letter on to home, & they therefore will be equally anxious to hear.

Good night dear old Fox | Yours Most Sincerely | C. D.—

I am quite knocked up

Postmark: Newcastle SE 9 1830
Christ's College Library, Cambridge (Fox 32)

From Fanny Owen [8 September 1830]

[Woodhouse]

Dear Charles—

Papa has been rather expecting you here this Week to slay some of his Partridges but I suppose you have had good sport at Maer & did not return so soon as you intended— I now write by Papa's desire to tell you that he started to *London* this evening he will not be absent more than five or six days & hopes to see you here when he returns— he is gone to Haleybury to see Arthur, who I am sorry to say poor fellow has been getting into scrapes—so Papa thought he had better go himself instead of writing any more— Only fancy poor Arthur's astonishment when he sees the Governor step out of the Coach! I wouldn't be him for *something* I know—

Of course the Beetles gave way to the Partridges at Maer but I hope you have had good sport in some way I have been riding every day lately. *Goldfinder* quite invaluable.— How does y.ʳ *Stupendous Animal* turn out? I hear M.ʳ Wynne was determin'd to make it up to you for waiting so long, & got the *largest* Shropshire afforded *18 hands & ½* high *without his Shoes*—is this true? I assure you I heard it as a fact—no horrid Woodhouse *joke indeed*—

Enough of this scribbling a bad pen thru'out—& I have no knife to mend it

Believe me d.ʳ Charles | very truly y.ʳˢ | Fanny Owen

Best love to the Sisterhood—

Wednesday night—

Postmark: Shrewsbury SE 9 18⟨3⟩0
DAR 204.4

From Fanny Owen   4 October [1830]

[Woodhouse]

Monday morning | 4th Oc^{ber}—

Dear Charles—

I send yr book by Francis who promises me to deliver it this evening so I hope you will get it in time— I suppose y.^r *principal* Horse has started before this, how nervous you will be till he gets to the end of his journey without *broken knees*— no *insinivations* I don't mean against the *superb* animal but accidents will happen we all know— Owen is at this moment scouring the Country on my Goldfinder the Hounds met near Nesscliff early this morning so I lent it him to hunt with them—if he should come home lame—"I shall be *bereft* indeed—' Owen leaves us today for Wrexham Races and is off to Norwich on Thursday— we shan't have him at *Dose Hall* (as he calls it) this winter again.— Of course Susan has given you a true and correct account of all the proceedings here last week we had excellent fun and Miss Ansons siege of "The Hill" astonished all our weak minds. I *could imagine* a *good* deal but the *forces* were brought on so well the *cannonade* so quick, and the barricades so *dexterously* managed, I'm sure no *General of the Age* could come near her in *skill* & *science*—

We have been in such a whirl here ever since I saw you, & promis'd to send you some scrawl of mine that ⟨I⟩ have really had not one moment to myself to do it, but lest you should think me a perfidious deceiver I send you one out of my Book it is very badly done but I will *really and truly* do you another so you must burn this when you get it, which shall be when you *next* honor the Forest with y^r presence— How long the Ansons will stay I can't guess but Owen says till **May**. Milady is going over to consult Dr Darwin & I really think the best advice he can give her is **change of air** give him a *hint* if you can— I must finish this scrawl as I am wanted to *mount guard* The Duty comes *very heavy* now— I shall write to Cath tomorrow tell her— believe me d.^r Charles | ever y^{rs} very truly | F. Owen

DAR 204.4

To W. D. Fox [8 October 1830]

Christ College [Cambridge]

Friday Evening

My dear Fox

When I received your last letter our house was full of company, & we had somebody there every day till I left home, and I always feel it quite hopeless to think of writing a letter if the house is not quite quiet.— We were all very glad to hear so good an account of Mrs Fox.—& I most sincerely hope she continues daily to gain strength.—

I arrived here in my most snug & comfortable rooms yesterday evening, after having had a most comfortless journey up here.— I left Shrewsbury on Tuesday & slept at Daventry, where I overtook my horse & rode him myself the two last days journey. The poor beast was so tired that he hardly knew whether he stood on his heels or his head.—& it will be some time before I undertake to ride a young horse a long journey again.— It will be very pleasant having horse up here. Moreover I think he will make a splendid hunter, from a specimen I had of him with Eytons hounds.—

There is not an individual up whom I know, & therefore I have had plenty of time to regret your absence.— How I wish you had been able to have stayed up here. we should have suited so well, each of us reading all morning & being idle all evening.— But it is not only when I am solitary that I regret your absence. Many many times do I think of our cozy breakfasts & even wish for you to give me a good scolding for swearing, & being out of temper or any other of my hundred faults— But is no use regretting; what cannot be altered.—

I have not seen Prof Henslow, but am going to a Party there to night; you have not told me half enough what you think about Mrs Henslow She is a devilish odd woman. I am always frightened whenever I speak to her, & yet I cannot help liking her.— I suppose you cannot at all tell me what your plans are, & whether there is any hope of your coming up to Cambridge.— but that of course entirely depends on your Mothers health; I trust you will write soon & tell me how she is going on as it is sometime since I last heard.—

This is a very stupid letter, but I will try not to write many such | & Believe me dear Fox | y^{rs} sincerely. C. D—

Postmark: OC 8 1830
Christ's College Library, Cambridge (Fox 33)

To W. D. Fox 5 November [1830]

[Cambridge]

My dear Fox

I have so little time at present, & am so disgusted by reading, that I have not the heart to write to any body. I have only written once home, since I came up.— This must excuse me for not having answered your three letters, for which I am really very much obliged.— In the last I received a 5£ note. I have not yet heard the end of the negotiations between Mrs Field & Graham.— I send your things per waggon as the box is very large, it would cost a fortune per coach.— In your insect box Henslow has put in a Cassida & Donacia.— I have taken out of your bottle 2 or 3 specimens of a black Cryptocephalus which you took in profusion.—

I have not stuck an insect this term & scarcely opened a case: if I had time I would have sent you the insects which I have so long promised, but really I have not spirits or time to do any thing. Reading makes me quite desperate, the plague of getting up all my subjects is next thing to intolerable.— Henslowe is my tutor, & a most *admirable* one he makes. the hour with him is the pleasantest in the whole day. I think he is quite the most perfect man I ever met with.— I have been to some very pleasant parties there this term.— his good nature is unbounded.—

I saw your friend M! Hey once or twice & liked him very much he appears to be a very sinsible man.— I have called on the young one,[1] but in this term our communication will go no further.—

I am sure you will be sorry to hear poor old Whitleys father is dead: in a worldly point of view it is of great consequence to him, as it will prevent him going to the Bar for some time.—

(Be sure answer this) What did you pay for the iron hoop you had made in Shrewsbury? Because I do not mean to pay the whole of the Cambridge mans bill.— You need not trouble yourself about the Phallus, as I have brought up both species.—

I have heard men say that Henslowe has some curious religious opinions; I never perceived anything of it. Have you?[2]

I am very glad to hear, after all your delays, that you have at last heard of a Curacey, where you may read all the commandments, without endangering your throat.— I am also still more glad to hear that your mother continues steadily to improve. I do trust you will have no further cause for uneasiness about her.—

with every wish your your happiness my dear old Fox | believe me Yours Most Sincerely | Chas Darwin

Friday Nov. 5.th.—

I heard from old Simpson the other day, he is going on very prosperously, & sends numerous kind speeches to you.— Do you in one of your letters call M! G. Jenyns[3] a good or a grand fellow? I am curious to know.—

Christ's College Library, Cambridge (Fox 34)

[1] William Hey.
[2] Apparently the rumour was groundless. In the *Autobiography*, pp. 64−5, CD's memory of John Stevens Henslow is that he was 'deeply religious, and so orthodox, that he told me one day, he should be grieved if a single word of the Thirty-nine Articles were altered.'
[3] The Rev. George Leonard Jenyns of Bottisham Hall, Cambridgeshire, father of Harriet Henslow and Leonard Jenyns.

To W. D. Fox [27? November 1830]

[Cambridge]

My dear Fox

After a good deal of hunting they found at Prof. Henslows house a box containing the various articles you mention, & if they will all go in, I will send them off to day. Your prints must be kept till some other time, also the beetle stick.—

Our accounts stand thus,

&c &c left unpaid at Henslows } including Lens		14 . 6
Brett. case & packing up		13 . 6
Mʳˢ Field	2 "	3 . 6
		£3 " 11 . 6
		5
Balance, on my side,		£1 " 8 " 6

I am reading very hard, & have spirits for nothing. I actually have not stuck a beetle all this term.— Young Hey does go to Henslow evening parties.—

I am very glad to hear that there is some chance of your coming up to Cambridge.— I shall be delighted to see you, but I hope it will be after examination is over, as I am far too much plagued to enjoy any thing at present.— I trust after your numerous delays that you really will be in orders by xˢᵗmas, & that I shall really pay you a visit in the Spring.—

I have nothing more to write about. You must put down my short letters to the right account.—stupidity | & believe me, My dear Fox | Yours most truly | C. Darwin

You have no occasion to make speeches about giving me trouble I shall be most ready at all times to do anything I can, for you.

Endorsement: C Darwin Novʳ 27 1830
Christ's College Library, Cambridge (Fox 35)

To W. D. Fox [23 January 1831]

[Cambridge]
Sunday

My dear Fox

I do hope you will excuse my not writing before I took my degree.—¹ I felt a quite inexplicable aversion to write to any body.— But now I do most heartily congratulate you upon passing your examination; & hope you find your curacey² comfortable; if it is my last shilling (I have not many) I will come & pay you a visit.—

I do not know why the degree should make one so miserable, both before & afterwards; I recollect you were sufficiently wretched before, & I can assure I am now; & what makes it the more ridiculous, is I know not what about.— I believe it is a beautiful provision of nature to make one regret the less leaving so pleasant a place as Cambridge.—& amongst all its pleasures, I say it for once & for all, none so great as my friendship with you.—

I sent you a Newspaper yesterday, in which you will see what a good place I have got in the Polls.—[3] As for Christ did you ever see such a college for producing Captains & Apostles.—[4] There are no men either at Emmanuel or Christ plucked.— Cameron[5] is gulfed,[6] together with other 3 Trinity scholars!— My plans are not at all settled, I think I shall keep this term, & then go & economize at Shrewsbury, return & take my degree.— A man may be excused, for writing so much about himself, when he has just passed the examination. So you must excuse.— And on the same principle do you write a letter brim-full of yourself & plans.— I want to know something about your examination: tell me about the state of your nerves.—: what books you got up, & how perfect? I take an interest about that sort of thing as the time will come, when I must suffer.— Your tutor Thompson begged to be remembered to you, & so does Whitley.— If you will answer this; I will send as many stupid answers as you can desire.—

Believe me dear Fox | Chas. Darwin.—

Postmark: Cambridge JA 23 1831
Christ's College Library, Cambridge (Fox 36)

[1] CD here refers to passing the examination for the B.A. Because he took residence only at the beginning of the second (Lent) term of 1828, he is officially listed among the B.A.s of 1832. See *Autobiography*, p. 68 and *LL* 1: 163.

[2] Fox had obtained a curacy at Epperstone, near Nottingham.

[3] '*The Poll*: those students who read for or obtain a "pass" degree' (*OED*). The examination consisted of six parts: Homer, Virgil, Euclid, Arithmetic and Algebra, Paley's *Evidences of Christianity* and *Principles of moral and political philosophy*, and Locke's *An essay concerning human understanding* (*Cambridge University calendar*, 1831). 'By answering well the examination questions in Paley, by doing Euclid well, and by not failing miserably in Classics, I gained a good place among the οἱ πολλοί, or crowd of men who do not go in for honours' (*Autobiography*, p. 59). CD was placed tenth on the list of 178 who passed (*Cambridge Chronicle*, 22 January 1831, 2d ed.).

[4] The 'Captain' headed the list of the 'Poll'; the last twelve in the Mathematical Tripos list were called the 'Apostles'—not to be confused with the famous Apostles or Cambridge Conversazione Society founded in 1820, of which CD's brother Erasmus was briefly a member in 1823 (Levy 1979, p. 301; Allen 1978, p. 27).

[5] Jonathan Henry Lovett Cameron. In a letter to Francis Darwin, written after CD's death, he refers to his 'bright and sunny' recollection of CD at school and university: 'At Shrewsbury we slept in the same room for some years & often beguiled the night with pleasant conversation. He was always cheerful & good tempered & much beloved by his school-fellows. He was not a great proficient in the school studies, but was always busy collecting beetles, butterflies &c. He spent some time, most evenings, with a blow-pipe at the gas-light in our bed-room At Cambridge I used to read Shakspere to him in his own rooms & he took great pleasure in these readings. He was also very fond of music, though not a performer & I generally got an order for him for Kings Coll. Chapel on Sunday evenings.' (DAR 112: 14).

[6] 'The position of those candidates for mathematical honours who fail to obtain a place in the list, but are allowed the ordinary degree' (*OED*).

From George Simpson [26] January [1831]

Feversham
Jan.ʸ

Dear Darwin

I write to thank you for the Paper, & at the same time to congratulate you on your *very very good degree*, tho I must say I should have been disapointed had you not been a leading man, knowing your predilection for Mathematicks.[1] I have just this moment returned from a Fox chase, we found in a nice place for a Tally, & ran very well til after the first check when Scent would not alow of more sport. I have witnessed a few very pretty things in the hunting way this year, but I begin to think of studying divinity which is more profitable to the soul than field sports. I am almost afraid to inquire whether Lumsley Hodgson[2] has passed as I can not find his name. I hope you will not get so very drunk but that you can find time to tell me a little of your future prospects, I suppose you will shortly look out for a partner for your future Vicarage, as well as a pretty patern for *nightcaps*. You have to keep another term I think, and as you will be to & fro to London do let me have the pleasure of seeing you before you return for good to Shrewsbury, which is an awful distance from Feversham.[3] The disturbances in our neighbouhood[4] are quite gone, however my Brother has thought fit to enlist in the Yeoman troops, in order to put some of them to the rout in case of necessity. I hope you did not forget to give my general invitation to Fox. I always was particularly partial to him, and hope at some time to view him again. What sort of a Master does Graham make[5]

I have no news to tell you, but I shall expect to hear from you at *some* time and with my best wishes for your future welfare, & good choice of a Wife believe me to | Remain sincerely Your's | Geo. Simpson

Postmark: A 26 JA 26 1831
DAR 204.3

[1] CD was notoriously weak in mathematics, except geometry. See letter to W. D. Fox, [29 July 1828], n. 1.
[2] Nathaniel Thomas Lumley Hodgson.
[3] Now called Faversham, in Kent.
[4] During the winter of 1830 field labourers in the counties south of the Thames demonstrated and rioted for higher wages. See Trevelyan 1942, p. 471.
[5] John Graham was elected Master of Christ's College in 1830.

From Henry Matthew [2 February 1831]

London

My dear Darwin

Though I have little or nothing to say which can be of any Interest, yet I am vain enough to think that you will not be sorry to hear from me—

Here I am in London, alone in a crowd, without a human being to exchange a word with, To me this is a situation totally new and I can not describe to you the horrors and depression of spirits to which it subjects me. I have hired Lodgings at 15 shillings per week for which I am furnished with a Study, Sitting Room Bedroom Kitchen and Dressing Room. Is not this a splendid establishment for a single man? There is one circumstance which *does* however slightly diminish the grandeur of the thing and that is the painful fact that this long enumeration of apartments is only one room with many names. I eat drink sleep study and partly dress my food in a garret about half as large as my Cambridge Rooms.— But remember, this is a mighty secret and I do not wish it to be known in Cambridge that I am in London at all much less in such a degraded condition.

I have this morning penned some sentimental ditties which I mean to get money for from some periodical Publication, if verse fails I shall try prose and not succeeding in either I shall pawn my coat and sell my Books

As if I had not curses enough to bear in my own proper person I am harassed with moaning supplicatory Letters from my wife, and another whom you know, full of entreaties to be allowed to join me and vehement assertions of their being willing to go through anything with me, even to living on potatoes and salt. All this is very fine but I have not even potatoes and salt to give them. My application at Royston[1] was answered by a Letter almost insulting. Cookesleys[2] Letter to Rivington[3] has not been answered at all. A happy state this for a man without a sous. Yet believe me I do not despair. Mergas profundo—etcet.[4] You love a quotation I know particularly when you do not understand it. I hope you have got your Book at Last. If not you will find it at Aikins. I forgot to add to my list of Blessings that I am going tomorrow before a Magistrate about my bastard, with one sovereign in my pocket to meet Law expences, arrears, and advance for a quarter— I suppose you guess by this where I shall Lodge tomorrow night— I do not date my letter because I do not mean to let you know where I am, but if I do not come down next Term I will see you here as you pass through, before which time you shall have my Direction— God bless you old Fellow | I am your sincere friend H Matthew

Upon second thoughts I almost wish that I had not sent you this wretched account of my self because I am sure that you Love me well enough to be concerned at it, but my dear Fellow I do assure you that after what I have gone through poverty is but a light misfortune— I am sure indeed that if I prosper in my Literary efforts, this Life of constant exertion, compelled by Necessity will be the only thing without wine, to save me from Madness.

By the way, I hear from Heaviside[5] that my marriage as it is termed is made known in Cambridge. Do not you ever allow it to be true—

Postmark: GB 2 FE 1831
Endorsement in CD's hand: Cecil St | Strand[6]
DAR 204.3

[1] A publishing firm founded by Richard Royston.

[2] Henry Parker Cookesley.

[3] Rivington was a leading theological publisher; since 1760 the publisher for the Society for Promoting Christian Knowledge (*DNB*).

[4] Possibly a reference to Horace, *Odes* IV. iv. 65 'merses profundo pulchrior evenit.' The Delphin text gloss reads 'si mari demergas.' The sense is 'you may plunge it into the deep—it will come up more beautiful.'

[5] In a letter to Francis Darwin (15 September 1882), James William Lucas Heaviside recalls that CD as an undergraduate was 'rather fascinated' by Matthew, an attractive but 'very intemperate' man (DAR 112: 56).

[6] Matthew's address (see letter from Henry Matthew, [14 February 1831]).

To W. D. Fox [9 February 1831]

[Cambridge]
Wednesday

My dear Fox

I should have answered your last letter earlier, as indeed was necessary to redeem my character.— I waited till I could hear from Baker; but the rogue has not yet been to my rooms; although I wrote to him.— I shall be very glad to be of any use in paying your bills & will make enquiries as you may direct me. NB. I owe you at present £1″4″6.— I paid Markham Bennet for Porterage, which makes a shilling or two difference from my last account to you.— I do not quite understand your last letter is Orridge & Aikens account a different affair? & now for giving you a scolding. I should like to know what you mean by such expressions as "begetting a horror of my handwriting" &c, is it a very refined species of irony? does it mean, if your very agreeable letters (I am forced to flatter in order to excuse myself) are horrible, that mine must be à fortiori, most intolerably horrible?— To use your own expression, no more humbug. I am always very gld to receive your letter & you know it, & in Johnsons language "there is an end of it."—

I shall leave Cambridge a little time after division, my duty draws me away; my inclination would keep me all the next terms, & *if* I possibly can, most certainly I shall stay up the greater part of next term.— I have so many friends up here (Henslow amongst the foremost) that it would make any place pleasant.— (NB you always spell Henslow with an e).—

Is there any hope of your coming up next term, if there is we will most certainly meet.— I will ⟨wr⟩ite again, when I have settled your affairs ⟨h⟩ere.— You ask after Eras ⟨ ⟩ we never correspond excepting on business.— He is in London, & now you know as much as I do.—

I could not write a long letter even to Charlotte Wedgwood or Fanny Owen, so must excuse this short one.— & Believe me dear old Fox | Your most sincerely | C. Darwin

Simpson's direction is Feversham Kent.— do write to him, he will be so glad to hear from you.

Postmark: Cambridge FE 9 1831
Christ's College Library, Cambridge (Fox 38)

From Henry Matthew [14 February 1831]

22 Cecil St Strand—

My dear Darwin,

We will meet again by God— The accusing Angel that flew up to Heavens Chancery with the oath, etcet. Vide Sterne.[1] But we must meet again yet alas not in Cambridge— Contrive a time and place and I will be there—

I answer your kind letter on the spirits engendered by a pint of Porter, The days of gin are over. I answer your generous remittance with a beggars gratitude with thanks, though I am not yet practised enough in the profession not to feel ashamed while I write. Do not think meanly of me. I assure you I had the hard choice of accepting your kindness or a Jail, for I had already pawned my watch. God bless you. Things will soon I trust be better with me. I have not yet heard from the reviewers, but I have shown my attempts to a man well versed in the profession, and he says all sorts of fine things concerning them I begin to think that I shall be the next Poet Laureate— I lie rather when I say that I feel shame at your kindness. I am perhaps rather ashamed of my necessities but I am proud indeed of the attachment which your kindness shows.— I must not be too grateful or you will cry out Damn the Fellow he never means to pay me I have written to Hamilton[2] but have received no answer so I begin to fear he is offended with me. I trust it is not so for in these times it is much harder to get a friend than to lose one and I believe Hamilton to be a thoroughly good fellow— You ask what my plan is? It is to stay here *certainly* one fortnight more when I shall know whether I can do any thing or not. If I fail I suppose I must go home, if I succeed I shall write to my Father and tell him I want nothing of him for the next two years except payment of Debts a List of which I shall transmit, and then live on in my garret scribbling and hoping for better things till the two years have crept away. Any thing rather than Somersetshire and old recollections— I hope you have supported Cookesley's character in my absence but I perceive that you have not convinced Cameron— And you dare to lift up your voice against the immortal Shelley, as if he was an Insect and to find fault with that Most perfect of lines Most Musical of Mourners weep again[3] I wish I had you here in my garret where there is no room to run away. I would persecute you for hours— Most humanised of Insect kill⟨ers⟩ blush again— I have just completed nine of the most sentimental stanzas ever edited for which I intend to get five guineas, so a sneer at Poetry touches at once my fruits and my fortunes

Write soon, like a gentleman as you are— I mean to have a fly at Tennysons Poems if I can, they have been lauded in the Westminster—[4]

Yours most sincerely H Matthew

Like is not an adverb but a preposition when used as you use it Vide Murray[5] Page 102 Lowth[6] page 21 Harris[7] page 60 Cobbet[8] passim—

God bless you old boy—and I do believe in God— Yours ever | Henry Matthew

Postmark: M FE 14 1831
DAR 204.3

[1] 'The ACCUSING SPIRIT which flew up to heaven's chancery with the oath, blush'd as he gave it in;—and the RECORDING ANGEL, as he wrote it down, dropp'd a tear upon the word, and blotted it out for ever.' (Laurence Sterne, *Tristram Shandy* (1759–67), vol. 6, ch. 8.)

[2] Possibly Edward William Terrick Hamilton.

[3] 'Adonais', 4. 1.

[4] The work referred to is Tennyson 1830, written while Tennyson was an undergraduate at Trinity College, Cambridge. The notice appeared in the *Westminster Review*, 'Tennyson's poems', vol. 14, January 1831, pp. 210–24. The anonymous author has been identified as William Johnson Fox, not Sir John Bowring, as Tennyson's biography in the *DNB* has it (*Wellesley Index* 3: 572).

[5] Lindley Murray's *English grammar* went through many editions and was used in schools to the exclusion of all others (*DNB*). A copy in Darwin Library—CUL of the two-volume edition of 1824 has the signature 'Robert FitzRoy 1831' on both covers and title-pages. In the margins of volume one some passages on the rules of syntax and the use of the comma are lightly scored in pencil.

[6] Robert Lowth or Louth wrote a short introduction to English grammar (1762) (*DNB*).

[7] Perhaps a reference to Harris 1751.

[8] William Cobbett wrote on grammar as well as on politics, economics, and agriculture in *Cobbett's Weekly Political Register* from 1802 until his death (*DNB*).

To W. D. Fox [15 February 1831]

[Cambridge]
Tuesday

My dear Fox

I am going out this evening & have only time to write about business—

I saw Baker this morning & told him your message about writing, which he did not seem to like, so I do it for him.— He has for you the following Birds, pair of Hen Harriers— 3 ash coloured Falcons.— Woodpecker— pair Bearded Titmice.— Shrike.— Swan. (I sent to him to Price, but have not heard whether he is up) The Swan is in very bad condition.— I have sent also a dusky grebe, which I procured sometime ago.— His bill amounts to 5£ "3ˢ "0. which includes a packing case 14ˢ & a bushel Ribston pipins 12ˢ— I have not yet paid Aiken but will see about it.— You will then be indebted to me to some small amount.— If you will give me directions I will tell Baker to send your birds off. (You had perhaps better have the engraving with them) Henslows former servant has left his service some time ago.—

Yours sincerely | Chas Darwin

I shall go away in the course of 10 days or there abouts.—
Adieu
If you have not read Herschel[1] in Lardners Cyclo[2]—read it directly.
1 " 6.—
3 " 17—

Postmark: Cambridge FE 15 1831
Christ's College Library, Cambridge (Fox 37)

[1] John Frederick William Herschel.
[2] Herschel's *Preliminary discourse on the study of natural philosophy* was published in 1831 in Dionysius Lardner's *Cabinet cyclopaedia*. It became an authoritative statement of the methods of scientific investigation, anticipating John Stuart Mill in the formulation of the famous four methods of scientific investigation. In the *Autobiography*, p. 68, CD says that Humboldt's *Personal narrative* and the *Preliminary discourse* 'stirred up in me a burning zeal to add even the most humble contribution to the noble structure of Natural Science. No one or a dozen other books influenced me nearly so much as these two.' The copy of the *Preliminary discourse* in Darwin Library — CUL has no annotations in CD's hand. Several passages are marked in the margin. These markings occur in section 19, the criterion of a true statement of a law of nature; section 129 and section 130, on naming and nomenclature; section 384, on the superiority of residents over travellers in scientific investigation; and section 385, on the importance of institutions and journals in promoting the spread of science.

From Henry Matthew [March or April 1831][1]

Kilve, Bridgwater

My dear Darwin,

Do not think me the most forgetful and the most ungrateful rascal in this world of heartlessness on account of my long silence. Till this last week I did not know where you were to be found and I am not at all certain that this will reach you. If it ever does come into your possession I hope you will believe the assurances of my undiminished regard which it brings with it. Our friendship (I hate the term it sounds so like cant) but I must use it for fault of a better, our friendship was the growth of a day but I trust it will bear fruit for years. Once for all I do love you and shall ever come what may to either of us.

I am now at home in the bosom of my family, (as the novel writers have it) and I know of no bosom which I had not rather lie in.— I came here like the prodigal son but was received more like a fatted calf. My Father is fierce my brother cold and my sisters in tears. Every post brings a Dun, and every Dun a scene. My Father abuses me for wasting my Talents, though he never discovered that I had such things till they were irrecoverably thrown away. But such is the trick of Governors. My reviewing scheme did not succeed to my wishes— I was too much bothered by my w—e[2] when in London to write much, and I have been too much bothered by my Father here to write at all. I sent some poetry as I told you written to *the* young Lady six years since to one of the Magazines and received a civil note in reply beginning with compliment and concluding with "Sorry that the poetical department was occupied. What a phrase poetical department. Who ever heard of the department of Apollo. I sent besides a humourous description of my own condition to another rascal but got no money for it. I received much praise indeed and an entreaty to write again with assurances of prompt attention et cet. But that would not feed me, so weighing the "solid pudding" of home against the "empty praise" of a garret, I pawned my clothes for my coach hire and here I am. This last week however I have taken more heart, and this morning I commenced writing again— Do you know of any one who is in want of a private tutor in your part of the world, a household drudge, I mean— This is the situation I am now looking out for Meanwhile, How do you go on old Fellow. Have you bottled any more beetles, or impaled any butterflies. Are you delighted at the prospect of reform,[3] or weeping over the pres⟨ent⟩ deca⟨de⟩ of abuses. Or are you still old quiet poco c⟨uran⟩te Darwin[4] caring for nought but your gin bottle and its constant accompaniment philosophy. Dost thou not see symptoms of a rejected Scribbler in this my letter? The artificial liveliness of a writer in a magazine the vain attempts of one who has much small talk on hand, because his stock will not sell— You think from all this that I am in uproarious spirits but you were never more mistaken in your life. I am a wretch.

I have been at H—— lately, and I have seen her who makes all others little worth seeing, Oh God what a woman; and then to feel myself tied to a fool. The chain drags heavily.

Are you yet become inspired by the Spirit of Shelley I say inspired, for to enjoy him is a sort of inspiration Read the Cenci and the Spirit of Solitude, and confess that Virgil is a driveller and Byron a copyist.

So the widow flung Cameron after all?

God bless you my dear fellow | Ever your most sincere friend | H Matthew[5]

DAR 204.3

[1] The date assigned is a probable one. The letter clearly was sent after 14 February 1831 (when Matthew wrote from London) and as it is directed to Shrewsbury Matthew may have known CD to be at home between terms. Easter term began 13 April 1831.

[2] Presumably 'wife' (see letter from Henry Matthew, [2 February 1831]).

[3] Lord John Russell had introduced the first Reform Bill on 1 March 1831. CD, a Whig, was pro-Reform.

[4] A reference to Samuel Butler's 'poco curante' censure of CD's extra-curricular interest in chemistry while at Shrewsbury School (see *Autobiography*, p. 46).

[5] William Makepeace Thackeray, who was also a friend of Matthew, visited him in July of 1831 and wrote to Edward Fitzgerald as follows: 'he is improved in mind, & appearance for he does not look the rake he used—& has met with some very sad & trying experience since last I saw him' (Ray 1945–6, 1: 151). In a later letter (17–19 May 1849) he refers to Matthew as 'that friend of my youth whom I used to think 20 years ago the most fascinating accomplished witty and delightful of men— I found an old man in a room smelling of brandy & water at 5 o'clock . . . grown coarser and stale somehow, like a piece of goods that has been hanging up in a shop window. He has had 15 years of a vulgar wife' (*ibid.*, 2: 541).

To W. D. Fox [7 April 1831]

My dear Fox

Do you mean to cut the connection; why do you not write? I sent the last letter; so by the laws of nations you ought to have written.— I was in such bustle, when I last wrote to you: that I really forget how our various money transactions go on.— I will state them.— I have in my possession your 5£: I have paid Aiken 2″13″6. But have not paid Bakers bill, (not having seen him) which amounts 5£″3ˢ″6 including 12ˢ for apples for Henslows— I shall start for Cambridge tomorrow week: but shall stay a few days in London to hear Operas &c &c.— Let me have a letter from you waiting at Cambridge, or before I go there: I will settle all your affairs for you.—

I expect to spend a very pleasant Spring term: walking & botanizing with Henslow: I suppose it is out of the question, your snatching a Parsons week[1] & running up to Cambridge. I think you would enjoy; I am sure I should;— Think of it.—

At present, I talk, think, & dream of a scheme I have almost hatched of going to the Canary Islands.— I have long had a wish of seeing Tropical scenery & vegetation: & according to Humboldt[2] Teneriffe is a very pretty specimen.—

Looking over your letter I find there is a bill Orridges, is it distinct from the 2£″13″6?

If you are not busy, you had better write to me before tomorrow week, & give me circumstantial account of every thing that you can think of.— How all your family are? &c &c.

Believe me dear old Fox | Most sincerely | Chas Darwin

Shrewsbury Thursday

PS. tell me how, where &c &c, you are living?

Endorsement: April 10/31
Christ's College Library, Cambridge (Fox 39)

---

[1] 'the time taken as a holiday by a clergyman who is excused a Sunday, lasting (usually) from Monday to the Saturday week following' (*OED*).

[2] See chapter two of Humboldt 1814–29; also *Autobiography*, p. 68, where CD writes that he copied out from Humboldt long passages about Tenerife. The English translation of *Personal narrative* is in Darwin Library—CUL in six volumes of various editions. Volumes one and two, in one, third edition, 1822, is inscribed 'J. S. Henslow to his friend C. Darwin on his departure from England upon a voyage around the World 21 Septʳ 1831'. All of the volumes have some marginal scoring of passages and occasional comments. Volume five has a list of page numbers on the end-paper; volumes one and two, three, and seven have notes by CD pinned in back.

From Fanny Owen [8 April 1831]

[Woodhouse]

My dear Charles—

I am quite ashamed of myself for being so troublesome to you—but *goodnatur'd people you know* always *do get imposed upon*—so I want to beg you to bring me something from Mr Whitneys on Monday—it is a bottle of *"Aspalterm"*[1] an oil color—and also half a dozen small brushes if he has any—

I can't go on with my *Dairymaid* without this color, so if you will bring it me on Monday—I can't say how obliged I shall be to you— it is a **very small** bottle—it goes down to the Governor's account there **of course**!— I got you an order for Halston[2] to day—so dont fail to come on Monday *betimes*, in spite of what all the *Sisterhood* may say or do to prevent you— the Black **Charger** is expected tomorrow—

Do bring me a **juicy** Book of some kind—any thing you can *purloin*—for I sadly want one— I dont care what it is—a *book* is a **book**—truism!

It is very late & if I dont finish the bag will be lock'd for the night and this wont get admittance—so excuse scribble & all the trouble I give you—believe me | ever y^{rs}— F. Owen—

Friday night—

Postmark: Shrewsbury AP 9 1831
DAR 204.4

[1] '*Asphaltum* is prepared from the bituminous substance of that name. When dissolved in oil of turpentine, it is semitransparent, and is used as a glaze.' (Bigelow 1831, p. 437.)
[2] Halston Hall, near Oswestry, home of Fanny's cousin John 'Mad Jack' Mytton, notorious for his wild, dissipated life (see Apperley 1837).

To Caroline Darwin [28 April 1831][1]

[Cambridge]
Thursday

My Dear Caroline,

I want to hear some Shropshire politics, & I write in the hopes of having an answer.— I will give you a short hand account of myself since I left Shrewsbury.— I spent a very pleasant week in London, but found it as I always do, very fatiguing (I expect Erasmus did also for we lived together). Through Tom's[2] assistance I got a ticket for the Antient music which was most admirable; but what I liked most in all London is the Zoolog: Gardens: on a hot day when the beasts look happy and the people gay it is most delightful.— Cambridge, I find, is one of the few places, where if you anticipate a great deal of pleasure you do not find yourself disappointed: every day in the Term that passes I feel a loss; and the days go so quietly that I never do half what I intend to in the morning.— I am very busy and work all morning till Henslow's lecture: in the evenings I generally go out somewhere, and occasionally dinner parties, where good-eating and good-talking make a most harmonious whole (I hope you are disgusted, I will excuse anybody till they have been to a Cam: dinner, & if they are there, and if they cry out "what a disgusting thing a good dinner is" I must give them up.) The Election here is a great bore, as Henslow is Lord Palmerston's right-hand man,[3] and he has no time for walks.— All the while I am writing now my head is running about the Tropics: in the morning I go and gaze at Palm trees in the hot-house and come home and read Humboldt: my enthusiasm is so great that I cannot hardly sit still on my chair. Henslow & other Dons give us great credit for our plan: Henslow promises to cram me in geology.— I never will be easy till I see the peak of Teneriffe and the great Dragon tree; sandy, dazzling, plains, and gloomy silent forest are alternately uppermost in my mind.— I am working regularly at Spanish; Erasmus advised me decidedly to give up Italian. I have written myself into a Tropical glow.

goodbye | C Darwin

tell me Gossip write soon

I took my Degree the other day:[4] it cost me £15: there is waste of money. Love to all you do not hear from me often, so you must excuse my being as Egotistical as Hope.— I begin to think Natural Hist: makes people Egotistical.

Copy
DAR 154

[1] The copyist noted 'Postmark Cambridge April 28th (?)'. The query apparently indicates that the year was illegible.
[2] Thomas Josiah Wedgwood.
[3] John Stevens Henslow, originally a Tory, followed Lord Palmerston when he changed parties in 1828. In the elections of 1831 Palmerston lost his seat as M.P. for Cambridge University because of his support of Parliamentary reform. (Jenyns 1862, p. 60.)
[4] Tuesday, 26 April 1831 (*LL* 1: 163 n.).

From [J. M. Herbert][1] [early May 1831][2]

If Mr Darwin will accept the accompanying Coddington's Microscope,[3] it will give peculiar gratification to one who has long doubted whether Mr Darwin's talents or his sincerity be the more worthy of admiration, and who hopes that the instrument may in some measure facilitate those researches which he has hitherto so fondly and so successfully prosecuted.

DAR 204.3

[1] In a letter to Herbert, 21 November 1872 (APS 425), CD says, 'Do you remember giving me anonymously a microscope? I can hardly call to mind any event in my life which surprised & gratified me more.'

[2] See letter to W. D. Fox, [11 May 1831], the source of the approximate date of this letter.

[3] Of three microscopes on display at Down House, one bearing the maker's name of 'Cary' corresponds in structure to that described by Henry Coddington in a paper read in 1830 to the Cambridge Philosophical Society (Coddington 1830). Professor Phillip Sloan of the University of Notre Dame, who has examined the instrument in connection with his study of CD's microscopy, has concluded that it is the one referred to in this letter. Herbert may have given Coddington's name to the instrument because it had been built by George and John Cary, London instrument makers, according to Coddington's specifications; it used an improved lens, which though not invented by Coddington was commonly called after him.

To W. D. Fox [11 May 1831]

Cambridge
Wednesday

My dear Fox

Cambridge has been in such a state of bustle & excitement for the last week,[1] that I have done nothing but go gossiping about the town.— But thank goodness we are once again quiet: & I have had time to think about my plans.— I am very, very sorry to say I cannot pay you a visit at present. I really have no right to travel so many miles (& ∴ cost so many shillings) for my own amusement: the Governor has given me a 200£ note to pay my debts, & I must be economical.— Independent of paying you a nice quiet, snug visit, I should have put a scheme into practice which I have long wished to do to see the Pictures at Stamford:[2] But both schemes must die the same death from inanition.— On the per contra side of the question Henslows lectures come into play, & I should have been sorry to have missed even one of them.—

And now for our eternal accounts.— I find I made a mistake in my last letter.— I have of your money, 6″4″6 Orridge's bill was 13^s & what ⟨I⟩ have paid for you amounts in toto, 3£″6″6 ∴ remains now 2£18^s— I have written to Baker & given him proper directions, & will pay hi⟨m⟩ his bill next time he comes to Cam.:—

Some goodnatured Cambridge man has made me a most magnificent anonymous present of a Microscope: did ever hear of such a delightful piece of luck? one would like to know who it was, just to feel obliged to him.—

My time here is very pleasant. I am very busy at 3 or 4 λογοι,[3] & see a great deal Henslow, whom I do not know, whether I love or respect most.— Mrs Henslow is hatching a young professor.—[4] She will be confined very soon.— As for my Canary scheme, it is rash of you to ask questions: My other friends most sincerely wish me there I plague them so with talking about tropical scenery &c &c.— Eyton will go next summer, & I am learning Spainish.—

How I wish we could meet. You would soon be tired of the subject

Good Bye | Chas Darwin

PS. Aiken is not ill John Day wants to know what to do with your wine as the Hampers are rotting & something must soon be done.

[1] General Election polling took place in Cambridge on 3, 4, 5, and 6 May 1831, when the Whig government went to the country over the first Reform Bill.
[2] The collection of the Marquis of Exeter at Burghley House, Stamford.
[3] Prose writings.
[4] Leonard Ramsay Henslow.

To W. D. Fox [9 July 1831]

Shrewsbury
Saturday

My dear Fox

I arrived at this stupid place about three weeks ago, but have had no quiet time for writing till now.— I am staying here, on exactly the same principle, that a person chooses to remain in the Kings bench.— Talking of poverty puts me in mind to give you a scolding: in your answer to my letter containing the reasons I could not come to Epperstone, you say you do not wonder at my not choosing to to come to so stupid a place. now treating the thing logically, 1st you *must* have know what you call stupid, is just what I like, & 2nd you *might* have know, that if I could, I would most assuredly have come if it were merely for the pleasure of seeing you ∴ you have no excuse, & are (as we say in Spanish) un grandisimo bribon.—[1]

I hope your prints arrived safe. I was in a perfect whirlwind of dust & confusion, when I sent them off, else I should have written with the box.— The Canary scheme goes on very prosperously. I am working like a tiger for it, at present Spanish & Geology,[2] the former I find as intensely stupid, as the latter most interesting. I am trying to make a map of Shrops: but dont find it so easy as I expected.—

How goes on Entomology with you? you are in a capital situation, that is if Sherwood forest is at all like the New forest.— Hope & Eyton did wonders the⟨re⟩ (I did not go propter pecuniam). Your imagination cannot fancy the number of red elater, melasis, Cerambycidous insects, without end.— I am just beginning Diptera.— L. Jenyns started me, what an excellent naturalist he is. I have seen a good deal of him lately, & the more I see the more I like him.— I feel just the same way towards another man, whom I used formerly to dislike, that is Ramsay of Jesus,[3] who is the most likely person (I dont know whether I told you before) to be my companion to the Canaries.— How much do you know of the particulars of our plans?

Shall you be at Epperstone in the Autumn, & if you are, would be convenient, if I could manage to pay you a visit—answer this sincerely.— Cannot you pay me a visit at Shrewsbury, is it impossible?

This letter is all about myself. Do thou likewise, good Bye | dear old Fox. C. Darwin

Postmark: Shrewsbury JY 10 18⟨3⟩1
Christ's College Library, Cambridge (Fox 41)

[1] A great rascal.
[2] CD's 'Journal' (see Appendix I), which he started to keep in 1838, has this retrospective entry for 1831: 'In the Spring Henslow persuaded me to think of Geology & introduced me to Sedgwick.' John Maurice Rodwell, writing about CD and their Cambridge days together, remembered talking over Sedgwick's lectures and CD saying: 'It strikes *me* that all our knowledge about the structure of our Earth is very much like what an old hen w^d know of the hundred-acre field in a corner of which she is scratching'. Later, speaking of Sedgwick's speculation about the probable antiquity of the world, CD exclaimed, 'What a capital hand is Sedgewick for drawing large cheques upon the Bank of Time!' (J. M. Rodwell to Francis Darwin, 8 July 1882, in DAR 112: 94v.).
[3] Marmaduke Ramsay.

To J. S. Henslow [11 July 1831]

Shrewsbury
Monday

My dear Sir

I should have written to you sometime ago, only I was determined to wait for the Clinometer: & I am very glad to say I think it will answer admirably: I put all the tables in my bedroom, at every conceivable angle & direction I will venture to say I have measured them as accurately as any Geologist going could do.— It cost 25^s made of wood, but the lid with plate of brass graduated.— Cary[1] did not approve of a bar for the plumb: so that I had *heavy* ball instead.— I have been working at so many things: that I have not got on much with Geology: I suspect, the first expedition I take, clinometer & hammer in hand, will send me back very little wiser & good deal more puzzled than when I started.— As yet I have only indulged in hypotheses; but they are such powerful ones, that I suppose, if they were put into action but for one day, the world would come to an end.— I have not heard from Prof: Sedgwick, so I am afraid he will not pay the seven formations a visit.— I hope & trust you did your best to urge him:— And now for the Canaries.— I wrote to Mr Ramsay, the little information which I got in town.— But as perhaps he had left Cam. I will rehearse it.— Passage 20£: ships touch & return, during the months of June to February.— But not seeing myself the Broker, the 2 most important questions remain unanswered, viz. whether it means June inclusive & how often they sail.— I will find this out before very long.— I hope you continue to fan your Canary ardor: I read & reread

Humboldt, do you do the same, & I am sure nothing will prevent us seeing the Great Dragon tree.— Would you tell L. Jenyns, that his magnificent present of Diptera has not been wasted on me Would you ask him how he manages Diptera when too small for a pin to go through.— I am very anxious to hear how M^{rs}. Henslow is.— I am afraid she will wish me at the bottom of the Bay of Biscay, for having been the first to think of the Canaries.— I am going now to trouble you with several questions.— Do you know A. Ways direction? Do you by any chance recollect the name of a fly that M^r. Bird sent through Downes.—[2] And now for a troublesome commission, would you be kind enough to exert your well-know judgment & discretion in choosing for me a Stilton Cheese; fit for eating pretty soon.— Would you have it directed to Shrewsbury & I will pay the man, when I come up in October.—

Excuse all the trouble I am giving you, & Believe me my dear Sir | Yours ever most sincerely | Chas. Darwin

Eyton begs to be most kindly remembered to you.— his mind is in a fine tropical glow.—

Postmark: Shrewsbury JY 12 1831
Royal Botanic Gardens, Kew

[1] George and John Cary, the London instrument makers.
[2] John Downes.

To W. D. Fox 1 August [1831]

[Shrewsbury]
August 1st

Dear Fox

I received your letter yesterday & sit down, according to your desire, to answer to the best of my power your questions.— Hope will not be in town during the time you mention, his direction is 37 Upper Seymour St. & I am sure he would be very glad to show you his cabinet, if you should at any time be in Town when he is.—

I am sorry to say I utterly forget both Stephens Christian name & direction but I should think you could find out his direction from his bookseller, which latter you can find from the covers of the British Entomology.— I have generally gone somewhere about 8 oclock: He is a very civil little man: there is one proper evening in the week, but I believe he has no objection to people out of the country calling at any time.—

I forget whether I mentioned to you that I have 2 or 3 insects from, either M^r. Rud, or Wailes of Newcastle,[1] Henslow gave them in my charge for you, & I cannot recollect for certain the Donor.— I will send these together with my long promised lot (which I am afraid will not be so numerous as you might expect) in the course of a few days.—

The Canary scheme does not take place till next June. I am sorry to hear that Henslows chance of coming is *very* remote. I had hoped it was daily growing less so.— I shall in probability to go to Cam to pay my bills somewhere about the end of October, but I do not know for certain.—

I cannot end this letter without adding, how *grieved* I am to find that you think me capable of telling base, hollow & deliberate falsehoods, in no other possible way can I interpret your letter.[2]

Yours sincerely, | Chas. Darwin

The man who made my cabinet is W. Edwards 29 Wilton St. Westminster. I advise to get one bigger than mine.— Mine cost 5$^£$ʺ10. & contained 6 drawers, depth, 1^{f}ʺ3. breadth 1^{f}ʺ7.—& whole cabinet stood in height 1^{f}ʺ4.—

Postmark: Shrewsbury AU 1 1831
Christ's College Library, Cambridge (Fox 42)

[1] L. Rudd, Esq., of Marton Lodge, Yorkshire, and George Wailes, Esq., of Newcastle-upon-Tyne, are frequently mentioned as contributors of specimens in Stephens 1827—46.
[2] Apparently this was meant seriously. See letter to W. D. Fox, 6 [September 1831]. In *LL* 1: 205, where an excerpt from that letter is printed, Francis Darwin says in a footnote: 'He [CD] had misunderstood a letter of Fox's as implying a charge of falsehood.' Fox's letter has not been found.

From George Peacock to J. S. Henslow [6 or 13 August 1831][1]

My dear Henslow

Captain Fitz Roy is going out to survey the southern coast of Terra del Fuego, & afterwards to visit many of the South Sea Islands & to return by the Indian Archipelago: the vessel is fitted out expressly for scientific purposes, combined with the survey,: it will furnish therefore a rare opportunity for a naturalist & it would be a great misfortune that it should be lost:

An offer has been made to me to recommend a proper person to go out as a naturalist with this expedition;[2] he will be treated with every consideration; the Captain is a young man of very pleasing manners (a nephew of the Duke of Grafton), of great zeal in his profession & who is very highly spoken of; if Leonard Jenyns could go, what treasures he might bring home with him, as the ship would be placed at his disposal, whenever his enquiries made it necessary or desirable;[3] in the absence of so accomplished a naturalist, is there any person whom you could strongly recommend: he must be such a person as would do credit to our recommendation

Do think on this subject: it would be a serious loss to the cause of natural science, if this fine opportunity was lost

The ship sails about the end of Sept.

Poor Ramsay! what a loss to us all & particularly to you

Write immediately & tell me what can be done

Believe me | My dear Henslow | Most truly yours | George Peacock

7. Suffolk Street | Pall Mall East

My dear Henslow

I wrote this letter on Saturday, but I was too late for the Post: What a glorious opportunity this would be for forming collections for our museums: do write to me immediately & take care that the opportunity is not lost

Believe me | My dear Henslow | Most truly yours | Geo Peacock

7. Suffolk St. | Monday

DAR 97 (ser.2): 1 − 3

[1] Internal evidence shows that this letter was written on a Saturday after Marmaduke Ramsay's death, which occurred on 31 July 1831. Letters from J. S. Henslow, 24 August 1831 and from George Peacock, [*c.* 26 August 1831] indicate that there was further correspondence before Henslow's letter of 24 August.

[2] Peacock, a Fellow of Trinity College and lecturer in mathematics, knew of the plans for the voyage from Francis Beaufort, Hydrographer of the Navy. It is not clear from Peacock's statement who made the offer to him. It may have been Beaufort rather than Robert FitzRoy. If so, it would explain some of the misunderstanding that arose later about the availability of the post for CD (see letter to Susan Darwin, [5 September 1831]).

[3] In a letter of 1 May 1882 to Francis Darwin (DAR 112: 67 − 8), Jenyns remembered that he took a day to think over the offer before deciding that he could not leave his parish. He and Henslow then thought of CD.

From J. S. Henslow[1] 24 August 1831

Cambridge
24 Aug 1831

My dear Darwin,

Before I enter upon the immediate business of this letter, let us condole together upon the loss of our inestimable friend poor Ramsay of whose death you have undoubtedly heard long before this. I will not now dwell upon this painful subject as I shall hope to see you shortly fully expecting that you will eagerly catch at the offer which is likely to be made you of a trip to Terra del Fuego & home by the East Indies— I have been asked by Peacock who will read & forward this to you from London to recommend him a naturalist as companion to Capt Fitzroy employed by Government to survey the S. extremity of America— I have stated that I consider you to be the best qualified person I know of who is likely

to undertake such a situation— I state this not on the supposition of y^r being a
finished Naturalist, but as amply qualified for collecting, observing, & noting any
thing worthy to be noted in Natural History. Peacock has the appointment at
his disposal & if he can not find a man willing to take the office, the
opportunity will probably be lost— Capt. F. wants a man (I understand) more
as a companion than a mere collector & would not take any one however good a
Naturalist who was not recommended to him likewise as a *gentleman*. Particulars
of salary &c I know nothing. The Voyage is to last 2 y^{rs} & if you take plenty
of Books with you, any thing you please may be done— You will have ample
opportunities at command— In short I suppose there never was a finer chance for
a man of zeal & spirit. Capt F. is a young man. What I wish you to do is
instantly to come to Town & consult with Peacock (at N° 7 Suffolk Street
Pall Mall East or else at the University Club) & learn further particulars.
Don't put on any modest doubts or fears about your disqualifications for I
assure you I think you are the very man they are in search of—so conceive
yourself to be tapped on the Shoulder by your Bum-Bailiff[2] & affect^e friend |
J. S. Henslow

(Turn over)
The exped^n is to sail on 25 Sept:[3] (at earliest) so there is no time to be lost

DAR 97(ser.2):4 − 5

[1] The cover is addressed to 'C. Darwin Esqr Shrewsbury To be forwarded or opened, if absent'.
[2] 'the bailiff that is close at the debtor's back, or that catches him in the rear' (*OED*).
[3] Since this date and the reference to Robert FitzRoy's wanting a companion are details that do not
appear in the letter from George Peacock to J. S. Henslow, [6 or 13 August 1831], Henslow must
have had at least one additional letter from Peacock.

From George Peacock [*c.* 26 August 1831]

My dear Sir
    I received Henslow's letter last night too late to forward it to you by the
post, a circumstance which I do not regret, as it has given me an opportunity of
seeing Captain Beaufort at the admiralty (the Hydrographer) & of stating to
him the offer which I have to make to you: he entirely approves of it & you may
consider the situation as at your absolute disposal: I trust that you will
accept it as it is an opportunity which should not be lost & I look forward with
great interest to the benefit which our collections of natural history may
receive from your labours
    The circumstances are these

Captain Fitzroy (a nephew of the Duke of Graftons) sails at the end of September in a ship to survey in the first instance the S. Coast of Terra del Fuego, afterwards to visit the South Sea Islands & to return by the Indian Archipelago to England: The expedition is entirely for scientific purposes & the ship will generally wait your leisure for researches in natural history &c: Captain Fitzroy is a public spirited & zealous officer, of delightful manners & greatly beloved by all his brother officers: he went with Captain Beechey[1] & spent 1500£ in bringing over & educating at his own charge 3 natives of Patagonia:[2] he engages at his own expense an artist at 200 a year to go with him: you may be sure therefore of having a very pleasant companion, who will enter heartily into all your views

The ship sails about the end of September & you must lose no time in making known your acceptance to Captain Beaufort, Admiralty hyd[r] I have had a good deal of correspondence about this matter, who[3] feels in common with myself the greatest anxiety that you should go. I hope that no other arrangements are likely to interfere with it

Captain will give you the rendezvous & all requisite information: I should recommend you to come up to London, in order to see him & to complete your arrangements I shall leave London on Monday: perhaps you will have the goodness to write to me at Denton, Darlington, to say that you will go

The Admiralty are not disposed to give a salary, though they will furnish you with an official appointment[4] & every accomodation: if a salary should be required however I am inclined to think that it would be granted

Believe me | My dear Sir | Very truly yours | Geo Peacock

If you are with Sedgwick I hope you will give my kind regards to him

DAR 97 (ser.2): 11 – 13

[1] An error for Captain Philip Parker King, commander of the *Adventure* and *Beagle* survey of the southern coasts of South America (1826 – 30). Robert FitzRoy was appointed commander of the *Beagle* in 1828 to replace Captain Pringle Stokes, who had committed suicide (*Narrative* 1: 188). King's account of the first expedition is in the first volume of *Narrative*.

[2] FitzRoy had brought four natives to England. One, Boat Memory, died of smallpox soon after arriving. The others were named York Minster, James (Jemmy) Button, and Fuegia Basket (a girl). FitzRoy's original plan was to fulfil a commitment to return the Fuegians to their native land; he had already chartered a small vessel for the voyage when an uncle obtained for him the appointment to survey the southern coasts of South America. For FitzRoy's account of the Fuegians, see *Narrative* 2: 1 – 16.

[3] Peacock has inadvertently omitted the name of his correspondent, who was almost certainly Henslow.

[4] In the event, CD's appointment was not official. Although CD lists himself on the title page of *Journal of researches* as 'Naturalist to the Beagle' and in the *Zoology* as 'Naturalist to the Expedition' this is not to be understood as an official title conferred by the Admiralty. The letters of the next month bear out the contention of J. W. Gruber 1969 and Burstyn 1975 that CD's situation was that of guest of Captain FitzRoy, who sought a 'well-educated and scientific person' as a companion (*Narrative* 2: 18).

To J. S. Henslow 30 [August 1831]

Shrewsbury
Tuesday 30th—

My dear Sir

Mr Peacocks letter arrived on Saturday, & I received it late yesterday evening.— As far as my own mind is concerned, I should I think, *certainly* most gladly have accepted the opportunity, which you so kindly have offered me.— But my Father, although he does not decidedly refuse me, gives such strong advice against going.—that I should not be comfortable, if I did not follow it.— My Fathers objections are these; the unfitting me to settle down as a clergyman.— my little habit of seafaring.— the *shortness of the time* & the chance of my not suiting Captain Fitzroy.— It is certainly a very serious objection, the very short time for all my preparations, as not only body but mind wants making up for such an undertaking.— But if it had not been for my Father, I would have taken all risks.—

What was the reason, that a Naturalist was not long ago fixed upon?— I am very much obliged for the trouble you have had about it—there certainly could not have been a better opportunity.— I shall come up in October to Cambridge, when I long to have some talk with you.— I will write to Mr Peacock at Denton, (in Durham?) but his direction is written so badly, that even with the assistance of the Post office, I am not certain about it— Would you therefore be so kind, if you know his or C. Fitzroys direction, would you send one line to the same effect.— My trip with Sedgwick answered most perfectly.— I did not hear of poor Mr Ramsays loss till a few days before your letter. I have been lucky hitherto, in never losing any person for whom I had any esteem or affection. My Acquaintance, although very short, was sufficient to give me those feelings in a great degree.— I can hardly make myself believe he is no more.— He was the finest character I ever knew.—

Yours most sincerely | my dear Sir. Chas. Darwin

I have written to Mr Peacock, & I mentioned that I have asked you to send one line in the chance of his not getting my letter.— I have also asked him to communicate with Cap. Fitzroy.— Even if I was to go my Father disliking would take away all energy, & I should want a good stock of that.— Again I must thank you; it adds a little to the heavy, but pleasant load of gratitude which I owe to you.—

Postmark: Shrewsbury AU 3⟨0⟩ 18⟨3⟩1
Royal Botanic Gardens, Kew

From R. W. Darwin to Josiah Wedgwood II 30 — 1 August 1831

Salop
30 August 31.

Dear Wedgwood

I am very glad you feel better, and that six of the turpentine pills answer.—
perhaps as they change the particular part of the bowel that is stimulated by
medicine, it is as well to still continue them for some time. You will be so good
as to attend to the buffy discharge from the bowels, because if that continues,
or increases, it may be a reason for giving up the use of these new pills, as they
may conduce to its formation.

Thank Frank[1] for his basket. I have not yet opened it.

Charles will tell you of the offer he has had made to him of going for a
voyage of discovery for 2 years.— I strongly object to it ⟨on var⟩ious
grounds, but I will not detail my reasons that he may have your unbiassed opinion
on the subject, & if you think differently from me I shall wish him to follow
your advice.

Dear Wedgwood yours affectionly | R W Darwin

Since writing the above Edward has opened the basket Henry left for me from
Frank & we find it is a parcel for your brother which they ought to have taken
on with his to the Hill.[2]

Wednesday 31.

Charles has quite given up the idea of the voyage.

Keele University Library (Wedgwood/Mosley 96)

[1] Francis Wedgwood, son of Josiah Wedgwood II.
[2] The Hill, Abergavenny, Wales, was the home of the John Wedgwoods.

To R. W. Darwin 31 August [1831]

[Maer]
August 31

My dear Father

I am afraid I am going to make you again very uncomfortable.— But upon
consideration, I think you will excuse me once again stating my opinions on the
offer of the Voyage.— My excuse & reason is, is the different way all the
Wedgwoods view the subject from what you & my sisters do.—

I have given Uncle Jos, what I fervently trust is an accurate & full list of
your objections, & he is kind enough to give his opinion on all.— The list & his
answers will be enclosed.— But may I beg of you one favor. it will be doing me

the greatest kindness, if you will send me a decided answer, yes or no.— If the latter, I should be most ungrateful if I did not implicitly yield to your better judgement & to the kindest indulgence which you have shown me all through my life.—& you may rely upon it I will never mention the subject again.— if your answer should be yes; I will go directly to Henslow & consult deliberately with him & then come to Shrewsbury.— The danger appears to me & all the Wedgwoods not great.— The expence can not be serious, & the time I do not think anyhow would be more thrown away, than if I staid at home.— But pray do not consider, that I am so bent on going, that I would for one *single moment* hesitate, if you thought, that after a short period, you should continue uncomfortable.—

I must again state I cannot think it would unfit me hereafter for a steady life.— I do hope this letter will not give you much uneasiness.— I send it by the Car tomorrow morning if you make up your mind directly will you send me an answer on the following day, by the same means.— If this letter should not find you at home, I hope you will answer as soon as you conveniently can.—

I do not know what to say about Uncle Jos.' kindness, I never can forget how he interests himself about me

Believe me my dear Father | Your affectionate son | Charles Darwin.

PS. Frank would be much obliged if you would forward the Crockery to the Hill.—

(1) Disreputable to my character as a Clergyman hereafter
(2) A wild scheme
(3) That they must have offered to many others before me, the place of Naturalist
(4) And from its not being accepted there must be some serious objection to the vessel or expedition
(5) That I should never settle down to a steady life hereafter
(6) That my accomodations would be most uncomfortable
(7). That you should consider it as again changing my profession
(8) That it would be a useless undertaking

DAR 223 and DAR 97 (ser.2): 10

From Josiah Wedgwood II to R. W. Darwin[1] 31 August 1831

Maer
31 August 1831

My dear Doctor

I feel the responsibility of your application to me on the offer that has been made to Charles as being weighty, but as you have desired Charles to consult me

I cannot refuse to give the result of such consideration as I have been able to give it. Charles has put down what he conceives to be your principal objections & I think the best course I can take will be to state what occurs to me upon each of them.

1— I should not think that it would be in any degree disreputable to his character as a clergyman. I should on the contrary think the offer honorable to him, and the pursuit of Natural History, though certainly not professional, is very suitable to a Clergyman

2— I hardly know how to meet this objection, but he would have definite objects upon which to employ himself and might acquire and strengthen, habits of application, and I should think would be as likely to do so in any way in which he is likely to pass the next two years at home.

3. The notion did not occur to me in reading the letters & on reading them again with that object in my mind I see no ground for it.

4. I cannot conceive that the Admiralty would send out a bad vessel on such a service. As to objections to the expedition, they will differ in each mans case & nothing would, I think, be inferred in Charles's case if it were known that others had objected.

5— You are a much better judge of Charles's character than I can be. If, on comparing this mode of spending the next two years, with the way in which he will probably spend them if he does not accept this offer, you think him more likely to be rendered unsteady & unable to settle, it is undoubtedly a weighty objection— Is it not the case that sailors are prone to settle in domestic and quiet habits.

6— I can form no opinion on this further than that, if appointed by the Admiralty, he will have a claim to be as well accommodated as the vessel will allow.

7— If I saw Charles now absorbed in professional studies I should probably think it would not be advisable to interrupt them, but this is not, and I think will not be, the case with him. His present pursuit of knowledge is in the same track as he would have to follow in the expedition.

8— The undertaking would be useless as regards his profession, but looking upon him as a man of enlarged curiosity, it affords him such an opportunity of seeing men and things as happens to few.

You will bear in mind that I have had very little time for consideration & that you & Charles are the persons who must decide.

I am | My dear Doctor | Affectionately yours | Josiah Wedgwood

DAR 97(ser.2): 6—8

[1] The letter was enclosed with letter to R. W. Darwin, 31 August [1831]. After the salutation CD wrote 'Read this last'. CD's account of the events up to 1 September, written on 16 December 1831,

reads as follows: 'I immediately said I would go; but the next morning finding my Father so much averse to the whole plan, I wrote to Mr Peacock to refuse his offer. On the last day of August I went to Maer, where everything soon bore a different appearance. I found every member of the family so strongly on my side, that I determined to make another effort. In the evening I drew up a list of my Father's objections, to which Uncle Jos wrote his opinion & answer. This we sent off to Shrewsbury early the next morning & I went out shooting. About 10 o'clock Uncle Jos sent me a message to say he intended going to Shrewsbury & offering to take me with him. When we arrived there, all things were settled, & my Father most kindly gave his consent.' ('*Beagle*' *diary*, p. 3).

From R. W. Darwin to Josiah Wedgwood II   1 September 1831

[Shrewsbury]
1 Sept 1831

Dear Wedgwood,

Charles is very grateful for your taking so much trouble & interest in his plans. I made up my mind to give up all objections, if you should not see it in the same view as I did.—

Charles has stated my objections quite fairly & fully—if he still continues in the same mind after further enquiry, I will give him all the assistance in my power.

Many thanks for your kindness— yours | affectionly | R W Darwin[1]

Keele University Library (Wedgwood/Mosley 96)

[1] Robert Waring Darwin must have written this note soon after receiving the foregoing letters and posted it before CD and his uncle arrived later in the day. This and Robert Waring's earlier letter to Josiah Wedgwood II (letter from R. W. Darwin to Josiah Wedgwood II, 30 – 1 August 1831) clearly call for some revision of the conception of CD's father as stern and unloving.

To Francis Beaufort   1 September [1831]

Shrewsbury
September the 1st

Sir

I take the liberty of writing to you according to Mr Peacocks desire to acquaint you with my acceptance of the offer of going with Capt Fitzroy. Perhaps you may have received a letter from Mr Peacock, stating my refusal; this was owing to my Father not at first approving of the plan, since which time he has reconsidered the subject: & has given his consent & therefore if the appointment is not already filled up,—I shall be very happy to have the honor of accepting it.— There has been some delay owing to my being in Wales, when the letter arrived.—[1] I set out for Cambridge tomorrow morning, to see Professor Henslow: & from thence will proceed immediately to London.—

I remain Sir | Your humble & obedient servant | Chas. Darwin

Stamped: SE 12 1831
Notation on letter by Robert FitzRoy: Sept 6th. Let him be borne on the books for victuals only
10 F.R
Public Record Office, Kew: Adm 1/4541 PRO D 262

[1] CD made a geological tour of North Wales with Adam Sedgwick from 5 to 20 August 1831 (see the
letter from Adam Sedgwick, 4 September 1831); he and Sedgwick, however, were together for no
more than about one week during this trip (see Barrett 1974, pp. 147–8). CD returned to
Shrewsbury on the 29th (see letter to W. D. Fox, 6 [September 1831]), after spending some time with
friends at Barmouth.

From Francis Beaufort to Robert FitzRoy   1 September [1831]

Sep 1st

My dear Sir

I believe my friend Mr Peacock of Trin^y College Camb^e has succeeded in
getting a "Savant" for you— A Mr Darwin grandson of the well known
philosopher and poet—full of zeal and enterprize and having contemplated a
voyage on his own account to S. America

Let me know how you like the idea that I may go or recede in time | F B

Copy
Hydrographer of the Navy, Ministry of Defence, Taunton, Letter book 3:233

To J. S. Henslow   [2 September 1831]

[Cambridge]

My dear Sir

I am just arrived: you will guess the reason. My Father has changed his
mind.— I trust the place is not given away.— I am very much fatigued & am
going to bed.— I daresay you have not yet got my second letter.—[1] How soon
shall I come to you in the morning. Send a verbal answer.

Good night | Yours. | C. Darwin

Red Lion.—

Royal Botanic Gardens, Kew

[1] The letter has not been found. It was evidently written at Maer on 31 August to inform Henslow that
CD and his uncle were attempting to persuade CD's father to change his mind. The letter was
delayed a day in the post. (See letter from Charlotte Wedgwood, 22 September [1831]).

Josiah Wedgwood II. Portrait by William Owen.

(Courtesy of the Trustees of the Wedgwood Museum, Barlaston.)

Robert FitzRoy, *c.* 1840.

(Courtesy of the Alexander Turnbull Library, Wellington, New Zealand.)

From Adam Sedgwick 4 September 1831

Tremadoc [Wales]
Sep.ʳ 4. 1831

Dear Darwin

I left Capel Curig the day before yesterday & the stupid red nosed waiter did not shew me your letter till a few hours before I started. Otherwise I should have endeavoured to profit by your information respecting Cwm Idwal.[1] I *ought* however to have seen the madrepores;[2] for last wednesday I went from Capel Curig to Cwm Idwal and thence clambered out at the top by the side of Twll Dy a very curious chasm which I suppose you have seen I then scaled the crests of Glider Bach & G. Fawr and zig-zagged down to the Inn. Your information did not however surprise me, as madrepores are quite as likely to be met with as terebratulæ, which seem to occur here and there thro' yᵉ Snowdonian chain. I found *terebratulæ* among the talcose slates of Foel Goch a precipice just to yᵉ west of Cwm Idwal— I found organic remains in yᵉ slate of Moel Shabod, but did not stumble on any bed in which they abounded. At first I found specimens about the middle of the great zone of slate, & afterwards in the stone walls above the wood (by the way stone walls are good localities for fossils and often tell us a good story wʰ it wᵈ be difficult to make out without 'em). I don't understand your puzzle about M. Shabod.[3] Why should not the rough beds at the bottom, on yᵉ N.W. side, pass under the blue slate with shells? I don't agree with you in thinking that the mass of trap on yᵉ crest of the hill is *under* the slate. It appears to me decidedly to be *over* it— And in yᵉ great Cwm with yᵉ small lake on yᵉ East side we see the slate under yᵉ trap.— Again the trap wraps round in a horse Shoe shape to yᵉ S.E.—twists round to yᵉ E. side of the great Cwm, & then runs in a mass about 200 yards wide in a direct.ⁿ about N.E. for a mile or two, between two great vertical masses of baked slate. E.G.,

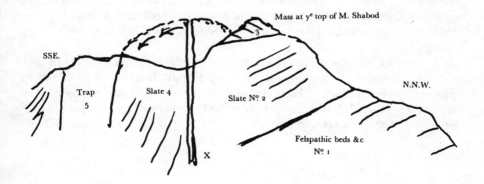

Mass at yᵉ top of M. Shabod

SSE.

N.N.W.

Trap
5

Slate 4

Slate Nº 2

Felspathic beds &c
Nº 1

X

My picture is detestable and out of all gentlemanlike proportions but you must suppose 4 & 5 to be pulled out considerably to ye left hand of X N°.1 & N°.2 are as in your letter; N°.3 ranges a mile to the S—. & then comes round to N°.5—the connexion is not *ideal*, as you may walk all the way on trap from N°.3 to N°.5. The slate N°.4 where in contact is as hard as a flint. I'm beginning to think that I shall not reach Barmouth this year. Therefore have the kindness to address me at Carnarvon, which I shall for some time consider as my head quarters—tho' I shall probably not often be there— To give you some outline of my progress I will send you a skeleton of my walk.

21st. (this day fortnight) from Carnarvon to Dolbadarn; having in ye morning heard a sermon an hour long & gone a geological Sunday walk towards the S.— 22^d. a hardworking day along the Crown, Shoulders & ribs of Snowdon 23^d. a still harder day, commencing with ye slate quarries. (*4 hours before Breakfast*, what w^d your man have said to that?)—& then scaling Lidir Mawr, & zigzagging along ye Crests of ye Chain to Twll Dy, & so home in ye dusk (*24th.* weather bound the whole day (*25th.* hammer my way to Cap⟨el⟩ Curig—lunch, & then hammer my way to Tinny Maes— making anot⟨her⟩ modest collection of geological specimens.) *26th.* Carnedd David, Carned Llewelyn, & so down the rough crags of Porphyry (we saw at a distance) down to Aber—& thence back to Tinny Maes. *27.* ascertain the nature of ye beds below the great slate zone by two or three traverses; & end at Bangor. *28.* In a great measure a day of rest—E.G. drive to Aber. *29th.* To Conway by the crests of the hill—thence across the water to ye ridges between great & little Orms head Ev^g drive to Llanrwst.— *30th.* Cross the mountains on foot by the line of y^e *lakes* to Capel Curig.: caught in a mist and deluge of rain and steer over ye mountains by Compass, reach the turnpike within a quarter of a mile of ye road turning off to ye Inn. *31* Lake Ogwen Glider Fawr & Bach, &c &c to C. Curig. 1st Sepr The crest & flanks of Moel Shabod— Evening excursion to ye knolls N. of the trap &c. 2^d Mountains between C. Curig and Llanrwst, & visit some of ye mines— 3^d. Bettws, Penmachno, Maentwrog, & Tremadoc, ascending one or two hills by ye way. 4th. Sunday, a day of rest—continual mist & rain.

My best regards to your friends at Shrewsbury. | Yours most truly A Sedgwick

P S. I saw no *basalt* at Lake Ogwen but a very black pyritous variety of rock something between Lydian stone & compact felspar. It differs from basalt in being extremely siliceous.— Perhaps I did not see the spot you mention. *I am going* as soon as wind & weather permit to make traverses between this place & Carnarvon. They will take 10 days

DAR 204.5

[1] CD's geological notes of the tour of North Wales are in DAR 5. They have been transcribed and annotated in Barrett 1974, in which this letter and letter from Adam Sedgwick, 18 September 1831 are also published. Professor Barrett's transcriptions of the many Welsh names in the letters have been adopted with his permission. Cwm Idwal is described in CD's notes (pp. 13–14), Barrett 1974, pp. 159–60.

[2] *Ibid.*, p. 160.

[3] *Ibid.*, pp. 160–1, but CD's notes make no mention of a 'puzzle'. The statement 'the Slate appears to overlie the Trap' appears on p. 161.

To Susan Darwin   [4 September 1831]

Cambridge
Sunday Morning

My dear Susan

As a letter would not have gone yesterday I put off writing till to day.— I had rather a wearisome journey, but got into Cambridge very fresh.— The whole of yesterday I spent with Henslow, thinking of what is to be done.—& that I find is great deal.— By great good luck, I know a man of the name of Wood,[1] nephew of Lord Londonderry; he is a great friend of C. Fitzroy & has written to him about me— I heard a part of C. Fs letter, dated sometime ago, in which he says "I have a right good set of officers & most of my men have been there before." it seems that he has been there for the last few years; he was then second in command, with the same vessel that he has now chosen.— He is only 23 years old;[2] but seen a deal of service, & won the gold medal at Portsmouth.[3] The admiralty say his maps are most perfect.— He had choice of two vessels, & he chose the smallest.—

Henslow will given me letters to all travellers in town whom he thinks may assist me.

Peacock has sole appointment of Naturalist the first person offered was Leonard Jenyns, who was so near accepting it, that he packed up his clothes.— But having two livings he did not think it right to leave them.—& to the great regret of all his family.— Henslow himself was not very far from accepting it: for Mrs Henslow, most generously & without being asked gave her consent, but she looked so miserable, that Henslow at once settled the point.—

Do not forward Henslows letter. you may open it, if you like.— & now for giving you some trouble.— Look in bedroom over the Edinburgh Journal of Science,[4] or some such title, & see whether the following papers are in it: 3 by Humboldt on isothermal lines:[5] 2 by Coldstream & Foggo.— on Metereology: Metereological observations:[6] Tell Edward to get all Shre. bills:

I should be obliged if my Father would place to my account here 100£ if at present convenient ditto at London.— what bank?

I am afraid there will be a good deal of expence at first.— Henslow is much against taking many things; it is mistake all young travellers fall into.— I write as if it was settled: but Henslow tells me, *by no means*, to make up my mind till I have had long conversations with C. Beaufort, & Fitzroy:

Good bye. You will hear from me constantly. direct 17 Spring Gardens *Tell nobody* in Shropshire yet.— Be sure not: C. Darwin

I was so tired that evening I was in Shrewsbury, that I thanked none of you for your kindness, half so much as I felt.

Love to my Father.

The reason I dont want people told in Shrops: in case I should not go, it will make it more flat.

Postmark: C.H 5 SE 1831 X[7]
DAR 223

[1] Alexander Charles Wood.
[2] Robert FitzRoy was 26. Wood may have told CD that FitzRoy was only 23 at the time he was given command of the *Beagle*, in 1828.
[3] In his examination for promotion to lieutenant, FitzRoy 'won the first medal . . . he did what has never been done before . . . he got full numbers' (Rev. James Inman, head of Royal Naval College, Portsmouth, to Bartholomew James Sulivan, in H. N. Sulivan 1896, p. 12).
[4] *Edinburgh Journal of Science*, edited by David Brewster and Robert Jameson, 1824–32.
[5] Humboldt 1817.
[6] Coldstream 1826; Foggo 1826 and 1827.
[7] The letter bears a London postmark. CD evidently carried it with him and posted it on Monday, 5 September.

To Susan Darwin [5 September 1831]

17 Spring Gardens London
Monday

I have so little time to spare that I have none to waste in rewriting letters so that you must excuse my bringing up the other with me & altering it.— The last letter was written in the morning. in middle of day Wood received a letter from C. Fitzroy, which I must say was *most* straightforward & **gentlemanlike**, but so much against my going, that I immediately gave up the scheme.—& Henslow did the same: saying that he thought Peacock has acted *very wrong* in misrepresenting things so much.— I scarcely thought of going to Town, but here I am & now for more details & much more promising ones.— Cap Fitzroy is town & I have seen him; it is no use attempting to praise him as much as I feel inclined to do, for you would not believe me.— One thing I am certain of nothing could be more open & kind than he was to me.— It seems he had promised to take a friend with him,[1] who is in office & cannot go.—& he only received the letter 5 minutes before I came in: & this makes things much better for me, as want of room of was one of Fs greatest objections.— He offers me to go share in every thing in his cabin, if I like to come; & every sort of accomodation than I can have but they will not be numerous.— He says that nothing would be so miserable for him as having me with him if I was unformfortable, as in small vessel we must be thrown together, & thought it his duty to state every thing in the worst point of view: I think I shall go on Sunday to Plymouth to see the Vessel.— There is something most extremely attractive in his manners, & way of coming straight to the point.— If I live with him he say I must live poorly, no wine & the plainest dinners.— The scheme is not certainly so good as Peacock describes: C F. advises me not make my mind quite yet: but that seriously, he thinks it will have much more pleasure than pain for me.—

The Vessel does not sail till the 10th of October.— it contains 60 men 5 or 6 officers &c.—but is a small vessel.— it will probably be out nearly 3 years.— I shall pay to mess the same as Captain does himself 30£ per annum, & Fitzroy says if I spend including my outfitting 500 it will be beyond the extreme.— But now for still worse news, the round the world is not *certain*, but the chance, most excellent: till that point is dicided I will not be so.— And you may believe after the many changes I have made, that nothing but my reason shall dicide me.—

Fitzroy says the stormy sea is exaggerated that if I do not chuse to remain with them, I can at any time get home to England, so many vessels sail that way & that during bad weather (probably 2 months) if I like, I shall be left in some healthy, safe & nice country: that I shall alway have assistance.— that he has many books, all instrument, guns, at my service.— that the fewer & cheaper clothes I take the better.—

The manner of proceeding will just suit me. they anchor the ship & then remain for a fortnight at a place.—

I have made Cap Beaufort perfectly understand me.: he says if I start & do not go round the world: I shall have good reason to think myself deceived.— I am to call the day after tomorrow, & if possible to receive more certain instructions.— The want of room is decidedly the most serious objection: but Cap Fitz. (probably owing to Woods letter) seems determined to make me comfortable as he possibly can.— I like his manner of proceeding.— He asked me at once.— "shall you bear being told that I want the cabin to myself? when I want to be alone.— if we treat each other this way, I hope we shall suit, if not probably we should wish each other at the Devil" We stop a week at the Madeira islands: & shall see most of big cities in S. America. C. Beaufort is drawing up the track through the South Sea.—

I am writing in great hurry: I do not know whether you take interest enough to excuse treble postage.— I hope I am judging reasonably, & not through prejudice about Cap. Fitz: if so I am sure we shall suit.— I dine with him to day.— I could write great deal more if I thought you liked it, & I had at present time.— There is indeed a tide in the affairs of men,[2] & I have experienced it, & I had *entirely* given it up till 1 to day:

Love to my Father, dearest Susan | good bye, Chas. Darwin

Postmark: C.H 5 SE 1831 X
DAR 223

[1] It is not clear when this promise was made, but neither Francis Beaufort nor George Peacock knew of it when they set out to find a naturalist. It may be that, on hearing about CD from Alexander Charles Wood and Beaufort, Robert FitzRoy had misgivings about sharing his quarters with a total stranger and that he then invited a friend whom he knew well and whose company he would enjoy.
[2] Shakespeare, *Julius Caesar*, 4. 3. 217 (Arden edition).

To J. S. Henslow [5 September 1831]

[17 Spring Gardens] London
Monday

My dear Sir,

Gloria in excelsis is the most moderate beginning I can think of.— Things are more prosperous than I should have thought possible.— Cap. Fitzroy is every thing that is delightful, if I was to praise half so much as I feel inclined, you would say it was absurd, only once seeing him.— I think he really wishes to have me.—[1] He offers me to mess with him & he will take care I have such room as is possible.— But about the cases he says I must limit myself: but then he thinks like a sailor about size: Cap. Beaufort says I shall be upon the boards & then it will only cost me like other officers.— Ship sails 10th of October: spends a week at Madeira islands: & then Rio de Janeiro.— They all think most extremely probable, home by the Indian Archipelago: but till that is decided, I will not be so.—

What has induced Cap. Fitzroy to take a better view of the case is; that Mr Chester,[2] who was going as a friend, cannot go: so that I shall have his place in every respect.— Cap Fitzroy has good stock of books,[3] many of which were in my list, & rifles &c So that the outfit will be much less expensive than I supposed.— The vessel will be out 3 years I do not object, so that my Father does not.— On Wednesday I have another interview with Cap. Beaufort, & on Sunday most likely go with Cap. Fitzroy to Plymouth.— So I hope you will keep on thinking on the subject, & just keep memoranda of what may strike you.— I will call most probably on Mr Burchill[4] & introduce myself.— I am in Lodgings at 17, Spring Gardens.—

You cannot imagine anything more pleasant, kind & open than Cap. Fitzroys manners were to me.— I am sure it will be my fault, if we do not suit.—

What changes I have had: till one to day I was building castles in the air about hunting Foxes in Shropshire, now Lamas in S America.— There is indeed a tide in the affairs of men.— If you see Mr Wood, remember me most kindly to him.—

Good bye, my dear Henslow | Yours most sincere friend | Chas Darwin
Excuse this letter in such a hurry.

Postmark: C.H 5 SE 1831 X
Royal Botanic Gardens, Kew

[1] The same evening Robert FitzRoy wrote his impressions of CD to Francis Beaufort:

> I have seen a good deal of Mr. Darwin, to-day having had nearly two hours' conversation in the morning and having since dined with him.
>
> I like what I see and hear of him, much, and I now request that you will apply for him to accompany me as a Naturalist. I can and will make him comfortable on board, more so perhaps than you or he would expect, and I will contrive to stow away his goods and chattels of all kinds and give him a place for a workshop.
>
> Upon consideration, I feel confident that he will have a much wider field for his exertions than I was inclined to anticipate on Friday last; and should we even be disappointed, by giving me the means of discharging him from the Books, he might at any time return to England or follow his own inclinations in South America or elsewhere.

(The original of this letter, once in the Hydrographer's Office of the Admiralty, has been lost. The letter is quoted in F. Darwin 1912, p. 547.)

[2] See letter to Susan Darwin, [5 September 1831], n. 1. H. F. Burstyn has suggested that the friend was Harry Chester, novelist and youngest son of Sir Robert Chester, who in 1831 was a clerk in the Privy Council Office (Burstyn 1975, p. 66). An inscription in volume one of a copy of Kirby and Spence 1828 in the possession of David Kohn tends to confirm this conjecture. It reads: 'Harry Chester | From his valued friend Robert FitzRoy'.

[3] See Appendix IV.

[4] William John Burchell. CD undoubtedly wanted to see him because Burchell had explored and collected in Brazil from 1825 to 1830. See *Journal of researches*, p. 101.

To Susan Darwin [6 September 1831]

17, Spring Gardens
Tuesday

My dear Susan

Again I am going to trouble you. I suspect, if I keep on at this rate, you will sincerely wish me at Terra de Fuego or any other Terra, but England.— First I will give my commissions.— Tell Nancy to make me soon 12 instead of 8 shirts: Tell Edward to send me up in my carpet bag, (he can slip the key in the bag tied to some string) my slippers, a pair of lightish walking shoes.—My Spanish books: my new microscope (about 6 inches long & 3 or 4 deep),[1] which must have cotton stuffed inside: my geological compass.—my Father knows that: A little book, if I have got it in bedroom, Taxidermy:[2] ask my Father if he thinks there would be any objection to my taking Arsenic for a little time, as my hands are not quite well—& I have always observed, that if I once get them well & change my manner of living about same time they will generally remain well.— What is the dose?— Tell Edward my gun is dirty: What is Erasmus direction, tell me if you think there is time to write & to receive an answer before I start: as I should like particularly to know what he thinks about it. I suppose you do not know Sir J. Macintosh direction?—

I write all this as if it was settled but it is not more than it was.—excepting that from Cap. FitzRoy wishing me so much to go, & from his kindness I feel a predestination I shall start.— I spent a very pleasant evening with him yesterday: he must be more than 23 old. he is of a slight figure, & a dark but handsome edition of Mᵣ Kynaston.—& according to my notions preeminently good manners: He is all for Economy excepting on one point, viz fire arms he recommends me strongly to get a case of pistols like his which cost 60£!!, & never to go on shore anywhere without loaded ones.— & he is doubting about a rifle.— he says I cannot appreciate the luxury of fresh meat here.— Of course I shall buy nothing till every thing is settled: but I work all day long at my lists, putting in & striking out articles.— This is the first really cheerful day I have spent since I received the letter, & it all is owing to the sort of involuntary confidence I place in my beau ideal of a Captain.—

We stop at Teneriffe. His object is to stop at as many places as possible. he takes out 20 Chronometers & it will be a "sin" not to settle the longitudes:³ he tells me to get it down on writing at yᵉ Admiralty that I have the free choice to leave, as soon & wherever I like:⁴ I daresay you expect I shall turn back at the Madeira: if I have a morsel of stomach left, I wont give up.— Excuse my so often troubling & writing, the one is of great utility, the other a great amusement to me.— Most likely I shall write tomorrow

Love to my Father.— Dearest Susan | C. Darwin

Answer by return of post

As my instruments want altering send my things by the Oxonian, yᵉ same night

Postmark: C.H 6 SE 1831 X
DAR 223

¹ See letter from [J. M. Herbert], [early May 1831], n. 3. The dimensions are close to those of the box of the Cary microscope at Down House (6.2 inches by 2.8 inches), which suggests that CD may have taken that instrument with him on the voyage. But, if so, it was not his only one. Another, also at Down House, bearing the name 'Bancks & Son, 119 New Bond Street', is, in the opinion of Professor Phillip Sloan, based on measurements of the focal lengths, the main one utilised by CD on board the *Beagle* (Phillip Sloan, personal communication). There is further evidence in the letter to W. D. Fox, 23 May 1833 that CD had the Bancks microscope on board. See also letter to J. S. Henslow, 28 [September 1831], n. 1.

² Swainson 1822 (viii + 72 pp.) fits this description, but no copy has been found in CD's library, nor has any mention of it, or of any other work on taxidermy, been found in his *Beagle* notes.

³ A memorandum from Captain Beaufort dated 11 November 1831 contains detailed instructions for chronometric observations to be made during the *Beagle* voyage in order to establish longitudes, such as that of Rio de Janeiro, about which authorities were in conflict (*Narrative* 2: 24−40). Eventually twenty-two chronometers were taken. See *Narrative* Appendix, p. 325 for a list with Robert FitzRoy's rating of their performance.

[4] No such statement has been found, though CD's *Beagle* letters make it clear that he felt free to leave whenever he chose.

To W. D. Fox 6 [September 1831]

17 Spring Gardens London
Tuesday 6th.—

My dear Fox

When you read this you will understand why I have not answered your letter earlier.— I returned from a geological trip with Prof: Sedgwick in N Wales on Monday 29th of August.— I found your letter there, & with a joint one from Henslow & Peacock of Trinity, offering me the place of Naturalist in a vessel fitted for going round the world.— This I at first, (owing to my Father not liking it I refused) but my Uncle, Mr Wedgwood, took every thing in such a different point of view, that we returned to Shrewsbury on 1st of September, & convinced my Father.— On 2d started for Cam: then again from a discouraging letter from my captain I again gave it up. But yesterday every thing was smoother.—& I think it most probable I shall go: but it is not certain, so do not mention it to any body.— I have had a most tremendous hard week of it.— Cap Fitzroy seems every thing I could wish: the most serious objections are the time (3 years), & the smallness of the vessel. Fitzroy is determined to get over the latter for me.— Our route is Madeira, Canary, Rio de Janeiro, 18 month in S America chiefly S. extremity, we shall see every principle city in it.—then through South sea islands, Australia, India, Home.— Of course nothing is quite certain. we sail on 10th October.— of course I will write to you again.—

Your letter gave me great pleasure.— You cannot imagine how much your former letter annoyed & hurt me.—[1] But thank heaven, I firmly believe that it was *my own entire* fault in so interpreting your letter.— I lost a friend the other day, & I doubt whether the moral death, (as I then wickedly supposed) of our friendship did not grieve me as much as the real & sudden death of poor Ramsay.— We have known each other too long to need I trust any more explanations.— But I will mention just one thing; that on my death bed, I think I could say I never uttered one insincere (which at the time I did not fully feel) expression about my regard for you— One thing more.— The sending *immediately* the insects on my honor was an unfortunate coincidence, I forgot how you naturally would take them.— When you look at them now, I hope no unkindly feelings will rise in your mind.— & that you will believe that you have always had in me a sincere & I will add, an obliged friend. The very many pleasant minutes that we spent together in Cambridge rose like departed spirits in judgement against me: May we have many more such, will be one of my last wishes in leaving England.—

God bless you dear old Fox May you always be happy.— | Yours truly | Chas
Darwin

I have left your letter behind so do not know whether I direct right

Postmark: C.H 6 SE 1831 X
Christ's College Library, Cambridge (Fox 43)

[1] See letter to W. D. Fox, 1 August [1831], n. 2.

To Susan Darwin [9 September 1831]

[17 Spring Gardens]
Friday Morning

My dear Susan

I have just received the parcel: I suppose it was not delivered yesterday
owing to the Coronation.—[1] I am very much obliged to my Father & every body
else.— Every thing is done quite right: I suppose by this time you have received
my letter written next day—& I hope will send off the things.—

My affairs remain in statu quo.— Cap Beaufort says I am on the books for
victuals, & he thinks I shall have no difficulty about my collections when I
come home.— But he is too deep a fish for me to make him out.— The only thing
that now prevents me finally making up my mind is the want of *certainty* about S S
Islands, although morally I have no doubt we should go there whether or no it is
put in the instructions: Cap. Fitz says I do good by plaguing Cap Beaufort: it
stirs him up with a long pole.— Cap Fitz. says he is sure he has interest
enough—(particularly if this administration is not everlasting: I shall soon turn
Tory.!) anyhow even when out to get the ship ordered home by whatever track
he likes.— From what Wood says I presume Dukes Grafton & Richmond interest
themselves about him.—[2] By the way Wood has been of the greatest use to me.—&
I am sure his personal introduction of me, inclined Cap Fitzroy to have me.—

To explain things from the very beginning; Cap Fitz first wished to have
naturalist & then he seems to have taken a sudden horror of the chances of having
somebody he should not like on board the Vessel: he confesses, his letter to
Cambridge, was to throw cold water on the scheme.— I dont think we shall
quarrell about politics although Wood (as might be expected from a Londonderry) solemnly warned Fitzroy that I was a whig.— Cap Fitz was before
Uncle Jos—he said "now your friends will tell you a sea Captain is the
greatest brute on the face of the creation; I do not know how to help you in this
case, except by hoping you will give me a trial."— How one does change.— I
actually now wish the voyage was longer before we touched Land. I feel my blood
run cold at the quantity I have do.— Every body seems ready to assist me. The
Zoological want to make me a corresponding member; all this I can construe
without crossing the Equator:— But one friend is quite invaluable, viz a M^r

Yarrell, a stationer & excellent naturalist: he goes to the shops with me & bullies about prices (not that I yet buy). hang me if I give 60£ for pistols.—

Yesterday all the shops were shut—so that I could do nothing.—& I was child enough to give 1 ' 1 for an excellent seat to see the procession— And it certainly was very well worth seeing.— I was surprised that any quantity of gold could make a long row of people quite glitter.— it was like only what one sees in picture books of Eastern processions.— The King looked very well, & seemed popular: but there was very little enthusiasm so little that I can hardly think there will be a coronation this time 50 years.—

The life Guards pleased me as much as anything: they are quite magnificent & it is beautiful to see them clear a crowd; you think that they must kill a score at least, & apparently they really hurt nobody, but most deucedly frighten them.— Wherever a crowd was so dense that the people were forced off the Causeway: one of these six feet gentleman, on a black horse, rode straight at the place, making his horse rear very high & fall on the thickets spot: you would suppose men were made of spong to see them shrink away.— In the evening there was an illumination, & much grander than the one on the Reform bill.—³ All the principal streets were crowded just like a Race ground.— Carriages generally being 6 abreast, & I will venture to say not going 1 mile an hour.— Duke of Northumberland learnt a lesson last time: for his house was very grand: much more so than the other great nobility: & in much better taste: every window in his house was full of perfectly straight lines of brilliant lights: & from their extreme regularity & number had a beautiful effect.— The paucity of invention was very striking, crowns anchors & W R.ˢ were repeated in endless succession.— The prettiest were gass pipes with small holes, they were almost painfully brilliant.— I have written so much about the Coronation, that I think you will have no occasion to read Morning Herald.— For about the first time in my life I find London very pleasant: hurry, bustle & noise are all in unison with my feelings.— And I have plenty to do in spare moments I work at Astronomy: as I suppose it would astound a sailor if one did not know how to find Lat & Long.—

I am now going to Cap Fitzroy, & will keep letter open till evening for any thing that may occur.— I will give you one proof of Fitzroy being a good officer, all officers are the same as before ²/₃ of his crew, & the eight marines, who went before all offered to come again: so the service cannot be so very bad: The admiralty have just issued orders for a large stock of Canister meat & Lemon juice & &c.—

I have just returned from spending a long day with Cap Fitz, driving about in his gig & shopping.— This letter is too late for to days post.— You may consider it settled that I go: yet there is room for change, if any untoward accident should happen: this I can see no reason to expect: I feel convinced nothing else will alter my wish of going.— I have begun to order things. I have procured case of good strong pistols & excellent rifle for 50£: there is a saving: good telescope, with compass 5£, & these are nearly the only expensive instruments I shall want.— Cap Fitz has every thing: I never saw so, (what I should call, he says not) extravagant a man as regard himself, but as economical towards me.— How he did order things. His fire arms will cost 400£ at least:— I found Carpet bag when I arrived; all right & much obliged.— I do not think I shall take any Arsenic: shall send partridges to Mr Yarrell, much obliged: Ask Edward to *bargain with* Clemson[4] to make for my gun: 2 *spare* hammers or cocks: 2 main spring: 2 sere springs: 4 nipples or plugs: I mean one for each barrell, except nipples of which there must be 2 for each: all of excellent quality & set about them immediately. tell Edward make enquiries about prices

I go on Sunday, per packet to Plymouth, shall stay 1 or 2 days then return, & hope to find letter from you.— few days in London: then Cam, Shrews, London, Plymouth, Madeira, is my route.— It is great bore my writing so much about Coronation. I could fill another sheet.—

I just been with Cap King, Fitzroy senior officer last expedition: he thinks that the expedition will suit me.— Unasked he said Fitzroys temper was perfect: He send his own son[5] with him as midshipman

The key of my microscope was forgotten it is of no consequence.

Love to all | Chas. Darwin

Postmark: 10 SE 1831 CX
DAR 223

[1] The coronation of King William IV on 8 September 1831.
[2] Robert FitzRoy was a nephew of George Henry FitzRoy, 4th Duke of Grafton. Charles Gordon Lennox, 5th Duke of Richmond, was a more distant relation.
[3] The Reform Bill did not become law until June 1832, but in March 1831 it passed its second reading in the House of Commons. This was the occasion of general jubilation.
[4] Shrewsbury gunsmith.
[5] Philip Gidley King.

To J. S. Henslow 9 [September 1831]

17 Spring Gardens
Friday 9th. evening

My dear Sir

You must have thought it very odd my not having written sooner.— I put it off yesterday & the day before owing to the Coronation & not seeing Cap. Fitz Roy & therefore not having anything particular to communicate.— To day I did not come home till too late for the post, having spent it with Cap Fitz going about the town & ordering things.— By this you will perceive it is all settled; that is to say I cannot possibly conceive any cause happening of sufficient weight to alter my determination.— I have ordered pistols & a rifle, both of which by Fitzroys account I shall have plenty of use for.— These really are nearly the only expensive things I shall want: Fitzroy has an immense stock of instruments & books.— viz takes out 5 Simpisometers,[1] 3 M Barometers.—[2] in books all travels, & many natural history books.— He does not appear to care for any expence as far as regards himself, but is very economical with respect to advice to me.— And now for my plans.— On Sunday I go packet to Plymouth stay there a few days: & then London: then Cam. where I shall finally settle things, pay bills &c & home to Shrewsbury: Then London again: Plymouth: Terra del Fuego.— The SS Islands are all but certain. I am on the books for Victuals.— but about my collections, Cap Beaufort said his first impression was, that they ought to be given to British Museum: but I think I convinced of the impropriety of this[3] & he finished by saying he thought I should have no difficulty so that I presented them to some public body, as Zoological & Geological &c.—

But I do not think the Admiralty would approve of my sending them to a Country collection, let it be ever so good,—& really I doubt myself, whether it is not more for the advancement of Nat. Hist. that new things should be presented to the largest & most central collection.— But we will talk of all this & many other things when we meet,—which I should think would be early the week after next.— Mr Yarrell has been quite invaluable to me; so very good natured & such very good advice: But ⟨the⟩y all say Cap. King will be of the greatest use.— The No^r of bottles is greatest puzzle.— Will you see about a iron net for shells.— Remember me most kindly to L Jenyns, & tell him I am to have a parcel to bring for him from Mr Yarrell: Would you enquire from him, in what Edinburgh Journal, there are some papers by Coldstream & Foggo?[4] send an answer to this question.— I must have a rain-gauge.— I hope when I return from Plymouth I shall find a letter from you: I received one to day from Prof: Sedgwick, but have not yet had time to read it.— You can have no idea how busy I am all day long.— & owing to my confidence in Cap. Fitzroy I am as happy as a king; if you were here to talk to, I should be a good deal happier.— I hope you will excuse all the trouble I give you, if you were like Liston in Paul Pry[5] to say you never would do a good natured thing again, I do not know what I should do.—

Good night my dear Henslow | Yours most sincerely | Chas Darwin

PS. All FitzRoy said about the letter of Peacock evidently from a very enthusiastic man, an elegant way of calling it inaccurate.—

Cary says your Clinometer is ready & he is working at the Camera obscura, it soon will be ready.—

I have just been with Cap King Fitzroy senior officer during last expedition & he has given me much good advice: but I am afraid he must have swept the Coast almost clear.—

I will write again before I come to Cambridge
Keep Syme on colours[6] in your mind.—

Postmark: 10 SE 1831 CX
Royal Botanic Gardens, Kew

[1] Sympiesometer, a form of barometer with gas instead of a vacuum in the tube above the liquid.
[2] Mountain barometers.
[3] CD may have come to this view because he had heard that the British Museum left undescribed so many of the specimens deposited there. Captain Philip Parker King's botanical specimens from the first voyage were a case in point.
[4] See letter to Susan Darwin, [4 September 1831], n. 6. Susan had apparently failed to locate the articles.
[5] Poole 1825, a popular farce.
[6] Syme 1814. The work contains plates of different tints for identifying the colours of specimens when they are taken by collectors. A copy of the second edition (1821) is in Darwin Library—CUL.

To Charles Whitley [9 September 1831]

17 Spring Gardens | London
Friday Evening

My dear Whitley

I daresay you will be surprised when you see the date of this letter, & perhaps you will be more so when you read it contents.—

When I arrived home, after having left Barmouth, I found letters from Peacock & Henslow offering me (from the Admiralty) the priviledge of going in a Kings ship on a surveying voyage round the world.— This I at first refused, owing to my Father not approving of the plan, but since then we have convinced him of the propriety of my going.— Accordingly after many doubts & difficulties I started for Cambridge, & then came on here, where I arrived on Monday.— And I believe now it is all finally settled.— Cap Fitz Roy, my captain, appears an uncommonly agreeable open sort of fellow—whom I liked at first sight: he is uncommonly civil: I am to live with him: the Vessel is very small, but it was his own choice.— It is such capital fun ordering things, to day I ordered a Rifle & 2 pair of pistols; for we shall have plenty of fighting with those d—— Cannibals: It would be something to shoot the King of the Cannibals Islands.—

Our route is Madeira, Canary Islands Rio de Janeiro. 18 months all about S America, chiefly Southern extremity.— South Sea Islands, (some new course) Australia India home.— I shall see a great number of places, as they take out 20 Chronometers to ascertain Longitudes—

Cap Fitzroy is very scientific & seems inclined to assist me to the utmost extent in my line.— I go on Sunday to Plymouth to see the Vessel. She sails 10th of next month.— So that I have not an idle moment.— I shot one partridge on the 1st devilish dear 3"13"6.[1] by 8 oclock I was off.— Remember me most kindly to the Lowes, I should like to hear their observations on my grand tour. tell Lowe Sen[2] that my things arrived quite safe, & I am very much obliged for all the trouble he took: There will be a paper published about the Fungus,[3] all my conjectures were right.— If any more can be got, & put into gin, & sent to Shrewsbury: it will be capital

I hope you will write to me. I am much obliged for your last note.— If I was see Lowe, I should think he would have a few questions to ask. I hope he will remain pretty easy in his mind.— Again remember me most kindly to the two Lowes I wish them all sorts of good luck, & Believe me dear old Whitley, Yours very truely | Chas Darwin

I saw poor old Herbert in Cam. he is pretty well tired of Cam poor old Fellow.—

Remember me most kindly to Beadon[4]

I added this postscript to the wrong letter.[5] Will you call at the Postoffice[6] & desire them to forward to Caernarvon a letter directed Prof: Sedgwick

I am quite ashamed to send such letters I am quite tired of writing.—

Postmark: 10 SE 1831 CX
American Philosophical Society 3

[1] The stamp duty on a game certificate. See Munsche 1981, p. 181.
[2] Henry Porter Lowe was at Barmouth in August 1831 with his younger brother, Robert Lowe. For Robert Lowe's memories of CD at Barmouth see Martin 1893, 1: 19–20; quoted in Barrett 1974, p. 149.
[3] CD apparently sent the fungi to Henslow (see letter to J. S. Henslow, 28 [September 1831]). A printed announcement of gifts received by the Botanical Museum and Library at Cambridge, dated 25 March 1832, lists 'Phallus impudicus, var ? . . . C. Darwin Esq.' No paper by Henslow has, however, been located.
[4] Probably Richard a'Court Beadon.
[5] Refers to deleted passage: 'The key of microscope was forgotten: never mind it; we soon opened it.— Good bye | Love to all.—| Chas Darwin' (see letter to Susan Darwin, [9 September 1831]).
[6] The letter is addressed to Whitley at Barmouth Post Office.

From John Coldstream 13 September 1831

My dear Darwin

I was no less surprized than delighted to see your handwriting once more on a letter addressed to me. It is indeed "a long time" since I received any accounts of you. You will be a most useful man in the situation which you have the prospect of filling in Captn Fitzroy's expedition, if your Zeal in the pursuit of Science, and your bodily strength remain the same as they were when I had the pleasure of seeing you here. I have no doubt that both are increased; and, accordingly, I anticipate much good from your labours in the South.— The paper by Dr Brewster to which you refer is that, I presume, on the series of hourly Meteorological observations made at Leith Fort in 1824 & −25. It is in the Xth Vol. of the Transactions of the Royal Soc^y. of Edin^r., published in 1826 or −27. The only other paper to which I can suppose you to refer is one "on the mean temperature of the globe". It is in the IXth Vol of the same Transactions.[1] As I have paid very little attention to Natural History of late, I feel myself but ill prepared to give you any information which might be of service to you:—but with regard to the collecting of marine animals, I may state, that I think a common oyster-trawl, of *the ordinary* size, would prove very serviceable. This you may readily procure in any of the fishing villages at the mouth of the Thames, (if not in London)—but, as you wish it, I shall sketch a figure of the dredge or trawl usually employed in the Firth of Forth.

Fig 1   Fig. 2.

The frame is made of iron, and measures from a to b, fig. 1. about 3 feet;—the bar which scrapes the ground has a blunt edge in front; the lower surface of the bag is formed of iron rings, the upper of strong netting. Fig. 1. is a front view with the bag hanging down. Fig 2 is a view of the implement in operation.

You might supply yourself also with a few lobster traps of various constructions. Many of the rarest of our Mollusca and Zoophytes are found adhering to the deep sea fishing lines; (such as are set for cod and haddock, and allowed to remain at sea for many hours together undisturbed). When at anchor, you might "shoot" some such lines, with small pieces of worm:eaten wood, or small baskets &c, as well as hooks, attached to them: by leaving these in the water over night, sunk to a considerable depth, you might obtain a rich supply in the morning.

I hope you will see Dr Grant before you sail, as he can give you many valuable hints. In determining upon the circumstances to which you should attend in making Meteorological observations, and the best mode of registering these, I think you can hardly set before yourself a better model than the excellent Journal contained in Cap^n. Beechey's Voyage to the Pacific.[2] It is truly valuable.

Allow me to suggest, also, that you should endeavour to obtain an interview with Professor Daniell, of King's College[3] before you leave. He is, I think, better qualified tha⟨n⟩ any other British philosopher to direct your attention to the points which you should chiefly attend to in observing Atmospherical Phenomena. If I can be of any further service to you in this matter I shall be very happy to hear from you. Have you heard of our mutual friend Glasspoole of late?

With the best wishes for your welfare & success, I remain, with much esteem, Yours most sincerely | John Coldstream

Leith 13th Sept 1831.

DAR 204.5

[1] Brewster 1826, 1823.
[2] Beechey 1831. A copy of another edition (Philadelphia 1832) is in Darwin Library – CUL. It is not annotated. The copy CD used during the *Beagle* voyage was probably FitzRoy's (see letter from Robert FitzRoy, 23 September 1831).
[3] John Frederic Daniell.

From Charles Whitley 13 September 1831

Barmouth
Tuesday Septr 13 1831

My dear Darwin,

Your letter, which arrived on my return from Snowdonia, certainly did surprise & on the whole please me. I congratulate you on the prospect of an employment after your own heart, & the opportunity it affords you of studying all the natural sciences at once, after your own taste. But as you are about to escape from my advice & reproaches for a considerable period I must just ask you whether you do not experience a few twinges of regret concerning your neglect of Mathematics especially of trigonometry & navigation—but this last you must learn, in the blindfold way of practice, by rules & tables. If however this Captn. Fitzroy is half such a good fellow as you take him to be he will not fail to teach you the rationale; & you will be foolish in my opinion if you don't give him a hint, as soon as the *sickness* has left you—& very foolish indeed if you don't study these things when you have an opportunity. As to studying any thing else unconnected with the business of the ship I fancy from what I have heard that Bishop Heber[1] is right when he asserts it to be no easy matter—but this is only hearsay. Now I dare say that you will think this advice much about as valuable as that which mammas usually give to good little boys when about to go to school. Never mind— It may be of use.

I will not disturb your "ordering" occupations or your cannibal shooting, fungus describing anticipations, by an account of a very pleasant ascent of

Snowdon undertaken by the Lowes & myself on a very clear Saturday or the view that we had or the pleasant walk home. We heard of Miller geologizing at Harlech, & I have despatched your letter to Sedgwick, at Caernarfon.[2] We are going on in the usual way & I hope to be at Shrewsbury by Oct.ʳ 1ˢᵗ Now I expect that you will write to me, if it be only a letter of three lines, to say if you shall be at Salop again & *when*? in as much as I desire greatly to see you again. You may be drowned, shot or feversmitten, or I may die from pure vexation & disappointment before you return, & I should certainly like to shake hands once more before we separate. We have jogged on amicably hitherto & there are few me⟨n⟩ I should miss more than yourself when the black day came to either of us. Thank you by the way for your early information. It was thoughtful enough for you—for you are a little given (inter nos) to be "uneasy in your mind" on occasions of "packing up." I shall also require you to tell me if there is any single thing on this earth which I can do for you either before you go or during your absence (for I shall expect an occasional letter). Now you must if you please write to me again on these subjects immediately. Once more I congratulate you on your luck (& I wish it were mine, for I have latterly had much vexation & am sick enough of England) as it will put it in your power to distinguish yourself in your favorite pursuits—& I flatter myself that you will not be slow to do so to your own satisfaction & the pleasure if not the envy of your friends of whom there are few more attached than

Your's very faithfully | Charles Whitley

Remember me most kindly to your brother

DAR 204.5

[1] Reginald Heber, who, as Bishop of Calcutta, travelled extensively in India (see Heber 1828).
[2] For Sedgwick's reply, see letter dated 18 September 1831.

To Susan Darwin　[14 September 1831]

Devonport.
Wednesday Evening

My dear Susan

I arrived here yesterday evening: after a very prosperous sail of three days from London.— I suppose breathing the same air as a sea Captain is a sort of a preventive: for I scarcely ever spent three pleasanter days.— of course there were a few moments of giddiness, as for sickness I utterly scorn the very name of it.— There were 5 or 6 very agreeable people on board, & we formed a table & stuck together, & most jolly dinners they were.— Cap. Fitz. took a little Midshipman (who by the way knows Sir F. Darwin, his name is Musters)[1] & you cannot imagine anything more kind & good humoured than the Captains manners were to him.— Perhaps you thought I admired my beau ideal of a Captain in my former letters: all that is quite a joke to what I now feel.— Every body praises him, (whether or no they know my connection with him) & indeed, judging from the little I have seen of him, he well deserves it.— Not that I suppose it is likely that such violent admiration—as I feel for him—can possibly last.— No man is a hero to his valet, as the old saying goes.—& I certainly shall be in much the same predicament as one.—

The vessel is a very small one; three masted; & carrying 10 guns: but every body says it is the best sort for our work, & of its class it is an excellent vessel: new, but well tried, & $\frac{1}{2}$ again the usual strength.— The want of room is very bad, but we must make the best of it.—[2] I like the officers, (as Cap. F. says they would not do for S! James, but they are evidently very intelligent, active determined set of young fellows.— I keep on ballancing accounts; there are several contra's, which I did not expect, but on the other hand the pro's far outweigh them.—

The time of sailing keeps on receding in a greater ratio, than the present time draws on: I do not believe we shall sail till the 20th of October.— I am exceedingly glad of this, as the number of things I have got to do is quite frightful.— I do not think I can stay in Shrewsbury more than 4 days.— I leave Plymouth on Friday and shall be in Cam: at the end of next week.—

I found the money at the Bank, & am much obliged to my Father for it.— My spirits about the voyage are like the tide, which runs one way & that is in favor of it, but it does so by a number of little waves, which may represent all the doubts & hopes that are continually changing in my mind. After such a wonderful high wrought simile I will write no more. So good bye, my dear Susan | Yours C. Darwin

Love to my Father.—

Postmark: Devonport SE 14 1831
DAR 223

[1] Charles Musters, listed as 'Volunteer 1st Class' by FitzRoy (*Narrative* 2: 20).
[2] For a detailed description of the *Beagle* see Darling 1978.

To Susan Darwin   17 [September 1831]

17 Spring Gardens
Saturday 17th

My dear Susan

I daresay you have received my letter from Plymouth.— I have nothing particular to write about, excepting to tell you on what day it is most likely I shall arrive in Shrewsbury.— I go on Monday night to Cam.—& most probably shall leave it on Wednesday or Friday, & shall arrive the following morning at 5 ocl in Shrews: have my bed ready accordingly.— What wonderful quick travelling it is.— I came from Plymouth 250 miles in 24 hours.—& arrived this morning. When I wrote last, I was in great alarm about my cabin: the cabins were not then marked out: but when I left they were & mine is a capital one, certainly next best to the Captains, & remarkably light.— My companion most luckily I think will turn out to be the officer whom I shall like best.—[1] Cap Fitz. says he will take care that one corner is so fitted up that I shall be comfortable in it & shall consider my home:— but that also I shall always have the run of his.— My cabin is the drawing one, & in the middle is a large table, over which we 2 sleep in hammocks, but for the first 2 months there will be no drawing to be done, so that it will be quite a luxurious room & good deal larger than the Captains cabin:

I dont care whom you now tell; for all is fixed & certain.—& I feel *well up it.*—not but what this has often been a difficult task & my reason has been the only power that was capable of it: for it is most painful whenever I think of leaving for so long a time so many people whom I love.—

But no more of this from myself or from any of you.— I have been in capital spirits ever since all was fixed, & if I go to the bottom, I shall go on this one point like a rational creature.— The use of the fire arms is most important. they almost lived for months, last voyage on the produce of them, so much so that Government allows powder & shot.— Not to mention it is never safe to go on shore without loaded arms, this is always sufficient to keep the natives pretty quiet

The object of the voyage is to make maps of Eastern side of terra del Fuego & Patagonia: likewise to settle Longitude of many places more accurately than they are at present: on the other side they named more than 50 new islands, so perfectly unknown is that part of the coast.—

I had intended writing to Maer & shall certainly do so— Have my shirts marked DARWIN.—& no number.—

Good bye love to my Father | Yours affectionately | Chas Darwin
I shall leave Shrewsbury on Friday 30th. or before.—

Postmark: C.H 19 SE 1831 X
DAR 223

[1] CD's cabin-mate was John Lort Stokes, Mate and Assistant Surveyor.

To J. S. Henslow   17 [September 1831]

17 Spring Gardens
Saturday 17th.—

My dear Sir
I arrived this morning from Plymouth & found your letter with six others on my table.— I mention this, as it will account for my writing to you a very short letter.— I called on your brother, but he was not at home, I heard there that you

left London yesterday: How very unfortunate it was my being detained in Plymouth: I should have much enjoyed taking a walk in London town with you.— I am much obliged for your asking me to take up my quarters with you: I will most gratefully accept it in every point but one, viz sleeping at your house.— I shall arrive in the middle of the night by the Mail,[1] & after 2 or 3 days shall start *very early* in the morning to Birmingham: So I cannot think of turning your house upside down for merely one night.— Will you be kind enough to order a bed for me at the Hoops[2] for Monday night: as it is almost certain I shall come to Cambridge then.— You may tremble at my arrival,—for I shall not give you a moments peace. I have so many things to ask about & talk about.— Every thing goes on very well.— The SS Islands daily become more probable.— My cabin is more comfortable than I expected: & my only difficulty is about the disposal of my collection when I come back.— I have seen this very morning Cap. Beaufort & had some talk on the subject.— There is one other disagreeable thing, but of this in future.— The ballance however is quite on the prosperous side.—

Excuse this hasty letter & believe me dear Sir, with my best thanks, Yours ever most sincerely | Chas. Darwin.—

Postmark: C.H 17 SE 1831 X
Royal Botanic Gardens, Kew

[1] The *Louth and Boston Mail* arrived at half-past two every morning from London; the *Rising Sun* coach to Birmingham left at 6.00 a.m. (*Cambridge University calendar*, 1831).
[2] The Hoop Inn, Cambridge.

From Adam Sedgwick 18 September 1831

Carnarvon.
Sep.[r] 18. 1831.

Dear Darwin

Before this you will have received a letter I addressed to you, some time since, at Shrewsbury. It contained a statement of what I was doing & had done— I have now resolved to confine myself to this county, & if I can finish it to my satisfaction I shall be well content to turn my back on these mountains for a season— I cannot but be glad at your appointment & I truly hope it will be a source of happiness & honor to you.— I really dont know what to say about books— N.[o] 1 Daubeny.[1] N.[o] 2. a book on Geology— D'aubuissions[2] work is one of the best tho' full of Wernerian nonsense.—[3] I dont think Bakewell a bad book for a beginner—[4] For *fossil shells* what is to be done?— Go to the Geological Society and introduce yourself to M.[r] Lonsdale[5] as my friend & fellow traveller & he will counsel you— Humboldts personal narrative you will of course get— He will at least show the right spirit with w.[h] a man should set to work— There is a small paper printed by the Geol. Soc.[y] containing directions for travellers &c—[6] Lonsdale will give you a copy: but it is a mere horn book[7] hardly worth your looking at— Study the *Geological Soc.[ys] collection* as well as you can—& *pay* them *back* in specimens— I am to *propose you* when the meetings begin.[8] I am in great hurry as my gig is at the door:—on my way to Clynnog from which place D.V. I hope in ten days to work my way round the great S. Western Promontory of Cardigan bay— I shall then return pack up—& start for Capel Curig—where I must halt again for yᵉ 3ᵈ time to make a traverse or two in yᵉ chain— But this depends on the weather— Should it fairly break up I must lodge

The Carnarvon Chain is very *troublesome* from the number of *anticlinal lines* wʰ I have to follow out from hill to hill & valley to valley upways, downways, & cross ways. I will try to give you a notion of one section

a. slate quarries on w. side of Mynydd Mawr
b. Mynydd Mawr, a *great anticlinal line*
b′.pass of Drws y Coed..
c. Drws-y-Coed ——Dº——
D. Moel Haebog E. Foel Ddu. an anticlinal li⟨ne⟩
f. Pass of Pont Aberglaslyn—
g. ⟨Cni⟩cht., k. h. hills toward Festiniog.

The prevailing dip in yᵉ Snowdonian chain is S.E with numberless great contortions; & the base of yᵉ series is near yᵉ line of the Slate quarries (a) on yᵉ west side of the Chain.— The *strike* of the beds in the chain is about N.N.E. with singular uniformity, till you reach the Eastern outskirts & then all is confusion.— The Merioneth chains are elevated in the same direction (as far as I have seen 'em) but yᵉ prevailing tilt seems to be to the N.W. I expect to find (*next year!*) a great central anticlinal axis in Merioneth.—[9] The place marked X in the secⁿ is the place where the two systems of elevation interfere with each other. But my picture is so detestable and out of all proportion that I fear you cannot comprehend it— I consider poor Ramsays death a grievous loss to the whole University— God bless you & preserve your health of mind & body. Most truly yours | A Sedgwick

I shall be happy to hear from you write to Carnarvon

DAR 204.5

[1] Sedgwick probably refers to Daubeny 1826 as particularly relevant to South American geology. A much spotted copy in Darwin Library—Down may be the one CD had on board the *Beagle* (see Appendix IV). A pencil drawing of volcanic islands in the section on the Azores (p. 26) and a note 'Covington Copy' next to a footnote on trachyte (p. 180) are more characteristic of notes CD made later while at work on the geology of the voyage.
[2] Jean François d'Aubuisson de Voisins.

[3] Abraham Gottlob Werner formulated the so-called Neptunian hypothesis that the rocks of the earth's crust were formed by depositions from a global ocean. It is not clear whether Sedgwick refers to Aubuisson's *An account of the basalts of Saxony* (1814) or to his two-volume more general geological work, the *Traité de géognosie* (1819), in which Aubuisson modified his Neptunism, though not his admiration of Werner as a founder of the science of geology. Both volumes are in Darwin Library — CUL. The *Traité*'s title-page bears the inscription 'C. Darwin HMS Beagle' and has many more annotations than the *Account*. CD refers to it frequently in his geological notes during the voyage. There is no evidence that CD also had the *Account* on board.

[4] Bakewell 1813.

[5] For a good account of William Lonsdale's contributions to geology and his early evolutionist ideas, see Tasch 1950.

[6] Geological Society, London 1808.

[7] A child's book of a single sheet in a frame, covered in horn for protection.

[8] CD was not elected a member of the Geological Society until 1836. See letter to J. S. Henslow, 9 July 1836, in which CD remembers this statement as an offer to nominate him.

[9] Sedgwick reported on the geology of North Wales to the Cambridge Philosophical Society on 11 March 1833. An abstract was published in the *Philosophical Magazine* 2 (1833): 381. His major finding was that 'the strata of that district are bent into *saddles* and *troughs*, of which the *anticlinal* and *synclinal* lines occur alternately, and are all nearly parallel to the "great Merionethshire anticlinal line." '

From Frederick Watkins [18 September 1831]

Barnbro. Rectory.
Sunday

My dear Darwin,

Never did I think so highly of our present Government, as when I heard they had selected Charles Darwin for Gt. naturalist & that he was to be trans-ported (with pleasure of course) for 3 years— Woe unto ye Beetles of South America, woe unto all tropical butterflies— So you another of ye old set, & may I not say a nearer & dearer than most of them, are going to launch on to ye Ocean of Life & leave behind ye little fortunatæ insulæ of fun & frolic & carelessness. By Jove, old Boy, truism tho' it may be, we shall never see again such days as we have seen, be half so happy as we have been, or eat half so much as we have ate; ye last at least is impossible.[1] In my humble opinion you are right in going & some of these ⟨   ⟩ ⟨wi⟩ll[2] see you ranked with Brognia⟨rt,⟩[3] ⟨d⟩e Candolle,[4] Henslow, Linnæus & Co.— Whilst I, luckless wretch, am rusticating in a country Parsonage & shewing people a road I dont know—to Heaven. One of our friends would say it was "a melancholy fact" that 3 years is a long time & in that long time much may happen both at home & abroad, sorrow sickness or ye grand finale—but if that time passes & finds us both on y[e] face of "this best of all possible worlds" why then, old boy, what a shake of y[e] hand we will have, what a bottle of Sherry what excursions, & what stories of wonders seen & dangers past. It may be as Cavendish[5] says he intends to find me on his return from Malta "in y[e] Snug Parsonage with M[rs]. W & some leetle Freddies" What a reunion of good & excellent fellows we might have he & you & Jack Venables[6]

&c &c. alas, alas we know not what is in yᵉ womb of time. But at least, old fellow, yᵉ worst fortune in yᵉ world cannot deprive us of many pleasant & sacred recollections (n.b. ⟨   ⟩)⁷ yᵉ note of yᵉ nightingale & yᵉ voice of yᵉ cherubim, yᵉ moonlight walk & yᵉ social glass (query, bottle?) yᵉ roll of yᵉ organ & yᵉ clash of knives & forks, with small-talk, arguments, billiard-balls & beetle hunting enough to furnish yᵉ most unfurnished head in Cambridge with ideas— We cannot expect quite yᵉ same hereafter but things perhaps higher & holier & believe me there is no one, I look forward to spending happy hours with, more sincerely than yourself— Our friends are scattering fast. Whitley prognosticates change in his vital barometer, Jem Turner⁸ & Jack Venab. are both preparing for orders & Curacies, Cavendish off next month to Malta for 3 or 4 years, Duncan dubbed a Viscount⁹ & of course on yᵉ wing, Grey¹⁰ looking out for something diplomati⟨qu⟩e, all that is good in Emmanuel Colbeck, Clutton & Co. rusticating,¹¹ worthy old Smith,¹² yᵉ only feather left in yᵉ wing. Its a heavy draft on yᵉ mental bank, & not cashed so easily. I dont think I shall return to Cambridge; this next week we are to be at Doncaster of course, Balls, Races, dinners & dissipation, but cares & blue-devils spoil all these & I'd rather be with you on those soft still evenings listening to yᵉ cries of strange birds, & admiring those wonderful forests.

It would be impertinent in an individual like myself to ask a naturalist to waste his time on me but if ever he has nothing to do & would kill a little time he may fancy how glad I should be to hear from him. Hang it I dont half like ending even this quasi-conversation, but it must be some time or other—so we shall dine together again in 3 years. What shall we have for dinner? My best wishes go with you on yᵉ sea & land in yᵉ Old World or New— do sometimes think of happy old times & remember that you've always one sincere friend in | Frederic Watkins¹³

Can I do any good for you in Cambridge or elsewhere. You know my ability is little but my will is great, if there is let me know & don't fear trouble F W

Postmark: Doncaster SE 19 1831
DAR 204.5

¹ Watkins was a member of CD's dining club at Cambridge, which was called the Gourmet or, by John Maurice Herbert, the Glutton Club. In a letter of reminiscences written to Francis Darwin, Watkins (then Archdeacon of York) described the Club as making 'a devouring raid on birds & beasts which were before unknown to human palate . . . I think the Club came to an untimely end by endeavouring to eat an old brown owl.' (DAR 112: 113v. — 114; see also *LL* 1: 168 — 70). Herbert lists the members of the Club, besides CD, Watkins, and himself, as [Charles Thomas] Whitley, [James William Lucas] Heaviside, [Jonathan Henry] Lovett Cameron, [Robert] Blane, and H[enry Porter] Lowe (DAR 112 (ser. 2): 70 — 1).

² Section of manuscript missing as a result of excision from verso. See n. 7.

³ Alexandre Brongniart.

⁴ Augustin-Pyramus de Candolle.

⁵ George Henry Cavendish.

6 Probably Richard Lister Venables, who was at Emmanuel, Watkins's College. No Jack or John Venables is listed in *Alum. Cantab.*

7 An excision, after an obliteration had earlier been made in heavy black ink.

8 James Farley Turner.

9 Adam Duncan, styled Viscount Duncan from September 1831, when his father was created Earl of Camperdown.

10 Possibly William Scurfield Grey.

11 William Royde Colbeck and Ralph Clutton, both Fellows of Emmanuel College.

12 Possibly the Rev. Thomas Smith, of Emmanuel College. He was a 'ten-year man'; i.e., an 'undergraduate, who had entered the University after having attained the age of twenty-four, and professed to have entirely devoted himself to the study of theology, was permitted, if he had performed the statutory exercises and ten years had elapsed since the date of his first admission, to graduate as a Bachelor of Divinity without having taken a previous degree.' (Winstanley 1940, p. 153).

13 The name is spelled 'Frederic' in this signature, but given as 'Frederick' in *Alum. Cantab.* and by Francis Darwin. Watkins's signature on a letter he wrote to Francis in 1882 is also 'Frederick'.

To Robert FitzRoy [19 September 1831]

17 Spring Gardens.

My dear Sir,

I have been hunting in several shops but have not succeeded in obtaining any pasteboard as thick as the sample you gave me. I send with this the thickest sort that they ever keep by them, and I hope it will answer your purpose.— I called on Watkins & Hill, and they promised they would use their best endeavours to hurry the glass house men. likewise they would try to get some colourless talc. I saw Captn Beaufort on Saturday & the result of the interview was that he could at any time take my name off the books; but that if the Admiralty were disposed to play the part of the wolf, it would not in his opinion make any difference my being or not being on the books— I mentioned that I believed the Surgeons collection[1] would be at the disposal of Government[2] and this he thought would make it much easier for me to retain the disposal of my collection amongst the different bodies in London. He advised me to reconsider and talk the subject over with you and to call on him before I ultimately left town. I have been very busy these two days in picking up information and everything goes on most prosperously. nothing less than a cable shall prevent me from seeing some time or another a Palm tree in its country; and what opportunity can possibly be better than the present— I do think it is the greatest piece of good fortune that ever happened to me— And I shall always recollect your kindness in helping me in every possible way to my end with the truest pleasure.— I start for Cam tonight, and from thence to Shrewsbury and shall again be in London by the 1st of October. *If you have time* and have anything to communicate, I shall be most grateful for a letter as I shall be anxious to hear how everything is going on. Captn Beaufort said he thought the

Surgeon could get apparatus free of expense from Sir W. Barnett[3] (or some such name.) but that I had of course better not as I should lose so much vantage ground over the Lords of the Admiralty

Have you Cap. Beecheys voyage to the Pacific? if you have not, I will buy it, as it contains some most excellent Meteorological Journals—

Believe me, dear Sir, | Your most sincerely obliged —| Cha[s] Darwin

Monday

Copy
DAR 144

[1] Robert McCormick was Ship's Surgeon and, as CD implies, it was normal for that officer to collect specimens on the voyage. When FitzRoy treated CD as the *de facto* naturalist, McCormick, who had reason to assume that this was his function, felt himself placed 'in a false position' and left the vessel at Rio de Janeiro to return to England (see letter to Caroline Darwin, 25 – 6 April [1832] and J. W. Gruber 1969). For a brief account of the naturalist tradition in the Royal Navy see Keevil 1957 – 63, vol. 4.

[2] The collections of the Ship's Surgeon and officers were considered government property. This was made explicit in the Admiralty instructions for the first voyage of the *Beagle*: 'You are to avail yourself of every opportunity of collecting and preserving Specimens of such objects of Natural History as may be new, rare, or interesting; and you are to instruct Captain Stokes, and all the other Officers, to use their best diligence in increasing the Collections in each ship: the whole of which must be understood to belong to the Public.' (*Narrative* 1: xvii). The instructions for the second voyage make no mention of collecting specimens. In a letter dated 16 November 1837 (DAR 164), FitzRoy states that the second *Beagle* voyage 'was the first employed in exploring and surveying whose Officers were not ordered to collect—and were therefore at liberty to keep the best of all—nay, all their specimens for themselves.' The Admiralty's policy seems to have varied with each voyage. In 1825 (two years before the first voyage of the *Beagle*), when Frederick William Beechey set out in the *Blossom*, the orders on collecting read: 'You are to cause it to be understood that two specimens, *at least*, of each article are to be reserved for the public museums after which the naturalist and officers will be at liberty to collect for themselves' (Beechey 1831, p. xiv, Admiralty instructions). The naturalist on this voyage was a 'Mr. Tradescant', a civilian, probably related to the famous seventeenth century father and son. His official appointment as lay naturalist is mentioned in the instructions.

[3] William Burnett, head of the Royal Navy Medical Department.

To W. D. Fox   19 [September 1831]

17 Spring Gardens (& here I shall remain till I start)
Monday 19th

My dear Fox

I returned from my expedition to see the Beagle at Plymouth on Saturday & found your most welcome letter on my table.— It is quite ridiculous what a very long period these last 20 days have appeared to me, certainly much more than as many weeks on ordinary occasions.— this will account for my not recollecting how much I told you of my plans, therefore I will begin a novo.— The expedition, under the command of Cap FitzRoy is fitted out principally for completing a survey of the S. parts of S America: The western shores of these parts have been well done by Cap King, under whom Fitzroy went out second in command.[1] We accordingly shall principally work on the Eastern coast of Patagonia from Rio de Plata to St[s] of Magellan.— The second object is to ascertain the longitudes of several places, more accurately than they are at present, & to carry a series of them round the world.— The expedition is entirely a government affair.

My appointment is not a very regular affair, as the only thing the Admiralty have done is putting me on the books for Victuals, value 40£ per annum.— I have some thoughts of having it taken off again. I should certainly do so, if I thought it would give me a more absolute disposal of my collection, when I return to England.—[2] But on the whole it is a grand & fortunate opportunity; there will be so many things to interest me.— fine scenery & an endless occupation & amusement in the different branches of Nat: History: then again navigation & metereology will amuse me on the voyage, joined to the grand requisite of there being a pleasant set of officers, & as far as I can judge this is certain.— On the other hand there is very considerable risk to ones life & health, & the leaving for so very long a time so many people whom I dearly love, is oftentimes a feeling so painful, that it requires all my resolution to overcome it— But every thing is now settled & before the 20th of Oct[r] I trust to be on the broad sea.— My objection to the vessel is its smallness, which cramps one so for room for packing my own body & all my cases &c &c.— As to its safety I hope the Admiralty are the best judges; to a landsmans eye she looks very small.— She is a 10 gun 3 masted brig.— but I believe an excellent vessel.—

So much for my future plans, & now for my present.— I go tonight by the mail to Cambridge, & from thence after settling my affairs proceed to Shrewsbury (most likely on Friday 23[d] or perhaps before): there I shall stay a few days & be in London by the 1st of October, & start for Plymouth on the 9th.— And now for the principal part of my letter.— I do not know how to tell you how very kind I feel your offer of coming to see me before I leave England.— Indeed I should like it very much; but I must tell you decidedly that I shall have very little time to spare, & that little time will be almost spoilt by my having so much to think about: & 2nd I can hardly think it worth your while to leave your Parish for such a cause.— But I shall never forget such generous kindness.— Now I know you will act, just as you think right, but do not come up for my sake. Any time is the same for me.— I think from this letter you will know as much of my plans as I do myself, & will judge accordingly the where & when to write to me.—

Every now & then I have moments of glorious enthusiasm, when I think of the date & cocoa trees, the palms & ferns so lofty & beautiful—every thing new everything sublime. And if I live to see years in after life how grand must such recollections be.— Do you know Humboldt? (if you dont, do so directly) with what intense pleasure he appears always to look back on the days spent in the tropical countries: I hope, when you next write to Osmaston, tell them my scheme, & give them my kindest regards & farewells.—

  Good bye my dear Fox. Yours ever sincerely | Chas Darwin

Postmark: C.H 19 SE 1831 X
Christ's College Library, Cambridge (Fox 44)

[1] See letter from George Peacock, [*c.* 26 August 1831], n. 1.
[2] On 15 September 1831, Robert FitzRoy had written to Francis Beaufort (F. Darwin 1912, p. 547): 'He [Darwin], Captain King and I *now* think that it would be better in many respects, that he should *not* be on the *Books*, but that he should go out in a strictly *private* capacity. I am, however, *equally* ready to receive him in *either* manner, and I have recommended his asking which plan meets *your* approbation. P.S.—He has seen his future dwelling and is satisfied with it.' In the end, CD apparently decided not to remove himself from the books. During the voyage, it is true, CD refers to paying FitzRoy for his mess (see letter to Caroline Darwin, [24 October — 24 November 1832] and letter to Catherine Darwin, 8 November 1834) but these statements refer to payments beyond the victuals supplied by the Admiralty. In his letter to J. S. Henslow, [5 September 1831] CD writes: 'Cap. Beaufort says I shall be upon the boards & then it will only cost me like other officers.—' In his letter to Susan Darwin, [5 September 1831] he says: 'I shall pay to mess the same as Captain does himself 30£ per annum', and FitzRoy was certainly on the books. This is further borne out by FitzRoy's statement that 'an offer was made to Mr. Darwin to be my guest on board, which he accepted conditionally; permission was obtained for his embarkation, and an order given by the Admiralty that he should be borne on the ship's books for provisions. The conditions asked by Mr. Darwin were, that he should be at liberty to leave the Beagle and retire from the Expedition when he thought proper, and that he should pay a fair share of the expenses of my table.' (*Narrative* 2: 19). Although this was written after the voyage was over, it is unlikely that FitzRoy would forget CD's having removed himself from the books.

From Charlotte Wedgwood   22 September [1831]

Maer
Sep^t 22^d

My dear Charles

  I congratulate you most sincerely on your fate being at last decided, & decided as you wished it. I wish you with all my heart all the enjoyment & improvement & beautiful scenes & life of interest that you are looking forward to, & above all a safe & happy return which will be the pleasantest of the whole. I was very glad to receive your letter. I had regretted that we had been so foolish as not to ask you to write when we were so particularly wishing to know what your fate would be, so that your letter coming unasked for gave me the more pleasure. For some time after you left Maer I was in a complete fidget thinking of the chances for or against your being cut out of the expedition by it's being offered in the mean time to somebody else which would have been the most vexatious way of losing it, & this was increased by our finding out that your letter to Mr Henslow, upon which it was possible that your fate might depend, was delayed at the Post office for a day—[1] Frank to make it more secure had sent a direction with it to be forwarded immediately, but most haste worst speed, for when he called at the Office in returning to inquire whether it was safe off, one of the first things he saw was the letter itself. We calculated however when you went so immediately to Cambridge that it would make no difference as you would we thought get there the same day that the letter ought to have arrived. I am delighted that you have fallen in with a Captain Wentworth—[2] such an extraordinary piece of good luck is a good omen for every else— I hope you will become real intimate friends which will double the pleasure of every thing. I wish you would not so completely set us down as your Lords of the Admiralty—when I think of your sisters my conscience is ill at ease & I shall feel guilty when I next see them—they will be very good natured if they do not bear us a grudge— I shall lay all the blame on my father & Hensleigh, & you can vouch for us that Hensleigh is the only that gave a strong opinion. I wish very much I could hear that Caroline will be returned time enough to see you or she will have more cause than any body to bear us a grudge. I am very sorry for the third year that has been added, but it is a great comfort that you can at any time if you wish it, quit the ship & return home when you meet with an opportunity. I do confess that that third year makes me tremble much more than I did before for the country parish & parsonage house where I should be very sorry not to see you established— I think it is the happiest kind of life & one which would almost oblige any one to be good, & something to oblige one to be good is what one feels the want of every day of one's life. That it will oblige you to work is I know one of the advantages that you think this expedition will give you— I wish it may but I am very much afraid that ship board is not a good place for working & that it will require a great deal of resolution & perseverance on your part to make it so. I have an earnest desire that you should prove that you have made a good choice, as well as that we have not done you an injury, for I cannot help remembering that but for that 1st of September your family would have had you safe at home & this you see makes me grave & preachy— I wish indeed that your time would have allowed you to see us before you go, but we would not have taken a day from your week at home on any account tho you must wish you had more time before you I really believe it is better for you not to have a great deal of time between taking your resolution & acting that there may not be time for all the objections to rise up which they always do with much more than their real weight when there is nothing more for them to do but to torment one, & as it is I hope you will be too hurried & busy for this disagreeable process. I suppose you are at Cambridge now— I wonder whether you are too busy & your head too full to observe the extreme beauty of today & yesterday, yesterday particularly Wednesday every thing looked supernaturally beautiful, even the larch plantations themselves— one of your tropical moonlight nights could hardly be more beautiful & it set me wondering what it could be that made the difference between it & other sunshiny days, but I think those days never come but in Autumn. Our Welsh party who returned the day before from the Menai bridge would have given a good deal for two or three such days, however they liked their tour in spite of clouds. Miss Julia Mainwaring came here today to intreat some of us to go & help her to entertain a party of officers today she being the only lady— nobody would go but Emma, who when she found she could not get Fanny to go & keep her in countenance had great scruples lest she should appear too Lydiaish,³ however by going rather early she hoped she should appear to be staying with Miss Julia, rather than come express to meet the officers. We are going to have the great honour of a visit from Dr Holland tomorrow think of that, however that we may not be too much puffed up it must be owned that Mary Holland & his children are here, which tho he did not know, he had a pretty good guess of, but it is something to be proud of notwithstanding & I hope we shall keep clear of the reform bill which I understand his temper cannot stand. Elizabeth & I are going next week to the Bent's at Derby for the music meeting—it will not rival Birmingham of famous memory. I shall like very much to hear from you again before you sail— tell me all little particulars about your arrangements on ship board &c. the Welsh party picked up some intelligence about you at Overton—that you will be able to go ashore for some time in stormy weather will be no small comfort I should think. All here desire to be most kindly remembered to you & have a great interest in all your plans & prospects. Once more warmly wishing you success I am ever dear Charles | Your affectionate cousin | Charlotte Wedgwood

I wonder where I shall direct to you next—

DAR 204.5

¹ See letter to J. S. Henslow, [2 September 1831].
² Captain Frederick Wentworth, hero of Jane Austen's *Persuasion* (1818).
³ Reference to the flirtatious Lydia Bennet, of Jane Austen's *Pride and prejudice* (1813).

From Fanny Owen [22 September − 2 October 1831]¹

Dr Postillion,

I *entreat* your acceptance of a *leetle* Purse which I hope you will *condescend* to use in remembrance of the *Housemaid* of the *Black Forest*—

I remain Dr C. yrs truly Fanny O

Pray remember me most kindly to *my friend*, Mr *Charles Mogg*—²

DAR 204.4

¹ Faint writing on the verso appears to read 'C.D. Shrewsbury'. The conjectured date refers to the time of CD's farewell visit to his father and sisters.
² Possibly Charles William Cumberland Mogg.

From Robert FitzRoy 23 September 1831

Devonport
Sept.ʳ 23ᵈ 1831

Dear Darwin

I read the first sentence of your letter—"Before you judge of my conduct" —and threw it away in a rage—saying "Damn these shoregoing fellows they never know their own minds,"—"well let's see what crotchet makes him refuse to go"—when upon reading further I found that so desperate a beginning only ushered in a simple request about a *Mid*!— I made certain you were *off* your bargain a **Lady** in the way, or something unforeseen!

I am sorry it is out of my power to take young Owen[1]—because the number of Mids allowed has been complete since the Vessel was commissioned.

There is no *chance* of a Vacancy. You were surely quite right to *ask*—I could but refuse—yet I would not have refused had I been *able* to oblige you.

I received the parcel from London & your letter—thanks to you.

I have Beechey's Voyage but not Head's Gallop—[2] You are of course welcome to take your Humboldt—as well as any other books you like—but, I cannot consent to leaving mine behind. *all* my *goods* go *with* me.

There will be *plenty* of room for Books. I have Daniells' Hyg.ʳ[3] but it would be well to have another as they are fragile. I have not a Pentagraph[4] because I do not think it of any material use to *me*.

Taking all things into consideration *I* think you had better be *on* the books, but do just as *you* like,—*you* are the person most interested.

I have arranged good & *dry* stowage for your things—and I think you will have small cause to complain of your Cabin—table—Drawers &cᵃ

The Dock Yard are making very slow progress—so that we shall not sail until the end of October You can remain away another week so as to be here on the 17ᵗʰ if you like.

Faithfully Your's | Rob.ᵗ FitzRoy

DAR 204.6.2

[1] Francis Owen, a younger brother of Sarah and Fanny Owen.
[2] Head 1826.
[3] A hygrometer invented by John Frederic Daniell in 1820 made possible precise measurement of the moisture in the atmosphere.
[4] An error for pantograph, a device for copying drawings on a different scale.

To Charles Whitley 23 [September 1831]

Shrewsbury
Friday 23ᵈ

My dear Whitley

I found your letter on my table, when I came from Plymouth; but I really I had not time to write to you from London or Cambridge, in neither of which places did I stay more than 2 days.— I saw Herbert at Cam. & am afraid this letter will not reach you.— I shall be very sorry for it, for I do not think I ever received a more kind letter than yours or one that gave me so much pleasure.— You ought to have in your mind, the prospect of leaving England for 3 or 4 years before you can understand how to enjoy such a letter from such a person as yourself.— I am afraid we shall not meet, I leave Shrewsbury this day week.— Herbert will find out from Henslow my direction, times of writing &c. & when I can (it must necessarily be very seldom) you shall hear of and from me.— I am rather tired this morning, & the post is soon going out; so that you will excuse this letter: I have nothing particular to say excepting that all is finally settled, & I have sealed away about half a chance of life.— If one lived merely to see how long one could spin out life.— I should repent of my choice.— As it is I do not.—

Again I thank you, my dear Whitley, for all your kind wishes, I hope they will come true, & that when I return, I shall find you a happy & useful man.

God bless you | My dear Whitley | Your affectionate friend | Chas. Darwin.—

Some Fungi have arrived much obliged for them.—

Ximenes Rare Books (December 1983)

From Fanny Owen [26 September 1831]

2. Northernhay Place, Exeter
Monday

My dear Charles,

I have this evening heard from Caroline that you leave home the end of this week—and that you wish to have a *good bye* from me before you go. I had not the **least idea** you were to go so soon, for they told me it was the end of October you sailed, so I **hoped** and fully expected I should have been at home in time to see you— I **cannot** *tell you* how *disappointed* & *vexed* I am that that cannot be. Little did I think the last time I saw you at the poor old Forest, that it would be **so long** before we should meet again!! This horrid Devonshire—fool that I was to come here— I shall just get home when you are gone I dare say— My dear Charles I do hope you will enjoy yourself & be the happiest of the happy, I would give any thing to see you once more before you go, for it does make me melancholy to think the time you are to be away—& Heaven knows what may have become of all of us by this time two years. at all events we **must** be grown **old** & steady— the pleasant days, and fun we have had at the Forest can never come over again— how I wish I was there this week to have one *last chat* with you I cannot bear to think you are really going *clear* away, without my saying one *good bye*!!

But I must drop this subject for I find I am getting prosy & melancholy & that wont do— They tell me you were at Plymouth about 10 days ago & so was I, how **very very** unlucky we never met, do you go there again? if you should perhaps you may pass through Exeter— I shall leave it on the 6th with the Hunts— I believe not come home direct but go with them to pay some visits— if possible I shall shirk and get the Gov— to meet me at Leamington or Birmingham for I think it will be awful *flat work*, dowagering about with the Hunts to unknown parts— I am sure I have been dull enough all this summer— hope I have expiated all my sins for a severe Penance I have had of it— I wont be *taken alive* again in that way when once I get home— *Home sweet home* you should hear me sing now—I assure you I do it **feelingly**—it would melt a heart of stone—or rather crack an **ear drum** of **Iron** to hear me—but here my powers have no scope I can never give vent to my feelings as I feel inclined— So poor old Williams is gone at last, a happy release for himself I should think—& certainly for every body related to him— a proper time being given up to becoming grief the awful ceremony will of course take place as soon as possible— how very provoking you should not be present—not even taste the **long** *expected* Plum Cake[1] how vexed I am you are going it is too selfish of me to say so, for I am sure it will be the very thing to suit you— did you throw yourself on the Governor's mercy, & confess your creditors, or what have you done? What a capital way of escaping *ungentlemanlike Tailors* &c— When you are *far from the Land* they may *whistle* for their cash for what *you care*! Well, dont be surprised if you hear I have *taken Ship* too and fled my duns— that **joyful** season Xmas is fast approaching— my heart sinks when I think of it—but there's nothing like putting a good *face* on it— I shall do so as long as I can— Pray write to me one last Farewell my dear Charles & tell me all your plans & prospects—where you are to go to—& all about it? And tell me too if I shall look out for a nice little Wife for the *Parsonage* by the time you return. tell me what you require and I will look about and get one in *my eye* by the time you want her—a proper knowledge of the *Beetle tribe* of course you require— bye the bye has *your faithless* Charlotte Salway[2] bee⟨n⟩ twined off yet—I have heard nothing of her As for all your Sisters I think they are gone crazy or *sulky* or sleepy or somethi⟨ng⟩ for not one line have I had from any of them these two months—they treat me with the most marked contempt.— I was much amused at Plymouth there is so much worth seeing— Mount Edgecombe[3] I dare say you saw—it is a beautiful Place.— I went on board the Adelaide and all over it—so can fancy you in your little Cabin—and I assure you you will not be forgotten, I shall often long to have you to laugh with and *scold* out of the Painting room— I wish I had made your Pincushions they might have been useful—and occasionally in taking out an *instrument of death for a Beetle* you would have called to mind the Manufacturer of the *useful article*—but it cant be helped now— this letter is *most prosy*, & duller than letter ever was before—but I cant help it you must take the *will for the deed*— write to me 2 Northernhay Place = I must now conclude—can only add—I most sincerely wish you every amusement & happiness possible—but only wish most heartily you were not going quite so soon that we might have one *more talk & laugh* first—but it is *not* to be— so good bye my dear Charles

Believe me always yours most sincerely and *affecty* | F O—

Burn this before *you sail for pitys sake*—

Postmark: Exeter SE 27 ⟨18⟩31
DAR 204.4

¹ Fanny's sister Sarah was soon to be married to Edward Hosier Williams (see letter from Fanny Owen, 2 [December 1831]).
² Charlotte Salwey.
³ Mount Edgcumbe, a sixteenth century mansion on the Cornish peninsular.

From Sarah Owen [27—30 September 1831]

My dear Charles,

You see I am as good as my word, or rather Mʳ Baker is, for I enclose the *promised* Pin,¹ the hair is genuine, & I am much flattered in the idea that it is destined to accompany you round the world—

We all felt very melancholy after your departure on Sunday, I do not know what Woodhouse will do without you for so long, but I hope & trust *we* may both meet with success in our respective new careers, & live to meet here again *very very* often; remember your promise about Nº 1, Belgrave Sᵗ² & pray think of me in the mean time, & write whenever you have an idle half hour. I assure you my parting promise to you shall be most religiously kept, & you may expect a true & correct account from the Pen of the Sufferer herself—

I am so glad you have a short reprieve for the sake of your Family, though perhaps *you* are not so well pleased with the delay—

God bless you, my dear Charles, believe that whenever I may change my *title*, I shall always remain your very sincere & affectionate Friend | Sarah—

DAR 204.4

¹ The pin, inserted in the letter, is preserved at the Cambridge University Library (DAR 204.4).
² Sarah's address after her marriage.

To J. S. Henslow 28 [September 1831]

My dear Henslow

I have received another parcel of the Phalli from *Barmouth*.— & another jar of them, which I gathered the day before yesterday in a very damp shady wood: I am more than ever convinced that they are different species.— The Shropshire ones are, whiter more conical & stiffer, than the Barmouth one: the ball more dark coloured & the cap has less jelly, & that not so dark coloured:

They are all preserved in gin & *brine* owing to the want of more spirit.—

I have sent some of the Leiodes.— Will you be kind enough, when you send my goods to London—you will enclose a piece of brick lapped up in the German fashion.— & mention likewise, what sort of Lens Mr Brown[1] recommended.— & lastly do not forget the introductions to Lowe[2] & Smith.—[3]

I heard from Cap Fitzroy yesterday he gives me a week more of respite, & therefore I do not leave this place till the end of this week, & London on the 16th of October.— I wish indeed that time was arrived, for I begin to be very anxious to start.—[4] My Father is getting much more reco⟨nci⟩led to the idea, as I knew he would, as soon as he became accustomed to it.—

Believe me dear Henslow | Yours most sincerely | Chas. Darwin

Wednesday 28th.
Shrewsbury

Royal Botanic Gardens, Kew

[1] Robert Brown had made important microscopical observations, among them the discovery of Brownian movement. See *Autobiography*, pp. 103–4, for CD's reminiscences of Brown. In a letter of 26 March 1848 to Richard Owen (New York Botanic Garden), CD compares a newly acquired microscope to 'the one, which I used on board the Beagle & which was recommended to me by *R. Brown*'. This is almost certainly the Bancks microscope now at Down House (see letter to Susan Darwin, [6 September 1831], n. 1). CD's is an improved version of Brown's own instrument, also made by Bancks, which is now on display at the Linnean Society.

[2] Richard Thomas Lowe, then residing at Madeira, was the author of a work on the flora and fauna of that island (Lowe 1833, read 15 November 1830). Stormy seas prevented the *Beagle* from putting in at Madeira, so CD did not meet Lowe.

[3] Andrew Smith.

[4] During this visit CD went to say farewell to the Wedgwoods at Maer, where his account of the prospective voyage apparently aroused some misgivings. On 29 September, his cousin Hensleigh Wedgwood wrote to his fiancée, Fanny Mackintosh: 'I wonder Charles is not damped in his ardour for the expedition. He says that Patagonia where they are going first to is the most detestable climate in the world, raining incessantly, & it is one vast peat bog without a tree to be seen. The natives will infallibly eat you if they can get an opportunity. They have got some tame Patagonians that they are going to take back & who promise to give up cannibalising but they do not believe a word of their promises. Then their mode of proceeding will be to anchor close to shore & remain there two or three weeks till they have surveyed all the country about & then go on to another place. It is very enterprising to go in spite of such discouraging accounts.' (B. and H. Wedgwood 1980, p. 215).

To Robert FitzRoy [4 or 11 October 1831]

17 Spring Gardens.

Dear FitzRoy,

I have nothing very particular to write about, except to assure you, that you shall have no occasion to "damn those shore-going fellows." It is much more likely that I should do the same to sea & shore fellows, if anything was to prevent my coming with you. All things go on most prosperously and everybody who knows me highly approves of my undertaking— I should be perfectly happy if it was not for the contest that is continually going on in my mind between the utility and the bulk of any intended object.— *I do assure you* I have *been as economical as I possibly could*, but my luggage is frightfully bullky—. I look forward with consternation to seeing M.ʳ Wickham—¹ if he grumbled merely at the number of my natural cubic inches, what he will do now I cannot imagine. If the worst comes and you cannot take my things there are two big cases that I can leave behind without very material injury— If you have time to send me one line will you inform me how I had better manage about my baggage when I arrive at Plymouth. I suppose the Beagle is not in readiness to receive it and if I recollect right, there is no possibility of getting near to my hotel in a boat.— I will now tell you what I have done about money— I have so arranged it, that Curtis & C.º will never dishonor my drafts, and that I suppose is all that is necessary. Have you any books on spherical trigonometry? as I hope & trust to read a little mathematics during the three years.— Mr.Earl² tells me to inform you that he is coming down the same day as I do—viz—Sunday 16.ᵗʰ ³ He would have started some days sooner than that period, only that he liked the opportunity of sailing with somebody whom he knew. I suppose the Beagle will not sail, till the beginning of November so that I shall have plenty of time to settle myself in my cabin. This will be a great advantage every thing is capital— I only hope as you say— it is not too much good luck for it to last.— If there is anything in London, which it is in my power to do for you, of course I shall be most happy to do it.

With many thanks for all the interest you have shewn in my affairs believe me, dear FitzRoy | Yours ever most sincerely | Cha.ˢ Darwin.

I saw George Cavendish who is in the Rifles & he gives a very poor character to our friend the Major.

Tuesday

Copy
DAR 144

---

¹ John Clements Wickham, First Lieutenant of the *Beagle*.

² Augustus Earle, an artist engaged by FitzRoy as draughtsman for the voyage. On 19 November 1831 FitzRoy reported to Beaufort: 'Messrs. Earle and Darwin are the very men, of all others, for their employment, and I assure you that Darwin has not yet shown *one* trait which has made me feel other than glad when I reflect how much we shall be together.' (F. Darwin 1912, p. 547).

³ CD did not arrive at Devonport until 24 October ('*Beagle*' diary, p. 5).

To J. S. Henslow [4 or 11 October 1831]

17 Spring Gardens
Tuesday

My dear Henslow

I called on your Brother yesterday & paid him the 7 " 12. which I owe you: & he told me that he was going to send a parcel to you: so that I seize the opportunity of writing to you on the subject of consingment.— I have talked to every body: & you are my only resourse; if you will take charge, it will be doing me the greatest kindness.— The land carriage to Cambridge will be as nothing compared to having some safe place to stow them; & what is more having somebody to see that they are safe.— I suppose plants & Bird skins are the only things that give trouble: but I know you will do what is proper for them.—

Will you give me as minute instructions about the directing, as if you were writing to an Otaheite[1] savage: or what will be better make a scetch of lid of box, & on it direction, *precisely & every letter*, as if it was one I was going to send it off from any remote place.— About paying for them, I should think the best plan will be, after the arrival of one or two cases, to write to my Father, & he will place the sum to your account at any bank in Cambridge you may choose:— I will write to him on the subject: I am so very busy, as never was anything like it before: I have hardly time to look about me: I suppose we do not sail till November, so that I will, of course write to you from Devonport.— M.r Brown has been of great use to me, & most exceedingly pleasant & goodnatured.—

Your Brother must think me a regular practised swindler, for most unluckily I gave him the money *sealed up*; which most properly he opened when lo & behold there was 6 instead of 7 " 10: I am sure he will think for the future you are very rash to trust me.—

Will you be kind enough to write to me, before I leave London on Sunday 16.th about consingements

Believe me, my dear Henslow | Yours ever sincerely obliged | Chas. Darwin.—

Royal Botanic Gardens, Kew

[1] Tahiti.

From Fanny Owen [6 October 1831]

Exeter
Thursday

My dear Charles.

Our letters must have crossed on the road— yours I received a few days ago; it was written indeed in a *Blue devilish* humour, and I'm sure imparted the same to me— I cannot bear to think my dear Charles that we are not to meet again for so long **three** years you say & I heard at first it was to be two— but that you will enjoy yourself I have not a doubt—& to remind you of the time you are to be absent, is nonsense & selvish— one last farewell—I cannot resist sending you— You say what changes will happen before you come back—"& you hope I shall not have quite forgotten you—" I doubt not you will find me in *status quo* at the Forest, only grown **old** & *sedate*—but wherever I may be whatever changes may have happen'd *none* there will **ever** be in my opinion of **you**—so do not my dear Charles talk of *forgetting*!! the many happy hours we have had together from the time we were **Housemaid & Postillion** together, are not to be forgotten—& would that there was not to be an end of them!! I dont know what we shall do at the Forest without you—& how sorry I shall be to have nobody to *scold out* of the Painting room— I have heard no particulars of your voyage from any body but that you expect to sail on the 15th—if you have a *little spare* half hour before you go do write to me again I should like so much to hear from you and direct Post Office Leamington— I shall be staying there for a few days with the Hunts on my way home— I leave Exeter next Monday the 10th—and am not sorry to do so— Caroline tells me you were actually walking about Plymouth the *very day I was*, I never did know any thing so unlucky we should not have crossed each *other's path!!* This letter is not worth its postage I know I am more dull & *matter of fact* than any body ever was—but take the *will* for the *deed* my dear Charles—as I cannot see you to say good bye it is a melancholy satisfaction to scribble to you once more—

God bless you my dear Charles & may you enjoy every possible happiness is the sincere wish of y^{rs} most affec^{tly} F. Owen—

If you have any time do write—but if you do not—I shall well know it is not your own fault— once more Farewell | my dear Charles—

Postmark: Exeter OC 6 1831
DAR 204.4

To Robert FitzRoy [10 October 1831]

17 Spring Gardens.

Dear FitzRoy,

Very many thanks for your letter; it has made me most comfortable, for it would have been heart breaking to have left anything quite behind & I never should have thought of sending things by some other vessel. This letter will, I trust accompany some talc.— I read your letter without attending to the name. But I have now procured some from Jones which appears very good—and I will send it this evening by the Mail. You will be surprised at not seeing me propriâ personâ instead of my handwriting But I had just found out that the large steam Packet did not intend to Sail on Sunday & I was picturing to myself a small dirty cabin with the proportion of $^{39}/_{40}$ of the passengers very sick—when Mr Earl came in and told me the Beagle would not sail till the beginning of November. This of course settled the point so that I remain in London one week more. I shall then send heavy goods by steamer, and start myself by the Coach on Sunday evening.

Have you a good set of mountain barometers—[1] Several great guns in the Scientific World have told me some points in geology to ascertain which entirely depend on their relative height. If you have not a good stock I will add one more to the list— I ought to be ashamed to trouble you so much but will you *send one line* to inform me? I am daily becoming more anxious to be off and if I am so you must be in a perfect fever. What a glorious day the 4th of November will be to me— My second life will then commence, and it shall be as a birthday for the rest of my life.

Believe me dear FitzRoy, | Yours most sincerely, | Chas Darwin.

Monday.

I hope I have not put you to much inconvenience by ordering the room in readiness—

Copy
DAR 144

[1] CD took with him a set of aneroid barometers. The tables for their use, with CD's notes, are in *Jones's companion to the mountain barometer & tables* in DAR 196.2.

From J. S. Henslow 25 October 1831

Cambridge
25 Octr 1831

My dear Darwin,

I have just received your letter about consignment with a statement of your attempt to cheat my Brother of 1£. Do look at the bill I see sent, for it runs in my head that you have read pounds for shillings & shillings for pounds & that you *ought* to have pd him 12"7"— instead of 7"12.— I can't be sure without searching after the Bills & putting the items together again—so that you may be right— As touching consignment I should think the best way is to do as Lowe does: direct as follows, Revd Profr Henslow, Cambridge to the care of J. W. Henslow Esqre 12 Clements Inn London, Antipodes or England.—the part underlined being optional.

I intend a great addition to my clinometer by giving it a sight for calculating <r distces when I think it will comprize all the Geologist wants. I have just met Watkins who is delighted to hear of your expedition. Downes is just returned from a short tour in Switzerland, but what is that to a Fuegian— The day after tomorrow our County election begins— Mr Jenyns[1] is chairman of Capt. Yorke's[2]

committee, proposed at one time to sleep at my house during the week, by which manouvre I had calculated upon getting my windows smashed— this would have been as good as foraging in the enemies quarters— He however thinks it better to sleep at the Inn where the Committee sit—so I suppose I shall remain in peace unless the Anties get the upper hand & think me too much of a radical— I presume however that these things begin to cease from interesting you & ∴ leave you to your better meditations on Mermaids & Flying fish.—

Y^s affectionately | J S. Henslow

DAR 204.6.2

[1] George Leonard Jenyns.
[2] Charles Philip Yorke, Tory anti-Reform Bill candidate in the Cambridgeshire elections. He was defeated by Richard Greaves Townley of Fulbourn, Cambs. (*Cambridge Chronicle*, 4 November 1831).

To J. S. Henslow 30 [October 1831]

4 Clarence Baths | Devonport
30th

My dear Henslow

Your letter has filled me with consternation.— I never knew anything so stupid as my making such a mistake.— I have lost your letter, but I have no doubt you are right.— If I merely trusted to recollection, I should yet think it was 7"12.— But after the little swindling affair with your brother, I will not trust my own self: It is too bad of me to give you so much unnecessary trouble, but perhaps you can find out the prices of the principal things such as paper & binding books, & that will be sufficient to know which of the sums it is.— I can easily through my brother contrive to pay you.— I am very much obliged for your directions about consignment.— I believe most of the things will first go to Falmouth (where I must get an agent), & then to Cambridge.— I will tell my Father that you will send him a note with an account of what you pay for me.—and I do not think you will find him as careless as I am.— I hope to be able to assist the Philosoph. Society when I come back.—but from all I hear, I suppose I shall be in honor bound to give largely to British Museum.— Every thing here goes on very prosperously. My beau ideal of a Captain is determined to make me as comf⟨ort⟩able as he possibly can.— But the corner of the cabin, which is my private property, is most wofully small.— I have just room to turn round & that is all.— My friend the Doctor[1] is an ass, but we jog on very amicably: at present he is in great tribulation, whether his cabin shall be painted French Grey or a dead white— I hear little excepting this subject from him.— The gun-room officers are a fine set of fellows, but rather rough, & their conversation is oftentimes so full of slang & sea phrases that it is as unintelligible as Hebrew to me.— Our Cabins are fitted most luxuriously with nothing except Mahogany: in short, every thing is going on as well possible. I only wish they were a little faster.— I am afraid we shall not bonâ fide sail till 20th of next Month.— I want your advice de Mathematicis. After looking at my 11 books of Euclid, & first part of Algebra (including binomial theorem?) I may then begin Trigonometry after which must I begin Spherical? are there any important parts in the 2^d & 3^d *parts* of Woods Algebra.— It is almost a shame to ask you, but I should be much obliged if you would write to me pretty soon.— You must be very busy; for if Messrs. Askew[2] & Darnell[3] have not got some fresh Brains in the vacation, they will give you some trouble:—

What an important Epoch 1831 will be in my life. taking one degree, & starting for Patagonia are each in their respective way memorable events.— And you have been most instrumental in getting them both.— Remember me most kindly to M^{rs}. Henslow.—Leonard Jenyns & all other friends.— I often think of your good advice of taking all uncomfortable moments as matters of course, & not to be compared with all the lasting & solid advantages:— Indeed I never can do better than when I think of you & your advice

Ever yours my dear Henslow | Most affectionately | Chas. Darwin

You give me your brother direction *12* Clements Inn. Is that right?

Postmarks: ⟨D⟩evonport OC 30 ⟨1831⟩; D 1 NO 1 1831
Royal Botanic Gardens, Kew

[1] Presumably CD refers to Robert McCormick, the *Beagle*'s surgeon, though not M.D. For a more sympathetic appreciation, see Keevil 1943.
[2] Henry William Askew.
[3] Daniel Darnell.

To Caroline Darwin [31?] October [1831][1]

4 Clarence Baths, | Devonport.
October 21st.—

My dear Caroline,

I received your letter the same day that I wrote to Susan— Since which I have had one from Katty, for which give her my thanks & love.— I want most particularly & directly to know about the Tutor's bill.— Was there somewhere about 30£. allowed for my furniture? In fact there could not have been, and it is too bad of M^r. Ash,[2] for I wrote on that subject solely to beg of him to subtract it from the bill, before sending it.— I can manage it through Henslow.— My Father must have thought me a regular swindler, as I said to him it would be about 8. or 16£.—

Everything goes on in such a regular manner, that I have nothing to write about.— I think Erasmus will come & see me start, so he will describe in how small a compass it is possible to pack a man.— Our ship will be, I fear, a regular raree show.—[3] And the poop cabin being a gay one it will be very troublesome for me.—

I shall begin very soon to stow away my things; but I do not suppose I shall sleep on board till about a week before starting.— It is very lucky for me that I think I shall like my fellow companion in the cabin the best of any of the officers.— Cap. Fitz. has given him a room in the house; so that we live together.— The advantage of living with the Captain is most decided, his quiet manners are quite delightful after the riot of the gun room.— The only drawback now, is the great deficiency of room to pack up things in.— I live in continual fear.— My room are 15^s a week, and I am now living as if we were on Ship board—viz 50£ per annum.— I shall pay (if my Father likes it) the Cap. 100£ for the 2 next years— For if his pocket is fathomable he must have found the bottom— From what I pick up I do not believe we shall sail till end of next month— nothing however can exceed the activity of the officers.—

Love to Susan and tell her I will not take Persuasion, as the Captain says he will not read it, & there is no danger of my forgetting it. More letters the better. Love to all: hurra, for the Valparaiso Volcanoes. | Yours affec— | C. Darwin.

Copy
DAR 154

---

[1] The copyist has 'October 21st', but CD did not arrive at Devonport until Monday, the 24th (*'Beagle' diary*, p. 5).

[2] Edward John Ash, Tutor and Steward of Christ's College.

[3] 'A show contained or carried about in a box; a peep-show.' 'This word [raree-show] is formed in imitation of the foreign way of pronouncing *rare show* (Johnson)' (*OED*).

To Caroline Darwin 12 November [1831]

November 12th.

My dear Caroline,

The tutor's bill is just as I expected—and I will contrive some plan through Henslow.— Most unfortunately Henslow has just lost his brother, so I do not like at present to trouble him.—

Everything here is most prosperous; the Beagle now looks something like a ship— They have just painted her and in a weeks time the men will live on board.— No Vessel has ever been fitted at all on so expensive a scale from Plymouth— I get into a fine naval fervour whenever I look at her. I suppose she is as good a ship as art can make her—and if I believe all I hear the Captain is as perfect as nature can make him— It is ridiculous to see how popular he is, ladies can hardly splutter out big enough words to express their big feelings—

I have been going out rather more lately than I wish. I dined yesterday at the Admirals Sir   Dixon[1] with Captain FitzRoy—where I met nobody but naval officers, the conversation would have been stupid to a Landsman,—but to me it was very interesting. I breakfasted yesterday with a M.ʳ Harris[2] whom I like more than anybody I have seen.— He has written a great deal on Electricity— This morning I did ditto with Col. H. Smith[3] a very clever old Gentleman.— Tomorrow I am going to Lord Morleys, and am going to ride over with Lord Borrington[4] to see the granite on Dartmoor.— So that I am quite gay & like the place very much.— I suspect from all I hear the sea-sickness is very much worse than I expected— More than half the naval officers feel uncomfortable at first starting.— I am sure, as soon as sea-sickness is over I shall soon fall into sea habits & like them.— I think I get accustomed to anything soon, and that will be half the battle won.— It is very lucky we did not sail earlier, for if we had started 6 weeks ago, I believe we should not, owing to S. W. Gale, have reached Madeira by this time.—

Tell Susan she need not be alarmed about my forgetting to give directions about writing. I presume Rio Janeiro will be principal place for some time.— I get letters for nothing— I fancy S. America will not detain us more than 18 months— What then nobody seems to know— It is certain that a new continent has been discovered somewhere far South.[5] Perhaps we may be sent in search.— I suppose you have received a letter from me since Susan's date.

Love to my Father & all others. C. Darwin.

Copy
In copyist's hand: Postmark 1831
DAR 154

[1] Admiral Sir Manley Dixon.
[2] William Snow Harris, known as 'Thunder and Lightning Harris' from his experiments with lightning conductors. For Robert FitzRoy's report on the efficiency of those installed in the *Beagle* see *Narrative* Appendix, p. 298.
[3] Charles Hamilton Smith.
[4] John Parker, 1st Earl of Morley, and his son, Edmund Parker, Viscount Boringdon.
[5] Probably a reference to the discovery of land in the Antarctic Circle by John Biscoe, in 1831. Biscoe explored the southern seas for the whaling and sealing interests of the firm of Enderby of London (*EB*, 'Polar regions').

To J. S. Henslow 15 [November 1831]

Devonport
15.ᵗʰ

My dear Henslow

The orders are come down from the Admiralty & every thing is finally settled.— We positively sail the last day of this month & I think before that time the Vessel will be ready.— She looks most beautiful, even a landsman must admire her. *we* all think her the most perfect vessel ever turned out of the Dock yard.— one thing is certain no vessel has been fitted out so expensively & with so much care.— Everything that can be made so is of Mahogany, & nothing can exceed the neatness & beauty of all the accomodations.—[1] The instructions[2] are very general & leave a great deal to the Captains discretion & judgement, thus paying a substantial, as well as many verbal compliments to him.— I will now give you an outline of the plans. 1st to Madeira or Canary (perhaps only the latter) Cape Verd, Fernando Noronha, Rio de Janeiro, Monte Video, then set to work at Patagonia Terra del, Falkland Islands, so as to consume about year & half. After this is completed to work our way Northward on W coast of S America, as far as Captain chooses, leaving time to take a good stretch across Pacific ocean—(taking some new course) New S. Wales, Van Diemens land.— Some of E Indian Island, Cape of good hope, So home.— I grieve to say time is unlimited, but yet I hope we shall not exceed the 4 years.— No vessel ever left England with such a set of Chronometers,[3] viz 24, all very good ones.— In short everything is well, & I have only now to pray for the sickness to moderate its fierceness, & I shall do very well.— Yet I should not call it one of the very best opportunities for Nat Hist. that has ever occurred.— The absolute want of room is an evil, that nothing can surmount.— I think L Jenyns did very wisely in not coming:, that is, judging from my own feelings, for I am sure if I had left College some few years, or been those years older, I *never* could have endured it.— The officers (excepting the Captain) are like the freshest freshmen— that is in their manners: in every thing else widely different.— Remember me most kindly to him, & tell him if ever he dreams in the night of Palm trees he may in the morning comfort himself with the assurance that the voyage would not have suited him.— I am much obliged for your advice, de Mathematicis. I suspect when I am struggling with a triangle I shall often wish myself in your room, & as for those wicked sulky surds, I do not know what I shall do without you to conjure them.—[4] My time passes away very pleasantly. I know one or two pleasant people, foremost of whom is Mr Thunder & Lightning Harris, whom I daresay you have heard of. My chief employment is to go on board the Beagle & try to look as much like a sailor as ever I can.— I have no evidence of having taken in man, woman or child.— I am going to ask you to do one more commission & I trust it will be the last. When I was in Cambridge, I wrote to Mr Ash, asking to send my college account to my Father after having subtracted about 30£ for my furniture This he has forgotten to do, & my Father has paid the bill, & I want to have the Furniture money transmitted to my Father. Perhaps you would be kind enough to speak to Mr Ash. I have cost my Father so much money.— I am quite ashamed of myself.—

I will write once again before sailing & perhaps you will write to me before then. Remember me to Prof Sedgwick & Mr Peacock.

Believe me Yours affectionately | Chas. Darwin

Postmarks: Devonport NO 15 1831; D 17 NO 17 1831
Royal Botanic Gardens, Kew

[1] The modifications ordered by Robert FitzRoy are described in *Narrative* 2: 17 – 18. For discussion of the reconstruction of the *Beagle* as equipped for surveying work, see Darling 1978, J. A. Sulivan 1979, and Stanbury 1979; for a history of her active service, see Thomson 1975.

[2] The Admiralty instructions, in a memorandum by Captain Beaufort, are reproduced in *Narrative* 2: 24 – 40.

[3] The chronometers (twenty-two, not twenty-four as CD says) and the measurements taken with them on the voyage are described in *Narrative* Appendix, pp. 325 – 31. To superintend and repair the instruments, FitzRoy engaged, on a private basis, George James Stebbing, son of a mathematical instrument maker at Portsmouth (*Narrative* 2: 19 and Appendix, p. 327).

[4] In his reminiscences of the Barmouth reading tour of 1828, John Maurice Herbert wrote to Francis Darwin: 'He had, I imagine, no natural turn for mathematics, for he gave up his mathematical reading before he had mastered the 1st part of Algebra, having had a special quarrel with Surds and the Binomial Theorem' (DAR 112 (ser. 2): 62).

To Charles Whitley 15 November [1831]

4 Clarence Baths | Devonport

My dear Whitley

I received your letter some days after date; & to my grief, you will see by my direction, it is impossible to pay Cambridge a visit for some years to come.— If I had earlier known how many repriefs I should have had, I certainly would have staid in London & then most assuredly I would have come & seen all the good old civilized Phys:ˢ once again in Cambridge.— Long indeed will it be, before I see such a set, as used to meet at your most classical Sunday evenings.—

My feelings overpower me when I think of the simple, the elegant, Glutton club & that day of victory & triumph & inward-glorying, which some call sublime, but the wise know it to be the full round feeling from a contented dinner.— Oh Lord what a jolly place Cambridge is.— But it is all over, so there is no use thinking about it. But I cannot help it; I suppose jolly old Herbert & F Watkins are up there.— I swear I would go without my dinner to sit by & see you three eat one. As for old Herbert, I will beat him in telling lies when I come back, if I dont may all men cry eternal shame on my soul.— I wish you three men the quickest the largest & the best digestions of any men in the united kingdoms.—

The man has just come for the letters & here I have been writing like a confounded fool.— But when I think of you & some few others, I must do one of two things.— either be bonâ fide melancholy & or talk like a fool.— The first would be 'too ridiculous' for after all what is four years, it is long to look forward to, but when once passed what a stage it will make in my life. I have no time for any more, give my very very best wishes to Herbert & Watkins

The scheme is a most magnificent one. We spend about 2 years in S America, the rest of time larking round the world

Remember me to Lowe & all others Especially to old Matthew if you see him.—

If you or any others have time for one line to tell me Cambridge; I shall be grateful for it. I shall be here till end of this month

My dear Whitley.— God bless you | Yours very affect! | C. Darwin

Nov. 15th

Where is Cameron

Postmark: Devonport NO 15 1831
T. H. W. Bower

To W. D. Fox   17 [November 1831]

4 Clarence Baths | Devonport

My dear Fox

I daresay you will be surprised to see my hand-writing, you I suppose thought that I was far on the high seas.— We have had delay after delay & now we do not sail till the 5th of next month.— I always had intended writing to you before making my final exit.— But as that period is so far distant.—I made up my mind to do it earlier.— I do not think I can do better than give you an outline of the Instructions, which came down yesterday.— They leave a great deal to the Captains judgement, & I am sure they could not leave it to a better one.—

We first go to Canary or Teneriffe, then to Cape Verd, Fernando Norunha, Rio de Janeiro, staying about a week in each of these places & rather longer at Rio.— I am very glad that, for I hear there is no view in the world at all equal to it.— Then Monte Video, which will be our headquarters for some time, or rather I should say, the point to which we often return for fresh provisions &c.— From Monte Video, we begin our regular work & go down the coast of Patagonia: parts of Terra del Fuego.—Falklands Islands:— After this done, we have some work about island of Chiloe, & then we are to proceed as far Northward as Captain likes (I daresay to California) so as to leave time to make a good traverse amongst Islands to New S Wales, after which take short cut amongst E Indian Islands, Cape of good Hope, so home.—

Every body, who can judge, says it is one of the grandest voyages that has almost ever been sent out.— Everything is on a grand scale.— 24 Chronometers. The whole ship is fitted up with Mahogany, she is the admiration of the whole place.— In short everything is as prosperous as human means can make it— Time, which no one can alter, is the only serious inconvenience.— Why, I shall be an old man, by the time I return, far too old to look out for a little wife. What a number of changes will have happened; I suppose you will be married & have at least six small children.— I shall very much enjoy seeing you attempting to nurse all six at once.— & I shall sit by the fire & tell such wondrous tales, as no man will believe.— When I think of all that I am going to see & undergo, it

really requires an effort of reasoning to persuade myself, that all is true.—
That I shall see the same land, that Captain Cook did. I almost doubt the truth
of the old truism, that man may do, what man has done.— when I think that I, an
unfortunate landsman, am going to undertake such a voyage,—I long for the time,
when sea sickness will drown all such feelings & that time I do suppose will be
the 5th of next month.— Will you write to me, & tell me, how you are getting
on.— I heard of you at the Music Meeting & that you sat by the incomparable
Charlotte. Report progress. I have hooked her into a correspondence.

Good bye | Yours affectionately | Chas Darwin

17th.

Postmark: Dev⟨onpo⟩rt NO 1⟨7⟩ 1831
Christ's College Library, Cambridge (Fox 45)

From J. S. Henslow 20 November 1831

Cambridge
20 Novr 1831

My dear Darwin,

As I have received the plates to Lowe's paper,[1] I thought it w^d be a pity not
to forward them to you, & so shall entrust them to L. Jenyns who goes to Town
tomorrow to send by some Plymouth Coach— They may be of service in directing
your attention whilst collecting land shells— In working your Surds remember
that you are operating merely on quantities with fractional indices & a little
practice will enable you to see nothing so formidable in them as you seem to
anticipate. Dress them up in this way & then compare them with the symbol $\sqrt{\ }$.

I have long since seen that the noble expedition upon which you are entering
would have been no way fitted for L. Jenyns. With a little self denial on your
part I am quite satisfied you must reap an abundant harvest of future
satisfaction— If I may say so, one of your foibles is to take offence at
rudeness of manners & any thing bordering upon ungentlemanlike behavior, & I
have observed such conduct often wound your feelings far more deeply than you
ought to allow it— I am no advocate for rudeness God forbid, & still less
for any thing dishonorable, but we must make abundant allowancances for
mal-education, early contamination, & vulgar feelings, if we really intend to
pass smoothly through life— & I therefore exhort you most sincerely & affectionately never to feel offended at any of the coarse or vulgar behavior
you will infallibly be subjected to among your comrades— Take St James's
advice & bridle your tongue when in burns with some merited rebuke, & the
impatient feelings which these evils must generate in your own polished mind
will gradually subside & you will be satisfied with the real & stirling worth
you will not fail to find beneath many a rough surface. I have all along preached

to you on the necessity of submitting to evils of various descriptions & when you come back I hope to have my lessons repeated to myself (if I should appear to need them) under the positive advanges of long & experienced trial. I am quite sure that you are the *right* man for the expedition you have undertaken, & that there is in you every thing that is wanted to make it turn out favorably— I am not yet very old, but I have a few more years experience over my head, & I have ever found the advantage of accommodating myself to circumstances. It is wonderful how soon a little submission conquers an evil & then all goes on smoothly.

Believe me ever Y^rs^ affect^y^ J. S. Henslow (Write again)

DAR 204.6.2

---

[1] Lowe 1833. There are six plates: 1—4 of Flora by Miles Joseph Berkeley; 5—6 of Shells by George Brettingham Sowerby Jr.

From Fanny Owen   2 [December 1831]

Woodhouse
Friday 2^d^—

My dear Charles—

In a letter I had from Catherine yesterday, she told me you were still waiting in suspense at Plymouth, & it was uncertain whether you sailed on the 5th or not— I hope *not* that this may reach you in time, for I want to have one bit more chat with you— I thought you were expecting every day to be *y^r^ last* at Plymouth so that if I wrote to you it would be useless or I assure you I should not have been silent all this time— my dear Charles how I do wish you had been *with* us on the awful **22^d^**.[1] I am sure you would *thoroughly* have enjoyed it all—from beginning to end it certainly (tho' I say it who should not) did go off most brilliantly— I was the **Undertaker** and managed the whole affair from *cutting up* of a *Ton* of cake to making *gallons* of *Rum Punch* for the evening's festivities— Susan & I of course you know were the Bridesmaids, and Mr Charles Jones the *Bridesmaid's Man*, about 10 carriages I think composed the Procession to Felton, the *dew Drops* **fell** about 11 o'clock, and I think really every body behaved with becoming fortitude & resignation—as for poor Mama she was wonderful The *Bridegroom* I think was the most *flabbergasted* of the whole party, poor thing I was quite sorry for him, he was as *white* as a *sheet* and as *I supported* him to the *halter* I really expected he w^d^ have fainted, no brandy was at hand tho' he entreated to have some— his stammering he was dreadfully nervous about but got through it all wonderfully the word *ch ch ch ch ch—erish* did stick some time but that was the only one— "As soon as the *ceremony* was over the happy Pair *stept* into their travelling carriage (*green* we are informed) and proceeded with all *possible speed* on a *romantic Tour* to the *Metropolis*, where it is understood they intend to attend all the

Theatres"— you may laugh at this announcement but it is *a* **fact**— did you ever hear of any thing half so unsentimental, the very first night they arrived in Town off they went to Covent Garden, the next to the Adelphi, & so on every night they have been in London— it was so like Sarah—determined to lose no time— but to return to our festivities at home— we had actually 37 people to dinner, two Tables, I President of a side table, and *didn't* I pass the Champagne I never allowed a glass to be empty a moment but before the Cloth was removed all **my** *gentlemen* became so much *more elevated* than those at the other table that I began to be in a fright lest they should expose themselves—but luckily nobody was too much elevated except Mr B. O. who rose to propose the *Ealth of our Ost & Ostess* and caused much fun— very numerous were the healths & toasts, drunk with 3 times 3—and several neat & appropriate speeches, also a beautiful song by that *wild Genius* Mr Crofton, composed by himself for the occasion, and it is much admired— I w^d send it you but think it not worth a double letter— after dinner, the dining room was cleared & prepared for dancing. A most brilliant Ball we had, kept up with the greatest spirit till 5 oclock on Wednesday morning, all the servants dancing. Papa opened the Ball to the tune of "come haste to the Wedding", with Mrs *Kenyon*, can you fancy them?— country dances were the order of the night, and excellent fun we had the only draw back was an occasional dreadful *kick*, from the *too well shod Fantastic* toes of some of the Beaux, but this was to be expected— altogether we had excellent fun, & I do **sincerely** regret my dear Charles, you were not of th⟨e⟩ Party, you might just as well have been as ⟨ ⟩ at Plymouth— How tiresome it is their keep⟨ing⟩ you so long in suspense— Pray my dear Charles do write me one last adieu if you have a spare half hour before you sail— I should like very much to send you some account of us now & then during y.^{r} absence if I knew where to direct to you? You cannot imagine how I have *missed* you already at the Forest, & how I do long to see you again— may every happiness & pleasure attend you dear Charles, and return to us as soon as *you can* I *selvishly* say!— I miss poor Sarah very much it seems like a dream I can hardly yet believe she is really gone!— I have no news really to tell you. I wish I had. Mrs Mytton is staying here at present she came for the wedding, having got her divorce business all settled before, and *her Squire* being safe at Calais, poor thing she is as happy as possible, & we are delighted to have her to keep up our spirits— I must tell you if you should not have heard of it that Edward Williams with all *proper Brother in Law good feelings* presented me the other day with a *beautiful Horse* at Mr Gore's sale— it is nearly thorough Bred, and *beautiful*, quite *perfect* I think & I am delighted with my present— all the fine things from Howell & James's would not have pleased me half as much— I have made the Governor give me a new London *saddle* & *bridle* for old Goldfinder, so you see I am quite a splendid turn out, on coming to my *Title* of Miss *Owen*— that name I must keep up with all *proper dignity*, & what a steady *old sober body* you will find me when you return from your Savage Islands— Miss Fanny Sparling, is going to marry *Dry* **Corbett**!!!² This is said every where as a positive *fact*— he has been living at Felton almost lately & we heard she said she did not know what people could find to laugh at in *Dry*— it is really *too good*—& it is *devoutly* to be hoped they will make a *match* of it for the *diversion* it will afford the Country—which is very flat I think at present—

I hear you like Plymouth very much, I thought it a delightful place when I was there for a few days, there is so much going on—

If you have time write to me my dear Charles— how I do wish you had not this horrible **Beettle** taste you might have staid "*asy*" with us here I cannot bear to part with you for so long—

God bless you my dear Charles excuse my dulness but believe me always | Yours most affec^{tly} | Fanny Owen

Postmark: Shrewsbury DE ⟨4⟩ 18⟨3⟩1
DAR 204.4

¹ The wedding day of Sarah Owen and Edward Hosier Williams.
² Dryden Robert Corbet of Sundorne Castle, Shropshire.

To J. S. Henslow 3 December [1831]

Devonport
December 3rd

My dear Henslow

It is now late in the evening, & to night I am going to sleep on board.— On Monday we most certainly sail, so you may guess in what a desperate state of confusion we are all in.— If you were to hear the various exclamations of the officers, you would suppose we had scarcely had a weeks notice.— I am just in the same way, taken all *aback*; & in such a bustle I hardly know what to do.— The number of things to be done is infinite. I look forward even to sea sickness with something like satisfaction, anything must be better than this state of anxiety.— I am very much obliged for your last kind & affectionate letter.— I always like advice from you; & no one whom I have the luck to know, is more capable of giving it than yourself.—

Recollect, when you write, that I am a sort of protegé of yours, & that it is your bounden duty to lecture me.— I will now give you my direction: it is, at first, Rio; but, if you will send me letter on first Tuesday (when packet sails) in February, directed to Monte Video, it will give me very great pleasure. I shall so much enjoy hearing a little Cambridge news.— Poor dear old Alma Mater. I am a very worthy son in as far as affection goes.— I have little more to write about. I shall be very glad to have some memorial of Ramsay.— My very short acquaintance with him appears like a dream,—which has left many melancholy yet pleasant recollections.—

I cannot end this without telling you h⟨ow⟩ cordially I feel grateful for the kindness you have shown me during my Cambridge life.— Much of the pleasure & utility which I may have derived from it is owing to you.— I long for the time when we shall again meet; &, till then, believe me, My dear Henslow | Yours affectionately & obliged friend | Chas. Darwin

Remember me most kindly to those who take any interest in me.—

Postmark: Devonport DE 4 1831
Royal Botanic Gardens, Kew

From Caroline, Catherine, and Susan Darwin 20 — 31 December [1831][1]

Wednesday December 20th.

My dear Charles—

It is now exactly a week since we rec^d your letter telling us you had been oblgd to put back to Plymouth after the stormy night & miserable 24 hours you had passed.[2] I was quite astonished how you could write so cheerfully after such suffering. I hope it served for a great piece of the seasoning you must go through before you are proof against sea sickness. Your account of the Captain was quite sublime— it was the Red Rover's[3] own still quiet manner & "low distinct tones" heard through all the uproar, & Papa's eyes were full of tears when he thought first of your miserable night & then of your goodnatured Captain in all the confusion paying you a visit & arranging your hammock— he must be quite a Captain Wentworth every thing you tell us of him makes him more & more perfect. Erasmus came home about an hour after we had your letter & was all astonishment to find we had later intelligence of you than he had. He said the Saturday night was quite tremendous at Devonport & his bed rocked almost as much as if he had been at sea— Eras was thoroughly wearied with his journey home & he contrived to see nothing the Coaches always getting into the towns, Bath, Worcester &c. just when dark. I was surprised he did not stay a day at Bath to look about him more particularly as Paganini was playing there so that he would have had a very nice amusemt for his evening. We made him tell us every thing he could about you, 'the Beagle' & your companions & I can fancy you very well seated in your snug corner to your table. he made us laugh too with his account of your delivery of a book & the run you were oblgd to take afterwards to escape the General— I will not go back beyond the time of Erasmus coming here as I hope you rec^d a letter I wrote telling you of the Wedgwoods being here & the birth of another little nephew to the great disappointment of Marianne & Doctor Parker. I have not yet seen the baby but they tell me (of course) it is a nice little creature & it is to be christened after you & D^r P's brother, making its appearance the very day of your sailing— This last week has felt to me very melancholy. I have been able to think of little besides you & though I hope & believe all you see will be great enjoyment, the selfish feeling of the great separation from you & the *long, long* time it must be before I can hope to see you again, is painful enough— in short, dear Charles you will be properly valued when we do see you again. I find Eras has hopes that you will return at the end of two years, & how we shall rejoice if his prophesy proves true— Papa is very well & I think every day gets more interest in the Hothouse which he is constantly going to see & I expect he will find it a very nice amusement and little occupation— Our days pass much as usual, cards in the eveng & after Papa is gone to bed Eras, Charlotte & we draw round the fire & have an hours cosy talk together & as I said before, generally about you— on (Monday Dec.ʳ 18.ᵗʰ)⁴ Susan and I sallied forth to pay our wedding call at Eaton,⁵ M.ʳ & M.ʳˢ Williams & M.ʳ and M.ʳˢ White having arrived there a few days previously— a few miles from Shrewsbury we met Sarah *alone in her carriage*. She was coming to see us, having for the first time she said "rung the bell" & ordered the carriage & she seemed quite proud of having ventured upon such a step of authority— we got into the carriage to her & had a very merry drive with her to Shrewsbury— she made us but a short call & asked us to dine with her the next Monday which Susan & I promised to do which we did, & got there about an hour before dinner. We found Fanny & Sarah alone in the Drawing Room. Fanny had we found insisted upon Sarah not running out to meet us in the Hall—but made her sit still "*to keep up her dignity*"— you cannot think how pretty & nice they looked when we came into the room, both looking so gay & happy— M.ʳ & M.ʳˢ Bruce, M.ʳ Edward Hanmer, & Henry Hill, dined there— as soon as dinner was over Sarah & I sat together alone more than an hour & she was so open & affectionate that I hardly ever felt fonder of her. Edward Williams was in very good looks, thanks Sarah thought to a beautiful black satin neck cloth she makes him wear, but I think owing to his looking so very happy. I never before had talked much to him & was quite surprised to find how pleasant he was, & moreover he never once stammered— Sarah looked particularly lady like at the head of her table & did the honors very nicely. I believe there has been but one jealous fit since the marriage & Fanny said Sarah behaved exceedingly well— She was half frightened after telling us of it whether she had not done wrong, & I am glad to find from different things that they do not intend to talk of his faults now he is married to Sarah— All the neighbourhood have been calling at Eaton, the Corbets of Sundurne & Acton Reynald,⁶ & many other families who never before would visit the Williams— I find M.ʳ Bruce's brother has been almost as great a traveller as you are going to be—having passed seven years in Arabia, Egypt, Nubia &c these Bruces are related to *the Bruce*⁷ which I did not know before. M.ʳ Bruce asked a great many particulars about you & begged I would tell you what happened to his brother who after 5 years travelling, & collecting & writing, was shipwrecked & all his papers destroyed & who never ceased regretting that he had not kept duplicates and sent off his journals & papers by every safe opportunity—he hoped you would profit by the hint. We finished the evening by a gambling game of cards— the Bruces have taken a house in London & all the brothers are gone away & I am happy to think Eaton is now clear of all the relations— the next morning Fanny came home with us & took Susan & Catherine on with her to Woodhouse for an Ellesmere ball & now I shall stop my letter for some days I never mean to make any apology for writing *Twaddle* for it is the only way you will know all or any thing about any body so goodnight dear Charles— in your first delight at all you are seeing at Rio you will hardly find time to read all I am writing—

Christmas day 25th— We have all been thinking of you & wishing you a happy Xmas, which to be sure is rather a farce *knowing* as we all do that you are pitching & tossing & as sick as may be— I was never more astonished when sitting painting quietly & believing you to be half way to the Madeiras to receive your two letters from Devonport & such long, interesting, agreeable letters. I am exceedingly glad you feel so much at home & comfortably established in the Beagle & if you are reconciled now to the confinement, so soon after going on board—I think with your Captain, & that you must fully deserve his compliment of being the "best of shore going fellows"— Dont be at all uneasy that you will not have letters enough— I dare say *every body* at home will *determine* they will write & I do not doubt they will—but *I* do not mean *to trust* to any one of them & unless I know for a certainty & have seen a letter to you put in the Post office in time for every Packet that sails I shall write myself if it is only half a page rather than let you miss having a letter & though I am careless in some respects, in any point of importance I can not reproach myself with *ever forgetting* what I have determined & promised to do, so you may be certain at any place you do not get a letter when you expect one that it will *not* be from our having forgotten to write, but from its having miscarried— Mr Towers tells us that Government Packets sail the first Tuesday in every month & I shall send of this letter by January Packet. remember how long the communication must be between us & give directions in proper time how long we are to continue writing to Monte Video & where afterwards— & now I want to beg a favor of you—to keep a letter written of three lines just to say you are well—ready to be sent off at any opportunity that may occur—for I suppose sometimes you may have an opportunity of sending a letter by some other vessel, when you would not have time to write & when we are feeling anxious & hopeless of a letter arriving I can hardly fancy greater pleasure than the very sight of your handwriting—though literally your letter only contained the words *you were well.*

Nothing has happened since I put my letter by— Charlotte is very agreeable but rather grave I think— Eras & she have a good deal of talking together but we think they both the poor creatures often look miserably shy—poor Charlotte blushes & colours to her fingers ends & Eras thinks she must have done so at Maer & *so innocently* have raised your never-to-be-forgotten *impertinence* which so petrified me— perhaps you will think this solution arises from Erasmus's jealousy & envy & I think it does seem rather suspicious— Fanny & Emma came here on Friday. Emma is looking very pretty & chats very pleasantly but I am still steady to my old friend Fanny—

A letter to my Father came from John Price, it ought to have been directed to Susan it was so full of messages to her & reproaches to her for having written only one side of note paper to him lately, when she was oblgd to write some message or communication for my Father to him. he supposed she thought it "highly improper to write 2 sides of note paper to a Bachelor &c". & very angry that she had not told him of your plans. Susan is furious at "his impertinence" Erasmus was ordered to write—he refused saying he would have nothing to do with "Lovers quarrels" you may guess how indignant Susan was & it has ended in my being victim—but I think I will so write that our correspondence may end where it begins—

Dec 29ᵗʰ. Dear Charles I have just had your letter of the 27ᵗʰ. on the point of sailing & I must say a few words to you for after reading any of your charming letters I never can read for very long afterwards & usually take a solo walk & think of you. I delight in your enthusiasm & can fancy no more vivid enjoyment than when you first land, see your "glorious sun", & exquisite vegetation— On Tuesday I got your dear affectionate letter by the early post & in the hopes of an answer reaching you wrote a short letter by that return of Post which I suppose after paying the dead letter office a visit will return to me in due time. I will not finish my letter till the day it must go that you may have the latest news.

Decʳ 31. I got your second letter on Thursday written when the "anchors were weighing, sails, unfurling"—& as my letter written on Tuesday returned this morning with (The Beagle has sailed) on its back—I suppose you are now really off. Eras, Charlotte Fanny & Emma all went off on Monday just filling the coach & a very pleasant merry party they must have made. they slept at Birmingham the first night. I have heard since from Eras who declares Dʳ Holland is very attentive to Charlotte & he is quite convinced has serious intentions. I feel very sure *Charlottes intentions* are equally serious & certainly will not have him—if she did but esteem & like I think he really would be a very affectionate husband, but that is match that will never be—

All Shropshire is Gossiping now at a quarrel between Dʳ. Dugard & Mʳˢ. Hill— Dʳ D. declaring Mʳˢ H promised him a certain sum (£1000 a year it is said) if he brought the marriage to bear between Sir Rowland & Miss Clegg,⁸ having made the promise before Mʳˢ Dugard & one of the young Hills, which she now denies & says "the finger of God" brought it to bear. Dʳ Dugard threatens to prosecute Mʳˢ. Hill in the Kings Bench for defamation for what she says & told the whole story a few days ago—[to] Sarah (Mʳˢ Williams), at a dinner party. On the other side, the Hills are furious say it is all false.

D.ʳ D. trying to get money—& they tell the story every where. every body are curious how this disgraceful affair will end.—

Susan and Cath want this flap to write a few lines to you, but their long letter will go by February packet directed to Rio unless in the mean time we have fresh direction— If you write to Eras, shall you direct to his Club? as he talks of leaving his Regent S.ᵗ lodging—but he has given you direction I dare say— I have been reading all morning Beecheys voyages & have at least double interest in his account of the South Sea islands from knowing you will visit them (first page)

I have read & *re*read so many times your letters that I know them by heart, *almost* I should say for you & I agree our memory is not our strongest point— I have just seen my Father he sends you his kindest love. I wish you had seen his pleased feeling look when I read him yʳ affectionate message.

Good bye my *very very* dear Charles | Yʳˢ very affect.ˡʸ | Caroline Darwin

Nancy was so pleased by your so kindly recollecting her She says she does not know what it would be proper for *her* to say *to you*

My dearest Charley. I have got this little scrap out of Caroline's enormous letter, just to tell you how often I think of you, and how very very much I hope you will have liked your voyage as much as can be expected, and when you get this, you will be beginning really to enjoy yourself, and see the wonders of the world. I can hardly believe I am really writing to you in South America. what will you think of this little speck, Shrewsbury, when you come back? You cannot think how interesting we shall find your letters—*we depend* on your writing as often as you have opportunity, it will be such a comfort to us to hear often from you. Your letters were very interesting even from Devonport; I long for your first account of Madeira. I see Beechey got to Teneriffe in 12 days. M.ʳ Owen talks so much of you, and with so much affection and feeling. I wish you could see his countenance about you. I have been several times at Woodhouse lately, for Balls. *I* will write more to you, I am determined, the 1.ˢᵗ of February. Goodbye, my dearest Charles, | Catherine.—

My dear Charley I cannot let this go without writing my own love & telling you how I long to have yʳ first Madeira letter it will be so very interesting. I am very sanguine about yʳ enjoyment as I think you are like myself & always look at the best side. I shall commence on a folio sheet in a very few days for the 1.ˢᵗ of Feb & now I must bid you Good bye wishing you a most happy prosperous New Year which begins tomorrow Ever Dear⟨est⟩ C yʳ most affectionate Susan Darwin

Postmark: Shrewsbury DE 31 1831
DAR 204.6.1

[1] The Wednesday before Christmas in 1831 was the 21st of December.

[2] Gale winds had forced the *Beagle* to return to Plymouth on 11 December. 'I suffered most dreadfully; such a night I never passed, on every side nothing but misery' (*'Beagle' diary*, p. 13).

[3] James Fenimore Cooper's novel of the sea, *The red rover, a tale* (1827).

[4] This must refer to Monday, 12 December, since another Monday (the 19th) intervened before the letter was begun.

[5] See letter from Sarah Owen, 31[December 1827], n. 3.

[6] The Corbet family of Sundorne Castle, Shropshire, (see *Burke's Landed Gentry 1879*) and the Corbet family of Acton Reynald, Shropshire (see *Burke's Peerage 1980*).

[7] Robert de Bruce VIII, King of Scotland.

[8] Dr Thomas Dugard was physician at the Shrewsbury Infirmary (*Gentleman's Magazine* n.s. 14 (1840): 556). Sir Rowland Hill and Anne Clegg, 'with whom he had a large fortune which he dissipated', were married 21 July 1831 (*Complete Peerage*). See also letter from William Mostyn Owen Sr, 1 March 1832.

From Catherine Darwin 8 January — 4 February 1832

Shrewsbury.
January 8th 1832.

My dearest Charles.

I think you will get Caroline's letter, which went off on New Year's day, the same time as this, but you must read her's first, that you may hear events in order. What a pleasure your first letter will be to us; you cannot think how I long for it— I assure you that I think of you, almost as much, as I am sure you must think of all of us. I feel like Ellen Tollet, that I will bear your long absence *if I can.*— I shall be very anxious to know how you continue to like all the Ship's Company, and especially the inimitable Captain, in short, every thing, I long to hear about you.—

I must now begin and tell you Owen news, of which there is some very surprising, and extraordinary. Caroline spent two days at Woodhouse this week; she thought she should find them quite alone, & quiet, and what was her surprise, on entering the room, to find Mr Biddulph[1] settled there.— You will be as much astonished as Caroline was, when Fanny took her out of the room, and told her that she was engaged to Mr Biddulph; he had proposed a few days before and been accepted, in the course of a secret ride, Fanny meeting him at the Queen's Head.— You may imagine how amazed we were, when Caroline came home, and told us; and I may add how grieved I was, when I thought of his dissipated, gambling character, though I am rather more reconciled now, as every body agree he is very affectionate, and *now* talks of spending great part of the year at Chirk Castle[2] quietly.— I do not think Fanny cares for him half as much as she did for John Hill;[3] but she is so exceedingly annoyed now, at the prospect of what Sarah will feel, when she hears it, that she thinks more about that, than she does about Mr Biddulph. M^r and Mrs Owen and all the family are very much alarmed about Sarah; they say she will be so dreadfully mortified, and so tremendously angry with Fanny, as she will of course fancy that Fanny was treacherous to her, and tried to attract Mr Biddulph, at the time of Sarah's flirtation with him. This is perfectly untrue, as Mr Biddulph declares, that his attachment to Fanny entirely arose from seeing Fanny's distress at the time of John Hill's desertion;—he was so charmed with her feeling and crying then, that he resolved he would try if he could not make her care as much for him; & from that time, his great anxiety was to shake off Sarah.— Mrs Owen is staying at Eaton now, to break it to Sarah; another and a great anxiety, as you may imagine, is that Mr Edward Williams should not find out that Sarah feels or cares about it. I think you will perhaps have a letter from Mr Owen by this Packet, as he said he would write to you, and I have no doubt, Fanny will write in it.— Your Portmanteau arrived safe the other day.— I must tell you that when Mr Biddulph first began to pay attention to Fanny, Mr Owen was so afraid that she should be "*blown upon*" again as Sarah was, that he woke Mrs Owen in the night, and declared, that if Mr Biddulph was only flirting again, he would call him out "and if I fall I shall leave my dying request to Owen, immediately to call him out again.—" Mrs Owen was much amused as you may imagine, at his midnight bloody thoughts.

January 29th I must go on with my letter, and tell you all that has happened since I wrote; I know how much you care for Owen news. Mr Owen was the person who broke it to Sarah, and between being so sorry for her feelings, & so frightened, he actually cried very much, when he told her; however Sarah bore it very much better than any body hoped, and though she did not speak to Fanny for a day afterwards, she then shook hands with her. Mr Williams is happily kept quite in the dark, and Sarah really seems exceedingly attached to him, and he violently in love with her.— I don't think Sarah feels much about this marriage, beyond anger and resentment at Mr Biddulph, having made her a "*cat's paw*" last Winter; and of course she will get over that in a short time.— Caroline and I have been staying with Sarah one night, and Sarah says she will fulfil her old promise, and write to you. I begged she would, and we will give her your direction. I wish you could know how much Mr Owen talks about you, and how nicely and affectionately he always speaks about you; I quite love him for his feeling about you.— Mr Owen positively refuses to take back the Cloak, so it must be kept here in the Laundry till you come back.— Fanny's marriage is to take place in March, I believe. They go first to Chirk Castle for a week, and then to London for the Spring.— I have been staying at Woodhouse some days, while Mr Biddulph was there, and certainly thought him very agreeable, and I cannot help hoping that with such an attaching wife as Fanny, he will reform, and become tolerably domestic, and I am in pretty good hopes about that dear Fanny's happiness. You will find her a *motherly old married woman* when you come back. I hope it won't be a great grief to you, dearest Charley, though I am afraid you little thought how true your prophecy of "marrying and giving in marriage" would prove.— You may be perfectly sure that Fanny will always continue as friendly and affectionate to you as ever, and as rejoiced to see you again, though I fear that will be but poor comfort to you, my dear Charles.—

Papa is quite well, and very fond of his Hothouse, which is finished, & very perfect, and some plants in it.— I do hope you will receive a good packet of Letters by this Packet. Erasmus and Charlotte will certainly write to you. Charlotte's letter will very much surprise you, as it has every body else. Only think of Charlotte's being going to be married after only a fortnight's acquaintance, with a man, who was a perfect stranger to all her family. Charlotte's letter too will give you an account of Hensleigh & Fanny Mackintosh's marriage on the 10th of this month.— England is gone mad, with marrying, you will think. It is Leap Year, you know when the Ladies take *their* turn of proposing.

I believe Caroline gave you an outline of the beginning of the Quarrel between Dr Dugard and the Hills, which has caused such a sensation in Shropshire.— Dr Dugard has finished the Comedy now, by signing a formal recantation of every thing he has said against the Hills. Whether he has eat his own words, under the bodily fear of the Horsewhips of Capt & Major Hill, or whether some idea of his own Character came over him, is not known.— Altogether it was an extraordinary story, that interested Papa most exceedingly.—

Do tell us whether you get the Papers, to tell you any Public news; you will see that we are to have a general Fast Day, though on what account, I don't exactly know, as the Cholera has almost completely died away.—[4] You must read the melancholy account of poor Colonel Brereton putting an end to the Court Martial on him, by shooting himself through the heart. It is said that he would certainly have been broke, if he had lived.— This Capt Warrington, whose Trial is going on now, runs a very bad chance; he might have got off his Trial, it is said; but he insisted upon standing it, since Col Brereton's death, as he fancied Col Brereton would have been the principal evidence It is said he will certainly lose his commission.—[5]

Harry & Jessie Wedgwood are here now; they came on Wednesday, and are very pleasant.— We heard the other day from the John Wedgwoods. Aunt Jane & Eliza begged us to send you their very best love and good wishes.— You will not much care, I guess, for their wishes.— If good wishes could be of use to you, you would have plenty of them; I am sure from many people in Shropshire, where every body liked you, and most loved you. What pleasure it will be to see you again, pleasure greater than anything else can give me, I am sure.—

I have been reading the account of the Mutiny of the Bounty, in the Family Library.[6] You will see Pitcairn's Island, perhaps. It gives such an excellent account of the goodness & religion of the people there, that I was very sorry to see, by an additional note, that the Missionaries had carried them off to Otaheite, where they will get depraved by those horrid Otaheitans. I was so much interested by the account Erasmus of the Sailor Missionary,[7] you have on board. It will be an extraordinary thing if his enthusiasm lasts, when he has seen the Country again.—

Goodbye, my dearest Charles. Papa's & every body's most affectionate love.— God bless you, and pray remember, if you love us, take every care of yourself, and your health.— | Ever, dear Charles your most affectionate | E. Catherine Darwin—

February 4th—

DAR 204.6.1

[1] Robert Myddelton Biddulph.
[2] Welsh border castle near Denbigh, North Wales. The seat of the Myddelton Biddulph family.
[3] On Hill's intention to propose to Fanny Owen see letter from Susan Darwin, 12 February [— 3 March] 1832.
[4] In 1831 the cholera had been confined mainly to North England and Scotland; in Newcastle alone there were 934 cases, of which 294 were fatal. As the epidemic spread, the King issued a proclamation directing that the 21st of March be observed throughout England as a day of fasting. For an account of the epidemic see *Annual register*, 1832, p. 47.
[5] Lieutenant-Colonel Thomas Brereton was tried for having failed in his duty to protect Bristol during the riots of 1831. Captain Warrington was found guilty of failing to order his troop out against the rioters. (*Gentleman's Magazine* 102.1 (1832): 84, 171).
[6] Barrow 1831.
[7] Richard Matthews, sent by the Church Missionary Society to accompany the Fuegians and to establish a mission at Tierra del Fuego.

From Charlotte Wedgwood 12 January — 1 February 1832

Dulwich
Jany 12— 1832

My dear Charles

I think it will be a very good opportunity to begin a letter to you, having a quiet hour to spend here before we return to that most idle & bustling of houses, Roehampton.[1] I believe this will be at Rio long enough before you without allowing for your being blown back two or three more times before our shores are fairly quit of you, but that is no matter & I cannot afford to let slip the opportunity of a wedding in the family as at the rate they have hitherto gone on at there is no chance of another occurring before you have finished your voyage round the world—as they are such very rare occurences it is a good plan that we are upon of doing up two members of the family at once tho on the other hand cousins marrying is a very humdrum affair & affords very little interest or entertainment— I am very much of Fanny M's maid's opinion, who being asked what she thought of her lady's marriage said "Well ma'am I think it wont make much difference". I am very glad it is over—they must feel so comfortable & at leisure now that the disagreeable interval since the time it was fixed is passed which was filled with nothing but tiresome settlings, moving of houses, & all sorts of plagues & Hensleigh between them & his new Magisterial duties was beginning to look worn out—[2] he had been very unwell all the week & was in bed Sunday & Monday morning being to be married on Tuesday & he thought himself so bad that he had written a note, to say that the marriage must be put off a couple of days when luckily his doctor arrived, told him he was quite well, recommended him some mutton chops & wine which so restored him that his note was burnt, & he appeared at dinner time in very good condition for his execution next day—the only serious consequence of his taking to his bed was his writing to Fanny that she must get the wedding ring, an indignity that I should suppose had never been put upon a bride before— however she was obliged to submit & sent out one of the Thorntons[3] for it— this was not the last indignity she was obliged to submit to neither, for first her gown did not arrive in time & she was obliged to strip one of her bridesmaids & be married in a borrowed one, & still worse she had to wait what seemed a long time in the church before the bridegroom made his appearance, & we began to be afraid he had taken to his bed again—however he appeared at last in very good case & accounted for the delay by his having a pair of hearse horses, 'a bad omen, and having to set down Judge Alderson[4] at his chambers, & bad as this beginning was, for the rest of the time he cut a very good figure & he & the bride both took off their spectacles for the ceremony. There were eight carriages the servants with enormous favours which brought us a rebuke from the superior taste of an old dirty woman in the crowd, who said, "Well if she had been going to be married she would have kept those things out of sight & not collected a crowd about her" There was a grand breakfast afterwards at which Lady Gifford presided 42 at table consisting besides all the branches of the family, of Thorntons innumerable besides a few other friends Before we went to church Sir James made me stuff a vol: of a new novel into his pocket I did not see what opportunity he had of reading it.

Jan 29th My letter has been lying by a little more than a fortnight— how little I thought when I put it by what would be the next piece of intelligence that I should add to it—nothing less than that I am engaged to be married— I am afraid you will think part of what I wrote at the beginning of this very deceitful, but I do assure you it was not— I had not the least notion then of what was going to happen to me & that I should ever be married seemed to me the most improbable thing possible. You will have most likely heard this news in some of your other letters but you will like to hear more about it from me. When Emma and I arrived at Roehampton we found Mr Charles Langton staying there— he is nephew of the Mr Langton[5] who married Marianne Drewe and is guardian to his son Bennet who always spends his holidays at Lady Giffords and it was to be with him during his holidays that Mr Langton spent this Christmas at Roehampton. He is a clergyman but has no living & has only a very small income now, but he was tutor to Lord Craven who has many livings in his gift &

he has no doubt that he shall have one of them—[6] he has also a rich grandmother so that he will be well off in future tho he is poor now. Some of Lord Craven's livings are in Shropshire & in very pretty parts of Shropshire—this will be delightful for me if M[r] Langton ever gets one of them—to fall by chance so near home & Shrewsbury would be high good luck. For the present we are going to take a house in Surrey near Guildford—it will be very pleasant to be within reach of London. Emma likes M[r] L almost as much as I do & was delighted when he proposed to me, which I tell you because you will think her a more impartial judge than me. I looked forward to seeing you established in your parsonage but now I suppose I shall receive you first in mine. I think it is the happiest life in the world & I hope dear Charles that we shall hereafter compare notes upon it when we have both tried it & found it as happy or nearly as happy as we expect. In looking forward to it myself & thinking of its advantages I feel more anxious that you should finish all your wanderings by settling down as a clergyman but it must be as a really good active religious clergyman, (you know you gave me leave to preach) in that only can the happiness consist, & if I did not think M[r] Langton would be all that, I think I would rather he were any thing but a clergyman. I feel a delightful trust in his high principles & kind nature which gives me a feeling of security that I have done what was wise as well as what was agreeable— it seems a very short time since I first saw him for me to judge so confidently of him & yet I do not feel the less secure for that. I am sure that in one respect being going to be married is very like going a voyage round the world—it makes one love all one's friends more than ever & it also makes one find out more affection in all one's friends than one ever knew of before— I have been receiving delightful letters from all mine, & from none more delightful than my dear Caroline— I shall hope to add one from you too to the list I like hearing from you very much, & a letter from you will always be a welcome sight tho with your many correspondents I do not expect it often. I forgot to say that our marriage will be in March. I am sorry I cannot send you any Maer news— My father has had nobody with him for some time but Jos & Frank— Frank sends us very pert acc⟨ounts⟩ of poor Papa's waterworks, & describes the cunning of the pipes and the wo⟨rks⟩ in always leading him on to fresh trouble & expense by a shew of success which regularly disappears as soon as their point is gained. My mother & Elizabeth return next week from Cresselly[7] to keep him company & very soon after we shall all return home. We have left Roehampton and are now staying partly with the Mackintoshes & partly with the Aldersons. The Sismondis & Fanny Allen are in town too so that we are a large family party. We had a very pleasant visit with Harriet, she is so thoroughly goodnatured & so eager to do every thing to make it pleasant. Erasmus is watching for a good play for us— we have been to one or two & all that we have done besides is dinners with the friends of the clan, particularly at D[r] Holland's who is risen exceedingly with us all & our conscience reproaches us for all that we have said against him— I think he really is improved & I am sure he is very friendly.

(Feb 1ˢᵗ) I hope it will not be long before there is some news from you from some of the islands you are to touch at— I shall like to hear your first impressions of tropical climates & that you are safe so far will be very pleasant to hear— I am surprised how little of an annoyance being cooped up in your little brig appears to be to you, dont take fire at my speaking of it in such terms but in spite of all your admiration at that lovely little vessel I cannot help feeling that it would be very difficult to get so completely rid of all one's notions of comfort as to be reconciled to all the confinement & discomfort of being on board her. I think you must feel that it does you a great deal of good to be made not to mind things, & by the account you give of the ship's company which Caroline shewed me it seems to have just the effect one should expect, of making the officers unselfish & good humoured Mʳ Langton was a midshipman for one year but did not like it— I am afraid you will certainly think less of him for being such a "shore going fellow"

It is pleasant to see Hensleigh & Fanny so happy as they seem—they find it much pleasanter than the hanging on state they were in so long. She received yesterday her first grand dinner party, & looked so elegantly dressed & received her company so gracefully that Hensleigh must have been proud of his wife. The lions of the party were two banished Poles, the Brahmin Ramohun Roy[8] & Sydney Smith. The Judge & Georgina with whom I am staying have been very cordial to me & G is risen again very much with me. Goodbye my dear Charles I hope I shall soon hear some news of you from Shrewsbury— God bless you & bring you safe home again Believe me yr very affc | cousin Charlotte Wewd

DAR 204.6.2

[1] The residence of Lady Gifford, née Harriet Drewe, was in Roehampton. She was related to the Wedgwoods, Mackintoshes, and Darwins through her mother, Caroline Allen.

[2] Hensleigh Wedgwood's appointment to a police magistracy in December 1831 made possible his marriage to Fanny Mackintosh (*Emma Darwin* 1: 242 − 3).

[3] Friends of Fanny Mackintosh (*ibid.*, 1: 186).

[4] Edward Hall Alderson, married to Georgina Drewe, Lady Gifford's sister.

[5] Algernon Langton.

[6] William, 2d Earl of Craven. In December 1832 Charles Langton obtained a living from him as Rector of Onibury, Shropshire (*Emma Darwin* 1: 254).

[7] Cresselly in Pembrokeshire, the family seat of the John Bartlett Allens.

[8] Ram Mohan Roy, an advocate of social and religious reforms in India, was lionised by English liberals.

From Bartholomew James Sulivan [17 January − 7 February 1832]

[Heights of various places on Quail Island (Cape Verde Islands).]

AmemS
DAR 32.1: 16a

CD ANNOTATIONS
at beginning: 'Quail Isl^d' *underlined*
at end: 'Another measurement 40'
on separate slip: 'N.B. All the beds dip in the direction of the diagonal line: so that the upper covering of
Lavas becomes much thicker in S end of the Island' *ink*

From J. S. Henslow 6 February 1832

Cambridge
6 Feby 1832

My dear Darwin,

As tomorrow is the first Tuesday in Feby I select today (my Birthday) for
keeping my promise, virtually made by your asking me to write to you on that
day. I heard of your adventurous departure thro' Mr Yarrel who was told of it
by Capt. King— You had a stout heart to resist the inclination which must
necessarily have come over you not to go on whilst you were in such a wretched
state of sickness as you are described to have suffered I trust however that it left
you soon afterwards & that you are now an experienced sailor— As you cannot
yet have quite lost all zest for Cambridge information I shall tell you a little
about the late examination, which seems to have been rather a merciless one, for
besides that some thirty men were frightened after the first day & cut & run,
there [were] 29 plucked, & among them Lord Sandwich & a nephew of L^d
Grey's at Trinity—& an Hon. DeGrey at St Johns— You see we are becoming
quite radical.— You will see by the papers which I suppose you get somehow or
other that Trin. had the Sr Wrangler (quite unexpectedly)— I was absent from
Cambridge myself on a visit to my Father's—spent two nights in London very
pleasantly with attending at the Geolog^l & Linnean—where I heard of you
from diff^t quarters among some of the members. Pray are you yet a Whig? for I
heard from Wood that your Brother told you it was impossible to touch pitch &
not be defiled.[1] Whatever you become I know it will be from honest conviction,
& therefore tho' I shan't change my principles myself I shall be quite content to
allow you to change yours without thinking the worse of you for so doing—not
that I suppose you will care if I should— Only as we used to agree in these
matters we will now agree not to let our disagreement (if it s^d turn out so)
trouble us— We have determined on erecting a mural Tablet to Ramsay's
memory in the Chapel of Jesus, It will be plain—with a medallion by Chantrey[2]
who has undertaken to assist us— I have a most admirable miniature likeness,

which impr⟨ess⟩es every one who sees it painted by Miss Jenyns from memory, having merely the outline to guide her from an old miniature taken by an instrument— I intend (at the request of his friends) to get it engraved for them, & as you may possibly like a print will select one which you may take or not as you like— If it is to be done by a first rate artist—will cost the subscribers about 10/6.—any surplus to be distributed to the poor— Worsley[3] has resigned the Tutorship of Downing— He was absent all this term & I acted for him as Chaplain & dinner-eater on Sundays & other days when Dawes was absent—[4] He thoroughly disliked the duties of a College life & has taken a wise step. Sedgwick is having his picture taken!! by Phillips[5]—to match a portrait of his of Buckland.[6] He makes the artist laugh so much that he can hardly get on with it— The two will be engraved as a pair— Whewell[7] has resigned his Professorship from want of time & I believe Miller[8] will succeed him— Whitley is a candidate for a Professorship of Mathematics in the new College about to be established in Durham— I can't recollect any other Cambridge news worth recording— My own household is flourishing— M⟨rs⟩ H. much improved in health & the Children blessed as usual— We had so amply discussed your prospects at T. del Fuego that I do not know what to say more till I shall have heard from you— I sometimes blame myself for having hinted to you rather plainly little pieces of advice, lest you should have thought me troublesome— but I am sure your good heart will ascribe my suggestions to the right motive of my being anxious for your happiness—which cannot be enjoyed in this troublesome world without daily restraint & submission to mortifications sometimes trifling sometimes grievous—always when patiently taken refreshing to the spirit— I feel the more anxious for you as I have been so mainly instrumental in your adopting the plans you have—& should your time pass unhappily shall never cease to regret my having recommended you to take the step you have of devoting yourself to the cause of science— Much therefore as [I] should like to see you return laden with the spoils of the Worl⟨d⟩ yet if you do not find yourself *content* I should much rather see you sooner than I hope to do— You must by the time you get this have had ample experience how far you are qualified to cope with difficulties & whether you can rise superior to them—whether you can enjoy yourself amidst them, & rejoice over them— If you have met with success hitherto be assured that you may go on safely & securely ⟨to⟩ the end—but if you have failed, then don't try any more—but come away— You will only be heaping up greater troubles— I shall endeavor to keep up our correspondence as you may be pleased to direct me how I am to succeed in getting my letters to you, & shall always write to you as freely as I can on the subject of your enterprize, judging from what I can learn from your letters may be the state of your wishes—

Believe me y⟨rs⟩ ever affectionately & sincerely J S. Henslow

[1] A reference to Robert FitzRoy's staunch Toryism.

[2] Francis Legatt Chantrey. The medallion of Marmaduke Ramsay was, however, executed by Joseph Theakston, who, besides producing work of his own, was employed by Chantrey.

[3] Thomas Worsley. He was elected Master of Downing in 1836 and served until his death (*Alum. Cantab.*).

[4] Richard Dawes.

[5] Thomas Phillips. The portrait is reproduced facing p. 499.

[6] William Buckland.

[7] William Whewell had been appointed Professor of Mineralogy in 1828 following John Stevens Henslow's resignation of the Chair in 1827 (see Winstanley 1940, pp. 40−1).

[8] William Hallowes Miller succeeded Whewell as Professor of Mineralogy and held the Chair until 1880.

To R. W. Darwin 8 February − 1 March 1832

(Brazils) | Bahia or St. Salvador

My dear Father

I am writing this on the 8th of February one days sail past St. Jago, (Cape De Verd), & intend taking the chance of meeting with a homeward bound vessel somewhere about the Equator.— The date however will tell this whenever the opportunity occurs.— I will now begin from the day of leaving England & give a short account of our progress.—

We sailed as you know on the 27th of December & have been fortunate enough to have had from that time to the present a fair & moderate breeze: It afterward proved that we escaped a heavy gale in the Channel, another at Madeira, & another on coast of Africa.— But in escaping the gale, we felt its consequence—a heavy sea: In the Bay of Biscay there was a long & continued swell & the misery I endured from sea-sickness is far far beyond what I ever guessed at.—[1] I believe you are curious about it. I will give all my dear-bought experience.— Nobody who has only been to sea for 24 hours has a right to say, that sea-sickness is even uncomfortable.— The real misery only begins when you are so exhausted—that a little exertion makes a feeling of faintness come on.— I found nothing but lying in my hammock did me any good.— I must especially except your receipt of raisins, which is the only food that the stomach will bear:— On the 4th of January we were not many miles from Madeira: but as there was a heavy sea running, & the Island lay to Wind ward it was not thought worth while to beat up to it.— It afterwards has turned out it was lucky we saved ourselves the trouble: I was much too sick even to get up to see the distant outline.— On the 6th in the evening we sailed into the harbour of Santa Cruz.— I now first felt even moderately well, & I was picturing to myself all the delights of fresh fruit growing in beautiful valleys, & reading Humboldts descriptions of the Islands glorious views.— When perhaps you may nearly guess at our disappointment, when a small pale man informed us we must perform a strict quarantine of 12 days. There was a death like stillness in the ship; till the Captain cried "Up Jib", & we left this long wished for place.—

We were becalmed for a day between Teneriffe & the grand Canary & here I first experienced any enjoyment: the view was glorious. The peak of Teneriffe. —was seen amongst the clouds like another world.— Our only drawback was the extreme wish of visiting this glorious island.— "*Tell Eyton, never to forget either Canary islands or S America*;—that I am sure it will well repay the necessary trouble but that he must make up his mind to find a good deal of the latter.— I feel certain, he will repent it, if he does not make the attempt".—

From Teneriffe to St Jago, the voyage was extremely pleasant.— I had a net astern the vessel, which caught great numbers of curious animals, & fully occupied my time in my cabin, & on deck the weather was so delightful, & clear, that the sky & water together made a picture.— On the 16th we arrived at Port Praya, the capital of the Cape de Verds, & there we remained 23 days viz till yesterday the 7th of February.— The time has flown away most delightfully, indeed nothing can be pleasanter; exceedingly busy, & that business both a duty & a great delight.— I do not believe, I have spent one half hour idly since leaving Teneriffe: St Jago has afforded me an exceedingly rich harvest in several branches of Nat: History.—[2] I find the descriptions scarcely worth anything of many of the commoner animals that inhabit the Tropic.— I allude of course to those of the lower classes.— Geologising in a Volcanic country is most delightful, besides the interest attached to itself it leads you into most beautiful & retired spots.—

Nobody but a person fond of Nat: history, can imagine the pleasure of strolling under Cocoa nuts in a thicket of Bananas & Coffee plants, & an endless number of wild flowers.— And this Island that has given me so much instruction & delight, is reckoned the most uninteresting place, that we perhaps shall touch at during our voyage.— It certainly is generally very barren.—but the valleys are more exquisitely beautiful from the very contrast:— It is utterly useless to say anything about the Scenery.— it would be as profitable to explain to a blind man colours, as to person, who has not been out of Europe, the total dissimilarity of a Tropical view.— Whenever I enjoy anything I always either look forward to writing it down either in my log Book (which increases in bulk) or in a letter.—[3] So you must excuse raptures & those raptures badly expressed.—

I find my collections are increasing wonderfully, & from Rio I think I shall be obliged to send a Cargo home.— All the endless delays, which we experienced at Plymouth, have been most fortunate, as I verily believe no person ever went out better provided for collecting & observing in the different branches of Natural hist.— In a multitude of counsellors I certainly found good.— I find to my great surprise that a ship is singularly comfortable for all sorts of work.— Everything is so close at hand, & being cramped, make one so methodical, that in the end I have been a gainer.—

I already have got to look at going to sea as a regular quiet place, like going back to home after staying away from it.— In short I find a ship a very comfortable house, with everything you want, & if it was not for sea-sickness the whole world would be sailors.— I do not think there is much danger of Erasmus setting the example, but in case there should be, he may rely upon it he does not know one tenth of the sufferings of sea-sickness.— I like the officers much more than I did at first.—especially Wickham & young King, & Stokes & indeed all of them.— The Captain continues steadily very kind & does everything in his power to assist me.— We see very little of each other when in harbour, our pursuits lead us in such different tracks..— I never in my life met with a man who could endure nearly so great a share of fatigue.— He works incessantly, & when apparently not employed, he is thinking.— If he does not kill himself he will during this voyage do a wonderful quantity of work.— I find I am very well & stand the little heat we have had as yet as well as any-body.— We shall soon have it in real ernest.— We are now sailing for Fernando Norunho off the coast of Brazil.—where we shall not stay very long, & then examine the shoals between there & Rio, touching perhaps at Bahia:— I will finish this letter, when an opportunity of sending it occurs.—

Feb 26th. about 280 miles from Bahia.— On the 10th we spoke the packet Lyra on her voyage to Rio. I sent a short letter by her to be sent to England on first opportunity.— We have been singularly unlucky in not meeting with any homeward bounds vessels, but I suppose Bahia we certainly shall be able to write to England.— Since writing the first part of letter nothing has occurred except crossing the Equator & being shaved.— This most disagreeable operation consists of having your face rubbed with paint & tar, which forms a lather for a saw which represents the razor & then being half drowned in a sail filled with salt water.— About 50 miles North of the line, we touched at the rocks of St Paul.— this little speck (about ¼ of a mile across) in the atlantic, has seldom been visited.— It is totally barren, but is covered by hosts of birds.— they were so unused to men that we found we could kill plenty with stones & sticks.— After remaining some hours on the island, we returned on board with the boat loaded with our prey.— From this we went to Fernando Noronha, a small island where the Brazilians send their exiles.— The landing there was attended with so much difficulty owing a heavy surf, that the Captain determined to sail the next day after arriving.— My one day on shore was exceedingly interesting. the whole island is one single wood so matted together by creepers, that it is very difficult to move out of beaten path.— I find the Nat: History of all these unfrequented spots most exceedingly interesting, especially the geology.

I have written this much in order to save time at Bahia.— Decidedly the most striking thing in the Tropics is the novelty of the vegetable forms.— Cocoa Nuts could well be imagined from drawings if you add to them a graceful lightness, which no European tree partakes of.— Bananas & Plantains, are exactly the same as those in hothouses: the acacias or tamarinds are striking from blueness of their foliage: but of the glorious orange trees no description no drawings, will give any just idea: instead of the sickly green of our oranges, the native ones exceed the portugal laurel in the darkness of their tint & infinitely exceed it in beauty of form.—

Cocoa-nuts, Papaws.—the light-green Bananas & oranges loaded with fruit generally surround the more luxuriant villages.— Whilst viewing such scenes, one feels the impossibility than any description should come near the mark,— much less be overdrawn.—

March 1ˢᵗ Bahia or St. Salvador.— I arrived at this place on the 28ᵗʰ of Feb & am now writing this letter after having in real earnest strolled in the forests of the new world.— "No person could imagine anything so beautiful as the antient town of Bahia; it is fairly embosomed in a luxuriant wood of beautiful trees.—& situated on a steep bank overlooks the calm waters of the great bay of All Saints.— The houses are white & lofty, & from the windows being narrow and long have a very light & elegant appearance Convents, porticos & public buildings vary the uniformity of the houses: the bay is scattered over with large ships. in short & what can be said more it is one of the finest views in the Brazils".— (copied from my journal) But the exquisite glorious pleasure of walking amongst such flowers, & such trees cannot be comprehended, but by those who have experienced it.— Although in so low a Latitude the weather is not disagreeably hot, but at present it is very damp, for it is the rainy season.— I find the climate as yet agrees admirably with me: it mak⟨es⟩ one long to live quietly for some time in suc⟨h⟩ a country.— If you really want to have a ⟨notion⟩ of tropical countries, *study* Humboldt.— Skip th⟨e⟩ scientific parts & commence after leaving Teneriffe.— My feelings amount to admiration the more I read him.— Tell Eyton (I find I am writing to my sisters!) how exceedingly I enjoy America & tha⟨t⟩ I am sure it will be a great pity if ⟨he⟩ does not make a start.— This letter will go ⟨on⟩ the 5ᵗʰ & I am afraid will be some time before it reaches you.— it must be a warning, how in other parts of the world, you may be a long time without hearing from.— A year might by accident thus pass.—

About the 12ᵗʰ we start for Rio, but remain some time on the way in sounding the Albrolhos shoals. Tell Eyton, as far as my experience goes let him study Spanish French, Drawing & Humboldt. I do sincerely hope to hear of (if not to see him), in S America.— I look forward to the letters in Rio. till each one is acknowledged mention its date in the next: We have beat all the ships in mæneuvering, so much so that commanding officer says we need not follow his example, because we do everything better than his great ship.— I begin to take great interest in naval points, more especially now, as I find they all say, we are the No 1 in South America.— I suppose the Captain is a most excellent officer.— It was quite glorious to day how we beat the Samarang in furling sails: It is quite a new thing for a "sounding ship" to beat a regular man of war.— And yet the Beagle is not at all a particular ship: Erasmus will clearly perceive it, when he hears that in the night I have actually sat down in the sacred precincts of the Quarter deck.— You must excuse these queer letters, & recollect they are generally written in the evening after my days work.— I take more pains over my Log Book.—so that eventually you will have a good account of all the places I visit.—

Hitherto the voyage has answered **admirably** to me, & yet I am now more fully aware of your wisdom in throwing cold water on the whole scheme: the chances are so numerous of it turning out quite the reverse.— to such an extent do I feel this that if my advice was asked by any person on a similar occasion I should be very cautious in encouraging him.— I have not time to write to any body else: so send to Maer to let them know that in the midst of the glorious tropical scenery I do not forget how instrumental they were in placing me there.— I will not rapturize again: but I give myself great credit in not being crazy out of pure delight.—

Give my love to every soul at home, & to the Owens I think ones affections, like other good things, flourish & increase in these tropical regions.—

The conviction that I am walking in the new world, is even yet marvellous in my own eyes, & I daresay it is little less so to you, the receiving a letter from a son of yours in such a quarter: Believe me, my dear Father Your most affectionate son | Charles Darwin

St Salvador, Brazils

I find after the first page I have been writing to my sisters

Endorsed in pencil on cover: 'Rec^d the 3^d of May—1832.—'
DAR 223

[1] In a personal letter from Bahia on 5 March 1832 (see F. Darwin 1912, p. 548), Robert FitzRoy wrote to Francis Beaufort at the Admiralty: 'He was terribly sick until we passed Teneriffe, and I sometimes doubted his fortitude holding out against such a beginning of the campaign. However, he was no sooner on his legs than anxious to set to work, and a child with a new toy could not have been more delighted than he was with St. Jago. It was odd to hear him say, after we left Porto Praya, "Well, I am *glad* we are *quietly* at *sea* again, for I shall be able to arrange my collections and set to work more methodically." He was sadly disappointed by not landing at Teneriffe and not seeing Madeira, but there was no alternative.' For other reminiscences of CD's seasickness by his ship-mates John Lort Stokes and Alexander Burns Usborne, see *LL* 1: 224.

[2] In his official report to Beaufort of 4 March 1832 (F. Darwin 1912, p. 548), FitzRoy wrote: 'Mr. Darwin has found abundant occupation already, both at sea and on shore; he has obtained numbers of curious though small inhabitants of the ocean, by means of a Net made of Bunting, which might be called a floating or surface Trawl, as well as by searching the shores and the Land. In Geology he has met with far much more interesting employment in Porto Praya than he had at all anticipated. From the manner in which he pursues his occupation, his good sense, inquiring disposition, and regular habits, I am certain that you will have good reason to feel much satisfaction in the reflection that such a person is on board the *Beagle*, and the certainty that he is taking the greatest pains to make the most of time and opportunity.'

[3] For the different kinds of records CD kept during the voyage see Appendix II.

To R. W. Darwin 10 February 1832

2 Days sail SW of S! Jago | Lat: 11 N.

Feb. 10th 1832

My dear Father

I have a long letter, all ready written, but the conveyance by which I send this is so uncertain.—that I will not hazard it, but rather wait for the chance of meeting a homeward bound vessel.— Indeed I only take this opportunity as perhaps you might be anxious, not having sooner heard from me.— All day long we have been in chace of a packet bound to Rio, & have this evening overtaken her, tomorrow a boat will go on board of her & this letter will be conveyed to Rio & from thence to Shrewsbury or to the fire.— We have had a most prosperous quick & pleasant voyage.— At first.—indeed till the Canary Islands.—I was unspeakably miserable from sea sickness & even now, a little motion makes me squeamish.— We did not stop at Madeira, owing to its blowing fresh, & at the Canary Islands, they wanted to put us in strict quarantine for 12 days— Sooner than submit to that, we sailed to Cape de Verds, & arrived at St. Jago on the 16th of January, having left England on the 27th of December.— The voyage from Teneriffe to S! Jago was very pleasant, & our three weeks at it have been quite delightful.— St. Jago although generally reckoned very uninteresting, was me most exciting.— Of course, the little Vegetation that there was, was purely tropical.— And my eyes have already feasted on the exquisite form & colours of Cocoa Nuts, Bananas & the beautiful orange trees. Hot houses give no idea of these forms, especially orange trees, which in their appearance are as widely different & superior to the English ones, as their fresh fruit is to the imported.—

Natural History goes on excellently & I am incessantly occupied by new & most interesting animals.— There is only one sorrowful drawback, the enormous period of time before I shall be back in England.— I am often quite frightened when I look forward.— As yet everything has answered brilliantly. I like every body about the ship, & many of them very much.— The Captain is as kind as he can be.— Wickham is a glorious fine fellow.— And what may appear quite paradoxical to you, is that I *literally* find a ship (when I am not sick) nearly as comfortable as a house.— It is an excellent place for working & reading, & already I look forward to going to sea, as a place of rest, in short my home.— I am throughily convinced, that such a good opportunity of seeing the world, might not again for a century.— I think, if I can so soon judge.—I shall be able to do some original work in Natural History.— I find there is so little known about many of the Tropical anima⟨ls.⟩

The effect of my sending this letter will ⟨be⟩ to spoil my longer one.—but I was determined not to lose any opportunity (at Cape Verds there was none) & it is doubtful how long it will be before we arrive at Rio.— The Albrolhos banks on coast of Brazil may last us some time.—

As yet I have not felt the heat more than in England.— In about a week it will be widely different.— You will always find my letters home very badly written, as I am exactly in case of having half an hours talk, & then it would be a struggle what should come out first.— This delay in letters will be a lesson not too soon to expect letters.

Give my very best love to everybody & believe me, my dearest Father, Yours | Most affectionate Son | Charles Darwin

DAR 223

From Susan Darwin 12 February [— 3 March] 1832

Shrewsbury
February 12th 1832

I must begin this folio by wishing you joy my dear Charley of being this day 23 years old; and I heartily hope it may find you happy, and that you may continue so for many and many a year to come.— Our plan as you may perceive is always for each of us to take you every three Months in turn & Feb^{ry} is my share.— Also another plan we have adopted by Caroline's suggestion is, never to go forwards, (I mean by that telling you what we have in contemplation) but on the contrary only to relate past events or else you would get confused & our letters would not answer the purpose of a Journal which I think is always most satisfactory in absences.— I am in spite of myself beginning to watch the Post in expectation of seeing an *outlandish* letter arrive: for if you have been lucky in meeting a Ship we might hear by this time I suppose from Madeira— Papa is more sensible & declares he will not begin to expect till March.— The only event which has happened this last week is poor old Mrs Darwin's death; who died at the Priory the 5th of this Month; we are all gone into Mourning this day being Sunday: but of course this does not matter for you— I believe she was 84 years old and was ill about a fortnight.— Sir Francis & Lady Darwin will I conclude leave their mountainous abode[1] & come to the Priory now, which the latter must prefer to the society of Eagles & Wild Boars.—

I repented last week that we had no intercourse with the Galtons[2] as Paganini has been performing several nights at Birmingham & I heard the Miss Galtons all attended the Theatre: If it had not been for Mrs Darwin death Papa most good naturedly proposed a scheme that we should have gone with Harry & Jessie to hear him which would have been delightful— That couple are spending a fortnight with us & we find them very agreeable inmates they fall into all our ways so nicely. They take to the Rubber of Whist as kindly as possible, & Harry for our benefit reads aloud a *corrected* copy of "Joseph Andrews"[3] which with plenty of *skipibus* we find very amusing— I think they will not bear living at Etruria much longer they both seem to hate the Potteries so cordially.—

I think there is great hopes that Charlotte *Langton* will settle hereafter in Shropshire Lord Craven has so many livings in this County at his disposal. The one I sh^d like them to have is Wistanstow very near the Craven Arms in a beautiful situation amongst the Clee Hills.—[4]

Hills put me in mind of Rocks & Stones &c & Papa bid me tell you with his love that he asked Mr Hughes to save any Shells for you in the Gravel Pit[5] that he might find: but he says there are none that you would care for.—

Catherine has been staying the last two days at Woodhouse to entertain Fanny whilst M^rs. Owen & her *one Chick* Caroline went to spend the week at Aqualate[6] for the Newport Hunt. M^r. Biddulph cannot come down f. London till the Reform question is settled,[7] & then the Marriage is to take place immediately & after spending the Honeymoon at Chirk they return to London. John Hill has just had a Living given him by one of his Uncles worth 600£ a year & M^rs. Drewe Corbet says he intended proposing again to Fanny & is very much vexed at just being too late. I cannot help being sorry for him tho' it is more than he deserves after his cold hearted behaviour.—

The more I hear of M^r. Biddulph the more I think he will make dear Fanny happy. he appears so very affectionate, not only to her, but to his Mother, which tells well for his character: I was nearly a week at Woodhouse whilst he was there, & tho' I thought him very agreeable I found him a very difficult person to get acquainted with & I feel rather a horror of going to visit at Chirk Castle.

Feb^ry 19th.

Another week has brought more news: for Frank Wedgwood is now engaged to be married to a Miss Mosley daughter of a Clergyman of good family in Derbyshire. he staid in the house with her a week at Loxley & some other place & then proposed to her. she w^d not accept him at once, but had him for trial another week at her father's house & now all is settled— She is very fat & not at all pretty, but exceedingly good tempered & a famous *scrattle*[8] so I daresay she will make him an excellent wife. None of his family have ever seen her except Harry, but she is to come & *shew off* at Maer, probably when Mr Langton makes his first appearance which will be very comical.— That good Charlotte has invited us all three to go to her Wedding & you shall have a long account of it from Caroline, for it will take place about the 15th. of March.— I foresee Maer will soon be qu⟨ite deserted⟩ for I expect Fanny & Emma to follow this mad m⟨arrying⟩ example before this year is over.—

I am now staying at Overton with Marianne and her 4 ⟨boys⟩ who are all very well & happy. Parky has been enquiring from me when Uncle Charles will come back, which is a question I cannot answer.— D^r. P. seems in very good spirits & ⟨   ⟩

I have just been reading the "Mutiny of the Bounty" in the Family Library it is a very old story but very interesting from Beecheys account of the happy state he found the Mutineers in at Pitcairn's Island.[9] I daresay you have Beecheys Voyage with you as Capt Fitzroy you said had such a large collection of Travels on board.— Catty & I dined at Onslow[10] last Thursday & I met there a Capt Meynard who put me to the blush for my Uncle Sir Francis Darwin by telling me he travelled with him in Greece & that Sir F's chief sport was *disfiguring* & *mutilating* all the Statues he came across.— This same Capt M. told me this was the time for all Ships to leave England & not return, so I fear yr letter will not meet with a conveyance so soon as I had hoped.— The Cholera has broken out in London so you have chosen yr time well for leaving England as now I suppose every town most likely will have its turn.— Mʳˢ Williams has just got into her new house in Belgrave St. but she cannot enjoy it much for she is ill with a very bad Cough, which she thinks is the Hooping Cough.— She has sent for yr direction & means to write to you.—

Poor Sam Beck has been getting worse & worse ever since you left home & now he is quite dying. Papa is just gone to see him. He is Dead—[11] Our Hot house is quite finished & we have got several Pines & plants in it. Papa sits there a great deal & it answers very well as a hobby for him. we have had pipes laid down in the Greenhouse & the regular warmth of the hot water makes the morning room very comfortable as it was apt to get very cold at night.—

I am come back to Shrewsbury now. Mariannes new little boy is to be christened Charles after his 2 Uncles of that name for Dʳ Parker's eldest brother in India is Charles.— You really must not blame me if Eras does not write to you for I regularly send him a Dun in every Postscript. he is now quite buried alive in a little Lab he has set up in his Lodgings which makes him quite forget times & seasons.— If he comes down at Easter I will stand over him & torment his heart out about writing to you.— I am afraid after all I shall be obliged to close this without hearing [yr pen] dear old Charley for March is come & this must go— All yr friends about are beginning to ask us if we have heard from you yet, & some say you got so much good wine at the Madeiras it was impossible you cᵈ write to us, but I know as soon as ever this is safe gone we shall all give a *scream* & Edward will present us with a Letter from you.—

I have one more Marriage to finish up with, but nothing very interesting: your charming Cousin Lucy Galton is engaged to marry Mʳ Moilliet: the eldest son of a *very fat* Mʳˢ Moilliet, who was once here, I forget if you were at home. The young Gentleman has a good fortune, so of course the match gives great satisfaction. I expect every day to hear Bessy & William Fox's marriage announced as I can't see what they are waiting for.— As I have no new Direction I must take my chance at Rio Janeiro.

All our affectionate Loves to you Dearest Charles & I am Ever Yʳˢ most *particularly* Susan E Darwin.

I have played a good deal of Music this winter for yr sake.—

Postmark: Shrewsbury MA 3 1832
DAR 204.6.1

1 Francis Sacheverel Darwin had married Jane Harriett Ryle in 1815; he was knighted in 1820. The 'mountainous abode' was at Sydnope, Derbyshire (see also letter to W. D. Fox, [1 April 1830]).

2 The Samuel Tertius Galtons.

3 The novel by Henry Fielding (1742).

4 A range of hills, the highest in the county, in south-west Shropshire.

5 CD mentions the Shrewsbury gravel pit in the *Autobiography*, p. 69, where he recalled Adam Sedgwick's reaction on being told that a tropical shell had been found there: 'if really embedded there it would be the greatest misfortune to geology, as it would overthrow all that we know about the superficial deposits of the midland counties.' CD continued: 'Nothing before had ever made me thoroughly realise . . . that science consists in grouping facts so that general laws or conclusions may be drawn from them.'

6 Aqualate Mere is a small lake on the western border of Staffordshire near the market town of Newport.

7 The House of Commons passed the Reform Bill on 23 March. It became law on 7 June 1832.

8 A dialect word. As used by the Darwins it usually has the sense of keeping accounts or being economical (see *Emma Darwin* 1: 139).

9 The following sentence, which had been written after 'Pitcairn's Island.' was deleted: 'I am afraid there is no chance of yr having it with you.'

10 Onslow, near Shrewsbury, was the seat of Colonel John Wingfield (*Burke's Landed Gentry 1952*).

11 'He is Dead' was added as an interlineation.

From William Mostyn Owen Sr 1 March 1832

Woodhouse
March 1st. | *1832*

My Dear Charles Darwin,

I have been much longer than I intended in commencing this Letter, & simply because, knowing I could do it any day, I have from day to day postponed it, perhaps expecting also that I might have something more interesting to communicate, & to render a stupid Letter somewhat more acceptable. Your own Family however will no doubt be before me in relating any news or Gossip this *barren* land produces, let me wait as long as I please to collect it— I will therefore delay no longer or flatter myself that I can write any thing that has not already been told you or that will make my Letter more welcome to day than it would have been some weeks ago. But this I am nevertheless happy & vain enough to believe that my Letter come when it will & whatever may be its contents will be as welcome as I assure you one from you will always be to me, provided it brings good tidings of you. I heard of you repeatedly before you left Plymouth & regretted exceedingly the bad Weather which for some time retarded your sailing & drove you back into Port, & hope your progress has not since been interrupted by any misfortune or disagreeable occurrence. I trust & pray too that you may be satisfied & pleased with your Situation & with the undertaking in which you have embarked, though at the same time I have the consolation of knowing that should you not be pleased or like to prosecute the Voyage throughout, you may halt wherever & whenever you please & can find your way home again— That you will do so however I do

not expect, for I heard you were most perfectly satisfied with your Capt &
companions & I dare say are by this time as much at home & at your ease in the
Ship as any of them. The first place you expected to touch at I think was
Madeira, & from thence I hope we may now very soon expect to hear from
you.— Instead of indulging in these conjectures further let me now endeavour to
recollect & to communicate all that has happen'd in which you are likely to feel
any Interest since you left us— And first if you will not call me a stupid Egotist I
will begin with my own Family— Sarah I believe you know was engaged, & I
think was married to Mr Williams just about the time you sail'd, & I hope &
believe is very happy— Now in London but pass'd about two Months at
Eaton.— Fanny who return'd out of Devonshire to attend her Sister's wedding
you will perhaps be surprised to hear has also found another admirer in the Man
who we last year rather thought admired her Sister & is now engaged to marry
Mr Biddulph, & I hope the event will have taken place before this reaches
you.— Though I am afraid he is now not very rich & indeed probably never will
be so, considering the large place &c he has to keep up & live at, it is certainly
what the world calls a very great match for her, & I know too it is quite as much
a love match on her side as on his, & I think there is every reason to hope she will
be happy— I don't know whether you ever met or are acquainted with him, but
he is very good looking & very Gentlemanlike & sensible, though like many
others he has at starting been guilty of some foolish acts; & amongst others, he
was very near being married to Miss Isabella Forester, & I am afraid lost some
money, though nothing like what was said, at Play. But now I hope & believe
his follies are at an end & his wild oats sown & that he will make a good
husband, for I think you will agree with me in saying, that tho' I say it who
perhaps ought not to say it, I am sure he will get as amiable & good a Wife as
ever was born. He has been introduced to your Father & I believe they are
mutually pleased with each other. He laughs & calls him Count Robert of
Chirk—I do not mean to his face, but joking with us.— This affair has been as
unexpected as unsought for by me or I believe I may add by Fanny herself—
who it seems won his heart chiefly by her conduct when her affair with J. Hill
was broke off, when he happen'd to be here & was made acquainted with all
that pass'd, and he now says, that he then determin'd if she ever gave up
J. H. he would try his luck, & this he did on the first opportunity after she returned
into this Country so all is well that ends well— And I am most heartily rejoiced
that she has nothing to do with such a Woman as Mrs Hill, between whom &
Dr Dugard you have I dare say been told there has been a grand explosion
lately occasion'd by his asserting that she had employ'd him to get Miss Cleg for
Sir Rowland with her 4 or 500,000£ for which & other services he made a
demand of 2000£ which was not acceded to & occasion'd a great blow up, &
though the Doctor has had the worst of the encounter & been obliged to recant
many People think he has been bought off—& the general Opinion is that if *his
hands are very dirty* Mrs Hills are not *quite clean*. A more disgraceful transaction

has certainly seldom taken place, & at all events poor silly Sir Rowland is as completely sold as any slave or beast of burthen, & whatever he gains in Money will lose, & ought to lose, in Character. You perhaps do not recollect that this is Leap Year or if you do may possibly have wisely made your escape for that reason, & if we may judge by the unusual number of Marriages that have already taken place in this Country & which are said to be in preparation I think the ladies must be exercising their Privilege without mercy— Amongst others Miss Parker has just carried off Sir Baldwin Leighton & as some say has caught a Tartar— Excepting these little varieties we are going on much in our usual hum drum style, a little hunting, a little shooting & now and then a little argument about the Reform Bill which is again undergoing a tedious debating in Committee of the Commons where however there is no doubt it will soon be passed by a vast Majority— In the Lords its success however still appears to be very doubtful but a large creation of Peers is talk'd of & I should think must be made if the Bill is to be carried, & why they are delay'd all well wishers to the Bill & to the present Ministers cannot understand— If Lord Grey has not courage & energy enough to carry it now, seeing what he has seen, *cout qu'il cout*, I think he deserves to be hanged by his own Friends and supporters, & is certainly quite unworthy his situation—but we will yet hope better things of him, though I confess I am half afraid of him, & should feel much more confident if he was out of the way & the business was left to Lords Brougham[1] and Althorp—[2] If it fails we must look I fear for something much worse & those who wish to do right will then share the Ruin which their infatuated opponents the Borough mongers & Bishops[3] will bring upon the Country—'Quos deus vult perdere prius dementat— But no more of Politics, for I suppose that the same conveyance that carries this Letter to you will also carry some English Newspapers.—

Arthur you will be glad to hear got through his Examinations at the India College at last with Credit, & if well will sail for Madras in May, where I hope you may meet in mutual good health— Francis is still with M[r] Meredith & what will be his Profession I am yet unable to determine, for in these times all are so full that it is difficult as well to chuse as to get employment & bread in any.— All the other Members of my numerous Family M[rs] Owen included are well & as you left them.— And so I am happy to say are all yours, of whom I should say more if I did not know that they will speak more to the purpose & to your satisfaction for themselves.— Your Father I think is particularly well, but still evidently not quite satisfied to lose sight of you for probably so long a period— God grant that you may return in health & find him so— At this distance I may say without blushing & I am sure without flattery that there is no Man living that I value and esteem more or that I am under so many obligations to— There is something about him so liberal, so high minded & yet so unassuming that it is quite impossible to know him well without respecting & loving him— You ought to be very proud of him, nor do I doubt but you are so— I think I have now exhausted all the stock of little & domestic events that have taken place in this Country since you left it, & am not aware that I have anything more to relate that can amuse or interest you—so shall conclude my stupid Epistle wishing you every possible success & gratification & with a most fervent hope that we may meet again in as good health as when you last visited this House— Fanny says that you must not suppose that the Lady of *Count Robert of Chirk* will ever forget the Friend of her youth & means to send you a few lines in the same envelope with this—

Believe me my dear Charles always most Sincerely & affectionately Yours | W^m. Owen

DAR 204.6.2

[1] Henry Peter Brougham, Baron Brougham and Vaux, made a famous speech on the Reform Bill's second reading.
[2] John Charles Spencer, Viscount Althorp. Leader of the Whigs in the Commons, he organised the passage of the Reform Bill in that House.
[3] Supporters of the 'rotten boroughs'. Over 200 (about half) of the so-called 'nomination' seats were eliminated by the Reform Bill. The Bishops feared that Reform would result in the disestablishment of the Church.

From Fanny Owen 1 March 1832

My dear Charles—
Your Sisters tell me they informed you in their last letter of the awful and important event that is going again to take place here— My fate is indeed decided, the die is cast—and my dear Charles I feel quite certain I have not a friend in the world more sincerely in my welfare than you are, or one that will be so truly glad to hear I have every prospect of Happiness before me in the lot I have chosen— I have known M^r. Biddulph a long time, and always liked him extremely, thought him **most** *agreable*, and good hearted, but on more intimate acquaintance I do feel convinced he possesses every quality to make me happy, is most sincerely & (*Heaven knows*) *disinterestedly* attached to me—in short I do think mine is a *happy* **end**, I have no misgivings about it—tho it is indeed an awful and a melancholy thing to leave ones quiet happy Home,—& all those people one has lived with & loved from childhood— I cannot help feeling low when I think of that, but to think I shall only be 10 miles from the poor dear, old Forest, is a delightful consideration— My dear Charles I will not prose any more on my own affairs, I know you are interested in all that concerns me or I should not have written so much— I would give a great deal to see you again & have one more merry chat, whilst I am still Fanny Owen—but alas that cannot be, but believe my dear Charles that no change of *name* or condition can ever alter or diminish the feelings of sincere regard & affection I have for *years* had for you, and as soon as you return from your wanderings, I shall be *much offended* if

one of your first rides is not to see *me* at Chirk Castle,—and find out what curious Beetles the place *produces*— I do not know, if you have ever seen Mr Biddulph—one thing I must tell you—that the *firm Catherine* had a most violent prejudice against him, but I have made her come here to meet him, and *even* **she** with her *well known firmness*, has now become a perfect convert, and declares *she* thinks him a *piece* of *Perfection*. I only mention this little anecdote as I think it may speak *for him*,— knowing as we *all do* how firm Cath—*can be*! Papa has been writing you a very long Budget and I fear told you all the news there is in these parts, which does not amount to much— Poor Arthur I suppose will sail for India about May or June— I shall be very sorry to see him go, poor fellow, he is a nice well disposed Boy, I do not think he is very happy in the prospect of going to India— Sarah & her *Hubby* seem to go on very happily. She has just got into her new House in Belgrave Square, seems to lead a very gay life— You have heard too of your favorite Charlotte Wedgewoods intended marriage, that seems to have been quite a *Bubble*, she went to London, met him, & all was settled *in a fortnight*. indeed the marriage fever seems to rage in Shropshire this year. Miss Parker has taken *to her care* that lively young Baronet Sir Baldwin Leighton. Miss Fanny Sparling they say is shortly to be united to Mr **Dry Corbet**— I dare say the infection will soon spread to your sisters— I should think poor *Forlorn Hope*[1] wished it had, when he tried his luck there some time ago— I hope you continue to like your Captain as much as on first acquaintance I hope too my dear Charles your expedition is answering all your most sanguine expectations, that you are enjoying yourself in every way as much as possible—and when you do return to the little Parsonage, and want *the little wife*, "pray give me a commission to look out for her—" Alas Miss C. Salway[2] is *lost to you*—but I have no doubt she may be replaced— I have not made much use of the old Painting Turret lately— What fun we *have had* in that poor old room—what happy days we have had together dear Charles— I cannot tell you how much we all miss you here, & how very often we talk about you— You are the Governor's *first favorite* I assure you he now seems quite happy in writing to you— I do not know at all when I shall be *twined off* as Owen calls it, but probably some time in April. Mr Biddulph is at present in London attending his parliamentary duties, *fighting* for Reform— I must now conclude this scrawl,—most anxiously my dear Charles shall I ever hope to hear of your happiness and prosperity, and remember you will always find me the same sincere friend I have been to you ever since we were *Housemaid* & *Postillion* together, and you must not forget your engagement to come & see me in the *old Castle* as soon as you reach again y.^r native place—

   Now adieu, and Heaven bless you my very dear Charles— | Believe me always yr | Sincerely attached | Fanny Owen

Woodhouse
March 1.st 1832—

Mama desires her kindest love to you she intends to send you a budget very soon— Caroline too sends her love—

DAR 204.4

[1] Probably a reference to Frederick William Hope, CD's entomologist friend.
[2] Charlotte Salwey.

From Caroline Darwin (with postscript by Marianne Parker) 12[−29] March [1832]

Maer—
March 12th | Monday.

My dear Charles—

Susan sent off her letter by the first Tuesday of this month directed to Rio as we have not yet had any fresh direction from you. We are in daily hopes of a letter from Madeira as it is high time we calculate for a letter & we are getting very impatient my dear Tactus to hear from you— I mean to fill this letter very much with Maer news. I am only afraid a letter without a new marriage will be very flat. I came here last Tuesday & found all the family at home & two Miss Tollets who however only staid one day to my great joy, for they are such talkers that I felt it vain to try to get a share & listening without talking oneself is dull work comparitivly—as *we* have often agreed.— On Thursday M[r] Baugh Allen arrived to stay *alas* a fortnight there never was such a tiresome, chattering, conceited, man. Every body here except Aunt Bessy are in despair at this visit & we all agree that when he does leave the room for a few minutes it is like the stilling of a storm (is not that beautifully affected?)— on Friday M[r] Langton arrived, & next Thursday week the marriage is to take place. Jessie also came here on Friday & now I have given you the outline I will tell you all particulars— Charlotte Fanny & Emma went to meet M[r] Langton who was to come per coach to Newcastle, which Charlotte was vy glad to do, to escape a public meeting with him: I returned to be in the room to see Uncle Jos's first greeting with him & to avoid a seperate introduction myself— We dined late & at 6[−] the coach arrived—all the family shook hands with him in a very nice & friendly manner— C. F & E all then went to dress & M[r] Langton & Uncle Jos & M[r] B. Allen stood talking of the weather an endless time. poor M[r] L saying it was "cold, remarkably cold," & then it had been "a fine day, particularly fine" & bright &[c] all repeating each other & contradicting each other in the most agonitic way till at length Uncle Jos took the unfortunate man to his room to dress. Uncle Jos said afterwards that he could not think "what bewitched M[r] L. to stand talking instead of going" & we all of course raised an outcry & said it was he himself who was to blame—& not M[r] Langton I suppose this long waiting for dinner did not agree with the constitution of the family, for when we did at length go to dinner at 7. oClock, Uncle Jos. *never* spoke, nor did Jos.— Mr L. was shy & constrained & not having had the advantage of a family sketch (such as you gave William Fox) I dare say thought this silence very strange & possibly an incivility to himself. he is like W. F a very polite ceremonious man & really very well bred & gentlemanlike— Frank I suppose thought the family *awe* struck by Mr L. & inclined to pay him too much attention— Frank but seldom addressed him & when he did, usually some ultra radical sentiment— When the toasted cheese was handed round Frank thought Mr Langton had not been offered any, so he took a plate as you would take a flat stone & made it skim across the table towards M.^r L & then asked him whether he chose to take any— After dinner Uncle Jos who is not quite well lay down on the sofa & read all evening & Jos lolled on the arm chair opposite & *literally* they neither of them spoke all evening or once addressed M.^r L which considering its being his first evening & the situation he was in I do think the most extraordinary piece of want of politeness I ever witnessed. The girls were all sadly annoyed particularly poor Charlotte the evening was very flat & constrained— Yesterday & Sunday went off much better, & dinner went off also very well. Harry came from the Stafford assizes & talked away. Frank was gone to pay a visit to Miss Mosely, so there were no plates flung across the table. One plate made a tremendous crash on the table slipping out of Jos's hands in such an odd manner that Harry afterwards asked if Jos was tossing the plate up to see whether Heads or tails came down— however he talked away very well to Mr L. & I still think when he *does* talk—he appears to much better advantage than most of his brothers. Mr Langton is I think a *very* agreeable man with the most pleasing countenance & manner I have often seen—his reading prayers & the bible last night was quite beautiful. Charlotte is looking so pretty & so happy & proud of M^r L it is a pleasure to see her & I do think when you get to know Mr Langton you will be as glad as we are that she has met him— they have taken a house for a year in Surey at Ripley— Franks Miss Mosly by all accounts has nothing very pleasing about her. She is good-humoured they say & very fond of Frank— it is settled they are to live at Etruria & Harry & Jessie to take some other house in the neighbrhd much to Jessies joy— Mr Paget Mosley—the brother, people think is rather taken with Fanny, he always singled her out to drink wine wine with & he watches her & remarks on all her little ways he did a very odd thing which if I can, without a grand prose I will tell you. He found out the day Mr Langton was expected who he had never seen— he got upon the Coach & went 5 miles with him, after a time he entered into conversation, made some allusion to who Mr L. was, by saying "you will be late for dinner &.^c—" then wished him joy, said he believed Charlottes white bonnet was not yet arrived—then praised her singing & after a pause said "Fanny sings very well also". Mr L. doubted, but Mr Paget persisted & afterwards they remembered Fanny had one night joined in some grand finale in Figaro— I think considering how little Mr Mosely knew of the Maer family & what a *slight connexion* he had with Mr Langton through this acquaintance this conversation was most extraordinary— On the 20th the party began to collect for the wedding. Catherine came. Mr Secker[1] with Harry from the assizes. Mrs Tollet & 3 daughters came to tea on 21st the young ladies in high spirits— Ellen Tollet made such a noise laughing & chattering that Uncle Jos grew quite cross, & left the room Charlotte looked remarkably pretty but very silent & seemed overcome & wished the party had not been so large. Mr Langton's manner was very nice & attentive w⟨ithou⟩t being disagreeably so— the next morning we all assemb⟨led⟩ in the drawing room between 9 & 10. there were two tab⟨les⟩ with tea & coffee for those who chose to t⟨ake⟩ any thing before going to church— Aunt Sarah & Miss Mos⟨le⟩y then arrived. the room looked so odd every body standing about in little groups, all gaily dressed. Charlotte in white silk bonnet & green pelisse— At 10 the glass doors were opened in the porch & we all set out Mr Langton & Mrs Tollet heading the procession & we all followed in pairs 7 couples then Uncle Jos & Charlotte & the bridesmaids. poor Emma was ill with a feverish attack & not able to attend. Robert[2] officiated & people said very well—the little church looked very full & gay. Uncle Jos said afterwards he thought Charlottes behaviour *quite perfect.*— When we got back to Maer we all went to breakfast in the dining room a long table covered with confectionery meats tea etc. it was a very noisy pleasant breakfast & a little before 12 M^r & Mrs Langton drove off. I have not heard from Charlotte since she got to Ripley— the next day Cath & I came home, & I enjoy the quiet and repose very much— Fanny Owens marriage is put off for a few weeks—the next letter you will hear those particulars. Have you been told of Lucy Galton's marriage to Mr James Moilliet. they were married yesterday & I am sure you must think Mr Moilliet a happy man—. I have not heard lately of the Foxes— One more marriage I must tell Yesterday there was a paragraph in the Shrewsbury paper saying the licence & ring bought & 2 persons coming to be married at St Chads when suddenly the bride groom changed his mind & positively refused the clergyman Mr Compson expostulated all in vain— *Mark* proved to be the Hero & our Laundry maid the heroine. Nobody had an idea they were going to be married & we have had no explanation of this odd behaviour of Marks. they were really we find married this morning—[3] I have no more family news, except poor Pincher has cut the sinew of his foot with a glass bottle & they fear will be lame for life— My Father is very well & takes great pleasure in the Hot house which answers very well & the green house is filled with pretty gay flowers from it— Marianne is here she will write a flap to you— We hear that in some paper the Beagle was mentioned, but stupid Mrs Sneyd[4] can neither tell us what was said nor where nor when. My dear Charles I do so long to hear of you every day I cannot help expecting & hoping for a letter—

poor Nancy nurses Mariannes baby Charles & I believe crys over your past baby days. ⟨   ⟩ Papa & we often and often talk of you & hope you have quite got over sea sickness & are well & enjoying yourself most thoroughly but I still build upon the hope you will be content with out staying the whole time with the expedition—

God bless you my very dear Charles Papas and all our kindest love | Ever yr affectionate | Caroline Darwin

March 29th | 1832.

My dear Charles

I have begged for a flap. I have often longed to write to you to tell you how often & how much I have thought of you, but I shall not undertake to be one of your correspondents, my letters would be too dull to send you & you will hear of any great events happening to me such as the birth of Children &c from here—& you must not think dear Charles that I forget you— Parky learns his Geography lessons by where Uncle Charles is going to— We long for a letter from you as you may well suppose— Caroline has told you all the news to be told—

Good bye & God bless you | My dear Charles— Y^{rs} ever & very affec | M P.

DAR 204.6.1

1 Isaac Onslow Secker.
2 Robert Wedgwood.
3 Mark Briggs and Anne Latham were married on 31 March 1832 (St Chad's, Shrewsbury, marriage register). Mark was the Darwin family's coachman (*Emma Darwin* (1904) 2: 13).
4 The Sneyds of Keele, Staffordshire, were friends of the Darwin family.

To Caroline Darwin   2−6 April 1832

My dear Caroline.—

We are now about a hundred miles East of Rio, & tomorrow the 3^d of April we expect to arrive at the capital of the Brazils.— My last letter was from Bahia, which place the Beagle sailed from on the 18th of last month.— On the whole I much enjoyed my first visit to S America.— I was however very unfortunate in being confined to my hammock for eight days by a prick on the knee, becoming much inflamed.— Bahia has one great disadvantage in being situated on so large a space: that it was impossible for us to walk but in one direction.— Luckily it was by far the most beautiful.— The scenery here chiefly owes its charms to the individual forms of the Vegetation: when this is united to lofty hills & a bold outline, I am quite sure the incapability of justly praising it, will be almost distressing.— I talk of *enjoying Bahia*, in order to be moderate: but this enjoyment, (weighted with 8 days confinement,) is well worth all the misery I endured between England & Teneriffe.— I am looking forward with great interest for letters, but with very little pleasure to answering them.— It is very odd, what a difficult job I find this same writing letters to be.— I suppose it is partly owing to my writing everything in my journal: but chiefly to the number of subjects; which is so bewildering that I am generally at a loss either how to begin or end a sentence. And this all hands must allow to be an objection.—

The *mean* temperature of Bahia was 80; being more accustomed to heat I suffered less from it there than at Praya, where mean temp was 73°.— The great difference of climate in the Tropics & colder zones consists in the higher temp: of the nights.— A mean of 84° for the whole year (at Guyara in Columbia) is the hottest place in the world.—[1] so certainly I have experienced a very considerable degree.— To me it is most enjoyable: I had expected to wish for the cold thawing days, which you have lately been shivering under: no give me the regions of Palms & Oranges & away with frost & snow.— It requires a little additional energy to set about anything, & a good deal more to resist a siesta after dinner: When having so indulged one wakes bathed in perspiration, but with the skin as cool as a young child.—

We shall in all probability stay more than a month at Rio.— I have some thoughts, if I can find tolerably cheap lodgings of living in a beautiful village about 4 miles from this town.— It would be excellent for my collections & for knowing the Tropics. Moreover I shall escape cauking & painting & various other bedevilments which Wickham is planning.— One part of my life as Sailor (& I am becoming one, ie. knowing ropes & how to put the ship about &c) is unexpectedly pleasant; it is liking the bare living on blue water, I am the only person on the ship who wishes for long passages: but of course I cautiously bargain with Æolus, when I pray to him, that with the winds he may keep the sea equally quiet.— Coming out of Bahia, my stomach was only just able to save its credit.— I will finish this letter full of I^s I^s I^s when at Rio.—

Rio de Janeiro. April 5th.— I this morning received your letter of Decr 31 & Catherines of Feb 4th.— We lay to during last night, as the Captain was determined we should see the harbor of Rio & be ourselves seen in broard daylight.— The view is magnificent & will improve on acquaintance; it is at present rather too novel to behold Mountains as rugged as those of Wales, clothed in an evergreen vegetation, & the tops ornamented by the light form of the Palm.— The city, gaudy with its towers & Cathedrals is situated at the base of these hills, & command a vast bay, studded with men of war the flags of which bespeak every nation.—

We came, in first rate style, alongside the Admirals ship, & we, to their astonishment, took in every inch of canvass & then immediately set it again: A sounding ship doing such a perfect mæneuovre with such certainty & rapidity, is an event hitherto unknown in that class.— It is a great satisfaction to know that we are in such beautiful order & discipline.—² In the midst of our Tactics the bundle of letters arrived.— "Send them below," thundered Wickham "every fool is looking at them & neglecting his duty" In about an hour I succeded in getting mine, the sun was bright & the view resplendent; our little ship was working like a fish; so I said to myself, I will only just look at the signatures:, it would not do; I sent wood & water, Palms & Cathedrals to old Nick & away I rushed below; there to feast over the thrilling enjoyment of reading about you all: at first the contrast of home, vividly brought before ones eyes, makes the present more exciting; but the feeling is soon divided & then absorbed by the wish of seeing those who make all associations dear.—

It is seldom that one individual has the power giving to another such a sum of pleasure, as you this day have granted me.— I know not whether the conviction of being loved, be more delightful or the corresponding one of loving in return.— I ought for I have experienced them both in excess.— With yours I received a letter from Charlotte, talking of parsonages in pretty countries & other celestial views.— I cannot fail to admire such a short sailor-like "splicing" match.— The style seems prevalent, Fanny seems to have done the business in a ride.— Well it may be all very delightful to those concerned, but as I like unmarried woman better than those in the blessed state, I vote it a bore: by the fates, at this pace I have no chance for the parsonage: I direct of course to you as Miss Darwin.— I own I am curious to know to whom I am writing.— Susan I suppose bears the honors of being M^rs J Price.— I want to write to Charlotte—& how & where to direct; I dont know: it positively is an inconvenient fashion this marrying: Maer wont be half the place it was, & as for Woodhouse, if Fanny was not perhaps at this time Mrs Biddulp, I would say poor dear Fanny till I fell to sleep.— I feel much inclined to philosophize but I am at a loss what to think or say; whilst really melting with tenderness I cry my dearest Fanny why I demand, should I distinctly see the sunny flower garden at Maer; on the other hand, but I find that my thought & feelings & sentences are in such a maze, that between crying & laughing I wish you all good night.—

April 6th.— A merchant in this town³ is going to visit a large estate, about 150 miles in the country.— He has allowed me to accompany him.— On the 8th we start & do not return for a fortnight.— It is an uncommon & most excellent opportunity,—and I shall thus see, what has been so long my ambition, virgin forest uncut by man & tenanted by wild beasts.— You will all be terrified at the thought of my combating with Alligators & Jaguars in the wilds of the Brazils: The expedition is really quite a safe one, else I will wager my life, my host &

companion, would not venture on it.— I believe a packet will sail before I return if so this letter will go.— I will of course write again from Rio.— When I return I shall live in a cottage at the village of Botofogo: Earl & King will be my companions; I look forward to living there as an Elysium,— The house & garden is overwhelmed by flowers & is situated close to a retired lake, or rather loch, as it is connected with the sea, but landlocked by lofty hills.— I suppose we shall be here for 5 weeks: & then to Monte Video which will be my direction for a very long time.— With your nice letters, I received a most kind & affectionate one from Henslow.— It is not impossible I shall have occassion to draw for some money.— Most certainly this is the most expensive place we shall perhaps ever again visit.— My time i⟨s so⟩ very much occupied, that my letters must ⟨do⟩ for the whole family.— Before leaving Rio I shall send a begging letter for some books (the ⟨enjoyment of which is immense⟩ & instruments.

I have had a great deal of plague in getting my passport: a revolution is expected tomorrow which made it more difficult.— I am very sleepy & hot. So my dearest Caroline & all of you | Good bye.— Yrs very affectionately | Chas. Darwin

My love to every body who cares for me.— I hope I shall hear from Mr Owen (& Fanny).— His so kindly talking of me I value more than almost anybody.—

DAR 223

[1] CD may have been thinking of Alexander von Humboldt's statement that La Guayra, Venezuela (now La Guaira), with a temperature at noon of 26.2° 'is one of the hottest places on earth' (Humboldt 1814–29, 1: 378).

[2] Philip Gidley King remembered the event many years later as follows: 'Though Mr Darwin knew little or nothing of nautical matters he one day volunteered his services to the First Lieutenant. The occasion was when the ship first entered Rio Janeiro. It was decided to make a display of smartness in shortening sail before the numerous Men-of-War at the anchorage . . . Mr Darwin was told off to hold to a main-royal sheet in each hand and a top-mast studding-tack in his teeth. At the order "Shorten sail" he was to let go and clap on to any rope he saw was short handed— this he did and enjoyed the fun of it often afterwards remarking "the feat could not have been performed without him".' (Notes made for John Murray's new edition (1890) of *Journal of researches*. Copy in DAR 106/7 (ser. 3): 16).

[3] Patrick Lennon (see *'Beagle' diary*, p. 49).

From J. M. Herbert 15−17 April 1832

St: John's College Cambridge
Sunday April 15th 1832.

My dear old Fellow

I now sit down to fulfil an engagement which I entered into with you before you left England; a task which has something so awful in it that it required no common exertion to break thro' my ordinary indolence to effect it. Dr Johnson says that writing a short letter to a distant friend is like making an old acquaintance a distant bow, (and this with respect to a letter for Milan): to what length then ought I to spin out this Trans-Atlantic, trans-equatoreal one of mine? Strange changes have already taken place in Cambridge; amongst others I am a Fellow of St. John's, and you must therefore imagine me a strict disciplinarian, and a good judge in Port. Whitley is pressing all sail to get elected Mathematical Professor at Durham new College; common reports have given him to Johnny Cameron's Widow: he himself professes to be looking forward to marriage and torpidity— Cameron is in his own native bogs, and the Gluttons are all dispersed save Watkins, Whitley & myself. By the way how do you like Patagonian feeding; Lowe thinks you may collect a bundle of curious facts with respect to Cookery. I am going to commence reading for the Law immediately, and have some idea of going to reside at Chester for the next two years, to read with a Law-Tutor there. I dined with our old friend Henslow yesterday, and met Cap! Ramsay there (a brother of Ramsay's); he has just returned from cruising on the African Coast; he is a very agreable man, and extraordinarily like his brother— After dinner we all went to Professor Smythes',[1] who gave a lecture on "Ladies" for the edification of Mrs Somerville,[2] and the Dublin Hamilton,[3] who have been lionizing here for the last week. It was very amusing; I had no idea the old fellow had so much in him; after his lecture we had some good music, which he of course accompanied with his obligato Foolery. Mrs Frere[4] sold him the other day beautifully. He was crying down poor Rossini in a merciless manner, saying that he had nothing but frippery & tinsel. She commenced playing what she pretended was a piece of Rossini's; he kept calling out tame! wretched! trash! &c!! She was really playing a noble bit of Handel. Henslow is going to take out this summer a party of Pupils either to Weymouth or the Isle of Wight, who wish to go out in the Poll (οἱ πόλλοὶ) and are at the same time attached to the study of Natural History; Mrs Henslow and family are going along, so they must spend a delightful summer; would you not jumped at such an opportunity some two or three years ago? I have heard nothing of Eyton since you left; plucking seems a family failing as C. Eyton has just been plucked for his Little-Go, and W. Eyton has left for the Army in despair.[5] As you might not perhaps have heard lately from Shrewsbury I will just tell you that Miss Owen of Woodhouse is married to — — .. & Miss Parker to Sir B. Leighton; and it is reported that Biddulph of Chirk is going to run off with Miss Fanny Owen, but I cannot in the least vouch for the authenticity of this last report; there has been a desperate case of Crim: Con: there between Offley Crewe & Mrs Broughton Streye; he has to pay £5000 damages.—[6] You will of course be anxious to hear how the Reform Bill is going on; the second Reading passed the Lords on Friday after a debate of four days; majority nine; Lords Harrowby, Wharncliffe, Haddington &c &c have come round; it is quite glorious to find how fast men are ratting;[7] you I think are amongst a Tory Crew;

just put one of them in Pickle as by the time you return home, he will be more valuable as a specimen for the Cabinet of the Antiquarian, than your Fungi & Coleoptera for that of the Naturalist; if you can get hold of one with Monboddo's Tail,[8] or with ears prolongated, it will be a doubly-interesting specimen. I expect that you will in addition to your book on Natural History, prepare one illustrative of the manners & customs of the Fuegans & Pata-gonians; I quite envy you the opportunities you have of collecting materials for lying; I, having long found it difficult to tell lies with good grace on the Home department, have by learning Spanish turned my attention to the Foreign; I expect great amusement from the discovery of America; do you mind and take advantage of 60º. S.L. The Cholera has been doing its work pretty effectually, but not to the extent that was expected; there have been in the last week 105 cases at Ely, & 49 deaths— It has not yet reached Cambridge— Of all places where it has raged furiously Paris seems to have been the worst; one day last week there were 1020 cases & between 300 & 400 deaths; Casimir Perier has been attacked with it, but has now nearly recovered; Heaviside's brother has had it there— Henslow wished me to tell you that he has got your 17 Vol: of "La Dictionnaire Classique &c" description of the Plates—[9] You will think this a very rambling, unintelligible letter, as I have paid no regard to order or, I fear, to intelligibility—going on the principle that all news must be agreable to one at such a distance from home— Tuesday Morning I was last night at a Quarterly Meeting of the Choral, which you used to patronize to such an extent; and if your old friend Keats was right, you will, as you used to, feel a thrill thro' your back-bone at the very mention of some of your old favourites; for "Heard melodies are sweet, But those unheard are sweeter—"[10] The Choral is improved to a pitch that you would hardly credit; the different parts are so well sustained, and so exactly in tune, that you might fancy each of them huge individual voices; I doubt whether you can understand my meaning. We had first the Overture to Esther; then a very judicious selection from the Dettingen Te Deum—[11] Then "O first created brow &c" out of Samson: Then "Let their celestial concerts all unite" which was sung so splendidly that you might as I said before believe it a Quartett by four tremendous voices— then that elegant chorus from Solomon "May no rash intruder disturb their soft hours" &c; lastly the "Great and Glorious" chorus of Haydn— The second part commenced with Handel's Overture to Rodelinda—very beautiful, grand, and simple.— Then a splendid new Mass of Hummel's. The whole concluded with the Saul Chorus "Gird on thy sword" &c— There was a concert at Huntingdon last November, at which the Choral sang; where the London Great Guns were highly delighted with the extreme precision with which they executed some very difficult pieces of Harmony. Cambridge has been Music Mad this spring, as we have had a succession of Concerts, tho' none, with the exception of those given by young Aspull, have been particularly brilliant. The

Septett thrives; and the Caucus is getting on very well— We have just had a severe contest for the Registraryship which has fallen vacant by the death of Hustler,[12] between Romilly of Trinity,[13] and Chevallier. It was made to a great extent a Party Question; at least, Romilly was chiefly supported by the Whigs, Chevallier by the Tories. I am happy to be able to tell you that the former succeeded, after a day's hard Polling— Miller has been appointed Professor of Mineralogy in the room of Whewell resigned. His lectures are tremendously stiff; he had a room-ful at first, but his audience gradually diminished down to five. Do you recollect Sharpe?[14] He is going to get the Travelling Bachelorship; he is at present busily engaged in illustrating an Archi⟨te⟩ctural work of Whewell's—[15] I have just been having my Prints valued; ⟨they are⟩ condemned to the Hammer, as I find nothing at pres⟨ent⟩ so ornamental as the Ready— I have succeeded in raising a considerable quantity of Wind,[16] and I now can take the wall of my Tailor[17] with the greatest satisfaction & nonchalance. Being out of debt, or even approaching to that blissful state, is truly an enviable feeling. I carried *my* principle, and almost every body's principle, "In for a penny in for a pound" rather too far, during my residence in the University, but I ought not to regret, as I have spent many happy days there, and not a few of them in your company. It is, I fear, a purely selfish feeling, when I say I wish you back; as we all confess to feeling somewhat uncomfortable on passing X^ts. Gate. Those who wish you well—and they are many, for you must not in your case believe the old Spanish proverb "Ahora que te veo me acuerdo"—[18] aint I bumptious?—ought to console themselves for your absence, by the reflexion that you are now engaged in collecting materials for future fame; that you are about to couple your name, already intimately connected with Science, with those of a Cuvier and a Humboldt. Don't think me guilty of Flattery—I know you will do great things, as it is impossible that your assiduity and talents should not succeed. When you do return, take compassion on the briefless barrister—

We are getting quite liberal at St: John's; I've just been asked to subscribe for Portraits of our two great Luminaries in Science & Literature—Herschel & Wordsworth—both of this College— Science is at present certainly on the Advance: when was there woman before M^rs. Somerville capable of abridging La Place's Mecanique Celeste. On being introduced to La Place, as having read his Book; he observed that there was only one woman before who had done so—a M^rs. Glegg—which was the name of *her* former husband.—[19] I have now told you all that my rambling wool-gathering head can think of, and my paper reminds me that it is time to bid you farewell; and, tho' it has not all the bitterness of a former one, this, believe me, is not without its sting. All friends here desire the kindest remembrance, among whom are Henslow, Whitley, & Watkins.— God be wi' you, and prosper all your efforts. I will not urge you to reply, as your time is necessarily much occupied: but let not the fear of writing a long letter deter you from writing, as a few lines to tell me that you are well and thriving will be most thankfully received by,

My dearest Darwin, your ever sincere friend | J. M. Herbert.

DAR 204.6.2

[1] William Smyth.
[2] Mary Somerville. Her adaptation, *Mechanism of the heavens* (1831), of Pierre-Simon, Marquis de Laplace's *Mécanique céleste* made her famous.
[3] William Rowan Hamilton, distinguished here from William Hamilton, the Scottish metaphysician and logician.
[4] Mary Frere, wife of William Frere, Master of Downing College.
[5] Charles James Eyton and William Archibald Eyton.
[6] An action was brought by Thomas Broughton Strey against John Offley Crewe 'for a criminal conversation with the plaintiff's wife' (*The Times*, 21 March 1832, p. 3).
[7] 'Desertion of one's party or principles' (*OED*).
[8] Lord Monboddo was the courtesy title of James Burnett, Scottish judge, who believed men originally had tails, which were gradually worn away by the habit of sitting.
[9] Bory de Saint-Vincent 1822 − 31. CD's set is preserved in Darwin Library − Down. Volume 17, *Atlas. Illustrations des planches* is without plates.
[10] John Keats, 'Ode on a Grecian urn' (1819).
[11] A setting of the 'Te Deum' written by Handel to celebrate the victory of Dettingen, 26 June 1743 (Grove 1980, 8: 121).
[12] William Hustler.
[13] Joseph Romilly.
[14] Edmund Sharpe. He studied architecture in France and Germany while on the Worts Travelling Fellowship (*DNB*).
[15] Whewell 1835.
[16] 'Saving money' (*OED*).
[17] To keep to the clear side of the walk, nearest the wall. 'I will take the wall of any man or maid of Montague's.' Shakespeare, *Romeo and Juliet*, I. I. 10 − 11 (Arden edition).
[18] 'Now that I see you I remember you'.
[19] Mary Somerville's first husband, who died in 1807, was Captain Samuel Greig (Somerville 1873).

To Caroline Darwin 25 − 6 April [1832]

Botofogo Bay
April 25th

My dear Caroline

I had sealed up the first letter, all ready to be sent off during my absence: but no good opportunity occurred so it & this will go together.— I take the opportunity of Maccormick returning to England, being invalided, ie. being disagreeable to the Captain & Wickham.— He is no loss.—[1] Derbyshire[2] is also discharged the survice, from his own desire not choosing his conduct which has been bad about money matters to be investigated.—

All this has been a long parenthesis.— My expedition lasted 15 days, most of which were ones of uncommon fatigue; I suppose for a civilized country travelling could not be worse.— the greatest difficulty in getting any thing to eat, & not undressing for the five first days.— I was very unwell for two days, & the misery of riding in a scorching sun for about 10 hours was extreme.— My horror of being left utterly destitute in a Venda will be better than any schoolmaster to make me learn Spanish, as soon as we get into those countries.— On the other side, there was a great interest & novelty in seeing the manner of living amongst the Brazilians, which rare of opportunity of doing during a few days in which I resided at a Fazenda, that is one of the most interior cleared estates.— Their habits of life were quite patriarchal.— Forest, & flowers & birds, I saw in great perfection, & the pleasure of beholding them is infinite.— I advice you to get an French engraving, Le Foret du Bresil: it is most true & clever.— This letter will be odds & ends, as really I have scarcely time for writing.— I send in a packet, my commonplace Journal.— I have taken a fit of disgust with it & want to get it out of my sight, any of you that like may read it.— a great deal is absolutely childish: Remember however this, that it is written solely to make me remember this voyage, & that it is not a record of facts but of my thoughts.— & in excuse recollect how tired I generally am when writing it.—

Earl & myself are now living in this most retired & beautiful spot.— I trust to spend a most delightful fortnight.— I have begun however with a bad omen.— whilst landing the boat was swamped; a heavy sea knocked me head over heels & filled the boat.— I never shall forget my agony, seeing all my useful books, papers,—instruments microscopes &c &c gun rifle all floating in the Salt Water: every thing is a little injured, but not much: I must harden myself to many such calamities.— It is very lucky I have such nice lodgings as the ship is turned inside out, a large party of the officers have gone up the river in the cutter.— I came just too late for this cruize.— I believe King is coming to live here: he is the most perfect, pleasant boy I ever met with & is my chief companion.— Wickham is a fine fellow.—& we are very good friends.— which in a selfish way is no common advantage.— And now for the Captain, as I daresay you feel some interest in him.— As far as I can judge: he is a very extraordinary person.— I never before came across a man whom I could **fancy** being a Napoleon or a Nelson.— I should not call him clever, yet I feel convinced nothing is too great or too high for him.— His ascendacy over every-body is quite curious: the extent to which every officer & man feels the slightest rebuke or praise would have been, before seeing him incomprehensible: It is very amusing to see all hands hauling at a rope they not supposing him on deck, & then observe the effect, when he utters a syllable: it is like a string of dray horses, when the waggoner gives one of his aweful smacks.— His candor & sincerity are to me unparralleled: & using his own words his "vanity & petulance" are nearly so.— I have felt the effects of the latter: but the bringing into play the former ones so forcibly makes one hardly regret them.— His greatest fault as a companion is his austere silence: produced from excessive thinking: his many good qualities are great & numerous: altogether he is the strongest marked character I ever fell in with.

Be sure you mention the receiving of my journal, as anyhow to me it will of considerable future interest as it an exact record of all my first impressions, & such a set of vivid ones they have been, must make this period of my life always one of interest to myself.— If you will speak quite sincerely,—I should be glad to have your criticisms. Only recollect the above mentioned apologies.—

I like this sort of life very much: I can laugh at the miseries of even Brazilian travelling.— I must except one morning when I did not get my breakfast till one oclock having ridden many miles over glaring sand.— Generally one is obliged to wait two hours before you can get anything to eat, be the time what it may.— Although I like this knocking about.—I find I steadily have a distant prospect of a very quiet parsonage, & I can see it even through a grove of Palms.—

Friday. The Captain has just paid us a visit & taken me to the Ministers, where I dine on Monday & meet the very few gentlemen there are in the place.— He has communicated to me an important piece of news; the Beagle on the 7th of March,[3] sails back to Bahia.— The reason is a most unexpected difference is found in the Longitudes it is a thing of great importance, & the Captain has written to the Admiralty accordingly.—[4] Most likely, I shall live quietly here, it will cost a little, but I am quite delighted at the thought of enjoying a little more of the Tropics: I am sorry the first part of this letter has already been sent to the Tyne: I must tell you for your instruction that the Captain says, Miss Austens novels are on every body table, which solely means the Jerseys Londonderrys &c.—

You shall hear from me again from Rio, how I wish I could do the same from you.—

Remember me most affectionately to every body, & to my Father, Susan & Catherine & Erasmus.— The latter must not forget to write to me.— I would write to each of you.—only it is in reality useless.— | Good bye & good night to all of you | Yours ever affectionately | Charles Darwin

April 26th Rio de Janeiro

DAR 223

[1] Robert McCormick stated his reason for leaving as follows: 'Having found myself in a false position on board a small and very uncomfortable vessel, and very much disappointed in my expectations of carrying out my natural history pursuits, every obstacle having been placed in the way of my getting on shore and making collections, I got permission from the admiral in command of the station here to be superseded and allowed a passage home in H.M.S. *Tyne*.' (McCormick 1884, 2: 222). Benjamin Bynoe, the Assistant Surgeon, served as Surgeon for the remainder of the voyage.

[2] Alexander Derbishire, Mate of the *Beagle* (*Narrative* 2: 19).

[3] A mistake for 'May'. The *Beagle* sailed for Bahia on 10 May (see '*Beagle' diary*, p. 60).

[4] *Narrative* 2: 75: 'As I found that a difference, exceeding four miles of longitude, existed between the meridian distance from Bahia to Rio, determined by the French expedition under Baron Roussin, and that measured by the Beagle; yet was unable to detect any mistake or oversight on my part; I resolved to return to Bahia, and ascertain whether the Beagle's measurement was incorrect.' Robert FitzRoy's measurement between Bahia and Rio was confirmed 'even to a second of time' (*ibid.*, p. 78).

From Catherine Darwin   26—27 April [1832]

Shrewsbury.
April 26th.

My dearest Charles.

We have been much disappointed not to have heard from you from Madeira, but suppose that either you have not touched there, or there was no conveyance to England. You cannot think how we all long and talk about your first letter from Rio; what pleasure it will be to hear that you are well and prosperous, and recovered from sea sickness, and are beginning to enjoy yourself. The last letter to you went the 1st of April directed to Monte Video, from Caroline,—and a letter went from Woodhouse by the 1st of March from Mr Owen and Fanny, directed to Rio Janeiro. If I am to judge of the pleasure letters from England will give you, by that which your's will give us, it will be great indeed, and I do suppose you must care even more, so mind to give us all possible directions about writing to you.— Papa and we all are well here, and just in statu quo you left us. We have had Erasmus down with us for Easter Week, and we have been often talking about you, my dearest old Charley. Erasmus came to us from Northamptonshire, where he had been paying a sentimental visit to his *friend* Mrs Whitworth, one of the many female friends he made abroad. We were rather *scandalized* at this, but were quite relieved by the first word he said about her; "that she was the horriddest brute that ever lived, and sang like a Barrel Organ". He found the house so intolerably stupid, that after spending two days there, he sent for a Post Chaise, and rattled off to catch the Wonder, promising to stay longer the *next* time he came to them.— We have not been able to make out the offence, except the poor woman's singing and her Husband's talking Toryism.— Fanny Owen's marriage is still delayed. Mr Biddulph has been staying at Woodhouse the last month, but he is so dawdling, and the Settlements are so long making, that it is the most tedious affair possible; and now there is another delay in Mr Owen's being obliged to go to London, to take Arthur up, and see him off; his Ship sails the middle of May. Caroline Owen is going up to London with them to stay with Mrs Williams, who has been exceedingly ill, and people say, is reduced to a perfect Ghost. You will hardly know your old friend Sarah, under her new name, and I am afraid Mrs Myddelton Biddulph will be still more strange to you.—

I wrote so far yesterday, and this morning poor Arthur called to wish us Goodbye on his road to London. It was most melancholy to see him, poor boy, hardly able to speak, and turning away his face from us to hide his tears. I never was so sorry for anybody. He had first had the parting with Mrs Owen and Fanny, and the children this morning, and this second leave taking to us seemed quite to overcome him. He looked the picture of sadness. Poor boy, how I hope he may live & be happy in India; it is a capital thing for him that he does not like wine.— Susan has been almost living at Woodhouse, lately, flirting alternately with the Captain, and Arthur.— You are not at all forgotten by Mr Owen, or by any of the Owens, I assure you; Mr Owen talks of you with as much affection and interest as ever. Sir Baldwin Leighton is another person who always enquires very much after you. He and his Bride called here the other day. I did not think that you knew much of him, and I suppose it is his own love of foreign travels gives him such an interest in you.—

This is the Wedding Day of Frank Wedgwood and Miss Mosley, and of Edward Holland, and Miss Isaac. Think what marrying times these are, that 2 relations should be married the same day. Erasmus gives such an account of Mrs Edward Holland. He says she is the reverse of Falstaff; she is not only stupid herself, but makes other people stupid, and that she is vulgar and sulky, and the state of their Country House he thinks will be alarming.— We know nothing of Miss Mosley yet, except that she is immensely fat, and the most excellent scrattle, talks of "cutting up pigs without waste". M^r and Mrs Langton are living at Ripley in Surrey, very comfortably.— It is a nice village, in a very pretty country, 23 miles from London. Charlotte seems to be very much in love by all accounts.— Do you remember your Prophecy you made to Erasmus? that you should find him tied neck and heels to E⟨mma⟩ Wedgwood, an⟨d⟩ heartily sick of her, in short in ⟨the⟩ same state that Harry & Jessie are supposed to be; I am much amused at your prophecy, and I think it may possibly have a good effect, and prevent its own fulfilment.—

Friday 27th of April. Erasmus left us today; he told me to send you his love, and to tell you that he did not write to you, as you and he had come to an understanding not to write to each other, and that *Brothers* never could write to each other. Erasmus talks with the greatest interest for your letters, and says "how grand it makes him feel, and how strange it is, actually to have a Brother in South America." Dearest Charley, how glad I shall be when our grandeur comes to an end, and we have you with us again,—and oh! that there was any chance of your returning from South America, before the Beagle's Course is finished. We are exceedingly interested to know whether your liking and admiration for your Captain continues, and how you like the rest of your company.

Pray remember how interested we are, and keep writing a Journal on and on for us, and tell us every thing. I am afraid you must have been much disappointed not to have seen Madeira, as we suppose, must have been the case, by not having heard from you.— I suppose you get the Papers, and will have seen how tremendous the Cholera has been in Paris. It is quite dying away in London now, but spreading over the Country fast. One case is said to have taken place in Whitchurch, so we shall have it directly in Shrewsbury, and Papa being the Head Doctor of the Board of Health is really awful. You will escape *this* danger at least; I wish I could be sure of your escaping all the innumerable other dangers you are exposed to. For Heaven's sake, take care of yourself, is all I entreat of you, and don't take any violent fatigues, and do your health great harm. That is what Papa is always so afraid of for you. I am sure *prudence* must do a great deal in saving people from risks and dangers, and my hope is in your sense saving you. God bless you, dearest Charles. I never knew how much I loved you before. Papa's and all our most affectionate loves. Will you send word the dates of the letters you receive, that we may know whether you receive all.—

Ever y^rs dearest Charles. E. C. D.

Postmark: Shrewsbury ⟨A⟩P 29 1832
DAR 204.6.1

To Catherine Darwin   May—June [1832]

Botofogo Bay, Rio de Janeiro
May—June

My dear Catherine

I have now altogether received three letters; your & Carolines together which latter I have answered & also sent my Journal by the Tyne, which was returning to England.— Susans (& one from Mr Owen) I received May 3^d.— The Beagle has not yet returned; so I am living quietly here & throughily enjoying so rare an opportunity of seeing the country & collecting in every branch of Nat. History.— I have just been rereading all your nice affectionate letters, & in consequence I have summoned resolution to begin a letter.— I am so wearied of writing letters & telling the same story; that if I stumble through this; it is almost more than I expect.— I have sent a list of commissions for poor Erasmus to execute; directed to Whyndam club tell my Father I am afraid some of them are expensive: but he cannot imagine the value such things are in a country, where even a watch never yet has been manufactured.— I am very glad to hear the hot-house is going on well; how when I return I shall enjoy seeing some of my old friends again.— do get a Banana plant, they are easily reared & the foliage is wonderfully beautiful.— I have not yet ceased marvelling at all the marriages: as for Maer & Woodhouse, they might as well be shut up. I received a very kind letter from Mr Owen & Fanny.— The former contained the warmest expressions of friendship to my Father.— (This letter will be odds & ends).— I suppose by this time you see how uncertain ship-letters must always be.— When we get to the South & have a 5 month cruize without seeing an homeward bound sail, together with the chances both before & after, the time might be almost indefinite between two of my letters.— The Admirals secretary here was under Cap: Maling; who seems to have had a great deal of duty, at a very precarious time: the secretary says that Mrs Maling entirely managed the political part.—[1]

June 6th.— The Beagle has returned from Bahia & brought most calamitous news.— a large party of our officers & 2 sailors before leaving Rio went a party in the Cutter for snipe shooting up the bay.— Most of them were slightly attacked with fever: but the two men & poor little Musters were seized violently & died in a few days.[2] The latter & one man were buried at Bahia.— The poor little fellow only two days before his illness heard of his mothers death.— What numbers snipe-shooting has killed, & how rapidly they drop off.— The Beagle will stay another 14 days at least & then we sail for Monte Video touching I hope at St Catherines.— She is getting in beautiful order; increased our compliment, got a new gun: put up boarding nettings, & rigged sweeps.—& now there is not a pirate a float, whom we care for & a thousand savages together, would be harmless.—

I have written letters to Charlotte, Mr Owen, Fox, & Henslow Herbert I mention it; being always anxious, when it is possible to know whether my letters have arrived safely.— I received a nice long one from Caroline, dated Maer.—& directed to Mr Darwin, HMS.— Am I a ship? or is his Majesty ship Beagle a dog? that you stick a *the* before it.— One would suppose she did not know the Jib boom from the Taffrill.—to see her direct in such a manner.— (odds & ends as I before said) Capt. Harding, brother of Mrs Hunt second Capt of the Warspite, is here & is very civil to me.— He sent to me to say he had 800 men under his command, & that I might have a boat for an hour or week, as I choose.— One of our officers lives at Falmouth, he gives the following direction for letters.— there are two packets sail every month one for Rio the other touching at Rio proceeds to M Video.— This one sails the Friday after the 3^d Tuesday in the Month & is of course the best way of sending my monthly letter.— The letter ought to be in a day before the Friday.— Whenever you should in doubt about direction, put, South American station.— Till I tell you stick to Monte Video You cannot imagine anything more calmly & delightfully than these weeks have passed by.— there never was a greater piece of good luck that the Beagle returning to Bahia.— Give my best love to Marianne & thank her for her postscript; & tell her to remember me most kindly to Dr Parker.— Remember me to all friends, especially Major Bayley & the Eytons.— tell Tom to keep his courage up for the Canaries, or Madeira would be very feasible. I drew 40£ (mentioned in my last letter) & I am afraid I shall be forced to draw 10£ more.— I really am very sorry.—but 12 weeks here instead of 4 has been a great increase.— My loudgings & board only cost 22 shillings per week.—

I ashamed to send so uninteresting a letter; but it will be to you unintelligible, how difficult I find writing letters.— At latter end of this month (June) we sail for M. Video.— Our first course will be I believe down the coast to Rio Negro, where there is a small settlement of Spaniards.— Our next will be to where man has never yet been; (that is as far as is known.)— How glad I am, the Beagle does not carry a years provisions; formerly it was like going into the grave for that time.— Living with the Captain, is a great advantage, in having what society there is, at my command.— I am only one in the ship who is regularly asked to the Admirals, Chargé d'affaires & other great men.—

With my very best love to every one. dear Katty | Yours most affectionately | Chas. Darwin

DAR 223

[1] See letter from Catherine & Susan Darwin, 4 December [1825].
[2] See Keynes 1979, pp. 55—6 (*Narrative* 2: 76—7) for Robert FitzRoy's account of the deaths.

To W. D. Fox   May 1832

Botofogo Bay, near Rio de Janeiro
May 1832.—

My dear Fox

I have delayed writing to you & all my other friends till I arrived here & had some little spare time.— My mind has been since leaving England in a perfect *hurricane* of delight & astonishment. And to this hour scarcely a minute has passed in idleness.— I will give you a very short outline of our voyage. We sailed from England after much difficulty on the 27th of December & arriv'd after a short passage to St Jago.— I suffered exceedingly all the first part, the snowy peak of Teneriffe by convincing me I was well on the road to see the world first put fresh life into me.— At St Jago my Natura Hist: & most delightful labours commenced.— during the 3 weeks I collected a host of marine animals, & enjoyed many a good geological walk.— Touching at some islands we sailed to Bahia, & from thence to Rio, where I have already been some weeks.—

My collections go on admirably in almost every branch. as for insects I trust I shall send an host of undescribed species to England.— I believe they have no small ones in the collections, & here this morning I have taken minute Hydropori, Noterus Colymbetes, Hydrophilus, Hydrobius, Gyrinus, Heterocerus Parnus, Helophorus Hygrotius, Hyphidrus, Berosus &c &c, as a specimen of fresh-water beetles.— I am entirely occupied with land animals, as the beach is only sand; Spiders & the adjoining tribes have perhaps given me from their novelty the most pleasure.— I think I have already taken several new genera.—[1] But Geology carries the day; it is like the pleasure of gambling, speculating on first arriving what the rocks may be; I often mentally cry out 3 to one Tertiary against primitive; but the latter have hitherto won all the bets.— So much for the grand end of my voyage; in other respects things are equally flourishing, my life when at sea, is so quiet, that to a person who can employ himself, nothing can be pleasanter.—the beauty of the sky & brilliancy of the ocean together make a picture.— But when on shore, & wandering in the sublime forests, surrounded by views more gorgeous than even Claude[2] ever imagined, I enjoy a delight which none but those who have experienced it can understand— If it is to be done, it must be by studying Humboldt.—

At our antient snug breakfasts at Cambridge, I little thought that the wide Atlantic would ever separate us; but it a rare priviledge, that with the body, the feelings & memory are not divided.— On the contrary the pleasantest scenes in my life, many of which have been in Cambridge, rise from the contrast of the present the more vividly in my imagination.— Do you think any diamond beetle will ever give me so much pleasure as our old friend Crux Major.— Can we ever forget our few days at Whittlesea Meer with little Albert?[3] It is one of my most constant amusements to draw pictures of the past, & in them I often see you & poor little Fan—Oh Lord, & then old Dash poor thing!— do you recollect how you all tormented me about his beautiful tail.— I am now living here by myself, as the Beaglle has returned to Bahia to settle a longitude question, & about the middle of next month we shall sail for Monte Video.— I rely upon your writing to me there (it will be our head quarters for a terrible long time) direct M⟨ ⟩ HMS. B⟨eagle⟩ Monte Video S. America: do as I have done, & tell me all about yourself how contrive to live on such a stationary, slow sailing craft as a Parsonage: what you are, have, & intend doing.— Remember minutiae become more not less interesting, as the distance increases.—

I suppose I shall remain through the whole voyage, but it is a sorrowful long fraction of ones life; especially as the greatest part of the pleasure is in anticipation.— I must however except that resulting from Natur— History; think when you are picking insects off a hawthorn hedge on a fine May day (wretchedly cold I have no doubt) think of me collecting amongst pineapples & orange trees; whilst staining your fingers with dirty blackberries, think & be envious of ripe oranges.— This is a proper piece of Bravado, for I would walk through many a mile of sleet, snow or rain to shake you by the hand, My dear old Fox. God Bless you.

Believe me | Yours very affectionately | Chas Darwin

Remember me most kindly to M^r & Mrs Fox & to all your family: Once more good night & good bye.

Christ's College Library, Cambridge (Fox 46)

---

[1] After the voyage CD's insect specimens were described by Charles Cardale Babington, Frederick William Hope, Francis Walker, George Robert Waterhouse, and Adam White. See *Collected papers* 2: 295–300.
[2] Claude Lorrain.
[3] Albert Way.

From Susan Darwin   12 May [− 2 June] 1832

Maer.

May 12th. 1832.

My dear Charles.—

On the 3^d of May we received your *last* written letter from St Salvador, which all the house rejoiced over most heartily: the happy account you give of yourself and all your enjoyments in the tropical world far exceeded what we most hoped for you.— Your letter has been read very often over to Papa (like M^{rs}. Bates)[1] and I think he never can again for shame make his old speech of *the gaol & the ship*: now he has heard what a comfortable home you find it.— Nancy & Edward were as much pleased as any of us to have tidings of you: Erasmus had been making some very bruttal jokes to the former a week before: "how that you were lost & we kept it from her" so the poor old soul very nearly cried when she heard you were really alive & well.— Papa was much interested by your miserable account of the sea sickness you had endured, & not a little proud of his prescription of the Raisins answering so well. I think he should publish such a discovery for the benefit of all such sufferers.—

The Letter you wrote by the Lyra came to us the day after we had your other letter which was very odd as the dates were so different but I suppose the Winds had kept all the Ships out together. We have sent your letters to Erasmus who is to shew them to Charlotte *Langton* & then the family here must see it for many enquiries they are making about you & many kind messages they desire me to send you.— The day after yr news came I had to write several despatches, first to Eyton copying out all your messages which I hope may have the desired effect of instantly starting him accross the Atlantic, which I fully expect he will do some time he talks so much about it: & whenever I have an opportunity I shall certainly urge him on.— Major Bayley & Woodhouse too I had to send the joyful intelligence, as they all begged to hear the first news of you, so I can promise you my dear Charley you are not att all forgotten amongst any of us, & I am very glad to hear the warm climate agrees so well with all your affectionate feelings.— We were very sorry for yr disappointment at the Canaries but I hope Cape Verd Islands comforted you: Master Parky went through the whole of yr voyage as a Geographical lesson: so Marianne made profit as well as pleasure fr your letter.—

Catherine wrote last month to Monte Video, where Caroline sent her March letter also.— I am in hopes we shall before very long hear from you again in answer to the letters you will find at Rio, & I hope then you will give us some more directions for our future letters as I think it is doubtful whether this ever reaches you.— I cannot help thinking how lucky it was you took that Tour with Professor Sedgwick as Geology seems to be so great a pleasure now.— Poor Arthur Owen sails this week I believe to Madras, he wrote me a little note f^r. London where he had been seeing his Ship very often It is the Abberton: there are only 2 young Ladies amongst the Passengers which I think he regrets as a very short allowance amongst all the *Writers*:² M.ʳ Owen is with him in London now preparing his Outfit: & Arthur expected he would be in such a bustle that by some mistake he would go in his place to India.— Francis wanted very much to go out with Arthur & I sh.ᵈ think it w.ᵈ be by far the best scheme if they can get any situation for such a pickle as he is.— Caroline Owen is staying in town with M.ʳˢ Williams: so poor Fanny who has very much lost her former *housemaid* spirits is left all alone with the M.ʳˢ in the Forest.— M.ʳ Biddulph has at last settled to have an Entail which has caused all this delay, & the marriage will take place as soon as M.ʳ Owen returns.— I am now spending a fortnight at Maer which I find sadly changed: Charlotte's loss is quite irreparable! & Frank being gone too, makes the party appear much smaller.— Uncle Jos talks much of poverty his Children having taken off so much money with their Marriages:— Fanny Wedgwood has a flirtation in han⟨d⟩ with M.ʳ Paget Moseley (brother to Franks wife) but as I hear he is a very dull man, I have no f⟨ear⟩ of Fanny accepting him.— Politics are much t⟨alked⟩ of here, as the Ministers are gone out, not being able to get the King to make Peers to carry the Reform, & of course that interests this public spirited family very deeply.— The week before I left home M.ʳ & M.ʳˢ Edward Holland on their bridal Tour came & spent 3 days with us.— It was rather strange their liking to come so directly after their marriage, particularly as little M.ʳˢ H. is very shy.— however Papa joked away with them famously & made them very merry & easy.— We heard such dismal accounts of the Bride from Erasmus who is a most fastidious mortal: that we were very agreeably surprised to find her rather pretty & a nice little creature all together & particularly well suited to Edward who seemed very proud of her.—

I am come home from Maer, & I find Eyton has been calling in consequence of our hearing from you: but he is going immediately into Germany for the next 3 months with M.ʳ F Hope on a Beetle expedition.— Professor Sedgwick dropped in here last week on his way into Wales: he talked much about you & sent his kind regards, & begged you might be told to examine the gravel banks of small rivers for animal remains.— We have seen to day in the papers a sad piece of news: that Sir James Mackintosh is dead: he had been ill about a week, & died the 30.ᵗʰ of May:— M.ʳ & M.ʳˢ Hensleigh Wedgwood will be obliged to retrench very much after losing £1200 a year.— Lord Grey is come into Office again so the Reform Bill will certainly pass triumphantly. It is a great pity poor Sir James has died before he has finished his History of England.—³ Papa desires I will give you his most affectionate love & many thanks for your very nice letter which has given him a great deal of pleasure.— When I told M.ʳ Owen how happy you were tears came into his eyes with pleasure I am sure he considers you one of his own children.— Fanny was married on the 31.ˢᵗ of May to M.ʳ Biddulph. I have not room here for any particulars, but Caroline will fill her next letter with them.

God bless you my very dear Charley. All our Loves & I am | Ever yr affecti |
Susan E Darwin

We have written once every Month since you left us. All directed to Rio Janeiro, till 1ˢᵗ of April, & then M Video Mʳ Owen has also written to Rio Janeiro.—

Postmark: Shrewsbury JU 2 1832
DAR 204.6.1

[1] Widow of the Vicar of Highbury, in Jane Austen's *Emma* (1816).
[2] Writer: 'A clerk in the service of the former East India Company' (*OED*).
[3] Mackintosh 1830−2.

To J. S. Henslow[1] 18 May − 16 June 1832

Rio de Janeiro.
May 18ᵗʰ 1832

My dear Henslow.—

I have delayed writing to you till this period as I was determined to have a fair trial of the voyage. I have so many things to write about, that my head is as full of oddly assorted ideas, as a bottle on the table is with animals.— You being my chief Lord of the Admiralty, must excuse this letter being full of my's & I's.— After our two attempts to put to sea in spite of the S.W.ˡʸ gales, the time at Plymouth passed away very unpleasantly.— I would have written, only I had nothing to say, excepting what had better be left unsaid: so that I only wrote to Shrewsbury.— At length we started on yᵉ 27ᵗʰ of December with a prosperous wind, which has lasted during our whole voyage:— The two little peeps at sea-sick misery gave me but a faint idea of what I was going to undergo.— Till arriving at Teneriffe (we did not touch at Madeira) I was scarcely out of my hammock & really suffered more than you could well imagine from such a cause.— At Santa Cruz, whilst looking amongst the clouds for the Peak & repeating to myself Humboldts sublime descriptions, it was announced we must perform 12 days strict quarantine.— We had made a short passage so "Up Jib" & away for St Jago.— You will say all this sounds very bad, & so it was: but from that to the present time it has been nearly one scene of continual enjoyment.— A net over the stern kept me at full work, till we arrived at St Jago: here we spent three most delightful weeks.— The geology was preeminently interesting & I believe quite new:[2] there are some facts on a large scale of upraised coast (which is an excellent epoch for all the Volcanic rocks to [be] dated from) that would interest Mʳ Lyell.—[3] One great source of perplexity to me is an utter ignorance whether I note the right facts & whether they are of sufficient importance to interest others.— In the one thing collecting, I cannot go wrong.— St Jago is singularly barren & produces few plants or insects.—so that my hammer was my usual companion, & in its company most delightful hours I spent.—

On the coast I collected many marine animals chiefly gasteropodous (I think some new).— I examined pretty accurately a Caryophyllea & if my eyes were not bewitched former descriptions have not the slightest resemblance to the animal.— I took several specimens of an Octopus, which possessed a most marvellous power of changing its colours; equalling any chamaelion, & evidently accommodating the changes to the colour of the ground which it passed over.—yellowish green, dark brown & red were the prevailing colours: this fact appears to be new, as far as I can find out.— Geology & the invertebrate animals will be my chief object of pursuit through the whole voyage.— We then sailed for Bahia, & touched at the rock of St Paul.— This is a Serpentine formation.— Is it not the only island in the Atlantic which is not *Volcanic?*—[4] We likewise staid a few hours at Fernando Noronha; a tremendous surf was running, so that a boat was swamped, & the Captain would not wait.— I find my life on board, when we are in blue water most delightful; so very comfortable & quiet: it is almost impossible to be idle, & that for me is saying a good deal.— Nobody could possibly be better fitted out in every respect for collecting than I am: many cooks have not spoiled the broth this time; Mr Brownes little hints about microscopes &c have been invaluable.— I am well off in books, the Dic: Class: is *most useful.*— If you should think of any thing or book that would be useful to me; if you would write one line E Darwin Whyndham Club[5] St James Sq^r.— He will procure them, & send them with some other things to *Monte Video*, which for the next year will be my head quarters.— Touching at the Abrolhos, we arrived here on April 4th; when amongst others I received your most kind letter: you may rely on it, during the evening, I thought of the many most happy hours I have spent with you in Cambridge.— I am now living at Botofogo, a village about a league from the city, & shall be able to remain a month longer.— The Beagle has gone back to Bahia, & will pick me up on its return.— There is a most important error in the longitude of S America, to settle which this second trip has been undertaken.— Our Chronometers at least 16 of them, are going superbly: none on record ever have gone at all like them.— A few days after arriving I started on an expedition of 150 miles to Rio Macaò, which lasted 18 days.— Here I first saw a Tropical forest in all its sublime grandeur.— Nothing, but the reality can give any idea, how wonderful, how magnificent the scene is.— If I was to specify any one thing I should give the preemenence to the host of parasitical plants.— Your engraving[6] is exactly true, but underates, rather than exagerates the luxuriance.— I never experienced such intense delight.— I formerly admired Humboldt, I now almost adore him; he alone gives any notion, of the feelings which are raised in the mind on first entering the Tropics.—

I am now collecting fresh-water & land animals: if what was told me in London is true, viz that there are no small insects in the collections from the Tropics.— I tell Entomologists to look out & have their pens ready for describing.— I have taken, as minute (if not more so) as in England, Hydropori,

Hygroti, Hydrobii, Pselaphi, Staphylini, Curculio, Bembididous insects &c &c.— It is exceedingly interesting observing the difference of genera & species from those which I know. it is however much less than I had expected I am at present red-hot with Spiders, they are very interesting, & if I am not mistaken, I have already taken some new genera.— I shall have a large box to send very soon to Cambridge, & with that I will mention some more Natural History particulars.

The Captain does every thing in his power to assist me, & we get on very well.—but I thank my better fortune he has not made me a renegade to Whig principles: I would not be a Tory, if it was merely on account of their cold hearts about that scandal to Christian Nations, Slavery.— I am very good friends with all the officers; & as for the Doctor he has gone back to England.—as he chose to make himself disagreeable to the Captain & to Wickham He was a philosopher of rather an antient date; at St Jago by his own account he made *general* remarks during the first fortnight & collected particular facts during the last.—

I have just returned from a walk, & as a specimen how little the insects are know.—Noterus, according to Dic Class. contains solely 3 European species, I, in one hawl of my net took five distinct species.— is this not quite extraordinary?.—

June 16th— I have determined not to send a box till we arrive at Monte Video.—it is too great a loss of time both for Carpenters & myself to pack up whilst in harbor.— I am afraid when I do send it, you will be disappointed, not having skins of birds & but very few plants, & geological specimens small: the rest of the things in bulk make very little show.—

I received a letter from Herbert, stating that you have a vol: of Dic Class— Will you send it to Whyndam Club.— I suppose you are at this moment in some sea-port, with your pupils.— I hope for their & your sake, that there will be but few rainy mathematical days.— How I should enjoy one week with you: quite as much as you would one in the glorious Tropics.—

We sail for Monte Video at the end of this month (June) so that I shall have been here nearly 3 months.— this has been very lucky for me.—as it will be some considerable period before we again cross the Tropic.— I am sometimes afraid I shall never be able to hold out for the whole voyage. I believe 5 years is the shortest period it will consume.— The mind requires a little case-hardening, before it can calmly look at such an interval of separation from all friends.— Remember me most kindly to M^{rs}. Henslow & the ⟨t⟩wo Signoritas; also to L. Jenyns, Mr Dawes ⟨   ⟩ Mr Peacock.— Tell Prof: Sedgwick he does not know how much I am indebted to him for the Welch expedition.— it has given me an interest in geology, which I would not give up for any consideration.— I do not think I ever spent a more delightful three weeks, than in pounding the NW mountains.— I look forward to the Geology about M. Video—as I hear there are slate there, so I presume in that district I shall find the junction of the Pampas of the enormous granite formation of Brazils.— At Bahia the Pegmatite

& gneiss in beds had same direction as observed by Humboldt prevailing over Columbia, distant 1300 miles: is it not wonderful?—

M Video will be for long time my direction:— I hope you will write again to me.— there is nobody, from whom I like receiving advice so much as from you.—

I shall be much obliged if you will get one of the engravings of poor Mr Ramsay & keep it for me.— Excuse this almost unintelligible letter & believe me dear Henslow—with the warmest feelings of respect & friendship | Yours affectionately | Chas Darwin

June. 16th.—

P.S. I found the other day a beautiful Hymenophallus, (but broke it to pieces in bringing home) & with it an accompanying Leiodes.—a most perfect copy of the Barmouth specimen.—^{7}

Royal Botanic Gardens, Kew

^{1} Henslow extracted passages from CD's letters and, without his knowledge, read them to the Cambridge Philosophical Society. The extracts were then published with some editorial changes, usually minor, in a pamphlet privately printed for the Society (Henslow 1835, *Collected papers* 1: 3 – 16). From this letter the following passages were extracted:
  1.8 'we started . . . December' 1.8
  1.18 'at St Jago . . . Mr Lyell.' 1.22
  1.25 'St Jago . . . companion' 1.26
  2.1 'On the coast . . . passed over.' 2.7
  2.10 'We then . . . formation.' 2.11
  2.23 '[After] Touching . . . April 4th '
  2.31 'A few days . . . 18 days.' 2.32
  3.1 'I am now . . . expected' 3.6
  5.1 'I have just . . . species.' 5.3
  8.14 'At Bahia . . . 1300 miles' 8.16
^{2} CD's first field notebook (no. 1.4, now at Down House) contains geological observations of the Cape Verde Islands (briefly excerpted in *Voyage*). More detailed notes are in the manuscript 'Diary of observations on the geology of the places visited during the voyage, Part 1' (DAR 32.1: 15 – 36). The Cape Verde mineralogical specimens are described in Harker 1907. The entire collection is now in the Department of Earth Sciences, Cambridge University.
^{3} 'I had brought with me the first volume of Lyell's *Principles of Geology* (Lyell 1830 – 3), which I studied attentively; and this book was of the highest service to me in many ways. The very first place which I examined, namely St. Jago in the Cape Verde islands, showed me clearly the wonderful superiority of Lyell's manner of treating geology, compared with that of any other author, whose works I had with me or ever afterwards read.' (*Autobiography*, p. 77). Henslow had recommended that CD take the first volume of Lyell's *Principles* on the voyage, 'but on no account to accept the views therein advocated' (*ibid.*, p. 101). CD's copy, preserved in Darwin Library – CUL, is inscribed 'From Capt FitzRoy'.
^{4} See *Darwin and Henslow*, p. 54 n. 1 for the modern view of the geology of St Paul Rocks.
^{5} Windham Club.
^{6} This apparently refers to the French engraving 'La Forêt du Brésil'. See letter to Caroline Darwin, 25 – 6 April [1832].
^{7} See letter to Charles Whitley, [9 September 1831] and letter to Caroline Darwin, [28 April 1831].

To J. M. Herbert [1—6] June 1832

Botofogo Bay, Rio de Janeiro
June, 1832

My dear old Herbert

Your letter arrived here, when I had given up all hopes of receiving another; it gave me therefore an additional degree of pleasure.— At such an interval of time & space one does learn how to feel truly obliged to those who do not forget one.— The memory, when recalling scenes past bye affords to us *exiles* one of the greatest pleasures.— often & often whilst wandering amongst these hills do I think of Barmouth, & I may add as often wish for such a companion.— What a contrast does a walk in these two places afford; here abrupt & stony peaks are to the very summit enclosed by luxuriant woods: the whole surface of the country, excepting where cleared by man, is one impenetrable forest.— How different from Wales, with its sloping hills covered with turf, & its open vallies.— I was not previously aware, how intimately, what may be called the moral part, is connected with the enjoyment of scenery.— I mean such ideas, as the history of the country, the utility of the produce, & more especially the happiness of the people, bring with them.— Change the English labourer into a poor slave, working for another, & you will hardly recognise the same view.—

I am sure you will be glad to hear, how very well every part (Heaven forefend except sea sickness) of the expedition has answered.— We have already seen Teneriffe & the great Canary; St Jago where I spent three most delightful weeks, revelling in the delights of first naturalizing a Tropical Volcanic island, & besides other islands the two celebrated ports in the Brazils, viz Bahia & Rio.— I was in my hammock till we arrived at the Canaries, & I shall never forget the sublime impression, the first view of Teneriffe made on my mind.— The first arriving into warm weather was most luxuriously pleasant; the clear blue skies of the Tropics was no common change after those accursed SW gales at Plymouth.— About the line it became sweltering hot.— we spent one day on St Pauls, a little group of rocks about $\frac{1}{4}$ of mile in circumference peeping up in the midst of the Atlantic.—there was such a scene here. Wickham (1st Lieut) & I were the only two who landed with guns & geological hammers, &c.— The birds by myriads were too close to shoot, we then tried stones, but at last, proh pudor!, my geological hammer was the instrument of death.—

We soon loaded the boat with birds & eggs.— Whilst we were so engaged, the men in the boat were fairly fighting with the Sharks for such magnificent fish, as you could not see in the London market.— Our boat would have made a fine subject for Sneyders;[1] such a medley of game it contained.— Tell Whitley, that I find my life on blue water not only very pleasant, but that it is an excellent time for reading; so quiet & comfortable, that you are not tempted to be idle.— We have been here 10 weeks, & shall now start for Monte Video.— where I look forward to many a gallop over the Pampas.—

I am ashamed of sending such a scrambling letter; but if you were to see the heap of letters on my table you w⟨ould⟩ understand the reason.— A short letter or a stupid one may be a hint for a cut amongst some people; but old gentleman, you might as well try to cut your tailor as me; so short or long do write to me again; a letter from you brings with it a thousand pleasant thoughts.— I fancy I can see you now in the two extreme cases, of the *dead march* to Dolgelley,[2] & the bogtrotting match with Selwyn.—[3] I am glad to hear music flourishes so well in Cambridge; but it as barbarous to talk to me of "Celestial concerts" as to a person in Arabia of cold water.— In a voyage of this sort if one gains many new & great pleasures, on the other side the loss is not inconsiderable.— How should you like to be suddenly debarred from seeing every person & place, which you have ever known & loved for five years? I do assure you I am occasionally "taken aback" by this reflection.— And then for man or ship it is not so easy to right again:— Remember me most sincerely to the remnant of most excellent fellows, whom I have the good luck to know in Cambridge. I mean Whitley & Watkins.— Tell Lowe I am *even beneath* his contempt I can eat Salt Beef & musty biscuits for dinner.— see what a fall man may come to.—

My direction for the next year & $\frac{1}{2}$ will be Monte Video.—

God bless you—my very dear old Herbert— May you always be happy & prosperous is my most cordial wish | Yours affectionately | Chas. Darwin.—

I have directed to you in a curious manner for fear of mistakes.[4]

American Philosophical Society 4

[1] Frans Snyders.
[2] Town in Gwynedd, Wales, a few miles inland from Barmouth.
[3] William Selwyn, or his younger brother George Augustus Selwyn. Both studied at St John's College with Herbert.
[4] The cover is addressed 'J. M. Herbert Esq.ʳ | Fellow of St. Johns Coll: | Cambridge | To be forwarded immediately.'

From Caroline Darwin 12 — 28 June [1832]

[Shrewsbury]
June 12.ᵗʰ

My dear Charles,

I cannot tell you the delight receiving your letter from S.ᵗ Salvador gave us—we had been so impatiently longing to hear from you & it was *such* pleasure to have such an interesting happy account of you We miss you & talk about you & think about you more even than I expected— by this time you will have had our packets of letters, we have written by the first of Tuesday of *every month* since you went to Rio & latterly to Monte Video— do not forget to give us notice when we are to change our direction— by counting months you will always tell if you receive all our letters for I again assure you till you return I take the responsibility of seeing that a letter goes to you every month— Susan had only room in her last letter to tell you Fanny Owen & Mr Biddulph were married on the 31st of May & promised I should add a more particular account— the party assembled was almost entirely female *all* the men being in London— nobody was staying in the house but the Humphreys my sisters & Mr Biddulps brother— I went early in the morning of the day with Miss Casteau as I had been staying a few days at Maer & did not get home till the 30th. We got to Woodhouse between 8 & 9 & Fanny soon sent for me to her room She was beautifully dressed in white of course, with her bonnet & veils on—all ready. She looked so odd & so much like a person going on the stage that we had a very merry laugh together. She was very nice & affectionate & I sat with her whilst the breakfast was going on— all the maids came in to have a look at the bride & after breakfast all the ladies who had arrived for breakfast came up stairs to see Fanny, which she found very disagreeable— those who came to breakfast were the Kenyons, Bridgemans, Mathews, Dymocks, Smythe Owens, & Cottons— at 10½ the procession began, 10 carriages— I went with Fanny Caroline & Emma poor Caroline was very low. Mr Biddulph looked very handsome & gentlemanlike & extremely nervous, during the ceremony he looked quite white & his hands blue. Fanny shook so much she could hardly write her name. they went off to Chirk from the church. Mr Hunt[1] I thought read the Service very well. We all returned to Woodhouse the day was long— the tenants assembled in front of the house playing games &^{c} the evening ended by a servants ball & all the farmers wives & daughters in the neighborhd came. the ball went off with great spirit & the dining room so crowded that the heat was intolerable. the next morning we all went home— Susan & Cath have since this met M^r & Mrs Biddulph for a day at Woodhouse, Fanny looking very handsome & happy. they went to London last week for a month— Mrs Owen & Caroline are now in London—staying with Mrs Williams. She (Mrs W.) has not got quite well from the accident by which she hurt her foot—& is still forced to use crutches— there has been a letter from Arthur Owen, he was more lucky than you being only sick one day— he wrote from Madeira in rapture with the flowers & climate— Nothing is yet settled about Francis— there were no Young ladies to Arthurs great disappointment— his Captain thinks his merchant ship the first in the world.— Eyton after all is not gone to Germany. Mr Hope did not wait for him & he would not follow— I shall mention all the people about: the Leightons we have seen a good deal of lately as Catherine & the girls have taken a drawing rage & go out sketching together most days— You have heard in Susan's last letter of Sir James Makintosh death, all his books are to be sold this spring. When the debts are paid it is supposed there will be about £3500 between all his children Hensleigh & Fanny will live near London so that Hensleigh can drive to his Police Office every day— they are coming to Maer next month, & the Langtons in September. Erasmus has I think rather a *spite* to Mr Langton he does not think him worthy of Charlotte, full of small talk & not very sensible but I do not feel sure that he is right in his opinion as he seemed determined not to like him. I have had many nice happy letters from Charlotte— Erasmus had intended going to Paris with Fred Hildyard in July but in a letter we had from him yesterday he says now that Paganini is in London he can not leave it— Marianne, Cath & I are going to the sea tomorrow, for a fortnight to the Rhyl near St Asaph— the 3 elder boys go with us & leave the baby at home.— I mention the number of children for fear like Erasmus you should forg⟨e⟩t all about your nephews— Papa is very well—he is going to have a new carriage built to hold two people instead of having a new Sulky— poor old Pincher is quite affectionate to me now that his master is away & I always pet him out of regard to his master. Nina is getting fat—

Frank Wedgwood & Mrs Frank are staying here. Mrs Frank is a very uninteresting person not at all agreeable, large & plain. She seems very fond of Frank a capital scrattle & amiable so I hope they will do well together & she is sensible— We are going this evening to the Circus to see a very good set of horses that are in Shrewsbury— Dr H. Johnson is come to live here & he dines with us & joins our party I forget whether you know him— I was surprised this morning to meet 3 Miss Hills of Berrington all going to ride at the Circus going alone without any gentleman or even a servant. Mr Everard Fielding was married last week to Miss Baughey[2]

I am ashamed of sending such an *abominably* stupid letter, but there never was a month with less to tell in it We have done nothing but garden & I think to write about *our* flowers would hardly do now that you are seeing tropical vegetation— Mr Sedgwick called for half an hour on his return from Wales & was very pleasant—what a very agreeable man he is & what an agreeable countenance he has. We heard a report that William Fox has been very ill, in a letter from Miss Bent to one of the Wedgwoods she says "he is now so much better that he walks out every day, but his Doctor bled him so violently that it will be long before he recovers his strength" Cath wrote to Julia to enquire of him but we have had no answer. Do be careful of yourself dear Charles. I can not help feeling afraid that you will make yourself ill by over exertion in some scheme which you enjoy. Papa & all join in love to you with mine my very dear Charles how I shall enjoy seeing your dear old face again.

Ever y^{rs} affly Caroline Darwin—

(Letter finished June 28th)

DAR 204.6.1

[1] Thomas Hunt.

[2] Anne Henrietta Boughey married Everard Robert Bruce Fielding, Rector of Stapleton, on 21 June 1832 (*Gentleman's Magazine* 102.2 (1832): 75–6).

From W. D. Fox 30 June 1832

Epperstone near Nottingham.
June 30. 1832

My dear Darwin

It is now I believe a month since your Sister was so kind as to send me word that you were at length heard of, and where I could write to you.— I commenced a letter at the time but was prevented finishing it, and ever since I have purposed writing from day to day, and as constantly put it off, sometimes owing to illness, sometimes idleness, & frequently from feeling that I had nothing in the world to tell you of that you will not hear from Shrewsbury; and I am now once more commencing (with a determination to finish it) merely that I may put you in mind of my existence & prevent your totally forgetting me in the midst of the wonders of Creation you are now surrounded by, & will behold previous to your return to England.— I can scarcely realize the idea sometimes of your being at such a distance, and revelling in the midst of scenes I have always longed intensely to see, and hope to have a sort of idea of sometime secondhand from your description.— I had often wondered where you were and how going on, and was very anxious to hear of you, when your sisters letter gave me the welcome news. From what she says of you, you seem most happily situated in every respect; your health, ship & Companions all remaining as perfect as you hoped they would prove previous to your departure. From all I hear of South America the Climate is very little to be feared with proper precautions. I cannot help a little fearing that the ardour (which I remember your shewing in Chase of Machaons in Bottisham Fen) may, to compare great things with small, lead you into difficulties, & into disregard of dangers of various kinds when in pursuit of Nat: History where all is new & all glorious to the last degree.— I often long so to be with you & join in your happiness, and think over the difference of our lots & the ridiculousness of my pursuits in Nat Hist: compared with yours. In consequence of a severe Inflammation of the Lungs I had early in April, I have been more or less Invalided ever since, and have amused myself in santering about the Fields on horseback studying the Small summer Birds of Passage, their nidification &c, and when thus employed the thoughts of you and your occupation most forcibly & frequently struck me. I pottering in a Hedge Rows to watch the proceedings of a Whitethroat & you surrounded by the Noble Trees of a S. American Forest with every luxury of vegitation & life around you.— You must have much regretted your not seeing the Madeira & Canary Islands, tho' perhaps the time thus saved will be abundantly recompenced hereafter and as they are pretty well explored, at least the former certainly, the harvest will be richer gathering where you now are & will be.— The extreme novelty of every thing around you, must now be rather wearing off, and you are becoming more used to the Intoxication of feelings, the Country you are now in, must produce.— I have often regretted one trait of your Character which will I fear prevent your making so great an advantage as you might do from your present travels, and which I regret also very much on my own account, as I might perhaps get the perusal of it;—I allude to your great dislike to writing & keeping a daily methodical account of passing events, which I fear (tho' I have also hopes the other way from the overwhelming influence of every surrounding object) will prevent you from keeping a Regular Journal.— If you do not do this, the vast crowd of Novelty which will surround you, will so jostle about ideas, that to say nothing of the many that will be lost altogether, the vivid reality & life which a memorandum taken at the moment gives to every passing event & thing, is done away with.— With this one exception (which I dare say you have overcome) I know of no one so fitted altogether for the expedition you are engaged in. We have had many extraordinary changes in England since you went, even in this short six months, what may occur before your return therefore in three years?— You have of course heard of the Incomparable Charlotte Wedgwood changing her name. From what I hear from all that know her husband or rather have seen him, her choice seems a very happy one, indeed she is not one that would readily be taken in. You would I think much regret to hear of Sir J. Mackintoshs death, as I have often heard you speak of him as one you much esteemed. Mrs Darwin of the Priorys Death will not much affect you. By the time of your return we shall be better judges of the happy effects of our Reform Bill, at least if it is allowed to have its natural course in the correction of the abuses of Church and state.— Party Spirit runs now very high indeed— The Tories are merely (to use their own words) endeavouring to prevent the vessel from altogether sinking, & the Whigs & Radicals all alive.— For some days we certainly were on the very verge of Revolution. The excitement in the Country was quite extraordinary among the lowest orders. All still but evidently all prepared for anything that might turn up— We have however now I trust safely passed this grand Corner, upon which so much hung & shall proceed steadily & prosperously, though much remains to be done that is formidable to look into.— The Cholera is now spread all over England, and tho' not the very dreadful scourge we were led to expect it to be, is very awful in many places.— It began I fear at Derby last week, & is now in many of the large Towns slowly making progress in England Scotland & Ireland.— I never remember such a season as the present. Every kind of crop promises to be most abundant. It has been an extraordinary year for Insects, but I have not been able to go in search of any. I have not seen any of your Family since you went, but hear very flattering accounts of all.— Your Father is uncommonly well.— All at Osmaston I am happy to say are much as usual.— My Father has been poorly but is much better.— Of Cambridge Friends I have not heard for some time.— I hope however very shortly to hear of Henslow. I ought to have taken my Masters degree as next week, & should have rather gloried in having my vote at the Commencement of a Reformed House of Commons, but I have been obliged to forego it.— Pulleine is to spend some days with me next week— Do you remember our excursion to Moncks Wood & Whittlesea at this time of year with Albertus Way At Whittlesea the Cholera has killed 48 & there are 130 new cases last report.— I have never heard of him, whether he is still at Leamington or not. Did you ever see Old M.ʳ Galton of Dudston.—[1] He is just dead after a lingering illness.— I must now give you a few lines about my own dear self.— I have as I told you before, been unwell which has incapacitated me from taking any duty for the last 3 months and I am only returned to Epperstone for a short time, as I fear I can be of no use at present I am now very much stronger than I have been, in fact comparatively well, but as is always the case with Chest Complaints, vary very much in health & spirits. I did at one time think I should never meet you again in this world, but trust now to do so & see you in full vigour after your wanderings are over.— I often look forward to the time of your return with great delight, and regret I did not see you before your departure. I had no idea that you had stayed so long in England— You scarcely left us in time to say it was 1831.—

I hope you will not be disgusted at my very stupid letter.— You who abound in novelty must not censure we plain housekeepers for having nothing to communicate. I do not ask you to write to me as you must have plenty to occupy your time, and I shall hear of you from Shrewsbury when you write there, as a few lines from them will give the information I want as to your welfare.—

And now my Dear Darwin with every wish for your welfare and success in all your undertakings & that I may again see you in health & happiness in Old England which after all is the prettiest & best Island in the world | Believe me your attached friend William D. Fox.—

DAR 204.6.2

[1] Samuel John Galton of Duddeston House who died in 1832.

To Catherine Darwin 5 July [1832]

Rio de Janeiro. | HMS Beagle
July 5.ᵗʰ—

My dear Catherine

I have only ¼ of an hour to write this—Sullivan[1] will put it in his parcel, so that it will only cost common postage.— I have received your letter directed Monte Video & previous to it one from Caroline from Maer.— Tomorrow we sail for Mon: Video.— If the wind is not directly against us, we shall touch at Cape Frio, the celebrated scene of diving for the Thetis wreck.—[2] They have fished up 900000 dollars.— If we are lucky enough (& it is very probable) to have a gale off St Catherines we shall run in there.— I expect to suffer terribly from sea-sickness—as we are certain to have bad weather.— After lying a short time at MV: we cruize to the South—but not I believe below Rio Negro— The geography of this country is as little known as interior of Africa.—

I long to put my foot, where man has never trod before— And am most impatient to leave civilized ports:— We are all very anxious about reform; the last news brought intelligence that Lord Grey would perhaps re-continue in.— Would ask Erasmus to add to the books—Pennants quadrupeds[3] (if not too late) in my bedroom.—& Humboldt tableaux de la nature.—[4] You cannot imagine what a miser-like value is attached to books, when incapable of procuring them.—

We have been 3 months here: & most undoubtedly I well know the glories of a Brazilian forest.— Commonly I ride some few miles, put my horse & start by some track into the impenetrable—mass of vegetation.— Whilst seated on a tree, & eating my luncheon in the sublime solitude of the forest, the pleasure I experience is unspeakable.— The number of undescribed animals I have taken is very great—& some to Naturalists, I am sure, very interesting.— I attempt class after class of animals, so that before very long I shall have notion of all.— so that if I gain no other end I shall never want an object of employment & amusement for the rest of my life.— (Sullivan only gives me 5 minutes more—).— I am now writing in my own snug corner.—& am as comfortable as man can be.— I am only obeying orders in thus writing a short letter.— When on the deserts coasts of Patagonia.—you will be a long time before hearing from me.— My journal is going on better; but I find it inconvenient having sent the first part home on account of dates—

Give my very best love to my Father & all others. | Most affection | Chas Darwin.—

DAR 223

[1] Bartholomew James Sulivan.
[2] H.M.S. *Thetis*, in which Robert FitzRoy had served for four years (1824−8) was wrecked on 5 December 1830. For FitzRoy's account, see *Narrative* 2: 67−72.
[3] Pennant 1781. A copy of the third edition (1793) is in Darwin Library−CUL. The title-page of volume one is missing; volume two has 'Charles Darwin 1826' on the flyleaf. The set is annotated; marginal scorings on pp. 236−42 of volume one, in the chapter on 'Dogs', are probably of a later date.
[4] *Tableaux de la nature* (Humboldt 1828), translation of Humboldt's *Ansichten der Natur*. There is no copy in the Darwin Library, nor has any reference to it been found in CD's *Beagle* notes.

To Susan Darwin 14 July − 7 August [1832]

[At sea; Monte Video]

My dear Susan

As in all probability we shall stay but a short time at M Video.—I take the opportunity of an idle evening at sea to begin a letter.— We are now (July 14th) about 300 miles from Rio: to day for the first time we have a fair wind:—before this calms & light contrary winds were only disturbed by squalls & gales.— For a week I suffered much from sickness, but am now nearly well again.— Every body is full of eagerness to commence our real work.— After laying in fresh water at M Video, we sail for Rio Negro.— Comparatively near as this is to the civilized world, yet the whole coast & interior country is *totally* unknown.— Falcners account,[1] inaccurate as it must be, is the only one.— I expect grand things in Natural History, but if that fails, the whole world, I suppose, does not produce so much game in any one spot.— I believe the Captain will proceed many miles up the river & I trust I shall be of the party.— I cannot imagine anything more interesting: the only thing unpropitious is the ferocity of the Indians.— But I would sooner go with the Captain with 10 men than with anybody else with 20.— He is so very prudent & watchful, as long as possible & so resolutely brave when pushed to it.—

As far as we are able to guess—the following is the rough outline for the future.— After coast of Patagonia return to M Video, & then proceed to Terra del & settle the Fuegians. Back to M Video— Afterwards to Valparayso.— From which point one more cruize will be to the South(?) & after that the wide world is open to us.— Even the prospect of walking, where European never before has, hardly recompenses for leaving the glorious regions of the Tropics; already is the change of weather perceptible.— Every one has put on cloth cloathes & preparing for still greater extremes our beards are all sprouting.— my face at presents looks of about the same tint as a half washed chimney sweeper.— With my pistols in my belt & geological hammer in hand, shall I not look like a grand barbarian?— Before leaving Rio we heard the news of Ld Greys minority, & are all most anxious to see how it will end.— It is not very likely that we shall receive letters before our return from the South: this will be a sad disappointment to me, as I there expect answer to my Bahia letter: for this gives to a correspondence an appearance of closer connection.—

I do not think I have ever given you an account of how the day passes.— We breakfast at eight oclock.— The invariable maxim is; to throw away all politeness.—that is never to wait for each other & bolt off the minute one has done eating, &c. At sea, when the weather is calm, I work at marine animals, with which the whole ocean abounds.— if there is any sea up.—I am either sick or contrive to read some voyage or Travels.— At one we dine. You shore-going people are lamentably mistaken, about the manner of living on board.— We have never yet (nor shall we) dined off salt meat.— Rice & Peas & Calavanses are excellent vegetables & with good bread, who could want more? Judge Alderson could not be more temperate, as nothing but water comes on the table.— At 5 we have tea.—

The Midshipmens birth have all their meals an hour before us, & the Gun-room an hour afterwards.

July 30th— Monte Video.— The packet will arrive here in a few days, so that I will make another attempt to fill my letter.— We arrived here on the 26th after a long & disagreeable passage.— The weather has been too heavy or too light.— I expect the further we proceed South, the more uncomfortable I shall find sea-life.— It was quite curious how much I felt the change of climate.— The thermometer has scarcely ever been below 50°, but yet with thick clothes I could not make myself warm.— Wherever we go, there is sure to be some disturbance.— as we passed the Frigate, she made signals to us.— "Clear for action" "& prepare to cover our boats". When shortly afterwards a heavy force in boats with Carronades ready mounted, passed by us to go to the Mole.— This merely turned out to be a substantial argument to convince the inhabitants they must not plunder British property.— I have only had one good walk on the turf plains, which one has so often read about.— There is something very delightful in the free expanse, where nothing guides or bounds your walk.— Yet I am disappointed in them, & as far as regards scenery, imagination could not paint anything m⟨ore⟩ dull & uninteresting.— How different from the Brazilan forest, where I could sit for hours together & find every minute fresh objects of admiration.— We certainly sail before another packet arrives from England. I am sorry for it. I quite long to hear from you after you have received a letter from me.— I cannot thank you all too much for writing so regularly to me.— The very regularity of time is a satisfaction, as it prevents unreasonable expectation.— My main object, Natural History goes on very well, & I certainly have taken many animals &c which would be interesting to Naturalists.— Independent of this satisfaction, I have begun so many branches, previously new to me; that even already I long to be in England to commence an attack upon several obscure little individuals.— I am going to draw 25£, which will make altogether since leaving England, 80£.— out of this at least Twenty has not been wasted, in as much as it has been spent about my collection.— For the next two months even with my ingenuity, I do not think I shall be able to spend a penny.—

I have just received intelligence we sail tomorrow for Buenos Ayres.— The Captain has heard some news about an old chart of the coast & he thinks it of sufficient importance to go there.— I am glad of it, the more places the merrier: when one is about one cannot see too many.— At last I shall deliver the letter to Mr Hughes from Mrs Haycock.—² The packet calls here on Wednesday so I leave this letter to be forwarded.—

Give my love to all at dear old Shrewsbury.— & dear old Granny I am & always shall be yours | very affectionately | Chas Darwin

July 31st.—— | Monte Video

Monte Video | August 7th.— I have procured my letter again in order to write some more.— We run up to Buenos Ayres, where a Guard Ship fired a shot close to us. This we took up as a great insult, & if our guns had been ready we should have returned it with interest.— We immediately made sail & returned here.— The Captain reported the circumstance to the frigate Druid lying at the Mount, & she has gone up to Buenos Ayres & obtained ample satisfaction for

the insult offered to us.—: Quarantine for the cholera was the excuse!.— We all thought we should at last be able to spend a quiet week, but alas the very morning after anchoring, a serious mutiny in some black troops endangered the safety of the town.— We immediately armed & manned all our boats, & at the request of the inhabitants, occupied the principal fort.—[3] It was something new to me to walk with Pistols & Cutlass through the streets of a Town.— It has all ended in smoke But the consequence is very disagreeable to us, since from the troubled state of country we cannot walk in the country.— The Packet will not sail yet for a week.— And now for a bit of business, in my letter to Erasmus, I tell him Lieut: Blanchard will transact the shipping of my Box.— We have just heard the news he has broke & gone to America.— I hope Erasmus enquired at the London agent.—Mrs Palsgrave No 3, Lyons Inn, Strand. If so he will have heard of this. If not a letter had better be sent to Falmouth I trust they are not lost.— What a loss it will be to me.— If they are regained there must be some means of forwarding them to M Video.— I am very sorry for all this trouble.— |
Yours affectionately.— | Chas Darwin

On the 17th we start for the Rio Negro.— Adieu.—

DAR 223

[1] Falkner 1774. No CD copy has been found; the copy used during the voyage was probably FitzRoy's.
[2] Charles Hughes, with Rodger, Breed & Co., Buenos Aires (note in CD's field notebook no. 1.12, Down House). Mrs John Hiram Haycock, mother of Edward Haycock, a prominent Shropshire architect (Hobbs 1960).
[3] On 5 August 1832, in Montevideo, Captain FitzRoy's assistance was requested to preserve order in the town against some mutinous black troops. 'The Beagle's crew were not on shore more than twenty-four hours, and were not called upon to act in any way' (*Narrative* 2: 95). CD was one of the fifty men involved, but he returned to the *Beagle* that night, as he 'had a bad headache' ('*Beagle*' *diary*, pp. 86 – 7).

To J. S. Henslow[1] [23 July —] 15 August [1832]

My dear Henslow

We are now beating up the Rio Plata, & I take the opportunity of beginning a letter to you.— I did not send off the specimens from R Janeiro; as I grudged the time it would take to pack them up.— They are now ready to be sent off, & most probably by the Packet.—[2] If so they go to Falmouth (where C. FitzRoy has made arrangements) & so will not trouble your Brothers agent in London.—

When I left England.—I was not fully aware how essential a kindness you offered me, when you undertook to receive my boxes.— I do not know what I should do without such head-quarters.— And now for an apologetical prose about my collection.— I am afraid you will say it is very small.—but I have not been idle & you must recollect that in lower tribes, what a very small show hundreds of species make.— The box contains a good many geological specimens.— I am well aware that the greater number are too small.— But I maintain that no person has a right to accuse me, till he has tried carrying rocks under a Tropical sun.— I have endeavoured to get specimens of every variety of rock, & have written notes upon all.— If you think it worth your while to examine any of them, I shall be *very* glad of some mineralogical information, especially in any numbers between 1 & 254, which include St Jago rocks.— By my Catalogue, I shall know which you may refer to.—[3] As for my Plants, "pudet pigetque mihi".[4] All I can say is that when objects are present which I can observe & particularize about, I cannot summon resolution to collect where I know nothing.—

It is positively distressing, to walk in the glorious forest, amidst such treasures, & feel they are all thrown away upon one.— My collection from the Abrolhos is interesting as I suspect it nearly contains the whole flowering Vegetation, & indeed from extreme sterility the same may almost be said of St Jago.— I have sent home 4 bottles with animals in spirits I have three more, but would not send them till I had a fourth.— I shall be anxious to know how they fare.— I made an enormous collection of Arachnidæ at Rio.— Also a good many small beetles in pill-boxes; but it is not the best time of year for the latter.— As I have only ¾ of a case of Diptera &c I have not sent them.— Amongst the lower animals, nothing has so much interested me as finding 2 species of elegantly coloured true Planariæ,[5] inhabiting the dry forest! The false relation they bear to Snails is the most extraordinary thing of the kind I have ever seen.— In the same genus (or more truly family) some of the marine species possess an organization so marvellous.—that I can scarcely credit my eyesight.— Every one has heard of the dislocoured streaks of water in the Equatorial regions.— One I examined was owing to the presence of such minute Oscillaria that in each square inch of surface there must have been at least one hundred thousand present.— After this I had better be silent.— for you will think me a Baron Munchausen[6] amongst Naturalists.— Most assuredly I might collect a far greater number of specimens of Invertebrate animals if I took less time over each: But I have come to the conclusion, that 2 animals with their original colour & shape noted down, will be more valuable to Naturalists than 6 with only dates & place.— I hope you will send me your criticisms about my collection; & it will be my endeavour that nothing you say shall be lost on me.—

I would send home my writing with my specimens, only I find I have so repeatedly occasion to refer back, that it would be a serious loss to me.— I cannot conclude about my collections, without adding that I implicitly trust in you keeping an exact account against all the expense of *boxes &c &c.*— At this present minute we are at anchor in the mouth of the river: & such a strange scene as it is.— Every thing is in flames,—the sky with lightning,—the water

with luminous particles, & even the very masts are pointed with a blue flame.—
I expect great interest in scouring over the plains of M Video, yet I look back
with regret to the Tropics, that magic line to all Naturalists.— The delight of
sitting on a decaying trunk amidst the quiet gloom of the forest is unspeakable &
never to be forgotten.— How often have I then wished for you.—when I see a
Banana, I well recollect admiring them with you in Cambridge.—little did I
then think how soon I should eat their fruit.—

August 15th. In a few days the Box will go by the Emulous Packet (Captn.
Cooke) to Falmouth & will be forwarded to you.— This letter goes the same
way so that if in course of due time you do not receive the box, will you be kind
enough to write to Falmouth.— We have been here (Monte Video) for some
time; but owing to bad weather & continual fighting on shore have scarcely ever
been able to walk in the country.— I have collected during the last month
nothing.— But to day I have been out & returned like Noahs ark.—with
animals of all sorts.— I have to day to my astonishment found 2 *Planariæ* living
under dry stones. Ask L Jenyns if he has ever heard of this fact. I also found a
most curious snail & Spiders, beetles, snakes, scorpions ad libitum And to
conclude shot a Cavia weighing a cwt:— On Friday we sail for the Rio Negro, &
then will commence our real wild work.— I look forward with dread to the wet
stormy regions of the South.— But after so much pleasure I must put up with
some sea-sickness & misery.—

Remember me most kindly to every body & believe me, my dear Henslow,
Yours affectionately | Chas. Darwin

Monte Video. August 15th.—

Royal Botanic Gardens, Kew

[1] The following passages from this letter were extracted by Henslow and published in the Cambridge
Philosophical Society pamphlet:
   3.2'My collection . . . Vegetation' 3.3
   3.6'I made . . . the latter.' 3.8
   3.9'Amongst . . . present.' 3.18
   3.19'I might . . . place.' 3.23
   4.4'At this present minute . . . flame.' 4.7
[2] On the same date as this letter Robert FitzRoy wrote to Francis Beaufort about CD's shipment: 'By
this Packet, the *Emulous*, he [CD] sends his first collection to the care of Prof. Henslow, at Cambridge,
there to await his return to England. I *fancy* that, though of small things, it is numerous and valuable,
and will convince the Cantabrigians that their envoy is no Idler.' (F. Darwin 1912, p. 548).
[3] CD's four catalogues of geological specimens are preserved in the Cambridge University Library, on
permanent loan from the Department of Earth Sciences, Cambridge University. In Harker 1907
they are described as 'a monument of patient labour. Under each number is a condensed description
of the rock, as seen by the eye and the lens, besides the necessary records of locality and occurrence.
On the opposite page are additional notes, also made during the voyage, giving the results of
examination with the blow-pipe, goniometer, magnet, and acid-bottle.'

[4] 'to my shame and disgust'.

[5] Henslow, in his excerpts from this letter, omitted 'true' and placed a question mark after 'Planariae'. He apparently doubted that the genus could exist out of water. (See letter from J. S. Henslow, 15 – 21 January 1833, in which he asks whether CD has mistaken species of the genus *Oncidium* for land *Planariae*, and CD's reply to J. S. Henslow, 24 July – 7 November 1834.) CD's 'Oscillaria' was changed to 'Oscillatoria'.

[6] An allusion to Rudolph Erich Raspe's book *Baron Münchausen's narrative of his marvellous travels and campaigns in Russia* (1785) which contains Raspe's personal reminiscences of Hieronymous von Münchausen.

From Catherine Darwin 25 July [– 3 August] 1832

> Shrewsbury.
> July 25th. 1832.

My dearest Charles

On the 31st of June we were delighted to receive your letter from Rio dated 6th of April, and on the 5th of July (this month) we received your second letter and Journal dated 25th of April, with the interesting account of your expedition into the Brazils. I cannot tell you how interesting and entertaining we find your letters and Journal, and what great joy it gives all the house when we have such happy accounts of you in every way. I run to tell Nancy and some of the other Servants, and the pleasure is universal over the house, as everybody loves you, and thinks of you, my dearest Charles. It is so delightful too to find how wonderfully your Voyage answers, and how excessively you have enjoyed yourself. I do not think any pleasure can be more vivid than your's must have been. I had no idea before I read your Journal of the extraordinary beauty of the Tropics. If you wish to have *my* Criticisms, I must say I think your descriptions most excellent, and gave me most lively pleasure in reading them. I was so interested I could not bear to stop reading it, till I came to the end, which was the case also with Marianne and Caroline, who both admired and liked it exceedingly. Your Journal & Letters were sent on to Marianne, Caroline and myself, at the Sea, where we have been the last three weeks. Susan read the Journal aloud to Papa, who was interested, and liked it very much. They want to see it at Maer, but we do not know whether you would choose that, and must wait till we hear from you, whether we may or not. It shall be kept most carefully for you.— The same Packet contained a note from Capt Beaufort, which I dare say you will be curious about, so I will copy it out. "Capt Beaufort presents his Compliments to Miss Darwin, with the enclosed letter, and perhaps she will pardon the liberty he takes in adding that Capt Fitzroy omits no opportunity of expressing the unqualified satisfaction he feels in Mr Darwin's society—and in his last dispatch he says "D. is equally liked and respected by every person in the Ship". Admiralty. June 29."—[1] Susan wrote to thank Capt Beaufort for his politeness.—

This is the 5th letter going to Monte Video, April, May, June, July & August. It is a pity that April was not directed to Rio, but we were obliged to follow your original directions.— We were rather puzzled by your writing by a slip of the pen, I suppose that the Beagle was to sail back to St Salvador on the 7th of *March*, but you must have meant to write *May*. It is a very nice scheme your remaining stationary at Botofogo Bay. I can conceive no thing more extraordinary and interesting than to be quietly living in a Brazilian Cottage,— but do not let the Cottage put the Parsonage out of your head, a far better thing, and which we were rejoiced to hear continued to be a vista to your prospects. I hope you will in all probability find Fanny Wedgwood *disengaged* and **sobered** into an excellent Clergman's Wife by the time you return, a nice little invaluable Wife she would be;[2] I will not quite promise though that you will find her disengaged, as another Clergman, M[r] Paget Moseley, Brother to M[rs] Frank Wedgwood, is said to be paying her very sedulous attention; but he is such a vulgar, fat, horrid man, I do not think it is possible she will have him. I must tell you a little scene that took place between them. They were admiring some Flowers in the Greenhouse, when M[r] Moseley declared he could show some far prettier flowers, and out of his pocket produced a scrap of paper, on which Fanny had scrawled some little flowers some weeks before. Emma was by, and was near choking with laughing at the man's odd manner, and Fanny's amazemen⟨t.⟩

Professor Sedgwicke has been so continually calling here of late at all the most unexpected times, tha⟨t⟩ I think the next piece of news you will hear, will be that Susan has turned into M[rs] Sedgwicke.—[3] The last time he called, he was on his road to Cader Idris.[4] When we were at the Sea (at Rhyl, in Flintshire, not far from Abergelley, an ugly place) we made the usual little Tour to Bangor and Conway, and also to your old Acquaintance, the Orme's Head, which Caroline is so delighted with, she quite longs to build a house there.— The Rhyl is a very ugly Sea Coast, and I found it quite a *Plas Edwardes*[5] and got heartily sick of it.— We have been reading a Review of M[r] Earle's Voyages in the South Seas,[6] in M[r] Lytton Bulwer's Monthly Review;[7] his book is much praised, and I think must be very entertaining.— I am very curious to know how often you get Newspapers, and how much of the Public news you know. The Papers will hardly tell you how much the Cholera has broken out again in London, and spread all over the Country; it is so strange its appearing this second time so much more among the higher classes. M[rs] Smith (Lord Forrester's Sister's) death was the most frightfully sudden thing; she was at the Opera on Saturday Night, was perfectly well till Luncheon time on Sunday, when she was suddenly seized, and was dead by 12 at night. She suffered agonies; every Physician in London was sent for, but they wasted 2 hours in disputing before they could do any thing.— We are free from it yet in Shrewsbury.— The County Members are canvassing about all Shropshire, for the Reformed Parliament. M[r] Pelham and M[r] Whitmore are undisputed for the South of Shropshire, and Sir Rowland

Hill, Mʳ Gore and Mʳ Coates for the North of Shropshire, which will be a disputed election. Mʳ Coates is the only Whig among them, and he is a very poor one. Mʳ Biddulph is to stand a Contested Election for Denbighshire with Lord Kenyon's Son, which is a very foolish thing, as he is almost certain to lose.[8] Poor Fanny will not have a very pleasant or easy life I am afraid; the old Mother, Mʳˢ Biddulph is so odious to her, and Mʳ Biddulph is such an exacting Husband.— It is too bad of Erasmus not writing to you, he is such an idle creature, and he is so engrossed with Paganini I suppose.— I long to hear from you again, my dearest Charles. You cannot think how it rejoices my heart, when we get a letter from you.

God bless you always & believe me, with every body's Love, yʳ most affectᵗᵉ | E. Catherine Darwin.

Postmark: Shrewsbury AU 3 ⟨1832⟩
DAR 204.6.1

[1] The original letter is in DAR 223.
[2] Frances (Fanny) Wedgwood 'died on August 20th, 1832, aged twenty-six, after a few days' illness from some inflammatory attack' (*Emma Darwin* 1: 250).
[3] Although Henrietta Litchfield reports (in *Emma Darwin* 1: 141 n.) that CD once told her that 'anything in coat and trousers from eight years to eighty was fair game to Susan', Susan was the only sister who never married.
[4] A mountain near Barmouth.
[5] Seaside town in North Wales. The 'Journal' (Appendix I) records that CD spent three weeks at Plas Edwards in 1819.
[6] Augustus Earle, the draughtsman of the *Beagle*, had lived in New Zealand and travelled in the Pacific in 1827. The work referred to was published in 1832 (Earle 1832). See letter to Caroline Darwin, 27 December 1835 for the reaction of Robert FitzRoy and CD to its account of missionary work.
[7] Edward George Earle Lytton Bulwer-Lytton edited the *New Monthly*, 1831–2.
[8] John Cressett Pelham was a Conservative candidate for Shrewsbury and was defeated; for South Shropshire, the Earl of Darlington and Robert Henry Clive, both Conservatives, were elected, defeating Thomas Whitmore, also a Conservative; for North Shropshire, Sir Rowland Hill (Conservative) and John Cotes (Liberal) defeated William Ormsby-Gore (Conservative) (Hanham 1972, pp. 283, 275). Robert Myddelton Biddulph stood as a Whig, for reform, and narrowly defeated Lloyd Kenyon (1805–69).

From Susan Darwin 15[−18] August 1832

Shrewsbury
August 15ᵗʰ | 1832

My dear Charles—

We received your last letter to Catherine with one for Mʳ Owen dated Rio June: on Saturday last and very great pleasure it gave us to have another happy account of yourself. Tho' you complain so much of the difficulty of writing we should not have found it out as we could not have had a nicer or more chatty letter than your last.— I think you have found out the way at last of making that idle old Dag write to you by sending him commissions to execute: he has sunk into such a Lethargy in London that it requires *three* letters from us before we can rouse him to send us a line in return.— I can't conceive why he has buried himself alive this lovely summer in the dirt of London for he has been talking of coming down & going abroad the last three months yet nothing comes of it.— Mʳ & Mʳˢ Hensleigh are at Maer & coming to see us early in September & I expect that by *some chance* Erasmus will then appear among us, for there is certainly great attraction in that quarter.— Your account of the fatal effects of Snipe Shooting is very melancholy especially poor little Musters death who was such a merry little animal: I suppose they must have died on Ship board? How fortunate it was you did not join that party which I think you regretted at the time. I hope my dear Charley this will be a warning to you to be *exceedingly careful* of not over tiring yourself lest you should bring on these fevers Papa sends his most affectionate love to you & bids me again repeat what we have all said continually how much we hope we may depend upon on your not allowing any false shame to prevent your returning whenever you feel inclined it would make us all so happy to have you back again. I am very much pleased to find the quiet Parsonage has still such charms in your eyes. it is so delightful to look forward & fancy you settled there.—and in spite of this marrying year I am sure you will find some nice little wife left for you.— Robert Wedgwood has been here & went on yesterday to the Hill from thence he proceeds to pay Mʳ & Mʳˢ Edward Holland a visit & inspect his own future abode at *Dumelton*[1] (which it is now called) He has taken to farming & keeps more than 40 Ducks & Fowls, also rents the Maer Pool from Uncle Jos. which I shᵈ think a doubtful speculation, as I believe he is to give 20 pounds a year for it.— Papa has not been as well as usual lately, he has had the Lumbago & altogether finds the least exertion too much for him: we are persuading him as much as possible to give up going out to see Patients, & then he wᵈ avoid any hurry or fatigue. The Hot house is a great pleasure & before yr letter came he had sent for a Banana—we eat our first Pine from the Hothouse on Monday last Uncle John being with us who pronounced it very good. Joseph's head is quite turned by this first production.— Papa desires me to tell you that the Bill for yr money at the Bank had not been presented on the 9ᵗʰ of this Month, but Papa has given directions to have it honored: & he gives consent to the commissions you have sent to ask Erasmus to get.—

Caroline and I set out the day after tomorrow into Derbyshire to pay our long promised visit at Osmaston. I can hardly beleive it is two years since William & Julia were here. I hope all the Sick Sisters will be pretty well as I shall like to get acquainted with them especially Frances Jane. William has not been well enough to do his duty lately so I hope he may be at home.— Caroline would like to meet Bessy Galton to see how the flirtation prospers, but it would put me in a rage to see them together so I hope she won't be there.— Mʳˢ Fox is our particular horror, such a turn for Mechanics alarms us. I have just got "Babbage on Machinery"[2] & shall certainly study it very diligently as a preparation.— Catherine & I went to Ness yes⟨terday⟩ to join a party on the Hill.[3] Owens of course the chief attraction. Woodhouse seems so altered & odd with little Caddy & Francis as props of that once gay house.— Poor Owen is very unwell in London with the Jaundice,—& Sarah has quite lost the Owen constitution she is composing a long letter to you which she has just written to ask me how to direct.— Fanny has been very busy lately *canvassing* Denbighshire with Mr Biddulph & very successfully. Ld Kenyons son has no chance against them.—

I daresay Caroline told you in her last letter what a pleasant day we spent at Major Bayleys & how very much he talked about his friend Charles. I have not seen Tom Eyton for an age but whenever I do will certainly urge him to go after you.—

The Cholera has at last reached Shrewsbury but there have not been 20 cases yet & I hope it may soon die away,—Catty I think is more alarmed about it thay anyone else in the household.— The present pet of the house is a young Cuckoo which was taken by some boys out of a Larks nest. I am afraid it will never live to say Cuckoo next Spring.—

We went to a Bowmeeting[4] at Pradoe on the 10th of this Month, & we all did nothing but stare at the unfortunate little Bride of Sir Rowland Hill's who made her first public appearance there. She looks very childish but rather pretty & pleasant looking. She was the great heiress Miss Clegg whose marriage had been kept secret more than a year by that old beldame Mrs Hill who of course contrived it.— After all our plans being settled to go to Osmaston we had a letter yesterday to tell us poor William was again so unwell they c^d not receive us I suppose it is Consumption they fear as his Chest is affected— He had got yr letter & was very much pleased by yr writing to him. What an unfortunate & sickly family the poor Foxes are.— This is a very dull epistle dear Charles but it is merely to tell you we are all well. pray let no opportunity pass without writing it is such a comfort & pleasure to hear from you my dearest C. & will you in future just mention the dates of our letters: So far not one has missed either way which is very comfortable.

God bless you & with all our affectionate Loves Believe me Ever Y^rs | Susan E Darwin

I know I have put H.M.S. wrong in this direction but in future I will do it right if you will pardon me this time.—

Postmark: Shrewsbury AU 18 1832
DAR 204.6.1

---

[1] Robert Wedgwood became Rector of Dumbleton, Gloucestershire, but not until 1850; the living was under the patronage of Edward Holland (*Alum. Cantab.*; *Clergy list* London 1852).

[2] Babbage 1832.

[3] Great Ness and Little Ness, villages about 7 miles north-west of Shrewsbury. The 'Hill' is probably Nesscliffe Hill, near Great Ness.

[4] A meeting to practise archery.

From E. A. Darwin   18 August [1832]

[London]
Aug. 18.

My Dear Charles.

I find by a letter from Catty that the packet sails on Friday, so I write this to tell you about your commissions, tho' I am afraid I shall hardly be able to get all your rattletraps in sufficient time to send them. Cuviers Mollusques[1] are not to be had in London and are very dear & scarce. of all the books of travels only one was to be had an imperfect copy without the Atlas for three guineas & a half so I did not get—of the others one was £40 & another £30, consisting of a vast number of plates in folio. In short I have got none of them. I have got Humboldt Fragmens de Geologie et de Climatologie Asiatique[2] which I suppose was the work you meant. The 8th vol of the Personal Narrative[3] was not published. Leopold Von Buch's Travels by Jamieson[4] were in Norway & not Sweden so I have got that in its place & hope it is right. Bohn[5] was very civil & thought he remembered something about the Linnæus[6] but as you did not mention the Edition & there are so many of them he is not certain that he shall be able to procure you the sheets. Mr Banks has promised me the mirrors &c but as you know he is a slow coach. If Bohn should fail, I will send you my Linnæus which I have ordered up from Shrewsbury in case. Scoresby's Arctic Regions[7] are not to be found at Shrewsbury, and as you did not seem very anxious about them I have not thought it worth while to buy them. I could not procure any lace *needles* (so you wrote it) but I have got some lace pins, & some bead needles as being the *finest* made. Everything else I have done.—

You would have had a very pleasant summer in Shropshire this year, Sedgwick & Murchison[8] were both geologizing there. Murchison went to examine all the country about the Ponsford Hills[9] which even I should like to have joined him in. I have been making a great many plans for the summer, but they have all broken through and I am living or rather vegetating in the quietest manner possible in London thinking it quite an exertion if I can get round St Jame's Park in the course of the day. I meant to have gone this summer with Sargeaunt[10] to see the Auvergne, but he has been voted by his Doctor in a delicate state of health and is ordered off to Italy for a couple of years.

I have established a very comfortable little lab in my lodgings, which has long been my great desideratum in my London life, and that and smoking fills up my day delightfully. Next month I mean to go and pay Charlotte a visit at Ripley—

Charlotte alas is very much deteriorated by her marriage contrary to the general rule of women being improved when they marry, and I fully expect that ⟨by⟩ the time you come home, she will have lea⟨rned⟩ to talk quite fluently about Lords & their pedigrees. I am quite convinced that is their subject tête à tête. I long to have a good groan with you over the incomparable throwing herself away. It certainly is the most wonderful event that ever happened in the family.

I am sorry to see in your last letter that you still look forward to the horrid little parsonage in the desert. I was beginning to hope I should have you set up in London in lodgings somewhere near the British Museum or some other learned place. My only chance is the Established Church being abolished, & in some places they are beginning to demand pledges to that effect. The question of pledges is now very much agitated, and I yesterday read an admirable letter from Tom Macaulay to the Electors of Leeds on the subject of pledges in which he refuses to give any direct ones merely stating his opinions openly & frankly. The abolition of taxes on newspapers, vote by ballot, abolition & commutation of tithes, abolition of slavery &c &c all of which he is in favor of, but which he will not pledge himself to vote for.[11] Now that we have got the Reform Bill people seem disinclined to make any use of it, as a vast proportion have either neglected thro' ignorance or else are unwilling to get themselves registered. This makes people more anxious than ever for vote by Ballot, and I have no doubt that will soon be carried. The poor old King is very unpopular now, going down to the house to prorogue parliament, he was received with groans, & at a great dinner of the National Political Union[12] when his health was proposed it was absolutely refused to be drunk. I have written you all this politics tho' I suppose you are too far from England to care much about it. Politics wont travel.

Good Bye My dear Charles, & write to me again when you want any more commissions and I shall have great pleasure in executing them and dont think of the trouble. I have great pleasure in reading your letters home and take the greatest interest in all your proceedings but do not wish that you should write to me, and if you do not hear very often from me it is you may be very sure not from want of love, but ⟨fr⟩om indolence, & not very well knowing what to write about London gossip will hardly carry to Shrewsbury much less across the Atlantic. Good Bye my best love & good wishes. E Darwin

Postmark: Paid 20 AU 1832 CX
DAR 204.6.1

[1] Cuvier 1817.
[2] Humboldt 1831. The flyleaf of the second volume is inscribed 'Chas Darwin Monte Video Novem: 1832'. The volumes are lightly annotated and scored, mainly in pencil. The subjects of interest suggest that the notes were made after the voyage. In volume one the half-title page has written on it CD's signature and 'Interesting parts begin P 84', 'The Andes P 143.' The facing page has the word 'Metaphysics'.

[3] The *Personal narrative* was complete in seven volumes. The eighth and ninth volumes of the *Voyage* were devoted to zoology and comparative anatomy; they were not translated.

[4] Buch 1813. The copy in Darwin Library — CUL is inscribed 'Chas Darwin M. Video Nov[r] 1832'. There are no marks or annotations.

[5] Henry George Bohn.

[6] The Darwin Library — Down has both the *Systema naturae*, Ed. 13a, Cura J. F. Gmelin (bound in 10 vols.) (Linnaeus 1789—96) and the *Systema vegetabilium*, Ed. 15a . . . (Linnaeus 1797). The latter is inscribed 'Erasmus Darwin Christ Coll 1825'. This may be the volume mentioned by Erasmus later in the letter. There was also at Shrewsbury an English translation of the *Systema vegetabilium* 'by a Botanical Society at Lichfield' (Linnaeus 1783a) (see letter from E. A. Darwin, [8 March 1825]). The Darwin Library — CUL has Linnaeus's *Philosophia Botanica* (Linnaeus 1783b), lightly annotated. There is no evidence that CD had it on board the *Beagle*. The notes on the end-paper relate to later botanical interests; one of them calls attention to p. 87: 'Maris fundus non destruit semina' ('The depth of the sea does not destroy seeds').

[7] Scoresby 1820.

[8] Roderick Impey Murchison.

[9] A Pontesford Hill is discussed and illustrated in Murchison 1839, pp. 81, 263—4.

[10] Possibly Frederick Thomas Sergeant.

[11] Thomas Babington Macaulay was M.P. for Leeds in 1831. The views listed by Erasmus are typical Whig causes of the 1830s. In his address to the electors of Leeds, Macaulay supported the Reform Bill as a moderate one which would prevent the present ministers being 'superseded by able, vicious and destructive radicals, who would trample on Whig and Tory alike' (*Annual register*, 1832, p. 47).

[12] Political body founded by Francis Place in 1831 and designed to put pressure on the House of Lords to pass the Reform Bill.

To Frederick Watkins 18 August 1832

Monte Video, Riv. Plata.
August 18th. 1832

My dear Watkins,

I do not feel very sure you will think a letter from one so far distant will be worth having I write therefore on the selfish principle of getting an answer.— In the different countries we visit the entire newness and difference from England only serves to make more keen the recollection of its scenes & delights. In consequence the pleasure of thinking of & hearing from ones former friends does indeed become great— Recollect this & some long winters evening sit down & send me a long account of yourself & our friends; both what you have, & what intend doing; otherwise in 3 or 4 more years when I return you will be all strangers to me Considering how many months have passed we have not in the Beagle made much way round the world. Hitherto every thing has well repaid the necessary trouble and loss of comfort. We staid three weeks at the Cape de Verds, it was no ordinary pleasure rambling over the plains of Lava under a Tropical sun but when I first entered on and beheld the luxuriant vegetation in Brazil it was realising the visions in the Arabian nights— The brilliancy of the Scenery throws one into a delirium of delight and a Beetle hunter is not likely soon to awaken from it, when whichever way he turns fresh treasures meet his eye. At Rio de Janeiro three months passed away like so many weeks— I made a most delightful excursion during this time of 150 miles into the country.— I staid at an estate which is the last of the cleared ground, behind is one vast impenetrable forest. It is almost impossible to imagine the quietude of such a life— Not a human being within some miles interrupts the solitude.— To seat oneself amidst the gloom of such a forest on a decaying trunk, and then think of home, is a pleasure worth taking some trouble for— We are at present in a much less interesting country— One single walk over the undulatory turf plain shows every thing which is to be seen. It is not at all unlike Cambridgeshire only that every hedge tree & hill must be levelled & arable land turned into pasture. All S. America is in such an unsettled state that we have not entered one port without some sort of disturbance— At Buenos Ayres, a shot came whistling over our heads; it is a noise I had never before heard, but I found I had an instinctive knowledge of what it meant. The other day we landed our men here & took possession at the request of the inhabitants of the central fort. We Philosophers do not bargain for this sort of work and I hope there will be no more. We sail in the course of a day or two, to survey the Coast of Patagonia as it is entirely unknown I expect a good deal of interest.— But already do I perceive the grievous difference between sailing on these seas and the Equinoctial ocean— In the "Ladies Gulf" as the Spaniards call it, it is so luxurious to sit on deck and enjoy the coolness of the night & admire the new constellations of the south— As for the old moon, she, nightingales, Jack Venables, & your jolly old self, form so pleasant a train of ideas that I never could want something to think about. I wonder where we shall ever meet again, be it when it may; few things will give me greater pleasure than to see you again and talk over the long time we have passed together

If you were to meet me at present I certainly should be looked at like a wild beast, a great grisly beard and flushing jacket would disfigure an angel.—

Believe me, my dear Watkins, with the warmest feelings of friendship, | Ever yours, | Charles Darwin.

Copy
DAR 148

From Sarah Williams 26[−31] August 1832

1. Belgrave Street.
Sunday 26. August. | 1832.

My dear Charles.

If you have ever thought of me I fear it has only been to upbraid me with forgetfulness of the promise I made to write to you. I assure you my own conscience has reproached me very very often, but I *know* I need not say that forgetfulness of you did not cause my silence, indeed, my dear Charles, no day has elapsed without your occupying some of my thoughts, & it was with very great pleasure I heard such good accounts had been received of you, Papa was *much pleased* with your letter, which he had about a fortnight ago, & I dare say has answered by this time. Susan tells me to direct to Monte Video, but I am afraid it may be very uncertain when this reaches your hands, whenever it does, I hope you will write & tell me so, without waiting till you are at *Terra del Fuego* from whence you know you promised me a letter— I often think what a wonderful number of changes have taken place amongst your Friends, since you saw them.— Fanny's marriage *must have surprised you* not a little, then the "incomparable Charlottes", & her Brother, Mʳ F. Wedgwood, I think your Cousin Hensleigh was *twined* off before you sailed & you must have been detained long enough at Portsmouth to hear that my *execution* had actually taken place, since which awful time, I have, I assure you, been *as happy as a Queen*, & have really nothing left to wish for being sadly spoilt & indulged by Edward, who is an Angel all but the Wings, (as Harry Wedgwood said of the "incomparable Charlotte's husband)— I do think there *never* was such a temper, *you* may imagine its excellence, when I tell you that even one of the *Warlike race* of Owen has never yet contrived to make a fight, or even a dispute, whether it is that the Owen blood is fast degenerating, I know not, but such is the fact.— I must try & give you some account of my life & adventures since we parted— When I married instead of going a shivering tour in the month of November, in search of the Picturesque, I declared I would instantly proceed on a romantic tour to London—where we arrived the day afterwards, that same evening I went to see Miss Fanny Kemble[1] (Heavens & Earth, I hear you exclaim,) & continued my Theatrical Tour for a week or more; till I had visited every play house then open— We soon afterwards returned to Eaton, where we remained till the end of January, & excellent fun we had, hunting almost every day with the Beagles—I mounted on a steady seasoned old Hunter, with as much sense as any Christian, he carried me most delightfully over most things, & never gave me a fall, he has since become my own property, he has been at grass all summer, & I hope to find him in hunting condition when the season begins— but to return to my story, we left Eaton the end of January & went to stay at *my Brother* Richard's,[2] in Pall Mall, whilst we were furnishing this house, which we took possession of the end of February, & I have never left it since, except for a few day's visit to the Bruces who have taken a place near Windsor— In March I had a very severe illness, which laid me up for some time, as soon as I was able, I thought a ride would do me good, & accordingly mounted a 4 year old Grey Mare which Edward had bought for me from Mʳ Gore ⟨&⟩ we went into the Park, (it was my sec⟨ond ti⟩me of riding her) & had just got opposite the Serpentine, when she started across the road at a Dog, turned short round, & down I came, her feet were unluckily entangled in my habit, & in trying to get away, she put her foot on my Ancle bone & the caulking of her shoe took a piece *clear* out, had it been half an inch lower, the Surgeons say I should have lost the joint— I jumped up directly, fancying I had only strained my ancle, & Edward put me into the house belonging to the *Humane* Society whilst he fetched the Carriage to take me home, that very evening Caroline & Arthur arrived, poor Arthur to prepare for his Voyage & the whole time he staid, I was never able to leave my bed the wound was so bad—it was 9 weeks before I was able to walk. I went about with a Crutch, & was lifted in & out of the Carriage— Edward would not consent to my riding this unlucky grey Mare any more, & I was obliged to sell her, but have supplied her place by a very nice Bay, which exactly suits me— Poor Arthur sailed the middle of May, he took with him a most tremendous looking Bull Dog, rather an odd Dog to choose, I thought but he was told it would be more useful to him than any other— We heard from him at Madeira, & he seemed to enjoy his voyage much poor boy, I hope he will be steady, when he arrives in India, & then there can be little doubt of his doing well— Caroline went down for Fanny's Wedding, & then returned here with Mama, they staid 5 weeks with us, came up & returned per *Wonder*, which they say is the most delightful of earthly conveyances— I feel quite odd at being left by myself so many hours in the day, for Edward goes to *Shop* about 11 & seldom returns before 5 or half past, it is very provoking we are detained in London so late this year, owing to M.ʳ Powell's absence, but I hope to see Eaton about the second week in Sep.ᵗʳ & then we shall remain there till after Xmas— I never was so long in London before, but I am such a *thorough Cockney* that I like it better than any other place, on the long run, & am glad it is my fate to spend so much time in it— Fanny has been in Town for 3 weeks, but is now returned to Chirk Castle, which I fear she will not find very pleasant for some time to come, as M.ʳˢ & Miss Biddulph are there for a *few* Months, & she is sadly in awe of them both. Who would have thought she would have married so soon after that unfortunate Citadel Affair!!—³ Caroline is now in possession of the title of *Miss Owen*, I assure you she is much *come forward* since she attained that dignity— Emma I suppose will make her *public* appearance this Winter, I hear she is much improved lately, but it is more than 6 months since I have seen her— That "rising Star of Ton" Matty Cotton is also very flourishing— Of course all the other Shropshire Marriages have been announced to you—Miss Boughey & M.ʳ E. Fielding, Miss Parker,—& Sir Baldwin—M.ʳ Mainwaring's *intended* marriage to Miss Salisbury, and M.ʳ Henry Lloyd's to a Bristol heiress, as soon as he gets a Living— If you do not make haste home, you will find nothing but *dead dogs* in Shropshire— Clare Leighton is still Clare Leighton, & I hear no *talk* of her changing her name, I am sure you would be sorry to hear of poor M.ʳˢ Mathew's sudden death. I do think nobody was ever so generally lamented, poor M.ʳ Mathew is not likely ever to recover it I fear— Erasmus dined with us not long ago, & I also met him at dinner at the Hollands—by the bye M.ʳ Holland is married & done for since you went— I cannot say I much admire his choice though I am not so violent as Erasmus, who declares i⟨f sh⟩e was to offer him

5£ to give her a kiss, he would not— Catherine is now on a visit to the Hollands, in Gloucestershire— I shall be *most delighted*, my dear Charles, to receive a letter from you, I have not forgotten the *solemn* promise you made me to come to N° 1 Belgrave S^t as soon as you arrived in London—dine with us, & *go to the* Play— I wonder when that day will come!! I hope you continue to like your *"Angel in the disguise of a Sea Captain"* as much as ever, have you made a large collection of Curiosities— I often laugh when I think of your very last *walk in the Forest*, when you discovered & rooted up those horrible funguses, & *bottled* them for the Professors at Cambridge— I should prefer *living* curiosities such as Monkeys or Parrots, which I have a great fancy for— I have now got 3 Pets—a Bull finch, & 2 nondescript little foreign birds. I had the ⟨   ⟩ misfortune to lose poor Beppo, who was taken by some wretch in human form when he had been 3 weeks in Town— I like this house very much, it is so airy & *quiet*—you will laugh at this being a recommendation to *me*, but you have no idea what a sober steady Matron I am become & strange to say, I have *almost* entirely lost my taste for gaiety & going out— You will hardly believe it is *Sarah Owen* who writes in this strain, but so it is— And now I fear I must bring this long scribble to a conclusion. Heaven knows *when* or *where* it may reach you, but whenever it does, I hope it will convince you that you are not & never will be forgotten by one of your oldest Friends, who now remains very sincerely, & affec^tely Yours, | S. H. Hosier Williams | (how does it look)

A letter directed Belgrave S^t. will always find me— God bless you, my dear Charles, I will write again ⟨very⟩ soon—

Postmark: 31 AU CX
DAR 204.6.2

[1] Frances Anne Kemble, actress then at the height of her success.
[2] Richard Williams, Sarah's brother-in-law (see letter from Fanny Owen, [January 1828]).
[3] An allusion to Fanny Owen's having been jilted by John Hill.

From W. D. Fox 29 August — 28 September 1832

Ryde. Isle of Wight.
August 29. 1832

My dear Darwin

The sight of your well known hand upon a letter back gave me no little pleasure as I began to doubt whether I should hear from you during your journeyings; it seemed so long since I had done so, I had however heard excellent accounts of you tho' not from you & that was next to it, tho' far from being so pleasurable as hearing of your happiness in your own words.— I wrote a very stupid letter some three months since which you have probably received ere this, & if not you have no loss, as I felt very stupid & owlish when I wrote it & almost repented sending such rubbish so far.— Your letter I got a fortnight since when I was just on the point of leaving home for this Island in consequence of another attack upon my Lungs, which tho' very much slighter than my former one in March, was still bad enough to make a change to Warmer Air desirable. I gained ground surprisingly for a week after leaving home, indeed was nearly well to my feelings, when an imprudent walk on a hot day gave me another attack which has lasted me now a week & will not depart while the very cold weather & rain we now have, last.— I have intended every day commencing this, but not felt equal to it, but trust now I have once begun, to go on with it & send it across the Atlantic tomorrow.— You tell me to give you a particular account of my life &c &c.— You would have a very monotonous description indeed, were I to give you one. Since my last letter I have vibrated between Epperstone & Osmaston, well enough to enjoy myself, but not sufficiently so to do any parochial Duty, or Entomologise, or in fact do any thing but amuse myself.— I shall not therefore inflict upon you the Diary of an Invalid, but sometime before Winter sets in I will write to you again & tell you how I then go on & how the World wags in these parts. I do not at present know where my Winter will be spent. I believe we (ie My Father, Mother & 2 younger sisters with my little niece[1]) stay here for about 6 weeks when it will depend upon my health whether I return to Osmaston or go somewhere where the Air is milder than Derbyshire to pass the Winter. As far as I understand myself, I have no actual affection of the Lungs but they are in such a delicate & excited state as to be extremely liable to all the Diseases liable to them, without great care &c. And unless I gain more strength & am very careful about them, Consumption would most probably ensue.— At present however I trust I am in a fair way of recovery & mean to leave no caution on my part wanting to ensure it.

September 18.— When I wrote the first page I fully intended sending this off immediately, but the next day I was so unwell that I could not finish it, & so continued for some days when finding I was losing strength very fast I applied to a D.r Barrett (here at present to recover from an attack of Cholera) who ordered me Blisters &c &c which kept me *very tame* for some time longer, and until lately I have had no inclination to perform the mechanical part of writing, tho' very often thinking of you. Indeed there is seldom half a day passes that I do not wonder what you are about & wish I was with you.— I am now very much better, as well or very nearly so, as before my last attack and enjoy myself much.— All my plans for the Winter are however altered and it is now fixed that I & my two younger sisters stay here during the Winter, and we have already got into our Winter Quarters, a very comfortable House which seems to be very warm & snug; I much wish you were likely to be an inmate of it for a few weeks during our stay; tho' I am a great Beast to wish you here when you seem so very delightfully occupied where you are. I saw by the Paper a few days since that a Ship from Rio left your little Beagle in safety there and of course you were then well or we should have heard; indeed from your account of yourself & from what I know of the climate of S. America from books & travellers, I should hope there is every chance of your having your health there as well as in England with common precautions.— September coming in must have brought England very much to your mind, & the Partridge shooting you used so much to enjoy here, & I trust will do again. I have now not heard from Shrewsbury for a month. At that time your Father & Sisters were in good health. Poor Fanny Wedgwoods death had just been a great shock to them, as it will be to you I am sure. Her death seems to have been very sudden, as she was not ill more than a week & then was not thought in any danger. How changed is that Family party since the last time (the only one) I saw them together.— Three sons & a daughter married & one gone.[2]

September 28.— Again was I stopped in my progress, and been so engaged since with one thing or another that I have never resumed my letter tho' every day resolving I would do so and send it off. I have often wished you were with me in my little walks here when I have seen Insects that in your infantine days of Entomology you would have thought much of. I have seen one Colias Hyale, and had I been able to go after them believe I might have seen all the species of that beautiful Genus in the Island. Not one capture have I made since I came except a few Coleoptera that have crossed me, & one Sirex juvenans, tho I every day see something new to me.— I often think when I see our minute crustacea or mollusca or any insect, of the gloriously beautiful & curious creatures that are lying in profusion round you, & fancy you revelling among them, with sky, ocean, Trees, in fact every thing grand & sublime. You must sometimes almost feel bursting with your feelings, tho' of course by this time the extreme novelty of Tropical Climes is somewhat softened down.— I met with a case of South American Insects here some days since which I was extravagant enough to buy merely because they came from where you are collecting.— There are among them some magnificent Cerambycidæ, Geotrupidæ, Cicadæ &c &c but I have at present no book by which I can make them out. There is also among them the Walking stick Mantis and a huge Nepa, or nearly allied to it.— From the list you give of the Water Beetles taken the day you wrote to me, your favorite Family seems to be very abundant.— I looked anxiously in Stephens for the Cryctocephalus Darwinii,[3] but it never appeared, nor do I think your name has ever been mentioned by him since your departure— You will however have your revenge in this I hope some time, as you will have plenty to tell the Lovers of Natural History when you return. I quite agree with you in what you say about minute insects not having been commonly collected abroad—I have very seldom seen any but those remarkable for size or beauty in any Cabinets.— Stephens is now quite out of my books. He has kept prevaricating so egregiously about his No[s] as to disgust many very much, and now he has crowned all but getting an injunction in Chancery against a book of Rennies[4] as infringing upon his work that has nothing to do with it, and has discontinued his N[os] for the last

2 months, because forsooth others may copy out of it, if he does so. He is sadly in want of cash I hear, which may extenuate his conduct. A new Entomological Monthly M⟨agaz⟩ine is just come out,[5] which I have not yet seen, but ⟨   ⟩ it is likely to be well supported by the first Entomologists and ⟨n⟩o contests are to allowed.— Are you aware that there is now no Duty upon Shells coming into this Country.—[6] I mention it as it may induce you to bring more than you otherwise might do over, as the Duty used to forbid a large collection.— You never told me whether you were to be allowed your own collection of Nat History or whether as you feared, our Government would swallow all. You surely will be allowed your duplicates, though after all the stores you will lay up in *the mind* is the great thing, in comparison with which, the Cabinet is of little moment.— Of course you see our Papers whenever a ship comes out and must devour them with no little interest.— Now the Reform Bill has passed away the Cholera & coming Election are the great topics.— The Cholera is now every where almost, but is not by any means so dreadful a visitor to those of our rank in Society as it at first threatened to be.— Among the poor & needy it is often very fearful in its ravages and sometimes also among the better classes, as in London where very many enjoying every comfort of life have been cut off.— In this Island there has been only one doubtful case though it has been at Portsmouth some weeks, and I believe many more have been attacked with it there, than is generally known. I say this from my own enquiries there as I often go over. To shew you how little is thought of it when it once has made its appearance, I yesterday went to the launch (with all our party of the Neptune 130 Gun ship, and I do not think an inhabitant of the country within 20 miles that could come, was left behind— It was a glorious sight to see that fine Harbour a mass of living souls all looking happy.— I do not know whether you ever were here. It is a beautiful place commanding Spithead and all that fine anchorage opposite Portsmouth, and generally enlivened, as at the present time, with Ships of war. We have now three English & one French there & are hourly expecting the French Fleet in, when they are going to give the last decision of their countries to Holland on the Belgic Question & if not acceded to, blockade all her ports.—[7] It is somewhat curious to see a French Frigate with the Tricolor, abreast a 74 taken from them, in our roads. I do hope that the antipathy of the 2 nations so long cherished is now about to cease, would that your favorite Text, "peace on Earth, & good will to men" generally might spread abroad over the whole earth. It has just struck me that from my maritime situation this winter, I might be able to be of some use to you in forwarding any stores you may want. If there is anything that I can do of any sort, do not hesitate to make use of me as nothing would give me more pleasure (as I hope you know) than to be of use to you. Of course you have agents in London for common things but I thought I might of some utility in sending any Nat. Hist: stores you may now want. I can easily clear out any packages for you here, get store cases, any thing. Pray use me if you can for my pleasure.— I have gone on scribbling nonsense till my paper is

gone. You will be glad to hear that I am now quite convalescent & hope a Winter here will quite restore me.— My Father Mother & sisters all in their best way and all wishing you every kind wish for success in all your undertakings, good health &c &c. I have just heard from Simpson, who tells me that Jeffry Hall[8] is actually settled in Canada. Simpson is gone into Church.— Of Henslow I have not heard some months. I conclude Erasmus is alive as I have never heard to the contrary but have not heard of him in any way for some months past, I might almost say years— I must now my dear Darwin bid you farewell for a time. I shall certainly torment you again before very long. If you can find time to write me 6 lines any time, they will be thankfully & joyfully received; just tell me how & where you are & what doing—no more— Goodby. May you have a prosperous voyage & both of us live the one to hear & the other relate Your Travellers Wonders

God Bless you & Believe me. Ever yours affectionately & faithfully W D. Fox

DAR 204.6.2

[1] Anna Maria Bristowe.

[2] Of the Wedgwood daughters, only the eldest, Sarah Elizabeth (Elizabeth), and the youngest, Emma, were still unmarried.

[3] Apparently a slip for *Cryptocephalus*. CD must have sent what he thought was a new species for inclusion in Stephens 1827–46.

[4] Rennie 1832.

[5] *The Entomological Magazine*, edited by Edward Newman (London 1833–8).

[6] The duty on shells and other natural history objects had been lifted in July 1825, 'An act to repeal certain duties and customs', 6 Geo 4. c. 104. See Swainson 1822 for an account of the various levies, and Lingwood 1984.

[7] In October–December 1832 the British and French cooperated in a blockade to force the Dutch to surrender Antwerp to newly independent Belgium.

[8] Jeffry Brock Hall emigrated to Canada and resided at Guelph, Ontario (*Alum. Cantab.*).

From Caroline Darwin   12[–18] September 1832

[Shrewsbury]
12th September 1832.

My dearest Charles,

Susan wrote by the vessel that went from Falmouth the 3[d] week of last month as you directed in your last letter dated June. I have it not by me to be accurate about the date, but as Susan has acknowledged it—it does not matter. poor Musters death was very melancholy. I wish he had been spared hearing of his mother's loss. I do hope you are very prudent & do consider the forlorn state you would be in a long bad illness with the miserable accommodation you would have in one of your scrambling expeditions, as indeed you had experience of from what your journal says— I know it is nonsense & "all foolishness" to use your own expression writing at this distance wishes & cautions but I must do it dear old Tactus as a relief to myself— You will I know feel very much for the sad loss the poor Maer family have had. About three weeks ago poor Fanny Wedgwood was taken ill with what they thought a sort of billious fever. She seemed very ill for two days with vomitings & pain & then appeared to get better, so much so that not one of the family had an idea she was in danger. 7 days after she became unwell Elizabeth sat up with her at night as *she* (Fanny) was too restless to sleep, towards morning she seemed cold & more uncomfortable & they sent for the apothecary. he came between 6 & 7 & thought her in gt danger & at 9 oClock she expired. only Elizabeth & Emma were with her as even after seeing the apothecary, from some misunderstanding none of the family had an idea her danger was so immediate. Uncle Jos was terribly over come & Aunt Bessy it was some time before Elizabeth could make her understand what had happened. My Father says mortification must have taken place in her bowels. at the time, the pain ceasing they all thought she was getting well. We had not heard of her illness till we had the letter with the account of her death. I have been very minute in telling all particulars as I know how much interested you are for all the Wedgwoods & you did poor Fanny justice in liking her & valuing her goodness & excellent qualities. She had no idea of her own danger, but as Elizabeth says in her letter to me, "this could not be needed, for one so pious, so humble & unselfish, & so good in all her feelings". the loss to Emma will be very great, hardly ever having been separated, all her associations of her pleasures & youth so intimately connected with her. M^r & Mrs Hensleigh were staying at Maer during the time which was fortunate as they have been a comfort to them all. Uncle Jos came over yesterday for a day to see Papa, he was cheerful & apparently much as usual, he says they are all better & cheerful at Maer except Aunt Bessy who they can not make go on with her usual little occupations, she sits by herself & looks very sad & dejected. Mr Baugh Allen is at Maer & has taken Aunt Bessy a little tour in North Wales in his Carriage for a few days which they think will do her good—and next Saturday Charlotte & Mr Langton come to Maer to make a long stay which will be a great comfort to them all. Charlotte had heard but a slight account of Fannys being unwell before the letter came to tell her what had happened— I had a letter just before from Charlotte speaking with pleasure of a letter she had had from you She was then writing to you but I suppose her letter was not finished before this & very likely you have not heard from her. She says Mr Langton has been most kind & sympathizing in this her first sorrow— I am very much puzzled what home news to tell you, one day passes after another with nothing worth telling to be had from it— however the very best news is that my Father seems quite as well as usual again. he had been as Susan told you unwell for several weeks. he is in very good spirits & amuses himself very well with the hothouse it has quite revived his old interest about flowers. he is going to get a Banana tree principally from your advice— he is giving up his business in great measure he still goes to some people, but it is so generally known that he declines going to any distance that he has now little difficulty in refusing going

to see people. A Dᵣ Goldie[1] is come to Shrewsbury from York highly recommended by Mʳ Vernon Harcourt[2] & Col. Gooch he seems a sensible man &
will be very likely to succeed. we have not yet seen Mʳˢ Goldie— Papa talks of
going for a day to Liverpool to see how he bears the fatigue of *Lionizing* as he has
a great wish to go to London which he has so long talked of.— Catherine is
staying at Overbury[3] with the Hollands. it is only 5 miles from Dumbleton. she
is enjoying her visit there extremely riding & driving about. they are sensible &
have nice horses & carriages— She is going a little tour in Derbyshire with them
& to pay a visit at Sir T. Denmans[4] & will not return home till the end of the
Month— she went to call or dine at the Edward Hollands & said they both
looked deadly dull— She suspects Mʳˢ E Hollands brother Mʳ John Isaac &
Charlotte will soon add another to this year of marriages— Susan goes in a few
days to the Hill to meet Jessie who is staying there she has been expecting to
go all this week but Harry who is to Chaperon her has been detained by
Bankruptcy business & it is uncertain which day he will be here— I hope you
still feel a proper interest about Nina & Pincher they are both invalids but
convalescent— Pincher cut his foot badly with broken glass but his lameness is
nearly cured. poor Nina has had a much worse misfortune—the ⟨   ⟩ old
Coach Horse seized hold of her by the hind leg lifted her up in to the manger &
would not let her go for a few seconds—her leg was badly broken & I am afraid
the joint injured but she is getting well & does not seem to suffer any pain now.
the *surgeons* who attended her were very goodnatured & I am sure every thing
possible was done— The Cholera has been in Shrewsbury now for some weeks
but it has been subsiding for the last week & it is now some days since we have
heard of any death—they say many people reported to have the Cholera really
had not, at least so Mʳ Wynn[5] says who of course is infallible—

I am glad to hear at last that idle Erasmus has written to you & I hope you
have received the books &c by this time. Captain Beaufort took charge of
them— William Fox has recᵈ your letter, he is still staying at the Isle of Wight
hoping to get better before winter

Mʳˢ Williams comes to Woodhouse next Monday. She is become very
delicate often unwell. the celebrated Owen constitution has quite failed with
her. I shall be very glad to see her again— I have only seen Mʳˢ Biddulph once
since her marriage I was staying at Woodhouse & she rode over & staid 3 hours
very pleasent & very pretty. Old Mʳˢ & Miss Biddulph are staying at Chirk
Castle & they are so formal & stupid Fanny says she is wearied to death. I think
she seems happy & attached to Mʳ B. I can't help thinking he is a *very* selfish
man though I have no great proof.— Francis is staying at Woodhouse as they
have not yet got any thing for him— I have written to you since reading your
journal which I liked *exceedingly*. I do hope when you have any safe opportunity
you will send us some more of it. it gives us all so much pleasure & interest
reading about you & it brings the Country &c in such a lively manner to one—
I have never told you dear Charles what great pleasure your most affectionate

dear letters give me & us all, you would be I am sure rewarded for the trouble of writing if you saw the delight a letter from you is received with— My Father bids me say y.ʳ money has not been drawn for at the Bank at least had not a fortnight ago—when he enquired. he sends his affectionate love. when you write he wishes you would mention the date of the letters you receive—I mean when ⟨ ⟩ they are written—as well as when rec.ᵈ

Harry & Susan went this morning to the Hill— My Father is so spirited that he has *settled* to go to London in a few weeks with Harry & Edward & he very much enjoys the thoughts of his grand Lark.

Good bye *very very* dear Charles Pray be prudent & careful I often make a day dream of seeing you so happy in your Parsonage, again Good bye & God bless you dear Charles. Yʳ affect | Caroline Darwin *Sept 25ᵗʰ*—6

I hope I have directed this letter right—

Postmark: Shrewsbury SE 19 1832
DAR 204.6.1

[1] George Goldie, M.D.
[2] William Venables Vernon Harcourt.
[3] Village in Worcestershire, on the border with Gloucestershire.
[4] Thomas Denman.
[5] Rice Wynne.
[6] The original provides no explanation of the discrepancy between the date at the end of the letter, which is certainly in Caroline's hand, and the post office stamp on the cover. One can only suppose that Caroline wrote on Tuesday, the 18th, and confused it with Tuesday, the 25th.

From Charlotte Langton 27 [September] 1832

Maer
October 27, 1832[1]

My dear Charles

I was very glad to receive your letter & to hear so good an account of the success of your voyage. I rejoice most cordially in the pleasure & benefit you receive from it & will continue to receive I hope. It appears to have answered much better than the most sanguine could have hoped when you not only enjoy the beautiful scenes you go into to the utmost but even get attached to your little cramped cabin in the Beagle, in which one thought you could meet with nothing but discomfort & inconvenience—that it was a good place for reading I never could have guessed & it gives me no little pleasure to hear that you profit by it in being so industrious—it will make this voyage a pleasure & advantage to you all your life instead of a mere present delight. I am glad too that you are not *too* fond of the sea so as to lose sight of the pleasures of a quiet domestic English country life, for I should be very very sorry if you continued to lead a wandering life which is I think bad for every body, to say nothing of the loss to their friends,

& not all the beauties of tropical climates would make up for the change. I have been putting off writing to you till I came to Maer—in a new place & among new people I thought I should be too much at a loss what to write to you about & that Maer would supply me with something to tell. The loss of our dear Fanny has changed it sadly since I left it in the Spring. the family seems diminished to such a small one compared with what it was then—for herself so good & innocent & unselfish as she was I can only feel that she is very happy to be taken out of the world before any distress or unhappiness came near her, her life was a very happy one & closed without knowing her danger or feeling the pain of separation from her family—poor Mama has borne her loss wonderfully well & all are very chearful. Hensleigh & Fanny were fortunately with them at the time & were the greatest possible comfort & support to them—I was very glad that we were able to come & take their places when they went away. It is the most beautiful September weather possible—shooting is utterly neglected Robert being away at the Hill, enjoying a holiday I should think very much, & all the more I suspect for having the good luck to meet Susan there. In the mean time Charles, Mr Langton that is, has undertaken to do his duty the two next Sundays. I remember the horror you used to express at the thoughts of doing duty at Maer— I believe he has a little bit of the same tho fortunately for Robert not quite so strong a one as your's. I do not feel like a true parson's wife yet & shall not till he has a living, which I wish might happen soon, it will be so much pleasanter to be settled down in a regular home, than as we now are not knowing how long we may stay where we are. I should be so glad for us both to begin to try to lead a useful life instead of feeling good for nothing & useless as I am sorry to say we do now—one feels it an excuse to do nothing when one is not fixed, tho one ought not I own. I am very much pleased with the beauty of Surrey—the village we live in not very pretty but we are in reach of very beautiful drives & have two ponies & a little carriage to take us about. There is a very pretty mixture of rich & highly cultivated country, with wild heaths commons & copse & wood which make a most charming riding country & I have often longed to have Caroline with her horse to shew it to. We have got acquainted with a good many people about but not many that I care about—I have no turn for forming new acquaintances or friendships which I think is rather a misfortun⟨e⟩ one loses much pleasure by it. Our chief dependence for society is on Hensleigh & Fanny, who are often inclined to come & refresh with us in the country, and Lady Gifford who is in our way to London. I have never yet ventured to ask Erasmus to come & see us, for he seems to find the country so fatiguing that I am afraid of having the mortification of seeing him dying with ennui in one day at Ripley. The last time I saw him he was making himself most useful to Fanny Hensleigh, looking at houses for her, for which he seems to have quite a taste luckily for her & Hensleigh, who is too busy at his office to have much time for house hunting. I think he must have the most extraordinary taste for London, to have been able to stay in it all this beautiful summer, but I suppose he is gone down to Shrewsbury by this time. Harry & Jessie are houseless now, & are staying at the Hill they think of taking a house at Keel, between Newcastle & Betley. Robert has been scheming in ducks & poultry this summer—the pool is covered with his ducks & I understand he made his eggs pay for all his butcher's meat, & now he is going to set up a cow—we used to think his genius lay most towards being a squire, but now I think he shews it very strongly in the life of a country clergyman, in which it must be said for him he has other merits than these farmyard pursuits. My father sends his love to you & desires me to say how glad he is that your undertaking appears to answer so well. I think Charles bears you some envy for so delightful an expedition, for his great passion is travelling, & if any thing makes him repent being married I think it will be being cut off from it—however his spirits have stood being obliged to refuse two such tempting offers this summer that I think I need not be afraid any more: one was a trip in a friends yatch to the Mediteranean, & the other one in another friend's yatch to America to coast & travel about there all at his friend's expence.

My mother & Elizabeth & Emma desire to be most kindly remembered to you— I need only wish you to continue to be as happy in your travels as you are now, which I do most warmly & that you will have a safe & happy return.

Believe me dear Charles your affectionate cousin | Charlotte Langton

Your friend Wilcox[2] is I believe going on pretty well—he has taken the Manor and gets what he can by selling the game, we buying what we want as we did before.

Postmark: SE 28 1832 Newcastle
DAR 204.6.2

[1] The postmark and internal evidence make clear that the letter was written on 27 September and dated 27 October in error.
[2] Wilcox has not been identified.

From Catherine Darwin   14 October [1832]

Shrewsbury.
October 14th.

My dear Charles,

We received your letter dated July 7th, sent by M[r] Sullivan's Parcel, the end of last month. It was an unusually long time on its voyage, nearly three months. We were exceedingly glad to have such a happy letter from you; your three months on shore must indeed have been as interesting & useful as it was possible to be. We were rather amused at your anxiety to leave civilized ports, and to hear the fate of Reform, two wishes not very compatible. As yet we have been very successful in correspondence; you have received all our letters up to the date of your last, and all your's also I think have safely arrived. You must tell us when to leave off directing to Monte Video & to put S. American Station instead. Charlotte desired me to tell you she wrote you a letter the end of last month (September) to Monte Video. It must be a melancholy letter I think, written so soon after poor Fanny's death. They seem to be all recovering their spirits at Maer, except poor Aunt Bessy, who feels it very much now, though she did not appear to do so at first. The Langtons are staying at Maer now; all the family seem to like Mr Langton very much; they say he is so merry & joking, and chatty, quite different from the sensible, grave man he was taken for before the marriage. All the London people, with Erasmus at their head have a great spite & prejudice to him, from his having been so much cried up at first I suppose. Erasmus is very audacious & wicked about him, and thinks him wearisome, & foolish & tiresome. Charlotte makes him the most devoted wife that ever was seen, perfectly wrapt up in Mr Langton, and talks & thinks of nothing else. This increases the spite of the London people, whose main subject of conversation seems to be finding fault with Mr Langton. Erasmus is here now; he has not stirred out of London all this Summer, till he has come down here now to breathe a little fresh air. He and Mrs Hensleigh seem to be thicker than ever; she is quite as much married to him as to Hensleigh, and Papa continually prophecies a fine paragraph in the Paper about them.— Papa is very well now, much better than he was in the Summer, and more occupied than ever with his pet, the Hot house; his Banana Tree is sent for, and a deep hole made for it in the highest part of the Hot house, that it may have room. Papa means to call it the *Don Carlos* Tree, in compliment to you.— Papa is also planning buying Audubon's Book on American Ornithology;[1] the author sells it himself, and will not allow any separate number to be sold, unless you take the whole which is 40 guineas in price. The Plates are magnificent, as they ought indeed to be. You will like to see some of the Plates of your old Friends again, when you come home.—

Your Books were all sent off, before your last letter arrived, mentioning the two others. We had the most extraordinarily hot weather in England from the 20th to the 27th of September; 70 in the shade; the common people attributed it to the Comet,[2] which first appeared, visible to Telescopes about that time.— The Cholera has died away in Shrewsbury now, after but few deaths; the last was that Mr Corbet of Ynsymanghwyn (near our old Plas Edwardes) whom perhaps you may remember in Shrewsbury. He had sunk to the lowest state, and died after a few hours of the greatest agony, so that his screams were heard in the adjoining houses.— I have very little to tell you about the Owens; I am going to Woodhouse this week, to meet Mrs Williams & Mrs Biddulph; it will be just like old times, as both the Husbands will be happily away. Mr Williams in London, and Mr Biddulph canvassing. Poor Sarah is looking very delicate I hear; she has been continually ill all this year, but is in capital spirits. Fanny is not said to live a very happy life at Chirk; she has had the horrid old Mother, Mrs M. Biddulph, & the Sister & Brother staying there a long time, and she dislikes them most cordially; they are very stiff & formal, and I should think there was a thorough hatred between them.— Fanny does not however at all beat under to them, but gives herself very proper airs.— They have a French Butler now at Woodhouse, an old Servant of their's in France, who the young Ladies shake hands with, and who chatters & talks all the time he is in the room, mimics the guests, &c &c. He will be *an addition indeed.* I tremble when I think of him.— They were very much pleased by your letter to Mr Owen.— I have been living the last 5 weeks with the Hollands; I paid them a long visit at their place in Worcestershire, Overbury, which is a very handsome house in a very nice country and then I went with the Hollands a Tour in Derbyshire to see the Lions there, and to pay a visit to Sir Thomas Denman's. It was very pleasant, but I got rather tired of the Hollands. I dislike Mrs Edward Holland, the Bride, nearly as much as Erasmus does. She is a disagreeable, little, dull, cold thing. Edward Holland has begun his new house, which is to be 4 years in building; it will be a nice large house, but Edward will be awfully pompous when he is master of it, for he can hardly contain his importance now. I was at the grand ceremony of laying the first stone.— We have heard to day a piece of news from Maer. Lord Craven has given Mr Langton a Living near Ludlow, between 3 & 400 a year, and in a very pretty situation. Mr Langton was Tutor to Lord Craven. It is very nice for Charlotte being settled within a short distance of Maer, and of here, and they are all exceedingly pleased at Maer.— My dear Charles, how I long for you to be settled in your nice Parsonage. I hope you retain that vision before your eyes.— People here think you will find cruizing in the South Seas such uninteresting work, that it gives us some hopes you will perhaps return before the Beagle.— Nancy begs I will tell you how very happy she is made, every time we hear from you.

God bless you, & take care of you, my dearest Charles. You cannot tell how often I think of you. | Papa's and all our best loves, to our dear Charles. | Y^{rs} | Catherine Darwin

Postmark: Shrewsbury OC 16 1832
DAR 204.6.1

[1] Audubon 1827−38.
[2] This was not the famous Halley's Comet, which appeared in 1835, but one which was visible during October and November 1832. *The Times* of 12 October 1832, p. 3, has a letter from John Herapath about comets in general and this one in particular. Many people attributed the unseasonably hot weather to its approach.

To Caroline Darwin 24 October — 24 November [1832]

[Monte Video]

My dear Caroline

We are now October 24th.—within a few legues of M: Video, & shall before morning drop our anchor there.— This first cruize has afforded very little matter for letters or for any other purpose.— You recollect the sand hillocks at Barmouth; we have sailed along 240 miles of coast, solely composed of such hillocks.— Instead of being as at Barmouth merely a border for the sea, here in Patagonia[1] they extend for some miles, till you reach the open plains, which are far less picturesque than the sand-hillocks.— Even with this & a good deal of bad-weather on our passage down, I have enjoyed the cruize.— Our furthest point South was Bahia Blanca, (a little N. of Rio Negro), where there is a small Spanish settlement or rather a fort against the Indians.— On entering the bay we met a little Schooner, in which was an Englishman, who is connected with two other small vessels (or rather covered boats) employed in sealing.— The man was tolerably ackquainted with the coast: the Captain thought this so fine a chance, that he has hired two of them & put two officers in each.—[2] They now are surveying the coast, which from the number of banks would have detained us a long time.— On our return from M:V: (which will be as soon as possible) we meet them at Rio Negro, & leaving them to work, push on for the South.—

This second cruize will be a very long one; during it we settle the Fuegians & probably survey the Falklands islands: After this is over (it is an aweful long time to talk about) we return to M Video: pick up our officers & then round the Horn & once more enter the glorious, delicious inter tropical seas.— I find the peep of Tropical scenery, has given me a tenfold wish to see more: it is no exaggeration to say, no one can know how beautiful the world, we inhabit is, who has only been in the colder climes.— The chief source of pleasure has been to me, during these two months, from Nat: History.— I have been wonderfully lucky, with fossil bones.— some of the animals must have been of great dimensions: I am almost sure that many of them are quite new; this is always pleasant, but with the antediluvian animals it is doubly so.—[3] I found parts of the curious osseous coat, which is attributed to the Megatherium;[4] as the only specimens in Europe are at Madrid (originally in 1798 from Buenos Ayres) this alone is enough to repay some wearisome minutes.— Amongst living animals I have not been less fortunate:— I also had in September some good sporting; I shot one day a fine buck & doe: but in this line, I never enjoyed anything so much as Ostrich hunting with the wild Soldiers, who are more than half Indians.— They catch them, by throwing two balls, which are attached to the ends of a thong, so as to entangle their legs: it was a fine animated chace.— They found the same day 64 of their eggs:—[5]

It is now nearly four months, since I have received a letter, so you may imagine how anxious I am for for tomorrow morning: We are all very curious about politicks; all that we know is that the bill is past; but whether there is a King or a republic according to the Captain, remains to be proved.—

Monte Video:— I have just received your letter of June 28, & Susans of May 12th.— Far from your letters *not* containing news; I am astonished at the wonderful number of events, which monthly takes place.—and I assure you no half famished wretch ever swallowed food more eagerly than I do letters.— I received one from Fox; who seems to have been suffering from much illness; but he now writes in good spirits.— Tell Susan her most elegant note of Tournure to Cap: Beaufort has travelled here.— Capt. Beaufort included it in a civil note to me "thinking that at the distance of 6000 miles, the hand-writing of those dear to us is gratifying".— The Captain is evidently a good hand at turning the Kaleideoscope of "thanks" "gratitude" "compliment" "&c &c" ".— If at any time you want to send me any large letters (including papers or double &c) &c put it under cover to Cap: B. & he says he will forward them.—

On Monday we run up to Buenos Ayres, as the Captain wants to communicate with the government.— we shall stay there for a week I intend to have some good gallops over the Pampas.— I suppose you all well know Heads book.— for *accuracy* & animation it is beyond praise. After returning here, we stay another week, & then for Terra del.— This second cruize will I suppose last between 6 & 9 months; so make up your minds for a gap in my correspondence but *not in yours*:— You need be in no fears about directions: till told to alter; merely put S America: all letters for HM ships pass through the Flag ship, which knows where to send to all on the station.— Although my letters do not tell much of my proceedings I continue steadily writing the journal; in proof of which the number on the page now is 250.—

We are now Novemb: 11. beating down the river to Monte Video.— We stayed a week at Buenos Ayres. I much enjoyed this long *cruize* on shore. The city is a fine large one: but the country beyond everything stupid.— I saw a good deal of Mr Hughes.— nothing could be more obliging than he was; he obtained a great deal of information for me & has undertaken several troublesome commissions, which otherwise I never could have managed.— When we winter in the Plata, I intend taking a long excursion to geologize the Uruguay country & shall see him again in B. Ayres.— I think I have infected him with a slight geological Mania, which I hope he will encourage.— We saw there also Colonel Vernon, a brother in law of Miss Gooch: he is a very agreeable person & has actually come all this distance as a Tour: he intends going by land to Lima, & so by Mexico back to Europe.— Very few fine gentlemen undertake such a tour as this.— I forget whether I mentioned that during our previous stay at M Video.—Mr Hamond[6] joined us.— He is a relation of poor little Musters & a very nice gentlemanlike person.— We were generally companions on shore: our chief amusement was riding about & admiring the Spanish Ladies.— After watching one of these angels gliding down the streets; involuntarily we groaned out, "how foolish English women are, they can neither walk nor dress".—

And then how ugly Miss sounds after Signorita; I am sorry for you all; it would do the whole tribe of you a great deal of good to come to Buenos Ayres.—

November 14th.— M: Video.— I have just been again delighted with an unexpected stock of letters.— One from Catherine July 25.—from Susan August 15th from Erasmus 18th.— These two last I owe to the change of time of sending them from the Tuesdays to the Fridays.— As it is a special favor, thank dear old Erasmus for writing to me & doing all my various commissions— I am sorry the books turn out so expensive & not to be procured.— I only knew them from references: of course any travels, by those employed in Nat: History are preeminently interesting to me.— I am become quite devoted to Nat: History— you cannot imagine what a fine miserlike pleasure I enjoy, when examining an animal differing widely from any known genus.— No schoolboy ever opened a box of plumcake so eagerly as I shall mine, but it is a pleasure, which will not come for the next 9 months.— I am glad the journal arrived safe; as for showing it, I leave that entirely in your hands.— I suspect the first part is abominaly childish, if so do not send it to Maer.— Also, do not send it by the Coach, (it may appear **ridiculous** to you) but I would as soon loose a piece of my memory as it.— I feel it is of such consequence to my preserving a just recollection of the different places we visit.— When I get another good opportunity I will send some more.— The Beagle is in a state of wonderful bustle & confusion.—there is not a corner, even to the officers cabins where food is not stowed.— The Captain seems determined, that this, at least shall not call us back.— I look forward with a good deal of interest to Terra del; there are plenty of good anchorages; so that it may blow great guns if it likes, & we can laugh at it.— Anything must be better, than this detestable Rio Plata.— I would much sooner live in a coal-barge in the Cam:—

Hurrah, (Nov 24th): have just received the box of valuable thank everybody who has had a finger in it, & Erasmus for packing them all up so well: Neither the Captain or myself have received (from some change in packets) any letters.— I should have like to have heard once again that you are all well & safe, before my long absence; I may say from, this world: at Buenos Ayres I drew 20£ for myself & here Cap FitzRoy asked me if I could pay an year in advance for my mess.— I did so, for I could not, although, perhaps I ought, refuse to a person who is so systematically munificent to every one who approaches him.— So that now, (one year being gone) am, as at first starting 2 years in advance.— Having drawn

AL incomplete
DAR 223

---

[1] CD is referring here to the country between Cape San Antonio and Bahia Blanca in the southern part of La Plata province. The province of Patagonia did not extend north beyond the Rio Colorado.

² Robert FitzRoy paid for the hire of the two boats, the *Liebre* and *Paz*, out of his own pocket. His hope that the Admiralty would reimburse him was disappointed. See Mellersh 1968, pp. 104, 130−2 and *Narrative* 2: 109−11.

³ At Punta Alta, near Bahia Blanca, CD had uncovered fossil bones of the *Megatherium*, a giant ground sloth, and several hitherto undescribed extinct mammals (see *'Beagle' diary*, pp. 102−7). They were later named and described by Richard Owen for *Fossil Mammalia*. Although CD was aware that many of his South American fossils were new, his identifications were inevitably vague and sometimes mistaken, as when he failed to distinguish *Megatherium* from other edentate forms later described by Owen as *Toxodon*, *Mylodon*, and *Glossotherium*.

⁴ The naturalists of the time thought *Megatherium* had dorsal armour—an error that apparently originated with Georges Cuvier, who had named and described it (in Cuvier 1812) from the Madrid bones referred to by CD (see Judd 1911, p. 9). The osseous coat belonged to *Glyptodon*, related to the modern armadillo.

⁵ At the time, CD wrote in his 'Zoological diary': 'In one days hunting 64 were found; 44 of these were in two nests—the other 20 [*interl*] scattered about.— It seems strange that so many *of the latter [*interl*] should be produced for no end.' (DAR 30.2: 112).

⁶ Lieutenant Robert Nicholas Hamond, who had been a ship-mate of FitzRoy's in the *Thetis*, was transferred to the *Beagle* at Montevideo from the *Druid*, of which he was Mate. Shortly after CD's death, Hamond wrote to Francis Darwin the following hitherto unpublished reminiscence: 'I have the most pleasant and happy recollections of your father during the short intercourse I had with him while in the Beagle, from the fact of his having joined with me in a request to the Chaplain of Buenos Ayres, where we were then staying to have the Sacrament of the Lords Supper administered to us, previous to going to Tierra del Fuego— We were both then young and looked on that Ordinance as many young did, and do, as I suppose they do now as a sort of vow to lead a better life. Our request met with so cold a response and the necessity put on us of engaging others to come with us; that our purpose was not carried out, but it shewed a disposition of mind I was glad to dwell on— Of course this was too delicate a passage in life to mention in public.' (DAR 112: 54).

To J. S. Henslow¹ [*c.* 26 October −] 24 November [1832]

Monte Video [Buenos Ayres]

My dear Henslow,

We arrived here on the 24ᵗʰ of Octob:² after our first cruize on the coast of Patagonia: North of the Rio Negro we fell in with some little Schooners employed in sealing; to save the loss of time in surveying the intricate mass of banks, Capt: FitzRoy has hired two of them & has put officers in them.— It took us nearly a month fitting them out; as soon as this was finished we came back here, & are now preparing for a long cruize to the South.— I expect to find the wild mountainous country of Terra del. very interesting; & after the coast of Patagonia I shall thoroughily enjoy it.— I had hoped for the credit of dame Nature, no such country as this last existed; in sad reality we coasted along 240 miles of sand hillocks; I never knew before, what a horrid ugly object a sand hillock is:— The famed country of the Rio Plata in my opinion is not much better; an enormous brackish river bounded by an interminable green plain, is enough to make any naturalist groan. So hurrah for Cape Horn & the land of storms.—

Now that I have had my growl out, which is a priviledge *sailors* take on all occasions, I will turn the tables & give an account of my doings in Nat: History.— I must have one more growl, by ill luck the French government has sent one of its Collectors[3] to the Rio Negro.—where he has been working for the last six month, & is now gone round the Horn.— So that I am very selfishly afraid he will get the cream of all the good things, before me.— As I have nobody to talk to about my luck & ill luck in collecting; I am determined to vent it all upon you.— I have been very lucky with fossil bones; I have fragments of at least 6 distinct animals; as many of them are teeth I trust, *shattered & rolled* as they have been, they will be recognised. I have paid *all the attention*, I am *capable* of, to their geological site, but of course it is too long a story for here.— 1st the Tarsi & Metatarsi very perfect of a Cavia: 2nd the upper jaw & head of some very large animal, with 4 square hollow molars.—& the head greatly produced in front.— I at first thought it belonged either to the Megalonyx or Megatherium.— In confirmation, of this, in the same formation I found a large surface of the osseous polygonal plates, which "late observations" (what are they?) show belong to the Megatherium.—[4] Immediately I saw them I thought they must belong to an enormous Armadillo, living species of which genus are so abundant here:[5] 3^d The lower jaw of some large animal, which from the molar teeth, I should think belonged to the Edentata: 4th some large molar teeth, which in some respects would seem to belong to an enormous Rodentia; 5th, also some smaller teeth belonging to the same order: &c &c.— If it interests you sufficiently to unpack them, I shall be *very curious* to hear something about them:— *Care must be taken*, in this case, not to confuse the tallies.—[6] They are mingled with marine shells, which appear to me identical with what now exist.—[7] But since they were deposited in their beds, several geological changes have taken place in the country.—

So much for the dead & now for the living.— there is a poor specimen of a bird, which to my unornithological eyes, appears to be a happy mixture of a lark pidgeon & snipe (N^r 710).— Mr Mac Leay himself never imagined such an inosculating creature.—[8] I suppose it will turn out to be some well-know bird although it has quite baffled me.— I have taken some interesting amphibia; a fine Bipes; a new Trigonocephalus beautifully connecting in its habits Crotalus & Viperus: & plenty of new (as far as my knowledge goes) Saurians.— As for one little toad; I hope it may be new, that it may be Christened "diabolicus".— Milton must allude to this very individual, when he talks of "squat like [a] toad",[9] its colours are by Werner,[10] *ink black, Vermilion red* & *buff orange.*— It has been a splendid cruize for me in Nat: History.— Amongst the pelagic Crustaceae, some new & curious genera.— In the Zoophites some interesting animals.— as for one Flustra, if I had not the specimen to back me up, nobody would believe in its most anomalous structure.— But as for novelty all this is nothing to a family of pelagic animals; which at first sight appear like Medusa, but are really highly organized.— I have examined them repeatedly,

& certainly from their structure, it would be impossible to place them in any existing order.— Perhaps Salpa is the nearest animal; although the transparency of the body is nearly the only character they have in common.— All this may be said of another animal, although of a much simpler structure.—

I think the dried plants nearly contain all which were then Bahia Blanca flowering. All the specimens will be packed in casks—I think there will be three: (before sending this letter I will specify dates &c &c).— I am afraid you will groan or rather the floor of the Lecture room will, when the casks arrive.— Without you I should be utterly undone.— The small cask contains fish; will you open it, to see how the spirit has stood the evaporation of the Tropics.—

On board the Ship, everything goes on as well as possible, the only drawback is the fearful length of time between this & day of our return.— I do not see any limits to it: one year is nearly completed & the second will be so before we even leave the East coast of S America.— And then our voyage may be said really to have commenced.— I know not, how I shall be able to endure it.— The frequency with which I think of all the happy hours I have spent at Shrewsbury & Cambridge, is rather ominous.— I trust everything to time & fate & will feel my way as I go on:— We have been at Buenos Ayres for a week.— Nov.ʳ 24ᵗʰ— It is a fine large city; but such a country; everything is mud; You can go no where, you can do nothing for mud.— In the city I obtained much information about the banks of the Uruguay.— I hear of Limestone with shells, & beds of shells in every direction.— I hope, when we winter in the Plata to have a most interesting Geological excursion in that country.— I purchased fragments (Noʳˢ: 837 & 8) of some enormous bones; which I was assured belonged to the former *giants*!!— I also procured some seeds.— I do not know whether they are worth your accepting; if you think so, I will get some more:— They are in the box: I have sent to you by the Duke of York Packet, commanded by Lieu: Snell to Falmouth.— two large casks, containing fossil bones.—a small cask with fish, & a box containing skins, spirit bottle &c & pill-boxes with beetles.— Would you be kind enough to open these latter, as they are apt to bec⟨ome⟩ mouldy.— With the exceptions of the bones, the rest of my collection looks very scanty. Recollect how great a proportion of time is spent at sea. I am always anxious to hear in what state my things come & any criticisms about quantity or kind of specimens.— In the smaller cask is part of a large head, the anterior portions of which are in the other large ones.— The packet has arrived & I am in a great bustle: You will not hear from me for some months:

Till then believe me, my dear Henslow, Yours very truly obliged, Chas Darwin.—

Remember me most kindly to Mʳˢ. Henslow.—

Royal Botanic Gardens, Kew

1 The following passages from this letter were extracted by Henslow and published in the Cambridge Philosophical Society pamphlet:

    1.1 'We arrived . . . Rio Negro' 1.2
    1.8 'I had hoped . . . groan.' 1.14
    2.8 'I have been . . . &c &c.' 2.22
    2.24 'They . . . country.' 2.27
    3.1 'there . . . creature.' 3.4
    3.5 'I have taken . . . toad" ' 3.10
    3.11 'Amongst . . . common.' 3.19
    5.8 'We have been . . . direction.' 5.12
    5.13 'I purchased . . . giants!!' 5.15

2 According to *'Beagle' diary* and Robert FitzRoy's meteorological log, the *Beagle* was still at sea on 24 and 25 October. The earliest they might have arrived at Montevideo was late on the 25th.

3 Alcide Charles Victor Dessalines d'Orbigny. From 1826 to 1833 he travelled throughout South America, collecting specimens for the Muséum d'Histoire Naturelle. He published the results in Orbigny 1835–47.

4 Described in *Fossil Mammalia*, pp. 63–73, by Richard Owen, who identified it as belonging to a distinct subgenus of Megatheroid Edentata, to which he gave the name *Mylodon darwinii*. The 'late observations' refer to English newspaper accounts of the *Megatherium* fossil found by Sir Woodbine Parish in 1831 (see letter to J. S. Henslow, 11 April 1833, n. 7).

5 After the voyage, in discussing the origin of his evolutionary views, CD frequently cited the relationship of living species like the armadillo to South American fossils as important in suggesting the possibility of transmutation (e.g., 'Journal', Appendix I, entry for 1837; *Autobiography*, pp. 118–19). Some of these references have led to the view that CD arrived at the hypothesis during the voyage. Most scholars, however, now hold that the 'conversion' to evolution came after CD had returned to London. See Sulloway 1982b for a discussion of the various views. Sulloway makes a convincing case that CD saw the evolutionary significance of his collections only after his ornithological and fossil specimens had been classified by John Gould and Richard Owen.

6 The tallies were tags with numbers corresponding to those in his catalogues of specimens.

7 Orbigny named twenty species of shells, all of living species, collected by CD from the Punta Alta formations (see *South America*, p. 83).

8 William Sharp Macleay. In his *Horæ entomologicæ* (Macleay 1819–21), he propounded the Quinary System of classification in which the five main animal groups are represented by 'circles of affinity'. To represent the continuity of forms the circles are arranged in a larger circle in which each is contiguous or 'inosculant' with two others. Loren Eiseley cited CD's use of 'inosculating' as evidence of an early, unacknowledged debt to Edward Blyth, but Macleay's system and its vocabulary were well known to CD long before he knew of Blyth. See Eiseley 1959 and Sydney Smith 1968. The inosculating bird is identified as *Tinochorus rumicivorus* in *Birds*, pp. 117–18.

9 *Paradise lost* 4. 799–800. CD had a copy of Milton's poems with him on the voyage.

10 Syme 1821.

From Charles Hughes 2 Nov [1832]

[Notes for a trip along the Rio Negro to Mercedes.]

AmemS
DAR 34.1: 14–15

From Susan Darwin   12 — 18 November 1832

Shrewsbury
November 12th. 1832

My dear Charles—

You will be surprised to hear that my Father has at last put his long talked of scheme into execution and set out to London two days ago with Caroline inside, & Edward & Harry the brother Antiquarians on the Seat behind, as upon our Lincolnshire tour.— We have not yet heard from them, and are very anxious for the first account of how he bears the fatigue They spend today (Sunday) at Oxford & get to London tomorrow night where they mean to spend about a week in seeing Sights, & a long List was drawn up Chiefly of all the buildings worth staring at.—

The new Carriage which my Father has had built chiefly for this occasion, most unluckily did not quite answer his expectations, as Hunt was careless about the measurement, & I am afraid the difficulty of getting in and out would very much spoil his pleasure. however it was very nearly as light as a Hack Chaise which was one great point.—

Catty & I feel very odd with the house, horses, Carriages & Servants all at our command without Papa sitting in his Arm chair by the fire.—

Sarah came as (Miss Owen) to spend 4 days with us about a fortnight ago before Mr Williams came down to Eaton, & we enjoyed having her exceedingly. even the philosopher Erasmus who was here at the time often exclaimed "It is a pleasure to look at her" and certainly she is looking more pretty than ever but very delicate.— She often talked about you & said "she depended upon y^r keeping y^r promise of letting her be the first person you dine with upon your return as Belgrave St was very convenient for that purpose". I wish I c^d foresee that before this time two years my dear old Charley you w^d fulfill that engagement.— Catherine & I are to go & spend two days this week at Eaton which will be an agreeable break in our solitude and mean to go to the Old Hunt Ball from Eaton with their party.—

I have been spending a very gay Autumn October at the Hill was very delightful. Tom & Robert were both there & both Masters of Gigs so with the Phaeton we had no ends of pretty drives and exploring parties. Also a family of Michells live close there & joined in our excursions. Miss Michell is rather good looking & very lively. we always used to set her down as Tom's Lady but now I have seen them together I see no symptoms on Tom's side.—

Capt Michell her brother was very pleasant & a famous Sportsman. when we made parties by the river side he used to catch Trout for our Dinner & cook them over a Stove which belonged to Tom on his Portuguese service.— They all enquired much about you & made me read some of your first Letters which interested them very much

Poor old Allen[1] absolutely talks of coming back to Maer to take care of the Parish next Christmas but I think his courage will fail when he sees Snow on the ground especially when he hears that Robert intends cutting the Parsonage Windows much larger which I am sure won't suit Allen's constitution.— Cath has told you in her last letter of Mr Langton having got the Living of Onnibury near Ludlow it is a charming place for us. he will have 70 acres of Glebe with it, but I am sorry to find he has no taste for farming, & declares he will keep niether Pigs or Cows, & Charlotte is a very bad Wife in that respect for she won't help to give him the proper country parsons tastes.—

I had never seen Mr. L. till just lately: he is very pleasant & sociable but not exactly the kind of man I sh^d have expected to have suited Charlottes fancy.— They come to their Living I believe next February so now we shall always see a great deal of them. My Father likes him extremely which is very fortunate.— I long for you to be acquainted with him & I am sure he feels friendly to you from his manner of speaking about you. I daresay partly from knowing what a Sea life is, he takes more interest in news of you. I must ask you a very ignorant question. Are you going to explore down towards the South Pole or not? or only the coasts of America? I must get Earles book to read, for that will put a little sense into my head I hope.— The Penny & Saturday Magazines[2] make the chief reading of the house at present, which we find *cheap* and profitable.— The two great Palm Trees are arrived & touch the top of the Hothouse now so I don't see how they ever can flourish.—

Charlotte Holland has announced in due form & with plenty of affectation, that she is engaged to marry Mr. John Isaac: he must have a strange taste to prefer her to Louisa, & I think the Bank which his Father and he belong to, must be breaking before he could take such a step.— They are to live close to Worcester.

Sunday the 18th. of Nov^{br}— I saw in the Paper last night that there had been an insurrection at Monte Video when 50 of the Crew of the Beagle were called upon & put it down— This account having reached England makes us hope we shall very soon hear of you my dear Charles & I wish the Admiralty would not be so long in sending out the Private Letters as I must finish this before I can expect to receive any news of you— The Newspaper says you arrived at Monte Video on the 9th. of August & in that case you have been a Month in getting there from Rio Janeiro which seems a long time.— We have heard twice fr London & they bring very agreeable accounts of my Father. he does not seem so much tired by sight seeing as Caroline expected & enjoys it all excessively. The Zoological Gardens he had been to once, but Caroline next Month will tell you all their annals.— We have been staying 3 days at Eaton & Mr Owen said he sh^{d.} write to you very soon. Mr Williams seems still over head & ears in love with Sarah tho' they will have been married a year next Thursday the 22^{d.} ⟨   ⟩ we are to go & celebrate the *anniversary* w⟨ith a⟩ *Flitch of Bacon.*

I saw Tom Eyton at the Hunt Ball he looks altered I think & not so pleasant as he was. They say that he is a gt deal with Mr Oakeleys who are very bad drunken young men, & I am sadly afraid he is spoiling. They say he is in Love with Miss Oakeley I am sure if you were in England he would not have fallen down hill in this way, but this in only *report* I am telling you & I wish it may not be true for it is a great pity.—

Catherine & I join in affecte Love to you my Dearest Charley & I am y^r most affecte Sister | Susan E Darwin.

Nancy talks about you morg, noon, & night.)

DAR 204.6.1

CD ANNOTATIONS
on cover: 'Mouse (1288) Inhabits the Sand dunes)'[3]

[1] John Allen Wedgwood.
[2] *Penny Magazine of the Society for the Diffusion of Useful Knowledge* (1832 – 45) and *The Saturday Magazine*.
[3] CD's catalogue of numbered specimens at Down House, on a page headed '1833 June Maldonado', reads: '1288. Mus, in camp. caught by bird: inhabiting especially sand dunes. Head in spirits (078).—'

To W. D. Fox [12 – 13] November 1832[1]

Rio Plata.
November 1832

My dear Fox.—

I am going to take you at your word, & send you a very short letter sooner than none.— It may appear very odd to you, but I never found so much difficulty in writing letters, as at present.—

There is so much to say, or might be said, that I am quite overwhelmed & generally finish by saying very little.— After leaving Rio de Janeiro, we sailed to M: Video.—from whence we had a surveying cruize to the South.— (at Rio I wrote to you).— All this part of S America is wonderfully stupid; sand hillocks & undulating plains makes a poor change after mountains & vallies glowing with the rich vegetation of the Tropics.— At M: Video we had not heard from anybody for four month: & as you may suppose, like a Vulture I devoured yours & other letters. I was very sorry, my dear old Fox, to hear from yourself & home, that you had been so unwell; I trust that when you receive this you will be sitting, well & cheerful, by a blazing fireside.— Eheu Eheu how long will it be before I enjoy that pleasure.— In about a weeks time, we sail down the coast for the Falkland islands & then for Cape Horn. I suspect there will be some difference between this & an English fire side: it would all be very tolerable, if there was some moderate limit steadily to look forward to.— But the Captains plans enlarge as the time advances, & I see no end to the voyage.— We return from Cape Horn & winter in the Plata; from thence we go to the other side & coast up to Panama, & then our voyage may be said again to commence.—

During these last month, the only source of enjoyment, & it has been a large one, has been from Nat: History.— I have principally been lucky in Geology & amongst pelagic animals.— An old piece of ambition of mine has been gratified, viz finding the remains of large extinct animals I think some of them are new; I have teeth & fragments of about 7 kinds. Even this does not reconcile me to leaving the golden regions of ye Tropics.— My peep at these climates has quite spoiled me for any other; I must however except the English autumnal day, the clearness of the atmosphere of which will stand comparison with anything.— Poor dear old England. I hope my wanderings will not unfit me for a quiet life, & that in some future day, I may be fortunate enough to be qualified to become, like you a country Clergyman. And then we will work together at Nat. History, & I will tell such prodigious stories, as no Baron Monchausen ever did before.— But the Captain says if I indulge in such visions, as green fields & nice little wives &c &c, I shall certainly make a bolt.— So that I must remain contented with sandy plains & great Megatheriums:—

At this present moment we are beating against a dead foul wind, on our return from B. Ayres to M Video.— Buenos Ayres is a fine large city & has an Europæan appearance, with the exception of a few wild Gauchos, with their bright coloured ponchos, riding through the streets.— Do you know Heads book? it gives an excellent account of the manners of this country.—

I hope you will write to me again: & recollect that all details about yourself gain in interest instead of losing from the distance: S. America, by itself will be the best direction for the Beagle: I shall be very busy next week in packing specimens. Henslow most kindly has undertaken to receive them. It is one more to the many many obligations I owe to him.— I always consider it one of the luckiest days in my life, when you introduced me to him.— The friendship of such a man, is indeed worth gaining:— Good bye, dear Fox. God bless you & keep you as happy, as you deserve to be, & Write to me again.

Yours very affectionately, Chas Darwin.—

Christ's College Library, Cambridge (Fox 46a)

[1] Dated from the time of the *Beagle*'s trip from Buenos Aires to Montevideo, according to '*Beagle*' *diary*, p. 113.

From J. M. Herbert 1[−4] December 1832

Manchester
1 Dec.ʳ 1832.

My dear Darwin,

I am indeed obliged to you for so soon meeting my demands, especially as the run on your bank of information must be so alarming. You will be surprised at seeing whence this letter of mine is dated I have now taken up a permanent abode at Chester, (at least for a year,) and as the distance from thence to Manchester is not very considerable, and my ball-room gazelle-like propensities so well known, I was very glad to gratify them at so little cost both of trouble & cash— I attended last night a meeting of the Philosophical Society[1] of this place, more for the sake of seeing old Dalton,[2] than of indulging any passion (can the word be used with propriety in relation to such a prude?) for science. D.ʳ Henry read a very interesting paper on the variation of temperature in fresh water lakes & the Ocean;[3] he gave us one fact which startled me very much; Lord Mulgrave found on his northern expedition[4] that at the depth of (I think) 4000 feet the sea had a temperature of only 26° Fahrenheit— How c.ᵈ it retain its fluidity? You have I suppose become a good practical mathematician long ere this; you only threatened Trigonometry, but I expect that the zeal which you always had & which you once gave me credit for, will have carried you into the regions of the "Mecanique Celeste", or that you will be at no assignable distance from it. As for myself I have long since put all my Mathematics, & for ever, on the shelf. I have now, only a mysterious recollection of Differential Coefficients, Polarisation &c &c. I hope however that I shall retain so much as will enable me to understand the (the $\kappa\alpha\tau'$ $\dot{\epsilon}\zeta o\chi\grave{\eta}\nu)$[5] book whenever it may make its appearance— I have already begun to picture myself its appearance & the nature of its contents—one of Murray's 4tos in Davidson's type?[6] How will it be entitled? "Observations physical, political & moral, made during a voyage r.ᵈ the world in the years 1831 − 1835 by C. Darwin F.R.S., F.L.S, &c &c". You will of course stay its publication, till these hieroglyphical characters be affixed to your name— I shall indeed revel in its fresh-cut pages. On your return you will find me a thin sallow hollow-cheeked lawyer, a wretched tenant of one of those uninhabitable abodes, (nick-named chambers) in or about Lincoln's Inn, with numberless bans settlements &c: heaped up around me; am I not sanguine enough to succeed? Things have taken a very strange turn since you left England in the political world— Denman has been appointed Chief Justice; who w.ᵈ ever have dreamed that the late Queens Attorney & solicitor General could be at the head of their profession?[7] yet such is the case. The Election is now very fast approaching, & every body is looking to it with anxiety; the Tories (poor souls!) are gone past recovery, the only question now is which of the Whigs or Radicals shall gain the ascendancy: Hobhouse is not thought sufficiently liberal for Westminster,[8] so they have started Col: Evans;[9] & he, it is said, is with Gordon to be nominated for our University![10] Henslow I see has just had a living, but whence, or of what value, I have not yet learned. His case has often convinced me of the necessity of reform in the Church; it was indeed a very humiliating consideration that one of her brightest ornaments sh^d be so long neglected— The House of Lords is not worth twenty years purchase—"Down with the Bishops" is I fear growing too general a cry. It is interesting to reflect on the very rapid march of liberal opinions— I was in company last night with a man whose father was in 1792 prosecuted for High Treason, merely because he advocated the repeal of the Test & Corporation Acts[11] & Reform in Parliament, and had his house nearly pulled about his ears by an infuriated Church & King mob. When shall we see the like again? In the literary world we have lost poor Macintosh, Scott, & Leslie,[12] three of Scotland's, nay of Britain's, most illustrious sons.—

I am extremely indebted to you for a most severe cut about my Tailor, yet tho' I cannot but congratulate you on the intention, I must condole with you on the effect produced, as (mirabile dictu) I am now absolved in that quarter— Of Whitley's schemes I know almost ⟨as⟩ little as you⟨rs⟩elf, as I have seen nothing of him for a very lon⟨g⟩ time, & he does not deign to answer my letters; you will conclude with me that he is carrying his lordship mightily in the 2^d court of St: John's, d——g the Whigs & lauding the Tories— I fear the Durham University has been deferred sine die; they speak of four years— The widow is married. Poor Heaviside! I sh^d like to hear him telling you his case, & exciting your commiseration— Of Watkins & Lowe I know nothing. I shall in future pay your judgment & opinions greater deference than I used to do; I have been for the last six months visited with those feelings which you & Henslow predicted, tho' I cannot think that the want of exercise contributed so much to them, as the irregularity of my college life. At one time I feared debility in its worst form, Lawrence has however relieved me by telling me that there is no cause for alarm. A journey, or rather a voyage, to the Pampas w^d be the best way of carrying his prescription into effect, as he has ordered me to take an unlimited quantity of exercise— I see by a late number of the Times that H.M.S. Beagle's services have been called for at Mont Video, & that she landed fifty men for the assistance of the place; you of course took some share in this brilliant affair. I have just rec^d a letter from Whitley; Durham he says will be decided this winter; Watkins & Cameron!! are both in Cambridge mad in their respective lines. I will now leave off sinning against Science & the public by trespassing on your valuable time, and permit you to return to your search for scolopendra & "ante-natal tombs" of moths &c. Write to me as soon as you *conveniently* can, & believe me ever, my dearest Darwin, | Y^{rs} most sincerely & faithfully, | J M Herbert

I have just seen a requisition to Lubbock,[13] signed by Sedgwick, Henslow, & all the tribe of worthy liberals that one really has a regard for, inviting him to stand for the University, with which he has complied. Whitley says it will have a good effect, as a few of the old Johnian incumbents will be brought from the nethermost parts of the land to support the first investigator of the Lunar Eccentric & the President of the Cockpit,[14] a summons which in this season of frost & fog must cause a few vacancies. Gordon has declined. Hobhouse I suppose will not be invited. You have I recollect a very singular way of reading letters, viz: a habit of devouring only one period at a time, & then putting by the remainder for luncheon the next day. In order to gratify this strange propensity of yours, & to cater for you for some length of time (tho' Heaven knows 'twill form but a sorry repast,) I have endeavoured to fill this sheet by crossing it till it flows over; if however you derive any either of pleasure or consequence from this strange Postscript, you are indebted to a young lady for the same, who suggested to me the impropriety (the shame I think she said) of sending so short a letter as I had written across the Atlantic. I contended that I knew your taste somewhat better than she could, and that you were an admirer of the brief & pithy, but my argument availed not, she pleaded so eloquently in your cause, that she could not fail of success. Yet do not attribute the stupidity of these latter lines to other than myself; I sh.^d be sorry that she should bear the onus. You will think it flattering to be supposed capable of contributing to the amusement of another by encroaching on his time; she can of course judge only from my conversational powers must they not be improved? When I was last in town I was lionized over a new Museum, called the N⟨a⟩val & Military, which I consider any thing but promising: it consists solely of donations, and appears to me utterly destitute of all arrangement. Herschel is to have it is said the Chair of Natural Philosophy at Edinburgh— They are now getting up a subscription to relieve Scott's family of the embarrassment in which he was involved by Constable's failure,[15] & to perpetuate the family estate of Abbotsford. Airy[16] has sent forth a manifesto declaratory of Lubbock's claims on the lovers of science, which I cannot think will contribute to his success;[17] he says that Lubbock's papers were the first that put us on a level with the rest of Europe as to investigations in the highest branches of mathematical philosophy.

God bless you. Addios, J. M. H.

Postmark: ⟨Manc⟩hester DE 4 18⟨32⟩
DAR 204.6.2

[1] The Literary and Philosophical Society of Manchester, founded in 1781. For a discussion of the cultural context of its founding see Thackray 1974.

[2] John Dalton was President of the Manchester Society, 1817−44.

[3] William Charles Henry. His paper was not printed in the Memoirs of the Society.

[4] Constantine John Phipps, 2d Baron Mulgrave, commanded a polar expedition in 1773.

[5] 'par excellence'; 'preeminent'.

[6] John Murray did become CD's publisher, beginning with the second edition of the *Journal of researches* (1845).

[7] Thomas Denman was Solicitor-General to Queen Caroline. Her Attorney-General, Henry Peter Brougham, 1st Baron Brougham and Vaux (1830), was Lord Chancellor, 1830−4.

[8] John Cam Hobhouse did, however, stand, and was elected for Westminster in December 1832.

⁹ George De Lacy Evans unsuccessfully contested Westminster in 1832.

¹⁰ Henry Percy Gordon. He was approached by the Whigs at Cambridge but when he did not pledge himself to vote for the ministers, 'it was agreed that Lubbock [see below] should be brought forward'. See Romilly 1967, p. 23.

¹¹ The Corporation Act (1661) extended by the Test Act (1673) made ineligible for public office Catholics and Protestant non-conformists.

¹² John Leslie.

¹³ John William Lubbock. During 1830–2 he published a series of papers on 'Researches on physical astronomy' (Lubbock 1830–2).

¹⁴ The office buildings containing the Treasury and Privy Council chambers in Westminster, so named because they were the former site of the cock-pit erected by Henry VIII.

¹⁵ Archibald Constable, Sir Walter Scott's publisher.

¹⁶ George Biddell Airy.

¹⁷ Lubbock withdrew just before the elections. Romilly commented in his diary: 'very wise of him— the Speaker [Charles Manners-Sutton] & Goulburn being sure' (Romilly 1967, p. 24).

From Robert FitzRoy [1833?]

[List of mountains and their heights.]

AmemS
DAR 40: 25

From Caroline Darwin   13 January 1833

Shrewsbury
Jan.ʸ 13ᵗʰ *1833*

My dear Charles,

You must forgive my sending you a very dull letter but I have been used rather unfairly— it is Catherines month & she did not tell me this moment that she has not even begun a letter to you & today being the last day I will not fail in my promise of writing myself when the others forget— My Father is very well, which I know is the main thing so it will not matter how little else I have to tell. I remember the anxious feeling I had even during my 3 months in France last year & it is impossible to shake it off when at a distance & for any reason not in good spirits. Your last letter which I answered the middle of December was the one to Susan just before you were to sail from Monte Video. I do hope my very dear Charles the cold & rains whilst coasting Patagonia have not made you ill— we are all impatient for your next letter, & if you find all these changes of Climate do not agree with your health come home & think of your snug parsonage— I finished my last letter at Overton. Parky & Henry feel proud in finding the place on the Map where their Uncle Charles is— Parky is growing very manly coming on nicely in his Latin & the most regular flirt I ever beheld. the week after I came home we had a visit from our new member Uncle Jos.¹ he seems very much pleased to have been returned with such a fine Majority & he says he has been at no expence & no trouble doing nothing but what his Committee ordered him & they were very merciful he even escaped being chaired—[2] he says he can not afford to bring his family with him to London but I think they expect to go at least for some months— Parliament meets the 31st of Jan^y—so he will soon have to leave Maer I hope the experiment will answer to him & that he will not repent— Jessie came with Uncle Jos & they staid two days.— it happened most unfortunately that we all three were going the very evening they arrived (unexpectedly) to Eaton to see a Play acted by the Family party of Williams's, Owens & Whites— poor Sarah herself did not act— The Owen constitution has quite failed & she is in a very delicate state of health. She looks very unwell & is out of spirits about herself thinking she shall never get strong & well again & can bear no exertion or fatigue

The piece they acted was The Irish Tutor a merry little bustling farce[3] & they all played their parts very well— there was afterwards a dance— All the rest of this month has passed perfectly quietly We play at Whist every evening with Papa & I think he seems quite as happy & comfortable as he used to be some years ago— Last week was the New Hunt week. We had nobody with us but the 2 Clives of Styche & Caroline Owen.— Mr Edward Williams died the beginning of the week so there was no party at Eaton— Mr Tom Pemberton[4] is also dead. that y⟨ou⟩ may know what is going on in this neighbourhood, there have been no marriages except Sir T. Boughy to Miss Louisa Giffard—[5] the Hunt ball I did not go to— Mrs Biddulph was there looking very handsome but I suspect she must find Mr Biddulph a tiresome person to live with— I am sure he is very selfish he has gained his election & is returned for Denbighshire. Your Pincher & Nina are both very well, I am about buying from Joseph your Grey horse I have tried him & I like him very much as far as I can judge—

Erasmus has been very gay at Parties at Lady Giffords & the Hensleighs &^{tc} he has not written very lately. Mrs Hensleigh is to be confined the beginning of next month so I suppose he can not be junkitting at her house now— Susan & Cath are going to two balls next week & I suppose these balls are Cattys excuse for not having written

they & Papa send their love to you and believe me my dear old fellow, with my own best love to you | Y^{rs} affec^{ly} Caroline Darwin

It is too late to wish you a happy new year—but I *do* wish it—

DAR 204.6.1

[1] Josiah Wedgwood II sat for Stoke-on-Trent, Staffordshire, which, as a result of the Reform Bill, had become a borough returning two members (Hanham 1972, p. 296).

[2] It was common practice in election celebrations to carry the victor about in a chair.

[3] *The Irish tutor; or, New lights: a comic piece in one act*, by Richard Butler, Earl of Glengall, first performed in 1822.

[4] Possibly the Mr Pemberton mentioned by CD in the *Autobiography*, p. 39.

[5] Sir Thomas Fletcher Fenton Boughey married Louisa Paulina Charlotte Giffard on 27 December 1832.

From J. S. Henslow 15 – 21 January 1833

Cambridge
15 Jan[y] 1832[1]

My dear Darwin,

I shall begin a letter to you lest something or other should persuade me to defer it till it becomes too late for the next packet— Wood & I had intended writing by the Dec[r] packet, but just as was about to do so your letter arrived stating that a Box was on its road, so I thought I had better delay till I had seen its contents. It is now here & every thing has travelled well. I shall however proceed by rule & answer your two letters first & then come to the Box. The 1st date of your first letter is May 18. & this I received at Cambridge in June, no, it was sent me from Cambridge in July, to Weymouth where I was spending the summer with my family and two pupils in exploring the geology &c &c of that neighbourhood, & a capable ramble we had. I stopped at Oxford in my way there, where the British Association had assembled for a weeks scientific discussion & a delightful time it was. Next summer this society is to meet in Cambridge. When at Oxford I received a letter from the L[d] Chancellor giving me a small living in Berksh: about 14 miles from Oxford. Of course I do not reside, as I never mean to quit Cambridge without something very extraordinary should happen.[2] I never mean to leave it for lucres sake. We returned to Cambridge in Oct[r] & have had the bustle of the Election to go thru'. We could make nothing of any attempt to squeeze a Whig in for the University so gave it up.[3] We have got 2 Whigs for the town and 2 Whigs & one Tory for the County— But the papers will tell you all this— At this moment I am examiner in Paley & in one hour have to attend in the Senate house. Now for a revision of your letters— I would not bother myself about whether I were right or wrong in noting such & such facts about Geology— note all that *may* be useful—most of all, the relative positions of rocks giving a little sketch thus. N[o]1. (specimen (a)) about 10 feet thick, pretty uniform in character— N[o]2 (specim. (b.c)) variable &c &c

When Sedgwick returns we will look over your specimens & I will send you our joint report—[4] they seem quite *large* enough!— I myself caught an Octopus at Weymouth this summer & observed the change of color whenever I opened the tin box in which I put it, but not in such great perfection as you seem to have done— The fact is not new, but any fresh observations will be highly important— Quere if a *serpentine rock* be not the produce of volcanic baking of a chloritic slate? The rock of St Paul *may* not be an exception to the usual character of the Isl^ds. of the Atlantic.[5] I have got the description of the plates to the Dict. Classique & will send it where you direct. Your account of the Tropical forest is delightful, I can't help envying you— So far from being disappointed with the Box—I think you have done wonders—as I know you do not confine yourself to collecting, but are careful to describe— Most of the plants are very desirable to *me*. Avoid sending *scraps*. Make the specimens as perfect as you can, *root*, *flowers* & *leaves* & you can't do wrong. In large ferns & leaves fold them back upon themselves on *one* side of the specimen & they will get into a proper sized paper. *Don't* trouble yourself to stitch them—for the really travel better without it— and a single label *per month* to

this side is folded back at the edges

those of the same place is enough except you have plenty of spare time or spare hands to write more. L. Jenyns does not know what to make of your land Planariæ. Do you mistake for such the curious Genus, "Oncidium" allied to ye slug, of which a fig. is given in Lin. Transact.[6] & are not the marine species also *mollusca*, perhaps Doris & other genera— Specimens & observations upon these w^d be highly interesting. If you could get hold of Cuvier's Anatomie des Mollusques,[7] you w^d find it very useful but I fear it is out of print— I will tell your Brother to enquire at Truttels.[8] Watkins has received your letter— And now for the Box— Lowe *underpacks* Darwin *overpacks*— The latter is in fault on the right side. You need not make quite so great a parade of tow & paper for the geolog^c. specimens, as they travel very well provided they be each wrapped up *German fashion* & closely stowed—but *above all things* don't put tow round *any thing* before you have first wrapped it up in a piece of thin paper— It is impossible to clear away the fibres of the tow from some of your specimens without injuring them— An excellent crab has lost all its legs, & an Echinus $\frac{1}{2}$ its spines by this error. I don't think however than any other specimens besides these 2 have been at all injured. Another caution I w^d give is to place the number on the specimen always inside & never outside the cover. The moisture & friction have rubbed off one or two—& I can't replace them. I shall thoroughly dry the different perishable commodities & then put them in pasteboard boxes with camphor & paste over the edges, & place them in my study or some very dry place. The heavy material I shall send to my lecture room, so soon as it is again habitable—for at present we are all in confusion—building a large Museum & lecture room & private rooms adjoining mine,[9] for Clark & Cumming— I must now leave off for the Senate house & put this bye till I can find a few more minutes to conclude it.—

Jany. 21. The Exam^n. is over & no X^ts. man plucked— I don't know whether you were acquainted with the men of this y^r. (except Downes who is N°. 26) or I w^d send you their names— The Capt: is Laffer of X^ts.—[10] I have just been putting bye the perishable articles in the way I said— *Birds*—several have no labels— the best way is to tie the label to their legs— One has its tail feathers crumpled by being bent from bad packing—the rest in good order— *Quad^s*. The large one capital, the 2 mice rather mouldy— Pack up an infinite quantity more of land & freshwater shells, they must be nearly all new— The minute Insects most excellent— what work you will have— You know better than I whether it is not dangerous to their antennæ & legs to pack them in cotton. I suppose if moistened by vapour they may be taken out quite safe.— The Lichens are *good things* as scarcely any one troubles himself to send them home— For goodness sake what is N°. 223[11] it looks like the remains of an electric explosion, a mere mass of soot—something very curious I daresay— W^d. it not be a good precautionary measure to transmit to England a copy of your memoranda, with your next packet? I know it is a dull job to copy out such matters—but it is highly expedient to avoid the chance of losing your notes by sending home a duplicate— Every individual specimen once arrived here becomes an object of great interest, & tho' you were to send home 10 times as much as you do, yet when you arrive you will often think & wish how you might & had have sent home 100 times as much! things which seemed such rubbish—but now so valuable— However no one can possibly say you have not been active—& that your box is not capital. I shall not wait for Sedgwicks return before I send this but must give you an account of the Geol^c spec^s in the next— I shall now forward this with the vol. of the Dict. Class. to your Brother & wish you a continuance of good success. I have no fears of your being tired of the expedition whilst you continue to meet with such as you have hitherto, & hope your spirits will not fail you in those dull moments which must occasionally intervene, during the progress of so long an undertaking. Downes & other friends have begged me to remember them to you most kindly & affectionately & Mrs Henslow adds her best wishes— Mine you well know are ever with you & I need not add that you s^d believe me | Most affect^ly. & sincerely y^rs | J S Henslow

My 3 children are well—& my boy is growing a very fine fellow— An increase expected next June— We are in Mourning for Mrs Henslow's Mother—

DAR 204.6.2

[1] The year was written as 1832, but should be 1833.

[2] The living, located at Cholsey-cum-Moulsford, was worth £340 a year. Henslow resided there only during the Cambridge Long Vacation (see *Darwin and Henslow*, p. 89 n. 1).

[3] This was the election from which John William Lubbock withdrew. Elected were Right Hon. Henry Goulburn (Conservative) and Right Hon. Sir Charles Manners-Sutton (Conservative), Speaker of the House of Commons. For these and other returns mentioned, see Hanham 1972, pp. 43–6.

[4] No such report has been found.

[5] See letter to J. S. Henslow, 18 May – 16 June 1832 and *Darwin and Henslow*, p. 54 n. 1.

[6] Guilding 1825.

[7] Cuvier 1817.

[8] Treuttel, Wurtz and Richter, foreign and classical booksellers, 30 Soho Square (*Post Office directory*, 1834).

[9] The Anatomy Museum and lecture rooms for the Anatomy and Chemistry Schools were built in 1832–3 in part of the former Botanic Garden, then located near Free School Lane.

[10] John Athanasius Herring Laffer.

[11] Specimen no. 223 in CD's 'Zoological diary' (DAR 30.1: 20) is identified as 'Mucor Linn.', a fungus.

From E. A. Darwin to J. S. Henslow 23 January [1833]

Jan 23.

Dear Sir

I have received your parcel containing a book &c and a letter for my Brother. I will forward them by the first opportunity, and will follow your suggestion in endeavouring to obtain the Anatomie des Mollusques which I do not think he at present has.

I feel very much obliged to you for the two letters you are so good as to send. I have had great pleasure in reading them, and will gladly make use of your permission to send them home, and will take care that they are returned to you.

I do not know whether I can be of any assistance to you either in receiving or forwarding the Boxes which my Brother may send to you, and can only beg that you will make any use of me that may be convenient

I remain yours, | Sincerely obliged | E. Darwin

24 Regent S!

P.S. My Brother mentions in his letter that his Box is to be forwarded through Capt. Fitz Roy's Agent at Falmouth The former agent of Capt Fitz Roy (I forget his name) failed and I had in consequence considerable difficulty in sending out some books. If you should happen to know the address of the present Agent who forwarded the Box, I should feel exceedingly obliged to you if you could send it to me. I should not venture to give you so much trouble if I were not so well acquainted with all your kindness to my Brother.

Postmark: V JA 24 1833
Royal Botanic Gardens, Kew

From W. D. Fox 23 January 1833

Ryde, Isle of Wight.
January 23. 1833.

My dear Darwin

As I remember promising that I would write to you again as Winter came on, I sit down to redeem it, fearing that you would almost rather I did not take the trouble, as I have nothing in the world to say that will interest you of any one excepting myself, and self is rather a dull subject to either write a letter upon, or receive one.— I had hoped to have heard a few lines from you in the last 5 months but I can easily imagine your time so taken up that your necessary letters home are quite sufficient for you, and perhaps in your case I should be as bad a correspondent. I have tried to instigate Julia for some weeks to write to your Sisters in order that I might hear what they knew of you, but hitherto without success, & if I do not soon have better luck, I feel half inclined to do it myself; Erasmus would not I conclude answer my letter if I wrote to him, nor do I know his direction.— I have seen so many vessels on the point of setting out to South America from Portsmouth & waiting for winds at the Mother Bank[1] that my erratic propensities have been often quite painfully excited and I have dreamt by the night that I was as busy as could be collecting with you, all around new, beautiful & strange. My destiny however is I fear quite fixed to the Continent at least, not to say (as perhaps may be much nearer the truth,) the country I was born in, and of tropical regions I must be content to hear from *Humboldt & Darwin.*— I hope your companion will be sufficiently great even for your enlarged ideas. But if I go on at this rate, I shall fill my paper without even telling you of myself, of whom I half flatter myself you will wish to hear, as I was in a very poor way when I last wrote to you. Since then I have become much stronger & better able to bear exertion, tho' I have had many attacks some of a more serious & others slight nature, and I still continue a good deal of an Invalid, and fear I shall do for some time to come. I did dread the Winter very much indeed but by great precautions & with the very mild climate of Ryde, I hope now to get it over pretty well, and then I trust that next Spring & Summer may do a great deal to take away the remaining affection of my Lungs. I often have great doubts whether I shall ever again be able to exert them for any continued length of time, as at present a few minutes quite oversets me without resting them. I must however hope for the best, at present I have very great cause for thankfulness that I am as I am.— I remain here with my two younger sisters & little Anna Maria, (who will have it that she has quite forgot you) for the Winter and most probably the Spring, when my health will determine what then becomes of me. You will be glad to hear that all at Osmaston are quite well, My Father and Mother only left here two days ago.— I was much pleased a few weeks since by finding out in this town, a splendid Case of Insects from Rio de Janeiro, furnished by a M[r.] Bescke Naturalist living on the Praca da Constituicao there—a Gentleman whom I daresay you visitted. What magnificent Lepidoptera there are there; There were several kinds of Mantis I never before saw & one of those Libellulæ which have such disproportioned long bodies. I fancied you in the height of Entomological Happiness & longed to be with you.

I have much enjoyed seeing our Navy constantly going & coming on account of this Dutch Blockade, and among all sizes have often fancied your little Beagle.

You will have seen in the Papers, that we have had a French & English Fleet lying together at Spithead, & since cruizing together & now lying in Downs. I rejoiced much at it & hope the National antipathies may be done away, but I have been much amused by the annoyance it has given many of the Officers in our Ships— In several I visitted they could scarcely find names sufficiently bad for the French Officers, & the older the Officers the more bitter their hatred. I went over the finest French Ship & was much pleased with her & her Officers & Crew, all picked for the purpose of showing Englishmen what Nick Frog can do in his Navy.— Erasmus Galtons Ship came here some months since & has just been paid off. He really is a very good specimen of a Midshipman & I am sure you would be much pleased with him. The whole Family came here & stayed some weeks while the ship was refitting for Holland. I wish much I could enter into your Geological Researches as well as your Ornitho^l: (which by the by you say nothing of, at which I marvel) Entomolo: and other Nat: Hist^y: Pursuits. I can easily imagine that the Spiders & adjoining tribes must be magnificent from the few I have seen. I have often thought that some parts of S. America would make a delightful Residence, & if one could get a few of ones most valued Friends to group together, I really think I could easily prevail upon myself to leave Good Old England, for a more sunny Clime. The Spartiate 74 is now lying in my sight just on point of coming to you with a New Admiral— I wish I could get to know some of the Officers that I might enquire when exactly they are going as it would be a great gratification to send you a *word of mouth* message.— There is something so freezing in sending stupid letters such a distance, & you scarcely having time to read them.— Our old Cambridge days often come over the mind like a dream.— They are hours gone by never to return I fear. I have been very busy here lately with the Pselaphidæ and Scydænidæ—I fancy your smile of contempt. I cannot help thinking that in other countries, larger genera allied to these will be found—perhaps it will be your lot, the S.A: is not the most likely place where you are at present.— I want much to know where you are likely to go after you leave S.A:— I fear I cannot again write to you, but that is of little consequence, you can sometimes send me a few lines, & I really hope you will. Mind I do not want more than 10 lines just to say how you are, where you are, & where going. You cannot think what pleasure a few lines of this kind would give me.— We all like Ryde very much—there are many pleasant sensible people here, & the climate of the Island is delightful. I had hoped to have found many Insects that I knew were taken here, but have not been able to look after them much. The Gulls have been a great source of amusement to me. I thought when I came I could tell them pretty well, & was rather mortified to find out my ignorance of them. I set to *studying* them (for really I worked hard) and with what dead specimens I could meet with, & 5 live tame ones I have picked up at different times, I have now made out all that come here, with their several changes of plumage from nest to 5 years old. I could not have done this without your old Friend Fleming.[2] It is a great pity he was so run away with by his fondness for new names, as he is decidedly the best Naturalist generally of any who publish.— I can make out Fish Mollusca and Birds by him when I cannot by any other book I possess guess at them. Since you went there is a very respectable Entomological Magazine set up, published every 2 months, and Rennie has just commenced "A Field Naturalists Mag". monthly at 1/.[3] & others are talked of. I forget whether Hewitson[4] had commenced his eggs before you left but I think he had. They are beautifully executed now, & sale of work encreases rapidly. I was amused with your seal of Cupid trimming the Sails of a vessel. If any love trimmed the sails of your Vessel to encircle the world, it must have been the love of Beetles, spiders, or Rocks.— If you continue (as you purpose doing you say) in the Beagle during her whole voyage, your opportunities as a Geologist will be very great indeed. Is there not some danger of your becoming like Waterton, so much attached to Wandering, that the itch will again become irresistable when you have been home for a year or two. I wish I knew whether there are any Books or things of any kind I could send you; do remember that any thing I can do for you in any way, will give me very great pleasure indeed. I think it very likely I shall be here till middle of Summer & here I am on spot for doing any thing— At all events I now know plenty here that will superintend any thing for me when I am away.— I have now set up here a Pony, a terrier, 5 gulls of various kinds, Julia a Cockatoo I bought her—& A M a Cat, so that our house begins to look very sociable. When I last heard of your Father—he was in excellent health & all your Family well & in good spirits, but you have probably heard from them. I must now conclude this long, rambling, nonsensical letter— I fear you will think among my other Ailments, I am somewhat Insane.— & now my Dear Darwin Believe one with the most grateful & pleasing recollections of former days at Cambridge & elsewhere, & ardent hopes for future long histories of men with tails & single eyes Ever your affectionate Friend | William Darwin Fox.

DAR 204.7

[1] Roadstead between Spithead and Isle of Wight.
[2] Probably Fleming 1828. An annotated copy is in Darwin Library—CUL.
[3] *The Field Naturalist* (London, 1833—5).
[4] William Chapman Hewitson. The first part of his *British oology* was published in 1833.

From R. W. Darwin and the Misses Darwin to J. S. Henslow
1 February 1833

Doctor & Miss Darwins present their Compts to Professor Henslow and beg to return a great many thanks for his kindness in allowing them to see the enclosed letters— the one written August 15ᵗʰ. is ten days later date than any they have received.—

Shrewsbury
February 1ˢᵗ 1833.

Royal Botanic Gardens, Kew

From Susan Darwin 3—6 March 1833

Shrewsbury
March 3ᵈ 1833.

Catherine who wrote last[1] will have told you my very dear Charles how glad we were to have your last letter dated November, and now it is gone to Erasmus— We have also seen a letter Mʳ. Hughes wrote to the Haycocks & it was speaking so much about you & the great pleasure he had in seeing you, that it was almost as good as hearing from yourself— We shall all treasure this last letter from you, more than any others, as you say we must not expect to hear again for six or nine months, which appears an endless time looking forward.— I congratulate you on your luck in finding those curious remains of the Monster M— I think Geology far the most interesting subject one can imagine & now I have found a very easy way of learning a little smattering of it. The penny Magazines give a few pages (which the most foolish person can understand) in every Number on the subject.— I think this clever penny work has come out since you left England we all *swear* by it as it contains every kind of knowledge written so pleasantly with prints.

The race of Wedgwoods is fast encreasing & I must give you the Annual Register.—

Frank's son & heir Master Godfrey was born on the 26ᵗʰ of January, & the Hensleighs have got a little Daughter[2] born the 6ᵗʰ. of Febʳʸ.— Uncle Jos has been attending Parliament now nearly a Month. We have not heard much about him except that he finds it fatiguing— His family go up to town after Easter when they will divide themselves between Charlotte & Hensleigh so will take no house in London.— The Radicals are getting so fierce & licentious in the Debates, that Papa gets more & more of a Tory every day.— This Government appears to be perfect for they let no abuses remain. Church Reform & Slavery will certainly be done this Session I shᵈ think.—[3] In short by the time you come back from your surveying expedition so many changes will have taken place that I can't imagine how you will ever learn them all for Newspapers 9 months old it wᵈ be impossible to read.—

I have been staying a great deal at Woodhouse lately to comfort poor Caroline who has *no* Sister now Emma being gone to stay with Sarah in London

& Fanny & Mr Biddulph have taken a small *nutshell* for 5 months whilst Parliament sits.— Francis is still at home they want to get him into a foot Regiment but find great difficulty. he seems to go on very well with Mr Owen which is fortunate. Mr Owen sometimes talks of packing him off to Canada but this is only a joke I suppose. They have heard f^t Arthur since he reached Madras. he wrote in great spirits & had gone through his first examination so well that they had augmented his salary.— They always talk and enquire much about you at Woodhouse & Mrs Owen still keeps her opinion of Charles Darwin being the happiest person she knows.— Caroline Owen laughed much at recollecting your walks in the wood with Sarah: & a⟨fter⟩ abusing us all for being such a *reserved* family, sa⟨id⟩ at least "*you* were an *exception* for she believed there was no family secret you would not tell to Sarah in the wood".— I am afraid this is all too true Master Charley.— I used to be surprised how you should like making such long visits at the *Forest*, but now I stay a fortnight at a time I find it much pleasanter, one becomes so completely one of the family.— It seems so odd to see them reduced to a party of *four* at Dinner. Sarah sent for yr direction the other day so I daresay you will find a letter on yr return.—

Charlotte Holland is married to Mr Isaac the 14th of this Month & Emma Wedgwood goes to the wedding.—

March 6th.

Captain Beaufort has very kindly written to tell us that if we send off this letter immediately to him it will catch the Buenos Ayres Mail & will most probably be the last news you will have for sometime I don't quite understand how this can be—but as you will like to hear as lately as possible that we are all well I pack this off tho' it is a mighty dull epistle

Dear old Charley I am afraid we shan't see yr hand writing till September but at least we have the comfort of knowing that your long silence means no harm & is unavoidable.— Catherine & I are just come in from a long walk with the (future Mrs Hope) alias Louisa Leighton[4] they talk of going abroad after their marriage which will suit the Gentleman much better than the Lady I guess.—

Papa desires his most affectionate love to you We all often talk about you & you are forgotten by nobody I assure you & I have nothing more to give but my best love & blessing | Dear old fellow y^r affectionate *Granny* D.

DAR 204.6.1

[1] Catherine's letter of 8 February has not been found, though it was received by CD the following June (see letter to Catherine Darwin, 22 May − 14 July 1833).

[2] Frances Julia Wedgwood.

[3] Church reform, which was widely anticipated in 1832, was not instituted. Slavery had been abolished in England in 1772 and slave trade in the colonies in 1806. In August 1833 a law emancipating the slaves in the colonies was passed.

[4] Louisa Leighton married Henry Hope, brother of Frederick William Hope.

From R. W. Darwin 7 March 1833

[Shrewsbury]

My dear Charles

As a packet of letters is going under cover to Capt Beaufort I must send you one line, tho' in fact I have not any thing to say besides expressing the pleasure we all feel at your still continuing to enjoy health & your voyage we all are very happy when we get a letter from you.

In consequence of the recommendation in your first letter I got a Banana tree, it flourishes so as to promise to fill the hothouse. I sit under it, and think of you in similar shade. You know I never write any thing besides answering questions about medicine and therefore as you are not a patient I must conclude

Your money accounts are all correct, the £20 = in November has appeared the other for Capt Fitzroy I have not yet received.

My dear Charles ever your affectionate | R W Darwin

Salop
7 March 1833.

DAR 204.6.1

From Caroline Darwin 7 March [1833]

[Shrewsbury]
March 7th.

My dear Charles

Though it is not my *turn* to write to you I cannot resist the opportunity of sending you my very best love— I was so glad to get your last letter of November 24th with such a happy account of yourself & I must congrat— you on being so successful and persevering in collecting I shall be very curious to read in your journal how you first heard of the Megatherium Coat & whether it is a part that has not before been sent to England— I thought I had heard Mr Sedgwick say some specimen of the coat was now in England—

We heard of you a few days later than your letter again through Mr. Hughes who seems to envy you your expedition his own life must be dull enough to be sure— I am so happy to be able to give you such a nice account of my Father he is looking very well & though since you went his powers of moving about, walking &tc are very much gone he is in every other respect quite like himself, he sees hardly any patients & refuses going to any distance but he amuses himself very well & his eyes from habit are not so soon tired by reading as they were formerly, we go on in the old way having our game of whist & Cassino every evening—

Charlotte Holland is to be married on the 14th of this month to Mr Isaac Emma Wedgwood is to be one of the Brides maids & go back with Louisa H. to

London— I expect Erasmus will be a very attentive Cavalier to her & nobody knows what will be the end of the drives in his Cab he will take her to & fro Clapham where the Hensleigh Wedgwoods live— there will be quite a new race sprung up before you return little Hensleighs & little Franks— Charlotte Langton says she almost rejoices to hear the expedition is to last 5 years longer as she thinks there is greater chance of your not staying all the time but I hope the report we heard was false of the time being lengthened— How anxious & impatient we shall be for your next letter after your cold Patagonian voyage my dear Charles it is such a happiness that your health bears what you go through— I wish I could think of any more home details— Mark has a very pretty little baby— Old Pincher is very well— I ride now the Grey you had from Joseph & like him very much

I am going to Maer tomorrow—such a small party—Uncle Jos being in London & Emma away. Uncle Jos stands the fatigue of Parliamt very well & I think writes in spirits as if he liked the life— he says 19 out of 20 of the speeches are very dull. he is a staunch supporter of Ministry & is very bitter against the Radicals All parties seem to agree that some strong measure in favor of Emancipation of the Slaves will be carried this session & I am sure that alone is enough to make one value the present Ministers—

Your journal is quite safe & we are far from thinking you over careful in being most anxious it should not be lost but pray be at ease it shall be taken the greatest care of— I had & so had my father & all of us so much pleasure & interest in reading it.

Good bye My very dear Charles and God bless you Sitting & writing in this old school room makes me feel so Motherly to you dear Tactus— Y^{rs} vy affecly | Caroline Darwin

DAR 204.6.1

To Caroline Darwin 30 March — 12 April 1833
                                        Falkland Island.— Berkeley Sound:
                                                    March 30th 1833
My dear Caroline
   The Beagle will sail in a few days for Monte Video, & as this sheet of paper is very large I have taken good time to begin my letter.— It is now four months since my last letter so I will write a sort of journal of everything which has since happened.— That we might not lose the long days we made a straight course for the South: my first introduction to the notorious Tierra del F was at Good Success Bay & the master of the ceremonies was a gale of wind.— This place was visited by Capt. Cook; when ascending the mountains, which caused so many disasters to Mr Banks I felt that I was treading on ground, which to me was classic.—[1] We here saw the native Fuegian; an untamed savage is I really think

one of the most extraordinary spectacles in the world.— the difference between a domesticated & wild animal is far more strikingly marked in man.—in the naked barbarian, with his body coated with paint, whose very gestures, whether they may be peacible or hostile are unintelligible, with difficulty we see a fellow-creature.—[2] No drawing or description will at all explain the extreme interest which is created by the first sight of savages.— It is an interest which almost repays one for a cruize in these latitudes; & this I assure you is saying a good deal.—

We doubled Cape Horn on a beautiful afternoon; it was however the last we were doomed to have for some time.— After trying to make head against the Westerly gales we put into a cove near the Cape.— Here we experienced some tremendous weather; the gusts of wind fairly tear up the water & carry clouds of spray.— We again put to sea, with no better success, gales succeed gales, with such short intervals, that a ship can do nothing.— After 23 days knocking about, we only reached false Cape Horn[3] a few miles distant.— This finale gale was worthy of the reputation which, this climate since Ansons times,[4] has possessed.— The Captain considers it the most severe one he was ever in.— We have already heard of two vessels which were wrecked at the very same period.— At Breakfast, I was remarking that a gale of wind, was nothing so very bad in a good sea-boat; the Captain told me to wait till we shipped a sea; it was prophetic; for at noon we shipped a great one; & it is a sight for a landsman to remember; one of our boats was knocked to pieces & was immediately cut away: the water being deep on the deck, it did me an infinity of harm, as it wetted a great deal of paper & dried plants..— I suffered also much from sea-sickness: & yet with all this I am becoming quite hardened; it makes me however, think with greater ectasy of the warm serene air & the beautiful forms of the Tropics.— No disciple of Mahomet ever looked to his seventh heaven, with greater zeal, than I do to those regions.—

Having found a good anchorage; we took the Fuegians & Matthews in a flotilla of boats to Jemmy Buttons country.— Jemmy's relations knew him, but having forgotten his language & being dressed in clothes, they paid no attention to him, & were much more earnest in begging for knives &c.— Having dug a garden & built houses.—the Captain went (taking me with him) on a long surveying cruize with two boats; when we returned to the Settlement, things were in a ruinous condition, almost every thing had been plundered, & the Fuegians had made such signs to Matthews that the Captain advised him not to stay with them. These Fuegians are Cannibals;[5] but we have good reason to suppose it carried on to an extent which hitherto has been unheard of in the world.— Jemmy Button told Matthews, a long time since, that in winter they sometimes eat the women.—certain it is the women are in a very small proportion.—yet we could not believe it.— But the other day a Sealing Captain said that a Fuegian boy, whom he had, said the same thing.— upon being asked, "why no eat dogs".— The boy answered "dog catch otter",—"woman good for nothing"—man very hungry".— He said they smothered, them: it is difficult to disbelieve two such distinct explicit accounts & given by boys.— Was ever any thing so atrocious heard of, to work them like slave to procure food in the summer & occasionally in winter to eat them.— I feel quite a disgust at the very sound of the voices of these miserable savages.— This boat expedition was exceedingly interesting: we went about 300 miles & were absent 23 days.— the worst part was the Fuegians being in such large bodies, that we were often obliged to find a quiet sleeping place after it was dark.— This often precluded us from the greatest luxury; a shingle beach for a bed: the greater part of the way was in the Beagle Channel, an arm of the sea, which connects the Atlantic & Pacific.— Some of the scenes from their retirement & others from their desolate air, were very grand.— Glaciers descend to the waters edge; the azure blue of the ice, contrasted with the white snow, & surrounded by dark green forests were views as beautiful, as they were novel to me.— An avalance falling into the water put us for a second in great peril.— Our boats were hauled up on the beach, but a great wave rushed onwards & nearly dashed them to pieces: our predicament, without food & surrounded by Savages would not have been comfortable.—

We arrived here in the Falkland Islands in the beginning of this month & after such a succession of gales, that a calm day is quite a phenomenon..— We found to our great surprise the English flag hoisted.—[6] I suppose the occupation of this place, has only just been noticed in the English paper; but we hear all the Southern part of America is in a ferment about. By the aweful language of Buenos Ayres one would suppose this great republic meant to declare war against England!— These islands have a miserable appearance; they do not possess a tree; yet from their local situation will be of great impo⟨rtance⟩ to shipping; from this Cause the Captain intends making an accurate survey.—[7] A great event has happened here in the history of the Beagle.— it is the purchase of a large Schooner 170 tuns, only 70 less than the Beagle: The Captain has bought it for himself, but intends writing to the Admiralty for men &c &c.—[8] Wickham will have the command; it will double our work, perhaps shorten our cruize, will carry water & provisions, & in the remote chance of fire or sticking on a Corall reef may save many of our lives.— It is the present intention to take the Schooner to the R Negro & then to refit, whilst the Beagles goes to M. Video: if so I shall stay at the former place; as it is a nice wild place, & the Rio Plata I detest.—

I have been very successful in geology; as I have found a number of fossil shells, in the very oldest rocks, which ever have organic remains.— This ⟨ha⟩s long been a great desideratum in geology, viz the comparison of animals of equally remote epocks at different stations in the globe.— As for living creatures, these wretched climates are very unfavourable; yet I have the great satisfaction to find my powers of examining & describing them have increased at a great pace.— As for our future plans I know nothing; circumstances alter them daily.— I believe we must have one more trip to the South, before finally going round the Horn, or rather passing the St^s of Magellan, for the Captain had enough of the great sea at the Cape to last him all his life.— I am quite astonished, to find I can endure this life; if it was not for the strong & increasing pleasure from Nat: History I never could.—

It is a tempting thought, to fancy you all round the fire, & I perhaps plaguing Granny for some music.— Such recollections are very vivid, when we are pitching bows under & I sea-sick & cold.— Yet if I was to return home now I should feel as if there had been no interval of time, I suppose it is from having so throughily made up my mind for a long absence.—

March⁹ 8th.— We have just had our usual luck, in a heavy gale of wind: but I wont write any more for I have not half got over my sea-sickness, & am ready to exclaim all is vanity & vexation of spirit.

April 12th Of this same vexation of spirit there is an abundance in a Ship: it is paying a heavy price, but not too dear, to see all, which we see; but such scenes it would be impossible to behold by any other means:—& for the zeal, which this voyage has given me, for every branch of Natural History I shall never cease being glad.—

Wickham will be a heavy loss to this vessel; there is not another ⟨in⟩ the ship worth half of him. Hamond also, who lately joined ⟨the⟩ Beagle, from stammering & disliking the Service intends leaving it altogether.— I have seen more of him, than any other one & like him accordingly.— I can very plainly see, there will not be much pleasure or contentment, till we get out of these detestable latitudes & are carrying on all sail to the land where Bananas grow. Oh those realms of peace & joy; I trust, by this time next year, we shall be under their blue sky & clear atmosphere.— At this instant we are shortening sail, as by the morning we expect to be in sight of the mouth of the Rio Negro.—¹⁰ I *send* by the Beagle (if I stay behind) a bill for 60£:.— I owe some little money & I hope to live on shore at the Rio Negro.— I shall get your letters in about a month's time; a pleasure which thanks to you all, never fails me.

With my most affectionate love to my Father & to all of you & may you all be happy: believe me dear Caroline, Y^{rs} very sincerely | Chas. Darwin

I have drawn a bill of Seventy Pounds: I shall stay nearly 2 months at the R. Negro

DAR 223

[1] Joseph Banks and Daniel Carl Solander sailed in the *Endeavour* with Captain Cook from 1768 to 1771. While climbing at Bay of Good Success on 16−17 January 1768, they were caught by a snowstorm and two members of their group perished. See Beaglehole, ed. 1962, 1: 218−22.

[2] In the '*Beagle' diary*, p. 119, CD noted: 'It [the difference] is greater than between a wild & domesticated animal, in as much as in man there is greater power of improvement.' Robert FitzRoy's reaction was very different: he remarked that Caesar had found the Britons painted and clothed in skins like these Fuegians (*Narrative* 2: 120−1).

[3] About 35 miles north-west of Cape Horn.

[4] George Anson, later Admiral, commander of a squadron sent to attack the Spanish colonies in South America in 1740, lost two ships in storms while rounding Cape Horn.

[5] This is denied by E. L. Bridges, son of Thomas Bridges, a missionary who settled in Tierra del Fuego in 1871. He says CD misunderstood his informants because the Fuegians gave the sort of answer they felt to be expected (Bridges 1948, pp. 31−6). Lothrop 1928, p. 118, corroborates Bridges's view that the Fuegians did not practise cannibalism.

[6] H.M.S. *Clio* and H.M.S. *Tyne* arrived in the Falklands early in January 1833 and the British flag was hoisted. Woodbine Parish, Chargé d'Affaires at Buenos Aires, who had earlier sent an official protest against the Argentine occupation, ordered that British sovereignty be reasserted. FitzRoy's account of the event and its historical background (*Narrative* 2: 228−40) presents the English view at the time. The Argentine protest of 17 June 1833 by Manuel Moreno, Minister Plenipotentiary to the Court of St James, is in *British and Foreign State Papers* 22 (1833−4): 1366−84; Lord Palmerston's response is in the same volume. FitzRoy reprinted it in *Narrative* Appendix, p. 150−62. For a later analysis of the problem, see Goebel 1927.

[7] On the economic and political background of the British voyages to chart the South American coasts in the period 1826−36, see Basalla 1963.

[8] See *Narrative* 2: 274−5. FitzRoy named the schooner *Adventure* after the *Beagle*'s companion vessel of the first voyage. The Admiralty strongly disapproved of FitzRoy's purchase (see *Voyage*, p. 184, n. 1 and Mellersh 1968, pp. 146−8).

[9] An error for April.

[10] Unfavourable sailing conditions changed this plan and the *Beagle* proceeded to Maldonado, 65 miles east of Montevideo.

To J. S. Henslow[1] 11 April 1833

April 11th.— 1833

My dear Henslow

We are now running up from the Falkland Islands to the Rio Negro (or Colorado).— The Beagle will proceed to M: Video; but if it can be managed I intend staying at the former place.— It is now some months since we have been at a civilized port, nearly all this time has been spent in the most Southern part of Tierra del Fuego.— It is a detestable place, gales succeed gales with such short intervals, that it is difficult to do anything.— We were 23 days off Cape Horn, & could by no means get to the Westward.— The last & finale gale, before we gave up the attempt was unusually severe. A sea stove one of the boats & there was so much water on the decks, that every place was afloat; nearly all the paper for drying plants is spoiled & half of this cruizes collection.— We at last run in to harbor & in the boats got to the West by the inland channels.— As I was one of this party, I was very glad of it: with two boats we went about 300 miles, & thus I had an excellent opportunity of geologising & seeing much of the Savages.— The Fuegians are in a more miserable state of barbarism, than I had expected ever to have seen a human being.— In this inclement country, they are absolutely naked, & their temporary houses are like what children make in summer, with boughs of trees.— I do not think any spectacle can be more interesting, than the first sight of Man in his primitive wildness.— It is an interest, which cannot well be imagined, untill it is experienced. I shall never forget, when entering Good Success Bay, the yell with which a party received us. They were seated on a rocky point, surrounded by the dark forest of beech; as they threw their arms wildly round their heads & their long hair streaming they seemed the troubled spirits of another world.— The climate in some respects, is a curious mixture of severity & mildness; as far as regards the animal kingdom the former character prevails; I have in consequence, not added much to my collections.— The geology of this part of Tierra del was, as indeed every place is, to me very interesting.— the country is non-fossiliferous & a common place succession of granitic rocks & Slates: attempting to make out the relation of cleavage, strata &c &c was my chief amusement.— The mineralogy however of some of the rocks, will I think be curious, from their resemblance to those of Volcanic origin.

In Zoology, during the whole cruize, I have done little; the Southern ocean is nearly as sterile as the continent it washes.— Crustaceæ have afforded me most work: it is an order most imperfectly known: I found a Zoëa, of most curious form, its body being only $1/6^{th}$ the length of the two spears.— I am convinced from its structure & other reasons it is a young Erichthus!—[2] I must mention part of the structure of a Decapod, it so very anomalous: the last pair of legs are small & dorsal, but instead of being terminated by a claw, as in all others, it has three curved bristle-like appendages, these are finely serrated & furnished with cups, somewhat resembling those of the Cephalopods.— The animal being pelagic it is a beautiful structure to enable it to hold on to light floating objects.— I have found out something about the propagation of that ambiguous tribe, the Corallinas.— And this makes up nearly the poor catalogue of rarities during this cruize. After leaving Tierra del we sailed to the Falklan⟨ds.⟩ I forgot to mention the fate of the Fuegians, w⟨hom⟩ we took back to their country.— They had beco⟨me⟩ entirely Europæan in their habits & wishes: so much so, that the younger one had forgotten his own language & their countrymen paid but very little attention to them.— We built houses for them & planted gardens, but by the time we return again on o⟨ur⟩ passage round the Horn, I think it will be very doubtful how much of their property will be left unstolen.—

On our arrival at the Falklands everyone was much surprised to find the English flag hoisted. This our new island, is but a desolate looking spot yet must eventually be of great importance to shipping.— I had here the high good fortune, to find amongst most primitive looking rocks, a bed of micaceous sandstone, abounding with Terebratula & its subgenera & Entrochitus. As this is so remote a locality from Europe I think the comparison of these impressions, with those of the oldest fossiliferous rocks of Europe will be preeminently interesting.[3] Of course there are only models & casts; but many of these are very perfect. I hope sufficiently so to identify species.— As I consider myself your pupil, nothing gives me more pleasure, than telling you my good luck.— I am very impatient to hear from you. When I am sea-sick & miserable, ⟨i⟩t is one of my highest consolations, to picture the future, ⟨w⟩hen we again shall be pacing together the roads round Cambridge. That day is a weary long way off: we have another cruize to make to Tierra del. next summer, & then our voyage round the world will really commence. Capt. FitzRoy has purchased a large Schooner of 170 tuns. In many respects it will be a great advantage having a consort: perhaps it may somewhat shorten our cruize: which I most cordially hope it may: I trust however that the Corall reefs[4] & various animals of the Pacific may keep up my resolution.—

Remember me most kindly to M⟨rs⟩ Henslow & all other friends; I am a true lover of Alma Mater, & all its inhabitants. Believe me My dear Henslow | Your affectionate & most obliged friend | Charles Darwin

Recollect, if should think of any books, scientific travels &c &c which would be useful to me do not let them pass out of y⟨r⟩ mind.

We are all very curious to to hear *something* about *some* great Comet, which is coming at *some* time: Do pump the learned & send us a report:[5]

I am convinced from talking to the finder,[6] that the Megatherium, sent to Geol: Soc: belongs to same formation which those bones I sent home do & that it was wa⟨she⟩d into the River from the cliffs which compose the banks:[7] Professor Sedgwick might like to know this: & tell him I have never ceased being thankful for that short tour in Wales

Royal Botanic Gardens, Kew

[1] The following passages from this letter were extracted by Henslow and published in the Cambridge Philosophical Society pamphlet:
  1.1 'We . . . Colorado).' 1.2
  1.3 'It is . . . Westward.—' 1.7
  1.10 'We at last . . . channels.—' 1.11
  1.12 'with two boats . . . trees.' 1.17
  1.23 'The climate . . . amusement.' 1.29; *omits* 1.26 'as indeed every place is,'
  2.1 'the Southern . . . work' 2.3
  2.3 'I found . . . Corallinas.' 2.12
  2.13 'After . . . Falklan⟨ds.⟩'
  3.3 'I had here . . . perfect.' 3.9
[2] The specimen (no. 485) is described in detail in the 'Zoological diary' (DAR 30.2: 128–33).
[3] Four genera of CD's Falkland Island fossil shell specimens are described as similar to Silurian and Devonian forms by Morris and Sharpe (1846). On the relationship of the fossils to those of European formations, the authors considered the number of species collected by CD to be 'too limited to justify any close comparison with the palæozoic fauna of other portions of the globe'.
[4] This is CD's first mention of an interest in coral reefs. It perhaps resulted from reading the second volume of Lyell's *Principles*, which he had recently received. His copy, inscribed 'M: Video. Novem⟨r⟩ 1832', is in Darwin Library—CUL. It is lightly annotated in pencil.
[5] See letter from J. S. Henslow, 31 August 1833.
[6] A reference to Woodbine Parish's agent, 'Mr Oakley, a gentleman of the United States' (Parish 1834, p. 404). In a field notebook entry of 3 November 1832 CD refers to him as 'Oakley, a Joiner with red hair, M. Video' (*Voyage*, p. 168). Later, under the same date is: 'Oakley's fossil—one scapula in true Tosca' (*ibid.*, p. 169). (See below, n. 7.) Parish presented the skeleton and other fossils

to the Royal College of Surgeons on 13 June 1832 after they had been exhibited at the Geological Society, London (see Clift 1835, p. 437 and n., and Parish 1838, pp. 175−7).

[7] On 20 November [1832] CD made the following note in his geological diary: 'In the Newspaper accounts the Megatherium lately presented by M.ʳ Parish to the Geological society, is stated to have been found in the *mud* in bed of the river Salado [*interl*].— Upon examining M.ʳ Oakley, who procured it for M.ʳ Parish, it seems the river flows through cliffs of the Tosca, & which doubtless is identical with that of Bahia Blanca & Buenos Ayres . . . M.ʳ Oakley clearly recollects that one of the Scapulas was imbedded in ['th' *del*] a mass of *Tosca*' (DAR 32.1: 71v.). Tosca is a limestone formation underlying the Pampean formation.

From Caroline Darwin   1 − 4 May 1833

May 1ˢᵗ | 1833.

My dear Charles—

I am afraid there is little chance of your getting this letter as Cap.ᵗ Beaufort in his note to Papa said he thought you would not receive any letters we wrote for some time after those the last we sent through the Admiralty office the middle of March— I hope you received them safe— Our last letter from you was dated 24ᵗʰ of November just before you began your southern expedition. My dear Charles how happy I shall be when we get your next letter with a good account of yourself after your stay in those cold & stormy regions. Very little has happened at home since we last wrote— I have been staying 3 weeks in Staffordshire at Maer & a few days at Betley—the John Wedgwoods were at Maer the greatest part of the time. Only Aunt Bessy Elizabeth & Jos were there of the family—it seemed very strange & melancholy to see so small a family party Uncle Jos finds he bears the fatigue & late hours of Parliament very well & is very much interested he is a strong ministerial supporter & can't bear the radicals—

Emma Wedgwood went to attend Charlotte Hollands marriage with M.ʳ John Isaac & then went to London with M.ʳˢ Holland & Louisa. She has since been visiting the Langtons at Ripley— the rest of the Maer family went to London this week where they will remain some months visiting their friends—

Susan & Cath have a very nice scheme of joining Harry & Jessie & taking lodgings for 3 weeks in London this spring—they set out the 21ˢᵗ of this month— We heard from Erasmus who seems very happy & seemingly leading a dissipated life for him. M.ʳ & M.ʳˢ Evans[1] of Portrane are in London & he sees a good deal of them. M.ʳ Evans is in Parliament for the County of Dublin— Erasmus is going to the meeting of the Philosophers at Cambridge in June—[2] he is going to visit Frederic Hilyard— Louisa Leighton was married to M.ʳ Hope last Tuesday, there was an immense crowd all S.ᵗ Chads place quite full the Leightons had a dance afterwards to which Catherine & Caroline Owen went. Susan was not well with a cold & stayed at home. Susan is gone this week with M.ʳ & M.ʳˢ Cotton to stay with W.ᵐ Clives at W. Pool—[3] Woodhouse is strangely altered since your days only that quiet Caroline at home & hardly ever any company. Francis is at home at present. They have had very pleasant letters from Arthur Robert Clive returned from India a few weeks ago, not to go back again. We have hardly seen him yet. he is grown very brown & looks much older from the last 3 years he has been away. Arthur Owen was staying in the same house with him in India— Robert Clive speaks very well of him Poor Fanny Biddulph is in daily expectation of her confinement— M^r B— quarreled with his Landlord & they had to move their lodgings last week— the night they got to them they were awoken by a smell of fire & it was discovered a beam was on fire connected with one of the chimneys— poor Fanny was in a gre⟨at⟩ fright & for some hours in momen⟨tar⟩y expectation of being obgd to leave the house, but before morning the fire was put out & she has not suffered from her fright.— Emma Owen is grown a very pretty girl— she is staying with Sarah Williams in London. My Father is grown quite larky— he is going next week to see York & I go with him. We go by Liverpool & return by Doncaster where we shall see Newstead Abbey which I shall like— My Fathers breathing is certainly much better this summer than it was last year. Capt Harding[4] called here a few days ago— I did not see him but my Father & Susan were very glad to hear about you from him— Capt Harding is looking himself very delicate, he wants to go at once to S^t. Helena & marry Miss Dona Dallas, but his friends will not allow him till he gets stronger—

Tom Wedgwood has been staying with us. he says Robert is becoming a *great* farmer he has taken several fields from Uncle Jos, keeps pigs, cows & did keep 17 ducks but they came to a lamentable end, as after they had begun to lay Uncle Jos said they eat the potatoes & must be taken from the Pool. Robert obeyed & is obgd to eat his ducks instead breeding from them as he intended—

Pincher & Nina are very well I am riding *your* Grey & every thing & every body are going on just as when the last monthly letter went so you must not mind such *twaddly* dull letter— the orchard is looking beautiful in full blow. I guess you see no such pretty sights in Patagonia

Good bye my very dear Charley | All join in kindest love | Ev^r y^r affc^tne | Caroline Darwin

May 4th.
Shrewsbury

DAR 204.6.1

[1] George Evans.

[2] The third meeting of the British Association took place in Cambridge, June 1833.

[3] Welshpool, Wales. William Clive was Vicar of Welshpool.

[4] Captain Francis Harding had recently returned from service in the South American station of the Fleet (O'Byrne 1849).

To Catherine Darwin 22 May — 14 July 1833

Maldonado. Rio Plata
May 22^d 1833

My dear Catherine

Thanks to my good fortune & my good sisters I have to acknowledge the following string of letters: (August I received many months ago:) September 12th Caroline: October 14th. Catherine: November 12th. Susan: December 15th. Caroline: & Jan 13th. Caroline:

My last folio letter was dated on the sea; after being disappointed at the Rio Negro.— the same foul winds & ill fate followed me to Maldonado; so that the Beagle proceeded direct to M: Video.— Here we remained only one night, when I received your four first letters: I really had not time to open & alter my letter, but sent it, as it was.— Leaving M: Video we came directly to Maldonado.— I the next day took up my residence on shore.— The Beagle has not yet returned (for she went again there) from M: Video, & I know nothing of our future plans: the purchase of the Schooner has so altered every thing. I have been living here for the last three weeks; it is quiet little village, surrounded on all sides by the endless succession of green turf hills & stony ridges.— I have had one little excursion which I enjoyed very much; I procured two trust-worthy men & a troop of horses & have had a 12 days ride into the interior.— the country continues very similar; so that one dreadfully misses the gorgeous views of Brazil.— I saw however a good deal of the Gauchos; a singular race of countrymen.— "Heads gallop" gives a most faithful picture; nothing can, I think, be more spirited & just than his remarks.—

Besides your letters I received several others.—one from Charlotte: 2 from Fox: also one of the very kindest I ever received in my life time, from M^{rs}. Williams.— I am very sorry to hear from your latter letters, that she has lost so much of the Owen constitution: I am very sure that with it, none of the Owen goodness has gone.—

I most devoutly trust that next summer (your winter) will be the last on this side of the Horn: for I am become throughily tired of these countries: a live Megatherium would hardly support my patience: the good people of Shropshire, who say I shall find cruizing in the South-seas stupid work, know very little of the numberless invertibrate animals, which abound in the intertropical ocean.— If it was not for these & still more for geology—I would in short time make a bolt across the Atlantic to good old Shropshire.— In for penny, in for pound.— I have worked very hard (at least for me) at Nat History & have collected many animals & observed many geological phenomena: & I think it would be a pity having gone so far, not to go on & do all in my power in this my favourite pursuit; & which I am sure, will remain so for the rest of my life.—

The following business piece is to my Father: having a servant of my own would be a really great addition to my comfort.—for these two reasons; as at present, the Captain has appointed one of the men always to be with me.—but I do not think it just thus to take a seaman out of the ship:—& 2^d when at sea, I am rather badly off for anyone to wait on me.— The man is willing to be my servant & **all** the expences would be under sixty £ per annum.— I have taught him to shoot & skin birds, so that in my main object he is very useful.— I have now left England nearly 1 & ½ years: & I find my expences are not above 200£ per annum:—so that it being hopeless from time to write for permission I have come to the conclusion you would allow me this expense.— But I have not yet resolved to ask the Captain: & the chances are even that he would not be willing to have an additional man in the ship.— I have mentioned this because for a long time I have been thinking about it.—[1]

June:— I have just received a bundle more letters.— I do not know how to thank you all sufficiently:—one from Catherine Feb 8th:—another from Susan, March 3^d; together with notes from Caroline & from my Father; give my best love to my Father; I almost cried for pleasure at receiving it.—it was very kind, thinking of writing to me.— My letters are both few, short, & stupid in return for all yours; but I always ease my conscience by considering the Journal as a long letter. If I can manage it, I will before doubling the Horn send the rest.— I am quite delighted to find, the hide of the Megatherium has given you all some little interest in my employments. These fragments are not however by any means the most valuable of the Geological relics. I trust & believe, that the time spent in this voyage, if thrown away for all other respects, will produce its full worth in Nat: History: And it appears to me, the doing what *little* one can to encrease the general stock of knowledge is as respectable an object of life, as one can in any likelihood pursue.— It is more the result of such reflections (as I have already said) than much immediate pleasure, which now makes me continue the voyage: together with the glorious prospect of the future, when passing the Straits of Magellan, we have in truth the world before us.— Think of the Andes; the luxuriant forest of the Guayquil; the islands of the South Sea & new South Wale. How many magnificent & characteristic views, how many & curious tribes of men we shall see.—what fine opportunities for geology & for studying the infinite host of living beings: is not this a prospect to keep up the most flagging spirit?— If I was to throw it away; I dont think I should ever rest quiet in my grave; I certainly should be a ghost & haunt the Brit: Museum.—

How famously the Ministers appear to be going on I always much enjoy political ⟨goss⟩ip, & what you, at home think will &c &c take place.— I steadily read up the weekly Paper: but it is not sufficient to guides one opinion: & I find it a very painful state not to be as obstinate as a pig in politicks. I have watched how steadily the general feeling, as shown at elections, has been rising against Slavery.— What a proud thing for England, if she is the first Europæan nation which utterly abolishes it.— I was told before leaving England, that after living in Slave countries: all my opinions would be altered; the only alteration I am aware of is forming a much higher estimate of the Negros character.— it is impossible to see a negro & not feel kindly towards him; such cheerful, open honest expressions & such fine muscular bodies; I never saw any of the diminutive Portuguese with their murderous countenances, without almost wishing for Brazil to follow the example of Hayti; & considering the enormous healthy looking black population, it will be wonderful if at some future day it does not take place.— There is at Rio, a man (I know not his titles) who has large salary to prevent (I believe) th⟨e⟩ landing of slaves: he lives at Botofogo, & yet that was the ⟨b⟩ay, where during my residence the greater number of smuggled slaves were landed.— Some of the Anti-Slavery people ought to question about his office: it was the subject of conversation at Rio amongst some of the lower English.—

June 19th. I write this letter by patches:— I have just spent a day on board to see old Wickham, who has returned from his little hired Schooner to be Captain of the new one.— This same Schooner will produce the greatest benefits to me.— The Captain always anxious to make every body comfortable, has given me all Stokes (who will be in the Schooner) drawers in the Poop Cabin, & for the future nobody will live there except myself.— I absolutely revel in room:— I would not change berths with anyone in the Ship.— The cause of our very long delay here, is coppering the Schooner; as soon as this is finished the Beagle will go for a month to R. Negro return to the R. Plata & take in provisions for the whole summer.— The Captain is anxious to then be able to pass on to Conception on the other side.— I am ready to bound for joy at the thoughts at it.— Volcanic plains: beds of Coal: lakes of Nitre, & the Lord only knows what more.— If this was certain I would hatch a grand plan, viz of now remaining behind, & posting up to B. Ayres; I heard of so many curious things there; per contrà at R. Negro cliffs almost built of fossil shells.— Was ever a Philosopher (my standard name on board) placed between two such bundles of Hay?— The worst of it is the B. A. bundle is rather expensive, & nearly all the 70£ is gone in paying what I owed & in my long residence here.— And then the mere reading the sum total from July 31 to 32 is enough to give one an indigestion: what it must have been to have paid it, I dont know:— I shall go on board in a weeks time & then I shall know more:—

June [July] 6th.— I am now living on board: The Packet has just come in; but no letters for poor me; I have no right to growl, for I suppose the Capt. Beaufort parcel has robbed this month. Farewell for the future to regular correspondence.— You must direct hereafter to Valparayso. Our plans are (always winds & waves permitting) to go for a month to the banks off the R. Negro, return to the Plata; find the Schooner, ready, take in at M Video one years provisions & Lark away to the land of storms, in the Autumn (your Spring) pass the Straits of Magellan: I am ready to bound for joy at this prospect. I long to bid adieu to the Atlantic. Already I almost fancy I see through a long vista of storms, the blue sky of the Tropics.— I wrote the other day to Mr Hughes' at B Ayres, & I am sorry to hear he has left that place, chiefly from ill health.— I have asked the Captain & obtained his consent respecting a servant.—but he has saved me much expence, by keeping him on the books for victuals & will write to Admiralty for permission.— So that it will not be much more than 30£ per annum.— I shall now make a fine collection, in birds & quadrupeds, which before took up far too much time. We here got 80 birds & 20 quadrupeds. Tell Caroline, to thank Charlotte very much for writing to me. When we are on the other side, I shall have more to say, & will then write to thank her.— I have lost all interest in this part of America, & I feel more inclined to growl, than write civilly to any-body.—

July 14th.— We have just had a trip to M: Video & in few days go Southward: I received letter of Caroline May 1st.— my last was the Beaufort parcel in March; the April one alas is lost: *Excepting* when the letters are sent from *home*, remember the 3ıı6 is temptation for any body to tear up the letter:[2] By the same packet which takes this the rest of my journal will arrive, through Capt. Beaufort.— so if it does not come, you will know where to enquire about it.— The journal latterly has not been flourishing, for there is nothing to write about in these well-known-uninteresting countries.— The letter ought to have made as it were two distinct ones: but when living on shore, I did not hear of conveyances to M. Video.— Once more I must thank you all for writing: it is so very delightful having a regular correspondence.—

Give my love to my Father & Erasmus & all of you: God bless you all.— | My dear Katty: Your most affectionately, | Chas. Darwin.—

P.S.— When you read this I am afraid you will think that I am like the Midshipman in Persuasion who never wrote home, excepting when he wanted to beg: it is chiefly for more books; those most valuable of all valuable things: "Flemings philosophy of Zoology" & Pennants Quadrupeds" these I have at home: "Davys consolation in Travel": "Scoresby Arctic regions": "Playfairs Hutton, theory of the earth" "Burchells travels" "Paul Scrope on Volcanoes" a pamphlet by "J. Dalyell Observations on the Planariæ, Edinburgh" Caldcleugh travels in S America.— If any of these books are expensive, strike them out: Tell Erasmus I shall be very much obliged if with my Fathers consent he will undertake this commission. If the 8th Vol of Humboldt or Sedgwick & Conybeares geological book is out I should like them both:[3] You people at home cannot appreciate the exceeding value of Books: Cary has 3sıı6 tape measure of about 12 feet.[4] I have lost mine: I have at present a double convex lens, fitted to the object glass, & about one inch in diameter: now I want one on a larger scale & with longer focal distance, for illuminating opake objects: it must be fixed on a stand with plenty of motions. I want to use it, by placing it near the Microscope, & thus have steady light on opake object.— I daresay an Optician must have made some such contrivance. Also another box of Promethians[5] (I blush like this red ink, when I ask for it) but the natives here are so much astounded at them, that I have wasted a great many:—& lastly 4 pair of very strong walking shoes from Howell if he has my measure; it is impossible to procure them in this country:

I guess, as the Yankys say, this a pretty considerable tarnation impudent Postcript: I have no doubt, Capt Beaufort will undertake to foreward the box to Valparaiso:—

DAR 223

[1] Syms Covington, 'Fiddler and Boy to the Poop cabin', became CD's servant and remained with him as assistant, secretary, and servant until 1839, when he migrated to Australia.

[2] 3*s.* 6*d.*, a considerable sum, was the postage for a letter to South America. Presumably CD means that a post office clerk—away from home, where the family was known—would be tempted to destroy the letter and pocket the fee.

[3] Burchell 1822−4, Caldcleugh 1825, Dalyell 1814, Davy 1830, Fleming 1822, Pennant 1781, Playfair 1802, Scoresby 1820, and Scrope 1825. No eighth volume of Humboldt's *Personal narrative* was ever published (see letter from E. A. Darwin, 18 August [1832], n. 3). No geological work by Sedgwick and Conybeare was published. Annotated copies of Fleming 1822 and Playfair 1802, and an unannotated Pennant (3d edition, 1793) are in the Darwin Library−CUL. Unannotated copies of Burchell 1822−4, Scoresby 1820, and Scrope 1825 are in the Darwin Library−Down. CD's copies of Caldcleugh 1825, Dalyell 1814, and Davy 1830 have not been found. Playfair and Caldcleugh were used by CD in the *Beagle*. Fleming and Pennant were sent to him from Shrewsbury and were probably used on board the *Beagle*, but there is no corroborating evidence, either in the books themselves or in CD's notes.

[4] CD's measurements were sometimes improvised and approximate. For weighing he balanced with his water flask and for more refined weights used bullets and pellets; e.g., 'Big rat weighs flask with water, without bottom 2 bullets, 4 pellets' (*Voyage*, p. 183).

[5] A kind of match. 'I carried with me some promethean matches, which I ignited by biting; it was thought so wonderful that a man should strike fire with his teeth, that it was usual to collect the whole family to see it: I was once offered a dollar for a single one' (*Journal of researches*, p. 47).

To W. D. Fox 23 May 1833

Maldonado. | Rio Plata.
May 23^d 1833

My dear Fox

Upon our return from a cruize amongst the islands of Tierra del Fuego,—I received your two letters dated at the wide of interval of August & January.—[1] I am very much obliged to you for writing; your letters never fail to throw me into a pleasant reverie of past times & this is one of the highest pleasures I now enjoy.— I find the loss of society a great one.— there is nothing on board the Beagle which can call to mind our evenings in Cambridge.— There is indeed a difference between one of your Coffee parties, with Whitmore &c &c & an evening spent in the gun-room.— But it will be all the same in fifty years, as a boy says, who is going to be well flogged, & thus I have brought myself not to care much for anything which does not interfere with Natural Hist.—

This summers cruize has not been a very profitable one; excepting some little in Geology.— I wish you would begin, like myself, to be a smatterer in this latter branch.— she will soon be the favourite mistress & one easy to be wooed.— I hope for better luck, when the happy day arrive of doubling the Horn & steering for warmer climes.—

This whole East side is totally devoid of all picturesque beauty; & the coast not being rocky & there being no forests, it is bad for the greater part of Zoology.— We are passing this winter in this vicinity; when we shall say farewell to the R. Plata, I know not: I trust that next summer will complete the whole of this part of S. America.— The voyage is an immense one; how different from the first proposed two years.— it is, as you say, a serious evil, so much time spent in wandering.— I often conjecture, what will become of me; my wishes certainly would make me a country clergyman.— You expect sadly more than I shall ever do in Nat: Hist:— I am only a sort of Jackall, a lions provider; but I wish I was sure there were lions enough.— Now this morning I have collected a *host* of minute beetles; who, I should like to know, in England is both capable & willing to describe them?— You ask me about Ornithology; my labours in it are very simple.— I have taught, my servant to shoot & skin birds, & I give him money.— I have only taken one bird, which has much interested me: I daresay it is as common as a cock sparrow, but it appears to me as if all the Orders had said, "let us go snacks[2] in making a specimen".— I collect reptiles, small quadrupeds, & fishes industriously, especially the first: The invertebrate marine animals, are however my delight; amongst them I have examined some, almost disagreeably new; for I can find no analogy between them & any described families.— Amongst the Crustaceæ I have taken many new & curious genera: The pleasure of working with the Microscope ranks second to geology.— I *strongly* advice you instanter to buy from Bancks in Bond St! a *simple* microscope, such as *the* Mr Browne recommends.—& then make out insects scientifically by which I mean separate, examine & describe the trophi: it is very easy & exceedingly interesting; I speak from experience, not in insects, but in most minute Crustaceæ.

I am very glad to hear in your last letter, that your health, after such struggles, is at last so much better, & that you are actually collecting the dear little beetles.— Your domestic arrangements at Ryde, sounded exquisitely ⟨com⟩fortable: it makes me envious to fancy them. I have told you nothing about our cruize to the South: because (you will say a very odd reason) I have too much to tell.— We had plenty of very severe gales of wind; one beating match of 3 weeks off the Horn; when it often blew so hard, you could scarcely look at it.— We shipped a sea—which spoiled all my paper for drying plants: oh the miseries of a real gale of wind! In Tierra del I first saw bona fide savages; & they are as savage as the most curious person would desire.— A wild man is indeed a miserable animal, but one well worth seeing.—

Will you write again? I make a poor return: but indeed letter writing is not my fort: if you were but in hail, I would talk you deaf on the spot.— Once more I must thank you for your most kind letters.— I assure you I well know how to

value & cordially be grateful for your friendship. Believe me, my dear old Fox |
Yours affectionately, Chas Darwin

Remember me most kindly to M^r & Mrs Fox & to every one at Osmaston.—
Tell Miss A. Maria I will remember her out of mere spite, because she wont
me.— Direct to Valparyso, pro futuro.— I have collected in this place about 70
species of birds & 19 Mammalia: Your question: what I did in Ornithology? has
done me good: I have watched the manners of the whole set:—[3]

Christ's College Library, Cambridge (Fox 46b)

[1] Letters from W. D. Fox, 29 August — 28 September 1832 and 23 January 1833.
[2] To take equal shares. The bird is probably the same 'inosculating creature' CD wrote of in his letter
to J. S. Henslow, [*c.* 26 October —] 24 November [1832].
[3] The words transcribed as 'watched' and 'set' are not clearly legible. If correct, the meaning appears
to be that CD observed the habits of the species collected. This is borne out by the Maldonado entries
in his 'Zoological diary' (DAR 30.2: 77—86v.), which contain detailed observations of flight,
nesting, feeding, and other habits. Most of these entries are reproduced in 'Ornithological notes', pp.
214—25.

From Catherine Darwin 29 May 1833

London Regent S^t
May 29th. 33.—

My dear Charles

You will be surprised to see the date of this, but Susan & I are enjoying
ourselves in London in lodgings with Harry & Jessie as Chaperons.— We came
up just a week ago, and have been exceedingly busy and gay ever since;— we
mean to be here three weeks altogether, if the money will hold out, for it goes at
an awful rate in London. It is very long now since we heard from you my dear
Charles, and I am afraid it will still be very long, before we can hope to hear
again. Caroline wrote to you last, the end of April; the Mails have changed back
again now to the first Tuesday in the month.— We have seen Capt Harding,
since Caroline wrote to you; he came to consult Papa, and looked terribly out of
health; he seemed to be a very gentlemanlike man, with very pleasing manners;
he gave us a long account of you, and your enthusiasm, and enjoyment, and I
enjoyed seeing a person, who had seen you in the last 8 months. Capt Harding
was going out to St Helena immediately to fetch his bride Miss Dona Dallas; it is
said that the Dallas are all going to return to England soon, as the Governorship
of St Helena is to be recalled.— I hope we shall soon see Mr Charles Hughes,
who is on his return home now; it is a great pity he is forced to leave Buenos
Ayres before you return to it.— We see a great deal of Erasmus now, as our
Lodgings are not very far from his, and he is goodnatured and pleasant, we
enjoy it very much. Driving in his Cab is one of the pleasantest things possible,
and he drives so very well, I am not afraid even in a London crowd. He seems to

be more in love than ever with Fanny Hensleigh, and almost lives at Clapham. Papa has long been *alarmed* for the consequences, & expects to see an *action* in the Papers. I think the real danger is with Emma Wedgwood, who I suspect Mr Erasmus to be more in love with, than appears, or than perhaps he knows himself. All the Maer Wedgwoods are now staying at Ripley, with the Langtons; Uncle Jos goes down there for his Sundays, but he is obliged to be a great deal in London, on account of Parliament; he has been very unwell and continually feverish, after this Influenza which has been so universal in London. About a month ago, people say that London looked as if it had the Plague; all the theatres stopped as 24 of the Singers were in bed, many of the Shops were closed; ninety clerks were in bed at the Bank of England, so that the business could hardly go on; it was not very fatal however, but much suffering while it lasted. We have all escaped it.— I am afraid you will find Maer still more sadly changed by the time you return, as poor Aunt Bessy's health is in a very precarious state; she had 3 fits in one day lately, which Papa thinks exceedingly dangerous. They do not seem to be aware at Maer, of the danger of these fits.— Papa and Caroline are alone at home now; we left dear Papa very well, and in very good spirits. He is become so spirited in touring, that he and Caroline went very lately another very beautiful Tour in Yorkshire; they went by Liverpool which Papa had never seen, saw the Rail Road, but did not go on it; then went on to the beautiful Yorkshire Abbeys, and so on to York Cathedral, and home by Litchfield; they were absent about ten days, and the Tour answered most perfectly to Papa, who enjoyed it most exceedingly and cam⟨e⟩ back in much better health than he st⟨arted.⟩

Do you remember the Evans of Portrane in Ireland? Mr Evans is Member for the county of Dublin, so they are in London, and are very nice friends for us. We are going with them and a large party down to Richmond by Steam on Saturday, dine there, and return in the Evening.— Mrs Evans enquired very much after you, and said that she could not conceive any thing she should enjoy more than your Voyage.— Sarah Williams is the most cordial, friendly person that ever lived; we find her invaluable in taking us about; she has a beautiful house in Belgrave Square, rather out of London;—but she is sadly out of health, and what is very odd, thinks so very much of her health herself, that Papa, and all her own family consider her as a regular hypochondriac. I suppose it is owing to Mr Williams petting her so exceedingly, and taking so much care of her.— Is it not the oddest change in the world? Sarah is just the very last Person I should have thought would have become full of health.— Poor dear Fanny Biddulph is very slowly recovering from her confinement; she had a little girl on the 7th of May; (think of Fanny as a Mother!!) Susan & I have been to see her continually; she looks deplorably ill & weak, and very lonely all alone in her London house; as for some extraordinary whim, Mr Owen would not let Mrs Owen come up to her confinement, so the poor thing was all alone. Mr Biddulph seems fond & affectionate to her, but he is a gay dissipated man, and desperately selfish also.

Goodbye my dearest Charles. Erasmus desires his best love to you, and with Susan's, believe me my dear old Charley whom I long to hear from again, believe me ever yrˢ | Catherine

Notes on cover, in Robert FitzRoy's hand: 'The *Snake* has *another* packet at Maldonado—not the *Snake* in the *grass*.'[1]
DAR 204.6.1

CD ANNOTATIONS
on cover: 'Steadman | Cangallo, 98.'[2]

[1] On 18 September, Robert FitzRoy had heard that H.M.S. *Snake* was at Maldonado with stores and letters for the *Beagle* (*Narrative* 2: 289).

[2] Steadman may have been a merchant in Buenos Aires. 'Steadman 30' occurs on a shopping list written in the field notebook CD kept at this time (Down House, *Beagle* notebook 1.13). There is no mention of Cangallo.

To J. M. Herbert 2 June 1833

Maldonado, Rio Plata
June 2ᵈ— 1833.—

My dear Herbert

I have been confined for the last three days to a miserable dark room in an old Spanish house from the torrents of rain.— I am not therefore in very good trim for writing, but defying the blue devils I will send you a few lines if it is merely to thank you very sincerely for writing to me.— I received your letter, dated December 1ˢᵗ a short time since.— We are now passing part of the winter in the R. Plata, after having had a hard summers work to the South.— Tierra del Fuego is indeed a miserable place; the ceaseless fury of the gales is quite tremendous.— One evening we saw old Cape Horn & three weeks afterwards, we were only 30 miles to Windward of it.— It is a grand spectacle to see all nature thus raging: but Heaven knows every one in the Beagle has seen enough in this one summer to last them their natural lives.— The first place we landed at was Good Success Bay, it was here Banks & Solander met such disasters on ascending one of the mountains. The weather was tolerably fine, & I enjoyed some walks in a wild country like that behind Barmouth The valleys are impenetrable from the entangled woods, but the higher parts, near the limits of perpetual snow, are bare.— From some of these hills the scenery from its savage, solitary character was most sublime.— the only inhabitant of these heights is the Guanaco, & with its shrill neighing it often breaks the glacial stillness.— The consciousness, that no European had ever trod much of this ground, added to the delight of these rambles.—

How often, & how vividly have many of the hours spent at Barmouth come before my mind. I look back to that time with no common pleasure; at this moment I can see you seated on the hill behind the Inn, almost as plainly as if we were really there.— It is necessary to be separated from all which one has been accustomed to, to know how properly to treasure up such recollections, & at this distance, I may add, how properly to esteem such as yourself.— My dear old Herbert I wonder when I shall ever see you again; I hope it may be as you say, surrounded with heaps of parchment; but then there must be sooner or later a dear little lady to take care of you & your house.— Such a delightful vision makes me quite envious.— This is a curious sort of life for a regular shore-going person, such as myself.— the worst part of it is its extreme length.— there is certainly a great deal of high enjoyment & on the contrary a tolerable share of vexation of spirit.— every thing however shall bend to the pleasure of grubbing up old bones & captivating new animals.— By the way you rank my Nat: Hist: labours far too high: I am nothing more than a lions provider; I do not feel at all sure, that they wi⟨ll⟩ not growl & finally destroy me.—

It does ones heart good to hear how things are going on in England.— Hurrah for the honest Whigs.— I trust they will soon attack that monstrous stain on our boasted liberty, Colonial Slavery.— I have seen enough of Slavery & the dispositions of the negros, to be thoroughly disgusted with the lies & nonsense one hears on the subject in England. Thank God the cold-hearted Tories, who as J Mackintosh used to say, have no enthusiasm except against enthusiasm, have for the present run their race.— I am sorry, by your letter, to hear you have not been well & that you partly attribute it to want of exercise.— I wish you were here amongst the green plains: we would take walks which would rival the Dolgelley ones: & you should tell stories which I would believe even to a cubic fathom of pudding: instead of this I must take my solitary ramble; think of Cambridge days & pick up Snakes, beetles & Toads.

Excuse this short letter; you know I never studied the complete letter-writer & believe me my dear Herbert | Your affectionate friend | Chas. Darwin.—

Pray, Write again: Remember me to all friends & Whitley. I shall never forget how many pleasant hours I have spent with the latter: Read Heads gallop, if you want an accurate account of this country:

Do you ever hear anything of F Watkins, Cameron or Matthews.— I wrote to the former many months ago, but he has never answered me.—

Direct pro futuro to Valparaiso

I have just met with the following quotation in the "*Sacred* history of the World", taken from the *Hereford*!! Journal. *November 1824.*[1] Carnations have been engrafted on Fennel & for the first two or three years the flowers were green: Likewise Peaches on a Mulberry, in which case the fruit will have a purple dye to the stone.—

Were you the original & ingenious experimentalist? I think I have heard you argue that White Lies do no harm.— Here are green Carnations & purple Peaches brought foreward to show the beneficence of Providence.— When such evidence is proved false, who will not become a Sceptic.— Reflect—, if the Author, what awful consequences may have been produced.—

American Philosophical Society 5

[1] Turner 1832 – 7, 2: 111 and n., quotes a 'Letter to the Editor' from 'Ethelbert', *Hereford Journal* . . . (24 November 1824). The quoted passage also appears in Bradley 1726, 2: 301. Herbert went to Hereford school, which explains CD's exclamation marks.

To J. S. Henslow[1] 18 July 1833

Rio de la Plata | H.M.S. Beagle
July 18th.— 1833.

My dear Henslow,

My last letter was dated on the sea.— I then expected to stop at the R. Negro in Patagonia; our domineering master, the wind, ordered otherwise; in consequence, the greater part of the winter has been passed in this river at Maldonado.— Amongst a heap of letters which awaited me, I was sadly disappointed not to see your hand-writing; for several months I had been looking forward with no little pleasure to hearing how you all are going on at Cambridge, & with a good deal of anxiety respecting the fate of my collections.— Our direction, for a long period hence, will be Valparaiso: I should be so much obliged if you would write to me:— You only know anything about my collections, & I feel as if all future satisfaction after this voyage will depend solely upon your approval. I am afraid you have thought them very scanty: but, as I have said before, you must recollect how much time is lost at sea, & that I make it a constant rule, to prefer the obscure & diminutive tribes of animals.— I have now got a servant of my own, whom I have taught to skin birds &c, so that for the future I trust, there will be rather a larger proportion of showy specimens.— We have got, almost every bird in this neighbourhead,[2] (Maldonado) about 80 in number & nearly 20 quadrupeds.— But, alas, excepting this there has not been much done.—

By the same packet, which takes this, there will come four barrells: the largest will require opening, as it contains skins, Plants &c &c, & cigar box with pill boxes: the two next in size, only Geological specimens need not be opened, without you like to see them, the smallest & flat barrell, contains fish; with a gimlet, you can easily ascertain how full it is of spirits.— Several of the pill-boxes are marked thus (X), they contain Coleoptera, & will require (as likewise the Case) airing & perhaps a little Essential oil.— This is not nearly all which I have collected this summer, but for several reasons I have deferred sending the other half.— It is useless attempting to thank you for taking charge of my collections: for as I know no other person who would; this voyage would then be useless & I would return home.—

Our future plans are, in a few days, to go to the R. Negro, to survey some banks.— I shall be put on shore: I wish we could remain there for a long time.— The geology must be very interesting— it is near the junction of the Megatherium & Patagonian cliffs.— From what I saw of the latter, in one ½ hour in St Josephs bay, they would be well worth a long examination.— above the great oyster bed, there is one of gravel, which fills up inequalities in its inferior;[3] & above this, & therefore high out of the water is one of such modern shells, that they retain their colour & emit bad smell when burnt. Patagonia must clearly have but lately risen from the water. After the Beagle returns from this short cruize, we take in 12 months provisions & in beginning of October proceed to Tierra del F.; then pass the Straits of Magellan & enter the glorious Pacific: The Beagle after proceeding to Conception or Valparaiso, will once more go Southward, (I however will not leave the warm weather) & upon her return we proceed up the coast, ultimately to cross the Pacific.— I am in great doubt whether to remain at Valparaiso or Conception: at the latter beds of Coal & shells, but at the former I could cross & recross the grand chain of the Andes.—

I am ready to bound for joy at the thoughts of leaving this stupid, unpicturesque side of America. When Tierra del F is over, it will all be Holidays. And then the very thoughts of the fine Coralls, the warm glowing weather, the blue sky of the Tropics is enough to make one wild with delight.— I am anxious to know, what has become of a large collection (I fancy ill assorted) of Geological specimens made in former voyage from Tierra del Fuego[4] I hope to see enough of this country to be able to make a rough sketch of it—& then of course specimens with localities marked on them, would be to me very valuable. Remember me most kindly to Prof. Sedgwick, perhaps he would enquire at Geolog: Soc: whether they are in existence.— Some body told me you had Volume of Dic Class: Explan: of Plates. My brother will in short time send me a parcel: by which it can come: his direction is Whyndham Club St. James Square. If you know of any book, which would be useful to me, you can mention it to him: I trust I shall find a letter (although it is a long time to look forward to) at Valparaiso: I shall be so glad to hear what you are doing.— Very often during your last Spring when the weather has been fine; I have been guessing whether it would do for Gamblingay[5] or whether at that very instant some revered Botanist was not anxiously looking at the *other* side of a fenny ditch.— The only piece of Cambridge news, which I have heard for a long time was a good one: it was that a Living has been given to you.— I hope it is true.—

Remember me most truly to Mrs Henslow & to Leonard Jennings.—[6] & believe me My dear Henslow.— Your most obliged & affectionate friend— Chas. Darwin.—

[1] The following passages from this letter were extracted by Henslow and published in the Cambridge Philosophical Society pamphlet:
>   1.3 'the greater . . . Maldonado.' 1.4
>   1.15 'We . . . quadrupeds.' 1.16
>   3.1 'in a few days . . . banks.' 3.2
>   3.3 'The geology . . . water.' 3.9; 3.7 'inferior' *changed to* 'interior'

[2] CD's idiosyncratic spelling of certain words, though not entirely consistent, continued almost to the end of the voyage. Frank J. Sulloway has tabulated the variations in CD's spelling of 'occasion', 'coral', and 'Pacific' in his letters, diaries, and other voyage documents (Sulloway 1982b) and has demonstrated the importance of the misspellings in helping to establish the time of writing of 'Ornithological notes' and other specimen catalogues as the last year of the voyage.

[3] Henslow transcribed the word 'inferior' as 'interior'; but the manuscript is clear and the meaning is confirmed by the description of the beds as 'underlying' in *South America*, pp. 5, 109.

[4] The specimens were sent to the Geological Society by Captain Philip Parker King. 'Of the specimens sent home by Capt. King from this remote quarter of the globe, it may be remarked, in general, that they agree perfectly with the rocks of Europe and other parts of the world;—the resemblance amounting, in several cases, to almost complete identity.' (*Proceedings of the Geological Society of London* 1 (1826−33): 29−31) A second collection was presented in 1831.

[5] Gamlingay, about 15 miles from Cambridge, a favourite entomological hunting ground. Leonard Jenyns described it as 'a locality so rich, from its geological position, in rare species of insects and plants not found in any other part of Cambridgeshire, and so well known to Cambridge men in later times, from the excursions yearly made there by Henslow with his botanical class' (Jenyns 1862, p. 24).

[6] CD's misspelling of Jenyns.

From Susan Darwin   22−31 July 1833

Osmaston [and Shrewsbury]
July 22^d 1833

My very dear Charles,

I hope you have not perceived the want of a letter for the Month of June, but if you did I am the one to blame as in the hurry of leaving London I quite forgot it was my turn to write. I daresay Caroline has been more vexed than you, because she did not know till too late to keep *her promise* of never letting you lose yr monthly letter.— We are all getting so impatient to hear again from you, & we have just got hopes of soon being gratified as in our paper it mentions The Beagle has been heard of on the 5th of April at the Falkland Islands & how very happy we shall be if this news brings us a letter fr you. I had determined not to expect one till September as you desired, but this paragraph in the paper has upset all my resolutions, & will make us watch the post very anxiously.—

You will see by the date of this that after vain attempts for 3 years Caroline & I at last are paying a visit at Osmaston We have been here nearly a week & are become very friendly with them all & like them exceedingly. Frances Jane is grown much stronger so that she comes down stairs & is able to sit up much more than formerly. I think she is very superior to the rest of the girls both in sense and agreeableness, besides being very handsome & interesting. Emma I like too extremely she really seems quite a perfect character of goodness. William is looking very delicate & they all seem low about him, he does not cough so I hope he is not yet consumptive. He is to spend the winter again at the I. of Wight which they say makes a gt difference in his health.—

They have all talked a gt deal about you & Julia was *boasting* the other day of her eyes being longer sighted than yours as she related the experiment that proved it, of William trying you both with some unknown book:— Mʳˢ Fox has tried hard to improve our minds with shewing us some manufactories, but so far we have escaped & as we only stay three days longer I think we shall go away as ignorant as we came.—

About a fortnight ago William was at the meeting of the Philosophers at Cambridge he said he heard yr things had arrived safe— P. Sedgwick made a most wonderfully eloquent speech upon resigning his office of President. P. Henslow could not give any of his parties because Mʳˢ H was lying in. It was a pity that idle Erasmus did not go to this meeting as Hildyard asked him to go to his house for it: he seems quite entranced by London for he has never once left it all this beautiful summer.

We were there 5 weeks exactly & did a wonderful quantity of amusement. Nancy paid the British Museum a visit & was quite delighted with beholding some stuffed animal with Capt Fitz roy's name on it as having discovered it, so that was quite nearly enough allied to *you* to make it the most interesting sight in London.— I fell in love with an Armadillo I saw at the Zoological Gardens & as they come fᵗ America I wish you wᵈ bring a pet one home, they trot along so pertinaciously, that I laugh whenever I think of it.—

Charlotte & Mʳ Langton are just come into Shropshire to take possession of their Living at Onnibury. They will be forced to build a new Parsonage house, but now they are living in a Farm house & seem very happy. it is very nice having them settled so very near us. I hope they will pay us a visit soon. I shall be curious to hear how Mʳ Langton likes his Clergymans life, for he felt shy at the thoughts of making acquaintance with his Parishioners— I am very sorry for poor William Fox being obliged to give up his Curacy at Epperstone I am sure he regrets it so much. Mʳ & Mʳˢ White his Rector have been staying here & we liked them very much they seem such excellent people.— I have been looking over some beautiful engravings of Birds by a Mʳ Selby. I suppose they have been published since you left England but I am sure you will get a copy when you see them, for they are far superior to Bewick's Birds.—[1]

I think Catherine must have told you that Fanny Biddulph has got a little Girl We saw a good deal of her when in London as she was slowly recovering & enjoyed having us to talk to very much when Mʳ Biddulph was at the House She has been very ill since her confinement & is now (altho' nearly 2 months since she lay in) too weak to travel.— It is very pretty to see how she doats upon her baby which she thinks a perfect beauty & it is christened Fanny Charlotte: Old Mʳˢ Biddulph has had a Paralytic stroke so I suppose the young ones will have Chirk Castle to themselves this Autumn.— I stayed 4 days at Woodhouse just lately. they see very little company there now & poor Caroline is rather dull: all the young ones were at home & the hay was making so we took novels, a bottle of Cider, & fruit, & spent 2 whole days upon the haycocks Francis undertaking to *hide the liquor* if the *Governor* approached.— It was high Strawberry season & Caroline Owen said *that* always put her in mind, of when you & Fanny used to *lie full length* upon the Strawberry beds grazing by the hour.— I think Mr Owen would enjoy having you to talk to not a little for he has nobody to listen to his campaigns in Flanders now you are away. They hear from Arthur very constantly but he complains very much of never receiving Letters f^t England which seems very strange, as you who are still farther off never miss any of ours.— Emma Owen is regularly come out & is a very pretty girl more like a *white* Mrs Owen than the rest.

About a month ago Papa & Caroline went a little tour into Yorkshire to see the Cathedral—& saw Liverpool with the Railway but did not go upon it: My Father enjoys these little tours very much & I hope next summer he will compass Edinburgh. His new Carriage answers very well being very light & he always takes Edward with him when he goes any distance. He still does not eat Bread at Breakfast which certainly makes his Breath much better.

Shrewsbury July 31st

The Langtons we expect to day A^t Bessy has been very dangerously ill from a severe attack of fits. Charlotte went up fr Onnibury to London hardly expecting to find her alive, but she has recovered in a wonderful manner, & Charlotte comes down to night p^r Wonder & we expect Mr Langton from his Living to meet her here.— The Biddulphs are now at Woodhouse poor Fanny & her little baby have both got the Hooping Cough. I am very glad she has left London. Papa went over to breakfast at Woodhouse yesterday in order to see some of the Hooping Cough patients as *Mr Owen's baby*[2] has it also

All our affectionate Loves to you Dearest Charley & Good bye from | yr very affecte Susan Darwin.

Nancy begs I will mention her lest you sh^d forget her.

Mr Charles Hughes has returned to England his health w^d not let him stay at Buenos Ayres any longer. He went to Canada first to see his vagabond Father. He is now with the Haycocks & is coming to dine here some day soon, when we shall hear a gt deal about you. I hope you keep quite well my dear Charles. Do your lips plague you now.

DAR 204.6.1

[1] Selby [1818—]34; Bewick 1797—1804.
[2] Sobieski Owen.

From Robert FitzRoy   24 [August 1833]

Beagle. | off M. Megatherii. [Punta Alta]

Saturday. 24th.

My dear Philos

Trusting that you are not entirely expended,—though half starved,—occasionally frozen, and at times half drowned—I wish you joy of your campaign with Gen[l.] Rosas—[1] and I do assure you that whenever the ship pitches (which is *very* often as you *well* know) I am extremely vexed to think how much *sea practice* you are losing;—and how unhappy you must feel upon the firm ground.

Your home (upon the waters) will remain at anchor near the Montem Megatherii until you return to assist in the parturition of a Megalonyx measuring seventy two feet from the end of his Snout to the tip of his tail—and an Ichthyosaurus somewhat larger than the Beagle.—

Our wise ones say that you are not enough of an Archimedes to accomplish the removal of this latter animalcule.

I have sent,— by Chaffers,—to the Commandant.— On *your* account,—and on behalf of *our* intestines,—which have a strange inclination to be interested by beef.[2]

If you have already departed for the Sierra Ventana— tanto mejor—I shall stay here,—at the old trade—"quarter-er-less four"—

Sancho goes with Chaffers in case you should require his right trusty service. Send word when *you* want a boat—*we* shall send, *once* in *four* days.

Take *your own time*—there is abundant occupation here for *all* the *Sounders,*—so we shall not growl at you when you return. | Yours very truly | Rob[t]: FitzRoy

P.S. I do not rejoice at your extraordinary and outrageous peregrinations because I am envious—jealous,—and extremely full of all uncharitableness. What will they think at home of "Master Charles" "I do think he be gone mad"— Prithee be *careful*—while there's *care* there's no *fear*—says the saw.

PS. 2[d.] (*Irish* fash)[3] Have you yet heard from Henslow—or about your collections sent to England?

DAR 204.7

[1] CD was waiting for the *Beagle* when she arrived at Bahia Blanca on 24 August. He and James Harris had ridden overland from the Rio Negro, having been given a passport and horses by General Juan Manuel de Rosas, commander of an expedition to exterminate the Indians. See *Journal of researches*, ch. 4.

[2] Edward Main Chaffers, Master of the *Beagle*, was sent by FitzRoy to obtain fresh meat.

[3] The second postscript precedes the first in the manuscript.

From J. S. Henslow 31 August 1833

Cambridge
31 Augt 1833

My dear Darwin,

I am afraid that I have been rather negligent in not writing sooner to announce the arrival of your last Cargo which came safe to hand excepting a few articles in the Cask of Spirits which are spoiled, owing to the spirit having escaped thro' the bung-hole— I am now in possession of your letter of last April, which has stirred me up to send you off a few books which I thought might interest you, & I have (or rather *shall*) write to your Brother to recommend one or two more— The fossil portions of the Megatherium turned out to be extremely interesting as serving to illustrate certain parts of the animal which the specimens formerly received in this country & in France had failed to do. Buckland & Clift exhibited them at the Geological Section[1] (what this means you will learn from the Report I send you)—[2] & I have just received a letter from Clift requesting me to forward the whole to him, that he may pick them out carefully repair them, get them figured, & return them to me with a description of what they are & how far they serve to illustrate the ostuology of the Great Beast— This I shall do in another week when I return again to Cambridge—for I am staying at present at Ely & am here merely on Saturday for L. Jenyn's duty tomorrow he having been unwell & advised not to take duty at present— I have popped the various animals that were in the Keg into fresh spirits in jars & placed them in my cellar— The more delicate things as insects, skins &c. I keep at my own house, with the precaution of putting camphor into the boxes— The plants delight me exceedingly, tho' I have not yet made them out—but with Hooker's work & help I hope to do so before long—[3] I never thought of putting your name down to a Tablet we have been erecting to poor Ramsay's memory in Jesus Chapel till lately— As the list has not yet appeared I have ventured to do so for 21/— I propose having an engraving (I think I told you) from an excellent likeness which Miss Jenyns made for me—& this I shall let the subscribers to the Tablet have at whatever the cost price may be, about 10/ or 12/— probably— I am sure from your respect for R's memory I have not done wrong in putting down your name— The comet you speak of is *expected* in 1835, according to calculation—but it seems very doubtful whether the calculation is correct— The papers of course talk nonsense about it, but it is really something out of the ordinary cometical occurrences—[4] Mrs Henslow produced me a fine girl on June 23, the day before the Association met— It proved quite a breeding week with the Cambridge Ladies Mrs Clark & Mrs Willis[5] being confined within a day or two of the same time— I long as much as you do to see the day when we shall be discussing the various events of your voyage together, but I hope also that there is much yet to arrive before you bend your way home again. Not but what I w^d have you return immediately if you are really tired out—but you remember how we used to talk of the certainty of many an annoyance that must arrive, & many a wish to be home again— If you propose returning before the whole period of the voyage expires, don't make up your mind in a hurry—but let it be a steady thought for at least a month without one single desire to continue—& if such an event should occur you may fairly conclude that you are sick of the expedition—but I suspect you will always find something to keep up your courage— Send home every scrap of Megatherium skull you can set your eyes upon.—& *all* fossils. Use your sweeping net well for I foresee that your minute insects will nearly all turn out new— (I must write on now to the end as I have transgressed the limits)—[6] I have turned Entomologist myself this summer for my little girls who have started a collection of Insects & Shells—& make me work for them— Poor Stephens has just lost 400£ in a Law suit & we are levying a subscription to help him on with his Illustrations— I delight in your descriptions of the few animals you now & then allude to—

    Believe me | affect.ʸ yʳˢ | J. S. Henslow

DAR 97(ser. 2): 14—15

[1] William Buckland was deputy chairman of the Geological Section of the 1833 British Association meeting; William Clift was Curator of the Hunterian Museum of the Royal College of Surgeons. The *Report* of the 1833 meeting makes no mention of CD's *Megatherium* bones.

[2] The *Report* of the British Association meeting of 1832 at Oxford. In 1832 Buckland addressed the General Meeting on the *Megatherium* bones found by Woodbine Parish (see Buckland 1832).

[3] William Jackson Hooker, then Regius Professor of Botany at Glasgow. Henslow described only some specimens collected later in the voyage: new species of Galápagos cacti (Henslow 1837) and the plants from the Keeling Islands (Henslow 1838). It was left to Joseph Dalton Hooker (William Jackson Hooker's son) to describe the collections from the Falkland Islands and Tierra del Fuego (Hooker 1844—7) and later the Galápagos flora (Hooker 1846a and 1846b). Most of the collection, however, remained unclassified until recently when Duncan M. Porter undertook to study it anew. For his account of the history of the work on CD's botanical specimens and the rediscovery of his notes, see Porter 1980a, 1981, and 1982.

[4] Halley's Comet appeared, as predicted, in 1835.

[5] Mrs William Clark and Mrs Robert Willis.

[6] Henslow had written on the part of the cover reserved for the address.

From Caroline Darwin 1 September 1833

Shrewsbury
September 1ˢᵗ | 1833.

My dear Charles

    We were *exceedingly* glad to receive your letter last week from Rio Negro dated 12 April and to hear you were well and happy I can easily think how impatient you must feel to be again in those delightful tropical climates of which I suppose one has very little idea from mere description. Do take care of yourself my dear Charles you were so apt at home to over exert yourself that we are all afraid when ever we read of your enjoying yourself. I will begin my letter with the very best news I can give. I mean my Father being *very well* & in good spirits and looking really much better than he did last year— Susan wrote last from Osmaston & she will have told you how thin & unwell poor William Fox was looking. he often talked about you and with great interest— seeing him put me so very much in mind of the pleasant rides I had with you and him when at Shrewsbury. it brought your voice & laugh so vividly before me & how I longed to hear it in reality— the eldest sister Eliza is I think the greatest bore I ever beheld & the poor lame girl Emma, the most perfect model of what a person should be,—I was much charmed by her, it is delightful I think to see a person so *very very* pleasing from goodness & religion acting upon every feeling as it does in her. We staid at Osmaston about 10 days & when we came back Erasmus came home & has been with us ever since—he is very nice & agreeable & we enjoy having him very much. I think he will go no expedition this summer but return to London when he leaves us. Mrs Evans of Portrane wanted to persuade him to go with her into Scotland, & I believe he would if he had not discovered there were to be some young ladies of the party who he disliked. he is very constant to Mrs Hensleigh Wedgwood & thinks her the nicest of women— She & Hensleigh were staying here last week with their baby who they are very fond of, & Erasmus with all his horror of babies plays with the little thing & watches it for any length of time—

Poor Aunt Bessy has been very dangerously ill—she has fits & in one of the last fell down & has lamed herself so that she cannot walk at all they had a tedious journey of 7 days from London with her, but she is at last safe arrived at Maer—

We have also had a little visit of 2 days from the Biddulphs. poor Fanny is very unwell with violent pains in her head at night & a good deal of fever— She has never been well a day since her confinement—& is looking so white, thin & refined you would scarcely know her, but very sweet & charming. Mr Biddulph seems very much in love & fond of her which is the best thing about him I think, though I like him far better than I thought I should before I knew him. he is clever and rather amusing occasionally. Old M^{rs} & Miss Biddulph are c⟨om⟩ing down to stay at Chirk Castle which ⟨is⟩ a great sh⟨o⟩ck to Fanny as it very much destroys he⟨r⟩ comfort there. I do not think she is at all altered by her marriage just as affectionate & unaffected as she used to be— I never see any of the Owens without them asking about you with the greatest kindness & interest Col. Mrs Leighton & Clare dined here to meet the Biddulphs one day & Clare was such a contrast to Fanny—

Mr George Maddocks has married his maid servant & since that has been flighty, so much so that he was put in restraint for a week— his madness first shewed itself by his stabbing 2 horses after driving them 60 miles because they could go no farther— I tell you all the gossip I can that you may know how the Shropshire world is going on— Susan's present hobby is work, as it was when you went—she is now doing a magnificent bunch of flowers in an enormous frame. *My* hobby is a new Infant School now finished & the children & Governess all properly established in it & Catherine has a little drawing rage, not pots & pans but old men and women which are certainly better than the former

We expect Marianne here tomorrow— she has your letter which I am sorry for as it is always pleasant to have a letter one is answering to refer to—tho' I wish it had brought the excellent news of the prospect of any time to expect you home. I am excessively glad it has answered to you so well & now two years of the dangers are over I am thankful to think & the Schooner must add to your safety

the draft for £70 was p^d on 31st July & my Father says you were quite right to mention in your letter when you draw for money. My Father sends his best love & so does Eras & we all dearest Charles | Ever y^rs affec^tely C S D.

Catherine has just been telling me how prettily & coquettishly Fanny Biddulph asked after you saying, "Has Charles quite forgotten me?" "does he ever mention me in his letters?" *I* have not at all forgotten our old Postillion & Housemaid days. Cath says she looked beautiful when saying all this—am I very immoral in repeating it?

DAR 204.6.1

To Caroline Darwin 20 September [1833]

Buenos Ayres.—
September 20th.

⟨My⟩ dear Caroline

I have just returned from a grand expedition; As a merchant vessel sails tomorrow for Liverpool, I will write, as much as I can before I go to bed.— The Beagle after leaving Maldonado, sailed for the R. Negro.— When ⟨   ⟩ I determined to go by land to Bahia Blanca ⟨&⟩ wait for the vessel; & subsequently having heard that the country was tolerably safe, I proceeded on to this city.—

It is a long journey between 500 & 600 miles, through a district, till very lately never penetrated except by the Indians & never by an Englishman.— There is now a bloody war of extermination against the Indians. The Christian army is encamped on the R. Colorado. in the progress, a few months since, from B. Ayres.—General Rosas left at every 10 or 15 leagues, 5 soldiers & a troop of horses.— When I was at the Colorado the General gave me an order for these horses.—¹ so fine an opportunity for Geology was not to be neglected, so that I determined to start at all hazards.— The horses &c were all gratis. My only expence (about 20£) was hiring a trusty companion.— I am become quite a Gaucho, drink my Mattee & smoke my cigar, & then lie down & sleep as comfortably with the Heavens for a Canopy as in a feather bed.— It is such a fine healthy life, on horse back all day, eating nothing but meat, & sleeping in a bracing air, one awakes as fresh as a lark.— From R. Negro to the Colorado, it is a dreary uninhabited camp with only two brackish springs.— from the latter place to B Blanca there are the *Postas*.— From Bahia Blanca to the Rio Salado the Postas are irregular, & excepting these, there is not an habitation.— There is sometimes a hovel & sometimes not & the soldiers live entirely on deer & Ostriches.— The wildness & novelty of this journey gave it great interest to me & the danger is not nearly so great as it appears, for the Indians are now all collecting in the Cordilleras for a great battle this summer.— I stopped two days to examine the Sierra de Ventana, a curious mountain which rises in the vast plain.— the ascent was excessively fatiguing, & there was but little to reward one for the trouble.— The plain merely resembles a sea, without its beautiful colour.—

At the Guardia del Monte, I found some more of the armour of the giant Megatherium, which was to me very interesting, as connecting the Geology of the different parts of the Pampas.— I likewise at Bahia Blanca found some more bones more perfect than those I formerly found, indeed one is nearly an entire skeleton.—[2]

The Beagle is now at Monte Video or Maldonado— I received a letter from the Captain, enclosing one from Catherine dated London May 29th.— As I have not my letter-case here I cannot say whether I received the April one.— I shall soon be on horse back again; there is a river to the North (the Carcaraña) the banks of which are so thickly strewed with great bones, that they build part of the Corrall with them.— Every person has observed them, so they must be very numerous.— I shall then return to M. Video & join the Beagle.— At the latter end of next month she sails for the Straits of Magellan & likewise pays the Falkland islands another visit.—

I am now living in the house of a most hospitable English merchant.—[3] it appears quite strange writing in an English furnished room, & still more strange to see a lady making tea.— I shall be obliged to draw rather largely for money.— I do it with more confidence, as I know for certain, after leaving the Plata, there will be 5 or 6 months of Southern economy.— I cannot at present say exactly what sum.— Travelling is very cheap in this country; the only expence is procuring a trusty companion, but in that depends your safety, for a more throat-cutting gentry do not exist than these Gauchos on the face of the world.— It is now the Spring of the year, & every thing is budding & fresh: but how great a difference between this & the beautiful scenes of England.— I often think of the Garden at home as a Paradise; on a fine summers evening, when the birds are singing how I should enjoy to appear, like a Ghost amongst you, whilst working with the flowers.— These are pleasures I have to view, through the long interval of the Pacific & Indian oceans.—

Good bye, God bless you all.— My dear Caroline, when shall we have a ride together. Yours most affectionately, Chas. Darwin. Give my very best love to my Father.—

DAR 223

[1] For CD's accounts of his meeting with General Rosas see '*Beagle*' *diary*, pp. 163—4, and *Journal of researches*, pp. 85—7.

[2] The nearly entire skeleton was of the *Scelidotherium*, a giant ground sloth (see *Fossil Mammalia*, pp. 73—99). In *South America*, pp. 84—5, CD discusses the fossil mammalia he found at Punta Alta, as later identified by Richard Owen.

[3] Edward Lumb, British merchant at Buenos Aires.

To J. S. Henslow [20—7] September 1833

Buenos Ayres
September 1833

My dear Henslow

A Spanish friend in Entre Rios has promised to send me a cargo of Bones; if they do arrive here: M[r] Lumb has kindly offered to forward them to you.— I leave this as a direction to him, & he will add the name of Ship, date, port &c or whatever is necessary.—

Believe me Yours most truly obliged | Charles Darwin

The Rev:[d] Professor Henslow
Cambridge University
England.—
Specimens of Natural History[1]

Royal Botanic Gardens, Kew

[1] CD left this letter with Edward Lumb to be forwarded with the cargo of bones to which it refers. The letter was not despatched until 2 May 1834, when Lumb wrote to Henslow telling him that a case of specimens with part of the head of a 'Megatherium' was being forwarded to him. That shipment, however, was not the one referred to in this letter, but another, containing a specimen found later in 1833. From Lumb's letter of 8 May 1834, it appears that the 'Spanish friend' is 'M[r] Hooker'. The cargo of bones has not been traced, though Lumb expected it to arrive soon after 2 May 1834.

From Catherine Darwin 27 September 1833

Shrewsbury.
September 27th 1833

My dear Charles,

The last letter we received from you was that dated April 12th, from your first landing at Rio Negro. We rec[d] it in August to our great pleasure, and Caroline has written to you since, for the 1st of this month, but she was in some doubt whether her letter went in time.

Erasmus was with us, when your letter arrived, as well as the Hensleigh Wedgwoods, and we were all very much interested by it. I am very glad indeed to hear of Capt Fitzroy's having bought the Schooner, it is a capital thing indeed. I cannot help being rather grieved when you speak so rapturously of the Tropics, as I am afraid it is a still stronger sign, how very long it will be, before we shall have you again, and I have great fears how far you will stand the quiet clerical life you used to say you would return to. Every body, who has heard of your beating about 23 days near Cape Horn, gives you unfeigned pity for it.— I saw Tom Eyton lately, who enquired much about you; he has been in Wales great part of the Summer he says, collecting; there has been a report that he has been paying great attention to Miss Slaney, and has been very much with the Slaneys, but how far that is true, I do not know. It seems very improbable that he should have the love of money so strong, as one can scarcely believe it to be any thing else.—

M^r & Mrs Henry Hope (Louisa Leighton that was) are gone a Tour up the Rhine & to Paris. The old Mrs Hope will not die, though the Leightons have been certain she could not live a month since the beginning of the year. Louisa must find it intolerable to live with the nasty old woman great part of her time.— I think Caroline wrote to you after the Biddulphs had been staying with us. They are gone to Chirk Castle now, and the detestable old M^{rs} & Miss Biddulph are come there, whom poor Fanny perfectly hates, & who are intolerably proud and disagreeable to her.— Poor Fanny is still very unwell; you would hardly know her, she is so changed, so delicate, pale, & thin; but so very charming and affectionate, quite like her dear old self. I wish you could have seen her pretty look, (when she was talking about you,) she turned to me, and said "I suppose he never mentions me"; with all her sweet old manner.— She talked a good deal about you very affectionately & warmly, & said how much she wished to see you again, & how very much she wished for your happiness.— Susan has been staying twice at Chirk, before the old Mrs Biddulph came there, and says Mr Biddulph appears to be as much in love with Fanny as possible, so that I hope he is worthy of her, in that respect.— Emma Owen is come out, & is exceedingly pretty; quite one of the prettiest of them. Francis is living idle at Woodhouse, and Mr Owen cannot tell at all what in the world to do with him. Mrs Williams is come down to Eaton— She has no children.— I must now tell you about your other old friend, Charlotte. We have all been staying with her at Onnibury, Mr Langton's living, 4 miles from Ludlow. It is an uncommonly pretty country about, & we had a very pleasant visit there.— We took your last letter to Charlotte to read to her, and she was much amused by it.— The present Parsonage house is such a miserable ruinous concern, that Mr Langton is going to build in a very pretty situation close. This is thought rather foolish of him, as he is very poor, but it would certainly be very provoking to spend the necessary 200 or 300£ ⟨on⟩ their present old affair.— Charlotte seems extremely happy, very full of scrattles & household cares, (so unlike her.) Mr Langton is rather a talking, visiting, chattering man; whom no one would ever expect Charlotte to have fallen in love with; but so it is. The worst thing about him is to my mind, that he governs most absolutely in all little trifling concerns, as well as in great matters. Susan attributes this to his having been one year on board Ship when a boy, seeing absolute authority; if this is the case, what will become of your poor wife, after so many years apprenticeship in the art of governing? She ought to be apprized of Susan's alarming theory.— Poor Charlotte has no brats, which I imagine to be a sore subject; however in spite of all she seems extremely happy, & very agreeable & nice, as of old.— Erasmus paid us a much longer visit than usual this summer; he went to Maer for a week, which he enjoyed extremely, and was very happy there, with Fanny Hensleigh, her Baby, *Miss Snow*, as it is called (short for Snowdrop) and Emma Wedgwood; all his favourites around him.—

I am afraid Erasmus is too idle to write to you; which is very naughty of him.—

Papa is very well, and is planning another little Journey, in the South of England to see the Cathedrals of Winchester, & Salisbury. Travelling does him a great deal of good.— He sends his most affectionate love to you, & with all our best of loves, believe me | ever, dearest Charles | Y^r very affecte Sister | E. Catherine Darwin

DAR 204.6.1

From Robert FitzRoy 4 October 1833

Beagle. Monte Video
4th October 1833

My dear Darwin

Two hours since, I received your epistle, dated 26th and most punctually and immediately am I about to answer your queries. (mirabile!!)

But firstly of the first—my good Philos why have you told me nothing of your hairbreadth scapes & moving accidents How many times did you flee from the Indians? How many precipices did you fall over? How many bogs did you fall into?— How often were you carried away by the floods? and how many times were you kilt?— that you were not kilt *dead* I have visible evidence in your handwriting,—as well as in a columnar paragraph in Mr Love's unamiable paper. You did not tell me whether you received the blank papers safely—you informal homo—how am I to feel certain that I have not signed what may blast my *immaculate* reputation? Harris[1] carried the Packet which contained them and promised to deliver them faithfully. How Sancho by Mr Hood's assistance,[2] contrived so to mismanage as to reach B.ˢ Ayres some days after Harris—Quien sabe? In it were 5 "Skimpy" lines, as Capt Beaufort would call them & a promise of better behaviour. Since the date of that note the Beagle has been two days at Maldonado—one day here and about a week between this & Cape Corrientes.— Not having any Stone pounders on board—nor any qualified person (the *Mate* being absent)—I could not think of landing,—so *you* have yet a *chance*,—"de verus" (it *blew* strong & prevented landing.) I believe you have heard from M.ʳ Parry and are aware of his loss.— If you have not heard from him—your *ally* (!! of bone stealing fame) will have informed you. Shocking as it was to him, and his family, but to him, most particularly,—I am in hopes that better times will be found by our good friend Parry,—in consequence of his being a single man.— Warm hearted and friendly as she was—and friendly to the utmost extent of her means—she had her share of woman's weakness and woman's failings. Robert Parry is gone to England in the "Mary Worrall", Merchant man,—to be placed at a school,— the young daughters are going to B.ˢ Ayres—also to school— M.ʳ P. intends to give up his house and turn "bachelor, in lodgings",— —a *wise* resolve, though painful indeed to the Father of a family,—think what a change in a domestic circle.

If M.ʳ P. has written as he intended you have heard of M.ʳ Martens³—Earle's Successor,—a *stone pounding artist*—who exclaims *in his sleep* "*think* of *me* standing upon a pinnacle of the Andes—or sketching a Fuegian Glacier!!!" By my faith in Bumpology, I am sure you will like him, and like him *much*—he is—or I am wofully mistaken—a "rara avis in navibus,— Carlo que Simillima Darwin".— Don't be jealous now for I only put in the last bit to make the line scan— you know very well your degree is "rarissima" and that *your* line runs thus— Est avis in navibus Carlos rarissima Darwin.— but you will think I am cracked so seriatim he is a gentlemanlike, well informed man.— his landscapes are *really* good (compared with London men) though perhaps in *figures* he cannot equal Earle— He is very industrious— and gentlemanlike in his *habits*,—(not a *small* recommendation).

Wickham gets on famously—really the "Lighter" will not merit *trifling* considerations— Mʳ Kent of the Pylades is at Gorriti—belonging to our Squad. We have plenty of men,—and *good* ones; and all is prospering— —

"*Well, but the conjunctions—the conjunctions*" I hear you saying—"*you have got to the end of a sheet of paper without telling me one thing that I wanted to know*".—

—This is the 4.ᵗʰ of October,—"*so the date of your letter tells me*"— —well— hum— —if—hum—but—we must consider—then—hum—tomorrow will be the sixth— "*Prodigious*"!! Do you know what I mean—"*to be sure*" so—and so & so—& hum hum hum & off goes the head!!—

I never will write another letter after tea—that green beverage makes one tipsy—besides it is such a luxury feeling that your epistle is not to go across the wide atlantick—and has only to cross the muddy Plata. It is so awful writing to a person thousands of miles off—when your conscience reproaches you with having been extremely negligent and tells you that six or eight or (oh—how awful) twelve months' "*History*" is due to your expectant and irate correspondent.

Still *you* get no *answer*— "*what is the Beagle going to do—will you tell me, or not?*"—

Philos—be not irate—have patience and I will tell thee all.

Tomorrow we shall sail, for Maldonado—there we shall remain until the middle of this month;—thence we shall return to Monte Video—to remain quietly, *if possible*, until the end of the month.— I will try all I can to get away from the River Plate the first week in November but there is much to do—and I shall not be surprised if we are detained even until the middle of November.— However—weather is of such consequence, that every long day gained will tell heavily—and I hope & will try hard to be off *Early* in November—[4] therefore do not delay your arrival *here* later than the *first few* days of November, at the *farthest*.

You say nothing about the "Journal of the expedition up the Rio Negro"—nor have you sent me the map of the province of Buenos Ayres— I pray you to *do the latter*—right speedily—and enquire about the *former*—from M.r Gore as well as the other man whose name I forget (Señor—Don—or Colonel Something, or somebody.)—but in writing to M.r Gore I mentioned it—so he will know it.—[5] I wish to compare the map with our charts—previous to sending them away—in order to "connive" a little; as your *friend* M.r Bathurst says.

Roberts[6] (of the Liebre) passed our bows *this morning* on board of the "*Paz*" bound to Rio Negro with a cargo of *tobacco*. he did not honor us with a visit—nor did he ask for Chico—respecting the former, he was somewhat rude, and as to the latter rather wise I think.—

Adios Philos—Ever very faithfully yours. Rob.t FitzRoy

DAR 204.7

[1] James Harris.
[2] Thomas Samuel Hood, British Consul-General in Montevideo (see *Narrative* 2: 293–4).
[3] Conrad Martens, who replaced Augustus Earle as draughtsman. Earle's poor health had forced him to leave the *Beagle* in August.
[4] The actual date was 6 December 1833. The delay was caused by the need to complete the charts of the surveys made by the *Liebre* and *Paz* ('*Beagle*' diary, pp. 191, 200).
[5] Philip Yorke Gore, Chargé d'Affaires in Buenos Aires.
[6] Mr Roberts was the pilot on board the *Liebre* (*Narrative* 2: 110).

From Susan Darwin 15 October 1833

Shrewsbury
Oct 15.th | 1833.

My dear Charles.

On October the 11.th we got your very nice letter dated Maldonado July

14ᵗʰ & two days afterwards your Journal arrived which I have only read a few pages of yet: but we mean to read it aloud comfortably in the Evenings— I think Papa particularly enjoyed your last letter & desires me to tell you most affectionately from him that he is *exceedingly* glad you have engaged a Servant as he is sure it will conduce very much to your comfort & only regrets you had not one sooner.— Pray tell us next time what countryman he is? if he had been a Negro, you wᵈ have said so I am sure: when you were praising their character— I have copied out what you say about the ill conduct of the man appointed to prevent Slaves Landing at Rio & I shall shew it Aᵗ Sarah who will I daresay take notice of it.— We have sent on yr Letter to Erasmus who will do yr London commissions & we will send up the two Books from here[1] with the Shoes fr Howell We shall be very anxious to hear that you get all your things safe, & Capt Beaufort is so civil that I have no doubt he will undertake to manage it.— I grudge very much for your sake, all this tiresome winter you are cruizing on this side of America as it appears so uninteresting that it is so much lost time to you,—but perhaps it will make you still more enjoy the warm climates if *that is possible.*—

Sarah who is now down at Eaton got your letter last week and I cannot tell you how much pleased & surprised she was to receive it. I think she is looking very well now & every body tells her she is grown "uncommon stout". She & Mʳ Williams seem most perfectly happy & that match certainly answered perfectly. Mʳ Owen was here yesterday looking very brisk, he desired me to ask if you had ever received his second letter.— Catherine bids me mention her last September letter was directed to Monte Video & she is afraid you won't get it.— Your message to Charlotte shall be given: we have been spending a few days at Onibury Mʳ Langtons Living: & I never did see so charming a country to live in: such lovely retired walks by the side of the river Onny. Mʳ Langton seems most extremely happy & finds himself excellent friends with his Parishioners I quite long for you to be settled in just the same kind of manner my dear Charley: I am sure I shall pitch my tent very near you in that case.— In two days time Papa, Harry, Edward, & I, are to set forth upon a small Architectural Tour to see the Cathedrals of Gloster, Winchester & Salisbury, & finish by spending a few days at the Hill where Harry will meet Jessie— It is unlucky that it is so late in the year as the 20ᵗʰ of Oct but Harry's business prevented our starting sooner we mean to take a pack of Cards and play Whist with Dumby[2] in order to pass the long Evenings at the Inns.—

I have just been reading an account of Ceylon in a kind of novel called "Cinnamon & Pearls"[3] the descriptions of the vegetation are so beautiful that I don't wonder you have a great desire to go there as of course you have read some more faithful history of it in Humboldt.— I have a much greater wish to see some tropical country than the old common place France & Italy—and I wonder people don't travel more to Madeira than sticking to Europe.— You will rejoice as much as we do over Slavery being abolished, but it is a pity the

Apprenticeship does not commence till next August as that is a great while for the poor Slaves to be at the mercy of the Planters who I sh^d. think w^d. treat them worse than ever.— I grudge too very much the 20 million compensation money: but perhaps it would never have been settled without this sum.—[4] The Poor Laws in Ireland will I suppose next Session be the great topic of interest. I have been reading some pamphlets which make me rather against the system.—

By the parcel that is going to you I think my Sisters will both write little notes: & I take the opportunity of sending you a *leetle* purse which I have been netting it is rather of the smallest so I hope your foreign coin is not very large,—& I know in old times you always used to tie up yr purse if it was of an ordinary length.

You will find Fanny Biddulph quite as charming as ever when you come back only more delicate looking. I have paid her 2 little visits at Chirk Castle this Autumn whilst old Madam & Miss Biddulph were absent, & you may suppose we were very merry in the old Castle without them. Fanny was exceedingly busy fitting up a few rooms she was to have separate for her use & so much in earnest over it that she actually went down on her knees to nail the Carpet whilst the carpenter went to his Dinner.— She has got a most lovely Poney carriage with two beautiful Grey ponies & used to take me delightful drives close to Llangollen which is only 6 miles distant fr. Chirk.— Caroline & William Owen were staying there with me the latter always talks much about you. he has become a great fisherman & wears a most absurd sailor's coat without tails.—

The present great friends of the Owens are the Boughey family who are come to live at Bicton near here— The Girls are handsome & uncommonly nice & unaffected very much of the Owen *genus*. Anastasia is quite a beautiful Girl.—

How fortunate we have been in never losing any of yr letters & I hope in future we shall have no such gaps as we had this Summer Nancy & Edward both partake of the *sensation* when a Letter comes from you & your absence does not make you at all less loved I can promise you by every one of us & all who know you. God bless you my dearest old fellow & Believe me always yr most affecte *Granny* | Susan Darwin.

Papa & All send their very best Loves to you. I wonder how we shall ⟨ev⟩er talk over every thing when you come bac⟨k.⟩ We read yr Letter to Uncle Jos & he remarked it was "very clear Capt Fitzroy must consider yr collections valuable or he could not apply to the Admiralty for leave to have yr Servant *on the Boards.*"

It is very delightful your Natural History pursuits answering so very well they must be such a never ending source of pleasure.

DAR 204.6.1

[1] Fleming 1822 and Pennant 1793.
[2] 'At *Whist*, An imaginary player represented by an exposed "hand" ' (*OED*).

[3] Martineau 1833.

[4] The bill abolishing slavery in the colonies became law on 28 August 1833. A system of apprenticeship for seven years, to serve as a transition to complete freedom, was established. £20,000,000 compensation was voted for the planters.

From Sarah Williams   21 October 1833

Eaton.

Monday 21ˢᵗ October | 1833.

I cannot tell you, my dear Charles, how *very glad* I was to see your handwriting once again, your nice long letter reached me about 10 days ago, & two or three days before *your Family* received any of your despatches. I was very proud to be able, *condescendingly* to assure them that their Brother was quite well, & had written to *me*— What a long time my letter took to travel to you, that *Valparaiso Post* is certainly a very slow one, & though you see I lose no time in answering your letter, how many months must elapse before it reaches you!!— I do not like to think of your distance from all of us, I will only look forward to your return, which according to your promises & my calculations cannot be delayed more than two years from this time, & about December *1835*, I shall expect every rap at the door of Nº. 1. Belgrave Street, will produce the celebrated South American Traveller & Naturalist, & that he will graciously condescend to dine; & then accompany the poor ignorant natives to the Play— You are very kind to make so many enquiries after my health, the Owen Constitution has at last shewn itself worthy of its former reputation, & I am now quite myself again, but indeed till within the last three months, I have never known what it was to be quite well. I have at last turned over a new leaf, & hope never to relapse into my old ways again— I wish I could say as much for Fanny, who is very far from well. She suffers sadly from Ague in her head, which she has had for many months, but as she is now under your Father's care, I have no doubt he will soon set her on her legs again— She is now at Chirk—where she leads but a melancholy life with her old Mother in Law— We remained in London this year till the beginning of September, & are now comfortably settled at Eaton till (I hope) the end of January. Emma was with me all the Year in London; she is now a Young Lady, *full fledged* so grown & improved, you would hardly know her— Woodhouse certainly is an altered place, though Caroline exerts herself wonderfully & has gained much spirit since *she came to her title*. Francis is still at home, waiting for a commission, & Charles & Henry at school at Mʳ Burd's— Baby is grown almost a big Girl, & quite the Governor's *right* hand. I do not know what would become of him without her, & she completely manages the Family, though she is wonderfully good, & does not presume on her influence— Of course Susan & Cath— have written you a circumstantial account of their London expedition this Spring with Mʳ & Mʳˢ Harry Wedgwood. Catherine's passion for *Pasta* still continues in full-force, & as Harry

Wedgwood said, they all agreed to act upon the *intensely selfish principle*, & each go their own way. I don't think they spent more than two Evenings at home during their stay in Town, & enjoyed every thing exceedingly— Has Susan let you into any of her Secrets respecting M!̣ Panting, who of course you know, she has certainly behaved very cruelly to him, in spite of all we *ventured* to say to her, for we thought him a very nice person, & I have no doubt she would have been very happy, but now it seems all at an end, & "bygones must be bygones"— if Susan has not told you anything herself, pray do not mention my having done so, but I cannot resist mentioning it to you, as you remember we have held many *confidential* conversations together, in our walks in the Forest, & *scrawls on the wall*— I very very often think of you, & the merry days we have passed together, & when you return, you will find *Sarah Owen unchanged* I assure you, though Catherine says "it's wonderful what excellent Wives those Owen's make"— I am as happy as possible, & am convinced that as long as I live, I shall never have reason to repent the *rash* step I took on the 22^d November 1831—now nearly 2 years ago— I have a "very proper influence", (which we used to talk of) though I think you will most likely hear me pronounce Shrewsbury like the *e* in Shr*e*w, this we settled was to be the criterion of my proper influence, & I own I have not shewn the Owen spirit in this one instance— Of course you have heard of Louisa Leighton's marriage, the Quarry Place party seem much pleased with it, but as you know I never *did* admire either of "those fond Hopes" tho' I think Louisa's better than the Beetle Hunter. the happy pair are now wintering in Italy, & I heard a very good & *true* story of them the other day— They were travelling in Germany, & arrived at a Town, which they walked about all day, & at last sat down to dinner at the Table d'Hote— in the course of conversation, M!̣ Hope observed to his next neighbour that he "hoped to get to Heidelburg the following day," when his Friend exclaimed in astonishment, "Why, you are *now there*," & it seems they had inspected the whole Town, without discovering *where* they were— This does not speak much for the *brightness* of the *Hopes*— Clare is very flourishing, I hear of no match for her— That "*rising star* of Ton" Matty Cotton is to astonish the County by her appearance at the Hunt. We are at present *at Daggers drawn* with her, as she chose lately *to abuse* all of us shamefully, & tell dreadful stories. I think the *Feud* will never end, though D!̣ Darwin strongly recommends a *Truce* being concluded for 2 years— I have not rode much lately, but I have a very nice horse of my own— I have taken violently to gardening, & work very hard in the Garden here, which is much improved & looks very pretty indeed— Do pray write to me again; my dear Charles. I am sure you would do so *by return of Post*, if you knew the pleasure it gives me to hear from you—always direct to Belgrave S!̣ as even if I am not there, the letter will be forwarded to me— If you capture any *Poll Parrots* or other bird or beast, you may forward them to me, even the *smallest monkey* will be thankfully received, but I suppose you despise such common things, & do not admit them into the Darwin Museum— I laughed much at your description of your long beard &c, but I think you seem tolerably comfortable, & I hope & trust you have never repented your voyage, or leaving the Land of your Fathers— I am going to send this dull effusion to Catherine to enclose in a packet she tells me she is to despatch to you tomorrow, I wish I could have sent you a more entertaining letter, but what can you expect from a stupid *old married* Woman like myself— I shall live in hopes of hearing again from you. Edward tells me to send his best remembrances to you, he wonders what I can find to tell you, to spin out so long a letter— God bless you, my dear Charles, I hope this may find you as well & happy as it leaves me. Believe me always your affectionate *old* Friend | Sarah—

I will write whenever I hear of an opportunity. I hope you will *do likewise*—

DAR 204.4

From Fanny Myddelton Biddulph [*c.* 21 October 1833]

[Chirk Castle]

My dear Charles,

Catherine writes to tell me she is going to send a packet to you— I cannot therefore lose the opportunity of adding my little scrawl— I do flatter myself you have not quite forgotten me altho' I am become a *steady stupid old matron*, the pleasant times we have passed together in the good old times of the *Housemaid* & *Postillion* at the Forest, *I* shall not easily forget, & steady & dull as I am I assure you I still look forward to some pleasant times again with you when you are cured of your *roving* turn, and settle quietly with the *little Wife* in the *little Parsonage*!! News I have little to tell you, innumerable letters you will receive with this from more able Pen's than mine— I have been laid up for the last 5 months, one illness after another, has pulled me down very much— Your good Father I think has now nearly set me up, for I am much better this last month & hope soon to be quite off the Invalid list which does not suit me at all— My *little Daughter* (how odd it sounds) is 6 months old, & a nice little creature she is— How I should like to see you here my dear Charles, I long for you to return, & I hope this next year will certainly bring you back— I have seen none of your letters, & know little or nothing of your proceedings. they tell me now & then where you are but I never can collect any particulars of your adventures— Susan has paid me two visits here & we were very merry, but now my Mother & Sister in Law have taken up their quarters for the Winter & as they are rather of a *serious* or *kill joy* turn of mind, they put a stop a little to our merrymakings— I like this place very much, & it is a delightful distance from the old Forest. I often pay it a morning visit— Emma is now a young Lady on her promotion, Matty Cotton is also turn'd loose a *blazing meteor*— Clare Leighton, now Miss Leighton much the same as when you left her, I think there is little new or amusing in Shropshire— Francis has just got an Ensigncy in the 63rd Regiment which is on its way to Madras but I believe he will go to the Depot in Ireland for the first year or two, & as the Reg! has not long to remain abroad he will only have a short time of India— Arthur is doing *very well* there—

How prosy I am my dear Charles I am really ashamed of sending such a dull little effusion feel more than half inclined to throw it into the fire—but my hope that you will take the *will for the deed* when you read it encourages me to let it go— I shall be very glad to hear from you but do not expect it I know you have enough to do without writing to me

Adieu my dear Charles. Believe me always | Yours most | affect^ly | F. Myddelton Biddulph

DAR 204.4

To Caroline Darwin 23 [October 1833]

Buenos Ayres
September 23^d—[1]

My dear Caroline

A vessel will sail in an hours time to Liverpool, & I will write as much as I can.— I have just returned from an adventurous tour.— I think I mentioned my intention of starting to the Northern parts of this Province.— I by chance procured Capt. Head's peon & arrived after a rapid gallop at St. Fe about 300 miles to the North— it was an interesting ride & good opportunity of seeing the real sea-like Pampas. at St Fe I was most unfortunately, rather unwell, so as to be unable to ride.— I crossed over to the Bajada the Capital of Entre Rios & there staid some days, but finding so much time lost I was obliged to embark on board a vessel down the Parana— This immense river, with its islands full of Tigers & Capinchos,[2] is so very great, as to appear only like an oblong lake.— When we arrived near Buenos Ayres, I left the vessel with the intention of riding into town.— The minute I landed I was almost a prisoner, for the city is closely blockaded by a furious cut-throat set of rebels.— By riding about (at a ruinous expence) amongst the different generals, I at last obtained leave to go on foot without passport into the city: I was thus obliged to leave my Peon & luggage behind; but I may thank kind providence I am here with an entire throat.— Such a set of misfortunes I have had this month, never before happened to poor mortal. My servant (Covington by name & most invaluable I find him) was sent to the Estancia of the Merchants whose house I am staying in.— he the other day nearly lost his life in a quicksand & my gun completely.—

We now here the house is ransacked (& probably his clothes all stolen!) Communication with the country is absolutely cut off, he cannot come into town, & the Beagle before long sails to the South.— Here is a pretty series of misfortunes, & there are plenty of smaller ones to fill up the gaps.—

I drew a bill a month ago for 80£. I am very sorry to say I shall be obliged from these great unexpected misfortunes to draw another one.— After my Fathers first great growl is over, he must recollect we shall be now 8 months to the South, where as last time I can neither spend or draw money.—the only security, I can give which will be trusted.—

Independent of all these uncommon mortifications & my illness at St Fe preventing my return by the Rio Uruguay, through a most interesting Geological country.—the tour answered well.— It is quite magnificent when I consider I have ridden nearly 800 miles in a North & South direction & the greater part through country most imperfectly known.—

We are in a pretty state in this nice city.— they think nothing of cutting the throats of 30 prisoners, whom they happened to take the other day:—and they are right; for what is it, to quietly stabbing all the Indian women above 20 years old or younger if ugly.— Oh these Creoles are such a detestably mean unprincipled set of men, as I hope this world does not contain the like.— There literally is only one Gentleman in Buenos Ayres, the English Minister.— He is has writt⟨en⟩ to or⟨der⟩ the Beagle up.— But we sail under such particular instructions I know not whether the Captain will come.— If he does all will be right about Covington.—otherwise I shall be obliged to send some small vessel or boat to smuggle him off the Coast.—[3]

In fact I am in a pretty pickle.— I wish the confounded revolution gentlemen would, like Kilkenny Cats, fight till nothing but the tails are left.— Some of the good people expect the town to be plundered.— Which will a very amusing episode to me.—

dear Caroline. Yours Chas. Darwin

I will write again.—

I sent home by the Capt: Beaufort about 2 or 3 months ago—some more of my journal. Be sure acknowledge it, & in more than one letter.

DAR 223

[1] 'September 23ᵈ—' was written by CD. 'September' has been deleted and 'October 1833' inserted, in what appears to be another hand, to correct CD's error.

[2] Jaguars (Spanish 'tigres') and capybaras.

[3] See *'Beagle' diary*, p. 191: 'These disturbances caused me much inconvenience; my servant was outside, I was obliged to bribe a man to smuggle him in through the belligerents. His clothese, my riding gear, collections from St Fe were outside, with no possibility of obtaining them. I was, however, lucky in having them all sent to me at M. Video.' See also letter from Edward Lumb, 13 November 1833.

To W. D. Fox 25 October 1833

Buenos Ayres
October 25th 1833

My dear Fox

In less than a fortnight we shall be on our course to Tierra del Fuego. All this summer we shall be buried alive amongst the Barbarians.— I send you this to wish you good night.— We shall not receive or write letters for the next six or eight months. I hope at Valparaiso (our future direction) to find one from you.— I have lately been a great wanderer. When the Beagle was at the R. Negro I left her & crossing the Colorado went to Bahia Blanca, from thence also by land, to this place.— It is a long & most dreary ride, & till very lately never performed.— The government here sent out a large army against the Indians in their route, they left at wide intervals a troop of horses & five men, forming a line of Postas to keep up some sort of communication with the Capital I obtained an order for these horses & was of course very glad to profit by such good fortune: it was rough work many days living on nothing but Ostriches & Deer & sleeping in the open camp.— When the weather is fine nothing can be pleasanter than the Gaucho fashion of travelling.— kill your game in the day, as soon as the sun sets, manacle your horses & with the cloths of the Recon[1] make your own bed.—

Falkner the old Jesuit who resided so many years with the Indians has given a most accurate account of this country.— One of the most interesting parts of this ride, was the ascent of the Sierra Ventana (or Casuahati of Falkner) a Mountain which rises in the campo like an island in the sea, to the height of between 3000 & 4000 feet.— In the greater part of the road the novelty & wildness were the chief charms, for one league did not differ an iota from another.—

After arriving here, in a weeks time I started to St Fe. This country is comparatively civilized & true Pampas with all its characteristic features, Thistles &c &c. My object in all this galloping was to understand the Geology of those beds which so remarkably abound with the bones of large & extinct quadrupeds. I hav⟨e⟩ partly succeeded in this; but the country is difficult to make out.— every thing in America is on such a grand scale, the same formations extend for 5 or 600 miles without the slightest change—for such geology one requires 6 league boots.— I thank Providence I have returned with an uncut throat; both Indians & these miscalled Christians, have most carotid-cutting faculties. We shall in a week leave for ever the muddy estuary of the Plata, & now the voyage may be said to have commenced.— I hope you will write to me, & as I do, give (but a longer) a history of yourself. Excepting my own family I have very few correspondents; & hear little about my friends.— Henslow even has never written to me.[2] I have sent several cargoes of Specimens & I know not whether one has arrived safely: it is indeed a mortification to me: if you should have happened to have heard, whether any have arrived at Cambridge, do mention it to me.— It is disheartening work to labour with zeal

& not even know whether I am going the right road.— How is Henslow's family & what is he himself doing? I should so enjoy receiving ever so short a letter from him.— But patience is a fine virtue & there does not lack the opportunity of exercising it.— Remember me most kindly to all at Osmaston; perhaps in five more years I may once again be there.—

I hope to hear you are reestablished at your Curacy, & leading a White of Selbourne life.—[3] Good bye, my dear Fox. As the Spaniard says, God protect you for many years | Yours affectionately | Chas. Darwin

Christ's College Library, Cambridge (Fox 46c)

[1] CD usually used 'recado' for 'saddle'; 'Recon' may be the gaucho term.

[2] Henslow had written on 6 February 1832 and again on 15−21 January 1833. The letter of 15−21 January 1833 arrived at Valparaiso on 24 July 1834 (see letter to J. S. Henslow, 24 July − 7 November 1834); there is no record of when the earlier letter was received.

[3] Gilbert White, *The natural history and antiquities of Selborne* (1789).

From Caroline Darwin 28 October [1833]

[Shrewsbury]
October 28th—

My dear Charles—

I have been reading with the greatest interest your journal & I found it **very** entertaining & interesting, your writing at the time gives such reality to your descriptions & brings every little incident before one with a force that no after account could do. I am very doubtful whether it is not *pert* in me to criticize, using merely my own judgment, for no one else of the family have yet read this last part—but I *will* say just what I think—I mean as to your style. I thought in the first part (of this last journal) that you had, probably from reading so much of Humboldt, got his phraseology, & occasionly made use of the kind of flowery french expressions which he uses, instead of your own simple straight forward & far more agreeable style. I have no doubt you have without perceiving it got to embody your ideas in his poetical language & from his being a foreigner it does not sound unnatural in him— Remember, this criticism only applies to parts of your journal, the greatest part I liked exceedingly & could find no fault, & all of it I had the greatest pleasure in reading—

Susan I dare say told you in her letter to Valparaiso dated 18th of October of M[r] Howels delight at receiving an order of shoes from you, this letter will go with them & with the books &[tc] that Erasmus sends. I have sent you a few little books which are talked about by every body at present—written by Miss Martineau who I think had been hardly heard of before you left England. She is now a great Lion in London, much patronized by L[d] Brougham who has set her to write stories on the poor Laws—[1] Erasmus knows her & is a very great admirer & every body reads her little books & if you have a dull hour you can, and then throw them overboard, that they may not take up your precious room— I was very glad you have more space now than at first & particularly glad you have a servant as that must be a comfort to you. I also send Whately (Archbishop of Dublin's) Scripture Revelations[2] I like it so very much & I think we often used to find we liked the same kind of books—

The papers now are very full of Captain Ross's safe return,[3] he seems to be quite satisfied with what he has done in finding there is no passage south of 74th, but I should have thought it very poor satisfaction for the misery & sufferings he has gone through— Catty & I have been passing a very quiet week together whilst Susan, Papa & Harry have been Cathedral hunting Papa has enjoyed very much all he has seen Susan says in her letter & tomorrow we expect them home again— I wish I could think of any home events— Cath I am sure will have told you all Owen gossip & I have been so busy the last month regulating my new Infant School that I could only give you a journal of the childrens progress in ba-be bi— I rejoice to find the more you see of the negro's the better you think of them & it is delightful to think in a few years we shall have no more slaves—that alone is enough to make one properly value this Parliament it is my turn to write next month & as in the interim I shall go to Maer, my next letter will be a Wedgwood one I dread the time you are looking forward to with such eagerness, for when once in the South seas there will be an end of regular correspondence & I wish you could see the happiness a letter from you gives to every one— dearest Charles good bye. Y^{rs} affly C D.

DAR 204.6.1

[1] Martineau 1833–4.
[2] Whately 1829.
[3] Captain John Ross had just returned from an Arctic surveying voyage, 1829–33, in search of the North-West Passage.

From Catherine Darwin 29 October 1833

Shrewsbury
October 29th | 1833.

My dearest Charles,

Your last letter was very interesting, & we were very much pleased to receive it & your Journal. I am very glad of this Parcel, as it gives Caroline an opportunity of sending you some little Books perhaps you will like. I *can* think of nothing to send you, except a little Chain; wch perhaps may be useful to hang an instrument to; it is with Susan's purse. Enclosed is another letter from Mrs Williams, which I know you like better than any thing else. What a nice letter your's was to her; She sent it here for us to read.— All your letters make one very melancholy in one respect, the *enormous* time that it seems your Voyage will keep you away. The original two years are over now; I wish to Heaven Capt Fitzroy had been strictly limited to time. We are going to read aloud your Journal to Papa—; it will do beautifully for Winter's Evenings.

I think as yet it has been quite wonderful how regularly our Correspondence has gone on. I do not think there was any April Letter lost; by the order in which we write to you, I think the April letter must have left here the end of March.—[1] I have really nothing in the world to tell you—so I must stop— Bless you dear Charley—ever y^rs affect^e | E. Catherine Darwin

DAR 204.6.1

[1] No letter from the family to CD dated in late March or in April has been found.

From Henry Stephen Fox[1] 31 October 1833

Rio de Janeiro
October 31st: 1833

Dear Sir

I have just learnt that the Packet Cockatrice is likely to find the Beagle still either at, or near to, Montevideo.— My object in troubling you with the present letter, is to urge you, in case you have not yet done so, to visit the little Island of Flores,[2] on which is the Lighthouse. You will find it a geological curiosity, and worthy I think of your inspection.— The whole island is a formation of Greenstone, different from any thing that occurs along the neighbouring continent, but similar to that (as far as I can recollect) of which I saw specimens, brought from Cape Horn, by Captain FitzRoy on his last voyage.— The formation is the more remarkable, as the island of Flores, though apparently out at sea as you go to it from Montevideo, is in fact considerably within the general line of coast.— You will find the finest ledges and cliffs of the Greenstone along the southeastern or seaward side of the island, much waterworn, and exceedingly hard and difficult to break.— All this history will however be quite superfluous, if you have already visited the place.— The island of Lobos I was not able to visit.— I found the islands of Martin-Garcia, the Dos Hermanas, and some other small islands in the middle of the channel of the Uruguay, above its confluence with the Paraná, to be small low knolls of granite, though surrounded, on both sides of the river, by interminable alluvial plains, pierced I believe by no rock whatever.—

I have taken the liberty of sending to you, by M^r: Rees of the Cockatrice, several specimens of rocks from Porto Alegre.[3] The main fundamental rock of that Country is granite, of remarkably white colour, from containing a superabundance of quartz. I found, lying over the granite, large formations, composing clusters of considerable hills, of the rock (specimen n°. 1)—perhaps you will have the goodness to tell me what it is. I also found large quantities, in the same neighbourhood, of the volcanic porphyry (n⁰. 2), not exactly in situ, but forming lines of disjointed masses, closely piled together, along the ridges of several hills. This rock very much resembles many of the ancient volcanic rocks of the north of Italy.— I again found large quantities of greenstone, not in situ, but in loose blocks, both in the island of St. Catherine's, and in the island of St. Sebastian; but none on the coast of the continent, opposite these islands.— In one place only, on the west side of the island of St. Sebastian, I found veins of greenstone, of a texture almost as close as basalt, in the granite.— The main rock along the whole length of coast, from Rio Grande up to this place, is granite.— I have thought you might like to hear these few particulars of the geology of the south of Brazil. The Province of Rio Grande would I think afford great interest to a geologist.—

I have not yet unpacked the greater part of my specimens; but if you should like to be furnished with any more specimens of the rocks of Porto Alegre, or of the greenstones above mentioned, I shall have great pleasure in sending them to you, with such statement of their site, &c, as my very slight knowledge may enable me to give.

I presume your head quarters, for the next cold season, will be Valparaiso. Perhaps you would in return oblige me, by sending me a specimen or two of the common rocks (not the rare minerals) of the coasts of the straits of Magellan, and of Chile, which you will visit.— When I had the pleasure of seeing you at Buenos Ayres, you were in expectation of discovering a volcanic formation in the neighbourhood of Bahia Blanca.—

I beg you to have the goodness to make my compliments to Captain FitzRoy, and to believe me | Your faithful and | obedient humble servant | H. S. Fox

DAR 39.1: 1—4

CD ANNOTATIONS
verso last page: 'November 1832'

[1] Fox had arrived at Rio de Janeiro as British Minister Plenipotentiary to Brazil in August 1833.
[2] CD preserved this letter among his geological notes (DAR 39.1: 1—4) for possible use in writing up the results of the voyage, but no mention is made in his later works of this or any of the other islands to which Fox refers. For CD's description of the geology of the Montevideo area see *South America*, pp. 145—7.
[3] The specimens are listed in DAR 39.1: 75.

To Frederick William Hope 1 November 1833

Buenos Ayres
November 1ˢᵗ 1833

My dear Hope

I have many times, since leaving England, intended writing to you.—but as many times put it off.— I believe the chief cause has been a conscience not quite free from shame.— I am not the worthy slayer of sufficient Hecatombs to venture to write to my old Instructor.— When I last saw you in London my promises were great, my performance I grieve to say does not equal them.— The Beagle for the last year has been cruizing either amongst the islands of Tierra del Fuego or on the barren coast of Patagonia.— Both these regions are most singularly unfavourable to the insect world:— In Tierra del I *captured several* Alpine Carabidous beetles, & one Carabus: & on the sandy desarts of the latter country, there are many of the Heteromeri.— But these in absolute numbers are not to be compared to the booty in one of your Achilles-like onsets. Before we came to these Southern regions inhospitable to Entomologists & Insects, I did pretty well amongst the Coleoptera.— I often thought of you, when sweeping the rich vegetation of the Tropics I *captured* the smaller Coleoptera by hundreds.— If, as I believe you told me, Europæan cabinets contain few of the minuter Beetles from tropical countries, I shall bring home a very great number of undescribed species both from Brazil & the Rio Plata.— It may be a foolish fear, but I often wonder, if any person will be found who will describe so many minute insects. This fear is rather a drawback in my collecting.— Excepting Coleoptera, pudet pigetque mihi, I have done scarcely anything. The incumbrance of a box & fly net is not trifling.—when I have to carry Geological tools fire-arms, spirit-bottle for reptiles &c &c. I hope however to improve, & be more diligent in this respect.— At Rio de Janeiro I took many water beetles, most exceedingly small Hydropori, Hyphidri, Hydrobii &c &c.— Also a fine species of that curious sculptured genus (I forget the name) which lives beneath stones in running water.— I was much interested by finding this; it could not fail vividly to recall some of our walks at Netley; in a like manner chacing Cicindela niveas amongst the burning san⟨d⟩ hills most forcibly reminded me of the Hybri⟨ ⟩ at Barmouth.— Judging from the Pamphlet, you gave me & which I have found very useful, the insects of the Rio Plata are tolerably well-known.[1] I regret therefore the less, not having worked as hard as I could have done.—

I took, the other day, a fine Leionotus.—

If you feel inclined I should much enjoy hearing from you. I know nothing of the scientific world of London.— The last thing I heard about you, was an age since, viz, that you were on your road to Germany & that Eyton failed to accompany you.— What is Eyton doing? He must be by this time a famous naturalist. remember me most kindly to him. I had hoped he would have by this time been wandering in some Terra incognita.— My direction is H.M.S. Beagle Valparaiso, if you will condescend to write to so recreant an Entomologist, I shall be very much obliged.— Remember me to the few friends I have amongst the Naturalists & Believe me dear Hope.— Your most obliged disciple. | Chas Darwin.—

Remember me most kindly to all your family, & my congratulations (although they arrive rather late) to your brother.—

Floreat Entomologia

I shall much enjoy some scientifico-entomologico-Gossip.— Once more Farewell.—

Hope Entomological Collections, University Museum, Oxford

[1] Probably Lacordaire 1830. A copy in the Darwin Library—CUL ('Philosophical tracts') is lightly annotated.

From Thomas Campbell Eyton   12 November 1833

Shrewsbury
Nov[r] 12th 1833

Dear Darwin

I have been writing to you for some time, and having just heard from your sister what your direction is have now set to work in good earnest to write to you. Hope has been very ill and is not now much better but I think that he fancies himself much worse than he really is. he has not done much in the insect way since he returned from Germany nor indeed have I, but I have been working very hard at the birds both English and foreign of English one I believe that I have one of the finest collections in this country. I have also made great progress in the anatomy particularly in that of fish birds & animals of the two latter I have now near a hundred skeletons some rather valuable ones. I had a letter from Yarrel the other day who gave me an account of some new birds discovered in England viz Alpine or White bellied swift, Carolina Cuckoo of Latham[1] and the red legged falcon F. rufipes of which I have been lucky enough to obtain a pair together with several other rare birds among which are the snowy & Eagle owl from Orkney. Sir Rowland Hill is making a collection of birds and animals both alive and dead at Hawkstone and as he spares no expence I should think that he would have a good one. I have several new and undescribed tracheæ and muscles of voice of birds of which I shall one of these first days publish some short account.[2] Jennings work on the vertebrates is not yet out and is not expected before next year it is to be on a much larger scale than was originally intended and to be called a manual.[3] Selby is going to publish a new work on birds[4] after the arrangement of Vigors.[5] A splendid work is going on by Gould, the birds of Europe folio coloured plates.[6] he has also published an excellent monograph on the Toucans[7] with a folio plate of each species which work I have been expensive enough to take but as I have been offered more than cost price for them since I am on the right side. I saw as I believe I told you before your sisters and father this morning the latter of whom has had a fit of the gout but has now recovered and the whole party are looking very well. I do not know that there is much gossip going on there now except that John Hill is to be married to Miss Kenyon of Pradoe which is I believe all settled or nearly so. I wish with all my heart that I was with you and had been with you all the time but next to that I wish to see you here in merry England again as I sandly want some body to talk to about nat his who cares about it as very few of those who profess to do so either know or in their hearts care one pin about it. I have trained some peregrine falcons to catch birds which is very good fun. I do not ask you to write to me because I think that you may be better engaged but should you have a few minutes to spare I think that I need scarcely tell you that I shall be most happy to hear from you if you collect any skins beware of the insects send them to the care of some body who will take care of them and open the cases otherwise you will have very little chance of finding any thing but a few feathers and beaks & legs remaining when you return I know of two or three collections which have been sent from India of which this has been the fate. Wishing you all health and happiness I remain | yours very truly | Thos C Eyton

DAR 204.7

[1] Latham 1781 – 1802.
[2] No separate publication on the trachea and voice muscles of birds has been identified; however, Eyton did employ his study of the trachea (and skeleton) as a basis for a new classification of ducks (see Eyton 1836, pp. 72 – 4). The concept involved was spelled out in more detail in Eyton 1838. Eyton's emphasis on internal organs was novel in bird taxonomy.
[3] Jenyns 1835.
[4] Probably a reference to volume two, *Water birds* (1833) of Selby [1818 –]34.
[5] Nicholas Aylward Vigors was an advocate of William Sharp Macleay's quinary system. Eyton probably refers to Vigors 1825.
[6] Gould 1832 – 7.
[7] Gould 1834.

To J. S. Henslow[1] 12 November 1833

Monte Video.
November 12th. 1833.

My dear Henslow.—

By the same packet, which takes this I send a cargo of specimens.— There are two boxes & a cask.— One of the former is lined with tin-plate & contains nearly 200 skins of birds & animals.—amongst others a fine collection of the mice of S America.— the other box contains spirit bottles, & will only require just looking at to see how the Spirit stands.— But the Bird-skins, if you will take the trouble will be much better for a little airing.— The Cask is divided into Compartments the upper contains a few skins.—the other a jar of fish, & *I am very anxious to hear how the Spirit withstands evaporation,*—an insect Case, which would require airing, a small box of stones.—which may be left in statu quo.—a bundle of seeds,[2] which I send as a most humble apology for my idleness in Botany.— They were collected in Port Alegra & in this country: the temperature of the former, must be that of a warm greenhouse.—& even plants of this country would require some protection (the olive & orange bear fruit here).— Also a bag of the sweepings of a Granary; it will be a Botanical problem to find out to what country the weeds belong: It might be curious to observe whether Europæan weeds have undergone any change by their residence in this country.— If they are like the men, I will answer for it they are not much improved.— I also send to the care of Dr Armstrong in Plymouth, an immense box of Bones & Geological specimens. I do this to avoid the long land-carriage: & as they do not want any care it does not much signify where kept.—another reason is, not feeling quite sure of the value of such bones as I before sent you.— I have one mutilated skeleton of the animal of which I sent the jaw with 4 ☍ ⌒ ○ ∘/ small teeth.—[3]

Since my last letter to you (middle of July, when I sent off some specimens) I have been, as they say here, un grande galopeador.— I left the Beagle at the R. Negro & crossed by land to B. Ayres. There is now carrying on a bloody war of extermination against the Indians, by which I was able to make this passage.— But at the best it is sufficiently dangerous, & till now very rarely travelled.— It is the most wild, dreary plain imaginable; without one settled inhabitant or head of cattle. There are military Postas, at wide intervals, by which means I travelled.— We lived for many days on deer & ostriches & had to sleep in the open camp.— I am quite charmed with the Gaucho life: my luggage consisted of a Hammer Pistol & shirt & the Recado (saddle) makes the bed: Wherever the horses tire, there is your house & home:— I had the satisfaction of ascending the Sierra de la Ventana, a chain of mountains between 3 & 4000 feet high;—the very existence of which is scarcely known beyond the Rio Plata.— After resting a week at Buenos Ayres, I started for St Fe ⟨   ⟩ on the road the Geology was interesting.— I foun⟨d⟩ two great groups of immense bones; but so very soft a⟨s⟩ to render it impossible to remove. I think from a fragment of one of the teeth they belonged to ye Mastodon: In ye R. Carcarana I got a tooth, which puzzles even my conjectures, it looks like an enormous gnawing one.—[4] At St Fe; not being well, I embarked & had a fine sail of 300 miles down that princely river the Parana.— When I returned to B. Ayres I found the country upside down with revolutions, which caused me much trouble. I last got away & joined the Beagle.— I am now going to have one more gallop to the Uruguay, & then we are off Tierra del Fuego.—

We shall for the future be much amongst Volcanic rocks, & I shall want more mineralogical knowledge.— Can you send me out any book, which with instructions from yourself will enable me to use my reflecting Goniometer. If you know of any, it would doing me a great favour to send it to Capt. Beaufort, who will forward it.— As I am very anxious to hear from you.—perhaps this will be the best manner of sending me a letter.— I want much to hear about your family.— L. Jenyns, your lectures excursions, & parties &c.—respecting all of which I have so very many pleasant recollections, that I cannot bear to know nothing.— We shall pass the St^r of Magellan in the Autumn & I hope stay some time in the Southern parts of Chili

There are two Volcanoes within 60 miles of Conception. I will run the risk of being eat up alive to see two real good burning Volcanoes. Oh the blue skys & the Bananas of the Tropics.— Life is not worth having in these miserable climates, after one peep within those magic lines.— Believe me my dear Henslow | Ever yours most truly obliged | Chas. Darwin.—

Would it not be a good plan to send sea-weeds in Spirits, having previously noted ye colour by Werner??[5]

Royal Botanic Gardens, Kew

[1] The following passages from this letter were extracted by Henslow and published in the Cambridge Philosophical Society pamphlet:
2.1 'I left . . . camp.' 3.7
2.10 'I had . . . Beagle.' 3.20
[2] In *Journal of researches*, p. 600 n., in a section devoted to advice to collectors, CD warns that: 'Seeds must not be sent home in the same case with skins prepared with poison, camphor, or essential oils; scarcely any of mine germinated, and Professor Henslow thinks they were thus killed.'
[3] Later identified by Richard Owen, who named it *Mylodon darwinii* (*Fossil Mammalia*, pp. 63 – 73).
[4] Richard Owen later identified it as belonging to *Toxodon platensis* (see *Fossil Mammalia*, p. 19, and *South America*, p. 88).
[5] Syme 1821.

To Caroline Darwin 13 November 1833

Monte Video
November 13th. 1833

My dear Caroline

I have to thank you for a letter dated September 1st & one from Susan July 22^d.— Since I wrote from B. Ayres, I have suffered a host of vexations, but at last every thing has ended prosperously. I with much trouble & by bribing got my servant in to the town & then started for this place, almost expecting the Beagle to have sailed.— I now find to my astonishment she will remain 3 weeks more in the river.— And here comes the whole purport of my letter, to announce more extravagance.— I have really now been struggling for a whole week, but there is a very interesting geological formation on the coast of the Uruguay, & every day I hear of more facts respecting it.— When I think I never shall be in this country again, I cannot bear to miss seeing one of the most curious pieces of Geology.— I wish any of you could enter into my feelings of excessive pleasure, which Geology gives, as soon as one *partly understands* the nature of a country.— I have drawn a bill for 50£.— I well know, that considering my outfit I have spent this year far more than I ought to do.— I should be very glad if my Father would make a real account against me, as he often says jokingly.— I hope he will not think I say this impertinently: The sort of interest I take in this voyage, is so different a feeling to any thing I ever knew before, that, as in this present instance I have made arrangements for starting, all the time knowing I have no business to do it. I wish the same feeling did not act so strongly with the Captain. He is eating an enormous hole into his capital, for sake of advancing all the objects of the voyage.— The Schooner, which will so very mainly be conducive to our safety, he entirely pays for.—

I have just packed up a Cargo of specimens. I send home nearly 200 skins of birds & the smaller quadrupeds & a fine set of fossil bones.— There is one skeleton, sufficiently mutilated, of an animal, of which I do not think there exists at present on the globe any relation.—[1] I am now living on shore in the house of an English merchant; as they are so busy, chart-making on board, that they would have nothing to say to me till this Packet sails. The whole coast of Patagonia is now completed, & please Providence, we trust by late in the Autumn to say the same of Tierra del Fuego. Poor Earl has never been well since leaving England & now his health is so entirely broken up, that he leaves us.— a Mr Martens, a pupil of C. Fielding[2] & excellent landscape drawer has joined us.— He is a pleasant person, & like all birds of that class.—full up to the mouth with enthusiasm.—

We are all beginning to long for "blue water" & I am sure I do, if it is merely to prevent my spending money.— My present scheme is not a very great one. I go to Colonia del Sacramiento, then up the coast of the Uruguay to the R. Negro, to the town of Mercedes.—from thence back in direct line to M: Video or perhaps to the lime-kilns at Paysandu,[3] 25 leagues up the Uruguay.— the whole round will be under 400 miles.—& the whole country inhabite⟨d.⟩ There is peace at last a Buenos Ayres, so that I have lost very little of my property.— Do you ever hear in England of these revolutions, which are considered as so important in this poor country?— It is late. I am not in a writing humour, so I will wish you good night.—

Give my love all & my thanks for all the long & very nice letters.— I will write again before we sail.— Yours very affectionately | Chas. Darwin.—

Love to Nancy

DAR 223

[1] Since CD does not give the location in which he found it, the fossil cannot be identified in *Fossil Mammalia*, but it may be the one, still unidentified in 1846, mentioned on p. 84 of *South America* as 'fragments of a head of some gigantic Edental quadruped' found at Punta Alta.

[2] Antony Vandyke Copley Fielding.

[3] CD does not record any visit to Paysandu but he saw a lime kiln on the way to Mercedes (see '*Beagle*' *diary*, p. 194).

From Edward Lumb   13 November 1833

Buenos Ayres
Novr 13 1833

My dear Sir

I am in due receipt of your's of the 5th. inst and am glad to find your stay delayed a little longer than you anticipated in Montevideo— You would receive f.^r Rosa all your goods & chattels & also the announcement of our reconstruction here so that now we are all quiet

I could not send you either the shot or powder both being prohibited by the law here and under present circumstances it was impossible to obtain permission— The Chart some of your Officers saw at Bahia Blanca was the original or a Copy of the one given by Capt Fitz Roy to the Government & borrowed by me for the use of the Captain of the Dolores— it is not published and I believe the Government do not intend publishing it: On your receiving your goods you will I think say that I fore told your wishes as they are packed as you require them I hope you will not lose any part of them I gave them in particular charge of the Captain and Owner who was also on board—

As the schooner "Dolores" had sailed for Bahia Blanca previous to your leaving it was impossible for me to obtain any further Information about the fish caught off Cape Corrientes than what Stewart the Owner has this day said to me viz that he had a number of fish on board some taken at New Bay, some in Belgrano bay and off Cape Corrientes on some Bank or other & he was not aware of other particulars & indeed had he been I do not expect he would have told me as he proposes sending the Schooner to fish there— Your information respecting the River Chupat[1] will be very interesting & should there be Neutrias please get me the correct Latitude and Longitude & any other "por memores"—[2] As you have been pleased to express yourself in so kind a manner for my poor services towards you during your staying here allow me to assure you that independently of your private worth which I consider intrinsic I do not consider I have done more than what any Englishman should do for the promotion of any scientific end which may tend to the aggrandisement of his Country & I shall be truly happy at any future period I can by any means in my power serve you or promote your views in a scientific way— pray write me when you arrive in Chili & also in England at some future period I may perhaps procure for you some specimen that may be interesting— M^{rs} L and family are well & unite with me in good Wishes. Should you go to Mercedes; you have my Letter to Keen which will answer every purpose— You would have been amused to see the ragged Reg^{ts} 7000 strong they say that alarmed the town a day or two— | I am Dr Sir | Yours truly | Edward Lumb

We have an American 18 gun Ship from Valparaiso homeward bound here; she arrived yesterday—

DAR 204.7

CD ANNOTATIONS
on cover: 17 | 11½ *[illeg]* | *[illeg]* *[Guide]* | 8½
M' | M

[1] The Indian name for the Chubut River in Patagonia. It had been explored by John Clements Wickham with the *Liebre* in February 1833 (see *Narrative* 2: 306–9). The *Beagle* did not, however, call there.

[2] According to Robert FitzRoy the location of the Chubut had been concealed for many years 'on account of the lucrative trade some individuals hoped to carry on by means of hides and tallow obtained from the herds of wild cattle' (*Narrative* 2: 307).

From Catherine Darwin   27 November 1833

Shrewsbury.
Nov^ber 27th. | 1833.

My dear Charles,

I believe your Parcel is not yet gone from London, but we think it safer to write to you by the Packet also, for fear you should not get the parcel.— There are several letters for you in the Parcel, from Mrs Williams & Mrs Biddulph, & from Caroline, so you must fish them out.— Tom Eyton has written to you also this month; he called here the other day, & enquired much about you.— Your letter to William Fox has arrived; we heard from them not long ago, at the Isle of Wight, where they are gone again to spend the Winter for William's health; he has been better lately.— We had two of your letters to Professor Henslow sent down here for us to see; one was dated May this year on the Sea, & the other July.— We are very much enjoying your Journal now, reading it aloud to Papa in the Evenings, and it meets with great success, and is pronounced exceedingly entertaining. Your account of the Brazilian people of the Patriarchal Vendas, and of the little graceful dancing Teresa, interested us very much.—[1] You never make any comments on your Companions on board, but that I suppose might not be safe, and does not come in the plan of your Journal.— I do not know whether you knew by sight a Capt Justice, a relation of the Clives, a naval man; he happened to dine with us yesterday, by ourselves, and we read aloud your Journal, by way of amusing him.— He was very useful to us in explaining some nautical terms, and he was perfectly charmed with your Journal, & would hardly let us stop reading it.— He thought you grown quite a Sailor by your language.—

Caroline's letter by the Parcel will have told you how unluckily Papa's Tour to Winchester & Salisbury Cathedrals was spoilt by the Gout; it seized him a little at the Inns, but not so badly as to prevent his getting on to the Hill, in Monmouthshire (the John Wedgwood's) where he was laid up 17 days with the worst fit of gout he ever had in his life. He & Susan came home about two weeks ago, when he was able to travel, and now he has quite lost it, and is as well as possible again. You may imagine how much it annoyed him, being so long ill, away from home.— His Tour answered perfectly to him in other respects; he admired the Architectural Beauties exceedingly & enjoyed Harry's company, as a fellow Antiquary. I hope he will make a custom of these Tours, now they answer so well to him.—

I have hardly any Owens news to tell you, except that Francis has got an Ensigncy, in the 63^d, which is quartered at Madras, where Arthur is. I do not know when Francis goes out, but I dare say, you will see both the Brothers, if you touch at Madras.— Poor Fanny Biddulph still continues very unwell; she is wretchedly altered in looks, so very thin & pale; she has never recovered her bad confinement. She is very happy otherwise apparently, and Mr Biddulph very devoted.— Mrs Williams keeps the gayest house possible at Eaton, continual dances & gaieties going on.

We are all in consternation at present at a piece of news Erasmus wrote us word a day or two ago: that Hensleigh Wedgwood has determined to resign his Police Magistrateship, worth 800£ a year, from some scruples about the system of administering oaths there, which he cannot reconcile to his conscience. He & his Wife & Child would only then have Hensleigh's private fortune £400 a year to live on; and Erasmus is horror struck at the sacrifice it is to Fanny Hensleigh and at the loss of her company to himself.— I cannot think what they will do, if Hensleigh persists, which he most probably will, though he has consented to hold an argument with Dr Holland, & with Sir Edward Alderson. I am afraid there is but little chance of their shaking his scruples, as he had them on his mind for some months.— It is most unfortunate to be sure, they will be so wretchedly poor; and Hensleigh once before refused a Fellowship at Cambridge, on ⟨   ⟩ of not subscribing to the 39 Articles.— The Maer Wedg⟨wood⟩s will take this business as calmly as any ⟨fa⟩mily could.— I have been staying at ⟨M⟩aer lately, which is a very pleasant house still, in spite of all its changes.— Poor Aunt Bessy is a melancholy sight; she is perfectly helpless, and cannot stir herself the least, owing to a pain in her leg, which makes her quite powerless.— Her fits are much more frequent than they used to be, and she is excessively altered since her dangerous illness this Summer, that one would not suppose she could last much longer.— I met Bessy Holland at Maer, Sister to Dr Holland, who talked a great deal about you, and *knew* more than I did about you; she hears all about you from their Friend C. Whitley who has the greatest interest in you. I can't conceive how Mr Whitley knows all about your plans, movements, discoveries &c, for I don't think you correspond with him. Whitley says that two letters of your's have passed through his hands, to other people (not to himself) and that he perfectly longed to break the seal of them, but of course did not.— I hear that your Theory of the Earth is supposed to be the same as what is contained in Lyell's 3^d Vol.[2] Some of your Friends or Whitley's Friends meant to send out the 3^d Vol to you; have you received it? Bessy Holland informed me also that you had sent a quantity of stones to Cambridge; is all this true? She says also that you have some intention of returning home before the Beagle—; how I wish this was true, my dearest Charles. What happiness it will be to see

you again, after such an absence!— Papa desires me to give you his best love, &
he is very glad that you are so happy and prosperous. He desires me also to tell
you how very much he is pleased with your Journal. Goodbye dearest Charles.
With all our loves believe me | ever y^rs affectionately | E. Cath^ine Darwin
   Pray, pray take care of yourself, and run no risks; and mind your health.

DAR 204.6.1

[1] The typical Brazilian *venda* is described in *'Beagle' diary*, pp. 51 – 2, and *Journal of researches*, p. 22 – 4.
   Theresa Price was the eight-year-old daughter of a merchant in Rio de Janeiro (*'Beagle' diary*, p. 71).
[2] The reference is probably to CD's adoption of Charles Lyell's uniformitarianism, which had
   impressed CD when he read the first volume of the *Principles of geology*. He did not receive the third
   volume until later in the journey (see letter to J. S. Henslow, 24 July – 7 November 1834).

To Susan Darwin   3 December [1833]

Monte Video.
December 3rd.

My dear Susan,
   Will you tell my Father I have been obliged to draw a Bill for 17 pounds—
This makes 217 in seven months. I can offer no excuse. But the Captain believes
that instead of passing the summer in Tierra del Fuego we shall also winter
there— I did not believe that anything could have made me look forward to so
miserable a prospect with joy— I do so for then I shall be able to make some
glorious excursions into the Andes with a better conscience than I have
undertaken my latter ones. We sail tomorrow, first up the River for fresh water
then for Port Desire on the Coast of Patagonia After that to East entrance of
Straits of Magellan where the Gales of Wind will decide which shall be the next
piece of work— It will be very interesting, but I am afraid likewise painful to see
poor Jemmy Button & the others— I expect to find them naked & half
starved—if indeed they have not been devoured during the past winter. My
gallop to the Uraguay was a very pleasant one— I went by Colonia del
Sacramento, so up the coast to the Rio Negro— I staid at an [estancia] a good
many leagues up the latter River, and from there returned in a direct line to
Monte Video— The heat of the Sun makes the fatigue of galloping excessive— I
could not for this reason go quite so far as I had hoped. Moreover the country in
many districts is shut up by the great thistle beds.— These thistles are from 8 to
10 feet high & form an impervious mass. The geology was very interesting to
me— I should very much have regretted not visiting this part of the province I
obtained many fragments of fossil bones, & a part of a head, which the Gauchos
had sadly mutilated but yet is in my eyes very valuable— For the last four
months I have not slept more than one night in the Beagle; to day I took all my
things on board meaning to stay— But I am writing this on shore; and what do
you think is the reason?— Proh Pudor—Sea sickness— Oh the next ten days will

be delightful! how I shall long for the green plain and its galloping horses. But it is the high Road to the Pacific, so I will not complain— We are in good state for the sea—12 months stores on board; & the Schooner well manned— The cause of our long delay has solely been owing to the Charts not being complete for sending home.— The Captain has been exerting himself to a degree which I thought no human being was capable of— The vast importance of the long days to the South was indeed a sufficient stimulant— The Admiral at Rio wrote to inform us that in two months he was going to send a ship to the Falkland Islands, with an Officer & party of soldiers to act as Governor. By this opportunity we shall receive our letters, and perhaps be able to write in answer.— This is a grand piece of good fortune— With the exception of this chance you may be a year before you again hear from me.— And this is a pretty sort of a stupid letter to send as a last But I am very tired with fighting on board against sea sickness—and at this present minute against a host of mosquitoes.— I now thank you all half sufficiently for writing so very regularly— No one in the Beagle has received so uniform a series of letters. I shall return to Shropshire quite au fait with the latest news. As we are now on the road (though not the shortest) to England—I can steadily look forward & count the time between this & the glorious moment of dropping Anchor in Plymouth Sound.— Till then & for ever God bless you all— No one ever had a better or dearer lot to say farewell to. | Yours etc. | Chaˢ Darwin.—

Copy
DAR 154: 123a — 123d

From Caroline Darwin 30 December [1833] — 3 January 1834

[Shrewsbury]
December 30ᵗʰ

My dear Charles

Your last letter was dated Sepᵗ 20ᵗʰ, Buenos Ayres, & it was a most agreeable surprise as we thought you had started for your endless Southern voyage— the account of your wild ride was extremely interesting— how strange it still seems to think of you as realy leading a Gaucho's life— a canter on the Oswestry road will feel rather flat I am afraid, though you may depend my dear Charles if I am not superannuated when you come home I *will* have many and many a happy ride with you— Erasmus is become quite a grand man he has a Cab & a few weeks ago my Father gave him a beautiful Grey horse a Hunter bred by Mʳ Wynne—& we hear from Eras that he (the horse) is as much at home in the streets of London as if he had led a town life all his days.— I think you will hardly know Eras he is become such a dissipated character & such a happy person. he seems always in good spirits & enjoys visiting about & liking & knowing many more people than he used— Hensleigh a short time ago had determined upon resigning his Police Magistracy from a scruple of conscience—he thought our Saviour's command "not to swear" was one which ought to be taken literally & that a judicial oath was consequently unlawful— I do not quite understand the reasons why he classed it among the commands to be taken literaly & not with a latitude— however Uncle Jos & his other friends have persuaded him to take time & study the subject more before he decides, which he has done & the arguments of some of his friends have I believe made him think differently, but this is not certain as he does not now talk on the subject— it w^d have been very melancholy to have had their happy household broken up & one knows of hardly any employment which Hensleigh could have taken, as in almost all an oath is a necessary form for being entered— I was staying at Maer whilst all this was in agitation & very anxious it made Uncle Jos & them all— Aunt Bessy is sadly changed since you saw her—her intellect *much* weakened & from a pain in her leg unable to stand or move herself in the least She sits or rather lies down in the big room upstairs which is now fitted up as a sitting room & makes a tolerably comfortable one— they are very much interested about you— Can you fancy any thing that the whole family would enjoy more than being transported to the banks of the River Carcarana (I think you call it) the banks of which you describe as being so thickly strewed with bones & fossil remains. I shall be very curious to hear the result of your expedition there— Charlotte & Mr Langton were at Maer for the winter—also Fanny Allen so altogether we mustered a large party, but dear old Maer is not what it used to be & never will again—

1834 Jany 3^d— I wish you my dear Charles a happy new year and *many* returns of it & very glad shall I be when I am able to wish you the same in person— I rec^d the day before yesterday your letter of the 23^d of October from Buenos Ayres, with the account of the disastrous ending to your ride— I see by the papers that on the 28th of Oct. trade was allowed to continue, so I trust you were soon released from your very disagreeable & alarming situation in that odious town— what a vilannous people they seem to be— We are excessively anxious for your next letter and my Father as you may believe sympathizes & felt very much for all your misfortunes & I am sadly afraid the great dangers you have gone through when you wrote were not over. I wish you were out of that town & safe again in the Beagle— My Father had not heard at the bank of the £80 you speak of having drawn a month ago, but it will be paid whenever it is asked for— the second part of your journal arrived quite safe I read it to myself & we all read it aloud— it was extremely interesting & I am very impatient for a third part—

The Cottons have been staying with us this week with Mattie who is come out & a very pretty merry girl— she has some of the Owen spirit— Also Robert Clive was here, he is exceedingly happy to be in England again & for life without the thoughts of a return to India to damp his pleasure— I like him very much he is the merriest & most taking of all the Clives & very pleasant— he seems exceedingly fond of Mattie & if she was a little nearer his own age I think he would not be long in trying his chance with her. the William Clives have had a sad disappointment in having a dead child— Marianne Clive is doing well now, but her life has been in the greatest danger from her confinement & it was a melancholy endin⟨g⟩ of he⟨r⟩ delight in the hope of a child— Poor Eliza Tollet is thou⟨gh⟩t to be in a consumption she has had a cough now for ⟨n⟩ine months & is getting gradually weaker— Frank Leighton has just been made Sub Warden of Magdalen College which the Leightons are much pleased at—

Next week we are going to a Play at Eaton an immense party in the house 23 the Biddulphs & all the Owens are to be there— Fanny Biddulph is still very delicate, so altered from what she used to look, but I think still prettier even than she used to be— Francis Owen goes to India next February— they have very good accounts from Arthur—

I think my Father is looking very well, all the better for the gout that he had when touring with Susan. leaving off his business in great measure, & not altogether, has answered remarkably well to him— giving him a little employment & not fatigue— I think when you come home you will be amused to see in the hothouse, his Banana, with its two leaves that we all admire & think so handsome. I cant say I do admire it now, for it is grown so tall that the glass prevents even the few leaves it has from appearing in their natural shape—

We have not seen Marianne Parker very lately She is educating her 4 little boys very nicely I think, they are very happy & tractable Dr. Parker & Marian are beginning to fret & puzzle about a school for Parky— Susan is very happy this week *scrattling* she is sitting now at the table I am writing from, with a long account book & innumerable bills— Catherine is now quite the junketting person in this house so eager for balls & visiting of all kinds— Pincher & Nina are both well— I wonder if Pincher will be very glad to see you again. Joseph has still your Grey horse. I am going to hire it this morning for a ride— poor old nurse Tante, we hear, is reduced to wear spectacles. She has never yet appeared to us in them, but when she is alone she can not resist them. She is deeply interested about you & we always tell her the news out of your letters, though I suppose the ideas she gains are but vague— You must take the will for the deed & not mind my writing such *very very* dull letters— living quietly at home it is really very difficult to find what to write about, but a letter to say all at home are well, you shall have every month till we see your dear old face again My Fathers, Susans, & Caths best & kindest of loves. Yrs most affcly dear Charles | Caroline Darwin

I open my letter for my Father who says he hopes I have said every thing kind from him to you. I did before I think give his love—

DAR 204.6.1

From F. W. Hope   15 January 1834

37 Upper Seymour St. | London.
Jan^y: 15th: 1834

Dear Darwin

Your letter of Nov 1st: 1833, has just come to hand, & I am glad to find the old Adage is true, "caelum non animum mutant qui transmare curraent"[1] I anxiously expected intelligence from you last year particularly so as Eyton I understand received a letter from you, "*better late than never*" is sufficient for me. Your description of Terra del Fuego with respect to Alpine Entomology I expected. I still think however it must have some peculiarities & particularly so if Volcanoes are in a state of activity there. It appears to me that the extremes of North & South will yield us the same forms the greater number of families of the Insect world are guided more by Vegetable than by Geological situation. With regard to temperature we want facts, to guide us; from some little attention to Geographical distribution I find it no certain rule at least very liable to exceptions. It chiefly holds good with respect to altitude which is variable in many places. In some places on Mountains at certain heights the same genera will be found, but here again we cannot fail to remark discrepancies, take for instance two Hills at the distance of 10 miles apart, at the same height & temperature you will find the forms differ considerably & often in toto if the Vegetation is different. You must however always observe on what side of a Mountain you capture your Insects as the South side has vegetation differing generally with north. It is probable also that the Strata have much to do in reference to certain families, some Rocks retain the Heat longer than others & drive Insects to the plains for want of moisture, many families delight in burning sands witness the Cicindelidae The Basaltic hills (if such) of Terra del Fuego are barren of Insects, so indeed are all Basaltic regions as far as I can learn. If you can send any facts respecting Insects in the vicinity of Volcanoes I shall be much obliged to you Several genera love volcanic situations, many others are not found in such places. You will find genera extend longitudinally to 3000 miles or more as to latitude, they vary considerably. In a word my good fellow if you are not tired with the above, I wish you to obtain & even draw up a Geographical chart of the Genera as far as it is in your power & as you are a Geologist & Botanist you can illumine our darkness by your attention to the Vegetable & Geological distribution of Insects—

You will be glad to hear that We have established an Entomological Society. We meet the first Monday in every Month, at present we have had two Meetings & we muster 110 Members without foreign Honorary Members, their number is limited to 10. Mr Children is President. Vigors, Horsefield Stephens Hope, Vice Presidents, Foreign Secretary Spence, Treasurer, Hope Secretary Gray, Curator, Waterhouse.[2] The objects of the Society, are the extension of Entomological Science & the formation of a Cabinet & Library to assist students &c. Pray attend to Larvae, & collect *Hymenoptera* from the Brazils &c all of which are valuable particularly the Chalcididæ i:e Insects from Galls, of Trees Shrubs &c.. The *Diptera* are also much wanted, do not fail to collect wholesale. I promise you all assistance in my power & I hope by the time you reach England Committees will be established for each Order.

My visit to the Continent enabled me to judge of the state of Science in Germany France & Holland. In our favourite pursuit, We are likely to take the lead, Our Cabinets are very rich but unfortunately not every One is so assiduous as a Darwin or a Stephens. The English in general will let foreigners describe for them what they are capable of doing themselves. Parasitic Insects are also much wanted. Pray do not be afraid of collecting for fear of having No One to describe your captures, attend to what I have said above; & I am sure I can find some of our Society who wd willingly describe the Hymenoptera In Coleoptera I will do my best. In the other Orders there are zealous persons to assist you therefore "nil desperandum". Before you left London Rennie had pirated Stephenss work, they both went to Law, the question was referred to Arbitration & it terminated as such cases generally do, in each person paying their own expenses Stephens's expenses amount to £400 there has been a Subscription set on foot for him in London which amounts at present to 80£. There is another at Cambridge which is nearly the same, it will close next September & I hope by that time the Sum will nearly reach the £400.

As to Geology from sending home the much desired bones of Megatherium your name is likely to be immortalized at the Cambridge Meeting of naturalists your name was in every mouth & Buckland applauded you as you deserved. I forgot to say Stephenss work was stopped it is now going on. The Coleoptera are nearly finished, only some few Staphilinidae to describe. There is chance when *they* are finished that it will stop I hope however the Public scientific world will then support him As to Science in general particularly Zoological, there never was a period when England had such leading men in all its branches as at present. Owen of the College of Surgeons has published some fine anatomical internal Structure of Animals.[3] your friend Dr Grant is working away at the Mollusca & Infusoria publishing at a great rate. He gave a Course of Lectures on the Regne Animal but did not elicit much novel matter As a Lecturer he is rather too grave, & rather too pedantic, too much given to *coin* hard words, at times He was eloquent & animated, generally verbose & lengthy. He is however what is more valuable a very amiable Man & strictly consciencious

Wishing you my dear Fellow all success in collecting & Health to enjoy yourself & a safe return to Old England with 1,0000,000,0000 Insects I remain Yours very truly & sincerely F W. Hope

P.S After consulting Eyton I took the liberty of putting your name down as a member of the Entomological Society

DAR 204.8

[1] 'They change their clime, not their mind, who rush across the sea'. Horace, *Epistles* I. 11. 27
(translated by H. R. Fairclough. Loeb Classical Library. 1978.).
[2] John George Children, Nicholas Aylward Vigors, Thomas Horsfield, James Francis Stephens,
Frederick William Hope, William Spence, George Robert Gray, George Robert Waterhouse.
[3] See the bibliography of Owen's papers in R. S. Owen 1894, 2: 335−6.

From Catherine Darwin 27−30 January 1834

Shrewsbury.

January 27th | 1834.

My dear Charles

I think Caroline has written to you since we received your letter of Oct 23^d from Buenos Ayres, when you were in an alarm at the chance of losing your Luggage & Servant. We were very sorry to hear of the unfortunate loss of your Gun, as every body who has heard of it, agree in its being such an irreparable loss in the Country you are in.— We are very anxious for your next letter to tell us that you are safe on Board again after all your Adventures, and your daring Rides & Exploits.— I can easily believe in the great interest of leading the Gaucho Life, and I only wish I could be quite easy as to the risk, among that villainous set of people.

Your Journal is safely packed up, now that we have all read it.— It is a very nice fashion you have got into of sending your Letters by Liverpool Packets; they come to us so much sooner than when they go to London. I am sure there is some delay at the Admiralty in forwarding them, for we see the news in the Paper from S. America, of the same date, a week sooner than we get your Letters.

Papa is very well, and manages to occupy himself with the small degree of Practise he keeps, as he has almost entirely given it up.— He desires his best love and to tell you how *very glad* he always is to hear from you. I have very little news indeed to tell you this time; the principal is that Caroline had a letter not long ago from William Fox in the Isle of Wight, announcing his intended marriage in the Spring to a Miss Harriet Fletcher, the daughter of a Sir Richard Fletcher who was killed at the siege of Zaragoza in the Peninsular War, and who resides near where the Fox's are, in the Isle of Wight.— He appears to be much in love with her, as far as one can tell by his letter, and what is a greater proof, is that it appears that his bad health has been partly caused by his anxiety about the success of his suit; so that now that is happily settled, I do hope he will become stronger & happier, and be able to resume his Curacy.— He hopes to find a Curacy in the South of England, as that climate agrees with him so much better; whenever he is able to undertake the duty, but I suppose he will not be strong enough to try it this Summer.— It is a very nice thing that so good a man should be restored to happiness & usefulness again, and I do hope Miss Fletcher will prove as nice a person as he deserves to have; any how she must be much better than that stupid Bessy Galton.— Another match has just been announced; but this, I am afraid will not interest you, and it must be owned, it is a desperately dull one.— Dr Holland is going to marry in second nuptials Sydney Smith's daughter;[1] she is old, and foolish by all accounts, so that it is not a match his friends particularly like. I do not know what his great friend Erasmus thinks of it, as he has not written to us of late.— Erasmus appears to live in a little round of visits about London; he goes the Circle of Clapham (the Hensleigh Wedgwoods) Dr Holland's, Lady Gifford's, and Mrs Marsh's.[2]

Poor Uncle Jos has to go up to London next week, to resume his Parliamentary duties; it will be a great bore for him to leave his family, & Maer, which he is so fond of, and go to London, which he dislikes so much.— Aunt Bessy continues just in the same state, quite helpless, but very comfortable otherwise.— Did Caroline tell you that when Robert Clive was here, not long ago, he talked of shooting with you, at Maer, the last time he had seen you? What a merry, pleasant goodnatured creature Robert Clive is.— I think he is so much the most agreeable of the Clives. The three Bachelors Henry, Edward & Robert Clive live together at Styche very comfortably.— You will be sorry to hear how ill poor Fanny Biddulph continues; she looks the ghost of herself, and Papa thinks her very seriously unwell; it has now continued so very long, ever since her confinement last May.— They talked of coming here this week, if she does not get better; Mr Biddulph is as attentive as possible to her.— Your other d⟨ear⟩ Friend Sarah Williams, has been leading ⟨the⟩ gayest and most rackety life this Winter; there never yet was such a lawless house as Eaton, nor such a merry one.— For about two months, there was an immense party collected there, about 20 in the house, and they went mad over Private Theatricals. They had regular Scenery, and Dresses, and they acted about three Plays, at different times.—

Thursday 30th.— I have got a piece of news to tell you now, which *ought* to interest you; the birth of Miss Louisa Jane Wedgwood,[3] Harry & Jessie's first & eldest daughter; she was born last Tuesday the 28th, and Mother & Daughter were both going on very well, when Harry wrote to us;— You will find a flourishing rising Generation when you return. This has been one of the most rainy Winters in England that has ever been remembered, we have had no frost or snow, but constant deluges of rain, pouring down. We have had *three* Floods, of the Severn, which is a most unusual number. I think it a very disagreeable kind of Winter, but people in general like the mildness of it very much.— I wish I could send you any Political news, but I am afraid I cannot— There have been constant rumours of Lord Grey's intending to resign, on account of his health, but with no truth in them, it is to be hoped.—[4] It is wonderful how many improvements were carried the last *long* Session; the great deed of the Emancipation of the Slaves, was upon the whole nearly as satisfactory as one could hope. How quietly the Slaves in the W. Indies have taken the certainty of their freedom; I hope they may go through their apprenticeships as quietly.—[5]

Goodbye, my dearest Charles.— We long to hear from you again and still more to get your first Letter from Valparaiso. How you will enjoy the hot

Climates again; all I entreat of you, is to take care of yourself, and not be rash in any thing. With all our most affectionate loves, believe me my dear dear old Charley, y^r very affe^{cte} Sister Catherine

Papa begs again I will give you his best love, and say every thing kind for him.—

DAR 204.6.1

[1] Saba Smith.

[2] Anne Marsh, a writer much admired by the Allens and Wedgwoods. She is frequently mentioned in *Emma Darwin*. Her sister Emma was Henry Holland's first wife.

[3] Louisa Frances Wedgwood.

[4] Lord Grey retired in 1834 as a result of a disagreement among the Cabinet about the renewal of the Irish Coercion Act of 1833.

[5] The term of apprenticeship, originally seven years, was reduced to five years when it became clear that the 'apprentices' would no longer work for their masters.

From Susan Darwin 12[−28] February 1834

Shrewsbury
February 12th | 1834

My dear Charles

This is your Birthday; so I must begin my letter to wish you joy, and many happy *returns* of it (but not abroad) mind that.

Papa who never forgets anniversarys remembered this day of course at Breakfast and sends you his best love & blessing on reaching 25 years. Poor old Nurse Nancy entertained me all the time I was dressing this mor^g with many lamentations over your absence on this day when you ought to be eating Plum pudding with us, & all the Servants say she has not failed to put them in mind of you; so as I have often told you before, you are not forgotten by the least of us.—

We are very anxious for your next letter to tell us how you escaped from Buenos Ayres & what is become of yr Luggage, but before this month is up I hope to hear news of you.— My Sisters have told you how very much we enjoyed your Journal and what a nice amusing book of travels it w^d make if printed, but there is one part of your Journal as your Granny I shall take in hand namely several little errors in orthography of which I shall send you a list that you may profit by my lectures tho' the world is between us.— so here goes.—

wrong	right according to sense.
loose. lan*s*cape. hi*ge*st	lose. landscape. highest.
profil. cann*a*bal	profile. cannibal. peaceable
peac*i*ble. quarre*ll*	quarrel.— I daresay these errors are the effect of haste, but as your Granny it is my duty to point them out.—

We have had the most surprising mild winter that ever was known or at least that I remember: not one days hard frost or anything like snow. I regret this for no reason except the Ice house being empty.— In consequence of this our Spring flowers in Feb^ry are in full blow but there is no pleasure with them so early.—

Parliament has just met & we read aloud the debates in the Eve^g— M^r O'Connell[1] still keeps the upperhand by talking & boring the House with Irelands wrongs till one is quite sick of him & his country.— I suppose Church & Corn Laws will be the great things this Session.—

We have not heard from Erasmus the last month he has taken a naughty fit and nothing can make him write. we have all three written in turns to reproach him in vain. Last year once he did the samething & then he confessed afterwards that he kept silent so long on purpose as he hoped to make us believe he was gone abroad.— He is quite a grand gentleman now with his own Cab and horse— we are expecting from London every day a new little Phaeton as the Car is pronounced quite *unseaworthy*.—

We had a very gay day yesterday 6 Owens came over to see a famous Conjuror perform at the Fox Inn which tempted Mr Owen to bring over Baby—& we were all much amused: his chief feat was shewing us how *to sit upon nothing*! which certainly might be a very useful accomplishment.

Francis Owen sails this Summer for the East Indies—how pleased poor Arthur will be to see him.—

Next time we have any opportunity of sending you books I ⟨sha⟩ll certainly send you "Peter Simple" the best Novel that has come out a long time, & will just suit you now as it is written by a Naval Officer Captain Maryatt:[2] and the sea terms which puzzled us you will understand & relish.— About a fortnight ago I went to Acton Burnell[3] to stay three days to meet the Owens. I enjoyed my visit extremely as I had never been there before & we walked a great deal about the Park which is very beautiful Emma Owen & I went to Mass one mor^g very early and I never saw anything like the mummery that was performed by the Priest and one of the Footmen in Livery who I suppose acted as Clerk— One of the Priests seemed to be quite captivated by Emma made her presents, and I am sure hopes to make a Convert of her— He has sent her a pair of Slippers worked by himself.— We have not seen Fanny Biddulph for some time the last thing we heard of was her sending poor *Bijou* the Poodle to Ireland with a Capt White who has begged it from her— The poor Dog latterly had been kept quite shut up at Chirk Castle so it is rather a happy change for it.—

Yesterday Robert Wedgwood came over here on his road to Welsh Pool to look at a Living which William Clive has offered to give him: but as it is only worth 120 pound a year he w^d not be better off than at Maer with his two Curacys so I suppose he will not accept it.— We are just returned from spending three days at Ness where Capt Cotton[4] one of Mr Cottons brothers was staying. he is a Navy Capt. just returned after 5 years absence & we heard a gt deal of naval talk. he is very handsome & agreeable, & remembers having once seen Capt Fitzroy as a Midshipman.—

It is in vain to keep my letter any longer open for the chance of hearing from you as I am afraid you must have sailed for Valparaiso without being able to write fr Buenos Ayres so Good bye my dearest old Charley from y^r very affectionate Susan Darwin

Postmark: Shrewsbury FE 28 1834
DAR 204.6.1

[1] Daniel O'Connell.
[2] Frederick Marryat.
[3] Acton Burnell, south-east of Shrewsbury, the seat of Sir Edward Joseph Smythe, Bart (Bagshaw 1851, p. 498).
[4] Francis Vere Cotton.

To J. S. Henslow[1]   March 1834

E. Falkland Is^d
March— 1834

My dear Henslow

Upon our arrival at this place I was delighted at receiving your letter dated Aug. 31.— Nothing for a long time has given me so much pleasure. Independent of this pleasure, your account of the safe arrival of my second cargo & that some of the Specimens were interesting, has been, as you may well suppose, most highly satisfactory to me.—

I am quite astonished that such miserable fragments of the Megatherium should have been worth all the trouble Mr Clift has bestowed on them. I have been alarmed by the expression cleaning all the bones, as I am afraid the printed numbers will be lost: the reason I am so anxious they should not be, is that a part were found in a gravel with recent shells, but others in a very different bed.— Now with these latter there were bones of an Agouti, a genus of animals I believe now peculiar to America & it would be curious to prove some one of the same genus coexisted with the Megatherium; such & *many other* points **entirely** depend on the numbers being carefully preserved.— My entire ignorance of comparative Anatomy makes me quite dependent on the numbers: so that you will see my geological notes will be useless without I am certain to what specimens I refer.— Since receiving these specimens, you ought to have received two others cargos, shipped from the Plata in July & November 1833.— With the latter there was a heavy box of fossil remains, which is now I suppose at Plymouth. I followed this plan from not liking to give you so much trouble: it contains another imperfect Megatherium head, & some part of the skeleton of an animal, of which I formerly sent the jaw, which had four teeth on each side in shape like this ⌒ ⌒ ○ ○ .— I am curious to know to what it belongs.—[2]

Shortly before I left M: Video I bought far up in the country for two shillings a head of a Megatherium which must have been when found quite perfect.— The Gauchos however broke the teeth & lost the lower jaw, but the lower & internal parts are tolerably perfect: It is now, I hope, on the high seas in pursuit of me.— It is a most flattering encouragement to find Men, like Mr Clift, who will take such interest, in what I send home.—

I am very glad the plants give you any pleasure; I do assure you I was so ashamed of them, I had a great mind to throw them away; but if they give you any pleasure I am indeed bound, & will pledge myself to collect whenever we are in parts not often visited by Ships & Collectors.— I collected all the plants, which were in flower on the coast of Patagonia at Port Desire & St. Julian; also on the Eastern parts of Tierra del Fuego, where the climate & features of T del Fuego & Patagonia are united. With them there are as many seeds, as I could find (you had better plant all ye rubbish which I send, for some of the seeds were very small).— The soil of Patagonia is *very* dry, *gravelly* & light.— in East Tierra, it is gravelly—peaty & damp.— Since leaving the R. Plata, I have had some opportunities of examining the great Southern Patagonian formation.— I have a good many shells; from the little I know of the subject it must be a Tertiary formation for some of the shells & (Corallines?) now exist in the sea.— others I believe do not.— This bed, which is chiefly characterised by a great Oyster is covered by a very curious bed of Porphyry pebbles, which I have traced for more than 700 miles.—but the most curious fact is that the whole of the East coast of South part of S. America has been elevated from the ocean, since a period during which Muscles have not lost their blue color.— At Port St Julian I found some very perfect bones of some large animal, I fancy a Mastodon.—[3] the bones of one hind extremity are very perfect & solid.— This is interesting as the Latitude is between 49° & 50° & the site is so far removed from the great Pampas, where bones of the narrow toothed Mastodon are so frequently found— By the way this Mastodon & the Megatherium, I have no doubt were fellow brethren in the ancient plains Relics of the Megatherium I have found at a distance of nearly 600 miles apart in a N & S. line.—

In Tierra del Fuego I have been interested in finding some sort of Ammonite (also I believe found by Capt King) in the Slate near Port Famine; on the Eastern coast there are some curious alluvial plains, by which the existence of certain quadrupeds in the islands can clearly be accounted for.— There is a sandstone, with the impression of the leaves of the common Beech tree also modern shells, &c &c.— On the surface of which table land there are, as usual, muscles with their blue color &c.— This is the *report* of my *geological section*! to you my President & Master.— I am quite charmed with Geology but like the wise animal between two bundles of hay, I do not know which to like the best, the old crystalline group of rocks or the softer & fossiliferous beds.— When puzzling about stratification &c, I feel inclined to cry a fig for your big oysters & your bigger Megatheriums.— But then when digging out some fine bones, I wonder how any man can tire his arms with hammering granite.— By the way I have not one clear idea about cleavage, stratification, lines of upheaval.— I have no books, which tell me much & what they do I cannot apply to what I see. In consequence I draw my own conclusions, & most gloriously ridiculous ones they are, I sometimes fancy I shall persuade myself there are no such things as mountains, which would be a very original discovery to make in Tierra del Fuego.— Can you throw any light into my mind, by telling me what relation cleavage & planes of deposition bear to each other?—

And now for my second *section* Zoology.— I have chiefly been employed in preparing myself for the South sea, by examining the Polypi of the smaller Corallines in these latitudes.— Many in themselves are very curious, & I think are quite undescribed, there was one appalling one, allied to a Flustra which I daresay I mentioned having found to the Northward, where the cells have a moveable organ (like a Vultures head, with a dilatable beak), fixed on the edge. But what is of more general interest is the unquestionable (as it appears to me) existence of another species of ostrich, besides the Struthio Rhea.— All the Gauchos & Indians state it is the case: & I place the greatest faith in their observations.— I have head, neck, piece of skin, feathers, & legs of one. The differences are chiefly in color of feathers & scales on legs, being feathered below the knees; nidification & geographical distribution.—[4]

So much for what I have lately done; the prospect before me is full of sunshine: fine weather, glorious scenery, the geology of the Andes; plains abounding with organic remains, (which perhaps I may have the good luck to catch in the very act of moving); and lastly an ocean & its shores abounding with life.— So that, if nothing unforeseen happens I will stick to the voyage; although, for what I can see, this may last till we return a fine set of white-headed old gentlemen.—

I have to thank you most cordially for sending me the Books.—[5] I am now reading the Oxford Report.—[6] the whole account of your proceedings is most glorious; you, remaining in England, cannot well imagine how excessively interesting I find the reports; I am sure, from my own thrilling sensations, when reading them, that they cannot fail to have an excellent effect upon all those residing in distant colonies, & who have little opportunity of seeing the Periodicals.— My hammer has flown with redoubled force on the devoted blocks; as I thought over the eloquence of the Cambridge President[7] I hit harder & harder blows. I hope, to give my arm strength for the Cordilleras, you will send me, through Capt. Beaufort, a copy of the Cambridge Report.—[8]

I have forgotten to mention, that for some time past & for the future I will put a pencil cross on the pill-boxes containing insects; as these alone will require being kept particularly dry, it may perhaps save you some trouble.—

When this letter will go, I do not know, as this little seat of discord has lately been embroiled by a dreadful scene of murder & at present there are more prisoners, than inhabitants.—[9] If a merchant vessel is chartered to take them to Rio I will send some specimens (especially my few plants & seeds).—

Remember me to all my Cambridge friends.— I love & treasure up every recollection of dear old Cambridge.—

I am much obliged to you for putting my name down to poor Rams⟨ay's⟩ Monument— I never think of him, without the warmest admiration.— Farewell my dear Henslow—believe my your most obliged & affectionate friend. Charles Darwin.—

N.B. What I have said about the numbers attached to the fossils, applies to every part of my collections.— Videlicet. Colors of all the Fish: habits of birds &c &&c

There is no opportunity of sending a Cargo: I only send this, with the seeds, some of which I hope may grow, & show the nature of the plants far better than my Herbarium. They go through Capt. Beaufort: Give Mʳ Whewell my best thanks for sending me his tide paper:[10] all on board are much interested by it.— Remember me most kindly to Mʳˢ Henslow & Leonard Jenyns.—

The Box of fossil remains, to which I have alluded is with Dʳ Armstrong at Plymouth: if you think it worth while, you can write to him (just stating the reasons) & he will (perhaps be too glad) to forward it. Capt. FitzRoy will mention in a letter the possibility of your writing for the Box.— It could be easily sent by water to London & from thence either by land or water to Cambridge.— | Once more dear Henslow, Farewell.—

Royal Botanic Gardens, Kew.

[1] The following passages from this letter were extracted by Henslow and published in the Cambridge Philosophical Society pamphlet:

> 2.2 'I have . . . preserved.' 2.9; 2.7 *omits* 'now' *before* 'peculiar'
> 4.4 'I collected . . . united.' 4.7
> 4.9 'The soil . . . color &c.' 5.7; 4.13 *omits* '?'; 5.5 'leaves of ' *changed to* 'leaves like'
> 6.1 'I have chiefly . . . distribution.' 6.12; 6.2 'by examining' *changed to* 'and examining'; 6.8
> 'Rhea' *changed to* 'ostrea'; 6.11 '& scales on legs, being feathered' *changed to* 'and scales; in the legs being feathered'

[2] See letter to J. S. Henslow, 12 November 1833, n. 3.

[3] It was not a Mastodon but a hitherto unknown extinct llama- or camel-like pachyderm, which Richard Owen named *Macrauchenia patachonica* (see *Fossil Mammalia*, pp. 35 – 56, and *South America*, pp. 95 – 6). The bones of CD's specimen are in the British Museum (Natural History).

[4] For CD's notes and observations on this species see 'Ornithological notes', pp. 273 – 4, and *Red notebook*, pp. 127, 130, 153. The new species was named *Rhea darwinii* by John Gould in 1837. See *Collected papers* 1: 38 – 40.

[5] One of them may have been the third volume of Lyell's *Principles of geology* (1833). CD first mentions having received it in his letter to Henslow of 24 July 1834, but the context suggests that it had arrived before the expedition up the Santa Cruz River, which was made following this visit to the Falklands. CD's copy in Darwin Library – CUL is inscribed only 'C. Darwin', with no date.

[6] The *Report* of the second meeting of the British Association at Oxford in 1832.

[7] Adam Sedgwick was elected President for the Cambridge meeting in 1833.

8 There is no record that CD received it, though it was almost certainly sent to him. In the Darwin Library — CUL there is a pamphlet of lithographed signatures of the members of the British Association who met at Cambridge, with a report of the proceedings of the public meetings, in 'Philosophical tracts', vol. 2 (a bound quarto volume of miscellaneous printed papers).

9 See *'Beagle' diary*, p. 209.

10 Whewell 1833.

From Caroline Darwin   9 — 28 March [1834]

[Shrewsbury]
March 9th.

My dear Charles

The very day Susan had sent her letter (Feb 27th.) to the Post yours from Monte Video November 13th. arrived— We had been anxiously wishing for an account of your difficulties having ended & I am exceedingly glad to find the newspaper report of Peace confirmed by your letter— My Father desires his kindest love to you & bids me say he did not growl or grumble at the last £50 you said you drew—& he says you must not fret about money—but be as good & prudent as you can. My Father does not suppose you can do any thing, but the next boxes & cargoes you send to England, he wishes you would talk or enquire a little what is charged for their conveyance, as my Father thinks that they make him pay too much—it may however be the conveyance in England to Cambridge that is so expensive.— We have not heard yet through Erasmus from Professor Henslow of the safe arrival of the last cargo of specimens skins of birds &tc but the Times newspaper says "several packages of specimens of fossils birds & quadrupeds skins & geological specimens have been collected by the naturalist M.r *Dawson* & sent to Professor *Hindon* at Cambridge by the last packet", so from this blundering account we may presume that they are safe, & I congratulate you, it would be very hard if they had been lost after all your zeal & industry. I am beginning to long to hear of the journey of 300 miles you were just setting out upon when last you wrote. Charlotte says in her last letter & I dare say very truly, that she supposes we have hardly any idea of the interest & delight of your adventurous travels, from our own jog trot tours— I think the wildest most adventurous *journey* I ever took was that with you over the Hills to Bala¹ & I am afraid now you will not be able to shudder over our perils of wind & water—poor old fellow how I wish after your fatigues you had the comfortable little parlour we had that night & tea & muffins to refresh your self—but seriously, I do most earnestly hope you may meet no dangers & that before this letter goes we may hear of your safe return to Monte Video.— My Father is very well he went last week for 3 nights to Camp Hill & had a very pleasant visit with Aunt Sarah one morning going to call at Keel & to see all the John Wedgwood family assembled there & the following morning he spent at Maer. Uncle Jos was not then attending his duties in Parliament of which I believe he is getting heartily tired— People say the power of the Radical party is gaining strength & that if this ministry should go out of which there have often been rumours & Parliament dissolved, there is great fear of a strong body of Radicals being admitted— It is very goodnatured of you remembering Nancy in your letters, she was so much pleased by the love you sent her in your last letter, I believe really she counts the months &c to your return with as much eagerness as we do—

Good Friday—March 28ᵗʰ— I dare say Erasmus will write to you himself, but for fear he should not I will copy a sentence in his last letter to me— "I called on Mʳ Clift the curator of the Museum at the College of Surgeons to read to him the passage in Charles's letter about the bones, and you never saw a little man so delighted, I have accordingly written to Plymouth to get them forwarded by sea, & I hope to make the College pay for the carriage. He has been working every spare hour for the last two months, & it certainly does appear a strange coincidence that the College should possess the front portion of the scull & that Charles should send home the remainder of it which enables them to complete the drawing tho' not the actual skeleton"—² I give you joy my dear Charles on having found these bones that delight the learned so much—& I do not doubt you will have long letters from Professor Henslow or some one who will tell you all particulars— Tom Eyton staid two days here the week before last & made most particular enquiries about you & wished he was with you— he said he would go to Cambridge & hear what people said about the specimens you have sent home & then write to you— he said he had many things to tell you which he was sure would interest you— we wanted him to stay another day to go with us to a party at Mʳˢ Leightons but he said no, he could not, it was impossible & at last told the reason, he was expecting by that days Coach a Chamelion & he must go & give it the meeting & some flies for its dinner— My Father was very much pleased by what he saw of him he thought he had so much accurate information on many subjects— I have two marriages to tell you of having taken place this month— William Fox is married to the Miss Fletch⟨er⟩ we told you of—& has taken a house in the Island of Wight⟨.⟩ Julia says he is looking very happy & well, but his che⟨st⟩ is not yet strong enough to undertake any duty—

Dʳ Holland is also married to a daughter of Sydney Smith it was a grand wedding, being married by the Archbishop of York³ Erasmus has not yet seen her so we do not know whether she will prove popular in his little London set—

We are just come in from hearing a most excellent sermon by Mʳ Harding—⁴ how happy it would make me dear Charles to feel sure I should some Good Friday be hearing you—or indeed I should be well content if at this moment I could tell where you were & to know that you were safe & well & happy—

Professor Sedgwick has published a most beautifully written discourse on the Studies of the University—⁵ I had no idea he was such a religious high minded man—from his book he must have very strong religious feeling—

My Father is very well & comfortable both in body & spirits— the methodical life he now leads suits him in every way so much better, than the bustling about as he was doing when you left home. I often look back to the dread I had of his leaving off business thinking he would be so flat & dull— I do not know of any Shropshire news—except that poor Mary Burton died a few days ago after a long painful illness—also Miss Kinaston[6] of Hardwick is dead— Susan is going to Woodhouse next Wednesday to stay with Caroline till she goes to London where she is going soon & Francis with her to pay Sarah a visit before he leaves England He believes that the ship he is to go in to India will sail the middle of April so he has but a short time before him— They have very good accounts from Arthur— Fanny Biddulph is better but looks so thin & pale you would think her sadly changed—she quite dreads going to London on account of the exertion, she feels so weak— Marianne comes to us next week she does not write to you, we always send on all news of you immediately to her— My Father is in the room & desires his kindest love, he asks me if I have told you of your fame & about the skull, which I have done. Susan's love also & Believe me my dear Charles Yʳˢ affecly and God bless you— Caroline Darwin—

Postmark: Shrewsbury MA 29 1834
DAR 204.6.1

[1] In 1826. See 'Journal' (Appendix I) entry for 30 October of that year.
[2] See Clift 1835, plate xliv for an outline drawing, made in 1832, of the *Megatherium*, showing the parts discovered by Woodbine Parish. CD's bones are figured in *Fossil Mammalia*, plate xxx.
[3] Edward Vernon Harcourt.
[4] John Harding, incumbent of St George's, Shrewsbury (*Alum. Oxon.*).
[5] Sedgwick 1833a.
[6] Letitia Kynaston.

From J. M. Herbert [28 March] 1834

Oxford & Camb: Univerʸ Club | St: James Square
Good Friday. 1834

My dear Darwin,

So long a time has elapsed since the receipt of your last kind letter, that I fear you will think I have forfeited by my silence all claim to another— It was I think in October that your letter arrivd; & now 'tis the very end of March. The delay is quite unpardonable, but I will not aggravate the fault by attempting a shuffling apology— Cambridge has been full of bustle during the last year. Henslow has of course told you all about the meeting of the British Association, how the Philosophers talked & ate & talked again; how many of them were to be found at Pot-Fair[1] instead of the Evening Meeting at the Senate House, the disgusting conduct of the Provost of Kings in shutting the Gates of the College against an immense crowd, after having allowed the use of Kings Piece for the exhibition of Deck's Fireworks; alleging as a cause of his so doing the mob outside, which consisted of Ld Fitzwilliam & Party, Sir John Herschel & Party, Airy & Party &c & all the élites; how he himself (Henslow) was knocked down by the Porter, in attempting to get the Key, & afterwards put into durance vile.[2] It w^d have done your heart good to have seen how Sedgwick expressed his disgust at the College & its miserable Provost. I was engaged nearly the whole of the meeting in the Commissariats Department so that I heard little of what was going on; in truth I cared little to hear it as having neglected all the little scientific knowledge which I once possessed (not mind from any disregard, but from mere idleness) I sh^d hardly have understood two consecutive sentences that were uttered. I hear that some Geological Specimens that you sent over were considered extremely valuable as supplying some desiderated link in the chain, & that you came in for a due share of κυδος. What a nice fellow young Lowe is, I mean the Albino. he just failed in getting a double first at Oxford, getting only a second in Mathematics—[3] We were together nearly the whole of the Meeting & took to each other amazingly. At the meeting of the B.A. the preceding year at Oxford, there was a discussion on the cause of the pinkness in the eyes of Albinos in his presence. After this philosophy came a series of excellent concerts, the diva Malibran & De Beriot[4] being the chief attractions. To ensure the expenses of this they had a guarantee fund, consisting chiefly of Members of the University; unluckily most of the names were Whig: & consequently the Tory Families in the County declined supporting a Concert under their superintendence! The poor fund had to pay up £450 to cover the deficiencies— The University has since been kept alive by an attack upon them by the redoutable Beverley, who has in his turn been demolished by Sedgwick.[5] Beverley's charges of Gambling &c ag^st the University led to some investigations into the "unhallowed nightly orgies" which terminated in the expulsion of Lord J. Murray & another who had played to the amount of £1200 in one night—[6] Old Chafy[7] the other day had an action brought ag^st him by a Fellow-Commoner of Jesus & an attorney for false imprisonment in confining them within the walls of Sidney, because they w^d not give their names— £25 damages— At the present moment everybody is talking about & the London Papers are full of, a petition from all the good men & true at Cambridge in favor of Dissenters being admitted to the University—[8] Toryism is evidently on its last legs, it is amusing to look upon their labored tho' abortive efforts; they are like swine—swimming upon the buoyant wave of sophistry they are cutting their own throats by their clumsy & ill-timed maneuvres— The University of Oxford have just elected the D. of Wellington (not one of their own body) their Chancellor in the room of the accomplished Ld Grenville—thus completing the Triad of Field-Marshals—Chancellors of the three Universities —Cumberland! Glouster!! Wellington!!![9] (Go to 1st page)[10]

L^d. Grey in presenting the Cambridge Petition made a most excellent appeal in behalf of the claims of the Dissenters; which Wellington attempted to answer— He spoke of the Dissenters as Atheists?? & designated our Articles the Articles of Christian Faith, thereby imputing Infidelity to any who declined to subscribe to them. Can you imagine any conduct more inane? They had already by their previous illiberality taught the Dissenters to look on them with suspicion, as members of a pseudo-tolerant church. Can insult or injury ever be forgiven?[11] Cumberland made an insignificant attempt; & Gloucester our learned Chancellor was too unwell to be in his place; Higgins c^d not screw him up to concert pitch.

I stayed at Cambridge till the end of July, living almost entirely with that clever amusing silly fellow Cameron. I never knew a more pitiable case than his, save perhaps Matthew's— There they live in worse than idleness; two men who with no great exertion might have cut no mean figure in the world. Cameron is a good deal sobered down, but Matthew is still I fear a most determined votary of Circe— He commenced a translation of Virgil for Poll-Men (by the way he is an excellent Poll-Tutor) & had proceeded to the end of the 2^d line, when Heaviside unluckily threw the M.S.S. into the fire, & poor Matthew has never since been able to renew the attack— I saw Watkins about six weeks back in Chester where he had gone to be examined for orders, but I have since heard that his title[12] was not good. Whitley is at Durham—a Reader in Natural Philosophy at the New College— He tried for the Mathem! Professorship, but was beaten by Carr,[13] who is since dead; I hope Whitley is now almost certain of succeeding. The Election comes on in July— He is very wroth with you for never having written to him— He talks very furiously of Matrimony— I have no notion who the lady is— Your friend Marindin[14] has just married Colville's sister— He is going into the Church— Last summer I was staying three Weeks at Dolgellau, which is I think a much nicer place than Barmouth—except that there are no delightful associations about the former like those of the latter; one day I started in the Pen-Maen boat to visit old Barmouth, absolutely feasting on anticipation of the joyous welcome I sh^d meet with from Squinney & my other old cronies— She however looked cold on me; asked me to take her lodgings. John Robert had forgotten me; the day set in to furious rain, Barnett gave me a bad dinner, the boat did not return that night to Pen-Maen so in a fit of bitter mortification I set out in the midst of the rain to walk to Dolgellau— On going down the hill from Llan— (I forget the name about two miles from Dolgellau)[15] I trod on a rolling stone with my sound foot, & so grievously sprained my ancle that I was confined to my sofa for a fortnight— There is surely some fatality about my connexion with Merionethshire— Just as I was thinking of the limping way in which I walked down the hill some 5 yrs before with you & Butler, & congratulating myself on my altered condition, this unlucky accident happened, & I had again once more to crawl from thence to Dolgellau, tho' now alone & unfriended— I am going to ffestiniog this next summer.

I have now left Chester & am settled in London— The change is a miserable one— I had formed such agreeable acquaintances in Chester— Here one is alone amidst a countless multitude. I ought to have been an Entomologist if the law of mutual Attraction hold good in all cases; for a certain species of the Cimex Genus has taken such an extreme fancy to me, that it does not leave through the whole night long— I fear the consciousness of having ill treated you has made me more than ordinarily stupid. When you get this letter over half of your pilgrimage will be completed. Go on & prosper. I look forward with pleasurable expectation to the uncut 4to. How I shall enjoy yr bursts of honest enthusiasm— God bless you— Every body here who knows you sufficiently to appreciate your worth wishes you the happiest success but none more than your affec^{ate} friend | J M Herbert.

Write when you have leisure—

DAR 204.8

[1] The Pot Fair or Midsummer Fair were popular names for the fair held at Barnwell, 3 miles northeast of Cambridge, on 23 June (Winstanley 1940, p. 129).

[2] 'That beast of a Provost [Rev. George Thackeray] shut the great gates—his Porter shut up Henslow for an hour in the Lodge for trying to keep the gates open: we had (after waiting an hour) to go down the Lane:—the fireworks good' (Romilly 1967, p. 36).

[3] Robert Lowe took a 1st class in classics, 2d in mathematics, at University College, Oxford, 1833 (*Alum. Oxon.*).

[4] Charles de Bériot.

[5] Robert Mackenzie Beverley, a graduate of Trinity, wrote a pamphlet in November 1833 on the corrupt state of the University. Adam Sedgwick wrote four letters in reply during the next two years. Beverley responded to the first, comparing Sedgwick to a bear dancing on hot iron. (Romilly 1967, p. 46; Winstanley 1940, pp. 86—7.)

[6] The Hon. George Augustus Frederick John Murray. On 11 February 1834, Joseph Romilly noted in his diary, 'we expelled Hon. G. A. Murray & Hunter for gambling (Murray won near £800 of Hunter, who confessed to Whewell)' (Romilly 1967, p. 48).

[7] William Chafy, Master of Sidney Sussex College (*ibid.*, pp. 49 n., 51).

[8] The petition to abolish religious tests for the degree was signed by sixty-three members of the University Senate, among them John Stevens Henslow, Adam Sedgwick, Charles Babbage, and George Peacock. A copy, made by Joseph Romilly, University Registrary, is in the Cambridge University Archives (CUR 118). See also Winstanley 1940, pp. 83—96.

[9] The third university was Trinity College, Dublin (Cumberland).

[10] Herbert continued the letter by writing vertically across the first page.

[11] The appeal to abolish religious tests in the universities failed. They were not repealed until 1871.

[12] 'A certificate of presentment to a benefice, or a guarantee of support, required (in ordinary cases) by the bishop from a candidate for ordination' (*OED*).

[13] John Carr.

[14] Samuel Marindin married Isabella Colvile on 13 March 1834.

[15] Llanelltyd.

To Edward Lumb[1] 30 March 1834

H.M.S. Beagle E. Falkland Island
March 30.ᵗʰ 1834

My dear Lumb,

There is a French Whaler lying here which sails to day for M. Video, & I take this opportunity of writing to you.— I am very anxious that the Megatherium head, which Mʳ Keen procured for me,[2] should not be lost. You will be, I am sure, be glad to hear, that the fossil relics of older days, which I found at B. Blanca, have been of preeminent interest to those few, who in England care about such things.— Prof: Henslow &c begs of me to collect every scrap of the bones of the head of the great monster; for this reason, the specimens which Mʳ Keen intended to forward from the R. Negro are the more valuable.— I dare say you have already sent them to the Admiral at Rio.— You will very much oblige me, by sending a line to Valparaiso to state time & Ship, by which they were conveyed there so that, if they fail to arrive, I can write to Sir Michael Seymour.—[3]

Since the Beagle left the Plata, we have had a pleasant cruize; we spent some time on the coast of Patagonia; the country swarms with Gaunaco several were killed, but besides these, there were few living creatures. For my own part, I found some interesting work for the Geological hammer.— I trust, with what I saw to the North, to be able to draw up a tolerable sketch of the geology of this Eastern side of S. America. We then entered the Straits & passed on to Famine & returned to survey the east coast of Tierra del F.— The entrance of the Straits are found to be much narrower than drawn in the charts.— After the coast was finished we ran down to near Cape Horn, & returned & beat up the Beagle Channel to J. Button's country. Poor Jemmy was quite naked, excepting a rag round his waist; he was however very happy, did not wish to return to England; had not forgotten his English & lastly, but not least, he had married a young; & for a Fuegian, a beautiful Squaw.— From thence, we scudded to this island; this seat of discord for the elements, as well as for Human affairs.— You will have heard of the murder of poor Brisbane &c &c; such scenes of fierce revenge, cold-blooded treachery, & villainy in every form, have been here transacted, as few can equal it.—[4]

I shall be curious to hear what the wise Gov⟨er⟩nment of B. Ayres says on the occasion I suppose "a just revolt," "their poor subjects groaning under the tyranny of England" &c &c. When you write you must tell me all the gossip. How goes on Mʳ Griffith & your new minister?—[5] How fare the Indians against the Cæsar-like Rosas!?— You must not forget to write to me under charge of British Consul, Valparaiso Remember me most kindly to Mʳˢ Lumb, & with my best thanks for all your kindness | Believe me Yours very truly | Charles Darwin

We shall soon Sail to the river of Santa Cruz: it must be from the account of the Indians an immense one: I will keep my eyes open for Nutrias.— I suppose what I wrote about the Chupat quite knocked on the head your scheme.[6] Do not forget to give my best remembrances to M^r & M^rs. Keen. I shall not soon forget what very pleasant days I spent at their Estancia.—

Armando Braun Menendez, Buenos Aires

[1] This letter is published, with a full discussion of its background, in Winslow 1975.
[2] Mr Keen was the owner of an estancia at which CD stayed on his trip to the Rio Negro in West Uruguay, 22–4 November 1833.
[3] Commander of the South American station of the British Navy.
[4] Matthew Brisbane.
[5] Charles Griffith was British Consul at Buenos Aires. The new Minister Plenipotentiary, Hamilton Charles James Hamilton, did not arrive until 11 October 1834 (*BDR*, p. 7).
[6] See letter from Edward Lumb, 13 November 1833.

To Catherine Darwin 6 April 1834

East Falkland Is^d.
April 6th.— 1834.

My dear Catherine

When this letter will reach you I know not—but probably some man of war will call here before in the common course of events I should have another opportunity of writing.— I have received your letter dated Sept 27th 1833, & Caroline's before that. Since leaving the Plata, we have had, pretty fine weather, & a very pleasant cruize. The gales have not been half so spiteful or so furious this year as last.— We reached Port Desire without one; & there we staid for about three weeks.— We also went to Port St Julian.— I was exceedingly glad to have these opportunities of seeing Patagonia: it is a miserable country, great sterile plains abounding with salt & inhabited by scarcely any animals but the Guanaco.— I was very lucky & managed to kill a couple of these animals: one of which gave us fresh meat for dinner on Christmas day.— The geology of this district abounds with interest; the recent elevation of this whole side of S. America can be most clearly proved— At Port St Julian, I had the good fortune to find some very perfect bones, of what I believe is some sort of Mastodon or Elephant

There is nothing like geology; the pleasure of the first days partridge shooting or first days hunting cannot be compared to finding a fine group of fossil bones, which tell their story of former times with almost a living tongue. After entering the St^s of Magellan; we had a very interesting interview with the Patagonians, the giants of the older navigators[1]; they are a very fine set of men, & from their large Guanaco mantles & long flowing hair, have a very imposing appearance.— Very few, however, were over 6 feet high, but broad across the shoulders in proportion to this.— They have so much intercourse with Sealers & Whalers, that they are semi-civilized: one of them, who dined with us eat with his knife & fork as well as any gentleman.— Many of them could talk a little Spanish.— For observations we ran on to P. Famine; justly so called from the terrible sufferings of Sarmiento's colony.—[2] Of this there is not now the least vestige; every thing is covered up by the deep entangled forest of Beech. We then returned to the outside coast & completed the Chart of the Eastern side: When this was finished after visiting some of the Southern islands we beat up through the magnificent scenery of the Beagle channel to Jemmy Buttons country. We could hardly recognize poor Jemmy; instead of the clean, well-dressed stout lad we left him, we found him a naked thin squalid savage. York & Fuegia had moved to their own country some months ago; the former having stolen all Jemmy's clothes: Now he had nothing, excepting a bit of blanket round his waist.— Poor Jemmy was very glad to see us & with his usual good feeling brought several presents (otter skins which are most valuable to themselves) for his old friends.— The Captain offered to take him to England, but this, to our surprise, he at once refused: in the evening his young wife came alongside & showed us the reason: He was quite contented; last year in the height of his indignation, he said "his country people no *sabe* nothing.— damned fools" now they were very good people, with *too* much to eat & all the luxuries of life—

Jemmy & his wife paddled away in their canoe loaded with presents & very happy.— The most curious thing is, that Jemmy instead of recovering his own language, has taught all his friends a little English: "J. Button's canoe & Jemmy's wife come".—"give me knife" &c was said by several of them.— We then bore away for this island,—this little miserable seat of discord.— We found that the Gauchos under pretence of a revolution had murdered & plundered all the Englishmen whom they could catch & some of their own country men.— All the economy at home makes the foreign movements of England most contemptible: how different from old Spain: Here we, dog-in the manger fashion seize an island & leave to protect it a Union jack; the possessor has been of course murdered: we now send a Lieutenant, with four sailors, without authority or instructions. A man of war however ventured to leave a party of marines, & by their assistance & the treachery of some of the party, the murderers have all been taken.—their being now as many prisoners as inhabitants.—

This island must some day become a very important halting place in the most turbulent sea in the world.—it is mid way between Australia & South sea to England. Between Chili Peru &c & the R. Plata & R. de Janeiro.— There are fine harbors, plenty of fresh water & good beef: it would doubtlessly produce the coarser vegetables. In other respects it is a wretched place: a little time since I rode across the island & returned, in four days: my excursion would have been longer: but during the whole time it blew a gale of wind with hail & snow; there is no fire wood bigger than Heath & the whole country is a more or less an elastic peat bog.— Sleeping out at night was too miserable work to endure it for all the rocks in S. America.—

We shall leave this scene of iniquity in two or three days & go to the Rio de la S.t Cruz: one of the objects is to look at the Ships bottom; we struck rather heavily on an unknown rock, off Port Desire, & some of her copper is torn off.— After this is repaired, the Captain has a glorious scheme: it is to go to the very head, that is probably to the Andes, of this river.— it is quite unknown. the Indians tell us it is two or 3 hundred yards broad & horses can no where ford it! I cannot imagine anything more interesting. Our plans then are to go to Port Famine & there we meet the Adventure, who is employed in making the chart of the Falklands. This will be in the middle of winter, so I shall see Tierra del in her white drapery.— We leave the Straits, to enter the Pacific by the Barbara channel, one very little known & which passes close to the foot of M. Sarmiento (the highest mountain in the South, excepting M. !!Darwin!!.).—[3] We then shall scud away for Concepcion in Chili.— I believe the Ship must once again steer Southward, but if any one catches me there again, I will give him leave to hang me up as scarecrow for all future naturalists.—

I long to be at work in the Cordilleras, the geology of this side, which I understand pretty well is so intimately connected with periods of violence in that great chain of mountains.— The future is indeed to me a brilliant prospect: you say its very brilliancy frightens you; but really I am very careful; I may mention as a proof, in all my rambles, I have never had any one accident or scrape.

And now for some queries.— Have you received a small square deal box, with part of my Journal, sent from the Plata in July 1833 (through Capt. Beaufort) Acknowledge it in more than one letter: recollect what a *bobbery* (a sea phrase)[4] I made about the other parcel.— I received a box with some delightful books & letter from Henslow: did Erasmus send it? there was not even a list of the books & I know not whom to thank. There is a Hon. Col. Walpole, consul-general at St Jago de Chili.—[5] Have I not heard of some such man at Walcot?— What sort of person is he?—

I do not recollect anything more to say: not having any apologetical messages about money, is nearly as odd a feature in my letters, as it would have ⟨been⟩ in Dick Musgrove's.—[6] I am afraid it will be, till we cross the Pacific, a solitary exception.

Remember me most affectionately to all the Owens tell dear Fanny I do not how to thank her, at this distance, for remembering me.— Continue in your good custom of writing plenty of gossip: I much like hearing all about all things: Remember me most kindly to Uncle Jos & to all the Wedgwoods. Tell Charlotte (their married names sound downright unnatural) I should like to have written to her; to have told her how well every thing is going on.— But it would only have been a transcript of this letter, & I have a host of animals, at this minute, surrounding me, which all require embalming & Numbering.—

I have not forgotten the comfort I received that day at Maer, when my mind was like a swinging pendulum.— Give my best love to my Father. I hope he will forgive all my extravagance—but not as a Christian—for then I suppose he would send me no more money.—

Good bye dear Katty to you & all ye goodly Sisterhood. | Your affectionate brother | Chas. Darwin.—

My love to Nancy. tell her if she was now to see me with my great beard, she would think I was some worthy Solomon come to sell the trinkets.—

I have enclosed a letter of my servants will you pay the postage & forward it: by being my servant, he looses the penny priviledge & his friends cannot afford ye 3^s ⁱⁱ 6^d —⁷

DAR 223

¹ Magellan is said to have named the people of this region Patagones when he observed gigantic footprints in the sand (*Narrative* 2: 133 – 4).

² Pedro Sarmiento had established a Spanish colony of 400 at the head of the Straits of Magellan in 1584. The name 'Port Famine' was given by Thomas Cavendish in 1587 when he found the starving colonists during his circumnavigation of the globe (1587 – 8) (see *DNB*, 'Thomas Cavendish').

³ In '*Beagle*' diary, p. 214, CD wrote: 'A mountain which the Captain has done me the honour to call by my name, has been determined by angular measurement to be the highest in Tierra del Fuego, above 7000 feet & therefore higher than M. Sarmiento.' Robert FitzRoy (*Narrative* 2: 215 – 16) is less certain, 'as the measurements obtained did not rest upon satisfactory data'. He gives its height (in a 'Table of remarkable heights', *ibid.* Appendix, pp. 301 – 3) as 6800 ft and Mt Sarmiento as 6910 ft. In the *Times atlas* the heights are given, respectively, as 2135 m and 2300 m.

⁴ 'Noise, noisy disturbance, "row" ' (*OED*).

⁵ *BDR*, p. 33, lists Lieutenant-Colonel Hon. John Walpole as Consul-General and Plenipotentiary at Santiago, Chile, but only for 1837 – 41. He was, however, certainly in residence when CD arrived in Santiago in August 1834 (see letter to Robert FitzRoy, [28 August 1834]).

⁶ See postscript of letter to Catherine Darwin, 22 May – 14 July 1833.

⁷ At the upper left corner of the first page, the following appears, in another hand: 'Mrs Hewtson Camelford Cornwall'—the address to which Covington's letter was to be forwarded.

From William Owen Sr   10 April – 1 May 1834

Woodhouse
April 10th 1834—

My Dear Charles,

Excuse this familiar address   It would hurt me & I believe you not less if I used any other, I will therefore offer no further Apology— Though I do not write often (this being I am afraid my third Letter only since you left us) & though I do not hear directly from you, I do not fail to make frequent enquiries about you from your Sisters, & have hitherto, as I trust I shall continue to be, been constantly gratified by the most satisfactory reports, though your travels have not been unattended with Perils & hardships, which, if I know you at all, I rather think will not make them less agreeable to you, even when they are taking place; & we can all enjoy the remembrance of dangers & difficulties when they are surmounted however little we have liked them when actually taking place or however ill we may have supported them— You have therefore yet a Pleasure to come which I trust I shall in some degree partake of, in hearing you recount some of your adventures; And as we have now enter'd upon the third year since you left England I shall soon begin to try to count, like a School boy before the Holidays, how long it will be before I may look for that Pleasure— And really my good fellow I do think that you have now indulged your rambling Propensity as much as is reasonable or good for any moderate Man, & I am anxiously hoping to hear you are thinking of returning home, not a word of which has yet been whisper'd— Habit is second Nature & I am afraid if you remain at Sea & on such an amusing expedition, (as it is to you at least,) you will acquire too great a fondness for rambling, & never again be content to sit down quietly amongst your Country Neighbours & be satisfied with the tame sport of Pheasant & Partridge Shooting—which I take this opportunity of telling you is by no means improved; on the Woodhouse domain at least, since you partook of it— The alteration in the Game act by which it is now permitted to be publicly sold has not answered the expectation formed of it in preventing Poaching, but on the contrary I am afraid tends to promote it— Besides this I believe I have had a d——d Rogue for a Keeper, & consequently have had very little Game the last season. Fortunately I have not had many Sportsmen with me, as William has been absent the whole Winter, either with his Reg^t at Dorchester or in Warwickshire & Leicestershire hunting—for which I am afraid his passion is rather upon the encrease. Poor Francis has been here alone, & by means of his Friends has had a great deal of hunting & shooting, & may now possibly cross you, as he has at last got a Commission in the 63rd. Reg^t which is at Madras, where he will proceed immediately to join it—& his brother being there makes that station more acceptable than it would have been otherwise. You will be pleased to hear that we continue to receive the most satisfactory accounts from Arthur, though he does not write so often or so fully as we should like to hear from him—still all he writes is good, his health is good, he makes no sort of complaints & seems perfectly satisfied with his Situation & the Country— which in one letter he says would do very well if it was not quite so far from England— This I think is as small an objection as he could well make & one which we should have been hurt if he had not discover'd. The meeting between him & Francis I can imagine will be cordial & delightful, & if you could only be of the Party it certainly would be most complete— When we last heard, which is not very lately, he was at Cuddalore, a very healthy good station ab^t 100 Miles south East I think of Madras, near the Coast & very near Ponticherry— One of the Cunliffes who is now in England was station'd there several Years & speaks very well of it, & amongst other things says it affords excellent shooting & Boar hunting, which you know will be no trifling recommendation to it in Arthurs Eyes— By the bye I was very sorry to hear from your Sister that you had lost your Gun which I fear is hardly to be well replaced where you are. If you ever see Cuddalore however I have little doubt but a Gun will be found, & that you will have some shooting again together— What would I not give to be of such a Party—but it is too late—I am getting very *hobbling* & unless you come here soon fear I shall never have the *Pleasure* of *rating* you again anywhere or shooting— Your Father is looking very well & I believe *is* very well, but like others gets older & considerably worse upon his legs than when you left, which is indeed the only difference worth noticing I observe in him— But your Sisters of course tell you every thing relating to your own Family better than I can & I hope frequently They also I dare say, for Ladies excel in that, (& indeed I might add that yours do in most things) amuse you with constant accounts of every event which occurs in this Country—& a Newspaper, which must be a great treat, now & then lets you into all the secrets in England worth knowing. Our Friends the Ministers do not go on quite as I could wish they would, i.e I think they do not do enough, & Lord Grey & Mr Stanley[1] think rather too much of **their order**. There was a fine row all through the Country as you no doubt have heard when the Elections took place, & in this Tory County it was pretty well to get one County Member who at least *calls him*self a Whig—[2] I voted agst him because I doubted his Professions.— We had a sharp contest also for Denbighshire, where Biddulph came in beating Ld Kenyon's Son in a canter— All this I am sorry to say did not pass without creating some ill blood, which however I think is cooling considerably. But a truce to a tale which you must have heard so often— And indeed I dont know to what I can turn without being pretty sure that your Sisters have anticipated me, & I believe I must be satisfied, though my letter is ever so barren & dull, if it tends to shew you my good intention & that I have not forgot you, & this at least I do flatter myself it will require little ability to convince you of— We have had a great Mortality in the Neighbourhood since you left— beginning with poor Mrs Mathew—since whom two of the Miss Sparlings, Miss Letitia Kynaston, Mrs Bourke, Young Edwards of Ness & last of all J. Mytton Esqr have fallen— The two latter in the course of last Week only— Edwards of a decline & Mytton worn out by debauchery of every kind—the miserable remains of an ill spent Life—& it is certainly rather a good thing as well for himself as for his Friends that he is gone, for all hope of any alteration in him for the better was quite out of the question, & he dragg'd on a miserable & disgraceful existence, first in one Gaol & then in another, & rarely sober enough to say that he was in his right Senses. Notwithstanding all this there are some few who pretend to regret him & who would think me very unfeeling for what I have said of him— I am glad to hear that his poor Wife has a very ample Jointure & that his Children are all pretty well provided for— No thanks to him, it was so settled when he married, or they would not have had a shilling & many of his debts I hear will now never be paid— Of Weddings I cannot recollect any fresh except Miss Charlotte Kenyon to J. Hill which I dare say has already been reported. Sarah & Fanny are now quite old steady Wives, & both, you will be glad to hear, are I believe as happy as any two wives in England— Sarah what they call pick'd her Calf³ a long time ago & I have not yet heard is likely to produce another— Poor Fan however was br! to bed of a Girl last Spring & suffer'd so severely after, that she has been in a very precarious state, but is now I think, thanks to your Father's care & skill, recovering fast— M^{rs} O— & my two Girls, for Miss Emma is now what they call come out, have lately been twice at your Father's for Ball⟨s⟩ the last for a Charity Ball for the Infirmar⟨y⟩ whose Funds are rather low— I was not o⟨f⟩ the Party, but I hear amongst others they me⟨t⟩ your Friend young Eyton there & liked him much— I also met him some time ago at Col Leighton's & was much pleased with him; but I dare say he is one of your correspondents so I will let him speak for himself—

I go on here in my little Forest much as usual but am rather more of a Fixture, & am now rather put out of my way to be obliged to take Francis to Town to equip & start him for India, which I suppose I must do in about a Week, as I have almost engaged his Passage to embark I believe at Gravesend before the end of the Month— How I do wish you may fall in with him somewhere— He is I am glad to say a good deal improved & steadier, & I know you will give him some good advice if you come across him. I do not feel that I have any single thing to tell you that can justify me for boring you with a longer Letter, nor indeed that what I have written can repay you for the trouble of the perusal, except in so far as it may accomplish my object by assuring you that nothing would give me more pleasure than to see you again under this roof—& that I remain most sincerely & affectionately Yours | Wm Owen

I had forgot what I think I hardly need have express'd the most anxious & sincere good wishes of all my Ladies old & Young for your welfare & safe return.

May the 1st All well at home so we do not write this month y^{r} affec^{te} C. Darwin⁴

DAR 204.8

¹ Edward George Geoffroy Smith Stanley became Colonial Secretary in Lord Grey's Ministry in March 1833. He resigned in 1834 on the question of Irish tithes.
² John Cotes, who narrowly defeated William Ormsby-Gore in the election of 1832.
³ Presumably suffered a miscarriage. To slip or cast one's calf was an eighteenth century expression meaning to miscarry (Partridge 1973).
⁴ Catherine Darwin.

From Edward Lumb to J. S. Henslow 2 May 1834

Bˢ Ayres
May 2ᵈ 18/34

Sir

I beg to enclose you bill of lading for a Case of Specimens of Natural History which by direction of Mʳ Charles Darwin I forward to you pʳ Brig "Basenthwaite" Mitchinson Martin for Liverpool— This Case contains part of the Head of the "Megatherium"—[1] I regret that on the passage down the River it should have been broken; previous to this accident the Snout or nose extended 1½ to two feet more than at present— These are not the bones referred to in the accompanying Letter—[2] I expect them down shortly when I shall feel proud in forwarding them to you— Permit me this opportunity of offering my Services to you & to assure you that I shall feel highly gratified if by any Information, or Specimens I can obtain in this Country I can contribute to the advancement of Science in my native land—

My last letter from Mʳ Darwin was from the Falkland Islands 30 March; at which time all was well— | I have the honour to be Sir | Your most obᵗ Serᵗ | Edward Lumb

Royal Botanic Gardens, Kew

[1] The cranium was described by Richard Owen in *Fossil Mammalia*, pp. 16—35, as that of a 'gigantic extinct mammiferous animal, referrible to the Order Pachydermata, but with affinities to the Rodentia, Edentata, and Herbivorous Cetacea'. Owen named it *Toxodon platensis*.
[2] See letter to J. S. Henslow, [20—7] September 1833.

From Edward Lumb 8 May 1834

Buenos Ayres
May 8 1834

My dear Sir—

I have delayed writing to you to the present period in order to give you as much News as possible from this side of the Andes I duly recᵈ your favour of the 30 March from the Falklands; which served to corroborate without details an unfortunate affair of which we had had several vague reports via Rio some days previous: Accounts recᵈ from a settler called Helsby who left the Falklands in the Challenger are all the particulars we have received; This affair is classed here in its true light and is not considered of any political tendency: This ill fated province is destined never to be quiet. 'Ere this leaves it is very probable that our present Government will have left office they have already tendered their resignations; they are not liked by the Restauradores or those Gentlemen who detained you on your route from Santa fe when you were last here— Rosas has returned from his glorious campaign and is now at the Monte; not having as yet made his entrée into the City: 3500 Indians have been put hors de combat according to his official accounts and an immense quantity of ground good & bad obtained in exchange for an Expenditure of 9 million of our "Volatiles"—

I observe what you say respecting the Bones: By the brig "Basenthwaite" Mitchinson Martin for Liverpool I forwarded the Bones and Specimens rec^d from Mr Keen carefully packed and addressed to the Revd Professor Henslow to whom I also forwarded bill of lading with your Letter & a few lines added by myself: and I shall forward a duplicate by the Packet— I expect when Mr Hooker[1] returns from his Estate, which will be shortly, that he will bring down a good collection, which shall be duly forwarded: I preferred sending the case by a Merchant Vessel to your plan of forwarding to Rio the expense is only 10/— for freight which will be paid in England. If I should not have intruded my services on the Revd Professor I would have requested my Agent at Liverpool to forward the Case; but I concluded he would have some friend to whose care he would recommend it: I am very glad that the Bahia Blanca relics have pleased your friends at home and as you appear so anxious for the safety of the head I think you will applaud my plan of forwarding by a Merchant Vessel instead of the Men of War via Rio where in transhipment it might probably be lost.

Our new Minister has not yet arrived; The Sparrow hawk Brig was spoken a few days ago to the northwest of Rio; he is reported to be on board this Vessel and will probably soon make his appearance; Mr Gore is quite tired of us I seldom see him in the Streets now; I formerly saw daily when making faces at the Wife of an English [Trainer] here but not being successful I think he has taken a dislike to our Neighbourhood— Mr Griffiths you will know is as though he did not exist for the majority of his Countrymen now we seldom hear any thing of him he is quite secluded— General Quiroga[2] from the Interior of whose exploits I presume you must have heard is at present here; with the money he brought down with him he has purchased One Million and a half of our new created Six per cents he is very rich and goes about here with a poncho wrapped round him— If you have not heard of him enquire his character in Chile— The last accounts from England are not interesting— Portugal and Spain were getting worse; The revolt of the Carlists in Spain was not put down altho' several say many battles had been fought in diff^t. parts of the Country: a number of Friars had been executed and others were in the field with pistols at their Belts & Cartridge Boxes on their shoulders: The English and French Squadrons in the Mediterranean had been recalled Russia having giving satisfactory accounts of her Views towards Turkey— An immense number of shipping had been lost on the coasts of England & France in the months of December & January last by a series of westerly Gales— A famine raged in some parts of Russia which caused great distress— You must not forget to write me and if I can be of any service to command me: When Hooker comes down I shall forward the bones; and I have charged M^r. Keen to procure any that he may hear of and I continually make enquiries myself it is therefore probable I may send some which may prove interesting— *[Zurillos]* has been down here and promised to procure me some from the Carcaraña—

As we are never to be perfectly settled here; a few fellows came into the town on the 29th ultim & fired several Shots into the house of the Canonized Vidal[3] who if you recollect during Balcarces[4] administration went to Santa fe to persuade Lopez[5] to come down with an Army to this province— They also fired into the house of the Minister Garcia[6] and killed a poor fellow who happened to enquire what they were about; the town was in great alarm it was only 8 °.Clock when the affair happened— Rivadavia[7] arrived on the 28th ul^t from France but was ordered on board again 3 hours after landing in order to prevent disturbance & the Government have consulted the Sala on the Subject— He still remains in the Roads— Two or three Vessels have arrived at Montevideo with Slaves or African colonists as they term them and on the representation of Mr Gore a vessel fitting out here has been detained— A Breed of dromedaries has been introduced in the Banda Oriental from the Canary Islands; it is supposed they will do well— Rivadavia has brought out a number of Mulberry plants and Silk worms with the intention of propagating the breed in this country— This Gentleman says he has been accused of attempting to monarchize the Republics of S. America and untill he is either found guilty or acquitted he will not leave the Country;— There have been a many discussions in the public prints relative to the unconstitucional proceeding of the Government in ordering him to leave the Country—

There have been many failures in America lately owing to the Removal of the Bank deposits belonging to the Gov^t and many angry discussions have taken place in the Senate & Congress on the subject: The presidents determination has caused a great deal of mercantile distress and this will in the end seriously affect the Yankee community at large— The Indians vanquished by Rosas at all points have lately made their appearance on the frontiers of Cordova & San Luis & have made terrible havoc so that they are not so completely destroyed as some anticipated— You must excuse this—heterogeneous mass of stuff which I hope will prove amusing; let me hear from you when convenient and with best wishes for your success & happiness, and kind remembrances from M^{rs} L and family believe me | My dear Sir | Yours very truly | Edward Lumb

P.S. To whom did you pay my little a/c at Montevideo J. G*[illeg]* has not *[illeg]* to me for it— L

Pray remember me kindly to Chaffers, & all enquiring. Lumb

DAR 204.8

[1] The owner of one of the estates at which CD stayed during his trip to West Uruguay. In his field notebook an entry reads, 'Hard rock at Mr Hookers Estancia' (Down House MS *Beagle* Notebook 1.13, 'Buenos Ayres, St. Fe and Parana', p. 4).

[2] Juan Facundo Quiroga, a regional *caudillo*.

[3] Pedro Pablo Vidal.
[4] Juan Ramón González Balcarce.
[5] Estanislao López.
[6] Manuel José Garcia.
[7] Bernadino Rivadavia.

From Susan Darwin [23] May 1834

Shrewsbury
May. 1834

My dear Charles—

Catherine very cleverly missed her turn of writing in the month of April, by inserting a few words in a Letter of Mr Owens, that he sent here to be directed: she had nothing to say, & so she just told you we were all well; & that she thought would be sufficient, which I hope you understood & don't think you have lost a letter.— I am afraid it will be very long before we can expect to hear again from you (except once by the Falkland Islands which you give us hopes of) It is very provoking the tiresome Charts keeping you in such an odious part of the world so long You must feel that such a waste of time & life. I have sent your last letter to Erasmus so I have not got it here to refer to: but I know it was dated sometime in November.— Caroline has told you of course how delighted the British Museum was to hear you had found part of the Skull of one of the unknown Animals[1] it certainly was a most extraordinary piece of luck that the bones sh^d just fit: it makes one think it really must belong to the identical same animal. We have just been reading a very clever book written by the famous Surgeon Sir Charles Bell "on the Hand" & there he laments very much that all the Antideluvian beasts should be given such uncouth names that no ignoramus can remember or pronounce them.—[2] Papa has long been talking of going to visit Tom Eyton's Cameleon, but unfortunately it died a fortnight ago. however he still went to Eyton & found only the old Squire at home who shewed him a number of curious water birds & he also inspected T. Eyton's room which was one mass of skeletons— Tom was staying with Major Bayley who always as you may be sure enquires most warmly about you.— Papa liked his visit very much he stayed there 2 hours & found Mr Eyton very agreeable tho a very strong Tory: as they talked over some of the new Ministerial plans with regard to the Poor Laws.—

Erasmus has behaved very shabbily this Spring & not paid us his usual Easter visit: he half lives with the Hensleighs, & Mrs H. has just had another child a boy:[3] How the Wedgwoods multiply!! Next week I am going to stay with Jessie & Harry at Keel where I shall be introduced to their new house as well as baby.— I am just come from Overton you cannot think what nice little boys all your 4 nephews are Marianne had serious thoughts of sending Parky to school this Summer as he will be 9 next Sept. however instead she has got a Schoolmaster of Overton to come & put Latin into him every day which is a very good thing: for he is so happy & good at home that I shall be very sorry when he is sent to school— Papa & Caroline are gone a short Tour into Wales chiefly to see Penrhyn Castle & the weather is so lovely I am sure they will enjoy it very much.—

We have sold our old crazy Car for 5 pounds & got from London a beautiful light little Phaeton which we can post with if we please so I hope we shall make good use of it this Summer.—

Poor Francis Owen sails to India next Tuesday he came to take leave of us with tears in his eyes & so low he c^d hardly speak: it is a great pity he c^d not get a commission to have gone out with Arthur for at that time he wished it, but now he was so happy at Woodhouse he could not endure "*the Black* Indies" as he always called them.— I think he fancied himself in love with Harriet Boughey which I suppose made his departure so painful.— Mr Owen is gone up to London to ship him off in the Broxbournebury.—

I am glad to say Mr Mytton is dead at last in the Kings bench, they made a most absurd pompous funeral,—which put Papa in a rage.—

Robert Wedgwood is going to leave Maer & has taken the Curacy of Mucklestone near Drayton which surprises every body as he will have to live in the house with that good for nothing old Mr Crewe which must be very disagreeable to him & he only gets 20 pounds more a year by the change.— Catty & I are living all alone in the Drawing room & with all the horses & Servants at our command we take our pleasure fr morg till night How I wish my dear old fellow you were here to make a third. I sometimes have a most violent longing to see you again but I hope & trust we shall have a happy meeting in two years & till then Good bye & God bless you my dearest C.

Ever y^r affecte | Susan Darwin

Postmark: Shrewsbury MY 23 1834
DAR 204.6.1

[1] See letter from Caroline Darwin, 9 – 28 March [1834]. Susan's mention of the British Museum is a mistake. She refers to the reception of the bones by William Clift and Richard Owen at the Museum of the Royal College of Surgeons.

[2] Bell 1833.

[3] James Mackintosh Wedgwood.

From Robert Edward Alison [June? 1834]

[Data to illustrate the recent elevation of the coast at Valparaiso.]

AmemS incomplete, diagram
DAR 36.1: 425 – 6

To Catherine Darwin   20 – 9 July 1834

a hundred miles South of Valparaiso.

Sunday.— July 20th.— 1834—

My dear Catherine.

Being at sea & the weather fine, I will begin a letter, which shall be finished when we arrive in Port.— I have received the whole series of letters up to yours of November, 1833.— I wrote last from the Falkland Is^{ds} (where the Conway left for us a letter Bag); in this I mention receiving a Box; which must have come from Henslow; The next Man of War that comes round the Horn will bring the one from you.— We left the Island of Chiloe a week since; for which place a succession of gales compelled us to bear up.— We staid there some days in order to refresh the men.— Pigs & potatoes are as plentiful as in Ireland. With the exception of this weighty advantage, Chiloe, from its climate is a miserable hole.— I forget whether you were at home, when my friend Mr Proctor[1] was there, & told us about the place, where his Uncle says it never ceases to rain; I am sure he must have meant Chiloe.—

Altogether the last six months since leaving the Plata, has been a most prosperous cruize.— Much as I detest the Southern Latitudes, I have been enabled, during this period to do so much in Geology & Natural History, that I look back to Tierra del Fuego with grateful & almost kindly feelings. You ask me about the specimens which I send to Cambridge I collect every living creature, which I have time to catch & preserve; also some plants.— Amongst Animals, on principle I have lately determined to work chiefly amongst the Zoophites or Coralls: it is an enormous branch of the organized world; very little known or arranged & abounding with most curious, yet simple, forms of structures.—

But to go on with our history; when I wrote from the Falklands we were on the point of sailing for the S. Cruz on the coast of Patagonia.— We there looked at ye Beagle's bottom, her false keel was found knocked off, but otherwise not damaged.— When this was done, the Captain & 25 hands in three boats proceeded to follow up the course of the river of S. Cruz.— The expedition lasted three weeks; from want of provisions we failed reaching as far as was expected, but we were within 20 miles of ye great snowy range of the Cordilleras: a view which has never before been seen by Europæan eyes.—[2] The river is a fine large body of water, it traverses wild desolate plains inhabited by scarcely anything but the Guanaco.— We saw in one place smoke & tracks of the horses of a party of Indians: I am sorry we did not see them, they would have been out & out wild Gentlemen. In June, in the depth of winter we beat through the Straits of Magellan; the great chain of mountains, in which Sarmiento stands presented a sublime spectacle of enormous piles of snow.— Scenery however, is not sufficient to make a man relish such a climate. We passed out by the Magdalen channel, an unfrequented & little known exit; on our passage up, before we were driven into Chiloe, Mr Rowlett, the purser, died; Having gradually sunk under a complication of diseases.—

So much for the past; our future plans are as yet very uncertain: After Valparaiso, we shall go to Coquimbo to refit.— Here the climate is fine, but every thing else bad; the desert of Peru may be said to extend so far South; where man-kind is only enticed to live by the richer metals.— Next summer there is a good deal of work to be done behind & around Chiloe; how far I shall accompany the vessels I do not yet know.—

Amongst all the things you & Susan have told me in the last letters; you do not ever mention Erasmus; I hope the good lazy old Gentleman is alive; tell him, I should like very much to have one more letter from him; perhaps the box will bring one: if he would write to me four letters during the whole voyage, I would not grumble at all.— As for all of you, you are the best correspondents a brother, 3000 miles off, ever had.— I wish you could inspire Erasmus with a little of the superabundance of your virtues.— I am afraid he thinks your stock is sufficient for the whole family.— I am much pleased to hear my Father likes my Journal: as is easy to be seen I have taken too little pains with it.— My geological notes & descriptions of animals I treat with far more attention: from knowing so little of Natural History, when I left England, I am constantly in doubt whether these will have any value.— I have however found the geology of these countries so different from what I read about Europe, & in consequence when compared with it so instructive to myself; that I cannot help hoping that even imperfect descriptions may be of some general utility.—

Of one thing, I am sure; that such pursuits, are sources of the very highest pleasures I am capable of enjoying.— Tell my Father also, how much obliged I am for the affectionate way he speaks about my having a servant. It has made a great difference in my comfort; there is a standing order, in the Ship, that no one, excepting in civilized ports, leaves the vessel by himself By thus having a constant companion, I am rendered much more independent, in that most dependent of all lives, a life on board.— My servant is an odd sort of person; I do not very much like him; but he is, perhaps from his very oddity, very well adapted to all my purposes.

July 29th, Valparaiso.— I have again to thank you all, for being such good sisters, as you are.— I have just received 3 letters, one from each of you, in due order the last being from Susan, Feb 12th.— Also the box of books, with sundry notes & letters.— I am much obliged for your chain, I wear Caroline's pencil case, suspended by it round my neck.— Thank Granny for her purse & tell her I plead guilty to some of her [*two words obliterated*], but the others are certainly only accidental errors— Moreover I am much obliged for Carolines criticisms (see how good I am becoming!) they are perfectly just, I even felt aware of the faults she points out, when writing my journal.—[3] The little political books are very popular on board; I have not had time yet to read any of them.— Everything came right in the box; the shoes are invaluable, tell Erasmus he is a very good old gentleman for doing all my commissions, but he would be still better if he would write once again.—four letters are too much, it will frighten him, so I will change my demand into two, & they may be as short as he likes, so that they really come from him.— One other message & I have done, it is to my Father, that I have drawn a bill of eighty pounds.— I must now hold out, as the only economical prospect, the time when we cross the South Seas.— I hope this will not be considered as a little "South Sea scheme".—[4]

Valparaiso is a sort of London or Paris, to any place we have been to.— it is most disagreeable to be obliged to shave & dress decently.— We shall stay here two months, instead of going North-ward, during which time the ship will be refitted & all hands refreshed. You cannot imagine how delightfull the climate feels to all of us, so dry, warm & cheerful: it is not here as in T. del Fuego where one fine day, makes one fear the next will be twice as bad as usual.— The scenery wears such a different aspect, I can sit on the hills & watch the setting sun brighten the Andes, as at Barmouth we used to look at Cader-Idris.— The time of year, being now winter, is very unfortunate for me, it is quite hopeless to penetrate the Cordilleras; There is a mountain, near here, at Quillota, 4700 feet high. I am going in a few days to try to ascend it; I fear however the snow will be too thick. R. Corfield is living here, I cannot tell you how very obliging & kind he is to me.—[5] He has a very nice house & before long I am going on shore to pay him a visit; he presses me most goodnaturedly to make his house my head-quarters.—

I have had some long & pleasant walks in the country; I am afraid it is not a very good place for Natural History; after my first ride I shall know more about it. I have received two letters from Henslow, he tells me my treasures have arrived safe & I am highly delighted at what he says about their value.— What work I shall have, when I return; there will be a glorious mass of what Wickham calls d——d beastly devilment. Although Wickham always was growling at my bringing more dirt on board than any ten men, he is a great loss to me in the Beagle.[6] He is far the most conversible being on board, I do not mean talks the most, for in that respect Sulivan quite bears away the palm. Our new artist, who joined us at M. Video, is a pleasant sort of person, rather too much of the drawing-master about him; he i⟨s⟩ very unlike to Earles eccentric character.—

We all jog on very well together, there is no quarrelling on board, which is something to say:— The Captain keeps all smooth by rowing every one in turn, which of course he has as much right to do, as a gamekeeper to shoot Partridges on the first of September.—[7] When I began this long straggling letter, I had intended to have sent it per Admiralty; but now it must be sent by Liverpool, so there will be double postage to pay.— Thank most affectionately those good dear ladies, Sarah W. & Fanny B: I am very sorry to find I have lost the second of Mr Owen's letters Remember me at Maer, Woodhouse & I believe those two houses will include every one, I shall care any thing about, when I return. How every thing will be altered by that time; looking at things from a distance, they appear to be undergoing changes far faster than when living amongst them.— Will Erasmus be married? all these gay doing with Cab & horses portend something eventful: Can he build a castle in the air, where he does not quarrel with his wife in the first week? If he has arrived at such a pitch, I know well I shall find him a well-broken-in subjected husband.

Give my best love to my Father, Erasmus & each of the Sisterhood.— Dear Katty, Your most affectionate brother | Charles Darwin

There are several good dear people, whom I should like much to write to, but at present I really have not the time. Thank Fanny for her nice, goodnatured note; I have just re-read it. The sight of her hand writing is enough alone to make me long for this voyage to come to some end.—

DAR 223

[1] George Proctor, a fellow student at Christ's College. His uncle, Robert Proctor, had lived and travelled in Peru and Chile in 1823 – 4. (See Proctor 1825, of which a copy, signed 'C. Darwin' on the inside front cover, is in Christ's College Library, Cambridge. There is no evidence that CD had it on board the *Beagle*.)

[2] CD's account is in Chapter X of *Journal of researches*. Robert FitzRoy read a paper about the expedition at the Royal Geographical Society on 8 May 1837 (FitzRoy 1837).

[3] Five lines of the manuscript ('tell her . . . journal') have been deleted in black ink, presumably by one of CD's sisters.

[4] A reference to the speculation and fraud of the famous 'South Sea Bubble' of the early eighteenth century.

[5] Richard Henry Corfield attended Shrewsbury School, 1816 – 19; CD entered in 1818 (*Shrewsbury School Register*). Corfield's letters to CD, 26 – 7 June 1835 and 14 – 18 July 1835, indicate that he was engaged in trading or shipping business at Valparaiso.

[6] John Clements Wickham had been placed in command of the *Adventure* by FitzRoy shortly after he bought her in March 1833.

[7] The opening day of the English shooting season.

From J. S. Henslow   22 July 1834

Cholsey Wallingford
22 July 1834

My dear Darwin,

It is now some months since I received your last letter, with the intention of answering it so soon as I should be able to give you an account of the safe arrival of your cargo of skins &c   These were delayed at Dr Armstrongs up to the time of my quitting Cambridge & I have only just heard that he has at length despatched them. He tells me however that every thing is safe, & that he had used the precaution of opening the cases & airing every thing for you— I recommended the fossils to be all sent to Mr Clift at Surgeons' Hall who has kindly undertaken to repair them & prepare them so that they shall be preserved without injury— Judging from what you sent before I did not hesitate to do this as they will be well worth the carriage to London, & could not possibly be in better hands than Clift's. I regret that I did not get the sweepings of the granary before I left Cambridge as I fear the delay will spoil most of the seeds which cannot now be sown before next Spring— Pray don't entirely neglect to dry plants— Those sent are **all** of the greatest interest— Send minute things, such the little ranunculus, & common weeds & grasses, not to the neglect of flowering shrubs of which you have sent some nice species of Berberry &c.—[1] I have not your letter bye me to answer your questions formally but I remember you enquire about a Goniometer— I would not advise you to bother yourself with one— It is an instrument of no use in the field, & of importance only in the hands of an experienced mineralogist in his *closet*. Phillips's book[2] must be quite as much as you *need* for the detection of the few ingredients which form rocks— Any that you can't make out you must describe conditionally & we will set you to rights 10 years hence when you return— Fox & his wife spent a day with us at commencement— He tells me that you are very irate at not having heard from me—which I don't exactly understand, as I should have thought that you ought to have received two letters at least from me by the time he heard from you— That I have not written so often as I ought I will readily admit for I never do any thing as I ought—but really & truly I have written & I trust that you have had positive proof of it before now. I don't know that I have much local news to tell you which is likely to be of any interest— You will see by the papers that we have been in various kinds of hot water,[3] in which however I am happy to say that I have escaped from scalding my own fingers, though I fear that the result has caused a few burning⟨s⟩ & cuttings among certain members of the University who ought to be abov⟨e⟩ such evils— Your Master,[4] I suppose you know, is married, & soon to be a Papa if all prospers— My own family is $3♀ + 1♂$, & if you delay your return much longer & I am equally fortunate as I have hitherto been you may be in time to stand Godfather to another.— I am at present rusticating for the Vacation at my living—& enjoy the change from a town to a country life most exceedingly— There are no immediate neighbours & I am not bothered by morning visits— My parish abounds in poor, & small farmers who leave every thing to the parson without attempting to assist him— However I am quite satisfied with my visit, the only drawback being the long distance which I have to bring my family—about 100 miles— I shall be very anxious as the time for your return approaches, to hear of you & look forward with the prospect of great satisfaction to the confabs we shall have together— Captn. W. Ramsay is about to start as commander of a Steam frigate for the W. Indies & if I had been a single or an independent man I should certainly have joined him for a few months cruise— How you would have stared to have seen me walking on the Quay at Monte Video.—

With kindest remembrances from all my family & most hearty good wishes from myself believe me ever | Your affectionate friend | J S Henslow

DAR 204.8

[1] For suggested identifications of these Tierra del Fuego specimens see *Darwin and Henslow*, p. 89 nn. 2, 3.

[2] Phillips 1816. Henslow's reference makes it likely that CD had the volume with him on the voyage. Unannotated copies of the third (1823) and fourth (1837) editions are, respectively, in the Darwin Library—Down and the Darwin Library—CUL.

[3] A reference to the controversy over the Dissenter question. See letter from J. M. Herbert, [28 March] 1834, n. 9.

[4] John Graham, Master of Christ's College.

To Charles Whitley 23 July 1834

Valparaiso
July, 23^{d} 1834

My dear Whitley

I have long intended writing just to put you in mind that there is a certain hunter of beetles & pounder of rocks, still in existence: Why I have not done so before, I know not; but it will serve me right, if you have quite forgotten me.— It is a very long time, since I have heard any Cambridge news.— I neither know, where you are living or what you are doing.— I saw your name down as one of the indefatigable guardians of the eighteen hundred philosophers.[1] I was delighted to see this, for when we last left Cambridge, you were at sad variance with poor science; you seemed to think her a public prostitute working for popularity.— If your opinions are the same as formerly; you would agree most admirably with Capt. FitzRoy, the object of his most devout abhorrence is one of the d——d scientific Whigs. As Captains of Men of wars are the greatest men going far greater than Kings or Schoolmasters; I am obliged to tell him every thing in my own favor; I have often said, I once had a very good friend an out & out Tory & we managed to get on very well together. But he is very much inclined to doubt if ever I really was so much honored.— At present none of us hear scarcely anything about politicks; this saves a great deal of trouble; for we all stick to our former opinions rather more obstinately than before & can give rather fewer reasons for doing so.—

I do hope you will write to me. ("H.M.S. Beagle, S. American Station" will find me); I should much like to hear in what state you are, both in body & mind.— ¿Quien sabe? as the people say here (& God knows they well may, for they do know little enough) if you are not a married man, & may be nursing, as Miss Austen says, little olive branches, little pledges of mutual affection.— Eheu Eheu, this puts me in mind, of former visions, of glimpses into futurity, where I fancied I saw, retirement, green cottages & white petticoats.— What will become of me hereafter, I know not; I feel, like a ruined man, who does not see or care how to extricate himself.— That this voyage must come to a conclusion, my reason tells me, but otherwise I see no end to it.— It is impossible not bitterly to regret the friends & other sources of pleasure, one leaves behind in England; in place of it, there is much solid enjoyment, some present, but more in anticipation, when the ideas gained during the voyage can be compared to fresh ones. I find in Geology a never failing interest, as ⟨it⟩ has been remarked, it creates the same gran⟨d⟩ ideas respecting this world, which Astronomy do⟨es⟩ for the universe.— We have seen much fine scenery, that of the Tropics in its glory & luxuriance, exceeds even the language of Humboldt to describe. A Persian writer could alone do justice to it, & if he succeded he would in England, be called the "grandfather of all liars".—

But, I have seen nothing, which more completely astonished me, than the first sight of a Savage; It was a naked Fuegian his long hair blowing about, his face besmeared with paint. There is in their countenances, an expression, which I believe to those who have not seen it, must be inconcevably wild. Standing on a rock he uttered tones & made gesticulations than which, the crys of domestic animals are far more intelligible.

When I return to England, you must take me in hand with respect to the fine arts. I yet recollect there was a man called Raffaelle Sanctus. How delightful it will be once again to see in the FitzWilliam, Titian's Venus; how much more than delightful to go to some good concert or fine opera. These recollections will not do. I shall not be able tomorrow to pick out the entrails of some small animal, with half my usual gusto.— Pray tell me some news about Cameron, Watkins, Marindin The two Thompsons of Trin:[2] Lowe, Heaviside, Matthews Herbert I have heard from: How is Henslow getting on? & all other good friends of dear Cambridge. Often & often do I think over those past hours so many of which have been passed in your company. Such can never return; but their recollection shall never die away.—

God Bless you, My dear Whitley. Believe Me, Your Most | Sincere Friend | Chas. Darwin

National Library of Australia, Canberra, MS 4260

[1] A reference to the British Association meeting at Cambridge in 1833. CD has doubled the attendance.
[2] Harry Stephen Thompson and Thomas Charles Thompson.

To J. S. Henslow[1] 24 July — 7 November 1834

Valparaiso
July 24th.—1834

My dear Henslow

A box has just arrived, in which are two of your most kind & affectionate letters; you do not know how happy they have made me.— One is dated Dec 12th. 1833 the other Jan: 15th of the *same year*!—[2] By what fatality it did not arrive sooner, I cannot conjecture: I regret it much; for it contains the information, I most wanted about manner of packing &c &c: roots, with specimens of plants &c &c: This I suppose was written after the reception of my *first* cargo of specimens.— Not having heard from you untill March of this year; I really began to think my collections were so poor, that you were puzzled what to say: the case is now quite on the opposite tack; for you are **guilty** of exciting all my vain feelings to a most comfortable pitch; if hard work will atone for these thoughts I vow it shall not be spared.—

It is rather late, but I will allude to some remarks in the Jan: letter: you advise me to send home duplicates of my notes; I have been aware of the advantage of doing so; but then at sea to this day, I am invariably sick, excepting on the finest days; at which times with pelagic animals around me, I could never bring myself to the task; on shore, the most prudent person, could hardly expect such a sacrifice of time.—

My notes are becoming bulky; I have about 600 small quarto pages full; about half of this is Geology, the other imperfect descriptions of animals: with the latter I make it a rule only to describe those parts, or facts, which cannot be seen, in specimens in spirits. I keep my private Journal distinct from the above.—[3] (N B this letter is a most untidy one, but my mind is untidy with joy; it is your *fault*, so you must take the consequence). With respect to the land Planariæ:[4] unquestionably they are not Molluscous animals: I read your letters last night, this morning I took a little walk; by a curious coincidence I found a new white species of Planaria & a (new to me) Vaginulus (3d species which I have found in S. America) of Cuv: I suppose this is the animal Leonard Jenyns alludes to.— The *true Onchidium* of **Cuv**:[5] I likewise know.— Amongst the marine Mollusques I have seen a good many genera & at Rio found one quite new one.— With respect to the December letter, I am very glad to hear, the four casks arrived safe; since which time you will have received another cargo, with the bird skins, about which you did not understand me.— Have any of the B. Ayrean seeds produced plants?—

From the Falklands, I acknowledged a box & letter from you; with the letter were a few seeds from Patagonia.— At present, I have specimens enough to make a heavy cargo, but shall wait as much longer as possible, because opportunities are not now so good as before.— I have just got scent of some fossil bones of a **Mammoth**!, what they may be, I do not know, but if gold or galloping will get them, they shall be mine. You tell me, you like hearing how I am going on & what doing; & you well may imagine how much I enjoy speaking to anyone upon subjects, which I am always thinking about, but never have any one to talk to with.—

After leaving the Falklands, we proceeded to the R. S. Cruz; followed up the river till within 20 miles of the Cordilleras: Unfortunately want of provisions compelled us to return. This expedition was most important to me, as it was a transverse section of the great Patagonian formation.— I conjecture (an accurate examination of fossils may possibly determine the point) that the main bed is somewhere about the Meiocene period, (using Mr Lyell's expression)[6] I judge from what I have seen of the present shells of Patagonia.— This bed contains an *enormous* field of Lava.— This is of some interest, as being a rude approximation to the age of the Volcanic part of the great range of the Andes.— Long before this it existed as a Slate & *Porphyritic* line of hills.— I have col- lected tolerable quantity of information respecting the period, (even numbers) & forms of elevations of these plains. I think these will be interesting to Mr Lyell.— I had deferred reading his third volume till my return, you may guess how much pleasure it gave me; some of his wood-cuts came so exactly into play, that I have only to refer to them, instead of redrawing similar ones.— I had my Barometer with me; I only wish I had used it more in these plains.—

The valley of S. Cruz appears to me a very curious one, at first it quite baffled me.— I believe I can show good reasons for supposing it to have been once a **Northern** St^{ts} like that of *Magellan*.— When I return to England, you will have some hard work in winnowing my Geology; what little I know, I have learnt in such a curious fashion, that I often feel very doubtful about the number of grains: Whatever number, they may turn out, I have enjoyed extreme pleasure in collecting them.—

In T. del Fuego I collected & examined some Corallines: I have observed one fact which quite startled me.— it is, that in the genus Sertularia, (taken in its most restricted form as by Lamouroux)[7] & in 2 species which, excluding comparative expressions, I should find much difficulty in describing as dif- ferent—the Polypi quite & essentially differed, in all their most important & evident parts of structure.— I have already seen enough to be convinced that the present families of Corallines, as arranged by Lamarck, Cuvier &c are highly artificial.— It appears they are in the same state, which shells were when Linnæus left them for Cuvier to rearrange.—

I do so wish I was a better hand at dissecting: I find I can do very little in the minute parts of structure; I am forced to take a very rough examination as a type for different classes of structure.—

It is most extraordinary I can no where see in my books one single description of the polypus of any one Corall (excepting Lobularia alcyonium of Savigny)[8] I found a curious little stony Cellaria (a new genus) each cell provided with long toothed bristles, these are capable of various & rapid motions,—this motion is often simultaneous & can be produced by irritation.— this fact, as far as I see, is quite isolated in the history (excepting by the Flustra with organ like Vultures Head) of Zoophites.— it points out a much more intimate relation between the Polypi, than Lamarck is willing to allow.— I forget, whether I mentioned, having seen something of the manner of propagation, in that most ambiguous family, the Corallinas: I feel pretty well convinced if they are not

Plants, they are not Zoophites:[9] the "gemmule" of a Halimeda contained several articulations united, & ready to burst their envelope & become attached to some basis.— I believe in Zoophites, universally the gemmule produces a single Polypus, which afterwards or at the same time grows with its cell or single articulation.— The Beagle left the St[s] of Magellan in the middle of winter; she found her road out by a wild unfrequented channel; well might Sir J. Narborough call the West coast South Desolation "because it is so desolate a land to behold".—[10] We were driven into Chiloe, by some very bad weather.— an Englishman gave me 3 specimens of that very fine *Lucanoidal* insect, which is described Camb: Phil. Trans: 2 males & one female.—[11] I find Chiloe is composed of Lava & recent deposits.— the Lavas are curious from abounding or rather being in parts composed of Pitchstone.— If we go to Chiloe in the summer I shall reap an Entomological harvest.— I suppose the Botany both there & in Chili is well known.—

I forgot to state, that in the four cargoes of specimens there have been sent 3 square boxes, each containing four glass bottles.— I mention this in case they should be stowed beneath geological specimens, & thus escape your notice perhaps some spirit may be wanted in them.— If a box arrives from B. Ayres, with Megatherium head & other *unnumbered* specimens: be kind enough to tell me; I have strong fears for its safety.—

We arrived here the day before yesterday; the views of the distant mountains are most sublime & the climate delightful; after our long cruize in the damp gloomy climates of the South, to breathe a clear, dry air, & feel honest warm sunshine, & eat good fresh roast beef must be the summum bonum of human life.— I do not like the looks of the rocks, half so much as the beef, there is too much of those rather insipid ingredients Mica, quartz & Feldspar.— Our plans are at present undecided.— there is a good deal of work to the South of Valparaiso & to the North an indefinite quantity.— I look forward to every part with interest. I have sent you in this letter a sad dose of egotism.—but recollect I look up to you as my father in Natural History, & a son may talk about himself, to his father.— In your paternal capacity, as pro-proctor what a great deal of trouble you appear to have had.— How turbulent Cambridge is become— Before this time it will have regained its tranquillity— I have a most school-boy like wish to be there, enjoying my Holidays.— It is a most comfortable reflection to me, that a ship being made of wood & iron, cannot last for ever & therefore this voyage must have an end.—

October 28th.— This letter has been lying in my port-folio ever since July: I did not send it away, because I did not think it worth the postage: it shall now go with a box of specimens: shortly after arriving here, I set out on a geological excursion, & had a very pleasant ramble about the base of the Andes.— The whole country appears composed of breccias, (& I imagine Slates) which universally have been modified, & oftentimes completely altered by the action of fire; the varieties of porphyry thus produced is endless, but no where have I yet met with rocks which have flowed in a stream; dykes of greenstone are very numerous: Modern Volcanic action is entirely shut up in the very central parts (which cannot now be reached on account of the snow) of the Cordilleras.— To the South of the R. Maypo I examined the Tertiary plains already partially described by M. Gay.[12] The fossil shells, appear to me, to be far more different from the recent ones, than in the great Patagonian formation; it will be curious if an Eocene & Meiocene (Recent there is abundance of) could be proved to exist in S. America as well as in Europe.— I have been much interested by finding abundance of recent shells at an elevation of 1300 feet; the country in many places is scattered over with shells, but these are *all littoral* ones. So that I suppose the 1300 feet elevation *must* be owing to a succession of small elevations such as in 1822. With these certain proofs of the recent residence of the ocean over all the lower parts of Chili; the outline of every view & the form of each valley possesses a high interest. Has the action of running water or the sea formed this deep ravine? Was a question which often arose in my mind, & generally was answered by finding a bed of recent shells at the bottom.— I have not sufficient arguments, but I do not believe that more than a small fraction of the height of the Andes has been formed within the Tertiary period.—

The conclusion of my excursion was very unfortunate, I became unwell & could hardly reach this place, I have been in bed for the last month, but am now rapidly getting well. I had hoped during this time to have made a good collection of insects &c but it has been impossible. I regret the less, because Chili fairly swarms with Collectors; there are more Naturalists in the country, than Carpenters or Shoemaker or any other honest trade.—

In my letter from the Falkland Is^d I said I had fears about a box with a Megatherium. I have since heard from B. Ayres, that it went to Liverpool by the Brig Basingwaithe.— If you have not received it—it is, I think, worth taking some trouble about. In October two casks & a jar were sent by H.M.S. Samarang viâ Portsmouth I have no doubt you have received them. With this letter, I send a good many bird skins; in the same box with them, there is a paper parcel, containing pill boxes with insects: the other pill-boxes require no particular care: You will see in two of these boxes, some dried terrestrial Planariæ, the only method I have found of preserving them (they are exceedingly brittle) By examining the white species I understand some little of the internal structure.— There are two small parcels of seeds.— There are some plants, which I hope may interest you, or at least those from Patagonia, where I collected every one in flower:— There is a bottle, clumsily, but I think securely corked, containing water & *gaz* from the hot Baths of Cauquenes, seated at foot of Andes & long celebrated for medicinal properties.— I took pains in filling & securing both water & gaz.— If you can find any one who likes to analyze them; I should think it would be worth the trouble.— I have not time at present to copy my few observations about the locality &c &c of these Springs.— Will you tell me, how the Arachnidæ, which I have sent home, for instance those from Rio appear to be preserved.— I have doubts whether it is worth while collecting them.—

We sail the day after tomorrow: our plans are at last limited & definite: I am delighted to say we have bid an eternal adieu to T. del Fuego.— The Beagle will not proceed further South than C. Tres Montes. From which point we survey to the North. The Chonos archipelago is delightfully unknown; fine deep inlets running into the Cordilleras, where we can steer by the light of a Volcano.— I do not know, which part of the voyage, now offers the most attractions.— This is a shamefully untidy letter, but you must forgive me & believe me | My dear Henslow | Yours most truly obliged | Charles Darwin

Nov.^b 7th.—

Royal Botanic Gardens, Kew

[1] The following passages from this letter were extracted by Henslow and published in the Cambridge Philosophical Society pamphlet:

    5.1 'After leaving . . . similar ones.' 5.15; 5.8 *CD's* 'field' *changed back to CD's del* 'mass'; 5.11 'period' *changed to* 'various periods' *and* '(even numbers)' *omitted by JSH*

    6.1 'The valley . . . *Magellan.*' 6.3

    7.1 'In T. del Fuego . . . rearrange.' 7.9; 7.8 'to me, that' *inserted after* 'appears' *by JSH*

    9.1 'It is most extraordinary . . . Pitchstone.' 9.22; 9.17 'Narborough' *incorrectly transcribed* 'Nasborough'; 9.22 'in parts' *omitted*

    11.1 'We arrived . . . Feldspar.' 11.6

    12.3 'shortly after . . . Tertiary period.—' 12.25; 12.12 'far more different' *changed to* 'differ more widely'; 12.13 'formation' *inserted after* 'Meiocene'; 12.22 'deep' *omitted*

[2] The December letter has not been found.

[3] On the records CD kept during the voyage see Appendix II.

[4] See letter from J. S. Henslow, 15–21 January 1833.

[5] Henslow marked this passage, 3.6 'With respect . . . Cuv: ' 3.11, for communication to the Cambridge Philosophical Society, but it was not reprinted in the pamphlet, possibly because he still felt CD to be mistaken about the land *Planaria*. CD described them in *Journal of researches*, pp. 30–1, and in more detail in a paper of 1844 (see *Collected papers* 1: 182–93).

[6] Lyell first used the term 'Miocene' in volume three of *Principles of geology*, p. 54.

[7] Jean Vincent Félix Lamouroux. CD had with him Lamouroux 1821, a lightly annotated copy of which is in Darwin Library–CUL, inscribed 'C. Darwin'.

[8] Marie-Jules-César Lelorgne de Savigny.

[9] Corallines were classified as animals by most of the authorities whose works CD had with him (Cuvier, Lamarck, and Lamouroux). Other naturalists held that they were plants. In 1826, when CD was a student at Edinburgh, his friend Robert Edmond Grant published a translation of a paper by August Friedrich Schweigger, which in Grant's view gave a 'great probability to the opinions formerly entertained by Pallas, Spallanzani, Cavolini, and Olivi, who, from their own observations on living corallines, regarded them as plants' (Schweigger 1826). A copy of the translation, bound with other reprints presented by Grant, is in Darwin Library–CUL.

[10] Sir John Narbrough (also spelled Narborough) visited the Straits of Magellan and the west coast of South America (Narbrough 1694). Since the *Beagle*'s survey covered much of Narbrough's route, Robert FitzRoy probably had a copy in the ship's library, but this is the only mention of the work. A version was reprinted in volume three (1813) of Burney 1803–17.

[11] Stephens 1833. CD may have received a copy of the paper from Henslow. It was read 16 May 1831 and described a new genus of beetle, *Chiasognathus*, which was sent to the Cambridge Philosophical Society by Dr Grant the surgeon of H.M.S. *Forte*. It too had been found on the island of Chiloé.

[12] Gay 1833. A lightly annotated copy is in 'Philosophical tracts', Darwin Library — CUL (see letter to Robert FitzRoy, [28 August 1834]).

From H. S. Fox   25 July 1834

Rio de Janeiro
July 25th: 1834.

Dear Sir

I avail myself of the opportunity of the Satellite sailing tomorrow for the Pacifick, to thank you very sincerely for your obliging letter of the 5th: of April, from East Falkland Island, which was brought to me by the Dublin a few days since, together with the very interesting specimens accompanying it. I had no idea that either those islands, or the coast of Patagonia, presented so much interest to a geologist, and you will have had the satisfaction of visiting regions entirely unknown to the rest of the world.— I will not fail to answer the queries contained in your letter, respecting the geology of Rio Grande, and other southern parts of Brazil which I examined, at least as far as my very limited knowledge of the subject permits of.— I have not time, however, to do so today.— Doctor Sellow, a Prussian naturalist, learned in all departments of natural history, passed I believe two years in exploring Rio Grande, and St. Pauls, and made extensive collections.[1] He died afterwards in the interior of Brazil, and I am ignorant whether he left any journal or account of his travels and observations. I heard frequently of him in that country, as also of M. Auguste St. Hilaire, a french traveller, but whose researches were confined to botany.[2] The Flora of the South of Brazil is little known, and extraordinarily beautiful, and as a far greater number of the plants would succeed in Europe, than of those collected in the tropical regions, it is a great pity that the Horticultural Society do not send out collectors,—and the more so, as the country is habitable and hospitable, and easy to travel about; indeed, the moment you pass from the oriental territory into the Brazilian, you emerge at once from a savage into a civilized region.— An English travelling naturalist, M^r: Andrew Matthews,[3] but whose objects are chiefly, I believe, entomology, and botany, is about to undertake an adventurous journey, directly across this continent, from Lima to Rio de Janeiro. I obtained passports and recommendations for him for this Government, and I wrote to Colonel Wilson[4] at Lima, to urge him not to neglect, if possible, collecting some specimens of rocks in the course of his journey. He will most probably have left Peru before you arrive there. At St. Iago de Chili, you will of course become acquainted with M^r: Caldcleugh,[5] who is residing there, whom you will find a very accomplished chemist and mineralogist, as well as a most agreable well informed and obliging person.— I hope you will continue your kindness, in sending me a few small fragments of the Andes, and of any of the islands that you visit.—

Our last political news from Europe bring word of the downfal, at last, of Dom Miguel, who, with his friend Don Carlos of Spain, embarked for England on the 30th: of May.— We have this intelligence from Lisbon, as the English June Packet is not yet arrived. I suppose this letter will find the Beagle somewhere on the coast of Chili or Peru.

I remain, My Dear Sir, | your faithful and obedient servant | H. S. Fox

DAR 204.8

[1] Friedrich Sellow. He concentrated on plant collecting in Brazil and Uruguay, 1814−31, but also made valuable observations on geology and native languages. 'Rio Grande and St. Pauls' are now the states of Rio Grande do Norte, Rio Grande do Sul and São Paulo.

[2] Augustin François César Prouvençal (also known as Auguste de) Saint-Hilaire travelled throughout Brazil for six years (1816−22) collecting in all fields of natural history.

[3] Andrew Mathews.

[4] Probably the 'M^r. Belford Wilson Consul General for H.B. Majesty to Peru' whom CD thanks for allowing him to copy an official statement of exports of nitrate of soda for 1831−4 (DAR 37.1: 685). Wilson served in the Bolivian army as A.D.C. to General Bolivar, 1823−30 (*Modern English Biography*).

[5] Alexander Caldcleugh.

To Caroline Darwin 9−12 August 1834

[Valparaiso]
August 9th.— 1834

My dear Caroline

A ship sails for Liverpool tomorrow. I will try to scribble this sheet full & if so send it off.— I received your letter, dated March 9th the day before yesterday & Mr Owen's long, one.— Give him my best thanks for writing so kindly to me; I will take an early opportunity of answering it.— I am much pleased, with what you have told me, respecting the fossil bones. I did not before understand in what particular way the head of the Megatherium came to be so much sought after.— I presume the big box, which Erasmus was going to send to Plymouth for, is one which I directed to be left at Dr Armstrong' (to save carriage) I am in great fear lest Mr Clift should remove the numbers or markers attached *to any of the Specimens*. Ask Erasmus to call on Mr Clift & state how anxious I am on this point. All the interest which I individually feel about these fossils, is their connection with the geology of the Pampas, & this entirely rests on the safety of the numbers.— Another point must clearly be explained to Mr Clift, it is with reference to the Coll: of Surgeons paying the expence of the carriage.— the ultimum destination of *all* my collections will of course be to wherever they may be of most service to Natural History. But **cæteris**

paribus the British Museum, has the first claims, owing to my being on board a King's Ship.— Mr Clift must understand that *at present* I cannot say, that any of the fossil Bones shall go to any particular Museum. As you may well believe, I am quite delighted that I should have had the good fortune (in spite of sundry sneers about seal & Whale bones) to have found fossil remains which can interest people such as Mr Clift.—

A small box has been forwarded from B. Ayres to Liverpool for Henslow, with part of a head, which I think will be more useful than any which I have sent.— With respect to the expence of the carriage it is entirely in England, everything *as yet* has been sent on the sea on "His Majesties Service". But they are very heavy & bulky.

Give my Father my best love & thanks for all his kindness about money, & tell him I can seriously say, that since leaving England I have spent none excepting in the furtherance of Natural History, & as little as I could in that so that my time should not be thrown away.— I am now living with Corfield; he is as hospitable & kind in deeds, as a Spaniard is in professions.—than which I can say no more. It is most plasant to meet with such a straitforward—thorough Englishman, as Corfield is, in these vile countries.— He has made his house so pleasant to me, that I have done less during the last fortnight, than in any time since leaving England.—

The day after tomorrow I start for a Geological excursion. Does it not *sound* awfully extravagant, when I say, I am going to buy a small troop of horses; with these I shall travel by a very round-about cours⟨e⟩ to St Iago, the gay Capital of Chili.— I shall there meet Corfield, who is going up to admire the beauties of nature, in the form of Signoritas, whilst I hope to admire them amongst the Andes.— I long to have a near view of this extraordinary & grand chain of mountains.— At this time of year however, it will not be possible to ascend to any height on account of the snow.—

This a very stupid letter to send.—but you have often told me you would rather have a short letter, than none.— So take the Consequences.— Give my best love to Marianne, we do not write to each other for the same reason, we are too busy with our children.— She with Master Robert & Henry &c, I with Master Megatherium & Mastodon: If I have a good opportunity, I will send home some more of my journal; which will give you some account of the Pampas galloping. I am ashamed of sending such a letter.—but take the will for the deed & Believe me my dearest Caroline | Yours most affectionately | Chas Darwin

My love to every body at home

August 12th.

DAR 223

From [Alexander Caldcleugh?] [28 August — 5 September 1834]

[A map of part of Chile between Santiago and San Fernando, with notes on people and places to visit.]

Amem incomplete
DAR 35.2: 405

To Robert FitzRoy [28 August 1834]
                                    My direction is the Fonda Inglese | S! Iago.
                                                                    Thursday
My dear Fitz Roy,
  I arrived at this gay city late last night, and am now most comfortably established at an English Hotel. My little circuit by Quellota[1] and Aconcagua was exceedingly pleasant The difficulty in ascending the Campana is most absurdly exaggerated We rode up 5/6[ths] of the height to a spring called the Aqua del Guanaco & there bivouacked for two nights in a beautiful little arbor of Bamboos. I spent one whole day on the very summit, the view is not so picturesque as interesting from giving so excellent a plan of the whole country from the Andes to the sea— I do not think I ever more thoroughly enjoyed a days rambling. From Quellota I went to some Copper Mines beyond Aconcagua situated in a Ravine in the Cordilleras The major domo is a good simple hearted Cornish Miner— It would do Sulivan good to hear his constant exclamation "As for London—what is London? they can do anything in my country." I enjoyed climbing about the mountains to my hearts content the snow however at present quite prevents the reaching any elevation— On Monday my Cornish friend and myself narrowly escaped being snowed in. we were involved in a multitude of snow banks, and a few hours afterwards there was a heavy snow-storm which would have completely puzzled us— The next morning I started for this place. I never saw anything so gloriously beautiful as the view of the mountains with their fresh and brilliant clothing of Snow— Altogether I am delighted with the Country of Chile— The country Chilenos themselves appear to me a very uninteresting race of people— They have lost much individual character in an *essay towards an approximation* to civilization My ride has enabled me to understand a little of the Geology—there is nothing of particular interest—all the rocks have been frizzled melted and bedevilled in every possible fashion. But here also the "confounded Frenchmen" have been at work. A M: Gay has given me to day a copy of a paper, with some interesting details about the Geology of this province published by himself in the Annales des Sciences—[2] I have been very busy all day, and have seen a host of people. I called on Col. Walpole, but he was in bed—or said so.— Corfield took me to dine with a M[r] Kennedy, who talks much about the Adventure & Beagle; he says he saw you at Chiloe— I have seen a

strange genius a Major Sutcliffe.[3] he tells me as soon as he heard there were two English Surveying Vessels at Valparaiso, he sent a Book of old Voyages in the Straits of Magellan to M[r] Caldcleugh to be forwarded to the Commanding Officer as they might prove of service— He has not heard, whether M[r] Caldcleugh has sent them to you— I told him I would mention the circumstance when I wrote.— The Major is inclined to be very civil— I do not know what to make of him. He is full of marvellous stories; and to the surprise of every one every now & then some of them are proved to be true— My head is full of schemes; I shall not remain long here, although from the little I have yet seen I feel much inclined to like it. How very striking & beautiful the situation of the city is— I sat for an hour gazing all round me, from the little hill of S[t] Lucia. I wish you could come here to readmire the glorious prospect— I can by no means procure any sort of Map.— you could most exceedingly oblige me if you would get King to trace from Miers[4] a little piece of the Country from Valparaiso to a degree south of R. Rapel—without any mountains. I do not think it will be more than $\frac{1}{2}$ an hours work— I have some intention of returning to Valparaiso by the Rapel.— If you would send me this *soon* and half a dozen lines, mentioning, if you should know anything about the Samarangs movements; it would assist me in my schemes very much—

Adios, dear Fitz Roy | y[r] faithful Philos. | C. D.

Copy
DAR 144

[1] Probably the copyist's mistaken transcription of 'Quillota'.
[2] See letter to J. S. Henslow, 24 July – 7 November 1834.
[3] Thomas Sutcliffe.
[4] Miers 1826.

From Frederick W. Eck [September 1834]

[Heights of various points in Chile.]

AmemS
DAR 35.1: 232

From Caroline Darwin 30 September 1834

[Shrewsbury]
1834 | Sept 30th

My dear Charles
We were delighted to receive your letter the week before last dated 6th April from the Falkland Isles— Your account of those barren Isles was very

interesting & of Patagonia. the Wedgwoods were surprised to find how much exagerated all former accounts of the heigth of the Patagonians had been, but my surprise was that they should really be such a fine race of men, most of them you said more than 6 feet— I hope by this time you will have had our letters saying we had received your journal, but for fear of their being lost, I will repeat we received the part of the journal ending July 1833, sent from Plata. I have been rereading with great pleasure both parts of it & shall be very glad when we get a *third* part. My Father does not know who the Hon Col Walpole was you mention—but it could not have been any one you have heard of at Walcot— Uncle Jos & Elizabeth were staying here when your letter arrived & they read it & saw your affectionate message & remembrance of them—& Charlotte like- wise—who was here with M.r Langton— I am sorry to say he is *very* delicate subject to cough & a delicate chest, so much so, that they intend wintering at the Isle of Wight—poor Charlotte is very anxious about him— they seem very happy & attached to each other— Uncle Jos franked your servants letter on to M.rs Hewston—

I do not find I have any thing to tell you that has happened during the last month, but no want of news shall prevent a letter being sent to you the end of every month to tell you my Father is well— I do not remember when I have seen him more pleased, than on receiving your last happy satisfactory letter— saying again, and again, how glad he was to have heard from you & that you were well & safe. it was a great comfort to know you had always been prudent & fortunate enough to get into no difficulties in your several ad- ventures. pray dear Charles do not let the having escaped so long make you careless & daring for the time to come. Sarah Williams dined here the day before yesterday— she is not at all changed by her marriage the same hearty friendly creature she ever was— when she heard that you had left Terra del Fuego and had vowed you would return there no more—she was *furious* "What a shame! & he promised I should have a letter from him *dated* from Terra del Fuego"— She desired her best love, & hopes you will receive a letter she wrote directed to Valparaiso—

The Biddulphs are returned from their Swiss Tour— Fanny looking exceed- ingly thin, but exceedingly pretty— Charles Owen is gone to Rugby & Henry who you left a good little boy is now aiming at being a Dandyfied young man— wearing a worked velvet waistcoat, white gloves & a cane! William Owen is beginning to recover from his accident, stands for a few moments every day & there are great hopes he will escape a stiff joint.

My sisters have been staying a fortnight in Monmouthshire with the John Wedgwoods but I know nothing worth telling about their visit & I have been for a few days to Overton—these have been the only outings during the last month— Marianne is very well but already beginning to fret a little at the prospect of losing Parky whose fate is at last determined, & to school he is to go next Easter— I dined at Onslow not long ago & met there a M.rs Murchison at least I cant remember the name but you will know, she being the wife of the Cambridge Mineralogist—[1] She was excessively interested about the fate of Jemmy Button & Fuegia—having known & seen a ⟨great⟩ deal of the latter— I am very glad you have seen poor Jemmy again & that he is happy though relapsed into his old uncivilized habits— Frederick Hope who is al⟨ways⟩ blundering, told us that the Admiralty had ordered the Beagle home—but this cannot have *any* truth for its foundation or I think we should certainly have heard from some of your friends if it had been in contemplation— I give you joy most heartily on the new animals you have found.— Erasmus & Susan told you in their last letter Aug 30th.[2] how delighted M.ʳ Clift was with your former specimens & that he thought the last boxes which had arrived contained remains of *much value*—

When I told Nancy how your beard was let to grow—& gave her your goodnatured message that she would take you for an old jew—she burst out crying. I really think poor Nancy looks forward with as much delight to seeing you again as ever my Father or any of us do.

I took Pincher & Nina a walk through some of the Berwick fields & Pincher still remembers your training so well, that though a Hare sprung up just before us & he looked as if he would have given worlds to follow he obeyed & walked close behind me without attempting to have a Hunt—

We have had workmen without end this summer about the House new slating &c, & latterly pulling down & rebuilding the green house on rather a larger scale than before— the wood was so bad that it was not supposed safe—so that we might have had a grand clash of glass some day— the Laboratory is turned into a Laundry is the other alteration Erasmus when he came home in the summer found every thing turned topsy turvy people ironing in his Lab, & a baby in his bed room, the Hensleigh children having taken possession of his room, we not expecting him so soon as he did come—

This is all sad twaddle to send across the world but it must be excused for one day passes so like another that nothing I have to tell you— So good bye my dear old Charles God bless you & I hope we may soon have some time to look forward ⟨to⟩ when we may hope to see you— Ever yʳˢ very af⟨ ⟩ Caroline Darwin.

My Fathers kindest lo⟨ve⟩ ⟨an⟩d my sisters—

DAR 204.6.1

[1] The Cambridge Professor of Mineralogy was William Hallowes Miller, but he did not marry until 1844 (*DNB*). 'Murchison' is likely to be the correct name, since the Roderick Impey Murchisons frequented fashionable circles in London, where the Fuegians were lionised, and Murchison was a friend of Colonel John Wingfield of Onslow (see Murchison 1839, pp. 39–40).

[2] This letter is missing, as are the other letters from the family for the months of June, July, and August 1834, though CD eventually received them (see letter to Caroline Darwin, [19] July – [12 August] 1835).

To J. S. Henslow   4 October 1834

Valparaiso
October 4th 1834[1]

My dear Henslow

I have been unwell & am not yet out of bed. I write to tell you that Capt. FitzRoy sent, a week ago, by H.M.S. Samarang through Portsmout^{h}.—2 Casks, containing bones & stones & a box with 6 small bottles, with very valuable specimens.— Besides these 2 Casks there is a large Jar.—

I will write soon again when I am well. Dear Henslow | Yours affectionately | Chas. Darwin

Valparaiso

Royal Botanic Gardens, Kew

[1] This was written before CD completed the letter begun on 24 July (letter to J. S. Henslow, 24 July — 7 November 1834).

To Caroline Darwin   13 October 1834

Valparaiso.
October 13th. 1834.

My dear Caroline

I have been unwell & in bed for the last fortnight, & am now only able to sit up for a short time. As I want occupation I will try & fill this letter.— Returning from my excursion into the country I staid a few days at some Gold-mines & whilst there I drank some Chichi a very weak, sour new made wine, this half poisoned me, I staid till I thought I was well; but my first days ride, which was a long one again disordered my stomach, & afterwards I could not get well; I quite lost my appetite & became very weak. I had a long distance to travel & I suffered very much; at last I arrived here quite exhausted. But Bynoe with a good deal of Calomel & rest has nearly put me right again & I am now only a little feeble.— I consider myself very lucky in having reached this place, without having tried it, I should have thought it not possible; a man has a great deal more strength in him, when he is unwell, than he is aware of. If it had not been for this accident, my ride would have been very pleasant. I made a circuit, taking in St Iago. I set out by the valley of Aconcagua I had some capital scrambling about the mountains. I slept two nights near the summit of the Bell of Quillota. This is the highest mountain out of the chain of the Andes, being 4700 ft high.[1] The view was very interesting, as it afforded a complete map of the Cordilleras & Chili.— From here I paid a visit to a Cornish miner who is working some mines in a ravine in the very Andes. I throughily enjoyed rambling about, hammer in hand, the bases of these great giants, as independently as I would the mountains in Wales. I reached the Snow but found it quite impossible to penetrate any higher.— I now struck down to the South, to St Iago the gay Capital of Chili.— I spent a very pleasant week there, receiving unbounded hospitality from the few English merchants who reside there.— Corfield was there also & we lived together at an inn.— St Iago is built on a plain; the basin of a former inland sea; the perfect levelness of this plain is contrasted in a strange & picturesque manner with great, snow topped mountains, which surround it.— From St Iago I proceeded to S. Fernando about 40 leagues to the South.— Every one in the city talked so much about the robbers & murderers, I was persuaded to take another man with me, this added very much to the expense; & now I do not think it was necessary. Altogether it has been the most expensive excursion, I ever made, & in return I have seen scarcely enough of the Geology to repay it.— I was however lucky in getting a good many fossil shells from the modern formation of Chili.—

On my road to S. Fernando, I had some more hammering at the Andes, as I staid a few days at the hot springs of Cauquenes, situated in one of the valleys.— From S. Fernando I cut across the country to the coast & then returned, as I have said very miserable to Corfields house here at Valparaiso. You will be sorry to hear, the Schooner, the Adventure is sold; the Captain received no sort of encouragement from the Admiralty & he found the expense ⟨of⟩ so large a vessel so immense he determined at once to ⟨give⟩ her up.—[2] We are now in the same state as when we left England with Wickham for 1st Lieut, which part of the business anyhow is a good job.— we shall all be very badly off for room; & I shall have trouble enough with stowing my collections. It is in every point of view a grievous affair in our little world; a sad tumbling down for some of the officers, from 1st Lieut of the Schooner to the miserable midshipmans birth.—& many similar degradations.— It is necessary also to leave our little painter, Martens, to wander about yᵉ world.—[3] Thank Heavens, however, the Captain positively asserts that this change shall not prolong the voyage.—that in less than 2 years we shall be at New S. Wales.—

I find being sick at stomach inclines one also to be home-sick. In about a fortnight the Beagle proceeds down the coast, touches at Concepcion & Valdivia & sets to work behind Chiloe. I suspect we shall pay T del Fuego another visit; but of this good Lord deliver us: it is kept very secret, lest the men should desert; every one so hates the confounded country. Our voyage sounded much more delightful in the instructions, than it really is; in fact it is a survey of S. America, & return by the C. of Good Hope instead of C. Horn. We shall see nothing of any country, excepting S. America. But I ought not to grumble, for the voyage is for this very reason, I believe, much better for my pursuits, although not nearly so agreeable as a tour.— I will write again before sailing. I am however at present deeply in debt with letters. I received shortly since a very kind long one from Mr Owen, which I will shortly answer.— Letter writing is a task, which I throughly dislike.— I do not mean writing to home: but to any body else, for really after such interval I have nothing to tell but my own history, & that is very tedious.—

I have picked up one very odd correspondent, it is Mr Fox the Minister at Rio. (it is the Mr Fox, who in one of Lord Byrons letters is said to be so altered after an illness that his *oldest Creditors* would not know him)

I forgot to thank Susan for her letter of May & Catherine for her pithy message *"We* do not write" because Mr Owen does.— I must previously have acknowledged your long letter for the foregoing month.—

We are all here in great anxiety to hear some political news. A Ship sailed from Liverpool just after Ld Greys resignation & we cannot guess who will succeed him.—[4]

Give my best love to my Father & all of you & Believe me my very dear Caroline | Yours affectionately | Charles Darwin.—

DAR 223

[1] Robert FitzRoy in *Narrative* Appendix, p. 303 records the height as 6200 feet. CD changed his figure to 6400 feet in *Journal of researches*, p. 312.

[2] In *Narrative* 2: 361−2, FitzRoy wrote: 'At this time I was made to feel and endure a bitter disappointment; the mortification it caused preyed deeply, and the regret is still vivid. I found that it would be impossible for me to maintain the Adventure much longer: my own means had been taxed, even to involving myself in difficulties, and as the Lords Commissioners of the Admiralty did not think it proper to give me any assistance, I saw that all my cherished hopes of examining many groups of islands in the Pacific, besides making a complete survey of the Chilian and Peruvian shores, must utterly fail. I had asked to be allowed to bear twenty additional seamen on the Beagle's books, whose pay and provisions would then be provided by Government, being willing to defray every other expense myself; but even this was refused. As soon as my mind was made up, after a most painful struggle, I discharged the Adventure's crew, took the officers back to the Beagle, and sold the vessel.'

[3] 'Mr Martens, the artist, has been obliged from want of room to leave the Beagle' ('*Beagle*' *diary*, p. 249). Conrad Martens emigrated to Australia, where CD visited him in 1836.

[4] Lord Melbourne succeeded Lord Grey.

From Catherine Darwin 29 October 1834

Shrewsbury
October 29th. 1834

My dear Charles

The last letter we received from you was dated April 6th of this year, from the Falkland Islands. Caroline has written to you since last month. We have been writing now a long time, directed to Valparaiso; I shall be so delighted when we hear from you from that place.— I was much interested by your account of Jemmy Button, & his young wife; I dare say it is happier for him to forget his

English Habits, as he is to remain there.— Caroline had a letter from William Fox lately, who wished to know where to direct to you, as he is thinking of writing to you again, and is very anxious to hear some news of you. He says that in July he heard some very flattering things said of your exertions in Nat History, by Professor Henslow; who was then regretting that a Bag of Seeds, which had long reached England, had not arrived at Cambridge. William Fox says that you seem to have added much to our Gigantic Fossil Remains.— William Fox & his Wife are now staying at Osmaston; they were to have paid us a visit this Summer, but his Wife has been ill, which prevented it. The only great piece of news I have to tell you, is that the Langtons are intending to spend this Winter either in Madeira, or in the West Indies; on account of Mr Langton's health being so delicate, there is such great fear of his going into a Consumption, & it is thought a warm Climate may be of use to him.— Who would have believed 4 years ago that you would have been in South America, & Charlotte in the West Indies!— Nothing is quite fixed yet, but they will certainly go somewhere, as they have parted with their Servants, and obtained the Bishop's Leave for two years absence from his Living,—and it will most probably end in the West Indies, as Mr Langton much prefers that to Madeira. They did not think of it till so late in the Autumn, that they will have a great hurry over their Preparations.— They will return next Summer of course, as I suppose nobody spends the Summer in those hot Climates, no invalid, I should say.— There is no other Wedgwood news; we have had Harry & Jessie staying here sometime, with their nice little Baby, who is a great beauty, Miss Louisa by name.— Uncle Jos is enjoying his holiday from Parliament, Aunt Bessy continues much in the same state;—her intellects are very much gone, and she is perfectly helpless, from her lameness, but otherwise she is very well, and very cheerful.— I suppose you will see in the Papers the account of the Houses of Parliament being so entirely burnt down; it was owing to carelessness entirely[1] Westminster Hall had a narrow escape, but fortunately was uninjured, and most of the valuable Papers also were saved.— Owing to the wood work being so very old, it blazed away with immense rapidity.— Erasmus has never sent us word whether he joined the crowd to see it; it must have been a splendid sight.— The Owens are all in just the same state; no news about them; poor William Owen hobbles about on crutches, and can now bend his knee seven inches, which is thought a great feat. He has sold all his Hunters, and is going to set up a Poney Carriage for the Winter; it is now 4 months since his accident; it is still doubtful whether he will ever get the entire use of it again.— The Biddulphs are returned from Switzerland; Fanny is ver⟨y⟩ well now.— I gave Mr Owen your message th⟨e⟩ last time I saw him; he always talks about you.— Do you remember Lloyd Kenyon, Mr Kenyon of Pradoe's eldest Son? I suppose we told you he went violently frantic in the Spring, and tried to kill a Man; he got quite well, after this attack, went out every where, and his family denied his madness; but about a fortnight ago, he went quite mad again, and after great difficulty, was taken up, and sent into confinement at Liverpool.— He never can be considered safe, after this second attack.— Susan and I returned yesterday from spending two days at your friend Mr Slaney's, in his new grand house at Walford.— It is a very large house, but not handsome, as he would be his own architect. Mr Slaney begged to be particularly remembered to you, and enquired much about you. Tom Eyton was staying in the house of course, with his love;[2] they are to be married in February, and to live with old Mr Eyton; the house at Eyton has been very much added to, and altered to receive the Bride.— Tom Eyton seemed to be very devoted; I think he will have a *very nice Wife*, and certainly a very handsome one; and they will suit so well both so fond of quiet pursuits.— I think you will find them a very nice couple to stay with, when you come back; Tom Eyton, I am sure, is very fond of you; he has written to you already *three times* at Valparaiso, and is going to write to you again.— He has got some Chinese Geese, which are very curious birds, and a variety of other creatures & birds at Eyton, which I daresay he has written you an account of.— You must write and congratulate him upon his marriage to be; he is very proud of his choice, I am sure.—

Nancy has been begging me to tell you how much she longs to see you again, and how much she thinks of you.

Goodbye dear old Charley. How I long to see your dear old Phiz again. Papa's & all our best of loves. | Yrs ever | Catherine Darwin

Papa is very well.—

DAR 204.6.1

[1] For an account of the fire see *Annual register*, 1834, pp. 155–69.
[2] Thomas Campbell Eyton was engaged to Elizabeth Frances Slaney, the daughter of Robert Aglionby Slaney.

From W. D. Fox 1 November 1834

Osmaston
November 1. 1834

My dear Darwin

It is now so long since I have heard from you that I thought you most probably were really in the South Seas, your long promised Land, and tho' I much wished to write to shew you that I had not forgotten you, I hesitated to do so thinking that a letter directed to South America would not reach you. I have however written to your Sister Caroline for instructions, & today having heard from her, that if I direct Valparaiso, you will certainly get it, down I sit determined that this shall forwith set sail after you.— My letters must all be sadly prosy I fear, but you say (& I believe you from my own feelings) that you like to hear from me, & therefore I write. I will commence by telling you, as it may be later information than you have when this reaches, that the Dr Erasmus & your Sisters are all quite well—the D^r particularly so & enjoying Gardening more than ever. You will be sorry to hear that Mr Langton is so unwell that he & his Wife are thinking of leaving England for the two next Winters—going to Madeira or the West Indies for the sake of the voyage. Your old friend Eyton is going to be married to Miss Slaney— Has not a love of Nat: Hist.^y been probably the means of this match, as Slaney is himself I know an Ornithologist.— This is the first thing I have heard of Eyton since I remember going to his Fathers with you to entomologise & see his Cygnus Bewickii.—

I must now tell you all about our Party here. We are all very much as when you were here. My Grandmother, Father, Mother, & Sisters all in the same health and following their avocations as then. Indeed every thing at Osmaston has gone on just as it has for years. Our Landlord Sir Robert Wilmot died after some weeks illness in July, & at the request of the Family was buried by myself—the first bit of duty I have done for two years. You will I dare say remember him well. I do not think that his death will make any difference as to our, or rather I should say, my Fathers living at Osmaston, as its present Possessor Wilmot Horton is Governor of Ceylon & so deeply in debt that he will most likely remain at present where he is.[1] I wrote to you about four months ago, and directed my letter to Valparaiso, so trust you have got it tho I dare say it was not worth much— It is always a comfort to me in writing foreign letters, that I pay the Postage. I think I wrote to you from the neighborhood of Doncaster where I was staying with my Wifes sister & her husband. After that I came here & spent a few weeks when we set out on a little tour into Yorkshire, where we were moving about the neighborhood of Harrogate York & Ripon till the middle of September. We were almost all the time at different friends & Relations Houses—amongst others, we were several days with Robert Pulleine, whom you must well remember at Cambridge. He is living in a most beautiful romantic part of Wensleydale;[2] in a most comfortable Parsonage House, surrounded by dogs, Pigs & Poultry. I never was with a more truly hospitable kind hearted fellow. He fed us upon Venison & Moor Game & wanted us very much to stay as many weeks as we did days. Among many of our old friends and former times Charles Darwin was not forgotten, & we talked you over thoroughly. I wish you could have been with us— It was just at the commencement of Grouse shooting, which is excellent in that neighborhood, and thro' Harriets friends I could have got as much shooting as we liked, & though I have never fired a Gun since I was in the Church, I should so much have enjoyed walking over this beautiful country with my much valued & old friend Darwin, and unless you are much changed, & have learnt to despise such small game since you took to Ostriches, you would I think have enjoyed it too. I hope we shall yet have this pleasure together, and as you bring down your Grouse or Partridge you can tell me prodigious tales of your sport in other climes.

We have this Summer travelled eleven hundred miles in a little Pony Carriage which with all our Luggage a stout Pony has conveyed. In this distance we have only had one break down, which was of no consequence as we soon got our

Carriage mended, & no one untoward event of any kind.— I cannot tell you how much we have enjoyed it, tho' I am not sure that our enjoyment would not have been as great if we had been quietly located in some snug little Parsonage—my wife being quite as homely & domestic in her tastes & pursuits as your humble servant.

We came here about six weeks since, intending to go to Ryde late last month in our little Carriage, and to stay there the Winter, and I had hoped that a Winter there in addition to my two last, would enable me to take a curacy in the South in the spring, as tho' very much better in my Lungs, I am still incapacitated from exerting them much: however things have fallen out otherwise—three weeks since my Wife was taken very suddenly extremely unwell with inflammation in the Peritonæum. We luckily had Medical Aid at hand immediately and after bleeding &c she was brought round, but requires extreme Care, as there is a great tendency to recurrence. This alone would have been sufficient to have made it extremely hazardous for her to have undertaken a long journey; but in addition to this, she expects to be confined at the end of December or beginning of the following month, and the two things together make it perfectly impossible for her to think of it. We therefore now purpose staying at Osmaston thro' the Winter, and in Spring I hope we shall set out a trio, instead of a Duet.

I begin now to think that your early prediction when you set off, of my having a family of children before you returned, may be realised, as at all events, unless something untoward occurs, there is a great probability of one child. I need not set forth to you the manifold advantages of Matrimony, as you were always a Philo⟨gyn⟩ist & purposed entering that state as soon as you could.— I will on⟨ly⟩ say that from my present experience, I warmly recommend it to you & all I wish well to. I love my dear little wife more dearly now if possible, than when we were married, & have every reason to hope that we shall pass thro' life together, each adding much to the others comforts. But you will say I am writing a most pretious heap of nonsense—& so I am—but the truth is, that I am always apt to write to you as I should speak.— I wish you would do the same— It is now so long since I saw your handwriting that I cannot tell you the pleasure it would give me. I feel however that your time is valuable & mine worth nothing, which makes a vast difference.—

My Uncle & Aunt Darwin[3] from Elston are now staying with us for a week. I was much amused a day or two since at my Uncles saying—"That he was quite sure no son of his would live to inherit Elston— That both his would die—that then the Estate would come to Erasmus, who he had understood never meant to marry—that you would be drownd in your present voyage—when it would pass to Sir Francis Darwin's children." Do you remember my being your first informant that Elston was entailed upon Erasmus & you after my Uncles sons?— His eldest Son is a dreadful invalid, having a complaint of the heart, but is a very nice dispositioned Boy. He has now had these attacks so long & always got over them, that I think he may be reared, but must always be an Invalid. His other son is a fine healthy Boy as ever was seen, so that I do not think your & Erasmus's chance is a very good one, but many more improbable events have happened.[4]

I heard a few days since a very poor account of Sir Francis Darwins health. He fancies that he has some mortal disease and doctors himself for it. He either has used strong Medicines which have much brought him down, or is become very low spirited. He does not go out any where to his old friends & is very uncomfortable about himself. He is just expecting an increase to his Family from Lady Darwin, who had I imagined long since ceased from such expectations. I had fully arranged to have accepted a very kind invitation I had from Shrewsbury this Summer to go there & spend some time at your Fathers, but time slipped on so fast while we were moving about the country, that at last we were obliged to give up the idea, fearing we could not get to the Isle of Wight before the Winter set in.

I wish much I knew where you were at this time. I find from Caroline that your last letter was in April from the Falkland Isles, and she mentions something of one Jemmy Button, thinking I know all about him, but she has quite mystified me as you never mentioned him to me. I hope you will experience as much delight in the South Sea as you anticipate— If you are not already on your way when this letter reaches Valparaiso, you must be on the point of setting off for that long promised land—

Pray do not think my dear Darwin from this very dull letter that my feelings of friendship are correspondently blunted— You are never a whole Day absent from my thoughts & I think there are few of your friends who will more heartily rejoice to welcome you to England again than myself. But you will find me a strange dull fellow then I fear. All here unite in sending their kind love & best wishes for you in every way & believe me to remain | Ever your faithfully attached friend | William Darwin Fox.

DAR 204.8

[1] Sir Robert John Wilmot Horton was Governor of Ceylon from 1831 to 1837.

[2] Spennithorne, North Yorkshire. Pulleine was curate from 1830 to 1845.

[3] The William Brown Darwins of Elston, the Darwin family seat near Newark, Nottinghamshire, since the mid-seventeenth century (see *LL* 1: 2 – 3).

[4] Of the two sons, one, William Waring Darwin, died the following year aged thirteen; the second, Robert Alvey Darwin, died unmarried in 1847 (*Darwin Pedigree*). He left Elston Hall to his sister Charlotte (Freeman 1978).

To Catherine Darwin 8 November 1834

Valparaiso.
November 8th. 1834

My dear Catherine
My last letter was rather a gloomy one, for I was not very well when I wrote

it— Now everything is as bright as sunshine. I am quite well again after being a second time in bed for a fortnight. Capt FitzRoy very generously has delayed the Ship 10 days on my account & without at the time telling me for what reason.— We have had some strange proceedings on board the Beagle, but which have ended most capitally for all hands.— Capt FitzRoy has for the last two months, been working **extremely** hard & at same time constantly annoyed by interruptions from officers of other ships: the selling the Schooner & its consequences were very vexatious: the cold manner the Admiralty (solely I believe because he is a Tory) have treated him, & a thousand other &c &c has made him very thin & unwell, This was accompanied by a morbid depression of spirits, & a loss of all decision & resolution The Captain was afraid that his mind was becoming deranged (being aware of his heredetary predisposition). all that Bynoe could say, that it was merely the effect of bodily health & exhaustion after such application, would not do; he invalided & Wickham was appointed to the command. By the instructions Wickham could only finish the survey of the Southern part & would then have been obliged to return direct to England.— The grief on board the Beagle about the Captains decision was universal & deeply felt.— One great source of his annoyment, was the feeling it impossible to fulfil the whole instructions; from his state of mind, it never occurred to him, that the very instructions order him to do as much of West coast, as *he has time* for & then proceed across the Pacific. Wickham (very disinterestedly, giving up his own promotion) urged this most strongly, stating that when he took the command, nothing should induce him to go to T. del Fuego again; & then asked the Captain, what would be gained by his resignation Why not do the more useful part & return, as commanded by the Pacific. The Captain, at last, to every ones joy consented & the resignation was withdrawn.—

Hurra Hurra it is fixed the Beagle shall not go one mile South of C. Tres Montes (about 200 miles South of Chiloe) & from that point to Valparaiso will be finished in about five months.— We shall examine the Chonos archipelago, entirely unknown & the curious inland sea behind Chiloe.— For me it is glorious C. T. Montes is the most Southern point where there is much geological interest, as there the modern beds end.— The Captain then talks of crossing the Pacific; but I think we shall persuade him to finish the coast of Peru: where the climate is delightful, the country hideously sterile but abounding with the highest interest to a Geologist. For the first time since leaving England I now see a clear & not so distant prospect of returning to you all: crossing the Pacific & from Sydney home will not take much time.—

As soon as the Captain invalided, I at once determined to leave the Beagle; but it was quite absurd, what a revolution in five minutes was effected in all my feelings. I have long been grieved & most sorry at the interminable length of the voyage (although I never would have quitted it).—but the minute it was all over, I could not make up my mind to return, I could not give up all the geological castles in the air, which I had been building for the last two years.— One whole night I tried to think over the pleasure of seeing Shrewsbury again, but the barren, plains of Peru gained the day. I made the following scheme. (I know you will abuse me, & perhaps if I had put it in execution my Father would have sent a mandamus after me), it was to examine the Cordilleras of Chili during this summer & in the winter go from Port to Port on the coast of Peru to Lima returning this time next year to Valparaiso, cross the Cordilleras to B. Ayres & take ship to England.— Would this not have been a fine excursion & in 16 months I should have been with you all. To have endured T. del F. & not seen the Pacific would have been miserable: As things are at present, they are perfect; the intended completion of *small* parts of the survey of S.W coast would have possessed no interest & the Coast is in fact frightfully dangerous, & the climate worse than about C. Horn.— When we are once at sea, I am sure the Captain will be all right again; he has already regained his cool inflexible manner, which he had quite lost.—

I go on board tomorrow; I have been for the last six weeks in Corfields house. You cannot imagine what a kind friend I have found him.— He is universally liked & respected by the Natives & Foreigners.— Several Chileno Signoritas are very obligingly anxious to become the Signoras of this house.— Tell my Father, I have kept my promise of being extravagant in Chili. I have drawn a bill of 100£ (Had it not better be notified to Mr Robarts & Co?).[1] 50£ goes to the Captain for ensuing year & 30 I take to sea for the small ports; so that bonâ fide I have not spent 180 during these last four months.— I hope not to draw another bill for 6 months. All the foregoing particulars were only settled yesterday: it has done me more good that a pint of Medicin; & I have not been so happy for the last year.— If it had not been for my illness, these four months in Chili, would have been very pleasant: I have had ill luck however in only one little earthquake having happened.— I was lying in bed, when there was a party at dinner, in the house; on a sudden I heard such a hubbub in the dining room; without a word being spoken, it was devil take the hind most who should get out first: at the same moment I felt my bed *slightly* vibrate in a lateral direction. The party were old stagers & heard, the noise, which always precedes a shock; & no old Stager looks at an earthquake with philosophical eyes.

Till you hear again, you may direct to Valparaiso. If however, it can be managed, far the best & cheapest mode is to get somebody in Liverpool to receive your letters & send them by the first ship, which sails for this port.— I shall thus receive them, very likely two months earlier than by the regular post. In this case they must be directed *to the care of R. Corfield Esqre*.—

I have written to Erasmus (directing Whyndam Club) to ask him to execute for me a commission.— if he is not London I daresay Hensleigh Wedgwood would be kind enough to do it, getting the letter to read from the Club.—

Good bye to you all, you will not have another letter for some time.—[2] My dear Catherine. Your affectionately | Chas. Darwin

My best love to my Father & all of you.— Love to Nancy.—

DAR 223

[1] The banking house of Robarts, Curtis & Co., 15 Lombard St, London.
[2] CD sailed for Chiloé in the *Beagle* on 10 November and did not return until 11 March 1835.

To J. S. Henslow 8 November 1834

Beagle. Valparaiso
November 8th 1834

My dear Henslow.—

This letter is merely to inform you that I send by H.M.S. Challenger two boxes, with Specimens.— She does not sail from this port till January, & will not arrive in England for at least 4 months afterwards.[1] This letter goes by the Challenger to England.— In one of the Cases, I have given you an account of all our proceedings & future prospects &c &c.— I have also sent a part of my Journal.— would you be kind enough to direct & book it by some Coach to Dr Darwin Shrewsbury.— I did not think of sending it till five minutes before closing the Box, otherwise I would have directed it.— Of course, if you are inclined, you can look at any part of my hum-drum letter-like journal There are three small parcels of seeds; the one in the oblong box I have labelled as coming from T. del Fuego. it comes from Chiloe: (Climate &c &c like T. del Fuego but considerably warmer).— I do not much expect, that any one seed will grow.—

Continue to direct Valparaiso: if you know any person in Liverpool who would post your letters to me by any of the numerous ships to this port I should receive them a couple of months sooner.— in this case, they must be directed *to the care of R. Corfield Esqr.*— Yours most truly obliged | Chas. Darwin

Royal Botanic Gardens, Kew

[1] The *Challenger* sailed from Coquimbo on 5 February 1835, but did not proceed to England. After arriving at Rio de Janeiro in March, she set out on 1 April on her ill-fated voyage to the west coast of South America (see 'Loss of His Majesty's frigate *Challenger*' 1835). CD's cases, if they were shipped in the *Challenger*, may have been transferred to another vessel at Rio de Janeiro.

From Susan Darwin [24] November 1834

[Shrewsbury]
November 1834

My dear Charles.—

You will be surprised to hear that the very Packet that takes this from Falmouth to you: conveys at the sametime the Langtons to Madeira where they have resolved to spend the winter on account of Mr Langton's health, as he has already had one attack on his chest which alarms them very naturally, having lost *nine* Brothers & Sisters by Consumption— It seems a great pity, that they should be forced to leave Onnibury just when they were so busy & happy making their house larger; & laying out their garden &c.—but as he is 33, very likely if he can get safe over another winter, there may be no longer any anxiety for him: so it is very wise of them to make this sacrifice.— They hesitated for a longtime between Madeira and the West Indies: both of them I think preferring the latter on account of the novelty & beauty of a tropical climate: but Uncle Jos decided the question by reminding them that most probably the West Indies would be in an unsettled state this first year after the freedom of the Slaves.— They were very anxious to find some good account of Madeira & Charlotte who searched yr favorite Humbolt was much disappointed to find he said nothing about it: however they have got some letters of recommendation to the Consul & Chaplain[1] so they won't feel quite cast away. The Pandora sails on the 5th of Decr. It is a new packet & built on some new & improved plan which makes it safer (Charlotte says) but I had not rather be the first for new experiments if I was them.— I daresay they will still proceed to the West Indies for Mr Langton has a gt wish to go there & when he is half way I don't believe he will turn back What a pity it is you can't meet them, but you will enjoy talking over tropical scenes some future time I hope. There could not possibly be a better Lady for taking such a voyage than Charlotte as she has no nervousness in her composition, & plenty of independance & carelessness about personal comforts. They mean to take no Servant with them which seems very rash considering the danger of illness in Mr Langton's case.—

The last letter from Erasmus was written in rantipole spirits—[2] he had been spending 5 days at Cambridge with the Hensleighs which they had all enjoyed prodigiously spending most part of their time with Professors Whewell & Sedgwick: and the day they enjoyed most was Sunday After going for the University Sermon to Kings, they went to Trinity where after chapel was over they had a most beautiful Concert. Sedgwick took them to his rooms in the Evg & Whewell met them there. Eras says Whewell took the lead in conversation which was of a religious turn: & Eras says he "is in despair he cannot write down his words for they were really *super human*" And in another part of his letter he says "the brilliancy & rapidity of Whewell's conversation with Fanny was such as I could hav⟨e⟩ formed no conception of— The two professors harmon⟨ising⟩ beautifully: Sedgwick's simplicity & good faith in all he says & his picturesque manner of conversation shewed off Whewell's, which is all speculative & generalizing always brilliant & so perfectly elegant I believe it would be impossible to change a single word" This extract from his letter is sufficient to shew you how delightful his visit at Cambridge must have been.— It will make you long to be amongst them. Sedgwick is just made Canon of Norwich.— The country is in a strange state at present, for quite unexpectedly the King has dismissed all the Whig Ministry &

made Duke of Wellington Premier, & how this will stand appears very doubtful, for they intend to have a Dissolution of Parliament in hopes to get more command over the House of Commons, & most people say they will certainly change for the worse & get more Radicals elected instead of Tories.

The last piece of news I have to tell you is a very sad one Poor Col Leighton was riding out last Wednesday the 19th with Clare quite alone having no Servant when he suddenly fell off his horse upon Coton Hill in a fit of Apoplexy and never spoke once but instantly expired.— Mr Wynne[3] went to the spot immediately but too late. he then galloped up here & begged Papa to go & break the shock to Mrs Leighton who was quite overwhelmed having just parted from him perfectly well & in good spirits—fortunately Louisa Hope's confinement was over & she has got a nice little boy so there was no danger in telling her & she with Clare exert themselves very much to comfort Mrs Leighton. Poor Frank was at Oxford & arrived the next night so now they are all together & the Funeral takes place tomorrow. He seems to be universally regretted as every one must respect him that knew him he was so conscientious & good in every way. Papa felt it very much as he was quite one of his truest friends.— My Sisters are gone to Maer to take leave of the Langtons or they would join Papa & me in very best love to you Dear Charley Ever y^r affecte | Susan Darwin

Old Nurse *Tanty & I* maundered over you for about an hour the other night. Our chief topic was wondering *when ever* you would *come back*

Postmark: Shrewsbury NO 24 1834
DAR 204.6.1

CD ANNOTATIONS
on cover: £60 Fox | Last letter received at Valparaiso

[1] The Chaplain was Richard Thomas Lowe (see letter to J. S. Henslow, 28 [September 1831]).
[2] Rantipole: 'Wild, disorderly, rakish' (*OED*).
[3] Rice Wynne.

From Caroline Darwin 29 December [1834]

24 Regent
December 29th

My dear Charles,

Your two letters were a delightful surprise to us & thank you much for them—the first dated July 29th to Cath from Valparaiso—and one to me August 9th— Mr Clift has been written to with all the directions you desired— Erasmus says a box with bones came to England by Liverpool in August, & he thinks from Buenos Ayres—a note he has just had from Pro.

Henslow dated Dec 22ᵈ. to say "owing to a mistake he had not recevᵈ your bones till now— he finds them all safe & has put them in a dry place— he also says he will write by the first parcel that goes to you— Erasmus says he does not know what bones Henslow refers to in the note I quote from— poor Eras. has been very ill but thank God is quite safe and well again— he was seized 3 weeks ago suddenly with gᵗ pain in his side which left him after a night but proved the beginning of a fever & inflammation upon his chest— for a few days he was extremely ill & reduced to such great weakness that he could not for many days move himself in bed— Most fortunately the Hensleighs found out by chance his illness & nursed him most kindly & tenderly. Dʳ Holland also has been most really kind. Susan came up last Sunday fortnight— I unfortunately was at Overton & did not hear of his illness in time to go with her but followed in a few days & most happily found the danger over & have had nothing but the comfort of seeing him eat sleep & gain flesh— he now sits up all day as merry & *pert* as possible— I go home tomorrow & Susan goes with him to Clapham for a few weeks before they go down to Shrewsbury. poor dear Papa had the gout flying about him so much that he was unable to travel to London, but he is very happy now— Eras sends his love to you & will write by a parcel which he hopes to be able to send out before long— as far as I can understand the delay is owing to part of one of the books not being printed which was to go in the parcel—. Erasmus seems to have enjoyed excessively a little visit of 5 days to Cambridge where he went with the Hensleighs— he said he felt as if he was in a dream walking about the gardens acting nursery maid to Snow—the eldest of the 2 Hensleigh children & the greatest of darlings with Erasmus— It is a very nice little entertaining thing— Eras called upon Pro— Henslow who apologized for not inviting him speaking about the state of health of his wife &ᶜ which at last Eras found out meant that Mʳˢ Henslow was on the point of her confinement— He came one eving & drank tea at the Inn with the Hensleighs & Eras— Also Sedgwick & Whewell came evy eving & were extremely agreeable— The Langtons sailed for Madeira 3 weeks ago— Mʳ Langton was pretty well with no fresh cold & in great spirits— He wants very much to go on to Rio & I do not think that Charlotte much objects but it is very doubtful whether they will put this scheme in execution— if they do go I can fancy nothing more provoking than that their visit did not take place when you were there How strange the very possibility of your meeting Charlotte at Rio seems—

We have heard a report that Capᵗ Fitzroy is promoted & that you are coming home by the Paci⟨fic⟩ I am afraid this is too good news to be true & even you I hope would be content if you see the South Sea Isᵈˢ— it will indeed be happy news to us when we have a certain time to look forward to seeing your dear face again— My Father is very well, rather fearful of gout but that is rather a good thing for him. Wishing you dearest Charles a happy new year Believe me Ever yʳ affᵗᵉ | Caroline Darwin—

I go home per Mail this eving & shall have a colder night than any you have known for some time

DAR 204.6.1

From Catherine Darwin 28 January [1835]

Shrewsbury.
Jan^{ry} 28th.

My dearest Charles

Caroline received your letter from Valparaiso of October 13th, about a week ago. I cannot tell you how sorry we were to hear that you have been ill, my dear Charley. It must have been so trying for you being ill when you were on an Expedition, and I am sure you must have suffered very much by forcing yourself to travel on, while so unfit for it.— We are in hopes of hearing from you soon again, as we are doubly anxious to hear of you now. Papa charges me to give you a message from him; he wishes to urge you to think of leaving the Beagle, and returning home, and to take warning by this one serious illness; Papa says that if once your health begins to fail, you will doubly feel the effect of any unhealthy climate, and he is very uneasy about you, and very much afraid of the fevers you are liable to incur in those Countries. Papa is *very much in earnest*, and desires me to beg you to recollect that it will soon be four years since you left us, which surely is a long portion of your life to give up to Natural History.— If you wait till the Beagle returns home, it will be as many years again; the time of its voyage goes on lengthening & lengthening every time we hear of it; we are quite in despair about it.— Do think of what Papa says, my dear Charles; his advice is *always* so sensible in the long run, and do be wise in time, & come away before your health is ruined; if you once lose that, you will never recover it again entirely.— I wish it was possible that anything we can say may have some effect on you; do not be entirely guided by those you are with, who of course, wish to keep you, & will do their utmost to that end, but do think in earnest of Papa's *strong* advice & opinion.— Caroline wrote to you last month, from London, when poor Erasmus was recovering from his dangerous fever; he has been with us now for about a fortnight, and is getting better and stronger every day, and I trust will soon be quite himself again; he will not stay with us much longer, but returns to London next week. He sends you his best love, & will write to you himself, when he goes back to London, with some Books that are going to you from Cambridge.— Eras did write you a bit of a letter in one of Susan's, that you must have had before you get this.— Eras is such a languid, indolent old fellow, that you must forgive him for not having written to you oftener.— I am very glad that you have now received all Mr Owen's letters, and I hope he will soon have your answer. Have you ever written to William Fox since his marriage? You will be sorry to hear that his Lady has had a dead baby, which was a great grief to them, and was most dangerously ill, soon afterwards with Inflammation of the Chest. Old Mrs William Darwin, Mrs Fox's Mother is dead, at the age of 90; I do not know whether you ever saw her, or knew of her existence. It is said they will leave Osmaston soon.— I can easily believe how very difficult and disagreeable you must find writing to friends, you have not seen for such a time now and I am sure *one* letter that you write, ought to count for *three* that you receive; but still you must every now & then write to your other Correspondents besides us, or they will not continue to write to you.— We rejoice exceedingly for you having such a kind useful friend in Mr Corfield; a friend in S. America must be invaluable.— I have only one extraordinary piece of news for you, which you will first laugh at, and then be sorry for.— I suppose you know Robert Wedgwood has been living the last half year as Curate at Muxton, in old Mr Crewe's house; he is a most disrespectable, horrible old man, and his son, Mr John Crewe also is a disgraced man, & banished from Shropshire;[1] all the John Wedgwoods were much vexed at Robert's going to live in that disrespectable house, and the⟨y⟩ are still more vexed now, when it comes out that Robert has fallen vehemently and desp⟨e⟩rately in love with Miss Crewe, who is *50* years old, and blind of one eye. The John Wedgwoods have tried in vain to break off this unlucky engagement, but all in vain; Robert is infatuated, and proud of his good fortune, and they will soon be married.— She is a clever woman, and must have entrapped him by her artifices; & she has the remains of great beauty to help her; it is said also that she has a violent temper, which is another bad point in this ill-starred match.— Robert is either 28 or 29, so there are either 21 or 22 years difference between them.— It is a regular case of Gobble Boy, I think.— The John Wedgwoods are here now; Aunt Jane desires her best love to you.— Uncle Jos' Parliamentary Days are over. he did not attempt to stand this last Election he would not have been returned if he had. Shropshire has actually returned 12 Tory Members who are called Lord Powis's Twelve Apostles.[2] Toryism rages in Shropshire more than ever, and there certainly has been a slight re-action in favor of the Tories over the Country; in general though, the Reformers are much stronger, and will, I trust soon rout out Sir Robert Peel, and his odious Ministry.—[3] Edward Holland is returned as Member for East Worcestershire.— I have not yet told you what nice accounts have been received of the Langtons; they arrived at Madeira, the 16th of December, after only 10 day's sail, which Charlotte speaks of enjoying very much, and she writes in great admiration of the beauty of the climate, which is hot for exercise (in Dec^ber) and our greenhouse flowers in full blow out of doors.— This will not sound so wonderful to you. Mr Langton was very well, when Charlotte wrote with no affection of his Chest.—

I have told you every thing now, my dear old Charley.— How I wish that your next letter might bring us the joyful news of your return; how happy that would make us.— Papa's kindest & best love; he is very well. Bless you, my dear old boy, and take care of yourself, at least. | Ever y^rs | Catherine Darwin

My dear Charles I was grieved to hear of your illness & your disappointment in again going to Patagonia— I hope & trust you will seriously consider whether it would not be wise in you to leave the Beagle & return home. you have now been longer than you originally intended & are not the slightest degree bound in honour to remain as long as the Beagle does. My Father & we shall be excessively happy to see you again & do think whether on account of your own happiness & health you had not better come back to us. | dearest Charles Goodbye & God bless' you. We are anxious for your next letter— | Caroline Darwin.

DAR 97 (ser. 2): 16—18

[1] See letter from J. M. Herbert, 15—17 April 1832, n. 6.
[2] Edward Clive, Earl of Powis.
[3] The Tory trend in Shropshire was not typical. Popular resentment at the dismissal of Lord Melbourne by King William IV and his appointment of Peel and Wellington caused a heavy Whig majority to be elected. Peel, repeatedly outvoted in the Commons, resigned in April 1835. Melbourne succeeded him and formed a Ministry that lasted six years.

From Charles Whitley 5 February 1835

College, Durham
Feb.y 5th 1835

My dear Darwin,

At length I have an opportunity of replying to your very welcome letter, & as a single sheet has limits, & those very soon attained I must begin my answer with a systematic account of my own proceedings. I think you were aware of my being a candidate for the office of Mathematical Professor in the University which was projected in this place about the time of your leaving England. Perhaps not,—but we will take that for granted, though you had enough to think of just at that time without plaguing your head about embryo Universities. I did not succeed in my wishes, but on the opening of the University in 1833 a subordinate Chair was created for my accommodation. To this has since been added the Mathematical Tutorship of the sole College erected within this University which with one or two other small offices, for I am a shocking pluralist, causes my pecuniary state to be a very comfortable one (though I am far from rich) even in our infant state. The Mathematical Professor too is dead, & I am looking anxiously forward to succeed him. If such should be my fortune I suppose I shall be planted in Durham for the rest of my days; and from what I can see at present,—& I have been here above a year—I shall have no reason to quarrel with my location. I am not yet married—I wish I were—or even engaged to enter upon that desirable state, but if I get this Professorship I think I shall not be very long in doing so. There is a goodly bevy of damsels round this city out of which a man may choose himself a helpmate. So that perhaps before you come home I may have 'settled' & be in a condition to offer you comfortable quarters—a hearty welcome & a warm fireside—which will I am sure be offered to no one more freely, or with a more earnest desire that they should be accepted, than to yourself. And I hope too that my new Alma Mater will by that time be firmly seated, & in the actual enjoyment of all the rights & privileges for which she is now contending. We have already an Act of Parliament, & we hope soon to have a Charter in addition, so that we shall be enabled to confer degrees. Our numbers at present are small enough, sixty undergraduates & six Dons, of whom Peile of Trinity is one. We cultivate the ornamental branches, such as Chemistry &c, a little, but for the encouragement of these there are Lecturers appointed who hold no appointments in the College, & are therefore to be considered as 'allied powers' rather than actual members of the Staff. And this may serve to give you some idea of my actual position. My books & prints have been duly transported hither & my household Gods may fairly be considered to have taken up their abode here. I have not been in Cambridge since the Autumn, when I left Heaviside there, flourishing. He is now Tutor of Sidney. I heard the other day from Cameron who is now living in Cambridge—much sobered by severe misfortune. Did you hear that his Father was dead, & in very indifferent circumstances? His Mother is now I believe living with him in a small house in Downing Terrace. He talks of going into orders shortly, if he can get a curacy. Marindin is married & has sold out of the guards. Of Mathew I know nothing. He was not in Cambridge at all during the summer when I was there. Watkins is now in orders & has got the curacy of Clyro in South Wales. It is the living of which Venables' Father is incumbent. Venables himself is married to a Russian Countess & has got a living in Herefordshire. Watkins seems contented & even happy in his humble occupation. I had a long letter from him the other day containing much rigmarole, some sense, & a great deal of amusement. He would I am sure set 'great store' by a letter from you. By the way I am more than half angry with you for not writing to me before. If you are not coming home shortly pray set yourself right again in my good opinion by giving me some further account of yourself—if possible *by return of post*. You think very often & very much, I daresay, of the friends you have left behind, but if all that your friends say & think of you, collectively, were to be put on the other side, my belief in the force of multiplication convinces me that the balance would be against you. I need not tell you much of Herbert as you have written to, & heard from him. He is still in London & will ere long be called to the bar where, *barring* laziness, he is very likely to shine. I spent a part of my Christmas Vacation, for Alma Mater Dunelmensis has, like her elder sisters, three Vacations, in Nottinghamshire viz: four days with Tom Butler, to whom as you have doubtless heard the *ex*-Lord-Chancellor[1] gave a living,[2] & three with the Lowes whose place of abode is only four miles distant from the said Tom's Rectory. The first was altogether a pleasant visit, & I need not say to *you* how pleasant the latter was. If there be any truth in old sayings your ears must have burnt most painfully during the whole time. Henry was unfortunately away, on his travels, but he was previously aware of my intention to visit Bingham & had sent a special request that I would join his brother Robert in writing a letter to him, which I did accordingly, & if it caused as much merriment to him at Bourdeaux as it did to us in the Vale of Belvoir I think his sides must have ached, as mine did, for a week afterwards. I need not of course retail to you any public news. The dismissal of the Whigs, the accession of Sir R. Peel & the Elections are of course to be found in newspapers even in your remote quarter of the world. We are at present a-gape on the subject of Church Reform, for the furthering of which a Royal Commission has just been issued.[3] I think myself of going into orders by & by & am therefore personally interested in the subject. And I hope that the integrity of Church property will be preserved. We are concerned also as a body in the distribution of the Church funds of the Diocese, especially of the Chapter, for a slice of which we naturally look. I saw Miss Holland at Newcastle the other day. She like every body else spoke of & enquired after you. You are aware that she has lost her post in Brook Street her brother having married a daughter of Sydney Smith the Reviewer[4] some ten months ago, who gave birth to a daughter herself within these last ten days. So the world goes round. Of Henslow I know little, save that the world goes more prosperously with him since he has got a living. There are plenty of new things for you to see & to hear in the way of music & painting. I have made some superb additions to my family of prints, and there are some glorious treasures added to the National Gallery. So let us look forward to intellectual pleasures yet to come. I have no more news for you, except that John Roberts of Barmouth has been drowned at sea in the 'Wellington'. Luckily for your friend Rhys Jones they had previously quarrelled & John had got another partner, who shared his fate. O. Gore is returned for N. Shropshire vice Cotes—retired.[5]

God bless you! & send you health & success! | Ever your's most faithfully | Charles Whitley.

DAR 204.9

[1] Henry Peter Brougham.

[2] Butler had become Rector of Langar with Barnston, Nottinghamshire (*Alum. Cantab.*).

[3] The Ecclesiastical Commission appointed by Sir Robert Peel was carried forward by the Whig government, which passed a number of reform acts designed to reduce the privileges and revenues of the Church.

[4] Sydney Smith was one of the founders of the *Edinburgh Review* in 1802.

[5] William Ormsby-Gore and Sir Rowland Hill, Bart, both Conservatives, were unopposed in the general election of January 1835.

From Susan Darwin 16 February 1835

[Shrewsbury]
February 16th: 1835

My dear Charles.—

We all hope very much before this letter is sent off we may hear again from you, as we do not feel quite happy and easy about you my dear Charley till we know you are quite got as strong as you were before your illness & we are so afraid you should have another return of the same attack.— Hearing of your being unwell as so great a distance is very uncomfortable and I wish with all my heart you would be sufficiently *home sick* as to proceed no farther on this endless expedition.— I met yesterday Capt Harding who has just brought home his Bride Miss Dona Dallas from St Helena, who looks quite as blooming as she did these 8 years ago when she lived above us.— Capt H. enquired much about you; and said that Capt Fitzroy was promoted to be Post Captain: but he did not think that would make any change in his plans or be likely to bring him back to England: which I was sorry to hear, as I had flattered myself, that he would quit the Brig Beagle for some larger vessel in that case.— It is a sad grievance Capt F. being forced to give up the little vessel he bought to accompany the Beagle, as besides running more hazard, you must be sadly crammed & crowded together.—

We could not keep Eras more than a fortnight here & one week at Maer before he would go back to his beloved London. however he writes to us in great spirits and seems to have caught no cold from his Journey so I hope he will soon be as strong as he was before his illness tho' that was not much to boast of.— If he is so naughty and lazy as not to write to you I am sure he has lost no affection for you, as he reads your letters often over and always talks with the greatest interest about you, and stands up warmly for the wisdom of your expedition as it has added so much to your happiness.— Though I have talked of his idleness: I know he means to write to you very soon, and I hope you have received part of a letter he wrote in one of mine to you last August or September.— Tom Eyton was here the other day and mentioned having got a letter from you since you have been at Valparaiso, but said that he had written you 3 Letters directed there & that you had never rec^d his first.—

We are daily expecting to hear of his marriage with Miss Slaney as it was to take place this month. I believe his house is not finished which is the cause of the delay, and he has not been well himself. We all like Miss Slaney very much & admire her beauty as much as you used to do in old times of Barmouth— Mr Slaney got turned out this last Election by Mr Pelham which must vex him.—

Catherine will have told you in her last letter of Robert Wedgwood's intended marriage with Miss Crewe of Muxton—just 20 years difference in their ages!— Robert came over here a fortnight ago to see Aunt Jane who is staying with us now: and talked over the matter they have given their consent much against their inclination as of course they dislike such an absurd match very much, besides all her family being very goodfornothing people.— I advised Robert to marry & go abroad as that would be much the best way of letting the *talk* subside, & he seemed very much inclined to follow my advice. People say she has a bad temper & it is impossible that she can be otherwise than very jealous of her young husband: really the Wedgwoods ought not to be allowed to chuse for themselves after Franks & Roberts specimens of wives.—

Caroline has had a very nice letter from Charlotte written from Madeira where I think she seems rather disappointed you will enjoy some future day talking over tropical scenery with her Mr Langton seems to have a great fancy for Rio Janeiro so I daresay they will proceed there what a pity it was not a year ago when you could have received them there.—

I am just returned from a short visit to Woodhouse where I went chiefly to see poor Owen since his accident— I was quite surprised to see how well he could move his Leg. he walks a great deal, but is obliged to strap it up that he may not bend the joint too much.— Fanny & Mr Biddulph were there and I found her quite as enchanting as ever. She is looking far more beautiful than ever I saw her before partly owing to her being so much more delicate looking.— Whilst we were walking round the Kitchen Garden she burst out laughing saying she could not help thinking how you & she in former times had stuffed yourselves over the strawberry beds, & from that she talked very affectionately about you & said how much she should enjoy seeing you again dear old Charley which is a wish I am sure we all have and so now Goodbye & bless you | my dear old fellow Ever y^r very affectionate | Susan E Darwin.—

Papa is very well & very anxious to have further news of you— All here send their affec^te love to you.

DAR 97 (ser. 2): 18–20

CD ANNOTATIONS
'Last letter. Coast of America.— Lima— dated February'

From Charles D. Douglas¹ 24 February 1835

Mr Charles Darwin | Dear Sir

I have sent you a dozen of the large beetles which I told you of having seen in Chiloé, & which I believe are not in your collection; I found them in the crutch of átenihue tree, thirty feet from the ground in a nest of moss; I was led to the spot by following one of them morning & evening for several days, & always lost sight of it near this tree, which is situated about 200yd^s from the church of Caucague, at last on the morning of the 15th curr^t I with some dificulty ascended the tree & found them in the crutch, as above refer'd. I killed them with heat in an earthen dish & put them in the box when cold.

Feb.[y] 20th at 11[h] 33[m] A.M. the Island of Caucague was Visited by an earthquaqe, which lasted seven minutes & forty seconds, it was not preceeded nor folowed by any subteraneous noise, the motion was horizontal from N.E. to S.W. slow & not strong, like the motion of a vessel in y[e] long gentle swell of the ocean, the trees were violently agitated, but very few were thrown down, the swell in the channel did not strike the shore more than two feet perpendicular height.

There is a block of fine grained granite, on the beach of Aucar, opposite the N.W. point of Caucague, length N. & S. 18f[t] 3in[s] hieght on the E. side 9f[t] 4in[s] circumference four feet above ground 54f[t] 10in[s]—the cross measure over the middle of the stone from ground to ground was N. & S. 31f[t] 3in. E. & W. 26f[t] 1in. there is a block of coarse grained granite on the point lobos of Caucahue, its elevation estimated 200 feet above the level of the sea—length 9 feet perpendicular thicknes 4f[t] Horiz[l] thicknes 6f[t] I have seen many blocks of granite on the surface of the ground in diffrent places & also on the beach, but never embeded in the cliffs, except in one instance, in point Chouan estimated length N. & S. 7 feet perp[r] thicknes 4f[t] Hor[l] thicknes unknown, estim[d] hieght above the sea 80 feet—

I have heard of a block of granite situated in the Potrero of the Reycaquines, 8 miles west of Lluco, said to be the largest stone on chiloe, a man on its top can over look the woods to a great distance, I know not, whether that circumstance is owing to the hieght of the stone, or to the elevation of ground it stands on. I expect the potrero will be surveyed shortly, should I be sent on the commision, I will measure the stone & send you the contents.[2]

I start this morning for Caylin & will return surveying the Indians lands as far as Rouca which will occupy me two months. If I should see any thing which I consider useful, I will send you word next Summer— Mr Robert Burr will forward this & the Box. | Chas D Douglas

Delcague Feby 24— 1835

DAR 35.1: 329–30

[1] Douglas was a surveyor, resident in Chiloé. He acted as pilot and interpreter to FitzRoy and provided him and CD with information about the country and its inhabitants. See *Narrative* 2: 364 and '*Beagle' diary*, pp. 252–3.

[2] CD entered Douglas's data on erratic blocks in his geological notes (DAR 35.1: 300). CD first observed them during the Santa Cruz expedition in April 1834 and thereafter noted instances of blocks at a distance remote from their original formations. In *Journal of researches* (pp. 288–90 and Addenda) he published his view that transportation by icebergs was the explanation, a conclusion presented in 1841 in more comprehensive form in his paper 'On the distribution of the erratic boulders . . . of South America' (*Collected papers* 1: 145–63). In his geological notes (DAR 34.2: 169) CD states that he had turned his attention to the transport of boulders as a result of 'some queries sent by M[r] Lyell to Capt. FitzRoy'.

To W. D. Fox [7—11] March 1835

Valparaiso.
March 1835

My dear Fox

Our correspondence seems to have died a natural death or rather I will say an unnatural death. I believe I wrote last to you, but it was before I heard the news of your marriage. You have my most sincere congratulations, mixed however with some little envy: I hope you are now stronger in your health; & then I am sure you will be as happy as you well deserve to be. How changed every body & every thing will be by the time I return. You a married clergyman, ave maria, how strange it sounds to my ears. I wonder when I shall see you: If you continue to reside in the Isle of Wight perhaps it will be in Portsmouth. If a dirty little vessel, with her old rigging worn to shreds, comes into harbor September 1836 you may know it is the Beagle. You will find us a respectable set of old Gentlemen, with hardly a coat to our backs. This same returning to dear old England is a glorious prospect; I wish it was rather nearer; but it is sufficient to make up for a thousand vexations. Five years is a sadly too long period to leave ones relations & friends; all common ideas must be lost & one returns a stranger, where one least expects or wishes to be so.— I hope at least it will not happen with you & me.— I think the recollections of the snug breakfasts & pleasant rambles at Cambridge, will make us remember each other. You are one of the indirect causes of my coming on this voyage: by taking me as your dog in the grand chace of Crux Major you made me an Entomologist & introduced me to Henslow. I am very glad I have come on this expedition, but like a Sailor I have learnt to growl at all the details— What I shall ultimately do with myself—Quien Sabe? But it is very un-Sailor-like to think of the Future & so I have done.—

We leave for ever the coasts of America in the beginning of September, our route lies by the Galapagos, Marquesas, Society, Friendly Iˢ, New Zealand (?), to Sydney. I hope, shortly after receiving this you will write to me at the latter place. I have heard nothing about you for a long time; excepting the one grand thing marriage; this certainly is a host in itself, but I should like to hear some more particulars, what doing, where living, & infuturity?— Can you drink to Hopes toast of "Entomologia floreat". in one of your letters you told me you had been collecting Pselaphidæ. In the damp forests of Chiloe & Chonos Archepelago, I had the satisfaction of taking many small English genera: amongst them Pselaphus, Corticari's, minute Staphylini, Phalacrus, Atomaria & Anaspis. (Remember the Fungi at Osmaston) &c &c & Elmis beneath a stone in a brook.—

Latterly however I have been paying more attention to Geology even to the neglect of marine Zoology. We are now making a passage from Concepciòn: you will probably have seen in the Newspapers an account of the dreadful earthquake.[1] We were at Valdivia at the time; the shock was not quite so strong there, but enough to be very interesting.— The ruins of Concepcion is a most awful spectacle of desolation. There absolutely is not one house standing.— I have thus had the satisfaction in this cruize both of seeing several Volcanoes & feeling their most terrible effects. It is certainly one of the very grandest phenomena to which this globe is subject.—

As soon as the Beagle reaches Valparaiso, I intend going on shore & shall reside there till 1st of June, when the Beagle will pick me up on her road to Guyaquil.— I am at present full of hope to be able to cross the Cordilleras & see the Pampas of Mendoza.— I am very anxious to connect the geology of the low country of Chili with the main range of the Andes.— I will keep this letter open for the chance of receiving one from you.— I should have written before sailing on the last cruize to the South; but I was very ill for 6 weeks & found all labor, even of writing too irksome. Farewell dear Fox. God bless you.— I hope you are both well & happy. | Your affectionate friend. | C. Darwin.—

Christ's College Library, Cambridge (Fox 47)

[1] CD described the earthquake at Concepción in his paper, read before the Geological Society in March 1838, 'On the connexion of certain volcanic phenomena in South America' (*Collected papers* 1: 53—86).

To Caroline Darwin 10—13 March 1835

[Off Valparaiso]
March 10th | 1835

My dear Caroline,

We now are becalmed some leagues off Valparaiso & instead of growling any longer at our ill fortune, I will begin this letter to you. The first & best news I have to tell, is that our voyage has at last a definite & certain end fixed to it. I was beginning to grow quite miserable & had determined to make a start, if the Captain had not come to his conclusion. I do not now care what happens. I know certainly we are on our road to England, although that road is not quite the shortest. On the 1st of June the Beagle sails from Valparaiso to Lima, touching only at one intermediate port—from Lima direct to Guayaquil—to the Galapagos, Marquesas so as to reach Otaheite middle of November, & Sydney, end of January of next year.—

This letter will be sent across land so will reach England soon: after receiving this you must direct till the middle of November to Sydney.—then till the middle of June to the C. of Good Hope.— We expect to arrive in England in September 1836.— The letters which come directed to S. America will not be lost for the Captain will write to the Admiral to forward them to Sydney.— I do so long to see you all again. I am beginning to plan the very coaches by which I shall be able to reach Shrewsbury in the shortest time. The voyage has been grievously too long; we shall hardly know each other again; independent of these consequences, I continue to suffer so much from sea-sickness, that nothing, not even geology itself can make up for the misery & vexation of spirit. But now that I know I shall see you all again in the glorious month of September, I will care for nothing; the very thoughts of that pleasure shall drive sea sickness & blue sea devils far away.—

We are now on our road from Concepciòn.— The papers will have told you about the great Earthquake of the 20th of February.— I suppose it certainly is the worst ever experienced in Chili.— It is no use attempting to describe the ruins—it is the most awful spectacle I ever beheld.— The town of Concepcion is now nothing more than piles & lines of bricks, tiles & timbers—it is absolutely true there is not one *house* left habitable; some little hovels built of sticks & reeds in the outskirts of the town have not been shaken down & these now are hired by the richest people. The force of the shock must have been immense, the ground is traversed by rents, the solid rocks are shivered, solid buttresses 6 — 10 feet thick are broken into fragments like so much biscuit.— How fortunate it happened at the time of day when many are out of their houses & all active: if the town had been over thrown in the night, very few would have escaped to tell the tale. We were at Valdivia at the time the shock there was considered very violent, but did no damage owing to the houses being built of wood.— I am very glad we happened to call at Concepcion so shortly afterwards: it is one of the three most interesting spectacles I have beheld since leaving England—A Fuegian savage.—Tropical Vegetation—& the ruins of Concepcion— It is indeed most wonderful to witness such desolation produced in three minutes of time. I wrote a short letter from Chiloe,[1] but forget at what date.— We had a remarkably pleasant boat expedition along the Eastern Coast. I am afraid it will be the last cruize of this sort. You cannot imagine what merry work such a wandering journey is; in the morning we never know where we shall sleep at night. Carrying, like snail, our houses with us we are always independent; when the day is over we sit round our fire & pity all you who are confined within houses.— I joined the Ship at the South extremity & proceeded with her amongst the Chonos Is[d] & Tres Montes. There was a good deal of rough water; & the geology not very interesting but upon the whole this cruize has been a very fair one.— Chiloe I have seen throughily having gone round it & crossed it on horseback in two directions. I am tired of the restraint of those gloomy forests of the South & shall enjoy the open country of Chili & Peru. Valdivia is a quiet little hamlet, just like those in Chiloe: We had an opportunity of seeing many of the famous tribe of Araucanian Indians. The only men in the Americas, who have successfully withstood for centuries the conquering arms of the Europæans.—

During this cruize, we have had the misfortune to loose 4 anchors; this is the cause of our now proceeding to Valparaiso—with only one anchor at the Bows it would not be safe to survey the coast. The Beagle will immediately return to

Concepcion, from there resume the survey & continue it to Coquimbo. Then she will return to Valparaiso take in provisions & start for Lima.— I shall leave the Ship for the present; & not join her till the beginning of June: the Captain most kindly has offered to run in to Coquimbo to pick me up, in his way up the coast to Lima.— I hope & trust it will not be too late to cross the Cordilleras; besides the interest of such a journey, I am most anxious to see a geological section of this grand range.— Two days after we get in port, I will be off for St Iago & cross the Andes by the bad pass, see Mendoza & return by the common one.—[2] I am much afraid of this cloudy weather, if snow falls early I may be detained a prisoner on the other side! I shall be obliged to spend a good deal of money; but I can most conscientiously say, I never spend a dollar, without thinking whether it is worth it. I am sure my Father will not grudge me a little more money than usual, for this is the last journey I shall be able take on shore; anyhow before we reach Sydney.— Oh the precious money wasted in Cambridge; I am ashamed to think of it.—

I am very glad of this spell on shore; my stomach, partly from sea sickness & partly from my illness in Valparaiso is not very strong. I expect some good rides will make another man of me.— And now our Voyage for many months will be in fine warm weather & the fair trade wind. Again I shall see palms, & eat Bananas & I look forward with pleasure to the very buzzing of the Mosquitos. The Captain is quite himself again, & thank Heavens as anxious to reach dear old England as all the rest of us.— The interval appears nothing—I can almost fancy we are running up the chops of the Channel & the look-out man has just hailed the "Lizard lights right ahead Sir"[3] There will be more men aloft that day, than on the deck.—

Valparaiso, 13th.— I am in all the delightful hurry of a quick march.— tomorrow morning at four oclock I start for St. Iago. I am yet very doubtful about the Andes, but hope for the best: A pretty thing if the Snow falls whilst I am in Mendoza! in that case I should have to beg my way up to Potosi.— I am now in Corfield's house, who is as hospitable & kind as he always is.— Tell my Father I have drawn a bill for sixty pounds.— When we arrived the day before yesterday, I only received two letters, (both most full of interesting news) from Katty Sept. & Caroline, October.— The June, July, August ones have mis-carried. I expect however they are in the Commodores ships, & Commodores are fully priviledged to forget the entire concerns of a ten gun Brig:— others are sufferers with me.— I am very sorry for this, because I actually suppose, that Erasmus has written; & it will indeed be hard if I lose this. Also it seems poor William Owen has badly hurt his leg.— I wish they had not met this fate. You allude to some of the fossil bones being of value, & this of course is the very best news to me, which I can hear.— See how much obliged I am to all of you for your faithful performance of the promise of monthly letters.— I might have been more than a year without hearing—it is now 10 months.— God bless you all, for the best Sisters anyone ever had.—

I cannot write more, for horse cloths stirrups pistols & spurs are lying on all sides of me.— Give my most affectionate love to my dear Father.— | Farewell. Chas. Darwin—

DAR 223

[1] This letter has not been found.

[2] CD left Santiago for Mendoza via the Portillo Pass on 18 March and returned via the Uspallata Pass, arriving on 10 April. See *'Beagle' diary*, pp. 288–306, and *Journal of researches*, ch. 17.

[3] The lights at Lizard Head, Cornwall, the southernmost point of the British mainland.

To J. S. Henslow[1] [10]—13 March 1835

Valp?
March 1835

My dear Henslow

We now are lying becalmed off Valparaiso & I will take the opportunity of writing a few lines to you. The termination of our voyage is at last decided on—we leave the coast of America in the beginning of September & hope to reach England in the same month of 1836. I am heartily glad of it, nothing should induce me to stay out any longer. As it is, it will be nearly as long as a seven years transportation. But now that I do clearly see England in the distance, I care for nothing, not even sea sickness. In October perhaps I shall be in Cambridge & who knows but taking a walk with you round by Shelford common.— You can hardly understand how I long to see you & all my friends again; & now there only wants a year & half to that time. We shall see a great many places in this interval, but I am afraid there will be but few opportunities for much Natural History. We are now making a passage from Concepcion.— You will have heard an account of the dreadful earthquake of the 20th of February. I wish some of the Geologists who think the Earthquakes of these times are trifling could see the way the solid rock is shivered In the town there is not one house habitable; the ruins remind me of the drawings of the desolated Eastern cities.— We were at Baldivia at the time & felt the shock very severely. The sensation is more like that of skating over very thin ice; that is distinct undulations were perceptible. The whole scene of Concepcion & Talcuana is one of the most interesting spectacles we have beheld since leaving England.—

Since leaving Valparaiso, during this cruize, I have done little excepting in Geology.— In the modern Tertiary strata I have examined 4 bands of disturbance, which reminded me on a small scale of the famous tract in the Isle of Wight.—[2] In one spot there were beautiful examples of 3 different forms of upheaval.— In two cases I think I can show, that the inclination is owing to the presence of a system of parallel dykes traversing the inferior Mica Slate. The whole of the coast from Chiloe to S. extreme of the Pen: of Tres Montes is composed of the latter rock; it is traversed by very numerous dykes, the mineralogical nature of which will I suspect turn out very curious. I examined one grand transverse chain of Granite, which has clearly burst up through the overlying Slate. At P. Tres Montes there has been an old Volcanic focus, which corresponds to another in the North part of Chiloe. I was much pleased at Chiloe by finding a thick bed of recent oysters shells, &c, capping the Tertiary plain, out of which grew large forest trees.— I can now prove that both sides of the Andes have risen in the recent period to a considerable height.— Here the shells were 350 ft above the sea.—

In Zoology I have done but very little; excepting a large collection of minute Diptera & Hymenoptera from Chiloe. I took in one day, *Pselaphus*, Anaspis, Latridius Leiodes, Cercyon, & *Elmis* & two beautiful true Carabi, I might almost have fancied myself collecting in England. A new & pretty genus of Nudibranch Mollusc: which cannot crawl on a flat surface: & a genus in the family of Balanidæ, which has not a true case, but lives in minute cavities in the shells of the Concholepas,[3] are nearly the only two novelties. You were surprised at hearing of land Planariæ; you will equally be so, when you see leaches, which live entirely out of water in the fore⟨sts⟩ of Chiloe & Valdivia.— Before the Beagle sails for Lima, I shall be obliged to send away one more box: this will be the last; with which I shall trouble you. I am afraid so many boxes must have been very much in your way. I trust they may turn out worth their stowage. I will write again, when this last Cargo is sent. You ought to have received about a month since 2 boxes by H.M.S. Challenger & before that 2 Casks & one jar by H.M.S. Samarang.— Will you write to me directed to Sydney, not long after receiving this letter.— I am very unreasonable in begging for so many letters; but bear with me for one year more.— If any come directed in the mean time to S. America, they will be forwarded to Sydney by the Admiral.—

Valparaiso March 13th.— I am on the point of starting to endeavour to pass the Cordilleras, but am very doubtful of the issue. Three month's letters are somewhere mislaid: but I hope they will be found.— Perhaps there may be a letter from you.— I am anxious to know whether the bird skins from the River Plate in a tinned box came safe.— I think that collection will be good, as I took much pains with them.— I am in a great hurry, so excuse this stupid shabby little letter. Oh the goodly month of September 1836.— To think that I shall again be actually living quietly in Cambridge.— It is too good a prospect, it will spoil the Cordilleras.

So my dear Henslow, good night | Your Most obliged & affectionate friend | Chas. Darwin.

Royal Botanic Gardens, Kew

[1] The following passages from this letter were extracted by Henslow and published in the Cambridge Philosophical Society pamphlet:

    1.1 'We now . . . 1836.' 1.4

    1.12 'You . . . novelties.' 3.7; 3.7 'Concholepas' *transcribed incorrectly as* 'concholapas'

[2] See *South America*, p. 124, for CD's description of the strata of the peninsula of Lacuy, Chiloé. The Isle of Wight formation is described in Conybeare and Phillips 1822, pp. 108–9, which CD had on board the *Beagle*. An unannotated copy is in Darwin Library–Down.

[3] The description of this specimen (*Cryptophialus minutus*) later led CD to his eight-year labour of Cirripedia classification. 'I had originally intended to have described only a single abnormal Cirripede, from the shores of South America, and was led, for the sake of comparison, to examine the internal parts of as many genera as I could procure' (*Living Cirripedia* 1: v; see also *Autobiography*, p. 117). The description is in *Living Cirripedia* 2: 566.

From Caroline Darwin 30 March 1835

Shrewsbury
March 30th | 1835

My dear Charles

Your last letter was to Cath dated Novr 8th and extremely interesting— How very kind Mr Corfield has been & what would have become of you if you had been unable to return to his house before you became so ill— poor dear Charles it is melancholy to think of you ill & suffering for a long month & I am sadly afraid it will be very long indeed before you are as strong & able to bear climate & dangers as you have done— We all cannot help feeling very sorry for your determination of remaining in the Beagle till the expedition is over & indeed it is not *only* the selfish wish we have to see you again— Do just think whether you are wise to encounter the danger & risk you must do—particularly now the schooner is given up. I do not doubt the enjoyment is extreme but I can not think the weeks of interest & pleasure can equal those of discomfort, danger & separation from all your friends— if there was any certain period for the end of the voyage it would be differt but the time has gone on lengthening & lengthening & it will end by your wasting the best years of your life on ship board. Capt Fitzroys health & spirits not being good certainly adds to ones uneasiness, & I can assure you not a friend you have feel there is the slightest reason for your continuing with the Beagle a day longer than you wish for your own pleasure— You have already made great collections & done much—& it will be a happy day when you are again in England to arrange them— I will teaze you no more dear Charles but do not decide without once more reflecting—

Mr Owen recevd your letter written on the 9th & was excessively pleased by it— he wrote a note to my Father quite overflowing with affection & as for Papa himself he was so much affected by thinking of you ill & forlorn that we hardly could mention your name to him all that day— he sends you his kindest most affectionate love I wish you could have heard all Papa said one day when we were talking about you— he wrote to Mr Corfield at Pitchford to say how grateful he felt for his sons kindness to you— I have had another letter from Charlotte— it is settled that she & Mr Langton go to Rio the end of April where they remain 4 or 5 months & then winter in one of the West Indian Isles. I am sure it would please Charlotte extremely if you would write a few lines to her at Rio. she always enquires & is so much interested about you— I am afraid poor thing she cannot like going to Rio instead of returning to Maer & seeing A^t Bessy— all at Maer are very well— A^t Bessy's understanding is fast failing b⟨ut⟩ her health much as usual. Elizabeth has ⟨been⟩ staying the last month at Clapham with the Hensleighs Erasmus is quite devoted to Snow (the eldest of Hensleighs children) he talks very vigourously of going to Switzerland this summer but I am sure he will not be able to leave his darling pet— I heard from Wm Fox, last week he talks of spending the summer at Barmouth or Beaumaris & paying us a visit on his road— his poor wife has been exceedingly ill all winter, she had a dead child & has never recovd her strength— what a pity in that sickly family that Mrs W. Fox should prove more delicate than any of them— he enquires most kindly after you—

Cath is at Overton gone to comfort Marianne for the approaching separation from Parky who goes to school at Oswestry at Easter— There is no family news to tell you Robert is not yet married to Miss Crewe & nobody knows exactly what they are waiting for. she flatters & coaxes his brothers—making flanel waistcoats & buying gingerbread for Allen & pressing Tom to stay at Muxton— when you last saw Robert you little thought you would find him married on your return to a woman more than old enough to be his Mother & such an odious disreputable family to marry into with herself having the reputation of a bad temper. her manner is so sweet & artificial that I fully believe she is not what she seems—& I pity poor Robert for it

I see by the Paper that Professor Henslow has a son.—[1] I hoped to have told you some political news but the debate is still going on upon the Irish Church by which Sir R Peel says he will stand or fall—[2]

Good bye, my very dear Charles all our kindest loves—you are an excellent correspondent & you may believe we do thoroughly value your letters— | Once more bless you & good bye—

DAR 97 (ser. 2): 20—1

[1] George Henslow was born on 23 March 1835 (*Alum. Cantab.*).

[2] The issue centred on the Irish tithes, paid mainly by Catholics, which provided the Church of England and its bishops with considerable funds. Lord John Russell proposed a resolution to inquire into the actual needs of the Church. Any surplus not required for the spiritual care of its members was to be applied to the education of all classes, regardless of their religious faith. Peel's ministry opposed the resolution and, when it was passed, resigned on 8 April 1835.

To J. S. Henslow[1] 18 April 1835

Valparaiso
April 18th—1835—

My dear Henslow.—

I have just returned from Mendoza, having crossed the Cordilleras by two passes. This trip has added much to my knowledge of the geology of the country.[2] Some of the facts, of the truth of which I in my own mind feel fully convinced, will appear to you quite absurd & incredible.— I will give a very short sketch of the structure of these huge mountains. In the Portillo pass (the more Southern one) travellers have described the Cordilleras to consist of a double chain of nearly equal altitude, separated by a considerable interval.— This is the case: & the same structure extends to the Northward to Uspallata; the little elevation of the Eastern line (here not more than 6000−7000 ft), has caused it almost to be overlooked. To begin with the Western & principal chain; we have where the sections are best seen, an enormous mass of a Porphyritic conglomerate resting on Granite. This latter rock, seems to form the nucleus of the whole mass & is seen in the deep lateral valleys, injected amongst, upheaving, overturning in the most extraordinary manner the overlying strata. On the bare sides of the mountains, the complicated dykes & wedges of variously coloured rocks are seen traversing in every possible form & shape the same formations, which by their intersections prove a succession of violences. The stratification in all the mountains is beautifully distinct & from a variety in the color can be seen at great distances. I cannot imagine any part of the world presenting a more extraordinary scene of the breaking up of the crust of the globe than the very central peaks of the Andes. The upheaval has taken place by by a great number of (nearly) N & S lines; which in most cases has formed as many anticlinal & synclinal ravines: The strata in the highest pinnacles are almost universally inclined at an angle from $70°−80°$.—

I cannot tell you how I enjoyed some of these views.— it is worth coming from England once to feel such intense delight. At an elevation from 10−12000 ft. there is a transparency in the air & a confusion of distances & a sort of stillness which gives the sensation of being in another world, & when to this is joined, the picture so plainly drawn of the great epochs of violence, it causes in the mind a most strange assemblage of ideas.

The formation I call Porph-Conglomerates, is the most important & most developed one in Chili; from a great number of sections, I find it a true coarse Conglomerate or Breccia, which by every step in a slow gradation passes into a fine Clay-stone Porphyry; the pebbles & cement becoming Porphyritic, till at last all is blended in one compact rock. The Porphyries are excessively abundant in this chain. I feel sure at least $4/5$ of them have been thus produced from sedimentary beds in situ.— There are Porphyries which have been *in*jected from below amongst Strata & others *e*jected which have flowed in streams: it is remarkable I could show specimens of this rock, produced in these three methods, which cannot be distinguished. It is a great mistake considering the Cordilleras (here) as composed of rocks which have flowed in streams. in **this** range I *no* where saw a fragment, which I believe to have thus originated, although the road passes at no great distance from the active Volcanoes.—

The Porphyries, Conglomerates, Sandstones & Quartzose Sandstones, Lime stones alternate & pass into each other many times (overlying, where not broken through by the Granite, Clay-Slate) In the upper parts the Sandstone begins to alternate with Gypsum, till at last we have this substance in a stupendous thickness. I really think the formation, is in some places (it varys much) nearly 2000 ft thick. it occurs often with a green (Epidote?) siliceous Sandstone & snow white marble: it resembles that found in the Alps in containing large concretions of a crystalline marble of a blackish grey color.— The upper beds, which form some of the higher pinnacles consist of layers of snow white gypsum & red, compact sandstone, from the thickness of paper to a few feet, alternating in an endless round.— The rock has a most curiously painted appearance.—

At the pass of the Puquenas in this formation, where however, a black rock, like Clay-Slate, without many laminæ occurring with a pale Limestone has replaced the red Sandstone, I found abundant impressions of shells.— The elevation must be between 12 — 13000 ft.— A Shell which I believe is a Gryphæa is the most abundant,—an Ostræa, Turritella, Ammonites, small Bivalves, Terebratula (?).— *Perhaps* some good Conchologist will be able to give a guess, to what grand division of the formations of Europe, these organic remains bear most resemblance.—[3] They are exceedingly imperfect & few. the Gryphites are most perfect.— It was *late* in the Season, & the situation particularly dangerous for Snow storms. I did not dare to delay, otherwise a grand harvest might have been reaped.—

So much for the Western line; in the Portillo pass, proceeding Eastward we meet an immense mass of a Conglomerate dipping to the West 45°, which rests on Micaceous Sandstones &c &c, upheaved, converted into quartz rock, penetrated by dykes, from the very grand mass of *Proto*gine[4] (large crystals of quartz, red Feldspar & occasional little Chlorite). Now this Conglomerate, which reposes on & dips from the Protogine <45°, consists of the peculiar rocks of the first described chain, pebbles of the black rock *with shells*, green sandstone &c &c: It is hence manifest, that the *upheaval* (& deposition at least of part) of the Grand Eastern chain is entirely posterior to the Western. To the North in the Uspallata pass we have also a fact of the same class.— Bear this in mind, it will help to make you believe what follows.—

I have said the Uspallata range is geologically, although only 6000 — 7000 ft a continuation of the grand Eastern chain.— It has its nucleus of granite, consist. of grand beds of various crystalline rocks, which I can feel no doubt are subaqueous lavas alternating with Sandstone, Conglomerates & white Aluminous beds (like decomposed feldspar) with many other curious varieties

of sedimentary deposits. These Lavas & Sandstones alternate very many times
& are quite conformable, one to the other. During two days of careful examination I said to myself at least 50 times, how exactly like, only rather
harder, these beds are to those of the upper Tertiary strata of Patagonia,
Chiloe, Concepcion, without the possible identity *ever* having occurred to me.—
At last there was no resisting the conclusion.— I could not expect shells for
they never occur in this formation; but Lignite or Carbonaceous shale ought to
be found. I had previously been exceedingly puzzled by meeting in the Sandstone
thin layers (few inches to feet thick) of a brecciated Pitchstone I strongly
suspect, the alteration, from the underlying Granite, has altered such beds into
this Pitchstone. The silicified wood, (particularly characteristic) was yet
absent. the conviction that I was on the Tertiary Strata was so strong, by this
time in my mind, that on the third day, in the midst of Lavas, & heaps of Granite
I began my apparently forlorn hunt.—

How do you think I succeeded? In an escarpment of compact greenish Sandstone I found a small wood of petrified trees in a vertical position, or rather
the strata were inclined about 20 − 30 to one point & the trees 70° to the opposite
one.— That is they were before the tilt truly vertical.— The Sandstone
consists of **many** layers & is marked by the concentric lines of the bark (I have
specimens)[5] 11 are perfectly silicified, & resemble the dicotyledonous wood
which I have found at Chiloe & Concepciòn: the others 30 − 40 I only know to be
trees from the analogy of form & position; they consist of snow white columns
Like Lots wife of coarsely crystall. Carb. of Lime. The longest shaft is 7
feet. They are all close together within 100 y^d & about same level; no where
else could I find any.— It cannot be doubted that the *layers* of fine Sandstone have
quietly been deposited between a clump of trees, which were fixed by their
roots.— The Sandstone rests on Lavas is covered by great bed, apparently about
1000 ft thick of black Augitic Lava, & over this, there are at least 5 grand
alternations of such rocks & aqueous sedimentary deposits; amounting in
thickness to several thousand feet.— I am quite afraid of the only conclusion
which I can draw from this fact, namely that there must have been a depression in
the surface of the land to that amount.— But neglecting this consideration it
was a most satisfactory support of my presumption of the Tertiary (I mean by
Tertiary, that the shells of the period were closely allied or some identical to
those which now live as in lower beds of Patagonia) age of this Eastern Chain.

A great part of the proof must remain upon my ipse dixit, of a mineralogical
resemblance, with those beds whose age is known, & the character of which
resemblance, is to be subject to infinite variation, passing from one variety to
others by a concretionary structure. I hardly expect you to believe me, when it is
a consequence of this view that Granite which forms peaks of a height probably
of 14000 ft has been fluid in the Tertiary period.—that strata of that period
are altered by its heat & are traversed by *dykes* from the mass: That these
Strata have also probably undergone an immense depression, that they are now

inclined at high angles & form regular or complicated anticlinal lines.— To complete the climax & seal your disbelief these same sedimentary Strata & Lavas are traversed by *very numerous* true metallic veins of Iron, Copper Arsenic, Silver & Gold, & that these can be traced to the underlying Granite.— A Gold mine has been worked close to the clump of silicified trees.—

If when you see my specimens, sections & account, you should think that there is pretty strong presumptive evidence of the above facts: It appears very important: for the structure, & size of this chain will bear comparison with any in the world. And that this all should have been produced in so very recent a period is indeed wonderful. In my own mind I am quite convinced of the reality of this. I can any how most conscientiously say, that no previously formed conjecture warped my judgement. As I have described, so did I actually observe the facts.— But I will have some mercy & end this most lengthy account of my geological trip.—

On some of the large patches of perpetual snow I found the famous Red Snow of the Arctic countries.— I send with this letter my observations & a piece of Paper on which I tried to dry some specimens. If the fact is new, & you think it worth while, either yourself examine them or send them to whoever has described the specimens from the North, & publish a notice in any of the periodicals.— I also send a small bottle with 2 Lizards: one of them is Viviparous, as you will see by the accompanying notice.— A M. Gay, a French Naturalist has already published in one of the Newspapers of this country a similar statement, & probably has forwarded to Paris some account:[6] as the fact appears singular, would it not be worth while to hand over the Specimens to some good Lizardologist & Comparative Anatomist to publish an account of their internal structure.— Do what you think fit.[7]

This letter will go with a cargo of Specimens from Coquimbo.— I shall write to let you know when they are sent off.— In the Box, there are two Bags of Seeds, one ticket, Valleys of Cordilleras 5000 — 10000 ft high; the soil & climate exceedingly dry; soil Very light & stony, extremes in temperature: the other chiefly from the dry sandy Traversia of Mendoza 3000 ft more or less.— If some of the bushes should grow but not be healthy try a **slight** sprinkling of Salt & Saltpetre.— The plain is saliferous.— All the flowers in the Cordilleras appear to be Autumnal flowerers,—they were all in blow & seed—many of them very pretty.— I gathered them as I rode along on the hills sides: if they will but choose to come up I have no doubt many would be great rarities.— In the Mendoza Bag, there are the seeds or berrys of what appears to be a small Potatoe plant with a whitish flower. They grow many leagues from where any habitation could ever have existed, owing to absence of water.— Amongst the Chonos dried plants, you will see a fine specimen of the wild Potatoe, growing under a most opposite climate & unquestionably a true wild Potatoe.— It must be a distinct species from that of the lower Cordilleras one.— Perhaps, as with the Banana, distinct species are now not to be distinguished in their varieties, produced by cultivation.— The Beagle is not at Valparaiso. So I cannot copy out the few remarks about the Chonos Potatoe.—⁸ With the Specimens, there is a bundle of old Papers & Note Books. Will you take care of them, in case I should loose my notes, these might be useful.— I do not send home any insects, because they must be troublesome to you & now so little more of the Voyage remains unfinished I can well take charge of them.—

In two or three days I set out for Coquimbo by Land, the Beagle calls for me in the beginning of June: So that I have 6 weeks more to enjoy geologizing over these curious mountains of Chili.— There is at present a bloody revolution in Peru: the Commodore has gone there & in the hurry has carried our letters with him; perhaps amongst them there will be one from you.— I wish I had the old Commodore here I would shake some consideration for others into his old body.— From Coquimbo you will again hear from me.— Till then Farewell. My dear Henslow— Yours very truly, C. Darwin

Our plans are altered. I have a ten weeks holiday & expect to reach as far as Copiapò & examine all that preeminently curious country abounding with mines:—

I shall not write to you till we reach *[left blank]* excepting half a dozen lines, just to inform you when my specimens leave this Port.— I am glad to say, that I believe this will be the last Cargo, with which you will be troubled.—

Royal Botanic Gardens, Kew

¹ The following passages from this letter were extracted by Henslow and published in the Cambridge Philosophical Society pamphlet:
    1.1 'I have . . . country.' 1.2
    1.4 'I will . . . facts.—' 10.8; 1.8 'Uspallata' *incorrectly transcribed* 'Uspellata'; 1.22 *Henslow inserted* 'Of dykes?' *as a footnote after* 'lines'; 3.8 'it is remarkable.' *omitted*; 3.11 'only' *inserted after* 'composed'; 5.7 'formations' *changed to* 'continents'; 6.4, 6.6 '*Protogine*' *transcribed* 'protogene'; 6.7 'pebbles' *transcribed* '*pebbles*'; 6.7 '*with shells*' *transcribed* 'with shells' 6.8 '*upheaval*' *transcribed* 'upheaval'; 6.10, 7.1 'Uspallata' *transcribed* 'Uspellata'; 8.5 '**many** layers' *transcribed* 'many horizontal layers' 8.7 '30—40' *transcribed* 'thirty to thirty-four'; 8.21 'live' *transcribed* 'lie'; 9.2—4 '& the . . . structure.' *omitted*; 9.4—5 'I hardly expect you to believe me, when it is a consequence of this view that' *altered to* 'According to this view'; 9.7 '*dykes*' *transcribed* 'dykes'; 9.8 'That these Strata have also probably undergone an immense depression,' *omitted*; 9.10 '& seal your disbelief' *omitted*; 10.1 'If' *omitted*
    11.1 'On . . . specimens.' 11.3
    11.5 'I also . . . account' 11.9; *Henslow here reprinted* 'an Extract from the Newspaper referred to by Mr Darwin' *as a footnote*
    12.2 'In . . . saliferous.' 12.7; 12.4 'stony' *transcribed* 'strong'
    12.11 'In . . . Cordilleras' 12.16
² CD's geological notes on this journey are in DAR 36.2. These notes, augmented by later reading, were used in writing *South America*. See ch. 7, ch. 8, pp. 237—43, and plates showing geological sections of the Peuquenes (Portillo) and Cumbre (Uspallata) Passes.

[3] The fossils were later named by Alcide d'Orbigny, with the corresponding European formations identified (see *South America*, p. 181).

[4] In CD's geological notes (DAR 36.1: 489v.) there is an entry which explains why he underlined 'Proto': 'Even if the Protogine is not posterior to the white Granite, it is so to the upheaval (owing to the Granite & Syenitic Greenstone in the West Cordillera) of the Puquenas chain.—Therefore its name of *Proto* is here very inapplicable.—'

[5] Robert Brown was much impressed by the specimens and identified them as 'coniferous, partaking of the characters of the Araucarian tribe, with some curious points of affinity with the Yew' (*South America*, p. 202).

[6] CD was right in assuming that Claude Gay would probably communicate the information to France (see Gay 1836).

[7] In the minutes of the Cambridge Philosophical Society for 14 December 1835, item 6 reads 'communications from C. Darwin, Esq., on Viviparous Lizards and on Red Snow'. CD describes the latter in *Journal of researches*, pp. 394–5. The lizards collected by CD, Claude Gay, and others are described by Thomas Bell in *Reptiles*. CD's field notes of 24 March 1835 contain his descriptions (see *Voyage*, p. 235). Henslow in his Cambridge Philosophical Society pamphlet, pp. 30–1, quotes the Chilean newspaper account referred to by CD.

[8] The notes in DAR 31.2: 314 are reproduced almost verbatim in *Journal of researches*, pp. 347–8.

To Susan Darwin 23 April 1835

Valparaiso
April 23ᵈ— 1835

My dear Susan

I received a few days since your letter of November: the three letters, which I before mentioned are yet missing: but I do not doubt they will come to life.— I returned a week ago from my excursion across the Andes to Mendoza. Since leaving England I have never made so successful a journey: it has however been very expensive: I am sure my Father would not regret it, if he could know how deeply I have enjoyed it.— it was something more than enjoyment: I cannot express the delight, which I felt at such a famous winding up of all my geology in S.— America.— I literally could hardly sleep at nights for thinking over my days work.— The scenery was so new & so majestic: every thing at an elevation of 12000 ft. bears so different an aspect, from that in a lower country.— I have seen many views more beautiful but none with so strongly marked a character. To a geologist also there are such manifest proofs of excessive violence, the strata of the highest pinnacles are tossed about like the crust of a broken pie. I crossed by the Portillo pass, which at this time of year is apt to be dangerous, so could not afford to delay there; after staying a day in the stupid town of Mendoza I began my return by Uspallata, which I did very leisurely.— My whole trip only took up 22 days.— I travelled with, for me, uncommon comfort, as I carried a *bed*!: my party consisted of two Peons & 10 mules, two of which were with baggage or rather food, in case of being snowed up.— Every thing however favoured me, not even a speck of this years Snow had fallen on the road.—

I do not suppose, any of you can be much interested in Geological details, but I will just mention my principal, results: beside understanding to a certain extent, the description & manner of the force, which has elevated this great line of mountains, I can clearly demonstrate, that one part of the double line is of a age long posterior to the other. In the more ancient line, which is the true chain of the Andes.—I can describe the sort & order of the rocks which compose it. These are chiefly remarkable by containing a bed of Gypsum nearly 2000 ft thick: a quantity of this substance I should think unparalleled in the world. What is of much greater consequence, I have procured fossil shells (from an elevation of 12000 ft) I think an examination of these will give an approximate age to these mountains as compared to the Strata of Europe: In the other line of the Cordilleras there is a strong presumption (in my own mind conviction) that the enormous mass of mountains, the peaks of which rise to 13 & 14000 ft are so very modern as to be contemporaneous with the plains of Patagonia (or about with *upper* strata of Isle of Wight): If this result shall be considered as proved it is a very important fact in the theory of the formation of the world.— Because if such wonderful changes have taken place so recently in the crust of the globe, there can be no reason for supposing former epochs of excessive violence.— These modern strata are very remarkable by being threaded with metallic veins of Silver, Gold, Copper &c: hitherto, these have been considered as appertaining to older formations. In these same beds (& close to a Gold mine) I found a clump of petrified trees, standing upright, with the layers of fine Sandstone deposited round them, bearing the impression of their bark. These trees are covered by other Sandstones & streams of Lava to the thickness of several thousand feet. These rocks have been deposited beneath water, yet it is clear the spot where the trees grew, must once have been above the level of the sea, so that it is certain the land must have been depressed by at least as many thousand feet, as the superincumbent subaqueous deposits are thick.— But I am afraid you will tell me, I am prosy with my geological descriptions & theories.—

You are aware, that plants of Arctic regions are frequently found in lower latitudes, at an elevation which produces an equal degree of cold.— I noticed a rather curious illustration of this law in finding on the patches of perpetual Snow, the famous Red Snow of the Northern Navigators.— I am going to send to Henslow, a description of this little Lichen, for him, if he thinks it worth while to publish in some of the Periodicals.—

I am getting ready my last Cargo of Specimens to send to England; This last trip has added half a mule's load; for without plenty of proof I do not expect a word of what I have above written to be believed.— I arrived at this place a week since, & am as before living with Corfield. I have found him as kind & good-natured a friend as he is a good man.— I staid also a week in St Iago, to rest after the Cordilleras, of which I stood in need & lived in the house of Mr Caldcleugh (the author of some bad travels in S. America): he is a very pleasant person & took an infinite degree of trouble for me.— It is quite surprising how kind & hospitable I have found all the English merchants.— Do mention to Mr Corfield of Pitchford, under what obligations I lie to his son.— Amongst the various pieces of news, of which your letter is full, I am indeed very sorry to hear of poor Col. Leighton's death. I can well believe how much he is regretted. It is a bitter reflection, when I think what changes will have taken place before I return. I pray to Heaven I may return to see all of you.—

When you write to the West Indies or Madeira, remember me most affectionately to Charlotte, I hope she will be happy there. When I enjoyed talking over all my schemes with her; how little did she expect to be so soon under a vertical sun & glowing atmosphere.— I am surprised at any Husband liking to take his Wife to such a country.—

The Beagle after leaving me here, returned to Concepcion: Capt Fitz Roy has investigated with admirable precision the relative level of land & Water, since the great Earthquake.— The rise is unequal & parts of the coast are now settling down again, probably at each little trembling which yet continue.— The Is^d of S. Maria has been elevated 10 feet: Capt Fitz Roy found a bed of Muscles with putrid fish that many feet above high water mark.— The Beagle passed this port yesterday. I hired a boat & pulled out to her. The Capt is very well; I was the first to communicate to him his promotion. He is fully determined, nothing shall induce him to delay the voyage a month: if time is lost in one place, something else shall be sacrificed.— Our voyage now will solely consist in carrying a chain of longitudes between important positions.[1]

My holidays extend till the middle of July: so that I have 10 weeks before me, & the Beagle will pick me up at any Port I choose. The day after tomorrow I start for Coquimbo. I have three horses & a baggage Mule, & a Peon whom I can trust, having now accompanied me on every excursion. The people moreover to the *North*, have a capital character for honesty, ie they are not cut-throats. The weather there also will not be hot & it never rains.— I shall extend my journey to Copiapo.—it is a great distance, but I feel certain I shall be most amply repaid. Everything which can interest a Geologist, is found in those districts, Mines of Rock-Salt, Gypsum, Saltpetre, Sulphur; the rocks threaded with metallic veins: old sea-beaches;—curious formed valleys; petrified shells, Volcanoes & strange scenery. The country geologically is entirely unknown (as indeed is the whole of South S. America), & I thus shall see the whole of Chili from the Desert of Atacama to the extreme point of Chiloe. All this is very brilliant, but now comes the black & dismal part of the Prospect.—that horrid phantom, money. The country where I am going to is very thinly inhabited & it will be impossible to draw bills.— I am therefore obliged to draw the money here & transmit it there.— Moreover it is necessary to be prepared for accidents: horses stolen.—I robbed.—Peon sick, a pretty state I should be 400 or 500 miles from where I could command money.— In short, I have drawn a bill for £100: : 0: : 0, & this so shortly after having spent 60 in crossing the Andes. In September we leave the coast of America: & my Father will believe, that I *will* not draw money in crossing the Pacific, because I *can* not.— I verily believe I could spend money in the very moon.— My travelling expences are nothing; but when I reach a point, as Coquimbo, whilst my horses are resting, I hear of something very wonderful 100 miles off. A muleteer offers to take me for so many dollars, & I cannot or rather never have resisted the Temptation.—

My Fathers patience must be exhausted: it will be patience smiling at his son, instead of at grief. I write about it as a good joke, but upon my honor I do not consider it so.— Corfield cashes the bill & sends it to his Father, who will bring it to the old Bank, where I suppose it can be transacted.—

I received a long & affectionate letter from Fox: he alludes to a letter which I have never received. I shall write to him from Lima; at present I have my hands full.— How strange it sounds to hear him talk of "his dear little wife". Thank providence he did not marry the simple charming Bessy.— I shall be very curious to hear a verdict concerning the merits of the Lady.— How the world goes round; Eyton married. I hope he will teach his wife to sit upright.— I have written to him: I am sure he deserves to be happy.— What are the two younger sons doing. I think, from what I saw at Cambridge, Tom is worth the pair.—

Your account of Erasmus' (does Erasmus live with the Hensleigh's for the last year their names have never in any letter been separated) visit to Cambridge has made me long to be back there. I cannot fancy anything more delightful than his Sunday round, of King's, Trinity & those talking giants, Whewell & Sedgwick: I hope your musical tastes continue in due force. I shall be ravenous for the Piano-forte. Do you recollect, poor old Granny, how I used to torment your quiet soul every evening?— I have not quite determined whether I will sleep at the Lion, the first night, when I arrive per Wonder or disturb you all in the dead of the night, everything short of that is absolutely planned.— Everything about Shrewsbury is growing in my mind bigger & more beautiful; I am certain the Acacia & Copper Beech are two superb trees: I shall know every bush, & I will trouble you young ladies, when each of you cut down your tree to spare a few. As for the view behind the house I have seen nothing like it. It is the same with North Wales. Snowden to my mind, looks much higher & much more beautiful than any peak in the Cordilleras. So you will say, with my benighted faculties, it is time to return, & so it is, & I long to be with you— Whatever the trees are, I know what I shall find all you.— I am writing nonsense—so Farewell.— My most affectionate love to all & I pray forgiveness from my Father. Yours most affectionately | Charles Darwin—

You send my letters to Marianne, so I do not send my particular love to her— I suppose her young gentlemen will be a small troop of Grenadiers by the time I return.— What a gang of little ones have come into the world, since I left England.—

DAR 223

[1] This was one of the major objectives of the voyage, as set forth in the Admiralty instructions. Robert FitzRoy summarised the principal results in *Narrative* Appendix, pp. 331 – 52. The chronometrical measurements are described as 'forming a connected chain of meridian distances around the globe, the first that has ever been completed, or even attempted, by means of chronometers alone.'

To Catherine Darwin 31 May [1835]

Coquimbo
May 31st

My dear Catherine

I have very little to write about; but as there will not be another opportunity for some time to send a letter, I will give an account of myself since leaving Valparaiso. My journey up here was rather tedious; I was obliged to travel so very slowly, that my animals might remain in good condition for the rest of their journey.— The country is very miserable; so burnt up & dry, that the mountains are as bare as turn-pike roads, with the exception of the great Cacti, covered with spines.— I visited very many mines; & since I have been here, I have made an excursion up the valley to see some famous ones of Silver. I reached the foot of the Cordilleras.— The geology goes on very prosperously; before I leave Chili, I shall have a very good general idea of its structure.—

The day after tomorrow, I start for Copiapò, passing through Guasko: on the 5th of July the Beagle calls for me at that place; from whence to Iquiqui & Lima.— This latter part of my journey, will be still less interesting than the former, as I understand nearly all the road is a desert. There is one Traversia of a day & half without a drop of water.— I shall be very glad, when once again settled on board the Beagle.— I am tired of this eternal rambling, without any rest.— Oh what a delightful reflection it is, that we are now on our road to England.— My method of travelling is very independent & in this respect as pleasant as possible. I take my bed & a Kettle, & a pot, a plate & basin. We buy food & cook for ourselves, always bivoucing in the open air, at some little distance from the house, where we buy Corn or grass for the horses.— It is impossible to sleep in the houses, on account of the fleas. Before I was fully aware of this, I have risen in the morning with my whole shirt punctured with little spots of blood, the skin of my body is quite freckled with their bites. I never formerly had any idea, what a torment, in these hot, dry climates, these ravenous little wretches could be.—

But gracias a dios one month more & farewell for ever to Chili; in two months more farewell South America.— I have lately been reading about the South Sea— I begin to suspect, there will not be much to see; that is, after any one group with its inhabitants, has been visited.— Everyone however must feel some curiosity to behold Otaheitè.— I am lucky in having plenty of occupation for the Sea part, in writing up my journal & Geological memoranda.— I have already got two books of rough notes.—

The Beagle is now in the Port, refitting before our long voyage: Everybody is living on shore in tents. Everything has been taken out of her even to the ballast.— She proceeds in a week's time to Valparaiso for 9 months provisions. I hope some vessel of war will come round, before she sail⟨s⟩; if not, I shall not receive any other letter from you, for the next 9 months, that i⟨s⟩ till we reach Sydney.— From Valparaiso I send a large cargo of specimens to Henslow; & these will be the last, for the rest I shall be able to carry, more especially as every month, my wardrobe becomes less & less bulky— By the time we reach England, I shall scarcely have a coat to my back.— And at present, as you may see, I have scarcely an idea in my head— So—

Farewell | Your affectionate Brother | Chas. Darwin

DAR 223

From R. E. Alison[1] 25 June 1835

Valparaiso
25th. June 1835

Chas. Darwin Esq.
Dear Darwin

By the arrival of the Beagle on the 14th. Inst. I recd. your much esteemed letter of the 29th. May, by which I was glad to learn that you had arrived safely at Coqbo—but I was sorry to hear that your rebel of a stomach had been annoying you again. I am pleased to find that you have had further evidence of a rising in the land of Chile; I have long thought that such was the case, but I am such a mere tyro in scientific matters, that I was ashamed to mention it, without it was confirmed by a more careful observation and by more capable persons—

A few days ago I went along the Coast from Playa-ancha towards Laguna, and in a ravine nearly parallel with that of Quebrada Verde and about 300 yards from the sea, I observed that it had intersected several strata of shells leaving them exposed to the right and left on both sides of the ravine, on one side they continued up a series of steps or beaches forming a little hill about 80 feet high from the brink of the ravine, and about 350 feet above the level of the sea—

The face of the hill was much covered with brush-wood, so that it was only by pulling it up and removing the earth that the shells could be found, and the steps were not well defined— The bottom of the ravine and the loose stones in it were gneiss of a very compact character with veins of feltspar; on digging a hole into one of the sides of the ravine about 3 feet from the edge I found the pelvis of some quadruped in a state of great decay. it was too small for a horse. I brt it in to show you the state of the bone but I do not think it worth sending you— The shells were the *large* concholepus, patella of various sizes some too small for the purposes of food, some turbos, and the metillus in a broken state, but I was not able to find some of the small concholepus.— I have sent you some for your inspection—

The situation is almost inaccessible from the sea, therefore it is not likely that they have been conveyed there for the purposes of food, nor in after times by the Spaniards to make lime as they would not have placed them round the sides of a hill—²

I ought to mention, that the shells higher up were much more decayed than those lower down, so if a rising has taken place, it has been per gradus & not per saltum—

I do not know whether you have examined the sea cliffs towards Viña de la Mar a little beyond the village of the Baron where you would observe the rocks about 14 feet above high water with numerous funnel shaped perforations, caused no doubt by lithophagi, & on a pointed rock which can only be reached by climbing with the hands & feet some balani are adhering to the surface on removing the dung of the gulls with which it is entirely covered— At the same level where you find the funnel shaped perforations, and where the rock is much decomposed, the Cactus tuna & other plants which require little soil and moisture are growing most luxuriantly a sign that they are always above high water mark— You will tell me that the sea may have retired as it has done in the Almendral from the detritus of the surrounding hills, but the bottom is all rocky and the water deep close to the shore, as I have frequently bathed there I am well able to judge— I am very possibly describing a spot which you have observed with more enquiring eyes than mine as I think you mentioned you had been round the rocks under the Castle of the Baron. if so, I am glad of it—

You are possibly aware that there is a general opinion amongst the natives that the sea is fallen from the great difference which has taken place in the depth of the Bay in the last 50 years, and when you wish to gain any information from them on that point, you must give in to that opinion, or they will think you mad if you ask them whether the earth is rising—

As at present we have no data to go by, it is difficult to know whether there are paroxysmal risings or a chronic impulsion however in Central America a high chain of hills were thrown up in the course of four years from 1824 to 1828, and a town which was in a plain became surrounded on one side by high hills—

The piece of fossil wood you allude to, and which I send was found in a ravine beyond Playa-ancha towards Quebrada verde, but I did not find it in situ, but in the water amongst the rocks. the sides of the ravine were a sandy conglomerate with rounded pieces of indurated clay similar to those in the road of the Alto del Puerto. the bottom of the ravine appeared a sort of grünsteinic rock—

You ask me what is my opinion respecting the direction of sounds accompanying earthquakes, from my own observation I have thought that it comes in the direction of the line of movement, but I am far from being certain that the opinion is correct, as some people have thought that the sound was in one direction & I thought in the opposite—

The shocks *generally* proceed from the North, and the great earthquake which happened here in Nov. 1822 came from that quarter, and the sea flowed in as "black as ink and the anchors & chains of ships anchored to windward shook dreadfully before they felt the shock." I merely mention this as having heard it from those who were here at the time without answering for the the truth of it, but I believe there is no doubt of the water coming in quite black from the North;— the Barometer in our store is a bent tube 49 Inches long, 19 of which is the utmost range, sunk down below the graduated part, equal to 20 Inches English previous to the earthquake of 1822—and whatever the season of the year may be rain almost always follows a heavy earthquake. the one in Nov 1822 was followed by torrents of rain— The natives say that when the volcanoes are active, there is no danger of earthquakes, but when they are quiet for a long period heavy shocks may be expected—after an earthquake they generally commence burning again— The people of the Country have various signs of a shock, such as the stars twinkling, rats making a noise, but there is not the slightest truth in it. Whenever the noise preceding a shock takes place all animals appear to be alarmed, the horse starts & snorts the fowls cackle & the dogs bark—

I wish much to hear of your report respecting the islands in the Pacific, and it will be curious if you find a sinking of the land there, & a rising here—[3]

I sincerely hope we may yet meet in the olden World and at all times it will give me much pleasure to hear from you and with every wish for your health & welfare | I remain | Dear Darwin | Yours sincerely | Rob.! Edw. Alison

DAR 36.1: 427—427a

CD ANNOTATIONS

1.1   By . . . persons— 1.8] *crossed*
3.3   The bottom . . . sending you— 3.7] *crossed*
4.1   it is not . . . hill— 4.4] *scored and crossed*
5.1   I ought . . . saltum— 5.3] *scored*
6.6   At the same . . . rock— 9.6] *crossed*
11.3   and the anchors . . . without 11.5] *crossed*
at end of letter: 'Elevation of Valparaiso'

[1] CD used information from this letter in his paper for the Geological Society of 4 January 1837 (*Collected papers* 1: 41—3) and in *South America*, pp. 31—5. Also included are data supplied by Alison in a memorandum on changes of land level from 1640 to 1834 (from R. E. Alison, [June? 1834]).

[2] One of the explanations advanced at the time for the existence of shells above sea level was that they were the remains of shellfish used for food at sites of earlier habitations. CD was especially cautious in his observation of shell deposits because he had recently seen heaps accumulated by the natives of Tierra del Fuego (see 'Observations of proofs of recent elevation on the coast of Chili', *Collected papers* 1: 41—3, and letter to [Alexander Burns Usborne], [*c.* 1—5 September 1835]). The correct names for Alison's shells are *Concholepas*, *Patellae*, and *Mytilus*. See *South America*, p. 32.

[3] CD had apparently discussed with Alison his hypothesis that a subsidence of land had taken place in the Pacific concomitant with the elevation of the South American continent.

From R. H. Corfield 26—7 June 1835

Valparaiso
June 26th 1835

My dear Darwin

It gave me infinite pleasure to receive your much wished for letter, which was brought by the Beagle. As I was very anxious to hear of your arrival at Coquimbo and though I am very glad to hear that you were well at the date of your letter, yet I was sorry to find my fears had been realized as what you would suffer on the road, for I was convinced you left this place too soon whereas by waiting another week you would have recovered so far as not to be so liable to a relapse. However I hope & trust you will not suffer from the weather we are experiencing having for the last 2 days more particularly suffered a severe norther, & heavy rain, but no ships lost, though if it had not been for the assistance rendered by the Beagle, some would have gone ashore.— Pray take care of yourself & let me have a line from you on your arrival at Lima just to say how you are, if it is only a few words I shall be satisfied I cannot express to you what I have felt since your arrival on the Coast,—I do not mean to compliment you or as the Irishman says give you any blarney—but I must candidly say, I do not recollect ever having experienced such pleasure, since I came to this Country as I have during your short stay with me a pleasure which I shall always remember with satisfaction & which has recalled to mind many associations of other days—which have much increased the desire I have of returning to England where I hope we shall again meet and I do not think you will have much difficulty in making me agree with you that Shropshire is in some respects, better than this Country. To me however individually Shropshire has few attractions as I am very little known there & never likely to live there yet still as one's Native County it has some charms—and I hope you will ere long be there to enjoy them I hope to be in England before this time 2 yrs., and shall endeavour to stop there, & fix myself in some business so as not to come out here again, *though there is many a worse country than this* although it is so barren— I cannot help laughing at your idea of seeing a farm in Shropshire watered by the honest rain of Heaven instead of the unnatural artificial streams of this Country—it is a very useful way of watering and do not you think *very original* However, I prefer mine own Country and I hope & trust God willing I shall be able to settle there— We are all here going on much as usual. Old White & his wife & Grandson, have sailed in the Conway All Ship news interesting to you Wickham will tell you, & no doubt you will be surprised to see him Captain of the Beagle until Capt Fitzroy rejoins you at Lima—[1] Your letter for England I shall send p Conway she being the first opportunity. your letter for Chiloe shall also go forward, but at present there is no vessel in the berth—[2]

I enclose a list of your clothes that you left for washing for which I paid 3$ 2. and the balance of 5$ 3 I paid Covington, who has paid 3cts for carting your things up to the mole, so that out of the 8$ 5 you gave me, he has to account to you for 5 dollars—

I do not send the letter you asked for, for Arica as I understand the vessel will not call there, but goes to Iquique where you have a letter for Mr Smith who lives there and I think you would receive little pleasure from an introduction to Gilman's House, at Lagua.—

Allison will write to you by the Beagle I dined with Capt Fitzroy on board the other day and a very pleasant comfortable evening I spent, so snug in his little cabin. We have got the Stove rigged up in our sitting room and it is really very useful and delightful, to sit round, on a wet night— Caldcleugh is down here he & Wickham dined with me the other day I have not been able to shew much attention to the officers of the Beagle for they have had a good deal to do, and I myself also busy at present preparing for the ships going to England and it being the end of the half year when there are always plenty of Accts to make out There is very little local news that I hear of and you will probably hear more from Wickham than I can tell you— We are daily expecting some vessels from Lpool, if I get any Salop papers I will send you some by Capt Fitzroy when he rejoins the Beagle at Callao & will then write again to you

I must conclude this stupid letter—and I trust it will find you in good health for the long voyage you have in prospect which I trust will be pleasant & prosperous and with every good wish for your happiness & welfare | Believe me my dear Fellow | very sincerely your friend | R H Corfield

Saturday June 27 1835

I merely add a few lines to ask you to apologise for me to Mr Stokes, whom I wished to have seen more of while staying here and told him I should send an invitation to ask him but some bad weather intervening I have been obliged to delay it until the last moment which being inconvenient for me on account of several little things & t.c I have been obliged to let him go away without seeing him—

A vessel arrived this morning in 90 days from Bourdeaux—. All quiet I understa⟨nd⟩ ⟨   ⟩, The French Ministry changed ⟨ ⟩el Soult[3] the new prime mini⟨ster.⟩ The Emperor of Austria—dead— Mr Abercomby[4] the New Speaker of the ⟨House⟩ of Commons & Sr Chas Sutton,[5] mad⟨e a⟩ peer I am told—

2 vessels sailed from Lpool for this place on 1st March— So I hope to be able to send you some more news soon—

Farewell—take care of yourself | Yours very truly | R H Corfield

[1] Robert FitzRoy had sailed with H.M.S. *Blonde* to rescue the crew of the *Challenger*, which had been wrecked on 19 May (see *Narrative* 2: 429 – 30 and letter to Caroline Darwin, [19] July – [12 August] 1835).

[2] This may have been a letter to Charles Douglas on Chiloé asking for his observations on the earthquake of 20 February. 'Being anxious to trace the effects of the earthquake to the south, I wrote, shortly after visiting Concepcion [March 1835], to Mr. Douglas, a very intelligent man, with whom I had become acquainted in the island of Chiloe' *Collected papers* 1: 55).

[3] Nicolas Jean de Dieu Soult, Duke of Dalmatia.

[4] James Abercromby.

[5] Charles Manners-Sutton.

From Mr Lambert[1] [*c.* July 1835]

You travel upwards in that dry ravine [Santandrés] for about 3 miles, where it ends, & on the summit of the hill which forms its boundary, you find large superficial beds of Sulphur & the Sulphate of Alumina.— In the ravine there are scattered about fragments of Pumice Lavas calcinèe Granite & Porphyry & other Volcanic productions.— The formation of all the country from Sant Andrès is generally of Granite & Porphyry covered by Volcanic eruptions. These substances present generally a red appearance, whilst the Granite & Porphyry formation on which they lay is black.— There are many points, where you can see the points of the Granitic hills rise in the middle of the Volcanic formation; the Lava not being in sufficient quantity to cover the whole.— All the ravines have been formed after the eruptions of the Lava.— This Volcanic formation extends from Copiapò to Atacama, forming in some places regular basaltic elevated plains. These detain the waters which flow from the Western side of the Cordilleras & cause those Salt lakes, where the inhabitants of Copiapò & Atacama provide themselves with that article.

Copy (in CD's hand)
DAR 37.1: 648

CD ANNOTATIONS
'It must be remarked, that Mr Lambert's statement that the Volcanic eruptions capped the hills prior to the excavation of the Valleys is in perfect accordance with my idea that the Tufa is submarine formation & not subaerial if subaerial, there is no possible reason for imagining why all the patches should have belonged to one level.—'

[1] Mr Lambert is mentioned in *South America*, 211. This letter is used on p. 230.

From R. H. Corfield 14 – 18 July 1835

Valparaiso
July 14 1835

My dear Darwin—

The Blonde with Capt Fitzroy on board being on the point of sailing I only write a few lines merely to ask you how you are and to express a hope that you have not suffered much from the fatigues of your journey but I trust this letter will find you at Lima in good health & spirits— I am anxious to hear from you as soon as you get to Lima We are going on much as usual, though at the present moment people are busy preparing for 3 vessels now on the point of sailing for England— On this account therefore you must excuse a short letter from me, and indeed to tell you the truth, I am not much in the humour for writing letters, my mind being rather burdened with some disagreably perplexing subjects relative to business, which are rather annoying me at this present moment— However there is little in our local news, that would interest you and you will hear from Capt Fitzroy the particulars of the Ship wreck of the Challenger, as well as any other news—relative to the Navy & shipping We have had several arrivals, viz the Sparrow Hawk Brig of War frm Rio de Janeiro, and 2 vessels from Lpool, with account of another having sailed before her but not yet arrived— Accounts are down to the end of March by a Belgian Vessel 82 days from Falmouth at which port she called on her way down Channel, by her we learn that the Tories are out of office I am sorry to say. I have received some Shrewsbury papers—and thinking they may be interesting to you, I send you some by this opportunity I have not read them so cannot tell their contents,—and they will be of more interest to you than to me—my acquaintance with Shropshire people being very slight and likely to be still less so—although I should be very glad to shake hands with you there again—for I do not know when I have experienced such pleasure as I did when you were here—

In the course of a few days I expect it will be decided whether I shall be in England before you, or, in all probability never as within the last 2 days a vessel has arrived from Lpool, with a new establishment, the partner of which is empowered to make proposals of partnership to me, and if I agree to accept them, I shall I imagine be bound down by the articles, to remain out here probably 10 years—in which case I feel more disposed to select a Señorita as you call them, & remain here altogether as I think a poor devil like myself will pass a happier or at all events, quite as agreable a life as in England—as I think *there is many a worse country than this* I hope however such may not be the case and at present I am not much inclined to accept any terms, & if I determine not I shall make a start for England before this time next year— Whether however we ever meet again or not, I shall always consider it to have been one of the happiest periods of my life, the short time you spent with me—

Wishing you therefore health & happiness, a safe & pleasant voyage and a joyful meeting with your friends in England— | Believe me to be | My dear Darwin, | ever sincerely yours | R H Corfield

July 18
Since writing the foregoing several more vessels have arrived from Lpool, one in 91 days brings accounts of the Tory ministry being thrown out, they having

been in two minorities of 33 & 27 on the Irish Tithe question, but you will hear all the best news from Capt Fitzroy I send you a few papers and one, the Morning Herald of Feby 21—containing some correspondence between a Mr Beaumont & Mr Townsend which are worth perusal[1] but Alison desired me by all means to send it you, as it also contains something relative to Cuvier which will be interesting to philosophers.[2] He desires to be remembered to you & if he has time will write—

I do not feel disposed to join the House I mentioned to you on the other side as I do not like them or their proceedings— So I hope yet to give you the meeting in England

Excuse this short & stupid letter, but believe me with every good wish | Yours ever | R H C

DAR 204.9

CD ANNOTATIONS
on cover: 'Gillman | *[*Cord*]* | M.. Gonzales[3] | Caldcleugh'

[1] The Rev. George Townsend's charge that Thomas Wentworth Beaumont, M.P. had expressed 'atrocious revolutionary sentiments against the Conservatives' led to an exchange of letters in *The Morning Herald* of Saturday, 21 February 1835, p. 3.
[2] The reference is to the summary of a eulogy of Georges Cuvier by Marie-Jean-Pierre Flourens at the French Academy of Science (*The Morning Herald*, 21 February 1835, p. 3).
[3] Mariano Gonzales, CD's guide on his Chilean tours (*'Beagle' diary*, p. 306).

To Caroline Darwin [19] July – [12 August] 1835[1]

Lima—
July— 1835—

My dear Caroline

My last letter was dated Coquimbo—I rejoice that I am now writing from Peru.— I have received the three months letters which were missing, & I know that in a few days I shall receive several more. In the mean time I will write an outline of our proceedings since the last letter. From Coquimbo I rode to Guasco, where in the valley I staid a few days; from that place to Copiapò, there is a complete desert of two & a half days journey, during which the poor horses had not one single mouthful to eat. The valley of Copiapo is a narrow little stripe of vegetation between districts utterly sterile.— Indeed the whole of Chili to the North of Coquimbo, I should think would rival Arabia in its desert appearance When in the valley of Copiapo I made two journeys to the Cordilleras & reached the divisions of the waters; it was most piercingly cold in those elevated regions, but the cloudless sky, from which rain does not fall more than once in several years, looked bright & cheerful.—

It is very hard & wearisome labor riding so much through such countries, as Chili, & I was quite glad when my trip came to a close. Excluding the interest arising from Geology, such travelling would be down right Martyrdom. But with this subject in your mind, there is food in the grand surrounding scenes, for constant meditation. When I reached the port of Copiapo, I found the Beagle there, but with Wickham as temporary Captain. Shortly after the Beagle got into Valparaiso, news arrived that H.M.S. Challenger was lost at Arauco, & that Capt Seymour a great friend of Fitz Roy & crew were badly off amongst the Indians.— The old Commodore in the Blonde was very slack in his motions, in short afraid of getting on that lee-shore in the winter; so that Capt Fitz Roy had to bully him & at last offered to go as Pilot.— We hear, that they have succeeded in saving nearly all hands, but that the Captain & Commodore have had a tremendous quarrel; the former having hinted some thing about a Court-Martial to the old Commodore for his slowness.— We suspect, that such a taught hand, as the Captain is, has opened the eyes of every one, fore & aft in the Blonde to a most surprising degree. We expect the Blonde will arrive here in a very few days & all are very curious to hear the news; no change in state politicks ever caused in its circle more conversation, that this wonderful quarrel between the Captain & the Commodore has with us.—

The Beagle after leaving the port of Copiapò, touched at Iquique, in Peru, a place famous for the exportation of Nitrate of Soda.— Here the country is an absolute desert, during a whole days ride, after leaving the Beach, I saw only one Vegetable production & this was a minute yellow Lichen attached to old Bones. The inhabitants send 40 miles for their water & firewood, & their provisions come from a greater distance.— From Iquique we came direct to this place, where we have been for the last week. The country is in such a state of Anarchy, that I am prevented from making any excursion.— The very little I have seen of this country, I do not like; The weather, now in the winter season is constantly cloudy & misty, & although it never rains; there is an abundance, of what the people are pleased to call Peruvian dew, but what in fact is a fine drizzle.— I am very anxious for the Galapagos Islands,—I think both the Geology & Zoology cannot fail to be very interesting.— With respect to Otaheite, that *fallen* paradise, I do not believe there will be much to see.— In short nothing will be very well worth seeing, during the remainder of this voyage, excepting the last & glorious view of the shores of England.—

This probably is the last letter, I shall write from S. America, I have written also to Mr Owen & Fox.— With the three months letters were two from Fox, the most kind & affectionate ones, which could be written.— He gives me a long account of his wife; I hope she is as nice a little lady, as he seems to think & assuredly deserves.— How very strange it will be, thus finding all my friends, old married men with families.—

July [August] 12th I have received three more letters making the chain complete from England to February 1835.— Capt Fitz Roy has arrived in good spirits & in a short time we sail for the *Galapagos*. He has just stated, five minutes ago on ye Quarter Deck that this time year we shall be very near to England. I am both pleased & grieved at all your affectionate messages, wishin⟨g⟩ me to return home.— If you think I do not long to see you again, you are indeed spurring a willing horse; but you can enter into my feelings of deep morti-fication, if any cause, ev⟨en⟩ ill-heath should have compelled me to have left the Beagle.— I say, should have, because you will agree with me, that it is hardly worth while, now to think of any such step.— Give my most affectionate love to poor dear old Erasmus, I am very glad, that ye same letter which brought an account of his illness, also told me of his recovery.— During my whole stay at Plymouth I have but one single recollection which is pleasant & that was his visit to me. Indeed, I do not know to what period of my life I can look back, without such thoughts coming to mind. I received his half letter & am grieved that I shall neither receive the letter & box which he is going to send till we reach the C. of Good Hope. What a good name that Cape has, indeed it with be one of good Hope when the Beagle passes its bluff Head.—

You will not hear from me for *upwards* of 10 month, nor I from you, in which time may God bless you all for being such kind dear relations to me. Farewell. Your affectionate brother | Charles Darwin.

NB. If you do not understand my former directions about letters you had better enclose them to Capt Beaufort.— Remember a letter too much (ie too late) is better than one too little

NB. 2^d.— Tell my Father I have drawn a bill for 30£, to take with me money for the Islands.—

DAR 223

[1] Dated from the arrival of the *Beagle* at Callao, the port for Lima (19 July 1835). Hence 'July 12th' later in the text is a mistake for August. The sentence preceding that date was also written in August, after the letter to W. D. Fox, [9–12 August] 1835.

To W. D. Fox [9–12 August] 1835

Lima
July,[1] 1835

My dear Fox,

I have *lately* received two of your letters, one dated June[2] & the other November 1834. (—They reached me however in an inverted order;—) I was very glad to receive a history of this the most important year in your life. Previously I had only heard the plain fact, that you were married.— You are a true Christian & return good for evil.—to send two such letters to so bad a

Correspondent, as I have been. God bless you for writing so kindly & affectionately; if it is a pleasure to have friends in England, it is doubly so, to think
& know that one is not forgotten, because absent.—

This voyage is terribly long.— I do so earnestly desire to return, yet I dare
hardly look forward to the future, for I do not know what will become of me.—
Your situation is above envy; I do not venture even to frame such happy
visions.— To a person fit to take the office, the life of a Clergyman is a type of all
that is respectable & happy: & if he is a Naturalist & has the "Diamond
Beetle",[3] ave Maria; I do not know what to say.— You tempt me by talking of
your fireside, whereas it is a sort of scene I never ought to think about— I saw
the other day a vessel sail for England, it was quite dangerous to know, how
easily I might turn deserter. As for an English lady, I have almost forgotten
what she is.—something very angelic & good. As for the women in these
countries they wear Caps & petticoats & a very few have pretty faces & then all
is said.—

But if we are not wrecked on some unlucky reef, I will sit by that same fireside
in Vale Cottage & tell some of the wonderful stories, which you seem to
anticipate & I presume are not very ready to believe. Gracias a dios, the
prospect of such times is rather shorter than formerly.—

From this most wretched "city of the Kings" we sail in a fortnight, from
thence to Guyaquil—Galapagos—Marquesas—Society Is^d, &c &c.—[4] I look
forward to the Galapagos, with more interest than any other part of the
voyage.— They abound with active Volcanoes[5] & I should hope contain
Tertiary strata.— I am glad to hear you have some thoughts of beginning
geology.— I hope you will, there is so much larger a field for thought, than in
the other branches of Nat: History.— I am become a zealous disciple of M^r
Lyells views, as known in his admirable book.— Geologizing in S. America, I
am tempted to carry parts to a greater extent, even than he does. Geology is a
capital science to begin, as it requires nothing but a little reading, thinking &
hammering.— I have a considerable body of notes together; but it is a constant
subject of perplexity to me, whether they are of sufficient value, for all the time I
have spent about them, or whether animals would not have been of more certain
value.—

I have lately had a long ride from Valparaiso to Copiapò; in the Northern
half the country is frightfully desert, & the sole source of interest was in the
Geology. The scarcity of fossil shells is very inconvenient, as it will render any
comparison of the formations with those of Europe nearly impossible. The
Andes, at the period when Ammonites lived, (which corresponds to the secondary rocks) must have been chain of Volcanic Islands, from which copious
stream⟨s of⟩ Lava were poured forth & subsequently covered with Conglomerates. Such beds form the Cordilleras of Chili.—

For the last months I have been shamefully negligent of all branches of
Zoology; I hope to make up a little in the Pacifick; but all our future visits will

indeed be flying ones.— The Captain talks about arriving in England September year. I doubt the possibility; but Heaven grant it may not be much after.— Will you write to me once again, soon after receiving this & direct to the C. of Good Hope, & in answer to it you will see me in Person; Till that joyful day arrives, I must wish you a long Farewell. I shall indeed be glad once again to see you & tell you how grateful I feel for your steady friendship.— God bless you. My very dear Fox. Believe me, | Yours affectionately | Chas. Darwin—

Christ's College Library, Cambridge (Fox 47A)

[1] A mistake for August (see letter to Caroline Darwin, [19] July — [12 August] 1835, n. 1).
[2] The letter has not been found.
[3] *Curculio imperialis* of Brazil, called the 'Diamond Beetle' because of its sparkling elytra.
[4] Of the four places named, only the Galápagos and the Society Islands were visited.
[5] Although some volcanic activity still occurs at the Galápagos, CD's expectations were disappointed. The only action he reports seeing is a small jet of steam issuing from a crater on Albemarle (Isabela) Island (*'Beagle' diary*, p. 338).

To J. S. Henslow[1] 12 [August] 1835

Lima
July[2] 12th.— 1835

My dear Henslow

This is the last letter, which I shall ever write to you from the shores of America.— and for this reason I send it.— In a few days time the Beagle will sail for the Galapagos Is^{ds}.— I look forward with joy & interest to this, both as being somewhat nearer to England, & for the sake of having a good look at an active Volcano.— Although we have seen Lava in abundance, I have never yet beheld the Crater.— I sent by H.M.S. Conway two large boxes of Specimens. The Conway sailed the latter end of June.— With them were letters for you.— Since that time I have travelled by land from Valparaiso to Copiapò & seen something more of the Cordilleras.— Some of my Geological views have been subsequently to the last letter altered.— I believe the upper mass of strata are not so very modern as I supposed.— This last journey has explained to me much of the ancient history of the Cordilleras.— I feel sure they formerly consisted of a chain of Volcanoes from which enormous streams of Lava were poured forth at the bottom of the sea.— These alternate with sedimentary beds to a vast thickness: at a subsequent period these Volcanoes must have formed Islands, from which have been produced strata several thousand feet thick of coarse Conglomerate.— These Islands were covered with fine trees; in the Conglomerate I found one 15 feet in circumference, perfectly silicified to the very centre.— The alternations of compact crystalline rocks (I cannot doubt subaqueous Lavas) & sedimentary beds, now upheaved, fractured & indurated form the main range of the Andes. The formation was produced at the time, when *Ammonites*, several Terebratulæ, Gryphites, Oysters, Pectens, Mytili &c &c lived.—[3]

In the central parts of Chili, the structure of the lower beds are rendered very obscure by the Metamorphic action, which has rendered even the coarsest Conglomerates, porphyritic.— The Cordilleras of the Andes so worthy of admiration from the grandeur of their dimensions, to rise in dignity when it is considered that since the period of Ammonites, they have formed a marked feature in the Geography of the Globe.— The geology of these Mountains pleased me in one respect; when reading Lyell, it had always struck me that if the crust of the world goes on changing in a Circle, there ought to be somewhere found formations which having the *age* of the great Europæan secondary beds, should possess the *structure* of Tertiary rocks, or those formed amidst Islands & in limited Basins. Now the alternations of Lava & coarse sediment, which form the upper parts of the Andes, correspond exactly to what would accumulate under such circumstances. In consequence of this I can only very *roughly* separate into three divisions the varying strata (perhaps 8000 ft thick) which compose these mountains. I am afraid you will tell me to learn my A.B.C.—to know quartz from Feldspar—before I indulge in such speculations.— I lately got hold of ⟨ ⟩ report on M. Dessalines D'Orbigny's labors in S. America. I experienced rather a debasing degree of vexation to find he has described the geology of the Pampas, & that I have had some hard riding for nothing; it was however gratifying that my conclusions are the same, as far as I can collect, with his results.—[4] It is also capital, that the whole of Bolivia will be described. I hope to be able to connect his Geology of that country, with mine of Chili.— After leaving Copiapò, we touched at Iquique. I visited, but do not quite understand the position of the Nitrate of Soda beds.— Here in Peru, from the state of Anarchy, I can make no expedition.—

I hear from Home, that my Brother is going to send me a box with Books & a letter from you.— It is very unfortunate that I cannot receive this before we reach Sydney, even if it ever gets safely so far.— I shall not have another opportunity for many months of again writing to you.— Will you have the charity to send me one more letter (as soon as this reaches you) directed to the C. of Good Hope Your letters besides affording me the greatest delight, always give me a fresh stimulus for exertion. Excuse this Geologico-prosy letter & Farewell till you hear from me at Sydney & see me in the Autumn of 1836. Believe me, dear Henslow, Yours affectionately obliged | Charles Darwin

Royal Botanic Gardens, Kew

[1] Henslow did not print excerpts from this letter, probably because it had not yet arrived when the Cambridge Philosophical Society pamphlet was published early in December 1835. CD's letter to Caroline, also written in August, has a postmark, 'Shrewsbury JA 4 1836'.

2 A mistake for August (see letter to Caroline Darwin, [19] July — [12 August] 1835, n. 1). The expected sailing 'In a few days time' mentioned in the first paragraph was delayed until 7 September.

3 See *South America*, Appendix, for descriptions of CD's specimens of fossil shells by George Brettingham Sowerby and Edward Forbes.

4 A report on the scientific results of Orbigny's voyage was published in 1834 by Blainville, Brongniart, and others (Blainville 1834). CD later found that he disagreed with the French naturalist on the age and origin of the Pampean formation (see *South America*, pp. 98 — 103).

To H. S. Fox 15 August 1835

Lima.

August 15th.— 1835

Dear Sir

The Beagle will sail in a few days from Callao.— Before leaving the shores of S. America, I am tempted to send you the accompanying specimens.— In themselves they have little value, but I hope you will more readily believe that mere forgetfulness is not the cause of the fewness of their numbers.— Generally speaking in the whole line of coast of Chili, where the Beagle for the last 12 months has been employed, the rocks are Granitic.— It is only in the interior, that the lower Crystalline formations are covered by the Porphyries & Breccias & only in the main Cordilleras, where these are again covered by the Gypseous, Sandstone, Limestones & ancient Lavas.— What I before stated about my manner of travelling, render the collecting even of sufficient specimens, much more duplicate ones extremely troublesome.—

In the latter end of March, I had the satisfaction of crossing the Cordilleras to Mendoza & returning by a different route. It is impossible to imagine more illustrative scenes of subterraneous violence, than these huge mountains present. The strata cracked & fissured by numberless dykes have been tossed about, like the ice on a running stream; there is however a degree of rough parallelism in the lines of disturbance. The scenery is on so grand a scale, & the atmosphere so clear & brilliant, that the whole was to me like entering on a new Planet.— After returning to Valparaiso, I travelled by land to Copiapò, whilst the Beagle was surveying the coast. It was a most dreary journey, in the deserts, which extend to the North of Coquimbo, there is no sort of interest excepting from Geology.— I hope now to be able to give some sort of outline of the superposition of the strata & the structure of the mountains in Chili. It is very certain that the general idea of the Cordilleras being composed solely of Volcanic rocks is quite incorrect. There is one point in the Geology of S. America, in which I am much interested, it is the recent elevation of the land.— That such has taken place & to a considerable amount on this coast I have abundant proofs. Have you ever noticed on land elevated from 30 to 200 ft above the sea, any *large* beds of marine shells, & which did not appear carried there by man? I think it probable that such might occur at R. Grande or South

Brazil; if you have any information on this head, I should be most grateful for such a communication.—[1] My direction in England is Shrewsbury or indeed to the C of Good Hope on my road there. I am afraid you will think me a very troublesome correspondent; I only wish I could send you instead of Geological questions & details, some specimens which would be worthy of your acceptance.—

The Beagle now proceeds direct to the Galapagos, from thence across the Pacifick to Sydney, C. of Good Hope to England.— I confess, I am so little accustomed to these long expeditions, that I look forward to this last stage, with more interest, than the whole of the voyage. I have the pleasure to remain, Your obliged & obedient servant | Chas. Darwin

Bodleian Library, Oxford, MS. Eng. lett. c, 235: 28

[1] The first two chapters of *South America* bring together the evidence of the recent elevation of both coasts of the continent. On the east coast the location of shells in formations observed by CD and Orbigny constitutes the main evidence. No evidence from Fox is mentioned.

To Alexander Burns Usborne[1] [*c.* 1 – 5 September 1835]

The recent elevation of the land above the level of the ocean is shown by beds of shells lying on the surface of the land.— There is often a difficulty in distinguishing between such as have been left by the sea & those at some remote period brought by man.— Where the shells or fragments are *in great numbers*, packed in layers, either with or without earth, & forming a *level* mass, the former cause may be generally assumed as certain.— Such shells are brittle & decayed & their color partially lost.— If amongst them there are many very small ones; if the situation is remote from the sea, or inaccessible from the immediate Beach,—if no water is found in the vicinity, of course there is a probability they have not been carried there by any former residents.— If old Barnacles, minute Corallines, Serpulæ, or impressions of such are found adhering on the *inner* surface of shells (this happens chiefly in Spiral Univalves such as Welks &c) it is manifest they must have been lying *dead* (& therefore of no use as food) for some time at the bottom of the sea.— I have frequently found such shells on the coast of Chili at an height from 20 to 400 ft.— If you should meet with shells, thus circumstanced in *any part* of your survey, & especially to the North of Lima; I should be much obliged, if you would mark on Paper the name of the Place, & estimate carefully the vertical height.— It would be well always to state the amount of (& reasons for) conviction which you feel respecting their origin.— Where an opportunity occurs, especially if the elevation should be great, the observation would be of infinitely more service if an angle of elevation could be taken.— The oftener you can observe & record this class of facts, in different places, so much the better; for the evidence respecting the rise of land becomes cumulative.— I may mention that the layers or beds of shells, or such loosely scattered on the surface sometimes occur in steps, like ancient *shingle* beaches; or in small step-form terraces (Sea).— If such should happen it should be mentioned.— Also, the state of the shells, whether very old in appearance & broken, whether decomposing into a white powder, whether chiefly of one or various sorts.— Also the extent, breadth & depth of the bed. Again, it would be well, where the height of the locality is considerable to collect a few specimens of the different sorts.—

The most likely situations to find such shells, are on *flat-topped* points near the Coast or in an island.—

Small specimens of the *prevailing* rocks (not loose fragments) wrapped up in paper with *attached label* of the *locality*, from any part of the coast will possess considerable interest.— I may except Cobija Iquique, (but the small neighbouring ports are **not** excluded) Arica, Islay, & Callao, from all of which I have specimens.—

I cannot state too strongly the value of all shells, whether petrified or not, extracted from stratified cliffs of sand, clay, or stone.— Their localities being carefully marked, & nature of imbedding stone.—

The first class of observations are less troublesome & to me even of more interest & value.—

Amem
DAR 207 (Letters)

[1] CD's memorandum was sent by Usborne to Henrietta Litchfield on 15 September 1882 with a note of reminiscences of CD as a shipmate (DAR 207). CD wrote the memorandum at the time Usborne and Charles Forsyth were left behind to survey the coast of Peru (see *Narrative* 2: 483). No mention is made in *South America* of any specimens collected or of observations made by Usborne during the survey.

To Susan Darwin 3 [September] 1835

Lima.
May 3ᵈ— 1835[1]

My dear Susan

I write to you again, chiefly for the purpose of telling my Father, that I have drawn a 50£ bill *instead of* 30£, which I mentioned in my last letter.— So that this must be notified to the Banker, otherwise he will be surprised at seeing the 50£.— Our prolonged stay in this place, has caused me to draw for the extra money.— This delay has been a grievous waste of time for *me*: the Captain discovered in Lima some old charts & Papers, which he thinks of considerable importance.— Two of the Midshipmen, Mʳˢ Usborne & Forsyth are to be left behind to survey in a small Schooner, the coast of Peru;[2] afterwards they will return in a Merchant man to England.— I wish indeed the last month had been spent at Guyaquil or the Galapagos: but as the Spaniard says "no hay remedio".— The Captain in a note which he sent me to day from Lima says "Growl not at all— Leeway will be made up.— Good has been done unaccompanied by evil—ergo—I am happier than usual"— So that, I am glad to say, that all this time will not be lopped off the period of our return.—

We shall go round the world, like a Flying Dutchman, & without doubt, if this was the third in stead of the fifth year the cruize would be delightful.— We shall arrive at Sydney just at the right time of year: the Captain intends going within the reefs through Iona St^{ts}.— We hear a famous account of this passage, smooth water, anchorage every night, beautiful scenery & splendid weather.— I am quite impatient to get into a glowing hot climate.— it sounds very odd to hear a person, in Latitude 12° wishing for warmth.— But really it is here uncomfortably chilly & damp with an eternally cloudy sky. When we reach the Galapagos, the sun will be vertically over our heads—& I suspect my wishes will be fulfilled to the uttermost.— Living quietly on board the Ship & eating good dinners have made me twice as fat & happy as I have been for some months previously. I trust & believe, that this month, next year, we shall be very close to if not in England. It is too delightful, to think, that I shall see the leaves fall & hear the Robin sing next Autumn at Shrewsbury.— My feelings are those of a School-boy to a the smallest point; I doubt whether ever boy longed for his holidays, as much as I do to see you all again.— I even at present, although nearly half the world is between me & home, begin to arrange what I shall do, where I shall go during the first week.— In truth I shall have a great deal to do, for a long time after we return. My geological notes are become very bulky, & before they can be of any use will require much overhauling & examination.— But sufficient for the day is the evil thereof.— We shall be in England next September & that is enough for me.—

Two men of war have lately arrived from Rio, but they brought no letters for the Beagle; so that the Admiral is forwarding them on to Sydney.— We all on board are looking forward to Sydney, as to a little England: it really will be very interesting to see the colony which must be the Empress of the South.— Capt King has a large farm, 200 miles in the interior.— I shall certainly take horse & start—I am afraid however there are not Gauchos, who understand the real art of travelling.—

I have scarcely stirred out of the Ship for the last fortnight: the country is in such a miserable state of misgovernment, that nothing can exceed it— The President is daily shooting & murdering any one who disobeys his orders.— One is that all property should be at the disposal of the state, & another, that every man from 15 to 40 should enroll himself, as ready to be his soldier.— Yesterday several young men were shot for neglecting to give in their names.— Is not this a precious state of things?—

Good bye, till I again write from Sydney.— Give my most affectionate love to my Father & to all at home.— My dear old Granny. Your affectionate brother | Charles Darwin—

Give my love also to Nancy

DAR 223

[1] CD's first sentence makes clear that 'May 3^d' is an error. The date was changed to 'September', probably by his sister Catherine (see letter from Catherine Darwin, 29 January 1836), and then to 'August'. August 3d is the date given by Nora Barlow (*Voyage*, p. 126) but the reference to Robert FitzRoy makes September the better date, since FitzRoy did not return from the *Challenger* rescue mission until 9 August (see *Narrative* 2: 482).

[2] Despite his earlier censure by the Admiralty, FitzRoy again purchased a small schooner, the *Constitución*, with his own funds and apparently again expected to be reimbursed by the Government (see *Narrative* 2: 482 – 3; for the Admiralty's negative response see *'Beagle' diary*, p. 333, n. 45).

From Catherine Darwin 30 October 1835

Shrewsbury.
October 30th. | 1835.

My dear Charles

We are very anxious to hear from you again, after your 2^d Land Expedition, which sounded very dangerous, and tremendously fatiguing, and we shall be delighted to hear that you are safe on board the Beagle again. I do hope you are careful not to over tire yourself very much, for fear of giving yourself a dangerous fever in those hot countries. What an invaluable friend Mr Corfield has been to you; we are all very grateful to him, I am sure.— I am so very glad to tell you that Erasmus has received your Journal safe, within the last two days, and also two Boxes very valuable; and there are also six large Boxes at Plymouth come from he does not know where. Erasmus had given up all expectation of receiving them, after the wreck of the Challenger, and hearing nothing from Capt Beaufort, about them; and his letter to day is written in much joy that they had safely arrived.— It will be very interesting to read your Journal, & I hope we shall soon have down here.— Erasmus says, that from the Court Martial, it appears the Challenger was wrecked in consequence of an alteration in the rate of the currents, as stated by Capt Beaufort, produced by the late Earthquakes, of which you gave us an account.—[1] I don't think I have written to you, my dearest Charles, since we had the capital news of your return next September; it is such a blessing to have a fixed time to look forward to, and one really not very far off now.— There is hardly any news to tell you, I am afraid, since Caroline last wrote to you; the Hensleighs have another Boy; (they have 3 children now, one girl, & two boys) and as Erasmus entirely adopts the children, you ought to be properly interested in having *another nephew*.— Are you aware that you are **really** to have a niece, a little Miss Parker born next January? if it should be Master Quintus Parker's appearance, he must be strangled, for that is the only thing left for us to do.—[2] I suppose you saw the Comet in a different month from what we did in England; the middle of October was the best time for seeing it with us, but it was so hazy all the month, that it could not be seen well at any time. Erasmus went to look at it with D[r] Holland, through Sir James South's great Telescope,[3] but there was such a mist, that it could be seen no better than with the naked eye.— We had your Friend Major Bayley to dine here not long ago, and he went out in the cold with me to look at the Comet through a Telescope; I saw it pretty well at last, but he could not. Major Bayley made many enquiries, about you, and talked much about you.— Frank Leighton also sent me a message to you a little while ago; he begs that you will remember your native Town, and keep any duplicates of curiosities, or specimens, for the Museum of Natural History; which has been begun this Summer in Shrewsbury.— There is a Society of Natural History formed,[4] and several people are very much interested about it; I am afraid you will look down upon the *specimens* in it, with sovereign contempt.— I saw you⟨r⟩ friend, Sarah Williams not long ago, a⟨nd⟩ she desired me to remind you particular⟨ly⟩ of your engagement to dine *& go to the Play* with her, the first evening you are in London. Sarah is afraid you cannot have received her last letter; it was sent with some of our's, and I am afraid it must have been among those that you did not receive in Valparaiso; I think you said they would be forwarded on to you, in the course of time.— There is no news of the Owen Family to tell you particularly. William Owen is gone to join his Regiment in Ireland, though he has a stiff knee still, and it is feared will always have it.— The poor little Biddulph Girl is still kept lying down on her back; though they say she is getting better; she is the most charming little child I ever saw, quite as charming as her Mother.— A[t] Sarah is staying with us now, and she desires I will give her love to you.— Nancy also begs I will tell you she is counting the months till your return.— Charlotte Langton also in a note we had to day from her says how pleasant it is to think that there is a fixed time for your return. Papa & Caroline desire their best of loves to you, my dearest old Charley. | God Bless you— | Y[r] ever affectionate | E. Catherine Darwin

Susan is at Overton.

DAR 97 (ser. 2): 22–3

[1] This was Robert FitzRoy's explanation (see *Narrative* 2: 479–80). FitzRoy's notes to this effect were read at the court-martial of Captain Michael Seymour of the *Challenger* at Portsmouth, 19 October 1835. Seymour was exonerated.

[2] The first four Parker children were all boys. As predicted, a niece, Mary Susan Parker, was born in January 1836.

[3] South's telescope had a 12-inch object glass, the second largest in the world.

[4] The Shropshire and North Wales Natural History and Antiquities Society. CD is listed as a member in 1835–6. In 1877 it was amalgamated with the Shropshire Archaeological Society.

From Susan Darwin 22 November 1835

Shrewsbury
Nov^{br}: 22^d. | 1835

My dear Charles—

I am happy to tell you that your Journal has arrived safe, there was some alarm about it; as the Challenger that you told us it was coming by; was wrecked owing to some new currents from the late Earthquake.— I don't know what Ship it did come in, but your heart w^d certainly have been broken if it had been lost.— Eras rec^d the Journal in London and lent it to the Hensleighs to read, who were exceedingly pleased with it, and think it will make a most interesting book of travels when you publish it.— We are now reading it aloud, and Papa enjoys it extremely except when the dangers you run makes him shudder: Indeed I think the escapes you have had of different dangers are quite providential! We never read anything so shocking as the murderous war upon the poor Indians—one can hardly believe anything so wicked at the present day as the conduct of General Rosas. Is he a Spaniard?— I cannot think how you c^d write such a collected account of your travels when you were Galloping so many miles every day.— When I have corrected the spelling it will be perfect, for instance *Ton* not *Tun, lose* instead of *loose.*— You see I am still your Granny— Since I began this Eras writes me word that your Journal he believes came on the Ship that brought the Crew of the Challenger so it was wonderfully lucky it was not lost.— Eras also says he hears that some of your Letters were read at the Geological Society in London & were thought very interesting,[1] and now I will copy another bonne bouche for you. Dr Butler sent Papa an extract from a Letter of Professor Sedgwick's to him which was as follows about you. "He is doing admirably in S. America, & has already sent home a Collection above all praise.— It was the best thing in the world for him that he went out on the Voyage of Discovery— There was some risk of his turning out an idle man: but his character will now be fixed, & if God spare his life, he will have a great name among the Naturalists of Europe."— I think this paragraph we ought to copy out in every Letter that goes to Sydney lest this sh^d miscarry. My dear Charley I am so happy you have this reward for all your excessive labour & exertions. I sometimes can hardly fancy you are my brother that I read of going through such hazardous enterprises.— I do long to hear you are safe out of South America for then I shall consider you comparatively safe. Catherine heard f^r you last week y^r Letter was dated Coquinbo the 31st of May and then you were going to undertake a month of land Journey I am sorry your inclination for the South Sea Islands has rather evaporated.—

You will be surprised to hear that Erasmus is turning into a busy Man. Government has appointed Commissioners for examing into all the Public Charities throughout England and Robert Mackintosh is made one of them and as he is obliged to have a Clerk Erasmus has taken that office & gets 150 p^r Annum. I don't expect Mr Eras will keep his place long, at least if it requires much work, and how he will manage with no Law knowledge seems a mystery:—[2] He sets out next Wednesday to undertake Berkshire.— Unfortunately Shropshire has been examined or we sh.^d have had him here which would have been very nice.—

Marianne has been suffering a good deal of anxiety this Month, owing to Parky having got the Scarlet fever at his School at Oswestry and he had it so severely that he could not be moved. Marianne went there to nurse him and was several nights without slee⟨p⟩ which q⟨ui⟩te knocked her up.— She expects to be con⟨fined⟩ in January so she was not at all in a fit state for so much exertion— She has however not caught the fever, and now as it is 3 weeks since we hope she will certainly escape. Parky is now at Overton & Caroline is there, whilst we have the three younger boys here and they are the best & nicest little men you ever saw, and often talk of Uncle Charles with awe & reverence.—

The John Wedgwoods are very busy transplanting themselves again from Monmouthshire into Staffordshire as they have taken a house about 4 miles from Maer in order to be near Jessie & also Allen.[3] Indeed they have now three children living in Staffordshire for M.^r & Mrs Robert reside at Muxton— I had a visit from them this Autumn & c.^d hardly make myself remember they were husband & wife they looked so much more like Mother and Son.—

I hope my dear Charles we shall have another Letter from you when you reach Lima as we shall be very glad you have done with that odious South America.— My Father Catherine & I send you our most affectionate Love & Good bye Ever y.^{rs} Susan E Darwin

Nancy begged I w.^d tell you how she counts the Months now with joy.—

DAR 97 (ser. 2): 24—5

[1] Extracts, taken from CD's letters to Henslow (attributed to 'F. Darwin Esq., of St. John's College, Cambridge'), were read to the Geological Society by Adam Sedgwick (see letter from Caroline Darwin, 29 December [1835]) on 18 November 1835 (*Proceedings of the Geological Society of London* 2 (1833—8): 210—12, *Collected papers* 1: 16—19). The geology of South America was so little known at the time that CD's letters excited much interest. Charles Lyell, who saw CD's reports on elevation as confirmation of his views, was particularly eager for more details. On 6 December 1835 he wrote to Sedgwick, 'How I long for the return of Darwin! I hope you do not mean to monopolise him at Cambridge.' (K. M. Lyell 1881, 1: 460—1; see also Wilson 1972, p. 425).

[2] Robert Mackintosh was Fanny Mackintosh Wedgwood's brother. On hearing the news that Erasmus had taken employment with him, Emma Wedgwood wrote on 29 November [1835] to her aunt, Jessie Sismondi: 'Erasmus is gone as his Clerk, which surprized us all that so idle a man should like to undertake it (viz. the Clerk), as it is supposed he will have a good deal to do. The girls at Shrewsbury tell him they are afraid the King will have a very bad bargain.' (*Emma Darwin* (1904) 1: 376).

[3] The John Wedgwoods took a cottage at Betley, Staffordshire. Their son, John Allen Wedgwood, was Vicar of Maer.

To Caroline Darwin 27 December 1835

Bay of Islands.— New Zealand.
Decemb 27th 1835.—

My dear Caroline,

My last letter was written from the Galapagos,[1] since which time I have had no opportunity of sending another. A Whaling Ship is now going direct to London & I gladly take the chance of a fine rainy Sunday evening of telling you how we are getting on.— You will see we have passed the Meridian of the Antipodes & are now on the right side of the world. For the last year, I have been wishing to return & have uttered my wishes in no gentle murmurs; But now I feel inclined to keep up one steady deep growl from morning to night.— I count & recount every stage in the journey homewards & an hour lost is reckoned of more consequence, than a week formerly. There is no more Geology, but plenty of sea-sickness; hitherto the pleasures & pains have balanced each other; of the latter there is yet an abundance, but the pleasures have all moved forwards & have reached Shrewsbury some eight months before I shall.—

If I can grumble in this style, now that I am sitting, after a very comfortable dinner of fresh pork & potatoes, quietly in my cabin, think how aimiable I must be when the Ship in a gloomy day is pitching her bows against a head Sea. Think, & pity me.— But everything is tolerable, when I recollect that this day eight months I probably shall be sitting by your fireside.— After leaving the Galapagos, that land of Craters, we enjoyed the prospect, which some people are pleased to term sublime, of the boundless ocean for five & twenty entire days. At Tahiti, we staid 10 days, & admired all the charms of this almost classical Island.— The kind simple manners of the half civilized natives are in harmony with the wild, & beautiful scenery.—

I made a little excursion of three days into the central mountains. At night we slept under a little house, made by my companions from the leaves of the wild Banana.— The woods cannot of course be compared to the forests of Brazil; but their kindred beauty was sufficient to awaken those most vivid impressions made in the early parts of this voyage.— I would not exchange the memory of the first six months, not for five times the length of anticipated pleasures.—

I hope & trust Charlotte will be enthusiastic about Tropical scenery, how I shall enjoy, hearing from her own lips, all her travels. I do not clearly understand from your last letters, whether she has actually gone to Rio, or only intended doing so.—

But I must return to Tahiti, which charming as it is, is stupid when I think about all of you.— The Captain & all on board (whose opinions are worth anything) have come to a very decided conclusion on the high merit of the Missionaries.— Ten days no doubt is a short time to observe any fact with accuracy, but I am sure we have seen that much good has been done & scarcely anyone pretends that harm has ever been effected. It was a striking thing to behold my guides in the mountain, before laying themselves down to sleep, fall on their knees & utter with apparent sincerity a prayer in their native tongue. In every respect we were delighted with Tahiti, & add ourselves as one more to the list of the admirers of the Queen of the Islands.—

Again we consumed three long weeks in crossing the Sea to New Zealand, where we shall stay about 10 days.— I am disappointed in New Zealand, both in the country & in its inhabitants. After the Tahitians, the natives, appear savages. The Missionaries have done much in improving their moral character & still more in teaching them the arts of civilization. It is something ⟨to⟩ boast of, that Europæans may here, amongst men who, so lately were the most ferocious savages probably on the face of the earth, walk with as much safety as in England. We are quite indignant with Earle's book, beside extreme injustice it shows ingratitude.—[2] Those very missionaries, who are accused of coldness, I know without doubt that they always treated him with far more civility, than his open licentiousness could have given reason to expect.— I walked to a country mission, 15 miles distant & spent as merry & pleasant an evening with these *austere* men, as ever I did in my life time.[3]

I have written thus much about the Missionaries, as I thought it would be a subject, which would interest you.— I am looking forward with more pleasure to seeing Sydney, than to any other part of the voyage.— our stay there will be very short, only a fortnight; I hope however to be able to take a ride some way into the country.— From Sydney, we proceed to King George's sound & so on as formerly planned. Be sure, not to forget to have a letter at Plymouth on or rather before the 1st of August.

Daylight is failing me, so I will wish you good bye,—how strange it is, to think, that perhaps at this very second Nancy is making a vain effort to rouse you all from your slumbers on a cold frosty morning.— How glad I shall be, when I can say, like that good old Quarter Master, who entering the Channel, on a gloomy November morning, exclaimed, "Ah here there are none of those d——d blue skys"

I forgot to mention, that by a string of extraordinary chances, the day before finally leaving the Galapagos, I received your letter of March. I am almost afraid, that at Sydney, we shall be too soon for our instructions respecting letters.

Give my most affectionate love to my Father, Erasmus Marianne & all of you. Goodbye my dear Caroline | Your's | C. Darwin

I have written to Charlotte. I also enclose a letter for Fanny will you forward it— I do not myself know the present direction.— I have also written to Sarah

DAR 223

[1] No letter by CD from the Galápagos has been found.

[2] Earle 1832, pp. 58–60.

[3] Six months later, on finding strong feelings at Cape Town against missionaries, Robert FitzRoy and CD wrote a defence of their work (see *Collected papers* 1: 19–38).

From R. W. Darwin to J. S. Henslow 28 December 1835

Shrewsbury
28 December 1835

Dear Sir

I am much obliged for the favour of your letter, for the flattering terms in which you speak of my son and for your kind attention in sending the copies of the extracts from his letters.[1]

We are all sensible how much Charles owes to you his success and the great advantage your friendship is to him. He feels and speaks of it.

I thought the voyage hazardous for his happiness but it seems to prove otherwise and it is highly gratifying to me to think he gains credit by his observation and exertion.

There is a natural good humored energy in his letters just like himself.

Dear Sir very faithfully | your obliged | R W Darwin

Professor Henslow

Royal Botanic Gardens, Kew

[1] The pamphlet, 'Extracts from letters addressed to Professor Henslow by C. Darwin, Esq', dated 1 December 1835, was privately printed for distribution to members of the Cambridge Philosophical Society (see *Collected papers* 1: 3 – 15).

From Caroline Darwin 29 December [1835]

[Shrewsbury]

My dear Charles

I received yesterday your dear affectionate letter from Lima, dated July 17th.—[1] it is delightful to think how very soon we shall have you at home again & now that it is possibly only six or 8 months before you will be England again I can with good heart say we are very glad you will finish the voyage to your own satisfaction & very thankful we all are that you have not been oblgd to give it up from ill health— you must now hear how your fame is spreading— a note came to my Father on Xmas day from Prof. Henslow speaking most kindly of you & rejoicing you would soon return "to reap the reward of your perseverance and take your position among the first Naturalist of the day" and with the note he sent my Father some copies of extracts from your letters to him printed for Private distribution the little preface to the extracts says they were printed for distribution "among the members of the Cambridge Philosophical society in consequence of the interest which has been excited by some of the Geological notices which they contain, & which were read at a meeting of the Society on the 16th of November 1835". My Father did not move from his seat till he had read every word of *your* book & he was very much gratified— he liked so much the simple clear way you gave your information Your frank unhacknied mode of writing was to him particularly agreeable— how very interesting, infinitely more so, than your former pursuit Geology must be. the accnt of some of your rides into the interior, particularly that when you found the wood of petrified trees is extremely interesting. I have not written to you since we receid your journal, but Susan has & will have told you the success it met with— I have read nothing that pleased me so much a long time—but for all this I do most heartily rejoice you can take no more such dangerous excursions— My Father has given away a few copies of the extracts to those friends who have all along felt the most constant interest about you—& first after Maer, Mr Owen, Fox, Eyton, the Leightons & Major Bayley who we saw a few days ago looking very well—

I have no home news, all being well & going on as usual. your little nephews, who are growing great tall boys, are gone back to Overton & next month I do hope a little neice may be added to the stock. Parky is quite well & recovd from the scarlet fever with out any one catching it from him— Erasmus has already given up his business & he says whatever people may please to say literary leisure is better than work— he was only 3 weeks with Robt & I think enjoyed the time taking it all in all, Mr Grant being an agreeable man & his fellow Clerk particularly so— he was obgd to give up as it was found that an Attorneys knowledge was necessary to do the business— he is talking of taking an unfurnished house & in his last letter says all but determined upon one with a nice garden & Balcony to smoke his Cigar, but he forgets to tell us in what part of London it is situated— What a happy meeting it will be for you both next Summer— Had you ever a message from Sarah Williams begging you to remember your old engagemt to her? She has just the same friendly hearty nature as ever— We were at her annual gayety this year of a Play & Ball which went off very well— Dr Butler has given up the schools & retires next summer he had intended continuing two more years but Mrs Butlers ill health has made give up— they are to live at the Hall with Mrs Floyd & her family— Mr Corfields daughter Mrs Servais (whose husband died in India) returned to Pitchford last week. Mr Corfield talked a great deal about you & we have again & again said how deeply grateful we feel to his son for his care & kindness to you when ill.— I think there is nothing going on at Maer. Elizabeth & the Langtons are staying here now. Mr Langton is a vy good kind man but so inferior to Charlotte in sense & ability—

My Father has been twice telling me not to forget his affectionate love to you & that he gives you joy with all his heart of all your laurels. My Father was reading the Athenaeum—a monthly Periodical—& he came upon the following passage— "Professor Sedgwick afterwards read extracts from letters addressed by Mr Darwin to Professor Henslow— They referred principally to the writers observations on the tertiary formation of Patagonia & Chili & on the changes of level between land & sea, which he noticed in these countries. The letters also contained an accnt of his discovery of the remains of the Megatherium over a district of 600 miles in extent to the Southwd of Buenos Ayres & a highly important description of the Geological structure of the Pass of Upsallata, in the Andes, where he discovered alternations of vast tertiary & igneous formations & the existence in the former, of veins of true granite, & of gold & other metals."[2] and now I think I must stop. I am very glad you have written to M.̲ Owen & Fox— I had last week such a kind letter from the former[3] making many enquiries about you— he expects to be a Papa next month:

Good bye my very dear Charles— all here join in kindest most affect.̲ᵃᵗᵉ love | C S Darwin

Dec.̲ 29.̲ᵗʰ

My dear Charles— I have begged for a corner to put in my love for myself & to tell you how warmly I rejoice in the comparatively near prospect of your return— it seems a great thing to be turned the corner of the last twelve month of your absence & I do hope that you will have had wandering enough to last you your whole life to come—except at least for such short absences as will not signif⟨y. I⟩ expect great pleasure from reading your journals. God bless you— Yr affect.̲ᵉ | friend C Langton

DAR 97 (ser. 2): 26—7

[1] This undoubtedly refers to the letter misdated 'July 12th' by CD (see letter to Caroline Darwin, [19] July — [12 August] 1835, n. 1).
[2] *Athenæum*, 21 November 1835, p. 8. The *Athenæum* was published weekly.
[3] Should read 'the latter', i.e., William Darwin Fox.

From C. D. Douglas[1] 5 January 1836

S.̲ⁿ Carlos [Chiloé]
5 January 1836.

Dear Sir.

I recieved your kind letter on the 27.̲ᵗʰ of July last year, & finding that you wish me to direct to you in England, I have taken mature time to investigate the questions on which you wish information, before returning my answer which I expect will arrive in England as soon as the Beaugle. I have been over the greatest part of the province since Feb.̲ʸ last, my information was verbal accuired on the spot where the observations were taken; & are the following.

There were three shocks felt at S.̲ⁿ Carlos on the 20.̲ᵗʰ Feb.̲ʸ the first at 10.̲ᵐ past 11.̲ʰ A.M. so weak as not to be generaly felt; no change was observed on the Sea, nor on the Volcanoes. It is worthy of remark that the Volcano of Osorno had been in moderate eruption, at least 48.̲ʰ before, that of Renigue[2] in moderate action much the same as it has been these 30 years past, the Corcovado in a state of inaction these twelve months past.

The second horizontal shock began at half past 11[h]. according to S[n]. Carlos town time & was said to last $7\frac{1}{2}$ minutes: it was described here & in every other place on this Island, the same as I felt it on Caucague, it was less felt in Calbuco than here, & on the Cordilleras not at all. people at work in the astilleras[3] of Mellipulli & Coyhuin, were not aware of it; & when told of it by those who felt it on the beach, they recolected that they were not able to strike fair with the ax for some time, some spoilt a board &c. by cutting too deep, while triming it down. The effect on the Sea in this harbour was instantaneous, not being quite low water, it fell ten inches during the shock & imediatly after it, began to flow with violence till two P.M. when it ebbed quicker than it had flowed; at half past two, it was low water: it then flowed with less violence than before till half past seven, having then attained four feet higer than common springs; empty water casks were washed over the punta Arena, to the westward: & it is afirmed by some that the tides were not quite regular the next day. I inclose you M[r]. Garrao's written acc[t].[4] it contains less information than I expected. The last shock began at a quarter past seven, P.M. was more perpendicular & lasted five seconds, & perhaps helped to raise the tide so high. Calbucanos on the beach near the entrance of the River Coyhuin, felt the middle shock, & directly after it passed, they saw the Sea advancing over the extensive flats, forming three waves & roaring as it aproached; it flowed three miles up the river & still'd the current on the first rapid; it then ebbed so strong that no boat could stem the current, till it left the flats dry their whole distance, which is nearly four miles; it flowed again in the evening, so high as to still the current on the fourth rapid which is the end of boat navigation. It is not known how high it rose above that, for being night, no rafts were plying. Note, there are twelve rapids between the beach & the road that leads up to the astillero. On the extensive shoals of Cheyhuau, off the Island Caylin, the ripples & swell were unusually violent all the afternoon. The effect on the tides were no where observed except in Caucague in S[n]. Carlos, & Coyhuin, although it must have been general all over the archipielago.

Great numbers of the poor inhabitants of this province, depend on the beach for their dayly food: these were all picking up shell fish when the shock happned, they were again on the beach in the afternoon, without observing any irregularity in the tides. What information can be expected from such people?

The Volcanos were as suddenly afected by the middle Shock as the Sea, that of Osorno threw up a thick colum of dark blue smoke during the shock, and directly that pass'd, a large crater was seen forming on the S.S.E. side of the Mountain, it boiled over melted lava & threw up burning stones to some height, but the smoke falling down, soon hid the Mountain in obscurity. And when seen again a few days after it showed very little smoke by day, but both craters continued to show a clear steady flame nightly, up to the date of my information, Sep[r]. 20th.

The action of Renigue was similar to that of Osorno, two curling pillars of white smoke had been seen all the morning, during the shock numerous small chimneys seemed to be smoking in the great Crater, lava & burning stones were thrown from a small crater on the S.W. side of the mountain, just above the verge of snow.

The thundering Corcovado, showed not the least Sign of activity, nor was it heard after the haze had hid the Cordilleras. So much from information: what follows is from Notes of my own observations, written in clear spaces of my poket work book little expecting they would be of any use. I shall remark here, that at the time, I had no idea of the earthquaqe having been felt beyond this province, also, that when I wrote you that hurried letter from Dalcague,[5] some twenty people were in the room, disputing loudly, different topics of private interest; so, as I do not recollect its contents, I shall continue my remarks the same as if it had not been written.

Caucague Friday 20th. Feby. watch set to true mean time by sun rise this morning. At 11^h 33^m A.M. felt an earthquake, motion horizontal & slow, similar to that of a ship at Sea going before a high regular swell, with three to five shocks in a minute somewhat stronger than the continued general motion, direction from N.E. to S.W. Forest trees nearly touched the ground in these directions, but none fell in our vicinity; pocket compas placed level on the ground, N. point set to lubbers point, remarked that it vibrated two points to westward, & only half a point to E. ward during the violent Shocks, & stood at N. when the motion was less violent, the Sea in the channel very smooth; 11^h 34^m the ripple caused by the ebb tide in midchannel disapeared (to be calculated at leisure whether low water or not) 11^h 37^m 10^s a shock more violent than any of the preceeding, a small wave advancing to each shore, & directly after the ripple in midchannel showed it had turned flood. Compas vibrated the same as in weaker shocks. 11^h 40^m 40^s another violent shock, two waves smaller than the preceeding. Compas vibrated as before the motion from this time, became gradually less distinct 11^h 40^m 45^s motion ceas'd entirely. Several bystanders imagined they felt it two minutes longer. After waiting a few minutes to be sure the earthquaqe had pass'd & a strong wind springing up at N.E. which set a swell into the channel, we resumed our occupations. 3^h 40^m P.M. tide more than half ebb, mark left by flood on the beach, afoot higher than yesterdays tide. I consider this tide contrary to the regular order of Nature (Rem^r. to inquire whether it has been general all over this Archipielago) 4^h 10^m boat in the offing met the flood tide strong. Wind fresh from N.N.E. & hazy weather, Cordilleras not seen. At 9.50. P.M. arrived at Tenaun, low water just turned flood. Feby 21. arrived at Delcague Mr Robt Barr thought the direction of the earthquake, from N.W. to S.E. which opinion was afterwards confirmed by observers in S^n. Carlos Castro & Quinched,[6] but he did not remark the efect on the Tides.

Quinched Feb.ʸ 27. Saw the volcanoes for the first time since the earthquake, Renigue threw up four colums of white smoke, the small crater formed outside the Mountain had gone out. The Corcovado was silent but the snow was melted round the N.W. crater. Osorno is not seen from here. On the Seven peaked Mountain South from the corcovado were three large black patches among the Snow, which had all the appearance of volcanic craters. I did not observe these spots when to the Southward in the Boats. Quinched Feb.ʸ 28. three colums of smoke proceed from Renigue the whole top of this Mountain which appears from the water like Table land, seen at this elevation seems to be the rim of a great Volcanic Crater; opning in a gully to N.W. & the smoke proceeds from small sugar loaf shaped hills situated within this, & their tops are seen over it at sunrise; during the night five small red flames are seen in a line, low at N.W. & highest at S.E. they are equidistant, show a steady light & appear like the street of a village illuminated. Quinched 1ˢᵗ. March Renigue has shown gradually les activity since the 28 ult.º & to day I can only distinguish one small colum of smoke I am told that it is never entirely inactive.

Frangui 16ᵗʰ. March had a fine view of the Corcovado this morning, from the N.E. point of this Island called Guechupicun, the curved ridge on its E. side is hid by this view, & the mountain appears a well shaped cone with two large craters, one open to N.W. & the other to S.W. & only seperated by a large rock that swells out N. & S. as it rises, & forms like a crown to the Mountain. The Snow appears to cover ¹/₅ of its perpendicular hieght.

March 26ᵗʰ. at 8ʰ. 13ᵐ. P.M. boat passing between the Island Lemuy & the village Chonchi, felt a smart perpendicular shock of an earthquake, which lasted ten seconds as near as I could guess, for before I could light a cigar & by its fire see my watch, the shock had passed. Five red fires seen in Renigue during the night. Corcovado silent.—

April 8. N.E. point of Island Quegui called peldén. The sun rose beautifully from behind the mountain Renigue, between two high curling colums of smoke. I saw the tops of fifteen conical hills within the great crater, & during the following night saw nine steady red fires, seven in a line & two straggling. An old Indian in whose house I lodged told me, that Renigue was formerly a very high three peaked Mountain, that two years before his marriage, its whole top fell in during an earthquake & it became a volcano, he never saw fire in it before that time, but it has been in constant action ever since

April 25ᵗʰ Boat passing between the Chengues & Quicavi. At 10ʰ P.M. Saw the volcano Osorno for the first time since Feb.ʸ 20ᵗʰ the low crater is larger than the high one, both show a steady white flame. At 11ʰ saw momentary flames issue from the side of the mountain, between the two craters: these flames first played round in a large circle, as if a new crater was about to be formed, but lastly they spread out in a straight line fliting up & down the side of the mountain between the two craters; these momentary flames all distant from each other & at least thirty in number, gave an idea of great distance between the craters, and also of this imense mountain being hollow perhaps only a thin shell which the flames can penetrate at pleasure, & some future earthquake may throw down like Renigue.

S.ⁿ Carlos Nov.ʳ 11 Strong gales from N.W. & heavy continued rain, the Tide rose to day 18 inches higher than comon springs, during the night the volcanoes of Osorno & the Corcovado were in violent action throwing up stones to great heights in the air, the thunder of both Volcanoes were distinctly heard. Renigue not seen from here. Nov.ʳ 20 arrivals from Calbuco state that the dark blue smoke thrown out of Osorno on the 11 & 12 inst. was in such quantity as to threaten darkness at midday. Dec.ʳ 1.ˢᵗ an arrival from the leeward coast, states that on the 11.ᵗʰ inst Talcahuano suffred a second ruin more dreadful than the first.

Caucague Dec.ʳ 5. at 10 P.M. emerging suddenly from a patch of wood land, through which my road lay, my attention was arrested by the grandest Volcanic spectacle I ever saw, the S.S.E. side of the Mountain Osorno had fallen in, uniting the two craters, & appeared like an imense river of fire, from the top rose an immense colum of dark blue smoke, ashes & lava, which the strong S.W. wind bent in an arch to N.E. & fell behind the mountain, a dense black cloud stood high aboe it, & discharged forked lightning towards it, three to seven flashes in a minute. At 10.ʰ 15.ᵐ a long square colum of burning matter was thrown very high from the Mountain top, & very large flash of lightning from the cloud struck it & stopt its ascent, it formed itself into a round globe & burst, scattring its fragments in every direction, some of which were followed & overtaken by lightning from the cloud & burst as before: Other smaller masses were thrown up succesively, many of which were struck by lightning & burst like the first. I appeared to be a tragic representation of Miltons battle of the Angels as descibed in Paradise Lost. At 10.ʰ 35.ᵐ an envious vapour propelled by the wind advanced from the Southward & hid the magnificent spectacle from my view, I waited till half past twelve expecting the screen would be raised as in the Theatre, but I was mistaken, for the vapour kept thickning into a dense black cloud, which remained stationary all night & next day, notwithstanding the wind continued fresh at S.W.

Chacaó Dec.ʳ 19. Volcano Osorno in violent action, the dark blue smoke which it iructated, setteled down on the gulf & appeared for several days like a new range of cordilleras, suffrining very little change in form or situation.

S.ⁿ Carlos 23 Dec.ʳ which was signalised by the strongest gale of wind that has been felt this year. it began at 8 P.M. at N.N.E. gradually veering to N.NW. & increasing till 12. & continued in lulls & violent squalls till 2 AM. of the 24.ᵗʰ when its violence abated & by five it was moderate; the rain was unusually heavy during the whole night. At half past one A.M. an alarm was given by some timid person, that the sea was coming in & had destroyed several houses on the beach: the panic became general in a few minutes, & great confusion ensued; the old & young of both sexes leapt from their warm beds & were soon drenched in the midnight rain, many were thrown down by the violence of the wind & rolled in the mud or bruised against the stones, & obliged to take refuge in the first house they could reach, where they waited in dreadful anxiety. I forced my way down the street with some difficulty & found, that the waters had only risen 14 inches higher than extreeme springs.

Before collecting these notes in form of a letter, I had intended to call your attention to several observations & querries in geology, Astronomy, Hydraulics &c, but considering the size of the packet, these remarks would swell my letter into, with the probable expence of postage, I have left them out: more willingly considering that all my observations & many others I have no idea of, must have occured to your own observation, & what are inexplicable querries to me, may be as plain as the alphabet to a man of your extensive knowledge.

I shall only trouble you with one more observation, diging in the E. cliff of the Island Caucague, I saw protubrences in the hardned sand, which had the appearance of the shells that you found bedded in rock: after many fruitles atempts, I succeded in cutting one out intire, it had the appearance of a thick clam shell, half wore by long beating on the beach, but so soft as only to be handled with the greatest care, breaking it with care it had some signs of concoidal fracture, the substance was like dried marl, & was ground into an impalpable powder between the fingers, the cliff where found was at least 200 feet above the present level of the sea. I will not determine whether this is a shell or not, till I find more perfect samples in other places. I intend to write to Cap.ᵗ Fitzroy during the insuing winter, when I shall have completed my observations on Indian & Spanish population. | Your humble & obedient Servant. | Cha.ˢ D. Douglas

DAR 39.1: 5—6

CD ANNOTATIONS
2.1   There . . . past. 2.6] *scored*
3.4   people . . . quicker 3.10] *scored*
5.1   The Volcanos . . . snow. 6.5] *scored*
6.4   just above . . . snow. 6.5] 'X' *in margin*
7.4   I shall . . . private interest; 7.7] *scored, ink*
8.3   of a ship . . . disapeared 8.10] *scored*
8.7   two . . . disapeared 8.10] *heavily scored*
8.14   another . . . distinct 8.16] *scored*
8.24   At 9.50. P.M. . . . Tides. 8.28] *heavily scored*
9.1   Quinched . . . inactive. 9.16] *scored*
9.3   The Corcovado . . . crater. 9.4] 'o' *in margin*
10.4   the other . . . hieght. 10.6] *scored, ink*
11.3   as near . . . silent.— 11.5] *scored*
12.3   tops . . . Mountain 12.6] *scored*

14.1 S^{n}. Carlos . . . Osorno 14.6] *scored*
14.6 on the 11 . . . first. 14.9] *triple scored, ink*
15.1 Caucague . . . lava, 15.5] *scored*
at end of letter: 'Compare. West Is^{d} of Scotland | coast of England | Volcan of *[*centrl*]* France'[7]

[1] For CD's extensive use of the information in this letter, see his 'On the connexion of certain volcanic phenomena in South America', *Collected papers* 1: 53 – 86.
[2] In his paper CD uses the Indian name 'Minchinmadom' for this volcano.
[3] Dockyards.
[4] The letter, in Spanish, signed *[*Humberto*]* Garrao, answers briefly and rather vaguely Douglas's questions about the level of the tides on 20 and 23 February 1835 (DAR 39.1: 6a).
[5] See letter from Charles D. Douglas, 24 February 1835.
[6] CD in *'Beagle' diary*, p. 253, calls this island 'Quinchao'. Philip Parker King (in *Narrative* 1: 271) refers to it as 'Achao, or Quinchao'.
[7] In his paper CD makes similar geographical references to give an idea of the extent of the volcanic action (*Collected papers* 1: 59).

To Philip Parker King [21 January 1836]

[Bathurst, New South Wales][1]

My dear Sir

I arrived here yesterday evening, certainly alive, but half roasted with the intense heat.— If my horses do not fail, I shall reach Dunheved[2] on Sunday evening & if you are at home, shall have much pleasure in staying with you the ensuing day.— I have seen nothing remarkable in the Geology or indeed I may add in anything else: It appears me, very singular, how very uniform the character of the scenery remains, in so many miles of country. At Mr Walker's Farm I staid one day, & went out Kangaroo hunting, but had not the good fortune even to see one. In the evening however, we went with a gun in pursuit of the Platypi & actually killed one.— I consider it a great feat, to be in at the death of so wonderful an animal.— I shall take advantage of your note of introduction to Mr Hughes & sleep there tomorrow night: if I should hear of anything remarkable in rocks of the neighbouring mountains I might be delayed there one day, in which case I should not reach Dunheved till Monday evening.—

Believe me, Dear Sir | Very sincerely Yours. | Charles Darwin.

Postmark: Bathurst NSW
Bathurst District Historical Society

[1] Bathurst is located on the banks of the Macquarie River about 100 miles inland from Sydney. CD had arrived there on 20 January (see *'Beagle' diary*, p. 383).
[2] King had retired to Dunheved, Penrith, N.S.W.

To Susan Darwin 28 January 1836

Sydney.

January 28th— 1836

My dear Susan

The day after tomorrow we shall sail from this place; but before I give any account of our proceedings, I will make an end with Business.— Will you tell my Father that I have drawn a bill for 100£, of which Fifty went to pay this present & last year's mess money. The remaining fifty is for current expenses; or rather I grieve to say it was for such expences: for all is nearly gone.— This is a most villainously dear place; & I stood in need of many articles. You will have received my letter some time ago, from New Zealand. Here we arrived on the 12th of this month.— On entering the harbor we were astounded with all the appearances of the outskirts of a great city:—numerous Windmills—Forts— large stone white houses, superb Villas &c &c.— On coming to an Anchor I was full of eager expectation; but a damp was soon thrown over the whole scene by the news there was not a single letter for the Beagle.— None of you at home, can imagine what a grief this is. There is no help for it: We did not formerly expect to have arrived here so soon, & so farewell letters.— The same fate will follow us to the C. of Good Hope; & probably when we reach England, I shall not have received a letter dated within the last 18 months. And now that I have told my pitiable story, I feel much inclined to sit down & have a good cry.

Two days after arriving here I started on a ride to Bathurst, a place about 130 in the interior, & the waters of which flow in to the vast unknown interior.— My object was partly for Geology, but chiefly to get an idea of the state of the colony, & see the country. Large towns, all over the world are nearly similar, & it is only by such excursions that the characteristic features can be perceived. This is really a wonderful Colony; ancient Rome, in her Imperial grandeur, would not have been ashamed of such an offspring. When my Grandfather wrote the lines of "Hope's visit to Sydney Cove" on Mr Wedgwood's medallion he prophecyed most truly.[1] Can a better proof of the extraordinary prosperity of this country be conceived, than the fact that 7/8th of an acre of land in the town sold by auction for 12000£ sterling? There are men now living, who came out as convicts (& one of whom has since been flogged at the Cart's tail round the town) who are said to possess without doubt an income from 12 to 15000 pounds per annum.— Yet with all this, I do not think this Colony ever can be like N. America: it never can be be an agricultural country. The climate is so dry & the soil light, that the aspect even of the better parts is very miserable. The scenery is singular from its uniformity.—every where open Forest land; the trees have all the same character of growth & their foliage is of one tint.— It is an admirable country to grow rich in; turn Sheep-herd & I believe with common care, you must grow wealthy: Formerly I had entertained Utopian ideas concerning it; but the state of society of the lower classes, from their convict origin, is so disgusting, that this & t⟨he⟩ sterile monotonous character of the scenery, hav⟨e⟩ driven Utopia & Australia into opposite sides of the World.—

In my return from my ride I staid a night with Capt King, who lives about 30 miles from Sydney.— With him, I called on some of his relations, a family of Mac Arthurs, who live in a beautiful very large country house. When we called I suppose there were twenty people sitting down to luncheon; There was such a bevy of pretty lady like Australian girls, & so deliciously English-like the whole party looked, that one might have fancied oneself actually in England. From Sydney we go to Hobart Town from thence to King George Sound & then adie⟨u⟩ to Australia. From Hobart town being superadded to the list of places I think we shall not reach England before September: But, thank God the Captain is as home sick as I am, & I trust he will rather grow worse than better.[2] He is busy in getting his account of the voyage in a forward state for publication. From those parts, which I have seen of it, I think it will be well written, but to my taste is rather defecient in energy or vividness of description. I have been for the last 12 months on very Cordial terms with him.— He is an extra ordinary, but noble character, unfortunately however affected with strong peculiarities of temper. Of this, no man is more aware than himself, as he shows by his attempts to conquer them. I often doubt what will be his end, under many circumstances I am sure, it would be a brilliant one, under others I fear a very unhappy one.

From K. George Sound to Isle of France, C. of Good Hope, St. Helena, Ascencion & omitting the C. Verd's on account of the unhealthy season, to the Azores & then England.— To this last stage I hourly look forward with more & more intense delight; I try to drive into my stupid head Maxims of patience & common sense, but that head is too full of affection for all of you to allow such dull personages to enter. My best love to my Father.— God bless you all. My dearest old Granny | Your most affectionate brother | Charles Darwin.

Tell my Father I really am afraid I shall be obliged to draw a small bill at Hobart. I know my Father will say that a hint from me on such subject is worthy of as much attention, as if it was foretold by a sacred revelation. But I do not feel in truth oracular on the subject. I have been extra⟨vag⟩ant & bought two water-color sketches, one of the S. Cruz river & & another in T. del Fuego; 3 guineas each, from Martens, who is established as an Artist at this place.[3] I would not have bought them if I could have guessed how expensive my ride to Bathurst turned out.

DAR 223

[1] The prophecy of the bright future for the new colony in Erasmus Darwin's poem (E. Darwin 1791, Canto 2: 315) was inspired by a Wedgwood medallion modelled from clay brought from Sydney soon after the colony was founded (see *Voyage*, p. 134).

[2] In a report to Captain Beaufort dated only two days earlier (26 January 1836), Robert FitzRoy commented: 'My messmate Mr. Darwin is so much the worse for a long voyage that I am most anxious to hasten as much as possible. Others are ailing and much require that rest which can only be obtained at home.' (F. Darwin 1912, p. 548). Francis Darwin observes that the most interesting point about this report 'is Captain FitzRoy's statement about the poor state of Darwin's health. I was quite unprepared for such a statement, and it seems probable that it was the beginning of the general breakdown in health which began so soon after his return to England.' (*ibid*). But CD's letters of this date make no mention of ill health and his twelve-day journey into the interior immediately before does not suggest any serious illness. One morning during the journey he did not feel well and 'thought it more prudent not to set out' ('*Beagle*' *diary*, p. 385), but he soon recovered. It is likely that FitzRoy was referring to CD's recurrent seasickness, as, a few days later (on 3 February 1836), he wrote: 'My messmate Mr. Darwin is *now* pretty well; but he is a martyr to confinement and sea-sickness when under way' (F. Darwin 1912, p. 548).

[3] The watercolours are listed in Keynes 1979 as No. 193 'Banks of Santa Cruz River' (owned by Mrs R. G. Barnet) and No. 150 '*Beagle* in Beagle Channel' (owned by George Pember Darwin). For excellent reproductions of Conrad Martens's watercolours of the voyage see Stanbury 1977 and Keynes 1979.

To J. S. Henslow [28–9] January 1836

Sydney—
January— 1836

My dear Henslow,

This is the last opportunity of communicating with you, before that joyful day when I shall reach Cambridge.— I have very little to say: But I must write if it was only to express my joy that the last year is concluded & that the present one, in which the Beagle will return, is gliding onwards.— We have all been disappointed here in not finding even a single letter; we are indeed rather before our expected time, otherwise I dare say I should have seen your handwriting.— I must feed upon the future & it is beyond bounds delightful to feel the certainty that within eight months I shall be residing once again most quietly, in Cambridge. Certainly I never was intended for a traveller; my thoughts are always rambling over past or future scenes; I cannot enjoy the present happiness, for anticipating the future; which is about as foolish as the dog who dropt the real bone for its' shadow.—

You see, we are now arrived at Australia: the new Continent really is a wonderful place. Ancient Rome might have boasted of such a Colony; it deserves to rank high amongst the 100 Wonders of the world, as showing the Giant force of the parent country. I travelled to Bathurst, a place, 130 miles in the interior, & thus saw a little of the country.— The system of communication is carried on in an admirable style; the roads are excellent, & on the Macadam principle; to form them vast masses of rock have been cut away. The following facts, I think, very forcibly show how rapid & extraordinary is the increase of wealth.— A fraction (I believe 7/8th) of an acre of land in Sydney, fetched by Auction twelve thousand pounds; the increase of public revenue during the last year has been 68,000£.— It is well known, that there are men, who came out convicts, who now possess an yearly income of 15,000£. Is not this all wonderful? But yet, I do not think this country can ever rise to be a second North America. The sterile aspect of the land, at once proclaims that Agriculture will never succeed.— Wool, Wool—is repeated & must ever be the cry from one end of the country to the other.— The scenery, from the extraordinary uniformity of its character, is very peculiar. Every where, trees of the same class & appearance are thinly scattered, with their upright trunks, over arid downs. The greatest change is that in some places the fire has been more recent & the stumps are black, whilst in others, their natural color is nearly regained.— On the whole I do not like new South Wales: it is without doubt an admirable place to accumulate pounds & shillings; but Heaven forfend that ever I should live, where every other man is sure to be somewhere between a petty rogue & bloodthirsty villain.—

In a short time we sail for Hobart town, then to K: Georges Sound, Isle of France, C. of Good Hope &c &c &c England.—

I last wrote to you from Lima, since which time I have done disgracefully little in Nat: History; or rather I should say since the Galapagos Islands, where I worked hard.— Amongst other things, I collected every plant, which I could see in flower, & as it was the flowering season I hope my collection may be of some interest to you.— I shall be very curious to know whether the Flora belongs to America, or is peculiar.[1] I paid also much attention to the Birds, which I suspect are very curious.—[2] The Geology to me personally was very instructive & amusing; Craters of all sizes & forms, were studded about in every direction; some were s⟨uch⟩ tiny ones, that they might be called quite Specim⟨en⟩ Craters.— There were however a few facts of interest, with respect of layers of Mud or Volcanic Sandstone, which must have flowed liked streams of Lava. Likewise respecting some grand fields of Trachytic Lava.— The Trachyte contained large Crystals of glassy fractured Feldspar & the streams were naked, bare & the surface rough, as if they had flowed a week before.— I was glad to examine a kind of Lava, which I believe in recent days has not in Europe been erupted.— In our passage across the Pacifick, we only touched at Tahiti & New Zealand: at neither of these places, or at sea had I much opportunity of working.— Tahiti is a most charming spot.— Every thing, which former Navigators have written is true: "A new Cytheræa has risen from the ocean".[3] Delicious scenery, climate, manners of the people, are all in harmony. It is moreover admirable to behold what the Missionaries both here & at New Zealand have effected.— I firmly believe they are good men working for the sake of a good cause. I much suspect that those who have abused or sneered at the Missionaries, have generally been such, as were not very anxious to find the Natives moral & intelligent beings.— During the remainder of our voyage, we shall only visit places generally acknowledged as civilized & nearly all under the British Flag. There will be a poor field for Nat: History & without it, I have lately discovered that the pleasure of seeing new places is as nothing. I must return to my old resource & think of the future, but that I may not become more prosy I will say Farewell, till the day arrives, when I shall see my Master in Natural History & can tell him, how grateful I feel for his kindness & friendship.

Believe me, Dear Henslow | Ever yours Most Faithfully | Chas. Darwin

Royal Botanic Gardens, Kew

[1] Henslow arranged the collection of CD's Galápagos plants, now in the Cambridge University Botany School Herbarium. Only a small part of this collection was described by Henslow; the remaining Galápagos plants were described by Joseph Dalton Hooker in 1846 (see letter from J. S. Henslow, 31 August 1833, n. 3).

[2] The Galápagos birds are famous for having aroused CD's first doubts about the permanence of species. For a discussion of their importance in CD's 'conversion' see Sulloway 1982b, and for a detailed analysis of the role of 'Darwin's finches' in particular see Sulloway 1982a. The entire *Beagle* collection of finches is now in the British Museum (Natural History) ornithological department at Tring. The specimens have been accurately identified and described for the first time in Sulloway 1982c.

[3] 'Rich crowned Cytherea', Homer, *Hymns*, 5: 1. The original, now called Cerigo or Kithira, is an island off the coast of the Peloponnesus.

From Catherine Darwin 29 January 1836

Shrewsbury.
29th | January | 1836.

My dearest Charles,

Your last letter was from Lima, in which you mentioned having drawn for 50 instead of 30£;— by some strange mistake it was dated *May 3[d]*, but we know you must have meant to write September 3[d], as the letter before, also from Lima, was dated July 20th.—[1] It is most delightful to think that we shall really have you home next Autumn;—we long to see you again, and to have the happiness of having you safely at home again, as much as you can wish to see us again. It gives universal joy to all your friends to have such a near time to look forward to; and I trust and hope that you may really keep to that time, as Capt Fitzroy seems to be as much in earnest as the rest of the Ship to be as expeditious as possible.— Papa was settling the other evening, what Bedroom you would have, when you come; and I shall much enjoy *turning out* of your room, to give it up to its dear old owner.— M[r] Owen sent us a very agreeable nice letter from you to read, dated also July 20th, from Lima; he was very much pleased with it, and talked a great deal about you, when I was lately at Woodhouse.— You will find very few changes for the worse, I hope, when you return, except poor Fanny Wedgwood's death; a great many marriages, and a whole host of children will have arisen, but, (with that one exception) nothing else of the melancholy kind.— Papa is very well, and walks surprisingly about the Town again. The Carriage drops him in the town, and then he walks a great deal about it, and the carriage goes to pick him up again.— I hoped to have been able to tell you of the birth of a Niece, as Marianne is expecting every day to be confined; and Caroline is gone to Overton to be with her; but as it has not yet happened, I am afraid I must send this off, before I can tell you if it is a little Girl, or that great misfortune, a 5th Master Parker.—

We have had a week's visit from Parky in his Christmas Holidays;—he is such a great, big, fine, spirited boy of 10 years old;—enjoying everything to the utmost, especially going on the Box of his Grand Papa's Carriage with Mark;— which was his great pleasure, hoping to be trusted with the Reins for a few minutes.— He rode over here on his Poney, and is a capital Horseman.

William Fox has a little daughter,[2] at last, I am happy to tell you:—he wrote to tell us of Mrs William Fox's safe confinement, the beginning of this month;—they are spending the winter at Ryde, in the Isle of Wight.— I dare say you will like his Lady better than we did; for one reason, she was in a bad state of health when we saw her, which I dare say accounted for much of her crossness.— We have sent William Fox one of the little books, with the Extracts from your Letters; every body is much pleased, with them, who has seen them; Professor Henslow sent half a dozen to Dr Butler; we sent one also to Tom Eyton;—he says he has written to you at Sydney, so you will have his opinion from himself of them.—

Harry and Jessie are staying with us now, and are both occupied in reading your Journal, which they find **very** entertaining, especially your descriptions of Tropical Scenery. My dearest Charley, we were so glad to hear in your last letter, that you had grown twice as fat, from living quietly in the S⟨hip.⟩ do tak⟨e⟩ care of yourself in those hot C⟨li⟩mates with the Sun, vertical over your head; pray take care of yourself, my dear old Charley, & come back to us in good health.— This new year will be the happiest that has been for a long time, if it brings you back to us:—I can hardly fancy you with us again, but it will be most delightful, when we really have you.—

Erasmus is busied with housekeeping cares at present; he has taken a house, belonging to an attorney in Marlborough S^t—Argyll Place—and is very busy now furnishing it, and setting himself up;—he finds the expence and difficulty of furniture and household cares, much more than he expected.—

I do not know whether you were acquainted with Mr Panting, of this town; we knew him very well, and saw a great deal of him at one time;[3] he died this week of a very malignant fever; Papa was only alarmed about him about three days, before his death;—the fever was so rapid & violent;—and it was a very melancholy death in many respects.— We have been reading the Wreck of the Challenger,[4] it is an interesting short account of their escape— there is no mention of who it is written by; how very generous, and how brave, & active Capt Fitzroy must have been in going to their assistance,—how much one admires him for it.

Goodbye, dearest Charley. Papa's & Susan's love,—and beg me to tell you how they long for September. | Yrs ever | Catherine | Darwin.

DAR 97 (ser. 2): 28—9

[1] CD's letter was misdated July (see letter to Caroline Darwin, [19] July — [12 August] 1835, n. 1).
[2] Eliza Ann Fox.
[3] See letter from Sarah Williams, 21 October 1833 in which she refers to Thomas Panting's interest in Susan Darwin.
[4] *A diary of the wreck of His Majesty's ship Challenger* . . . 1836.

From Susan Darwin   12 February 1836

Shrewsbury
February 12th *1836*

My dear Charles,

It is always my fate to write to wish you joy upon your Birthday; but thank goodness this is the last I can do so, into foreign parts.— You are today 27—and I hope all the rest of your life you may spend very happily amongst us.— we often speculate whether you will have had sufficient travelling to serve you for life: & I think the Yes's Yes's generally carry it.— Our two last letters have been full of your fame & glory so I will say nothing about it except that I am as much delighted as any of them at your present success & future prospect of distinguishing yourself in Geology— I was reading the other day part of your early Journal just before you left Plymouth when you made yourself an outline of how you meant to pass your time,[1] & amongst your studies I was surprised to find no mention of Geology but this must have been an oversight, because just after your tour with Professor Sedgwick you must have been hot on the subject.—

You will I know be very glad to hear that Marianne had a little Girl on the 31st of Jan[ry] which has made her excessively happy— she is a very nice little thing they say, & is to be called *Mary Susan*.— Parky is grown a famous fine lad and gets on capitally at school being now at the head of his class— he says the only thing he remembers of Uncle Charles, is his teaching him to say "*Oh berry.*" Tom Eyton is also turned Author, & in return for yr Geological work, which we sent him; has sent us, two Numbers of a continuation of Bewicks Birds; which he means to proceed with. they have tale pieces also, but these are inferior to Bewicks.— I had no idea he could draw but the birds are most of them very well done.[2]

Erasmus is no longer in noisy lodgings in Regent St. but has taken a roomy house in Marlborough St & has set up housekeeping in earnest, how much he will be cheated remains to be proved,—however it is very nice for his friends as now we can visit him comfortably whenever we like, & you too will probably find it very useful to have good quarters always ready in town. Sarah Williams this winter when I was staying at Eaton, desired me to remind you with her love of your old engagement to come & take your first Dinner with her in Belgrave St: but tho' I repeat this as I promised, *we* shall not allow you to go anywhere before you come to us at Shrewsbury: my dear Charley how delightful it will be to see you. I very often dream about you, besides thinking of you continually, and pity you in these horrid stormy equinoctial gales. I hope you won't go exploring too boldly in New Holland as I think land dangers are more to be dreaded than sea one's, and I am sorry to hear you are thinking of visiting Capt King if he lives far off Sidney.—

Poor Eras's troubles about housekeeping are quite pathetic, first of all he was excessively puzzled how to get an old woman into his house without furniture, or furniture into his house without an old woman; & then after he had accomplished that difficulty he had to carry 13 Cab loads of Glass bottles &c from his Lab.— Still each letter is full of the *"eternal botherations"* as he calls them & now we are busy breaking in a horse to send up to London for him; as his beautiful grey horse is dead who I suppose lived & died since your days.—

Papa & we often cogitate over the fire what you will do when you return, as I fear there are but small hopes of your still going into the Church:— I think you must turn Professor at Cambridge & marry a Miss Jenner if there is one to be had.—

Old Nancy is very much pleased at your kind mention of her in yr Letters, & bids me tell you a day never passes without her thinking of y^r return, which I fully believe poor old soul.

I wish we could hope to see y^r hand writing before May. I long for some account of the Galapageos for I never read any thing about those frying hot Islands. God bless you my dear Charley may I see yr dear face in 6 months time & Ever believe me | Y^r very affec^{te} | Granny | S E Darwin

DAR 97(ser. 2): 30—1

[1] See *'Beagle' diary* entry for 13 December 1831, p. 14.
[2] Eyton 1836.

To Catherine Darwin 14 February 1836

Hobart Town. Van Diemen's Land
February 14th.— 1836.

My dear Catherine

I am determined to begin a letter to you, although I am sadly puzzled, as you may see by the length of the date, to know what to write about. I presume you will have received, some few days before this, my letter from Sydney.— We arrived here after a six days passage, & have now been here 10.— Tomorrow morning we Sail for King George Sound.—1800 miles of most Stormy Sea.— Heaven protect & fortify my poor Stomach.— All on board like this place better than Sydney— the uncultivated parts here have the same aspect as there; but from the climate being damper, the Gardens, full of luxuriant vegetables & fine corn fields, delightfully resemble England.—

To a person not particularly attached to any particular kind, (such as literary, scientific &c,) of society, & bringing out his family, it is a most admirable place of emigration. With care & a very small capital, he is sure soon to gain a competence, & may, if he likes, die Wealthy.— No doubt in New S. Wales, a man will sooner be possessed of an income of thousands per annum. But I do not think he would be a gainer in comfort.— There is a better class of Society. Here, there are no Convicts driving in their carriages, & revelling in Wealth.— Really the system of emigration is excellent for poor Gentleman.— You would be astonished to know what pleasant society there is here. I dined yesterday at the Attorneys General, where, amongst a small party of his most intimate friends he got up an excellent concert of first rate Italian Music. The house large, beautifully furnished; dinner most elegant with *respectable*! (although of course all Convicts) Servants.— A Short time before, they gave a fancy Ball, at which 113 people were present..— At another very pleasant house, where I dined, they told me, at their last dancing party, 96 was the number.— Is not this astonishing in so remote a part of the world?—

It is necessary to leave England, & see distant Colonies, of various nations, to know what wonderful people the English are.— It is rather an interesting feature in our Voyage, seeing so many of the distant English Colonies.— Falklands Island, (the lowest in the scale), 3 parts of Australia: Is^d of France, the Cape.—St Helena, & Ascencion— My reason tells me, I ought to enjoy all this; but I confess I never see a Merchant vessel start for England, without a most dangerous inclination to bolt.— It is a most true & grievous fact, that the last four months appear to me ⟨as⟩ long, as the two previous years, at which ra⟨te⟩ I have yet to remain out four years longer.— There never was a Ship, so full of home-sick heroes, as the Beagle.— We ought all to be ashamed of ourselves: What is five years, compared to the Soldier's & Civilian's, whom I most heartily pity, life in India?— If a person is obliged to leave friends & country, he had much better come out to these countries & turn farmer. He will not then return home, on half pay, & with a pallid face.— Several of our Officers are seriously considering the all important subject, which sounds from one end of the Colony to the other, of Wool.

My Father will be glad to hear, that my prophetic warning in my last letter, has turned out false.— Not making any expedition, I have not required any money.—

Give my love to my dear Father I often think of his kindness to me in allowing me to come this voyage—indeed, in what part of my life can I think otherwise.—

Good bye my dear Katty. I have nothing worth writing about, as you may see,— Thank Heaven, it is an unquestioned fact that months weeks & days will pass away, although they may travel like most arrant Sluggards. If we all live, we shall meet in Autumn. | Your affectionate Brother | Charles Darwin.—

DAR 223

To W. D. Fox 15 February 1836

Hobart Town.— Van Diemen's land—
February 15th 1836

My dear Fox

On our arrival at Sydney, we all on board the Beagle were bitterly disappointed in not finding a single letter.— For the first occasion, the Beagle was before her appointed time; & hence the cause of our grief. I daresay otherwise I should have received a letter from you.— It is now a long time since I heard any news.— the last was, from home, of Mrs Fox's ill health.— You have had much to endure in your own bodily suffering & if to this is superadded unhappiness from another & deeper source you will indeed have a heavy burthen to support.— But I sincerely hope, my dear Fox, I am croaking about calamities, which have passed away & that you are as happy as you ought to be from the bright picture you drew in your last letter.—

I presume you heard from me at Lima; since that period time has hung rather heavily on hand.— Not that the present is absolutely disagreeable, but I cannot refrain from thinking of the future.— I am sure, if a long voyage may have some injurious tendencies to a person's character, it has the one good one of teaching him to appreciate & dearly love his friends & relations—

Now that the object of our voyage is reduced simply to Chronometrical Measurements, a large portion of our time is spent in making passages.— This is to me, so much existence obliterated from the page of life.— I hate every wave of the ocean, with a fervor, which you, who have only seen the green waters of the shore, can never understand. It appears to me, I am not singular in this hatred.— I believe there are very few contented Sailors.— They are caught young & broken in before they have reached years of discretion. Those who are employed, sigh after the delights of the shore, & those on shore, complain they are forgotten & overlooked: All think themselves hardly used, that they are not sooner promoted, I thank my good stars I was not born a Sailor.— I will take good care no one shall shall ever persuade me again to volunteer as Philosopher (my accustomed title) even to a line of Battle Ship.— Not but what I am very glad I have come on the expedition; but only that I am still gladder it is drawing to a close.— I have had little opportunity, for some time past of doing anything in Natural History.— I draw up very imperfect sketches of the Geology of all the places, to which we pay flying visits; but they cannot be of much use.[1] Leaving

America, all connected & therefore interesting, series of observations have come to an end.— I look forward with a comical mixture of dread & satisfaction to the amount of work, which remains for me in England. I suppose my chief ⟨place⟩ of residence will at first be Cambridge & then London.— The latter, I fear, will in every respect turn out most convenient. I grieve to think of it; for a good walk in the true country is the greatest delight, which I can imagine.— I shall find the different societies of the greatest use; judging from occassional glimpses of their periodical reports &c, there appears to be a rapidly growing zeal for Nat: Hist.— F. Hope informs me, he has put my name down as a member of the Entomological Soc:— I do not know, whether you are one.— Formerly, when collecting at Cambridge, how very useful such a central Society would have been to us Beetle Capturers. The banks of the Cam, the Willow trees, Panagæus Crux Major & Badister, which was not cephalotes, all form parts of one picture in my mind. To this day, Panagæus is to me a sacred genus.— I look at the Orange Cross, as the emblem of Entomological Knighthood. At Sydney I took a fine species, & long did I look at it, as compared to any other insect.— Poor little Albert Way, I wonder, what has become of him. I wish I could think he was well.—

I do not understand where you are now residing, in the last letters from home, (which was several months ago) nothing was mentioned. Probably I shall not receive another letter, before reaching England, if it turns out so, there will be then a space of 18 months,[2] of the events of which I shall be entirely ignorant.— God grant they may not be unfortunate.— I think it will be on a September night when we shall first make the Lizard lights. On such an occassion I feel it will be quite necessary to commit some act of uncommon folly & extravagance. School boys are quite right in breaking the binding of their books at the end of the half year & likewise Man of Wars men, when they throw guineas into the sea or light their tobacco pipes with Pound notes, to testify their joy.— The time is now so short, before, I trust, we shall meet, that I feel it is almost useless to describe imperfectly, what we shall have opportunities of talking over. Visiting Australia, which one day will rise the Empress of the South, was interesting. It has given me a grand idea of the power & efficiency of the English nation. To see Colonies which in age, bear the proportion of tens of years to hundreds, so far outstepping in Civilization those of S. America, is really most astonishing.— Although full of wonder & & admiration at this Spectacle, I should be very loth to emigrate. The moral state of the lower orders is of course detestable; the society of the higher is rancorously divided by party feelings & the country itself is not to me pleasing. But with respect to money-making it is a very paradise to the Worshippers of Mammon.— It is an undisputed fact that there are Emancipists now living worth 15,000! pounds per annum.—

After touching at King Georges Sound we proceed to the Isle of France.— It will clearly be necessary to procure a small stock of sentiment on the occassion; Imagine what a fine opportunity for writing love letters.— Oh that I had a sweet Virginia to send an inspired Epistle to.—[3] A person not in love will have no right to wander amongst the glowing bewitching scenes.— I am writing most glorious nonsense, so that I had better wish you good night, although at this present moment you probably are just awaking on a cold frosty morning. We are on opposite sides of the World & everything is topsy turvy: but I thank Heaven, my memory is in its right place & I can bring close to me, the faces of many of my friends.

Farewell, my dear Fox, till that day arrives, when we shall really once again shake hands. God bless you.— | Your affectionate friend | Chas. Darwin.

Christ's College Library, Cambridge (Fox 48)

[1] For an account of CD's geological observations while in Tasmania, see Banks 1971.
[2] In the event, CD received letters at the Cape of Good Hope, after thirteen months without one, and at Bahia, Brazil (see letters to Catherine Darwin, 3 June [1836] and to Susan Darwin, 4 August [1836]).
[3] A reference to Jacques Henri Bernardin de Saint-Pierre's *Paul et Virginie* (1787).

From Caroline Darwin 28 March 1836

[Shrewsbury]
1836 | March 28th.

My Dear Charles,

Susan & Cath have written since we had your last letter from Lima Augt 4th. but for fear of accidents I just mention it— I can hardly believe how near home you will be when you get this & that these *long long* five years are really just ended— many an hour we now spend in talking about you & pitying you for the confined feeling our little spot of a house & garden will seem to you when you come among us again—it will be as if you were awakened from a dream when you find every thing & every body just as you left them except all 6 of us being pretty considerably aged—Pincher & Nina inclusive— We saw one of your friends a few days ago who made many enquiries—Mr. Herbert. he goes the Oxford circuit now & met Cath at the Ball & called here the following day. his hair is quite grey I dont whether that is the effect of these 5 years or whether it was so when you last saw him, but it gives him a venerable appearance. Sir E Alderson was on the circuit this time & we had a half hour visit from him which would have been very merry & pleasant if he had not brought his brother Judge with him—but it was rather aweful to have to entertain two live Judges at once— My Father did not see them, he has had a slight attack of Gout in his hand which obliged him to keep his bed for a few days, he is however well again now & no bad effects remaining but his hand being swelled— he still talks of going in May a little tour to see Edinburgh again with Catherine & Harry—

We expect Erasmus down next week, that is to say if these cold winds will but go & he takes Susan back with him to introduce her to his new House. You will find the comfort & pleasure of his having a House to receive you instead of Lodgings as before—& this house has so many spare rooms that you will have space for some of your numerous boxes &^{c} I will give his direction as you may chance miss our future letters *43 G^t Marlborough St* it is very doubtful if you will receive this but I will send a few lines to tell you all is well— Susan is staying at Woodhouse— Fanny Biddulph was here yesterday with her little girl, it is oblgd still to lie down & gets rather worse than better— poor little thing it has such pretty engaging ways & its Mama is so fond of it— She (I mean Mrs Biddulph, not the child) is looking herself wretchedly ill & thin, you would hardly know her she is so altered from the robust girl she was formerly— there have been letters a few days ago from Francis & Arthur Owen from India both well & in good spirits— Mr Owen dined & slept here last week he quite charmed me by his look when we were talking about you, so full of affection & feeling. we were saying how very glad we were that you & your Captain had continued such good friends through your long voyage & Mr Owen said "Yes, but who could quarrel with Charles?" the words arn't much but his look said a great deal— dear old fellow I dont think you will find any of your friends love you less from their separation from you & it will not last much longer now— Catherine has been staying at Maer & having a very agreeable visit there which is really surprising considering how much the House is altered—poor A^t Bessys memory very much gone & confined altogether to the sofa & Charlotte the main prop in former times of the conversation gone— I am afraid you will not think her improved by her marriage she is certainly graver & more silent

As I have little expectation of your getting this letter I will finish with all our best loves to you dearest Charles My Fathers in particular— Good bye & God bless you. August will soon be here Ever y^r affec^{te} C. S Darwin

DAR 97 (ser. 2): 32–3

To Caroline Darwin   29 April 1836

Port Lewis, Mauritius.
April 29th 1836.

My dear Caroline,

We arrived here this morning; as a Ship sails for England tomorrow, I will not let escape the opportunity of writing. But as I am both tired & stupid, my letter will be equally dull. I wrote from Sydney & Hobart town, after leaving the latter place, we proceeded to King Georges Sound. I did not feel much affection for any part of Australia; & certainly, nothing could be better adapted, than our last visit, to put the finishing stroke to such feelings.—

We then proceeded to the Keeling Is^{ds}.—[1] These are low lagoon Is^{ds} about 500 miles from the coast of Sumatra.— I am very glad we called there, as it has been our only opportunity of seeing one of those wonderful productions of the Coral polypi.— The subject of Coral formation has for the last half year, been a point of particular interest to me. I hope to be able to put some of the facts in a more simple & connected point of view, than that in which they have hitherto been considered. The idea of a lagoon Island, 30 miles in diameter being based on a submarine crater of equal dimensions,[2] has alway appeared to me a monstrous hypothesis.

From the Keeling Id we came direct to this place. All which we have yet seen is very pleasing. The scenery cannot boast of the charms of Tahiti & still less of the grand luxuriance of Brazil; but yet it is a complete & very beautiful picture. But, there is no country which has now any attractions for us, without it is seen right astern, & the more distant & indistinct the better. We are all utterly home sick; I feel sure there is a wide difference between leaving one's home to reside for five years in some foreign country, & in wandering for the same time. There is nothing, which I so much long for, as to see any spot & any object, which I have seen before & can say I will see again.— Our heads are giddy, with such a constant whirl. The Capt, continues to push along with a slack rein & an armed heel.— thank Heaven not an hour has lately been lost, or will again be lost.

It is probable, if we escape the heavy gales off the Cape, we may reach England 8 weeks after you receive this letter. Our course beyond the Cape & St Helena is not certain; I think it will end in touching at Bahia on the coast of Brazil. With what different sensations I shall now view that splendid scene, from formerly. Then I thought an hour of such existence would have been cheaply purchased with an year of ordinary life, but now one glimpse of my dear home, would be better than the united kingdoms, of all the glorious Tropics. Whilst we are at sea, & the weather is fine, my time passes smoothly, because I am very busy. My occupation consists in rearranging old geological notes: the rearranging generally consists in totally rewriting them. I am just now beginning to discover the difficulty of expressing one's ideas on paper. As long as it consists solely of description it is pretty easy; but where reasoning comes into play, to make a proper connection, a clearness & a moderate fluency, is to me, as I have said, a difficulty of which I had no idea.—

I am in high spirits about my geology.—& even aspire to the hope that, my observations will be considered of some utility by real geologists. I see very clearly, it will be necessary to live in London for a year, by which time with hard work, the greater part, I trust, of my materials will be exhausted. Will you ask Erasmus to put down my name to the Whyndam or any other club; if, afterwards, it should be advisable not to enter it, there is no harm done. The Captain has a cousin in the Whyndam, whom he thinks, will be able to get me in.— Tell Erasmus to turn in his mind, for some lodgings with good big rooms in some vulgar part of London.— Now that I am planning about England, I really believe, she is not at so hopeless a distance.— Will you tell my Father I have drawn a bill of 30£.— The Captain is daily becoming a happier man, he now looks forward with cheerfulness to the work which is before him. He, like myself, is busy all day in writing, but instead of geology, it is the account of the Voyage. I sometimes fear his "Book" will be rather diffuse, but in most other respects it certainly will be good: his style is very simple & excellent. He has proposed to me, to join him in publishing the account, that is, for him to have the disposal & arranging of my journal & to mingle it with his own. Of course I have said I am perfectly willing, if he wants materials; or thinks the chit-chat details of my journal are any ways worth publishing. He has read over the part, I have on board, & likes it.—

I shall be anxious to hear your opinions, for it is a most dangerous task, in these days, to publish accounts of parts of the world, which have so frequently been visited. It is a rare piece of good fortune for me, that of the many errant (in ships) Naturalists, there have been few or rather no geologists. I shall enter the field unopposed.— I assure you I look forward with no little anxiety to the time when Henslow, putting on a grave face, shall decide on the merits of my notes. If he shakes his head in a disapproving manner: I shall then know that I had better at once give up science, for science will have given up me.— For I have worked with every grain of energy I possess.— But what a horridly egotistical letter, I am writing; I am so tired, that nothing short of the pleasant stimulus of vanity & writing about one's own dear self would have sufficed.— I have the excuse, if I write about my self, Heaven knows I think enough about all of you.—

We shall leave this Isld. in 6 days time; if there is any opportunity, I will write from the C. of Good Hope & that letter possibly may be the last you will receive, before you see me arrive, converted into an ancient, brown-colored Gentleman. The minute the Ship drops her anchor in the mud of old England, I will start for ⟨Shr⟩ewsbury.— I trust we shall find letters ⟨at⟩ the Cape; but I have many fears; the date of the last letter I received was 13 months ago: This is a grievous period, to be entirely ignorant, about all, one care's most for.— It is probable we shall arrive ⟨ear⟩ly in September; you must recollect the possibility ⟨of⟩ my not having received letters for 18 months, so retell me any thing important; if I do not come by the 14th of Septemb. write again to Plymouth post-office. So that when, I start for home—I may travel with a certain mind.

God bless you all. May you be well & happy. Forgive such a letter; I am sure, you would sooner have it, than nothing.— So once again farewell to you all.— give my most affectionate love to my Father & all | My dearest Caroline | Your affectionate brother | Chas. Darwin.

DAR 223

[1] A group of coral islands in the Indian Ocean. They are discussed at length in *Coral reefs*.

[2] This was the generally held view, shared by Charles Lyell (*Principles of geology* 2: 290 – 1). In the *Autobiography*, p. 98, CD says that his theory, that coral reefs were formed by the upward growth of coral during the gradual subsidence of the sea-bed, 'was thought out on the west coast of S. America'. No statement of the theory, written at that time, has been found. The records of CD's early thinking on the subject are discussed in Appendix V. The first exposition of the general outline of the theory is in a manuscript headed 'Coral islands', dated 1835 and probably written at sea between Tahiti and New Zealand, 3 – 21 December 1835. It forms part of CD's 'Geological notes' (DAR 41: 1 – 22). It has been transcribed and published in Stoddart 1962, with an account of the probable chronology of the development of the hypothesis. (See also Stoddart 1976.)

To Catherine Darwin 3 June [1836]

Cape of Good Hope.
June 3^{d} 1836

My dear Catherine,

We arrived here the day before yesterday; the first part of our passage from Mauritius was very favourable, and the latter as execrably bad. We encountered a heavy gale of wind, which strongly reminded us of the old days, near Cape Horn; It is a lucky thing for me, that the voyage is drawing to its close, for I positively suffer more from sea sickness, now, than three years ago.— All hands, having been disappointed in letters at Sydney & Mauritius made up their minds for a grand pile at this place.— The mountain of letters, alas, has dwindled into ⟨a⟩ small packet of about a dozen: amongs⟨t⟩ them I had the good fortune of receiving yours of Jan. *1836*!.— Nine months' letters are wandering over the wide ocean, which we shall not receive till some time after reaching England; But if you knew the glowing unspeakable delight, which I felt at being certain that my Father & all of you were well, only four months ago, you would not grudge, the labor lost in keeping up the regular series of letters.—& it has only happened by such order that I have received this last letter.—

When I wrote from Mauritius, I begged, that the Plymouth letter might contain a short abstract of the last 18 months; now it need only go back as far as January. Pray do not disappoint me this; for otherwise I shall be uncomfortable in my journey instead of enjoying, the sight of the most glorious & the most beautiful of countries. I believe I have at home, a leathern Portmanteau, great coat, & cloth leggings: if so, will you have them sent, by the 1st of September directed to "Lieut. Sulivan, to the care of Mr Elliot Royal Hotel, Devonport.—(to be kept till *H.M.S.* Beagle arrives)".— We go from hence to St Helena; between which place & England, our stages are not yet determined.—

The Beagle is now lying at Simons Bay, more than 20 miles from Cape town, where I now am— This ⟨is⟩ a pretty & singular town; it lies at the foot of an enormous wall, (the Table mountain), which reaches to the clouds, & makes a most imposing barrier.— Cape town is a great inn, on the great highway to the east; an extraordinary number of houses are occupied as boarding houses, in one of which I am now settled: the first day I got amongst a set of Nabobs, who certainly, poor fellows, all together could not have produced a Liver as good as the hero in Beppo.[1] They were heavy prosers. I was quite bewildered with Cawnpoor & so many "poors," & with rushing from Calcutta to Bombay, backwards & forwards.— in despair—I effected a most precipitate retreat; & deliver me in future from the Nabobs.—

Tomorrow morning I am going to call with Capt. F.R. on the Sir J. Herschel. I have already seen the house which he has purchased; it is six miles from the town & in a most retired charming situation. I have heard so much about his eccentric but very aimiable manners, that I have a high curiosity to see the great Man.—

The day after tomorrow, I hope to set out on a short ride of 3 or 4 days, to get a few glimpses of African landscape, or rather I should say, African deserts.— Having seen so much of that sort of country in Patagonia Chili & Peru, I feel myself to a certain degree a connoiseur in a desert, & am very anxious to see these. Every country has its peculiar character; & every country is well worth seeing. But oh the country of countries; the nice undulating green fields & shady lanes. Oh if you young ladies have been cutting down many of the trees (& I shall recollect every one), I never will forgive you.—

I am quite delighted at hearing Erasmus is turned house holder; I hope I shall be able to get lodgings at no great distance, & then London will be a very pleasant place. I often however think Cambridge would be better, I can not make myself cockney enough to give up thoughts of a quiet walks on an Autumnal morning, in the real country.—

I have been a good deal horrified by a sentence in your letter where you talk of "the little books with the extracts from your letters". I can only suppose they refer to a few geological details. But I have always written to Henslow in the same careless manner as to you; & to print what has been written without care & accuracy, is indeed playing with edge tools.[2] But as the Spaniard says, "No hay remedio".—

Farewell for the present & God bless you all.— I have a strong suspicion that my Father will hear of me again before the time of sailing, which will happen in 10 days time.— Give my love to the young Miss Parker; for I hope I have a little niece, instead of a fifth nephew. My dear Catherine | Your affectionate Brother. C. D.

N.B. I find I am forced, after all to draw a Bill of 30£ at once.—it is not that I am at all sure I shall want the money here, but if on my return from the country my funds fail, I shall not at the moment not know what to do.—

DAR 223

[1] Nabob was a term applied to persons who returned from overseas, usually India, having acquired great wealth. CD refers to Beppo, the eponymous hero of Lord Byron's satirical poem (1818).

William Mostyn Owen Sr. (Courtesy of William Mostyn-Owen.)

Erasmus Alvey Darwin. Portrait by George Richmond, 1841.
(Photograph courtesy of Hensleigh Wedgwood.)

John Stevens Henslow. Lithograph by T. H. Maguire, 1849.

(Courtesy of the Syndics of the Fitzwilliam Museum, Cambridge.)

Adam Sedgwick. After the portrait by Thomas Phillips, 1832.

(Courtesy of the Trustees of the British Museum.)

[2] Henslow had entered the following caveat in his prefatory remarks to the Cambridge Philosophical Society pamphlet of extracts: 'The opinions here expressed must be viewed in no other light than as the first thoughts which occur to a traveller respecting what he sees, before he has had time to collate his Notes, and examine his Collections, with the attention necessary for scientific accuracy.'

To *South African Christian Recorder*[1] 28 June 1836

On the whole, balancing all that we have heard, and all that we ourselves have seen concerning the missionaries in the Pacific, we are very much satisfied that they thoroughly deserve the warmest support, not only of individuals, but of the British Government.
Robt. FitzRoy
Charles Darwin

At sea, 28th June, 1836

South African Christian Recorder 2(1836): 238
Reprint in Darwin Library—CUL: 'Philosophical tracts'

[1] The original communication, published under the title 'A Letter, Containing Remarks on the Moral State of Tahiti, New Zealand, &c.' was in two parts: the first, longer section, written in the first person except for the three introductory paragraphs, and signed by FitzRoy alone, contained excerpts from CD's journal. Both parts are published in *Collected papers* 1: 19 – 38.

To J. S. Henslow 9 July 1836

St. Helena.
July 9th.—1836
My dear Henslow

I am going to ask you to do me a favor. I am very anxious to belong to the Geolog: Society. I do not know, but I suppose, it is necessary to be proposed some time before being balloted for, if such is the case, would you be good enough to take the proper preparatory steps. Professor Sedgwick very kindly offered to propose me, before leaving England: if he should happen to be in London, I daresay he would yet do so.—[1] I have very little to write about.— We have neither seen, done, or heard of anything particular, for a long time past: & indeed if, at present, the wonders of another planet could be displayed before us, I believe we should unanimously exclaim, what a consummate plague. No schoolboys ever sung the half sentimental & half jovial strain of "dulce domum" with more fervour, than we all feel inclined to do.— But the whole subject of dulce domum, & the delight of seeing one's friends is most dangerous; it must infallibly make one very prosy or very boisterous— Oh the degree to which I

long to be once again living quietly, with not one single novel object near me.—
No one can imagine it, till he has been whirled round the world, during five long
years, in a ten Gun-Brig.—

I am at present living in a small house (amongst the clouds) in the centre of
the Isl^d & within stone's throw of Napoleon's tomb. It is blowing a gale of
wind, with heavy rain, & wretchedly cold: if Napoleon's ghost haunts his dreary
place of confinement, this would be a most excellent night for such wandering
Spirits.—

If the weather chooses to permit me, I hope to see a little of the Geology, (so
often partially described) of this Isl^d— I suspect, that differently from most
Volcanic Isl^ds its structure is rather complicated. It seems strange, that this little
centre of a distinct creation should, as is asserted, bear marks of recent elevation.

The Beagle proceeds from this place to Ascencion, thence to C. Verds (What
miserable places!) to the Azores, to Plymouth & then to Home. That most
glorious of all days in my life will not however arrive till the middle of October.
Some time in that month, you will see me at Cambridge, when I must directly
come to report myself to you, as my first Lord of the Admiralty.— At the C. of
Good Hope, we all on board suffered a bitter disappointment in missing nine
months' letters, which are chasing us from one side of the globe to the other. I
daresay, amongst them there was a letter from you; it is long since I have seen
your hand writing, but I shall soon see you yourself, which is far better. As I am
your pupil, you are bound to undertake the task of criticizing & scolding me for
all the things ill done & not done at all, which I fear I shall need much; but I
hope for the best, & I am sure I have a good, if not too easy, task master.—

At the Cape, Capt Fitz Roy, & myself enjoyed a memorable piece of good
fortune in meeting Sir J. Herschel.— We dined at his house & saw him a few
times besides. He was exceedingly good natured, but his manners, at first,
appeared to me, rather awful. He is living in a very comfortable country house,
surrounded by fir & oak trees, which alone, in so open a country, give a most
charming air of seclusion & comfort. He appears to find time for every thing; he
shewed us a pretty garden full of Cape Bulbs of his own collecting; & I
afterwards understood, that every thing was the work of his own hands. What a
very nice person Lady Herschel appears to be,—in short we were quite charmed
with every thing in & about the house.— There are many pleasant people at the
Cape.— Mr Maclear,[2] the astronomer, was most kind & hospitable.— I became
also acquainted with Dr A. Smith, who had just returned from his expedition
beyond the Tropic of Capricorn.— He is a cap⟨ital⟩ person & most indefatigable observer: he has brought back an immense collection, & amongst
other things a new species of Rhinoceros.—[3] If you had heard him describe his
system of travelling & mode of defence, it would have recalled the days of
enthusiasm, which you have told me, you felt on first reading Le Vaillant.—[4]
Dr Smith shortly goes to England, he will soon return & recommence his travels

& either succeed in penetrating far into the interior, or, as he says, leave his bones in Africa.—[5]

I am very stupid, & I have nothing more to say; the wind is whistling so mournfully over the bleak hills, that I shall go to bed & dream of England.— Good night, My dear Henslow | Yours most truly obliged | & affectionately | Chas. Darwin.—

Royal Botanic Gardens, Kew

[1] See letter from Adam Sedgwick, 18 September 1831.
[2] Sir Thomas Maclear.
[3] See Andrew Smith 1838—49 for the zoology of the expedition. Smith's diary has been published in Andrew Smith 1939—40.
[4] Levaillant 1790, a popular book of travels in Africa.
[5] Andrew Smith left for England in 1837 but did not return to South Africa.

To Caroline Darwin 18 July 1836

[On board *Beagle*, bound for Ascencion]
July 18th.— 1836.—

My dear Caroline

We are at this present moment driving onwards with a most glorious tradewind towards Ascencion. I am determined to pay the debt of your most excellent correspondence; by at least writing to you all, as often as I can. I will leave this letter at Ascencion to take its chance of being forwarded. Before attempting to say anything else, I must disburthen my mind, of the bad news that our expected arrival in England, recedes, as we travel onwards. The best judges in the Ship entertain little hopes of it, till the end of October. The next three months appear infinitely tedious, & long, & I daresay the last three weeks, will be worse, as for the three closing days, they, by the same rule, ought to be intolerable. I feel inclined to write about nothing else, but to tell you over & over again, how I long to be quietly seated amongst you.— How beautiful Shropshire will look, if we can but cross the wide Atlantic before the end of October. You cannot imagine how curious I am to behold some of the old views, & to compare former with new impressions. I am determined & feel sure, that the scenery of England is ten times more beautiful than any we have seen.— What reasonable person can wish for great ill proportioned mountains, two & three miles high? No, no; give me the Brython or some such compact little hill.— And then as for your boundless plains & impenetrable forests, who would compare them with the green fields & oak woods of England?— People are pleased to talk of the ever smiling sky of the Tropics: must not this be precious nonsense? Who admires a lady's face who is always smiling? England is not one of your insipid beauties; she can cry, & frown, & smile, all by turns.— In short I am convinced it is a most ridiculous thing to go round the world, when by staying quietly, the world will go round with you.—

But I will turn back to the past, for if I look forward, I lose my wits, & talk nonsense. The Beagle staid at St Helena five days, during which time I lived in the clouds in the centre of the Is^d.— It is a curious little world within itself; the habitable part is surrounded by a broad band of black desolate rocks, as if the wide barrier of the ocean was not sufficient to guard the precious spot. From my central position, I wandered on foot nearly over the whole Island; I enjoyed these rambles, more than I have done any thing for a long time past. The structure of the Is^d is complicated & its geological history rather curious.— I have discovered a monstrous mistake, which has been handed from one book to the other, without examination. It has been said, that Sea shells are found on the surface of the land, at an elevation little short of 2000 ft. & hence that, this Is^d. though possessing an entirely unique Flora, must have been raised, within a late period, from beneath the Ocean.— These shells turn out land shells! But what is very singular, they have ceased to exist, in a living state on the Is^d.—[1]

I heard much of old General Dallas & his daughters.—[2] People speak very well of him—(as a well intentioned old goose).— He took much pains in improving the road & other public works, was most hospitable, magnificent, & popular.— The young ladies were the gayest of the gay.— Finally he was the last of the E. Indian Company's Governors, with an income more than quadruple the present.— Hence perhaps the lamentations at his departure.—

From St Helena, I wrote to Erasmus a long & a heavy letter all about myself, it was directed to the Wyndham Club.—[3] I most earnestly hope Erasmus will not be wandering on the continent about the time of the Beagle's return; I am delighted he has taken a house, as he will more probably now be a fixture.— I shall really have so much to say, that I fear I shall annihilate some of my friends.— I s⟨hall⟩ put myself under your hands; & you must undertake the task of scolding, as in years long gone past, & of civilizing me.— Oh for the time when we shall take a ride together on the Oswestry road.—

My dear Caroline I do long to see you, & all the rest of you, & my dear Father.— God bless you all— Your most affectionate | brother. Chas. Darwin.

P.S. I have kept this flap open in case of receiving any letters tomorrow when we reach Ascencion.—

[*Written in pencil on outer flap of cover:*] There is a Ship in the offing & this must go.— There are letters, but the bundle has not been opened.

DAR 223

[1] See *Journal of researches*, p. 582.

[2] General Charles Dallas. His daughter Davidona married Captain Francis Harding, a friend of the Darwin family (see letters from Caroline Darwin, 1–4 May 1833 and from Susan Darwin, 16 February 1835).

[3] This letter has not been found.

To Susan Darwin 4 August [1836]

Bahia, Brazil
August 4th.

My dear Susan

I will just write a few lines to explain the cause of this letter being dated on the coast of S. America.— Some singular disagreements in the Longitudes, made Capt. F. R. anxious to complete the circle in the Southern hemisphere, & then retrace our steps by our first line to England.— This zig-zag manner of proceeding is very grievous; it has put the finishing stroke to my feelings. I loathe, I abhor the sea, & all ships which sail on it. But I yet believe we shall reach England in the latter half of October.— At Ascension I received Catherines letter of October & yours of November; the letter at the Cape was of a later date; but letters of all sorts are inestimable treasures, & I thank you both for them.—

The desert Volcanic rocks & wild sea of Ascension, as soon as I knew there was news from home, suddenly wore a pleasing aspect; & I set to work, with a good will at my old work of Geology. You would be surprised to know, how entirely, the pleasure in arriving at a new place depends on letters.— We only staid four days at Ascension & then made a very good passage to Bahia.— I little thought ever to have put my foot on a S. American coast again.— It has been almost painful to find how much, good enthusiasm has been evaporated during the last four years. I can now walk soberly through a Brazilian forest; not but what it is exquisitely beautiful, but now, instead of seeking for splendid contrasts; I compare the stately Mango trees with the Horse Chesnuts of England. Although this zigzag has lost us at least a fortnight, in some respect I am glad of it.— I think I shall be able to carry away one vivid picture of inter-tropical scenery. We go from hence to the C. de Verds, that is if the winds or the Equatorial calms will allow us.— I have some faint hopes, that a steady foul wind might induce the Captain to proceed direct to the Azores.— For which most untoward event I heartily pray.—

Both your letters were full of good news:— Especially the expressions, which you tell me Prof: Sedgwick used about my collections.— I confess they are deeply gratifying.— I trust one part at least will turn out true, & that I shall act, as I now think.—that a man who dares to waste one hour of time, has not discovered the value of life.— Prof. Sedgwick men⟨tionin⟩g my name at all gives me hopes that he will assist me with his advice; of which in many geological questions, I stand much in need.— It is useless to tell you, from the shameful state of this scribble that I am writing against time; having been out all morning—& now there are some strangers on board to whom I must go down & talk civility.— Moreover, as this letter goes by a foreign ship, it is doubtful whether it ever will arrive.— Farewell, my very dear Susan & all of you.. Goodbye | C. Darwin—

Endorsement: 'The last Letter.—'
DAR 223

To Josiah Wedgwood II [5 October 1836]

[Shrewsbury]

My dear Uncle

The Beagle arrived at Falmouth on Sunday evening, & I reached home late last night. My head is quite confused with so much delight, but I cannot allow my sisters to tell you first, how happy I am to see all my dear friends again. I am obliged to return in three or four days to London, where the Beagle will be paid off, & then I shall pay Shrewsbury a longer visit. I am most anxious once again to see Maer, & all its inhabitants, so that in the course of two or three weeks, I hope in person to thank you, as being my first Lord of the Admiralty. I am so very happy I hardly know what I am writing.

Believe me, Your most affectionate nephew | Chas. Darwin

Remember me most kindly to Aunt Bessy & all at dear Maer.—

DAR 185

Caroline Darwin to Sarah Elizabeth (Elizabeth) Wedgwood [5 October 1836]

[Shrewsbury]
Wednesday

My dear Elizabeth

Charles is come home—so little altered in looks from what he was five years ago & not a bit changed his own dear self—he had landed at Falmouth on Sunday evening & travelled night & day till he came to Shrewsbury late last night— We heard nothing of him till this morning when he walked in just before breakfast— We have had the very happiest morning—poor Charles so full of affection & delight at seeing my Father looking so well & being with us all again— his hatred of the sea is as intense as even I can wish—arriving at its climax by a storm on the Bay of Biscay. He is looking very thin[1] —but well— he was so much pleased by finding your & Charlottes kind notes ready to receive him I shall indeed enjoy my dear Eliz going to Maer with him, how happy he will be to see you all again— When I began this letter I did not know he would feel tranquil enough to write himself, but he said he *must* be the first to tell Uncle Jos of his arrival— he feels so very grateful to Uncle Jos & you all & has been asking about every one of you— he must go to London to be able to be on the spot when his things are taken out of the Beagle so I am afraid he will certainly be gone before A^t Sarah comes to us—which I am sorry for. Now we have him really again at home I intend to begin to be glad he went this expedition & now I can allow he has gained happiness & interest for the rest of his life

Good bye dear Eliz it is pleasant to wr⟨ite to⟩ those who sympathize so entirely with us Y^r aff C S D

DAR 185

[1] A 'Weighing Account' book kept by CD's father and later by Josiah Wedgwood III records CD's height and weight as of 7 October 1836 as '5 ft, 1 13/8 in., 10 stone, 8^1/$_4$ lbs.' By December 1836 he weighed '11 stone, 12 lbs.' (Down House MS).

From William Owen Sr 5 October [1836]

My Dear Charles,

I cannot express the Pleasure your kind Note received a few hours ago has given me, & I most Sincerely congratulate you as well as all your Family upon your safe return, after so long an absence, to your *dear* home. I am sure your Father & indeed all your Family must have been quite overcome with delight, & I hope, nay I am certain they will suffer me to participate in it—but words are foolish things & as often used to flatter or deceive as to express the honest Feelings of our hearts, which I will therefore not attempt—but the long & the short, as they say in this Country, is that I must see you as soon as possible, & if I was not unfortunately engaged to meet a Man here tomorrow by appointment I should certainly be with you at breakfast— Unfortunately I have the same engagement on Friday— Perhaps you could come here on Saturday & if you can & will give us your Company as long as you can afford I will not say how happy it will make me as well as the M^{rs} but pray bring your Gun with you, for I have not forgot the amusement we used to have together, & I am anxious to see whether you are *improved by your travels* or whether I am again to be your Instructor.

We expect some Friends here on Monday to stay a few days & hope you will be able to make one of the Party, & if any of your Sisters will come with you we shall be still more pleased— I know it is in vain to ask your Father.— But you talk of riding over— If you mean for a call & one night only I will not have you—& therefore if you cannot make it convenient to come & stay a few days, give me a line by return of Post & I will try to come over on Saturday Morning just to shake you by the hand & to satisfy myself that it is the same Charles Darwin I formerly knew & valued so much.— I am no flatterer but in truth & Sincerity I do assure you that I feel the same so⟨rt⟩ of Pleasure & Joy in your return that I think I should do to see one of my poor Boys again from India—of whom by the bye we have just received very good accounts— | Yours most Sincerely | W^m. Owen

Woodhouse
Wed^y. Night— | Octr. 5th.

Postmark: Shrewsbury OC 7 1836
DAR 204.10

To Robert FitzRoy 6 October [1836]

Thursday morning. Oct 6th

My dear FitzRoy,

I arrived here yesterday morning at Breakfast time, & thank God, found all my dear good sisters & father quite well— My father appears more cheerful and very little older than when I left My sisters assure me I do not look the least different, & I am able to return the compliment— Indeed all England appears changed, excepting the good old Town of Shrewsbury & its inhabitants— which for all I can see to the contrary may go on as they now are to Doomsday— I wish with all my heart, I was writing to you, amongst your friends instead of at that horrid Plymouth. But the day will soon come and you will be as happy as I now am— I do assure you I am a very great man at home— the five years voyage has certainly raised me a hundred per cent. I fear such greatness must experience a fall—

I am thoroughly ashamed of myself, in what a dead and half alive state, I spent the few last days on board, my only excuse is, that certainly I was not quite well.— The first day in the mail tired me but as I drew nearer to Shrewsbury everything looked more beautiful & cheerful— In passing Gloucestershire & Worcestershire I wished much for you to admire the fields woods & orchards.— The stupid people on the coach did not seem to think the fields one bit greener than usual but I am sure, we should have thoroughly agreed, that the wide world does not contain so happy a prospect as the rich cultivated land of England—

I hope you will not forget to send me a note telling me how you go on.— I do indeed hope all your vexations and trouble with respect to our voyage which we now *know* **has** an end, have come to a close.— If you do not receive much satisfaction for all the mental and bodily energy, you have expended in His Majesty's Service, you will be most hardly treated— I put my radical sisters into an uproar at some of the *prudent* (if they were not *honest* whigs, I *would* say shabby), proceedings of our Government. By the way I must tell you for the honor & glory of the family, that my father has a large engraving of King George the IV. put up in his sitting Room. But I am no renegade, and by the time we meet, my politics will be as firmly fixed and as wisely founded as ever they were—

I thought when I began this letter I would convince you what a steady & sober frame of mind I was in. But I find I am writing most precious nonsense. Two or three of our labourers yesterday immediately set to work, and got most excessively drunk in honour of the arrival of Master Charles.— Who then shall gainsay if Master Charles himself chooses to make himself a fool.

Good bye— God bless you— I hope you are as happy, but much wiser than your most sincere but unworthy Philos. | Chas. Darwin.

Shrewsbury.

Copy
DAR 144

To J. S. Henslow 6 October [1836]

> Shrewsbury.
> Octob.ʳ 6ᵗʰ.—

My dear Henslow

I am sure you will congratulate me on the delight of once again being home. The Beagle arrived at Falmouth on Sunday evening, & I reached Shrewsbury yesterday morning.— I am exceedingly anxious to see you, and as it will be necessary in four or five days to return to London to get my goods & chattels out of the Beagle, it appears to me my best plan to pass through Cambridge. I want your advice on many points, indeed I am in the clouds & neither know what to do, or where to go. My chief puzzle is about the geological specimens, who will have the charity to help me in describing their mineralogical nature?— Will you be kind enough to write to me one line by *return of post* saying whether you are now at Cambridge.— I am doubtful, till I hear from Capt. F. R. whether I shall not be obliged to start before the answer can arrive, but pray try the chance.— My dear Henslow, I do long to see you; you have been the kindest friend to me, that ever Man possessed.— I can write no more for I am giddy with joy & confusion.— Farewell for the present.— | Yours most truly obliged | Chas. Darwin

Thursday Morning.—

Postmark: Shrewsbury OC 6 1836
Royal Botanic Gardens, Kew

From Robert FitzRoy [19 –]20 October [1836]

Dearest Philos

What you will say to me for not having written before I know not—but really I have *not* been idle or forgetful.

I trusted to Fuller[1] for all immediately necessary information and I will now try to give you the rest.

Captain Beaufort was out of town when my letters & papers reached London (from Falmouth) and the Chart duster put them *away* in a *corner* (excepting *one* private one) to await Capt. B's *return*!!! Those papers related to the Chronometric results &c &c—upon which the necessity for our going to Woolwich was to be founded—[2]

Orders had been sent to Plymouth for the Beagle to pay off there—but Ld Amelius Beauclerk had civility & sense enough to stay proceedings and approve of my going to London to see the Lords & Masters myself— I boarded Sir John Barrow, and then made a stalking horse of *him* while attacking the *others*.

All was satisfactorily settled in a very short time—and they acceded civilly to my proposal of calling at *Portsmouth*. I was delighted to see that the Valp^o cargo of charts had not only *arrived* but that they were mostly *Engraved*—or in the Engraver's hands—and on a *large* scale. They have given much satisfaction at the Hyd^l Office.[3]

I have promised to give them a short paper for the Geog^l Society—[4] a slight *sketch* of our voyage— I will do what I can—according to time and ask you to add, and correct.

Rice Trevor[5] & Alex^r Wood crossed me on the road—they in one mail—I in another—but I was soon down again & with them at Devonport— Fuller told me you looked very well and had on a *good hat*!

Who the deuce was *my cousin* in a *broad brimmed* hat?

I was delighted by your letter.— The account of your family—& the joy tipsy style of the whole letter were *very* pleasing. Indeed Charles Darwin I have *also* been *very* happy—even at that horrid place Plymouth—for that horrid place contains a *treasure* to *me* which even *you* were ignorant of!! Now guess—and think

& guess again. Believe it, or not,—the news is *true*—I am going to be *married*!!!!!!! to Mary OBrien—[6]

Now you may know that I had decided on this step, long *very long* ago.— All is settled & we shall be married in December. Rice Trevor Alex! Wood & Talbot like her *much*. Pray call on my sister in Stratton Street—she longs to see you,—& ask to see the *children*.—

Money matters are better than *you* think. Your's most sincerely | Rob! FitzRoy

Beagle. Portsmouth 20ᵗʰ Oct. (night)— arrived this morning—shall sail tomorrow morning for Woolwich Sights are *taken*.

DAR 204.10

[1] Fuller was FitzRoy's steward in the *Beagle* ('*Beagle*' *diary*, p. 81).

[2] The plan for the chronometric readings called for the final reading to be made at Greenwich. This was done on 28 October, after which the *Beagle* proceeded to Woolwich to be paid off (*Narrative* 2: 638).

[3] FitzRoy's outstanding ability as a surveyor was cited in a report to the House of Commons, 10 February 1848. Captain Beaufort is quoted as saying, 'From the Equator to Cape Horn, and from thence round to the river Plata on the eastern side of America, all that is immediately wanted has been already achieved by the splendid survey of Captain Robert FitzRoy' (*DNB*, 'Robert FitzRoy').

[4] FitzRoy 1836.

[5] George Rice Trevor, married to FitzRoy's sister Frances.

[6] Mary Henrietta O'Brien, daughter of a country gentleman and Major General (Mellersh 1968, p. 172).

To Caroline Darwin 24 October [1836]

[43 Great Marlborough Street]
October 24ᵗʰ—

My dear Caroline,

I send you these few lines, just to prove I have not quite forgotten you in all the bustle of London— These four days since leaving Cambridge have entirely been passed in calling on various naturalist people, but my plans only become more perplexed instead of any clearer— I think the Beagle will reach Woolwich in one or two days— I have heard from the Captain; & what do you think, he is going to be *married*!! to a Mary O'Brien, with whom formerly he had a long flirtation. I cannot tell you particulars, but will keep the letter to gratify your little propensity of curiosity— They are to be married in December.— I believe the Lady is a good one, but it is a most inconvenient time to marry— I mean to send all my property from the Beagle to Cambridge, where I have determined to reside for at least some months.— But I fear London must be the final quarters.— I find the geologists all willing to render all assistance & exceedingly kind,—but the Zoologists seem to think a number of undescribed creatures rather a nuisance.—

I do not think mortal man ever talked more than I have done during the last three days. I am quite tired & long to be living quietly with dear old Henslow, at Cambridge— This skipping about is part of the five years voyage.— The voyage is not yet come to an end.— I do not see any prospect of being able to leave London under a fortnight. I think your plan of my going direct to Maer a very good one— But I will write again in about a weeks time, when I shall see a little more of the way— I hope & trust Charlotte will be there,—Maer will not be itself without her.—

The letter from E. Holland contained no message.— It was a mere letter of civility, indeed I may say of patronage.— If you make out this letter you will be both very clever, & patient.— I wish I was with you at this moment at Maer.— I foresee my next visit to Shrewsbury will be one constant struggle of much can be done in a certain number of days.— Pray write a gossiping untidy letter to me.— I find Erasmus most agreeable, & his house most comfortable.

Your most affec^ate | Chas. Darwin.

Copy
DAR 154

From Emma Wedgwood to Frances Mackintosh Wedgwood
[24 October 1836]

[They are impatient for CD's arrival. Emma is reading Head (1826) 'to get up a little knowledge for him'. CD has nearly settled in favour of living in Cambridge.]

ALS 4pp
Keele University Library (Wedgwood/Mosley 233)

To Charles Whitley 24 October [1836]

London
October 24th:—

My dear Whitley

I was very glad to receive your letter, which I did the day previous to my leaving Shrewsbury. It was a piece of high virtue on your part, being able to write a single line on such an occasion.

I most heartily congratulate you. Long may you live in your now perfect state. We poor bachelors are only half men,—creeping like caterpillars through the world, without fulfilling our destination. Herbert did not call on the Saturday, as you told me he would. As it was impossible for me, from previous arrangements to have staid till that day, I left a letter for him, but which he has never received.— I am anxious to communicate with him, and have not an idea how to manage it. I am at present waiting in London till the Beagle arrives at Woolwich to be paid off. After packing up my all my goods, I shall pay Shrewsbury a visit of a few days longer than my former one, and then return to Cambridge, where I suppose I shall take up my residence for several months.— Ultimately I shall migrate to London.— I spent a few days with Henslow, before coming up here, & received a most cordial welcome from the few friends, still living in Cambridge. Amongst them were old Heaviside & Matthews, & when I return, we are to have a quiet little dinner. How I wish you were there to join us—but you, you lucky man, are enjoying far more cosy meals.— So much for my plans.. When I shall be enabled to visit you at Durham, Heaven only knows. I am at present at an utter loss to know how to begin, the arrangement of specimens and observations collected during the five long years All I know is, that I must work far harder, than poor shoulders have ever been accustomed to do.—

If you can muster time, do write to me and direct Shrewsbury (if in a fortnight's time, or 43 Great Marlborough St.[1] if earlier). In your last letter you told me, nothing beyond the main grand event, pray write in more detail. Do not forget to give me old Herbert's direction. I wish I could see some early prospect of shaking you by the hand, and talking over old Cambridge days. Do you remember our very long walks, how pleasant they were, such nice quiet tranquil days never can return.

Farewell my dear Whitley long may you enjoy your present happiness. Once again Farewell | Yours ever most truly | Charles Darwin

I fear my direction to you is not a very correct one, tell me if it is right[2]

Postmark: EX 2⟨5 OC⟩ 25 1836
British Library MS Dept. Add. 41567 fols. 248−9

[1] The address of E. A. Darwin, with whom CD stayed while in London.
[2] The letter is addressed to 'Charles Whitley Esqre Durham University Durham'.

From Emma Wedgwood to F. Mackintosh Wedgwood [28 October 1836]

[CD will not get to Maer that week. The Langtons are leaving and will meet him at Shrewsbury.]

ALS 4pp
Keele University Library (Wedgwood/Mosley 233)

To J. S. Henslow [30 — 1 October 1836]

43. Great Marlborough St
Sunday.

My dear Henslow

I have delayed writing, as I daily expected the Beagle would arrive, and I should be better able to tell you how my prospects go on.— I spent yesterday on board at Greenwich, & brought back with me the Galapagos plants; they do not appear numerous, but are I hope in tolerable preservation.— Tomorrow I will procure a box & will send them to Cambridge.— I will keep this letter till I do so.— I called on your brother, but he was not at home, I left a card asking him to send anything he might happen to have, to my brothers, where I am now staying.— I have not made much progress with the great men, I find, as you told me, that they are all overwhelmed, with their own business. Mr Lyell, has entered in the *most* goodnatured manner, & almost without being asked, into all my plans. He tells me, however, the same story, namely that I must do all myself.— Mr Owen seems anxious to dissect some of the animals in spirits;[1] & besides these two I have scarcely met anyone who seems to wish to possess any of my specimens.— I must except Dr Grant, who is willing to examine some of the corallines.— I see it is quite unreasonable to hope for a minute, that any man will undertake the examination of an whole order.— It is clear the collectors so much outnumber the real naturalists, that the latter have no time to spare.— I do not even find that the collections care for receiving the unnamed specimens.— The Zoological Museum[2] is nearly full & upward of a thousand specimens remain unmounted. I daresay the British Museum, would receive them, but I cannot feel, from all I hear, any great respect even for the present state of that establishment. Your plan will be not only the best, but the only one, namely to come down to Cambridge, arrange & group together the different families & then wait till people, who are already working in different branches may want specimens.— But it appears to me, to do this, it will be almost necessary to reside in London.— As far as I can yet see, my best plan will be to spend several months in Cambridge, & then, when by your assistance, I know on what grounds I stand, to emigrate to London, where I can complete my geology, & try *to push on the Zoology.*— I assure you I grieve to find how many things make me see the necessity of living for some time in this dirty odious London.— For even in Geology, I suspect, much assistance & communication will be necessary in this quarter: for instance in fossil bones of which none, excepting the fragments of Megatherium, have been looked at. And I clearly see without my presence never would be.— Fossil shells.—charts, maps, communication with FitzRoy. Mr Lyell says also, that the power of taking any odd specimen, to the different societies, where many Naturalists are met together is a very essential point.— However true this may be, I am very sure the assistance I shall get in Cambridge will be infinitely more, than I ever should receive in London.— Prof. Sedgwick very kindly hunted out my quarters, & I breakfasted with him.— I am sorry to find he will leave Cambridge so soon.—[3] From the delay in the Beagle's arrival, I do not know whether I shall be able to come down before the end of the month, for I have yet to visit Shrewsbury.— If you have opportunities, talk a little with him, on those points you said his advice would be most valuable; namely the form of publication. I think from what I hear, that a volume, would be less troublesome & pleasanter, than detached papers.— Also about fossil shells. Is Sowerby[4] a good man? I understand his assistance can be purchased.—

Mr Clift says he will ask Prof. Buckland to look at the bones;[5] I should think he would rather like it, as Mr Clift says some belong to forms which he himself does not at all know.— I am anxious to know, whether Prof. Sedgwick recommends any particular nomenclature for the rocks.— I have often thought of your really most kind offer of talking with M^rs. Henslow about my taking up my quarters with you. Few things could give me more happiness, and at the same time do me more real good. But I fear I should me much in the way, & I have been thinking of another plan which would be better for the work; that is to take lodgings with two sitting rooms & a bedroom, (which I daresay could be procured), in one of which my servant could work & it would at the same time serve for a warehouse for the skins &c &c &c.— Perhaps my servant might live in the house.— In College I should only have one room, which, although a large one, would be inconvenient; and as it may turn out more advisable not to remain a whole year, it is a great expence to buy furniture crockery &c &c &c for any shorter period. If I subsequently live in London, I shall follow my brothers plan take the whole of an unfurnished house, excepting the shop or office, then furnish two rooms, & keep the others for lumber.— Such a house can be got for less than 100£ per annum. I believe Mr Ash clearly understood it was quite a chance, whether I intended coming to reside the whole yea⟨r.⟩ If you should happen to meet him (but otherwise not) just mention that I shall probably not reside for such a time, & he will then understand, that I should not trouble him about rooms.—

I find this letter, which is a most unmerciful long one all about myself, extending over so much paper, that I will send it with the plants, & write another just to forewarn you of the Box. Perhaps also I shall be able by that time to announce the heavy cases with geolog. & other specimen⟨s.⟩ Would it not be the best plan, *if it can be so managed*, to leave the heavy geolog: boxes at the warehouse, so that when I come down & the room is ready to take them direct there? I have forgotten to mention one bad bit of news, namely that Cuming was at the Galapagos.—[6] Did he collect plants, I doubt it, because the far greater part of the plants only live near the summit of the mountains, some miles from the coast? I shall grieve, if you lose your tiny botanical feast,— I only wish I had known the Botanists cared so much for specimens & the Zoologists so little; the proportional number of specimens in the two branches should have had a very different appearance. I am out of patience with the Zoologists, not because they are overworked, but for their mean quarrelsome spirit. I went the other evening to the Zoological Soc. where the speakers were snarling at each other, in a manner anything but like that of gentlemen.

Thank Heavens, as long as I remain in Cambridge there will not be any danger of falling into any such contemptible quarrels, whilst in London I do not see how it is to be avoided. Of the Naturalists; F. Hope is out of London, Westwood,[7] I have not seen; so about my insects I know nothing.— I have seen Mʳ Yarrel, twice, but he is so evidently so oppressed with business, that it is too selfish to plague him with my concerns.— He has asked me to dine with the Linnæan on Tuesday; and on Wednesday I dine with Geolog: so that I shall ⟨   ⟩ all the great men.— Mʳ Bell I ⟨   ⟩ is so much occupied that there is no chance of his wishing for specimens of reptiles.—[8] I have forgotten to mention Mʳ Lonsdale,[9] who gave me a most cordial reception, & with whom I had much most interesting conversation.— If I was not much more inclined for geology, than the other branches of Natural History, I am sure Mʳ Lyell's & Lonsdale kindness ought to fix me.— You cannot conceive anything more thoroughily goodnatured, than the heart & soul manner, in which he put himself in my place & thought what would be best to do.— At first he was all for London versus Cambridge, but at last I made him confess that for some time at least the latter would be for me much the best. There is not another soul, whom I could ask, excepting yourself, to wade through & criticize some of those papers which I have left with you.— Mʳ Lyell owned that second to London, there was no place in ⟨Engl⟩and, so good for a naturalist as Cambridge. ⟨U⟩pon my word I am ashamed of writing so many foolish details; no young lady ever described her first ball with more, particularity.— With respect to the Mathematical instruments, I told the Captain, that I was sure you would allow him to leave the boxes for a week or two longer till he was established & knew where to have them directed to.

Monday evening I have determined to send the plants by wagon with the bird skins, for reason which you will know by the letter which announces the boxes.

Yours ever most sincerely | Chas Darwin—

Again I have been compelled to change my plans.— I send the plants, per coach, & this letter with them.— I will also write one line by the post in case of any accident. On Thursday the four boxes will arrive at Camb⟨rid⟩ge by Marsh's Wagon.—

Royal Botanic Gardens, Kew

[1] CD had met Richard Owen at dinner with the Lyells the day before (R. S. Owen 1894, 1: 102).

[2] The Museum of the Zoological Society, then at 33 Bruton Street. The collection was dispersed some years later (*LL* 1: 273).

[3] Adam Sedgwick's duties as a Prebendary of Norwich Cathedral required that he be in residence in December and January of each year (Clark and Hughes 1890, 1: 433–4).

[4] George Brettingham Sowerby described CD's specimens of fossil shells in appendixes to *Volcanic islands* and *South America*.

[5] William Buckland added a 'Supplementary note' about CD's fossil bones to the second edition of his *Bridgewater treatise* (Buckland 1837, p. 603).

[6] Hugh Cuming, famous for his collections of shells, visited the Galápagos Islands in 1829. He collected only a few plants. They were described, together with CD's specimens, in Hooker 1846a and 1846b.

[7] John Obadiah Westwood.

[8] Thomas Bell did agree to describe CD's specimens for *Zoology* and, after much delay, *Reptiles* was published (1842–3).

[9] William Lonsdale described some of CD's coral specimens for *Volcanic islands*.

To J. S. Henslow [1 November 1836]

Tuesday Evening

My dear Henslow

You will have probably received, before this letter reaches you, the box with the Galapagos plants, which also contains a longer letter. The box starts tomorrow morning by the Fly Coach.— Four boxes were also sent by Marsh's Wagon to day.— they will reach Cambridge on Thursday morning.— The two very heavy & largest ones contain geological specimens; if it could be so arranged would it not be better to leave them in the warehouse, in such case pray see that they are so placed as not to be moved, & that they are housed dry.—

The smallest box contains spirit cases; the square one bird skins & insects. These perhaps had better be carried away into safety.—

I have just returned from dining at the Linnæan club, & attending a meeting to hear a couple of intensely stupid papers.—[1] I became acquainted with Mr Bell, who, to my surprise, expressed a good deal of interest about my crustaceæ & reptiles & seems willing to work at them.— I also heard, that Mr Broderip would be glad to look over the S. American shells.—[2] So that things flourish well with me.— You have made me known amongst the botanists; but I felt very foolish, when Mr Don[3] remarked on the beautiful appearance of some plant with an astoundingly long name, & asked me about its habitation. Some one else seem quite surprised that I knew nothing about a carex from ⟨I⟩ do not know where. I w⟨as⟩ at last forced to plead most intire innocence, & that I knew no more about the plants, which I had collected, than the Man in the Moon.— Pray write me a line to say how the Cambridge world is going on; & whether my boxes arrive safely.— Tell me whether you are disappointed with the Galapagos plants. I have some fears.—

Pray remember me most kindly to Mrs Henslow.— I long to be again in Cambridge.— Believe me | Yours ever most truly obliged | Chas Darwin

43. Great Marlborough.—

Postmark: EX 2 NO 2 1836
Royal Botanic Gardens, Kew

[1] 'Read [were] Observations on the *Esula Major Germanica* of Lobel by Edward Forster Esq. V.P.L.S., F.R.S. and a paper on the Wooraby or Curary Poison of the Indo-Americans, by Robert H. Shomburgk . . .' (Minute book, Linnean Society, London, General Meeting of 1 November 1836). CD is listed as a visitor introduced by Mr Yarrell.
[2] William John Broderip.
[3] David Don.

To W. D. Fox 6 November [1836]

43 Great Marlborough St^t
Novr 6th

My dear Fox,

I have taken a shamefully long time in answering your letter.— But the busiest time of the whole voyage has been tranquillity itself to this last month. After paying Henslow a short but very pleasant visit, I came up to town to wait for the Beagle's arrival. At last I have removed all my property from on board, & sent the specimens of Natural History to Cambridge, so that I am now a free man.— My London visit has been quite idle, as far as Nat: History goes but has been passed in most exciting dissipation amongst the Dons in science.— All my affairs indeed are most prosperous; I find there are plenty, who will undertake the description of whole tribes of animals, of which I know nothing. So that about this day month, I hope to set to work: tooth and nail at the Geology, which I shall publish by itself.—

I do not know how I shall be able to manage a visit to the Isle of Wight.— My plans are on Thursday to go to Shrewsbury by the way of Maer & Overton staying at each place about a couple of days.— For Shrewsbury I hope to spare ten days, but I foresee, it will be a most uncomfortable visit,—a mere struggle how many people can be visited, to whom I am bound, & indeed, (if time, just at present was not so precious) am most anxious to see.—

I shall return to London for two or three days & then go and reside in Cambridge for some months, & try if I can settle my jolted brains into some kind of order.— Have you during the winter any sort of business, or anything which will tempt you to pay Cambridge a visit.— It is too long a journey in the depth of winter, and I fear for your health, to take for any one motive; though you have offered to come up all the way to London, merely for a flying visit. When I leave Shrewsbury, I will write again, as by that time, I shall know more certainly about the future. At present I am exceedingly doubtful between the merits of London and Cambridge.—

It is quite ridiculous, what an immensely long period it appears to me, since landing at Falmouth. The fact is I have talked and laughed enough for years instead of weeks, so my memory is quite confounded with the noise.—

I am delighted to hear you are turned Geologist. when I pay the Isle of Wight, a visit, which I am determined shall somehow come to pass, you will be capital cicerone to the famous line of dislocation.— I really suppose there are few parts of the world more interesting to a Geologist than your island.—

Amongst the great scientific men, no one has been nearly so friendly & kind, as Lyell.— I have seen him several times, & feel inclined to like him much. You cannot imagine how good-naturedly he entered into all my plans. I speak now only of the London men, for Henslow was just like his former self, & therefore a most cordial and affectionate friend.— When you pay London a visit I shall be very proud to take you to the Geological Society, for be it known, I was proposed to be a F.G.S. last Tuesday. It is, however a great pity that these & the other letters, especially F.R.S. are so very expensive.—

I do not scruple to ask you to write to me in a week's time in Shrewsbury, for you are a good letter writer, & if people will have such good characters they must pay the penalty.

I wish I had the pleasure of being acquainted with Mrs Fox, that I might beg to be remembered to her, but I hope the time is not far distant, when I shall be so.—

Pray tell me what relation I am your daughter, for I have been puzzling my brains & can come to no conclusion. When formerly at Cambridge you explained to me my various relations, you little thought the next time you would be genealogist for your own daughter.

Good Bye | Dear Fox Yours C. D.

Postmark: D No 7 1⟨83⟩6
Christ's College Library, Cambridge (Fox 49)

To Charles Wilkes[1] [7 November 1836]

43. Gr^t Marlborough St

Dear Sir

I am going into the country for a few weeks on Thursday.— As I am anxious to have the pleasure of conversing a short time with you, concerning your long & most interesting voyage, I have thought the best plan of meeting would be to fix some hour.— Shall you be disengaged on Wednesday between 12 and 1 o'clock? If this hour is inconvenient, will you send me one line, & fixing some other.— In case I do not hear, I will call at Long's[2] on Wednesday.

Yours very faithfully | Charles Darwin

Monday

American Philosophical Society 6

[1] Lieutenant (later Admiral) Charles Wilkes, American naval officer and explorer, was in London to obtain instruments for an exploring expedition to the Antarctic, the islands of the Pacific, and the Pacific Northwest, 1838—42 (see Borthwick 1965).
[2] Long's Hotel, 13 Clifford St, Bond St, London (*Boyle's court and country guide*, London 1836).

To Caroline Darwin [9 November 1836]

[43 Great Marlborough Street]
Wednesday Evening.

My dear Caroline,

I did not write yesterday Evening because my plans were so unsettled. They are not quite so yet. I hope to start by the mail on Friday Evening. Will you ask if the Phaeton can be sent for me on the Saturday to Stone,[1] that is provided you do not hear from me tomorrow to say the Contrary. My fossil bones are all unpacked at the College of Surgeons, and I have some doubt whether I shall anyhow be able to finish them tomorrow. Some of them are turning out great treasures. One animal, of which I have nearly all the bones is very closely allied to the Ant Eaters, but of the extraordinary size of a small horse. There is another head, as large as a Rhinoceros which as far as they can guess, must have been a gnawing animal.[2] Conceive a Rat or a Hare of such a size— What famous Cats they ought to have had in those days! Do not think me very unfeeling, but I almost wish I was not coming down to Shropshire, for it will be such a hurry skurry visit— I think I ought to be back in London. I think I ought to be back in London in 14 or 16 days. So that besides a couple of days at Maer and Overton, I have Woodhouse, Eyton, & other visits to pay, which I should grieve to omit— Anyhow we shall have a nice quiet drive together. Whilst in Shropshire I have some Geology to do. Mr Murchison has lent me a manuscript map, and has asked me to look at a part of the Country, which he has been describing— It is near Minsterley,[3] a very beautiful scene of very ancient true volcanic phenomenæ.—

My London visit has been very pleasant, but too much visiting, although only to people whom I am delighted to become acquainted with. Erasmus is just returned from driving out Miss Martineau.— Our only protection from so admirable a sister-in-law is in her working him too hard. He begins to perceive, (to use his own expression) he shall be not much better than her "nigger".—

Imagine poor Erasmus a nigger to so philosophical & energetic a lady.— How pale & woe begone he will look.— She already takes him to task about his idleness— She is going some day to explain to him her notions about marriage— Perfect equality of rights is part of her doctrine. I much doubt whether it will be equality in practice. We must pray for our poor "nigger".—

Remember me to all at Maer. I long to have a good walk in all the old haunts, and round the pool.

Good bye, | C. Darwin—

Copy
DAR 154

[1] A village about 7 miles east of Maer, on the Mersey Trent Canal.

[2] The two fossils were later described and named by Richard Owen as *Scelidotherium leptocephalum* and *Toxodon platensis* (*Fossil Mammalia*, pp. 16–35, 73–99).

[3] A village 9 miles south-west of Shrewsbury. Murchison 1839, p. 276, describes greenstone rocks near Minsterley. The map may be the one Murchison was preparing for that work. CD's annotated copy, in three volumes, is in Darwin Library – CUL. The map is at Down House.

From S. E. (Elizabeth) Wedgwood to Hensleigh Wedgwood
[16] November [1836][1]

Maer
Wednesday Nov

My dear Hensleigh

Charles's visit has been exceedingly pleasant. Caroline went in the phaeton to meet him on Saturday, and as he did not come by the mail she waited till several more coaches came in & then came back—and we quite gave him, when in a quarter of an hour after her, he appeared. He looks very much thinner, but it has improved his looks, and his countenance is so pleasant that his plainness does not signify— his manners are uncommonly pleasant it is impossible not to be quite fond of him— it is very pleasant to see him so ready to enjoy himself. We had to distribute proper shares of his company to everybody. Eliza came here on Saturday to stay over Sunday & Harry came both those days & went home in the evening—& on Monday again he appeared directly after breakfast but diverged from Madeley (where they all walked to see my father's Yew-tree, that he has never been able to make anybody curious about before, but now Charles is going actually to do it the honour to describe it to Mr Henslow) & we saw him no more. Frank & Fanny came on Sunday, & we had Aunt Sarah—

We were particularly anxious that Aunt Sarah's day should go off well, and by great good luck he got upon just the subjects that would suit her—& was remarkably pleasant all the evening—(we had begun to be low at dinner for he was shy, & we could not get on) and the next morning after breakfast we fell again into such a pleasant fit of talk that she actually ran to the door when the phaeton *was* there, and ordered it away again for an hour! Charles gave us Capt Fitzroy's letter to read, which Sarah was very much pleased with—indeed we are all duly impressed with a proper profundity of admiration for the Captain— Harry enjoyed his cross questioning of Charles thoroughly— We got several little geological lectures— Harry was rather anxious to know whether we are going up or going down— I learn that the learned world are coming fast round to Mr Lyell, only that earth-quakes were bigger in those days— It is very lucky that Charles's bones are turning out so valuable—it was so great a chance, for of course before he went he could not have had much opportunity of knowing what was already known. I perceive that he rather skips over his residence at Cambridge and looks forward to settling in London with the most pleasure— It will be very pleasing for his sisters if he does— There is a certain German pamphlet on the corals of the red sea,[2] that he amused us with his account of Eras translating for him, & how pale he looked after his labours— I offered to finish translating it for him but he refused. I dont know whether from not caring about it or not to give me the trouble— If Eras thinks he would really like to have it & that I could do it (though I dont know how he is to know that—but I have read a great deal of Kozebue)[3] ask him to send it me—it is a job I should like. Charles feels he must learn German, but I dont see how he can get time— we gave him a vocab—and looked for Robinson der Junger—[4] but we could not find it— has Fanny taken it for Snow?

They were to have gone yesterday morning—but by great good ⟨lu⟩ck Caroline wrote for a chaise by post to Tern-hill & the chaise did not arrive here till two o'clock—and we persuad⟨ed⟩ them it was quite too late to go that day, and would only disturb the Dr when once he had given them up— so instead we set out for a good walk with Caroline on the pony & had the pleasure to see the empty chaise rattle away past us back again— They are gone in good earnest today—& Emma to Betley court—[5] I have luckily had Ettore Fieramosca[6] to console me, and I have done no duty & not gone out, but sat in an arm-chair by the fire reading it, till I heroically tore myself from it to write you this letter—but it was in a dull part— It is a pleasing & interesting book—on the model of P. Sposi[7] & Walter Scott—but not so charming as that—but shewing as that does good & nice feelings in the author—everywhere but about some asses that he throws head over heels in a cruel manner. I was very glad indeed to get Caroline to stay here for Charles— she enjoyed his visit so much— I have not seen her a long time in such good spirits—only the longer she stay⟨s⟩ the more sorry she is to part with him. Charles told us his boat story,[8] & very striking it was—from his own feeling about it—and a dreadful story of Miss Martineau's about two young women, that made Sarah look quite as if she was suffering acutely for some time after— it was an inconceivable thing to happen amongst civilized Christians— In the last Emancipa⟨tor⟩[9] there is the most barefaced thing one ever heard of yet—a pamphlet advertised with a full account of the trial & execution of 10 men, 5 black & 5 white by Lynch law! & it appeared as if the negroes were examined by flogging before they were hanged— The names date and all the particulars are given⟨.⟩

Clem Broughton[10] has offered Charles the curacy of Norbury—but I have no doubt Frank has employed some of his leis⟨ure⟩ which is likely to last for the next 3 months in writing to you & will have told you all about it. Robert is delighted with it, and Eliza has taken hold of the occasion to write to Mrs R. a friendly letter which is a very kind thing of her. Jessie is going on quite well. We are expecting Uncle Allen & Emma tomorrow for a fortnight & Mrs Holland & Louisa ought to arrive tomorrow too, as they have never written to tell us they have altered the day but I dont expect them. Charlotte has been giving her first dinner party & suffering most severely—but I have no room to give you Charles's account of her troubles

Dont you think one of our large family of lamps would do for Fanny? & then perhaps she would let you get your books.

We have been keeping an eye on the sky but all in vain— I was up from $\frac{1}{2}$ past 2 to 4 on *the* night—but nothing was to be seen—not that I got up for that but my mother was not comfortable & was sitting up in bed.

AL incomplete?
Postmark: 18 NO 18 1836
Keele University Library, Wedgwood/Mosley 129

[1] CD arrived from London on 12 November, left on the 16th (see Appendix I).
[2] Ehrenberg 1834, an important source for CD's *Coral reefs*. A heavily annotated copy is in 'Philosophical tracts', Darwin Library—CUL.
[3] Otto von Kotzebue 1821, 1830.
[4] [J. C. L.] H[aken] 1802, *Robinsons des Jüngern*, an elementary German reader. Two German dictionaries are in the Darwin Library—Down: *Rabenhorst's pocket dictionary* (1829), inscribed 'Erasmus Darwin', and volume one of a two-volume set edited by Johann Gottfried Flügel, Leipzig (1838).
[5] The home of the Tollet family.
[6] Tapparelli 1833.
[7] Manzoni 1827.
[8] CD may have told the story of a narrow escape in Tierra del Fuego, when an exploring party from the *Beagle* was nearly marooned. Robert FitzRoy, in *Narrative* 2: 217, states 'had not Mr. Darwin, and two or three of the men, run to them [the boats] instantly, they would have been swept away from us irrecoverably.'
[9] Abolitionist weeklies of that name were published in Boston and New York.
[10] Clement Francis Broughton. It is not clear to whom the offer was made. Possibly Charles Langton is meant, but he was rector of Onibury at the time, and the next sentences suggest that Elizabeth intended to write 'Robert'. In any case, the offer was not accepted. In the next *Clergy list* (London 1841) no curate is listed for Norbury.

From J. M. Herbert [19 November 1836]

9 Old Square Lincoln's Inn
Saturday Night.

My dear Darwin,

Welcome, right welcome, was the sight of that good round unmistakeable hand of yours, as your letter lay on my breakfast table this morning: I began to fear that there must be some fatal obstacle in the way of our ever again meeting or corresponding— I did not pass thro' Shrewsbury until the Saturday after I had intended, which will excuse the servant from the charge of stupidity in not giving me the letter; besides, I was in such a desperate hurry that the man had no time to recollect y^r message— The first thing I did after I got to town, which was some 2 or 3 days before you left, was to endeavour to beat up your quarters; but alas vain were all my enquiries; nobody that we knew seemed to be conscious of y^r return to England. The Court Guide told me that y^r brothers lodgings were in Regent St:; I went there, whence I was directed to 34 Marlborough St:[1] but the engaging fair one, who opened to my knock, c^d give me no clue to the object of my enquiries— I then went to Yarrell's; he had seen you the night before, & knew that you were going to stay in town 'till the Saturday, but had no notion where you were to be met with. My slow brain never suggested Mr Hope to me 'till it was too late, and 'till Secker told me that he had heard from y^r cousin (Wedgwood) that you had actually left town. Mind you must soothe my disappointment next week, or when you next come to town, by *exhibiting* (for you are, as you s^d you s^{hd} be, a Lion, or at all events a Puma) at my Rooms, N^o. *9. Old Square, Lincoln's Inn*—

I am right glad to hear of y^r intention of going down to Cambridge, as I have serious thoughts of spending the Xmas vacation there myself, being at length in a condition to satisfy the unliquidated claims of those gentlemen from whom even you did not on all occasions think it prudent to take the Wall.[2] I say I am right glad of y^r determination, as we will have sundry reunions, to remind us of those by-gone days, which most of us can recur to with pleasure, but of the happiness of which I for one doubt the return—

You ask me about my future plans; Heaven only knows; I am in the most perplexing state of indecision poor mortal was ever in; one moment I am bent on turning a country conveyancer, another—a circuit lawyer, another—Chancery has its charms,[3] then I have serious thoughts of going into orders, & putting myself under the tuition of some country curate to grind me in divinity. Don't however suppose that I am turned grumbler in my old age, nor expect me to be a philosopher; I am just what I was, the same mixture of absurdity, egotism, & imperturbability, (so long as the electricity is not too violently excited by stroking against the hair,) with a strong dash of the chiaro oscuro, producing however no effect.

I can't admit the charge of having defamed y^r musical taste, tho' I have often spoken of y^r want of musical memory, the coexistence of which with y^r musical taste & longings has been often the subject of my most curious speculations. Yet I can't understand how or when or to whom I even mentioned it, so as to come to the ears of y^r fair cousins, whom I do not remember ever to have met or seen. 'Faith I think I see & hear you now, on my having tried you with "God save the King" or "Cherry Ripe"—"By G—— old fellow, I don't know, but isn't it the Hallelujah I heard in King's last Sunday?" At times I imagine y^r fondness for music must have been the result of reasoning, tho' the intensity of y^r feeling it, as evinced by y^r *back-bone* criticisms on Handel's Choruses, w^d lead one to infer a very large developement of the organ of Tune.[4]

Whitley's wedding was rather a grand affair, what with a Bishop officiating as Minister, & myself as Paranymph.[5] On the night after his wedding the Company staying in the Bride's Sisters' house got up an Amateur Play, & I was obliged to personate Bob Acres!![6] Mrs Whitley is a very ladylike accomplished & agreable person, with more pretentions to intellectual than personal powers of fascination; she is just the person one w^d wish, tho' I think not expect, him to marry. You must not be offended at his not answering y^r letter, as that is his custom, he having the unfeelingness to treat me, his cousin ex parte maternâ, in this manner. Have you seen Cameron yet? He has just been ordained to the Curacy of Whitchurch. Watkins is stationed at Clirow[7] in Radnorshire, as Venables's Father's Curate. We must have a meeting of the Club[8] to welcome you on y^r return, when you must bring up y^r reports of TransAtlantic Cookery— What is the Story between you & Buckland?[9]

God bless you. I shall be on the most anxious look out for you— | Y^{rs} most aff^{ly} J M Herbert

Postmark: C NO 21 1836
DAR 204.10

[1] E. A. Darwin lived at 43 Great Marlborough Street.
[2] See letter from J. M. Herbert, 15 – 17 April 1832.
[3] Herbert was called to the bar in May 1835. He became a County Court Judge in 1847 (*Alum. Cantab.*).
[4] See *Autobiography*, pp. 61 – 2, for CD's memories of music at Cambridge.
[5] Best man.
[6] A character in Richard Brinsley Sheridan's *The Rivals* (1775).
[7] Now spelled 'Clyro'.
[8] The Glutton (or Gourmet) Club (see letter from Frederick Watkins, [18 September 1831], n. 1).
[9] Herbert may have heard that William Buckland intended to discuss CD's fossils in the second edition of his *Bridgewater treatise* (see letter to J. S. Henslow, [30 – 1 October 1836], n. 5).

From Emma Wedgwood to F. M. Wedgwood [21 November 1836]

[Tells of the pleasure that CD's visit gave the family.]

ALS 4pp
Keele University Library (Wedgwood/Mosley 233)

To Caroline Darwin [7 December 1836]

[43 Great Marlborough Street]
Wednesday.

My dear Caroline,

I should have written to you earlier, but I had little to say and little time to say that little. I will give you my annals. On Sunday I dined at Clapham,[1] and had a very pleasant party.— I enjoyed the satisfaction of making Erasmus jealous by Snow flirting with me and sitting on my knee, and at the same time Hensleigh & Fanny were very jealous at Bro's coqueting with Erasmus. Fanny & Hensleigh have agreed to look over my Journal and give me some detailed criticisms. My plans have, since being here, become more perplexed, with respect to the Journal part. I am becoming rather inclined to the plan of mixing up long passages with Capt Fitzroy. Dr Holland looked over a few pages, and evidently thought that it would not be worth while to publish it alone, as it would be partly going over the same ground with the Captain. The little D^r talked much good sense, and, what was far more surprising much sincerity. I shall go on with the geology and let the journal take care of itself. I have been very busily engaged almost every hour, but have been to very few parties, on Friday however we go to the Marshes.[2] I called as in duty bound, on Miss Martineau, and sat there nearly an hour. She was very agreeable and managed to talk on a most wonderful number of subjects, considering the limited time. I was astonished to find how little ugly she is, but as it appears to me, she is overwhelmed with her own projects, her own thoughts and own abilities. Erasmus palliated all this, by maintaining one ought not to look at her as a woman. Miss Martineau has asked me to meet on Saturday week Fanny Butler,[3] who has suddenly appeared from the other side of the Atlantic. I should have much liked this, but on Saturday or at furthest on Monday, I go to Cambridge, and the invitation is for Saturday week. Nothing has happened very particular, with respect to my affairs. The fossil bones are at a higher premium for curiosity than ever. I have seen Lyell, who was exceedingly friendly, & gave me much sensible advice and told me how he managed to make the most of his time; an art which no one understands better.

Will you tell my Father I have not yet called on the Bank, but intend doing so.

I will write from Cambridge; excuse this abominable letter. I really have not industry after a hard day of excitement to write a more tidy one. | Yours C Darwin

Copy
DAR 154

[1] The Hensleigh Wedgwoods lived in Clapham.
[2] See letter from Catherine Darwin, 27–30 January 1834, n. 2.
[3] The actress Fanny Kemble had married Pierce Butler in 1834.

To W. D. Fox 15 December [1836]

Cambridge—
December 15

My dear Fox

It is now sometime since I wrote to you at Ryde, since that time, I have been living a life of busy enjoyment, but now, at last, I see (at least) a short prospect of quiet tranquillity.

I am at present staying in Henslow's house,[1] (which by the bye, must be new since you were in Cambridge) it is not very large, but very comfortable. Henslow and Mrs Henslow are so kind and affectionate, that I quite feel to them as to my nearest relations. Their comfortable ways of life however do not suit hard work, and, in consequence, tomorrow I migrate into solitary lodgings in Fitz William St. (not far from where Hoare lived). In the mornings I have to arrange my specimens into groups, and to examine all of the geological fragments one by one, which will be a most tedious task, and in the evenings, I shall write. As it is now Christmas vacation I hope to extract a good many solid hours out of each day.— I do not think excepting Henslow, that we have one single friend in common, now at the university.— It appears to me, most strange to stand in the court of Christ, and not to know one undergraduate: It was however some kind of satisfaction to find all the old "gyps".— John Dowly enquired much after you, as did some others.—

After packing up all my goods on board the Beagle, I went to Maer, where I staid two most uncommonly pleasant days. I found Mr Wedgwood much aged, perhaps more changed, than any other individual. as for poor Aunt Bessy, she is only the wreck of her former self.— With all these changes, and the absence of Mrs Langton, yet the family party was most agreeable.— I returned home with Caroline, and staid at Shrewsbury about a fortnight.— My time was most grievously destroyed by visits to stupid people, who neither cared for me, nor I for them. My father told me, next time I wrote, to remember him most kindly to Mrs Fox and yourself—and to say how very anxious he always is to hear how you are going on.— I do hope, when you write, you will be able to send good accounts. My direction for the next month will be Christ Coll: after that period or a little later, I fear I again shall be compelled to change my place of residence to London.— This will be necessary if Capt Fitz Roy & myself publish any joint stock concern, and indeed it will be necessary for all zoological information.—

To complete my annals; from Shrewsbury I went to London, where I staid a week with my good dear old brother Erasmus, and the day before yesterday arrived here.— Whilst in London I disposed of the most important part of my collections, by giving all the fossil bones to the College of Surgeons.— Casts of them will be distributed, and descriptions published.— They are very curious and valuable: one head belonged to some gnawing animal, but of the size of a Hippopotamus! another to an Ant-Eater of the size of a horse!—

When shall we ever meet. I cannot really spare time at present to go to the Isle of Wight. Already several people have attacked me, for not having sooner commenced.— And it is out of the question your taking the trouble of crossing so many miles of country, and leaving M^rs. Fox. I do not see any earlier prospect than the summer, without some good accidental circumstance should draw you up to town. But pray write to me; though we may not meet, let tell to each other our respective annals. I truly hope your next annals will be good and comfortable ones.—

Good Night. I am alone Henslow and Madam having gone to a Ball which I shirked. | Yours ever most truly | C. D.

Postmark: Cambridge DE 16 1836
Christ's College Library, Cambridge (Fox 50)

[1] CD had arrived in Cambridge on the 13th. That afternoon, Joseph Romilly recorded in his diary, 'Drank tea with Mrs Henslow to meet Marchesa &c: here also I met Mr Darwin (g.s. of Bot^c Garden) who is just returned from his travels round the world: he declares that in "terra del fuego" whenever a scarcity occurs (wch is every 5 or 6 years) they kill the old women as the *most useless* living creatures: in conseq. when a famine begins the old women run away into the woods & many of them perish miserably' (Romilly 1967, p. 110).

From Emma Wedgwood to F. M. Wedgwood   17 December 1836

[The Darwin family are anxious for the opinions of Fanny and Hensleigh Wedgwood on CD's journal. Emma is convinced that Henry Holland is wrong if he thinks that it is not worth publishing.]

ALS 4pp
Keele University Library (Wedgwood/Mosley 233)

To Richard Owen   19 December [1836]

Decemb. 19th.—

My dear Sir

I have just written and will send at the same time with this, a letter to Sir Ant: Carlisle.—[1] I have done exactly as you recommended me.— I thought myself compelled to fix on the British Museum in preference to that of Paris,[2] because I was carried on board a King's Ship; and the public collection of the country certainly has claims on me.— If the collection had been made entirely at my own expense, that is on board a Merchant vessel, then I should not have hesitated in making a different choice.— I quite agree with you that the British Museum ought to make returns when it has the power.— I suppose you could not venture to propose another set for Paris. Their value would be so much more in that collection than in the British Museum.— I ought to make up my mind to give my own set to Paris; but I confess I should be grieved to lose my trophies. I should feel like a knight who had lost his armorial bearings.— If the Council should not choose to go to the expence necessary for making all the casts; it was suggested to me here, that the College might pay the price of forming the casts, and the public bodies purchase the models, but I think you will agree with me, that if this can be avoided, it will be better.—

With respect to great head of the Rodent,[3] I certainly feel inclined to run the risk of taking a cast, because the models will be more generally useful, even in case the head itself should be injured or destroyed. But I am sure after the kind and effectual manner with which you have entered on this affair, I cannot do better than follow your advice.— I, at one time, began to think that the fossil bones would be as troublesome to me, and as of little service, as some other branches of my collection are likely to be.— But now I look back to the trouble I took in procuring them with great satisfaction. I do assure you, I feel very grateful to you, for having given me such good assistance.

I have scarcely begun to unpack my cases; in the course of a week I shall have every thing open, and I already know of one very large bone (of a Mastodon??) which I will forward to the College.— When separating the animals in Spirits, I will put by any that, I think, will interest you.—[4] And, it will be a great pleasure to me, if I chance to possess any thing which will be of use to you in your numberless investigations.

Believe me, my dear Sir, | Your very truly obliged | Chas. Darwin

Christ Coll: | Cambridge

Postmark: A 19 DE 19 1836
Fitzwilliam Museum, Cambridge

[1] Sir Anthony Carlisle, Vice-President of the Council and Chairman of the Board of Curators, Royal College of Surgeons. CD's letter to him has not been found, but the substance is given in the minutes of the RCS Board of Curators meeting of 21 December 1836. (See next letter.)

[2] The Muséum d'Histoire Naturelle, Paris, which contained Georges Cuvier's collection of extinct Mammalia.

[3] The *Toxodon platensis*. (See *Fossil Mammalia*, pp. 16–35.)

[4] Minutes of the Museum Committee of the Royal College of Surgeons, as abstracted by Sir Arthur Keith in 1908, list a donation from CD on 4 January 1837: '28 mammalian skeletons in spirit, all rare, and 20 specimens of Birds'. These may be the specimens on two lists in DAR 29.3, in CD's hand, with accompanying notes, bearing the stamp of Richard Owen's estate. The numbers are not, however, identical; the DAR 29.3 lists contain 26 mammals and 24 birds.

To Anthony Carlisle [19 December 1836]

[Cambridge]

[Read a Letter of the 19th Instant from Mr Charles Darwin of Christs College, Cambridge stating that understanding from the Conservators that a Series of fossil Bones collected during the voyage of H:M: Surveying Vessel Beagle possesses a peculiar Interest as connected with Specimens already in the Museum of this College that it had always been his intention to present such Bones to some public collection on the condition that Casts thereof should be given to the leading Public Bodies for the sake of making them more generally useful, specifying the British Museum the Geological Society and the Universities of Cambridge and Oxford, and one set for himself: and that under such Conditions he should be most happy to present the entire series to the Museum of this College.][1]

Minute book of Board of Curators, The Royal College of Surgeons of England

[1] The minute continues:

'And Mr Owen stating that such fossil remains (which were found in the Pampas to the Southward of Rio de la Plata) are highly desirable for the Collection more especially the Cranium of a hitherto unknown gigantic rodent Animal and parts of the skeleton of a Gigantic Armadillo which would be a most important addition to the Series of Osteology which already enriches the Museum of this College.

Resolved: That this Board recommend to the Council the acceptance of Mr Darwin's liberal offer upon the conditions proposed.'

CD's specimens remained at the Royal College of Surgeons until it was damaged by bombing in World War II. Specimens that survived were later removed to the British Museum (Natural History).

From William Owen Sr   19 December 1836

My Dear Sir,[1]

I have been some time in search of a Tutor for my Son Charles, who if nothing better offers I intend for the Church, & as you are now at the seat of learning I take the liberty, which I know you will excuse, of asking if you can assist me in the matter; & I do so, not only because I feel assured you will be happy to do anything you can for me either in this or any other way, but because I have that confidence in your good Judgment that would make me better satisfied by your recommendation of such a Person than by that of almost any Man I know.

I must however tell you that, (as you already know,) I have a large Family & am a poor Man, & therefore cannot afford to pay extravagantly for such a Person—& on this acc^t I have already refused two offers—the one requiring £200 per Annum & the other 150 with extras which with travelling expences, Pocket Money & Clothes would make the total expense not much less than 250—& I am told that perfectly competent Tutors & eligible Situations are to be met with for from 100 to 130—which would suit my Finances much better—

Charles is I think more than 17 Years of Age but both in appearance & manner is much less— He is a very honorable, & well disposed Boy & has always maintain'd a good Character & a good place at Rugby, & I removed him a few Month's ago because, though he was doing well, his younger Brother[2] who is a quicker Boy, having got three removes in one Year, got into the same Form with him, & I did not like to expose him to the chance of being pass'd by his Junior; an event which might have created a Jealousy or ill blood between them which I thought it better for both should be avoided— And though the younger Boy has the greater Facility in learning the elder I believe has quite as good common sense & will probably make as good & as useful a Man. Whether he is now qualified to commence his Career at the University or not I am not able to say—but he is so Boyish both in manner & appearance that I think it will be for his advantage to pass another Year at least with a good Tutor, & as I feel more inclined to send him to Cambridge than to Oxford I have a further reason for asking your assistance. I need not tell you that Boys at home *in the Country* will not pay much attention to their Books & I am therefore very anxious to find a good Tutor for him as soon as possible; & I think if the Tutor had two or three other Boys who were pursuing the same studies & were nearly of the same Age it would be a recommendation.—

I suppose you receive frequent accounts from your own Family of whom I shall therefore say no more than that I believe they are all well— Your Sister Katharine was here two days ago & I narrowly miss'd seeing your Father Yesterday at the Lyth.—[3]

I am almost sorry you are so well & I hope so agreeably employ'd at Cambridge for I fear there is now no chance of my having an opportunity this Winter of renewing my instructions to you in the *Science* of Shooting—& I am getting so old & so bad on my Legs that I fear I shall not be able much longer to attend to or to *instruct you*, & you really were so promising a Pupil that I was rather proud of you.— We are all well & all unite in kind regards to you with | Yours most sincerely | W[m]. Owen

Woodhouse
Dec[r] 19th— *1836*

DAR 204.10

CD ANNOTATIONS
on cover: 'Rev[d] Edw[d] Heawood | Sevenoaks | Kent'

[1] The unusual formal salutation and style of the letter suggest that William Owen Sr initially intended it to be an impersonal one that CD could pass on to someone else.
[2] Henry Owen.
[3] A seat near Ellesmere, about 5 miles north-east of Woodhouse.

To The Master & Fellows, Caius College
[19 December 1836 — 6 March 1837]

[Cambridge]

Mr Darwin presents his compliments to the Master & Fellows of Caius Coll.
and is extremely sorry he is prevented by a previous engagement the honor of
dining with them on Thursday.—

Monday Evening

American Philosophical Society 7

From Hensleigh Wedgwood [20 December 1836]

Clapham
Tuesday.

My dear Charles

I hope by this time you have received your manuscript which we entrusted on
Sunday to Erasmus to send to you It is difficult to discriminate how much of the
interest we felt in it belongs to yourself, from what it would have possessed if it
had been written by an indifferent person. I think however that it would have
been a very interesting journal even in that case. I am not in general a good
reader of travels, but I found no part of yours tedious. We read a great deal of it
aloud too, which is a more severe test. I am afraid we have done but little
towards the end for which you lent it us, having only marked a few of the
passages that pleased us most, & that not throughout. I was especially interested
with your account of Otaheite & New Zealand. I think that of New Holland to
which you called my attention is well worth insertion. In short there is more
variety and a greater number of interesting portions than in 99.100^{ths} of the
travels that are published I should not have the least doubt of it's success & I
think the less it is mixed up with the Captains the better. If Dr Holland read
your journey from the Rio Negro & thought it would not do for publication it
only affects my opinion of his taste & not the least in the world the merits of the
thing itself.[1] I liked your account of Keeling island, but your theory of the
lagoon islands seemed to us not quite clearly enough explained.[2]

Will you remember me very kindly to Henslow. | Yours affect^y | H Wedgwood.

Postmark: CX 20 DE 20 1836
DAR 204.10

[1] See letter to Caroline Darwin, [7 December 1836].

[2] The manuscript of the *'Beagle' diary* has a marginal comment at the end of CD's explanation of the
formation of lagoon islands: 'Good, but the 1st pt not quite clear'. This notation, and several others,
were probably made by Hensleigh Wedgwood and not, as Nora Barlow conjectured, by Robert
FitzRoy (see *'Beagle' diary*, pp. 304 n., 317 n., 328 n., 400 n.).

From Anthony Carlisle 23 December 1836

London College of Surgeons
Decr 23^d 1836.

Dear Sir

I have had the satisfaction to submit your Letter of the 19th. to the Board of Curators who expressed their intire approval of the terms and conditions annexed to your highly liberal offer of your valuable specimens of South American fossil remains—

The Board have recommended their acceptance to the next Council of the College, and I hope that you will soon receive their decision—[1]

Dear Sir | Your much obliged Servt | Anthony Carlisle

Charles Darwin Es^r
&c &c &c

DAR 204.10

[1] The Council, at its Quarterly Meeting on 12 January 1837, resolved: 'that Mr Darwin's Donations to the Museum be accepted upon the conditions annexed thereto: and that accordingly the Board of Curators be authorized and requested to direct the provision of the necessary number of Casts by Mr Jones, Modeller to Sir Francis Chantry' (RCS Council minutes). The cost of preparation of the casts for distribution is recorded as £60 (minutes of the Board of Curators, January 1837).

From Sarah Elizabeth (Sarah) Wedgwood 23 December [1836]

Camp Hill near Woore | Stafford
Decr 23^d

My dear Charles

As I like to take time by the forelock I have during your absence been presenting my wedding gifts to my nieces & nephews without too curiously inquiring whether they were ready for the same, I wish now to close my account by begging you will do me the kindness to accept the 40 pounds enclosed & to consider it as a wedding offering.

I was sorry to hear that you were very unwell after you left Maer, I hope you are well again and stronger than you were. Caroline rather puzzled me by saying that you have taken a lecture room at Cambridge, but I suppose you mean to employ it not exactly in its own line.

All our people hereabouts are pretty well, the Seabridgians[1] are spending a week at Maer: they expect Susan next week at Seabridge. I was surprised to see Miss Martineaus book advertised,[2] I think that must be a sign of its coming out in less than a year.

Good bye my dear Charles | ever yours affectionately. | S Wedgwood

DAR 204.12

[1] Harry and Jessie Wedgwood.
[2] Probably a reference to Martineau 1837.

From Charles Lyell 26 December 1836

16 Hart Street, Bloomsbury:
December 26, 1836.

My dear Sir,—

I have read your paper[1] with the greatest pleasure, and should like to point out several passages which require explanation, and must have a word or two altered, but it would be impossible in a letter. I have made notes on them, and hope you will call here before you read the paper. Will you come up on Monday, January 2, and come and dine with us at half-past five o'clock, or come at five, and I will go over the paper before dinner? No one dines with us but Mr. and Mrs. Horner and one daughter, and Mr. Horner will be glad to renew his acquaintance with you.[2]

We dine early, because we have one of our small early tea parties, and one or two are to be here, to whom I should like to introduce you, besides a few whom you know already. If you cannot get here to dinner, you must if possible join the evening party.

The idea of the Pampas going up, at the rate of an inch in a century, while the Western Coast and Andes rise many feet and unequally, has long been a dream of mine.[3] What a splendid field you have to write upon!

I have spent the last week entirely in comparing recent shells with fossil Eocene species, identified by Deshayes.[4] When some great principle is at stake, all the dryness of minute specific comparisons vanishes, but I heartily long for some one here with a collection of shells, and leisure to talk on these matters with. Lonsdale is overpowered with work. Don't accept any official scientific place, if you can avoid it, and tell no one that I gave you this advice, as they would all cry out against me as the preacher of anti-patriotic principles. I fought against the calamity of being President[5] as long as I could. All has gone on smoothly, and it has not cost me more time than I anticipated; but my question is, whether the time annihilated by learned bodies ('par les affaires administratives') is balanced by any good they do. Fancy exchanging Herschel at the Cape, for Herschel as President of the Royal Society, which he so narrowly escaped being,[6] and I voting for him too! I hope to be forgiven for that.[7] At least, work as I did, exclusively for yourself and for science for many years, and do not prematurely incur the honour or penalty of official dignities. There are people who may be profitably employed in such duties, because they would not work if not so engaged.

Whenever you come up, you must be here on Wednesday, and if you like to dine at the club do so.[8] There is no vacancy, but you stand the first of those who are knocking at the door for admission.

Yours very truly, | Charles Lyell.

Published in K. M. Lyell (1881) 1: 474−5

[1] CD's paper 'Observations of proofs of recent elevation on the coast of Chili' was read at the Geological Society on 4 January 1837 (*Collected papers* 1: 41−3).

[2] CD had met Leonard Horner while a student at Edinburgh. Horner was Lyell's father-in-law.

[3] This specific comparison does not occur in the paper as printed in the *Proceedings of the Geological Society*, vol. 2, 1838 (*Collected papers* 1: 41−3), though the difference in rates of elevation is clear. The point that CD and Lyell were interested in, however, is that the entire continent is rising, and that the more rapid elevations caused by volcanic action on the west coast are 'irregularities of action in some more widely extended phenomenon' (*Collected papers* 1: 43).

[4] Gérard Paul Deshayes amassed an enormous collection of recent and fossil molluscs. Statistical tables based on Deshayes's collection provided important support for Lyell's division of the Tertiary into Eocene, Miocene, and Pliocene. (Lyell 1830−3, 3: 49−60 and Wilson 1972, pp. 259−60, 301−7.)

[5] Lyell served as President of the Geological Society during 1835−6 and again in 1849−50.

[6] In 1830 John Herschel was unsuccessfully nominated for the presidency of the Royal Society as part of an effort to increase the influence of scientists within the organisation. From 1834 to 1838 he carried out important astronomical observations at the Cape of Good Hope.

[7] Lyell actively solicited votes for Herschel (Wilson 1972, p. 304).

[8] Lyell's club was the Athenaeum. CD became a member on 21 June 1838 (Ward 1926).

From Catherine Darwin   27 [December 1836]

Shrewsbury.
Tuesday. | 27th.

My dear Charley,

We want to hear from you again, to tell us how you are going on, and whether you have settled yourself comfortably by this time.— I suppose you received the Box all safe.— We have been living as quiet as mice lately,—and I hope you also have at last begun to be really quiet, which I know is what you most wish for.— It is so horridly cold, that one is nearly frozen up by it, and you must long for your Patagonian Dresses again.— Erasmus wrote us an account that Miss Martineau's soirée turned out most brilliant;—the principal Lions were Mrs Butler, and a Mr Carlyle and his Wife; *he* writes all the articles on German Literature, in the Foreign Quarterly,[1] and *she* is a little woman, with a profusion of hair, one of Erasmus' married loves.— Neither Erasmus nor Mr Empson[2] appear to have at all admired Mrs Butler;—and say she looked as if she was acting Juliet.— It was very odd the Hensleighs were not at that Party.— I heard from Fanny Hensleigh the other day, who says that she & Hensleigh find your Journal so interesting, that it is quite difficult to stop to criticize.— You have settled nothing more on that subject of course; and by the bye, have you ever heard from the Capt lately? I do believe he has jilted Mary O'Brien after all, as I have never seen his marriage in the Paper.— We heard through Maer the other day, that Hensleigh had been to see your Bones at the College of Surgeons, and that the unique head was lying in a room with workmen; which certainly ought not to be;— can you do any thing about it?—

I think Caroline wrote last to you; did she tell you what a capital Concert we had at the Theatre here— How I wish you had been at it, for I am sure you would have liked it. Caradori's[3] singing was beautiful, but the best performer was Ole Bull[4] on the Violin, who I think very superior to Paganini; I met young Corfield[5] there, and he told me to tell you that the Mr Thompson, who married Lady Fitzwilliam is *not* your friend, as he is *not* a Fellow Commoner, and his name is Henry, not Leonard.—

I was at Woodhouse, about ten days ago, for an Ellesmere Ball, and met Fanny Biddulph there; Mr Biddulph was out on a shooting party, so I saw her, and her charming little girl quite comfortably; she is better than when you saw her; you cannot think how much she admired your beautiful flowers; I wish you could have heard how thoroughly she & all the young Ladies there admired their beauty;—and I am sure dear Fanny was *very* much pleased by your remembrance of her.— She is going to write to you soon, and will send the note here for me to forward on to you; & she declares she will begin "my dear Mr Charles" as she says, you wrote her such a *formal* note, beginning, "my dear Mrs Biddulph".— I wore my pretty flowers also one night at Woodhouse, which were very much admired indeed.

You will envy me, when you hear that I brought back your Beauty, Miss Boughey with me in the Phaeton, to Bicton.— Mr Smythe was at Woodhouse also, who is supposed to be desperately in love with her, but I did not see much signs of it.—[6]

Goodbye, dear Charley— Pray write to us, as we long to hear from you. Yrs ever, Catherine.—

Papa will place to your account next week, 1st of January, 50£, half year's allowance, and 51£, half year's interest due.—

Postmark: Shrewsbury DE 28 1836
DAR 204.10

[1] Thomas Carlyle wrote no articles on German literature-after 1832 for *Foreign Quarterly Review*. Hannibal Evans Lloyd reviewed German books and John Stuart Blackie wrote some articles on German literature. Blackie was a Carlyle disciple and his style was similar; since all contributions were unsigned, his articles may have been mistaken for Carlyle's. (See *DNB* and *Wellesley Index* vol. 2.)

[2] William Empson.

[3] Maria Caradori-Allan.

[4] Ole Borneman Bull.
[5] Probably William Corfield.
[6] Anastasia Elizabeth Boughey married Edward Joseph Smythe Jr in 1840.

From Robert FitzRoy 30 December 1836

Bromham | Bedford
30. Dec.[r] /36

Dear Darwin

While in London a few days since I consulted M[r] Broderip about Capt King's Journal. He recommended a joint publication such as is sketched in the accompanying paper One volume might be for King—another for you—and a third for me. The *profits* if *any*, to be divided into three equal portions—

What think you of such a plan?—

Shall I accept the offer of M[r] Colburn or wait and talk the matter over with you when we meet? I shall be in London and *remain* there after the 15th of January.

I told M[r] Colburn I should give him an answer in *January*.

My **Wife** and I are *snowed up* at Bromham—my sister's place.— Pray return the proposed agreement under cover to my brother in law *Hon* G. R. Trevor—at this place.

Where are you? and where are you *going* to be? I shall send this letter to your brother—as the surest *direction*.

I wish you a happy Christmas my good Philos.—

Ever sincerely | your friend | Rob[t] FitzRoy

My wife though *personally* unacquainted desires her kindest wishes for your happiness—

DAR 204.10

APPENDIX I
Chronology 1821–36

In August 1838, Charles Darwin began his 'Journal' in which he recorded various events in his life beginning with his birth and ending with an entry dated December 1881, four months before his death. This small notebook, which measures 3 inches by $4\frac{1}{2}$ inches, tells us the periods Darwin was away from home, the progress and publication of his work, and important events in his domestic life.

This appendix includes a transcription of Darwin's manuscript 'Journal' for the period 1809–36. Initially Darwin appears to have kept alternate pages of the 'Journal' blank for comments about particular entries on the opposite page and for additional entries. All the entries on these pages are printed within double angle brackets (« »). In many cases it is impossible to tell whether these entries were made later than those on the opposite page and therefore the angle brackets simply indicate that the enclosed text occupies a different physical position in the manuscript without presupposing when the entry was made. All alterations, interlineations, and any additions made in a different ink or in pencil are footnoted. The version published by Sir Gavin de Beer as 'Darwin's Journal' (1959), was edited before the original had been found and used a transcription made by an unknown copyist. The original 'Journal', which was located in 1962, is now in the Darwin Archive, Cambridge University Library (DAR 158) and reveals that the copyist did not clearly distinguish between the various types of entries it contains.

For the period covered by this volume all the entries in the 'Journal' are retrospective. Whether these early entries were made from notes or from memory is not known. The entries for the years of the voyage of H.M.S. *Beagle* are very sparse, presumably because Darwin felt that this period was adequately covered by the journal he wrote on board the *Beagle* and which, in a reorganised form, was published as *Journal of Researches*.

The editors have added relevant information throughout this transcription of the 'Journal', particularly in the years of the *Beagle* voyage. These interpolations are enclosed in square brackets to distinguish them from Darwin's own entries.

Journal | Charles Darwin | August 1838[1]

1809   Feb 12[th.] Born at Shrewsbury Parish of St. Chad[2]

1813   summer. Went to Gros near[3] Abegele for sea-bathing, some of my earliest recollections dates from this.

1817.   Went to Mr Case's[4] school in the Spring. (8 years old)
——   July My mother[5] died.—

1818   July[6] Went with Erasmus on party of pleasure to Liverpool.
——   Midsummer.— Went to Dr Butler's school.
——   September. Ill with scarlet Fever

1819   July. Went to sea at Plas Edwards & staid there three weeks.

1820   July. Went tour with Erasmus to Pistol Rheyadur.—[7] on horseback

1822   June. Tour with Caroline to Downton.— «My first recollection of having
         some pleasure in scenery date as far back as this.— 13 years old.—»
         July— to Montgomery & Bishops Castle, with Elizabeth.[8]
         September.— To old Mr Cotton.—
         [October. Began to assist Erasmus with chemistry.]

[1823   'Mineralogy, Botany, Chemistry and Entomology'.] [9]

1824.   November.— Marianne was married to Dr Parker

1825   June 17th. Left Shrewsbury school for ever.— 16 years old
——   October. Went with Erasmus to Edinburgh.—

[1826   February. Taxidermy lessons.]
1826   June 15. Walking tour into North Wales with Hubbersty
——   Oct^10 30th. Riding tour with Caroline Vaynor Park, Bala &
——   November 6th. Went to Edinburgh by myself
[1826   10 November. Elected to Plinian Society.]

«1826⎫ winter. Examined marine animals on shores of Frith of Forth with D^r
 1827⎭ Grant & Coldstream.—»
         [Lectures in natural history, Robert Jameson.]
         ['much interesting natural-history talk' with William Macgillivray.][11]
         [Attended meetings of Wernerian Society with Dr Grant.]
         [Zoological walk to Portobello with William Kay.][12]

1827   [16 March − 23 April. Began notes on marine animals.][13]
         [27 March. Read papers at Plinian Society on 'Ova of *Flustra*' and
         'Pontobdella muricata'.][14]
         «In Spring went tour. Dundee St. Andrews Sterling. Afterwards
         Glasgow. Belfast Dublin.— Then London. Paris.—»

1827 In Spring went to Dublin & Portran[15], & then to London & Paris [in May] with Uncle Jo[16]

1827. In the Autumn paid many visits to Woodhouse

—— September. Maer & Sir J. Mackintosh[17]

—— Christmas. Went to Cambridge

« 1827 }
  1828 } Became acquainted with Fox & Way & so commenced Entomology»

1828 Summer. Went to Barmouth (with Herbert & Butler) to read with Butterton

«1828. Very keen in collecting insects in the spring at Cambridge & at Barmouth in the Autumn.»

1828. Sept. Maer. & thence to Osmaston Hall for Music Meeting.

[1829 19—24 February. In London with Erasmus. Entomological visits to F. W. Hope and J. F. Stephens.][18]

1829. Summer. Paid Barmouth a visit.— [on entomological expedition with F. W. Hope].

—— October. Bermingham. Music Meeting with Wedgwoods.
[Michaelmas term, Cambridge. Entomology with Leonard Jenyns; Friday evenings at J. S. Henslow's.]
[December. Three weeks in London with Erasmus. Entomological visits to F. W. Hope and J. F. Stephens.][19]

[1830 January. Cambridge. Reading for 'Little Go'.]
[March. Passed 'Little Go'.]
[August. Entomological tour, North Wales, with F. W. Hope and T. C. Eyton.]

1830 Christmas vacation spent at Cambridge (?)

«1830. Continued to collect insects: to hunt, shoot & be *quite* idle.»

1831. Christmas passed my examination for B.A. degree [22 January 1831], & kept the two following terms.
«During these months lived much with Prof. Henslow, often dining with him, & walking with. became slightly acquainted with several of the learned men in Cambridge.—which much quickened the little zeal, which dinner parties & hunting had not destroyed.—»
[February. Read J. F. W. Herschel's *Preliminary discourse* and reread A. von Humboldt's *Personal narrative*.][20]
«In the Spring, paid Mr Dawes a visit, with Ramsay, Kirby[21] & Henslow[22] & talked over an excursion to Teneriffe.— In the Spring, Henslow persuaded me to think of Geology & introduced me to Sedgwick.— During midsummer geologized a little in Shropshire»

[July. Studied Spanish and geology.]

—— August. Went on Geological tour by Llangollen, Ruthven Conway, Bangor & Kapel Curig. where I left the Professor Sedgwick[23] & crossed the mountains to Barmouth. Returned to Shrewsbury at end of August.— refused offer of Voyage—

Sept. Went to Maer— returned with Uncle Jos to Shrewsbury.— [Father's approval of voyage won] thence to Cambridge & London.

—— Sept. 11. Went with Capt. FitzRoy in steamer to Plymouth to see the Beagle

—— 22^d returned to Shrewsbury, passing through Cambridge

1831 Oct. 2^d. Took leave of my home Staid in London.

Oct 24th. Reached Plymouth

«October & November. These months very miserable.—»

—— December 10th Sailed. but were obliged to put back

—— 21st Put to sea again & were driven back

—— 27th. Sailed from England on our circumnavigation.—

1832. Jan. 16th. First landed on a tropical shore St. Jago[24]

[15 Feburary. Landed on St Paul Rocks.]

Feb 29th Landed at [Bahia] Brazil

[27—8 March. Abrolhos.]

[5 April. Landed at Rio de Janeiro.]

[25 April — 25 June. Lived at Botafogo.]

[26 July — 19 August. Montevideo.]

[6 September — 17 October. Bahia Blanca.]

[26—30 October. Montevideo.]

[2—10 November. Buenos Aires.]

[14—26 November. Montevideo.]

Dec. 2^d.— Tierra del Fuego [The *Beagle* anchored off the coast of Tierra del Fuego on 16 December 1832 and left for the Falkland Islands on 26 February 1833.]

[1833 1 March — 6 April. East Falkland Island.]

[28 April — 3 July. Lived at Maldonado.]

[June. Fossil bones exhibited to British Association in Cambridge.][25]

[3—11 August. Rio Negro.]

[11 August — 20 September. Overland trip from Rio Negro to Bahia Blanca and Buenos Aires.]

[27 September — 4 November. Expedition to Santa Fé and return to Montevideo.]

[4 November — 6 December. Montevideo.]

[14—28 November. Expedition to Mercedes and return.]

[December. Inscribed as a founding member of the Entomological Society by F. W. Hope.][26]

1833   Dec 6th Sailed for last time from Rio Plata
[23 December — 4 January 1834. Port Desire.]

[1834   26 January. Strait of Magellan.]
[2 February. Port Famine.]
[5 March. Sailed from Tierra del Fuego.]
[10 March — 7 April. East Falkland Island.]
[18 April — 8 May. Expedition up the Santa Cruz river.]
[1—8 June. Port Famine.]

1834   June 10th Sailed for last time from Tierra del Fuego [Arrived west coast via Magdalen channel.]
[28 June — 13 July. Chiloé Island.]
[23 July. Arrived at Valparaiso.]
[2 August — 10 November. Resided in Valparaiso with R. H. Corfield.]
[14 — 27 August. Geological excursion to base of Andes.]
[28 August — 6 September. Santiago.]
[6—26 September. Travelled back to Valparaiso. Ill in bed from 27 September to end of October.]
[10 November. Sailed for Chiloé Island.]
[21 November — 4 February 1835. Chiloé Island.]

[1835   22—8 January. Expedition from San Carlos to Cucao, Chiloé Island, and return.]
[8—22 February. Valdivia.]
[20 February. Earthquake at Valdivia.]
[4—6 March. Concepción. Saw effects of earthquake of 20 February.]
[11 March. Arrived at Valparaiso.]
[12—14 March. Resided with R. H. Corfield]
[14—18 March. Santiago.]
[18 March — 10 April. Crossed Cordillera via Portillo Pass to Mendoza, returned to Santiago via Uspallata Pass.]
[10—15 April. Santiago.]
[15—17 April. Travelled to Valparaiso.]
[17—27 April. resided with R. H. Corfield.]
[27 April — 5 July. Journey from Valparaiso to Coquimbo and Copiapó.]
[5 July. Rejoined the *Beagle*.]
[19 July — 6 September. Callao and Lima.]

1835   Sept. 5th Sailed from West[27] shores of South America [The *Beagle* sailed for the Galápagos Islands on 7 September.]
[15 September — 20 October. Galápagos Islands.]

[15 – 26 November. Tahiti.]
[22 November. Visited coral reef off Matavai.][28]
[21 – 30 December. Bay of Islands, New Zealand.]
[23 – 4 December. Waimate, New Zealand.]

[1836   12 January. Arrived at Sydney Cove, New South Wales, Australia.]
[16 – 27 January. Expedition from Sydney to Blue Mountains, Bathurst,
and Dunheved.]
[28 – 9 January. Sydney.]
[5 – 17 February. Tasmania.]
[6 – 14 March. King George's Sound, Australia.]
[1 – 12 April. Keeling Islands.]
[29 April – 9 May. Isle of France (Mauritius).]

1836   May 31. Anchored at the Cape of Good Hope
[8 – 15 June. Visited Dr Andrew Smith, Sir John Herschel, and others.]
[8 – 14 July. St Helena.]
[19 – 23 July. Ascension Island.]
[1 – 6 August. Bahia, Brazil.]
[12 – 17 August. Pernambuco, Brazil.]
[4 September. St Jago.]
[20 – 4 September. Azores.]

1836   October 2^{d} Anchored at Falmouth.
Oct.— 4th Reached Shrewsbury after absence of 5 years & 2 days.
Oct— 14th London
——— 15th Cambridge
——— 20th London
——— 28. Beagle arrived at Woolwich. was paid off November 17th—
[29 October. Dined with Charles Lyell and Richard Owen.]
[November. Elected to Geological Society of London.]
November 12th Maer
——— 16th Shrewsbury
December 2^{d} London
——— 13th Cambridge

[1836   16 December – 6 March 1837. Living in lodgings in Fitzwilliam Street,
Cambridge.]

1 'Journal . . . 1838' written on the cover of the notebook.
2 'Parish . . . Chad' *interl.*
3 'near' *over* 'at'.
4 Rev. G. Case. See *Autobiography*, p. 22.
5 Susannah Darwin.
6 'July' *interl.*
7 Pistyll Rhaiadr, falls on the Rhaiadr river, Denbighshire.
8 Possibly Sarah Elizabeth (Elizabeth) Wedgwood.

9   CD's interest in these subjects is mentioned in the letter from Catherine Darwin, *c.* June 1823.

10 'Oct' *over* 'Nov'.

11 *Autobiography*, p. 53.

12 Fellow undergraduate at Edinburgh. CD's account of the walk is in DAR 5: 49−51.

13 DAR 118; printed in *Collected papers* 2: 285−91.

14 In 1873 CD inserted opposite the first page of his Journal and also opposite his entries for 1827 the following extract from *Nature*, 20 November 1873, p. 38: 'The first paper contributed by him [CD], entitled "On the Ova of the *Flustra*", in which he announces that he has discovered organs of motion, and, secondly, that the small black body hitherto mistaken for the young of *Fucus loreus* is in reality the ovum of *Pontobdella muricata*, exhibits his early habits of minute investigation.' After the first insertion CD added the comment: 'My first paper 2^d year in Edinburgh'; and after the second insertion: 'These 2 papers were read by me at close of 1826 or early in 1827 before Plinian Soc. in Edinburgh; & were first read by me. I was then 18 years old'.

15 Portraine, County Dublin.

16 '1827 In Spring . . . Uncle Jo' *added in blue-black ink.*

17 'September . . . Mackintosh' *interl.*

18 See letter to W. D. Fox, [26 February 1829].

19 See letter to W. D. Fox, [3 January 1830].

20 *Autobiography*, pp. 67−8.

21 Possibly Henry Kirby.

22 'Henslow' *deleted in pencil.*

23 'Sedgwick' *interl.*

24 'St. Jago' *added in pencil.*

25 See letter from J. S. Henslow, 31 August 1833.

26 See letter from F. W. Hope, 15 January 1834.

27 'West' *interl.*

28 *'Beagle' diary*, p. 357.

APPENDIX II
Darwin's *Beagle* records[1]

During the voyage CD kept extensive records of his observations and activities. His notes form a tiered system, beginning with field notes, which were later expanded when CD was arrranging and describing the specimens he had collected. Later in the voyage, during the longer crossings of the Pacific and Atlantic, CD reworked some of his notes into memoranda that summarised his observations on specific subjects.

The records are of five kinds: field notebooks, a personal diary, notes of zoological and geological observations, and specimen catalogues.

1. The 'field notebooks'.

When travelling ashore, CD used a small pocket notebook for making rough notes of his observations in the field. Nora Barlow has described the eighteen field notebooks that were preserved at Down House, Downe, Kent and has published extracts from them in *Voyage*. One of the eighteen notebooks (No. 1.17) is now missing, but the text is available on a microfilm made during the 1970s. It is probable that originally there were more than eighteen, since no notes of CD's visits to New Zealand and part of Australia are recorded in any of those at Down House. Notebook No. 1.2, the next to last one in the series, has been transcribed and published in a critical edition by Sandra Herbert (*Red notebook*). It was begun during the closing months of the voyage and continued in England until the spring of 1837. It contains the first clear assertions of CD's belief in the transmutation of species. The last notebook at Down (No. 1.5) is not strictly a field notebook since it was begun after CD's arrival in England.

The notebooks, which were not usually filled in chronological order, are listed here in the time order of their first headings and entries, with their Down House catalogue numbers:

1.1	Unlabelled (single page only)
1.4	Cape de Verds; Fernando Noronha; Bahia; Rio de Janeiro
1.10	Rio de Janeiro excursion; M. Video; Bahia Blanca
1.12	Buenos Ayres (city); Beagle Channel; ascent of Port Desire creek
1.14	Falkland; Maldonado (excursion); Rio Negro to Bahia Blanca
1.11	Bahia Blanca to Buenos Ayres

1.13	Buenos Ayres; St Fe & Parana; Cordillera of Chile
1.9	Banda Oriental
1.8	Port Desire; Port Famine; Tierra del Fuego; E. Falk- lands; Chiloe; Chonos Archipelago
1.15	Valparaiso; up Aconcagua; to St Jago
1.18	Santiago book
1.17	Santiago, Lima, Galápagos, Tahiti
1.16	Valparaiso to Coquimbo
1.7	Coquimbo to Copiapò
1.6	Copiapò; Despoblado; Isle of France; Cape of Good Hope; St Helena
1.3	Sydney; Mauritius; Bathurst
1.2	RN [Red notebook]
1.5	St Helena Model

2. The *Beagle* diary.

During the voyage CD called his personal chronicle of activities and events his 'Journal', but to avoid confusion with his *Journal of researches* and the pocket diary started later it is referred to in the *Correspondence* as the *Beagle* diary; its published version, edited by Nora Barlow, is referred to as *'Beagle' diary*. CD began it on 24 October 1831, the day he arrived at Devonport. From time to time during the voyage, sections were sent home, and CD relied on it as well as his letters to keep his family informed. The manuscript, which is preserved at Down House, has been published in facsimile by Genesis (Guildford, 1979). Soon after his arrival in England CD prepared it for publication in 1839 as the third volume of the *Narrative*, under the title *Journal and remarks*. Also in 1839, the same work was published separately as *Journal of researches into the geology and natural history of the various countries visited by H.M.S. Beagle*.

3. The zoological diary.

The chronological record of the zoological observations that CD kept through-out the voyage has been bound into four volumes in the Darwin Archive at Cambridge University Library (DAR 30.1, 30.2, 31.1, and 31.2), titled 'Diary of observations on zoology of the places visited during the voyage'. Information from the zoological diary and notes, as well as from the geological notes on fossils, was later made available to the naturalists who described the specimens in the *Zoology*. CD edited the five volumes and used his *Beagle* notes to write the introductions and relevant sections on geology, geography, and on the habits of the animals. The descriptions were written by Richard Owen (*Fossil Mammalia*), George Robert Waterhouse (*Mammalia*), John Gould (*Birds*), Leonard Jenyns (*Fish*), and Thomas Bell (*Reptiles*). CD's numerous observations on marine invertebrates in the zoological diary have never been published, although a prospectus for the *Zoology* announced that CD planned to include descriptions of them (Freeman 1977, p. 26). The plan was probably given up because the £1,000 grant from the Treasury for publication of the *Zoology* was exhausted.

4. The geological diary and notes.

The geological notes are the most extensive of the records CD kept on the voyage. They comprise three bound volumes of 'a diary of observations on the geology of the places visited during the voyage' (DAR 32.1, 32.2, and 33). In addition, there are ten bound volumes of 'notes on the geology of places visited on the voyage' (DAR 34 to DAR 38, each in two volumes). Portions of DAR 39.2, 41, and 42 contain a mixture of rough notes and essays of varying length on the geology of particular localities and on topics such as cleavage, fossils, and coral islands. The original manuscript of 'Coral islands', one of the essays in DAR 41, has been transcribed and edited by David Stoddart (1962).

From the geological notes, augmented by his later readings, CD composed his articles and books on the geology of South America. These works, published between 1837 and 1846, won him a place among the leading geologists in England. (For the articles, see *Collected papers* 1: 41 – 145; the books were *Coral reefs*, *Volcanic islands*, and *South America*.)

5. Catalogues of specimens.

CD's master catalogue of animals and plants, compiled on board the *Beagle*, is at Down House. It is in six parts: three labelled 'Catalogue for specimens in spirits of wine' (Nos. 1 – 660, 661 – 1346, 1347 – 1529), and three labelled 'Printed numbers' (Nos. 1 – 1425, 1426 – 3344, 3345 – 3907). The numbers correspond to those given to the specimens described in the 'Zoological diary' (DAR 30 and 31) and refer to the numbers on paper labels attached to dried and solid specimens.

From this master catalogue and the zoological diary twelve separate classified annotated catalogues were prepared during the summer of 1836 so that they could be turned over to naturalists who would be engaged in naming and describing the specimens in their special fields (Sulloway 1982b). Some of them were returned to CD when the work was completed, and they are now in the Darwin Archive: DAR 29.1 contains 'Animals', i.e., mammals; 'Fish in Spirits of Wine'; 'Insecta'; and 'Shells in Spirits of Wine'; DAR 29.2 contains the manuscript that was edited by Nora Barlow and published as 'Ornithological notes'. These notes contain the famous passages on the Galápagos tortoises and birds, in which CD's first stirrings of doubt about the immutability of species are recorded. Recent work indicates that they were written after leaving the Galápagos, probably between mid-June and mid-August of 1836 (Sulloway 1982b, p. 337). DAR 29.3 includes 'Shells', 'Insects in Spirits of Wine', 'Mammalia in Spirits of Wine', and 'Birds &c &c in Spirits of Wine'. The last two are notes on specimens sent to Richard Owen late in 1836 or early in 1837 and referred to in Darwin's letter of 19 December 1836. DAR 29.3 also contains nine manuscript pages of figures and plates of drawings of specimens, some from microscopical examinations made by CD during the voyage, and an alphabetical index to the zoological specimens.

Two notebooks containing lists, in John Stevens Henslow's hand, of the dried plants collected in South America were found in 1980 in a cupboard at the Cambridge University Herbarium. With the notebooks were eleven pages of notes on plants, on the same paper and in the same format as the notes in DAR 29 but in the handwriting of CD's amanuensis, Syms Covington, with insertions by CD. (See Porter 1981, 1982.) These notes and lists are now on deposit at the Cambridge University Library. Two additional catalogues are located in the British Museum (Natural History): 'Reptiles in Spirits of Wine' in the Zoological Library and 'Insects' in the Entomological Library (Porter 1983). Finally, four catalogues of CD's mineralogical specimens have been deposited at the Cambridge University Library by the Department of Earth Sciences of Cambridge University, which has custody of the collection.

From CD's letters it is clear that there were other notes and lists, though it is not certain that they were compiled during the voyage: notes on *Fungia* were sent to Charles Stokes (see *Calendar*, Letter 478) and Henry Denny was sent a set on *Pediculi* (*Calendar*, Letter 770). Francis Walker's descriptions of CD's large collection of Chalcididae (Walker 1839) were also probably based on a classified catalogue with notes by CD.

Porter (in press) summarises the current disposition of CD's lists and *Beagle* specimens. A useful listing of works that contain printed descriptions of species collected by CD has been compiled by Paul H. Barrett as Bibliography I of *Collected papers* 2: 295 – 300. Since that list was published the editors have located the following additional works containing descriptions:

Babington 1841 – 3
Harker 1907
J. D. Hooker 1844 – 7
W. J. Hooker and Arnott 1836, 1841
Morris and Sharpe 1846
R. Owen 1853
Porter 1980b
Ramsay *et al.* 1862
Sulloway 1982c
C. O. Waterhouse 1875
G. R. Waterhouse 1837 – 40

---

[1] Short titles are used for references to CD's published writings. The full titles are listed in the bibliography.

APPENDIX III
Persons on board the *Beagle*, 1831–6

Robert FitzRoy in *Narrative* 2: 19–20 states that the established complement of the *Beagle* was sixty-five officers and men and, with the supernumeraries, the total on board at sailing time was seventy-four. His list follows:

Robert Fitz-Roy	Commander and Surveyor.
John Clements Wickham	Lieutenant.
Bartholomew James Sulivan	Lieutenant.
Edward Main Chaffers	Master.
Robert Mac-Cormick	Surgeon.
George Rowlett	Purser.
Alexander Derbishire	Mate.
Peter Benson Stewart	Mate.
John Lort Stokes	Mate and Assistant Surveyor.
Benjamin Bynoe	Assistant Surgeon.
Arthur Mellersh	Midshipman.
Philip Gidley King	Midshipman.
Alexander Burns Usborne	Master's Assistant.
Charles Musters	Volunteer 1st Class.
Jonathan May	Carpenter.
Edward H. Hellyer	Clerk.

Acting boatswain: sergeant of marines and seven privates: thirty-four seamen and six boys.

On the List of supernumeraries were—

Charles Darwin	Naturalist.
Augustus Earle	Draughtsman.
George James Stebbing	Instrument Maker.

Richard Matthews and three Fuegians [York Minster, Fuegia Basket, and Jemmy Button]: my own steward [Fuller]: and Mr. Darwin's servant [Syms Covington].[1]

On 24 July 1832, during the sail from Rio de Janeiro to Montevideo, CD copied from the ship's watch-bill the names of the officers and crew (*'Beagle' diary*, p. 81). His list, tabulated for clarity, follows:

Boatswains mates, J. Smith & W. Williams;
Quarter-Masters, J. Peterson, White, Bennett & Henderson;
Forecastle men, J. Davis, Heard, Bosworthick (ropemaker), Tanner, Harper (sailmaker), Wills (armourer);

Foretop men, Evans, Rensfrey, Door, Wright, Robinson, MacCurdy, Hare, Clarke;

Maintop men, Phipps, J. Blight, Moore, Hughes, D. Johns, Sloane, Chadwick, Johns, Williams, B. Blight, Childs;

Carpenter's crew, Rogers, Rowe, J. May, James;

Idlers,[2] Stebbing (instrument mender); Ash, gunroom steward; Fuller, Captain's do: R. Davis, boy do; Matthews, missionary; E. Davis, Officers' cook; G. Phillips, ships' cook; Lester, cooper; Covington, fiddler & boy to Poop-cabin; Billet, gunroom boy;

Royal Marines,—Beareley, serjeant; Williams, Jones, Burgess, Bute, Doyle, Martin, Middleton, Prior, (midshipmen's steward);—

Boatswain, Mr. Sorrell;

Carpenter, Mr. May;

Midshipmen, Mrss. Stewart, Usborne, Johnson, Stokes, Mellersh, King, Forsyth.

Hellyar, Captain's clerk;—

Mr Bino, [Bynoe] acting surgeon;

Mr Rowlett, purser;—

Mr Chaffers, Master;—

Mr Sulivan, 2nd Lieutenant;

Mr Wickham, 1st Lieutenant;

R. FitzRoy, Commander.

There are (including Earl, the Fuegians & myself) 76 souls on board the Beagle.[3]

FitzRoy records the following changes that took place during the voyage:

In April 1832, Mr. Mac-Cormick and Mr. Derbishire returned to England. Mr. Bynoe was appointed to act as Surgeon. Mr. Mellersh received a Mate's warrant; and Mr. Johnson joined the Beagle as Midshipman. In May Mr. Musters fell a victim to fever, caught in the harbour of Rio de Janeiro:—Mr. Forsyth took his place.

Mr. Earle suffered so much from continual ill health, that he could not remain on board the Beagle after August 1832; but he lived at Monte Video several months previously to his return to England. The disappointment caused by losing his services was diminished by meeting Mr Martens at Monte Video, and engaging him to embark with me as my draughtsman.

In March 1833, Mr. Hellyer was drowned at the Falkland Islands, in attempting to get a bird he had shot. In September 1833, Mr. Kent joined as Assistant Surgeon. In June 1834, Mr. Rowlett died, at sea, of a complaint under which he had laboured for years: and the vacancy caused by his lamented decease was filled by Mr. Dring.

Mr. Martens left me, at Valparaiso, in 1834; and Mr. King remained with his father, at Sydney, in Australia, in February 1836.

(*Narrative* 2: 20).

A change FitzRoy does not record is that Robert Hamond, a midshipman, joined the *Beagle* from H.M.S. *Druid* on 31 July 1832. He transferred to H.M.S. *Pylades* on 24 May 1833 (*'Beagle' diary*, pp. 84, 152). He and CD became friends (see letter to Caroline Darwin, [24 October — 24 November 1832]).

After the mission at Tierra del Fuego was abandoned as too dangerous, Richard Matthews continued with the *Beagle* to New Zealand, where he joined his brother Joseph in missionary work. Fitzroy continues:

After these changes, and at our return to England in October 1836, the list stood thus—

Robert Fitz-Roy	Captain and Surveyor.
John Clements Wickham	Lieutenant.
Bartholomew James Sulivan	Lieutenant.
Edward Main Chaffers	Master.
Benjamin Bynoe	Surgeon (Acting.)
John Edward Dring	Purser (Acting.)
Peter Benson Stewart	Mate.
John Lort Stokes	Mate and Assistant Surveyor.
Arthur Mellersh	Mate.
Charles Richardson Johnson	Mate.
William Kent	Assistant Surgeon.
Charles Forsyth	Midshipman.
Alexander Burns Usborne	Master's Assistant.
Thomas Sorrell	Boatswain (Acting.)
Jonathan May	Carpenter.

And on the List of supernumeraries were Mr. Darwin : George J. Stebbing : my steward : and Mr. Darwin's servant.

Our complement of seamen, marines, and boys was complete at our return, and generally during the voyage; because, although many changes happened, we had always a choice of volunteers to fill vacant places.

(*Narrative* 2: 20 − 1).

---

[1] Covington did not become CD's servant until July 1833. Until then he was one of the six boys (see CD's list). FitzRoy has apparently counted him twice. If so, the complement at sailing was seventy-three.

[2] ' "A general designation for all those on board a ship-of-war, who, from being liable to constant day duty, are not subjected to keep the night-watch" (Smyth Sailor's Word-bk.)' (*OED*).

[3] The total on CD's list comes to only seventy-five, and May, the carpenter, has apparently been counted twice. Allowing for this oversight reduces the total to seventy-four.

APPENDIX IV
The books on board the *Beagle*

'Considering the limited disposable space in so very small a ship, we contrived to carry more instruments and books than one would readily suppose could be stowed away in dry and secure places'. So wrote Captain FitzRoy in the *Narrative* (2: 18).

CD, in his letter to Henslow, 9 [September 1831], discussing the preparations for the voyage, refers to FitzRoy's 'immense stock of instruments & books . . . in books all travels, & many natural history books.' FitzRoy encouraged CD to bring any books he felt he would need, even if it meant duplicating some of FitzRoy's own: 'You are of course welcome to take your Humboldt—as well as any other books you like—but, I cannot consent to leaving mine behind . . . There will be *plenty* of room for Books.' (Letter from Robert FitzRoy, 23 September 1831).

On board, FitzRoy states, 'Our books . . . were collected in one cabin, under Mr. Stebbing's charge, and lent to the officers, without reserve, under certain regulations.' (*Narrative* Appendix, p. 327).

The books were kept in the poop cabin where CD worked and slept. John Lort Stokes and Philip Gidley King shared the working space with CD. According to Keith Thompson (1975), the cabin measured 10 feet by 11 feet.

Plan of the poop cabin of H.M.S. *Beagle* where CD slept and worked. The annotations are in CD's hand. (DAR 44).

A copy of the regulations has survived in the papers of John Lort Stokes (National Maritime Museum STK Papers B.4054). They read as follows:

1. The books in the Poop Cabin are at the Service of all the Officers of the Beagle who will comply with the following regulations: —

1) Books are to be taken from, and returned to their places by the Person appointed for that purpose.

2) Every Book, whether *Old* or *New, bound* or *Unbound*, is to be covered, temporarily, by the person who has it in use.

3) Books are not to [be] transfered from one Officer to another without the knowledge of the person who has it in charge.

4) Two Catalogues will be kept, one for general use, the other for the Cabin.

5) The names of those who take Books are to be written in a list kept for that Purpose.

6) Any Officers who have books which they think will be generally useful and of which there are not already Duplicates in the Catalogue will confer a general benefit by lending them in a similar manner, inserting their names in the Catalogue, and if more convenient keeping them in the Poop Cabin.

7) Books are to be taken, or returned in morning before ½ past 8.

8) Books are never on any account to be taken out of the Vessel.[1]

The two catalogues referred to in the regulations have unfortunately been lost and FitzRoy, so far as is known, left no records from which the library could be reconstructed. However, from the *Beagle* correspondence, CD's diary, field notebooks, and the extensive zoological and geological notes it is possible to compile a list of works used by CD during the voyage. Some of the books CD refers to were his own copies, but the majority were probably part of the *Beagle* library. Some notion of the size of the 'immense stock' which CD mentions may be had from a letter FitzRoy wrote to his sister during an earlier voyage (16 March 1826): 'I flatter myself I have a complete library in miniature, upwards of 400 volumes!' This in a cabin 6 feet 6 inches square![2]

Since most of the works on the list are derived from the unpublished zoological and geological notes in the Darwin Archive (DAR 29–38), a brief description of those records may be helpful in understanding the nature of the evidence they provide. For the most part the notes follow in chronological sequence on the rectos of the folio pages. The versos were used by CD for comments in amplification of his observations. Many of these were made later in the voyage. After his return to England, when CD used the notes for writing his *Journal of researches* and the later works based on the results of the voyage, he made numerous references to books in the margins or at the end of the verso entries. Fortunately, these later notes can almost always be distinguished—they are usually in pencil or in a different ink.

During the voyage pencil was used almost exclusively in the field notebooks when travelling ashore. On board, or when he was in residence ashore, the observations are almost always in ink, usually written with CD's favourite Brahma pens. References to books in pencil, or in ink of a different shade from that normal to the *Beagle* notes, cannot therefore be considered reliable evidence that the books were on board unless there is independent corroboration elsewhere in the notes.

Certain other references to books have also been eliminated from consideration, even though they occur in the main texts of the folio rectos or in dated entries on the versos. These are of three kinds:

1. References to which CD prefixed a 'V[ide]' or 'Mem'. For example, 'V. Sabine Horticultural Trans' or 'Mem. Audubon in Wern. trans' are interpreted to mean that CD intended to consult them after the voyage unless they follow a specific reference, or a quoted passage, or unless corroborative references exist elsewhere in the notes.

2. References that may only be remembered from a work with which CD was familiar. Obvious examples are references to Bernardin de Saint Pierre's *Paul et Virginie* and to characters in novels by Jane Austen.

3. Secondary sources. CD usually was careful to indicate when he was using a secondary source, probably because he anticipated using his notes for later publication, but occasionally he failed to do so. For example, two references to Felix Azara's works in notes made during 1833 cite secondary sources (DAR 33: 254 and DAR 30.2: 112) but several later ones have been found (DAR 42: 73) that are taken from Griffith's edition of Cuvier's *The animal kingdom* (1827–35) without citation of the source. Since no other reference occurs that clearly and unambiguously comes directly from Azara one must presume that his works were not on board.

The *Dictionnaire classique d'histoire naturelle* was a much cited secondary source during the voyage. Any reference to an author who contributed articles to the *Dictionnaire* is considered to be a reference to that work unless it fits a separate work by that author. For example, at least one of the references to Lamouroux (DAR 30.1: 48 v.) contains material that is not in the *Dictionnaire* and could only have come from that author's *Exposition méthodique* (1821).

The elimination of these three categories leaves a considerable number of works which are likely to have been on board. The first list that follows includes those for which the evidence that CD used them during the voyage is considered reliable. The second list includes those books for which there are some, but only inconclusive, indications that they were on board. The evidence for items on the first list is of four kinds:

1. There are volumes now in the Darwin Library in Cambridge that contain inscriptions or annotations that can be identified as having been written during the voyage. These titles are designated by symbol *. The nature of the evidence is noted in each entry.

2. CD sometimes mentions in his letters, diary, and notebooks that he has received, or is reading, certain books, or that he has them in his possession. FitzRoy, in his *Narrative*, also mentions many books, and as one might expect many of his references coincide with CD's. However, since FitzRoy's account was written after his return to England and no manuscript survives from the voyage itself, it has been used as a source for only three works which he explicitly states he used while on board (see Kotzebue, Krusenstern, and Whewell). The works in this second category are designated by symbol §.

3. Some works, like the *Dictionnaire classique* are referred to so frequently throughout the voyage, and the references occur in so many different contexts, that there can be no doubt that they were part of the *Beagle* collection. These titles are designated by symbol ‡.

4. The remaining works on the list are mentioned less frequently, some of them only once, but the references are clearly in a text written during the voyage and of such a nature (e.g., passages quoted or paraphrased, facts stated, or page references given) that CD must have had the work beside him as he wrote. A few books are mentioned in letters from the family as having been sent to CD (e.g., Fleming, Martineau, Pennant); although they were not acknowledged or referred to later by CD, it is considered highly probable that they were received, since letters in the same mail arrived safely. (Considering the difficulties and hazards of communicating with the *Beagle*, it is a remarkable fact that all of the letters sent to CD by his sisters are accounted for, though not all have been preserved.)

The possibility exists, of course, that some of the titles in the last category are references to unacknowledged secondary sources, even though the citations are extended or specific. Only the more closely related works on board have been searched for such cases. Considering CD's normal practice of citing his sources, it is not thought likely that further search would eliminate more than a very small number of titles, if any.

Another possibility that has been considered is that a work might have been borrowed from someone with whom CD became acquainted while living on shore. It is possible, for example, that Nuñez's book was borrowed from Edward Lumb, the merchant with whom he stayed in Buenos Aires, but the subject of the book makes it equally plausible that CD or FitzRoy thought it would be useful and acquired it in preparation for the voyage. References to works in notes made during periods ashore at Santiago, Cape Town, and Sydney have been examined with this problem in mind, but no reason has been found that would tip the balance in favour of elimination of any of the books; the possibility nevertheless remains that a small proportion of those used by CD during the voyage were not actually in the collection on the shelves of the poop cabin.

Despite the size of the list, it is clear that other works must have been on board. CD would not have had occasion to consult the technical books used by FitzRoy in his survey work, and there were probably other works about voyages in

FitzRoy's collection which he does not mention. Other books are referred to in a way that makes it impossible to identify them, as for example, CD notes that he is reading a 'Spanish story book' (*Voyage*, p. 196). In another field notebook, at Cape Town in May 1836, he lists, probably with the intention of purchasing them, a spelling dictionary and German books (*Voyage*, p. 252). Although there is no specific mention that these works were used, there is evidence that CD corrected some of his idiosyncratic spelling during the summer of 1836 (Sulloway 1982b, pp. 331–2, n. 13). Finally, there are two works that CD does not refer to at all, but which we can be certain were on board the *Beagle*: a Spanish–English dictionary and the Authorised Version of the Bible.

In the list the most frequently cited sources are the zoological and geological notes made by CD during the voyage. They are in the Darwin Archive in the Cambridge University Library in bound volumes DAR 29–38. Other sources used are the manuscripts at Down House of the field notebooks and the journal CD kept on board. If a published version of a manuscript source exists references are to it rather than the original. Except for *Voyage*, however, the page references are to the manuscript page numbers.

The following abbreviations and symbols are used:

DAR — Darwin Archive
CUL — Cambridge University Library
Down — Down House, Downe, Kent
* — CD's copy contains evidence of use on board the *Beagle*
§ — mentioned in a letter or other source as being on board
‡ — works frequently cited during the voyage
† — CD's copy annotated (†† indicates numerous annotations)

Thus, the entry below for Aubuisson de Voisins, *Traité de géognosie*, conveys the following information: CD's copy, now in Darwin Libary–CUL, was used on board. The inscription that provides evidence of this is quoted and other bracketed data gives the location of a reference to the book on folio 61 (recto) in part 1 of volume 32 of CD's geological diary (DAR 32.1) in the Darwin Archive. The copy in the Darwin Library at CUL is heavily annotated. (N.B. Works cited during the voyage are often in the Darwin library, but they are not considered to be the same copies unless definite evidence that they were on board is to be found in them.)

§Anson, George. *A voyage round the world, in the years MDCCXL, I, II, III, IV by George Anson Esq . . . compiled from papers . . . of . . . Lord Anson . . . by Richard Walter*. London, 1748. (*'Beagle' diary* p. 62).

*Arrowsmith, Aaron. *A new general atlas, constructed from the latest authorities*. London and Edinburgh, 1823. (Contemporary annotations by CD and Wickham, see *Voyage*, p. 128 n.). CUL, Adv.b.79.18 †.

*Aubuisson de Voisins, Jean François d'. *Traité de géognosie*. 2 vols. Strasbourg, 1819. (Inscription in vol. 1: 'C. Darwin HMS Beagle'; DAR 32.1: 61). Darwin Library – CUL ††.

‡Beechey, Frederick William. *Narrative of a voyage to the Pacific and Beering's Strait . . . 1825, 26, 27, 28*. London, 1831. (DAR 31.1: 276v.; 33: 253v.). Darwin Library – CUL, 1832 Philadelphia edition ††.

Bible (Authorised version)

‡Blainville, Henri Marie Ducrotay de. *Conchyliologie et malacologie* in *Dictionnaire des sciences naturelles. Planches. 2e Partie, Zoologie*. Paris, 1816 – 30. (DAR 30.1: 6, 12v.). Darwin Library – CUL.

§——et al. Rapport sur les résultats scientifiques du voyage de M. Alcide d'Orbigny. . . *Nouvelles Annales du Muséum d'Histoire Naturelle* 3 (1834): 84 – 115. (DAR 37.1: 677v.; letter to J. S. Henslow, 12 July 1835).

*Boitard, Pierre. *Manuel d'entomologie*. 2 vols. Paris, 1828. (Inscription in vol. 2: 'Charles Darwin Rio Plata Aug 7th 1832'). Darwin Library – CUL †.

‡Bory de Saint-Vincent, Jean Baptiste Genevieve Marcellin, ed. *Dictionnaire classique d'histoire naturelle*. 17 vols. Paris, 1822 – 31. (Letter from J. S. Henslow, 15 – 21 January [1833]). Darwin Library – CUL.

§Bougainville, Louis Antoine de. *A voyage round the world*. Translated by John Reinhold Forster. London, 1772. (*Voyage*, p. 178; DAR 34.2: 153).

§British Association for the Advancement of Science. *Report of the 2d meeting . . . Oxford, 1832*. London, 1833. (Letter to J. S. Henslow, March 1834 and letter from J. S. Henslow, 31 August 1833).

——. *Lithographed signatures of the members . . . who met at Cambridge, June 1833. With a report of the proceedings . . .* Cambridge, 1833. (Letter to Charles Whitley, 23 July 1834). 'Philosophical tracts', Darwin Library – CUL.

Brongniart, Alexandre. Rapport fait à l'Académie Royale des Sciences, sur les travaux de M. Gay. *Annales des Sciences Naturelles* 28 (1833): 26 – 35. (DAR 35.2: 396). 'Philosophical tracts', Darwin Library – CUL ††.

‡Buch, Leopold von. *Travels through Norway and Lapland . . . Translated . . . by John Black. With Notes . . . by Robert Jameson*. London, 1813. (DAR 30.2: 154). Darwin Library – CUL.

Bulkeley, John and Cummins, John. *A voyage to the south-seas, in the years 1740 – 1. Containing a faithful narrative of the loss of his Majesty's ship the Wager*, etc. London, 1743. (DAR 36.1: 447).

Burchell, William John. *Travels in the interior of Southern Africa*. 2 vols. London, 1822 – 4. (*Voyage*, p. 182; *Red notebook*, p. 86). Darwin Library – Down †.

Byron, George Anson, 7th Baron. *Voyage of H.M.S. Blonde to the Sandwich Islands, in the years 1824–25.* London, 1826. (DAR 31.2: 333; Stoddart 1962, p.4).

Byron, John. *The narrative of the Honourable John Byron Commodore in a late expedition around the world* . . . Aberdeen, [1786], 1822. (DAR 31.1: 206; 35.1: 235v.).

‡Caldcleugh, Alexander. *Travels in South America, during the years 1819* . . . *21.* 2 vols. London, 1825. (DAR 31.2: 319; 32.1: 60v.).

Candolle, Augustin Pyramus de and Sprengel, Kurt Polycarp Joachim. *Elements of the philosophy of plants.* Edinburgh, 1821. (DAR 30.2: 162v.).

Colnett, James. *A voyage to the South Atlantic and round Cape Horn* . . . London, 1798. (DAR 31.2: 318v.).

Conybeare, William Daniel and Phillips, William. *Outlines of the geology of England and Wales.* Pt 1. London, 1822. (DAR 35.1: 317). Darwin Library – Down.

Cook, James. *Voyages* (editions unidentified; see also Hawkesworth, John). (DAR 32.2: 89v.; Robert FitzRoy's letter to the *South African Christian Recorder*, 28 June 1836, *Collected papers* 1: 20).

‡Cuvier, Georges. *Le règne animal.* 4 vols. Paris, 1817. (DAR 30.1: 29v.). Darwin Library – CUL, 2d edition, 5 vols., Paris, 1829–30, vols. 4, 5, *Crustacés*, etc. par M. Latreille †.

‡——. *The animal kingdom arranged in conformity with its organization . . . with additional descriptions of all the species hitherto named, and of many not before noticed.* By Edward Griffith and others. 16 vols. Edinburgh, 1827–35. (DAR 33: 255v.; number of volumes on board unknown).

Dampier, William. *A new voyage round the world.* London, 1697. (*Red notebook*, pp. 8e, 10; '*Beagle*' *diary*, p. 407).

Daniell, John Frederic. *Meteorological essays and observations.* London, 1823. (DAR 30.1: 41).

‡Daubeny, Charles Giles Bridle. *A description of active and extinct volcanoes* . . . London, 1826. (DAR 32.1: 23v., 26). Darwin Library – Down.

‡De la Beche, Henry Thomas. *A selection of the geological memoirs contained in the Annales des Mines* . . . Trans. with Notes. London, 1824. (DAR 36.2: 522; Stoddart 1962, p. 14). Darwin Library – Down.

‡——. *A geological manual.* London, 1831. (DAR 32.1: 53).

Desaulses de Freycinet, L. C. see Freycinet, L. C. Desaulses de

Desmarest, Anselm-Gäetan. Plates for Crustacea, Zoea, etc. in *Dictionnaire des sciences naturelles.* Paris, 1816–30. (DAR 30.2: 122).

Dictionnaire classique d'histoire naturelle. See Bory de Saint-Vincent, J. B. G. M., ed.

Dictionnaire des sciences naturelles. Par plusieurs Professeurs du Jardin du Roi. (Number of volumes on board unknown; references are to plates and descriptions by Blainville and Desmarest. See Blainville, H. M. D. de and Desmarest, A. G.).

Duperrey, Louis Isidore. *Voyage autour du monde . . . sur la corvette . . . La Coquille 1822–5. Zoologie* par MM. [R. P.] Lesson et [P.] Garnot. 2 vols., atlas. Paris, 1826–30. (DAR 30.2: 184; 31.1: 241).

§Earle, Augustus. *A narrative of a nine months' residence in New Zealand in 1827* . . . London, 1832. (Letter to Caroline Darwin, 27 December 1835).

Ellis, William. *Polynesian researches, during a residence of nearly six years on the South Sea Islands* . . . 2 vols. London, 1829. (DAR 37.2: 798; Stoddart 1962, pp. 6, 8, 11).

Encyclopædia Britannica. 20 vols., 1 vol. *Supplement.* 6th ed. Edinburgh, 1823. (DAR 30.2: 198v.; 33: 254).

§Euclid. *Elements of geometry.* (Letter to J. S. Henslow, 30 October 1831).

‡Falkner, Thomas. *A description of Patagonia, and the adjoining parts of South America* . . . Hereford, 1774. (DAR 31.1: 207v.).

Fitton, W. See King, P. P. *Narrative of a survey of the intertropical and western coasts of Australia.*

Fleming, John. *The philosophy of zoology* . . . 2 vols. Edinburgh, 1822. (Letter from Susan Darwin, 15 October 1833). Darwin Library—CUL ††.

Flinders, Matthew. *A voyage to Terra Australis.* 2 vols., atlas. London, 1814. (DAR 30.1: 30v.).

Forster, John Reinhold. *Observations made during a voyage round the world* . . . London, 1778. (Stoddart 1962, p. 20).

Freycinet, Louis Claude Desaulses de. *Voyage autour du monde* . . . *1817—20.* 9 vols., 4 vols. plates, Paris, 1824—44. (DAR 32.1: 52v.).

Frézier, Amédée François. *A voyage to the south-sea and along the coasts of Chili and Peru in* . . . *1712—14.* London, 1717. (DAR 31.1: 206v.).

Gay, Claude. Aperçu sur les recherches d'histoire naturelle faites dans l'Amérique du Sud . . . 1830 et 1831. *Annales des Sciences Naturelles* 28 (1833): 369—93. (DAR 35.2: 396). 'Philosophical tracts', Darwin Library—CUL †.

Greenough, George Bellas. *A critical examination of the first principles of geology in a series of essays.* London, 1819. (DAR 32.2: 77)

——. Anniversary address (1834). *Proceedings of the Geological Society of London* 2 (1833—8): 44—70. (DAR 37.2: 797).

Griffith, Edward. see Cuvier, Georges

Hall, Basil. *Extracts from a journal written on the coasts of Chili, Peru and Mexico for the years 1820, 1821, 1822.* 2 vols. Edinburgh, 1824. (DAR 37.1: 662).

Hawkesworth, John. *An account of the voyages undertaken . . . for making discoveries in the Southern Hemisphere . . . performed by Commodore Byron, Captain Wallis, Captain Carteret, and Captain Cook in the Dolphin, the Swallow and the Endeavour drawn from the journals which were kept by the several commanders and from the papers of Joseph Banks* . . . 3 vols. London, 1773. (DAR 31.2: 349v.).

§Head, Francis Bond. *Rough notes taken during some rapid journeys across the Pampas and among the Andes.* London, 1826. (DAR 36.1: 469v.). Darwin Library—Down.

‡Henslow, John Stevens. Geological description of Anglesea. *Transactions of the Cambridge Philosophical Society* 1, pt 2 (18): 359—452. (DAR 33: 217—22; *Red notebook,* pp. 5e, 7e, etc.). Darwin Library—CUL ††.

Horsburgh, James. *Directions for sailing to and from the East Indies, China, New Holland, Cape of Good Hope and the interjacent ports.* 2 pts. London, 1809–11. (DAR 30.1: 2v.).

‡Humboldt, Alexander von. *Essai géognostique sur le gisement des roches dans les deux hémisphères.* 2d ed. Paris, Strasbourg, 1826. (DAR 32.1: 56v.). Darwin Library – CUL †.

——. *Fragmens de géologie et de climatologie asiatiques.* 2 vols. Paris, 1831. (DAR 35.2: 401; Stoddart 1962, p. 22a). Darwin Library – CUL.

*——. *Personal narrative of travels to the equinoctial regions of the new continent . . . 1799 – 1804 . . . translated into English by Henrietta Maria Williams.* 7 vols. London. 1814–29. (Vols. 1 and 2, in one, 3d edition, inscribed from J. S. Henslow to CD 'on his departure', September 1831). Darwin Library – CUL ††.

*——. *Political essay on the kingdom of New Spain.* Translated by John Black. 2 vols. New York, 1811. (Inscription, both volumes: 'Chas Darwin Buenos Ayres'). Darwin Library – CUL.

Jones, Thomas. *A companion to the mountain barometer.* 2d ed. London, n.d. [1802]. (Letter to Robert FitzRoy, [10 October 1831]). DAR 196.2 †.

*Juan, George and Ulloa, Antonio de. *A voyage to South America . . .* Translated by John Adams. 2 vols. 4th ed. London, 1806. (Inscription in vol. 1: 'Robt FitzRoy to Charles Darwin'; *Red notebook,* pp. 75, 105e, 106e). Darwin Library – CUL, vol. 1 only.

King, Philip Parker. *Narrative of a survey of the intertropical and western coasts of Australia* (includes: Fitton, William. An account of some geological specimens . . .). 2 vols. London, 1827. (DAR 38.2: 867v.; 'Geology', by W. Fitton, *Red notebook,* p. 6e).

*——. Description of the Cirrhipedia, Conchifera and Mollusca in a collection formed . . . 1826–30. *Zoological Journal* 5 (1832–4). (Inscribed 'Charles Darwin Esq from the Author Dunheved Jan 26 1836'). 'Philosophical tracts', Darwin Library – CUL.

——. Manuscript journal of voyages of *Adventure* and *Beagle,* later published as vol. 1 of *Narrative.* (DAR 30.1: 66; 30.2: 153).

——. Some observations upon the geography of the southern extremity of South America. *Journal of the Royal Geographical Society* 1 (1832): 155–175. (DAR 32.1: 120v.). 'Philosophical tracts', Darwin Library – CUL †.

Kirby, William and Spence, William. *An introduction to entomology.* 4 vols. London, 1815–26. (DAR 30.2: 119; 30.1: 54v.). Darwin Library – CUL, vol. 1, 3d ed.; vol.2, 2d ed.; vols. 3 and 4, 1st ed. ††.

§Kotzebue, Otto von. *A voyage of discovery, into the South Sea and Beering's Straits . . .* Translated by H. E. Lloyd. 3 vols. London, 1821. (*Narrative* 2: 523).

§Krusenstern, Adam Johann von (Kruzenshtern, Ivan Fédorovich). *Atlas de l'Océan Pacifique dressé par M. de Krusenstern.* St Petersburg, 1824–7. (*Narrative* 2: 506).

§Krusenstern, Adam Johann von. *Recueil de mémoires hydrographiques pour servir d'analyse et d'explication à l'Atlas de l'océan Pacifique.* 2 vols. St Petersburg, 1827. (*Narrative* 2: 506).

Labillardière, Jacques Julien Houton de. *Relation du voyage à la recherche de La Pérouse* . . . 2 vols. Paris, [1799−1800]. (*Red notebook*, pp. 5e, 12). Darwin Library − Down.

Lacordaire, Jean Théodore. Mémoire sur les habitudes des coléoptères de l'Amérique méridionale. *Annales des Sciences Naturelles* 20 (1830): 185−291; 21 (1830): 149−194. (DAR 30.1: 59). 'Philosophical tracts', Darwin Library − CUL †.

‡Lamarck, Jean Baptiste Pierre Antoine de Monet de. *Histoire naturelle des animaux sans vertèbres.* 7 vols. Paris, 1815−22. (DAR 30.1: 45v.). Darwin Library − CUL †.

‡Lamouroux, Jean Vincent Félix. *Exposition méthodique des genres de l'ordre des polypiers.* Paris, 1821. (DAR 30.1: 13v.; letter to J. S. Henslow, 24 July − 7 November 1834). Darwin Library − CUL †.

La Pérouse, Jean François de Galaup de. *A voyage round the world performed in the years 1785, 1786, 1787, and 1788, by the Boussole and Astrolabe* . . . 2 vols., atlas. London, 1798−9. (*Red notebook*, p. 15e).

‡Latreille, Pierre André. Vols. 4 and 5, *Crustacés*, etc., of Cuvier, Georges, *Le règne animal.* (DAR 30.1: 48v.). Darwin Library − CUL, 2d edition.

Lesson, René-Primevère. See also Duperrey, L. I.

——.*Manuel de mammalogie.* Paris, 1827. (DAR 31.1: 236, where it is confused with Rang, see below). Darwin Library − Down.

——. *Manuel d'ornithologie.* 2 vols. Paris, 1828. (DAR 30.2: 182v., 184). Darwin Library − Down.

*Lyell, Charles. *Principles of geology* . . . 3 vols. London, 1830−3. (Inscriptions: vol. 1 (1830), 'Given me by Capt. F.R C. Darwin'; vol.2 (1832), 'Charles Darwin M: Video. Novem.ʳ 1832'; vol. 3 (1833): 'C. Darwin'; letter to J. S. Henslow, 24 July − 7 November 1834). Darwin Library − CUL † (vols. 1 and 2).

Mackintosh, James. *The history of England.* Volume one. London, 1830. (Robert FitzRoy's letter to the South African Christian Recorder, 28 June 1836, *Collected papers* 1: 20). Darwin Library − CUL.

Martineau, Harriet. 'A few little books written by Miss Martineau'. (Letter from Caroline Darwin, 28 October [1833]).

Maskelyne, Nevil. *Tables requisite to be used with the nautical ephemeris.* 3d ed. London, 1802. (Notes on fly-leaf: 'Taken from Jones' (see Jones, T.)). Darwin Library − CUL †.

*Mawe, John. *Travels in the gold and diamond districts of Brazil.* London, 1825. (Inscription: 'Chas. Darwin Octob: 1832 Buenos Ayres'; DAR 33: 160). Darwin Library − CUL.

Maximilian, Prince of Wied-Neuwied. *Travels in Brazil in the years 1815, 1816, and 1817.* 3 vols. London, 1820. (DAR 32.1: 51v.).

Michell, John. Conjectures concerning the cause . . . of earthquakes. *Philosophical Transactions of the Royal Society of London* 51 (1760): 566–634. (*Red notebook*, p. 80). Darwin Library – CUL †.

Miers, John. *Travels in Chile and La Plata* . . . 2 vols. London, 1826. (DAR 31.2: 319; letter to Robert Fitzroy, 28 August 1834). Darwin Library – CUL †.

Milton, John. *Paradise lost.* ('*Beagle*' *diary*, p. 107; letter to J. S. Henslow, 24 November 1832).

*Molina, Juan Ignacio. *Compendio de la historia geografica natural y civil del Reyno de Chile.* Part 1. Madrid, 1794; (*Compendio de la historia civil del Reyno de Chile*). Part 2. Madrid, 1795. (Inscription: 'Charles Darwin Valparaiso 1834'). Darwin Library – CUL ††.

*Murray, Lindley. *An English grammar.* 2 vols. 5th ed. York, 1824. (Inscription: 'Rob! FitzRoy 1831'). Darwin Library – CUL †.

§Narborough, John. *An account of several late voyages.* 2 parts. London, 1694. (Letter to J. S. Henslow, 24 July – 7 November 1834).

§*New Testament* (Greek). ('*Beagle*' *diary*, p. 14).

*—— (German). (Signed 'C. Darwin H.M.S. Beagle'. Copy examined by Sydney Smith *c.* 1968. Present owner unknown).

Nuñez, Ignacio Benito. *An account* . . . *of the United Provinces of Rio de la Plata.* London, 1825. (DAR 33: 269v.).

Owen, William Fitz William. *Narrative of voyages to explore the shores of Africa, Arabia, and Madagascar* . . . 2 vols. London, 1833. (*Red notebook*, p. 64).

Pernety, Antoine Joseph. *Journal historique d'un voyage fait aux Îles Malouines en 1763–4.* 2 vols. Paris, 1770. (DAR 32.2: 132v.). Darwin Library – CUL, Berlin 1769 ed. †† (vol. 2).

Phillips, William. *Elementary introducton to* . . . *mineralogy* . . . London, 1816. (Letter from J. S. Henslow, 22 July 1834; *Red notebook*, p. 89). Darwin Library – CUL, 4th ed., 1837.

‡Playfair, John. *Illustrations of the Huttonian theory of the earth.* Edinburgh and London, 1802. (DAR 34.1: 92v.). Darwin Library – CUL ††.

Quoy, Jean René Constant and Gaimard, Joseph Paul. *Zoologie,* vol. 1 of Freycinet, Louis Claude Desaulses de, *Voyage* . . . (Stoddart 1962, p. 14).

Rang, Sander. *Manuel de l'histoire naturelle des mollusques et leurs coquilles.* Paris, 1829. (DAR 30.1: 30). Darwin Library – CUL.

*Richardson, Samuel. *The history of Sir Charles Grandison* . . . 7 vols. London, 1781. (Vols. 3–7 inscribed 'Chas. Darwin Buenos Ayres Sept. 1833'). CUL Adv.d.79.22–6.

§Rivero, Mariano de. Memoria sobre el rico mineral de Pasco, pp. 70–101 of *Memoriales des ciencias naturales* (DAR 37.1: 690). 'Philosophical tracts', Darwin Library – CUL.

Roussin, Albin-René. *Le pilote du Brésil, ou description des côtes de l'Amérique méridionale* . . . Paris, 1826. (DAR 32.1: 1, 2).

[Saint Pierre, Jacques Henri Bernardin de]. *Voyage à l'Isle de France,* . . . *par un officier du Roi.* Amsterdam, 1773. (DAR 31.2: 362v.; 38.1: 887).

Scrope, George Julius Poulett. *Considerations on volcanos* . . . London, 1825. (DAR 35.2: 381). Darwin Library — Down.

Spanish — English Dictionary. (Probably Neuman and Baretti, *Dictionary of the Spanish and English languages.* 2 vols. 5th ed. London, 1831. Darwin Library — CUL, vol. 1 only.).

*Spix, Johann Baptist von and Martius, Carl Frederick Philip von. *Travels in Brazil in the years 1817 — 1820.* 2 vols. London, 1824. (Inscription in vol. 2: 'Chas. Darwin Octob: 1832 Buenos Ayres'; DAR 32.1: 7). Darwin Library — CUL † (vol.2).

Stephens, James Francis. Description of *Chiasognathus Grantii* a new lucanideous insect . . . *Transactions of the Cambridge Philosophical Society* 4 (1833): 209 — 17. (Letter to J. S. Henslow, 24 July — 7 November 1834).

‡Syme, Patrick. *Werner's nomenclature of colours with additions, arranged so as to render it highly useful to the arts and sciences* . . . 2d ed. Edinburgh, 1821. (DAR 30.1: 30). Darwin Library — CUL.

Thompson, John Vaughan. *Zoological researches and illustrations.* Vol. 1, Pt 1. Cork, 1828 — 30. (DAR 30.2: 133v.).

Turnbull, John. *A voyage round the world in the years 1800, 1801, 1802, 1803, and 1804* . . . London, 1805. (Robert FitzRoy's letter to the *South African Christian Recorder*, 28 June 1836, *Collected papers* 1: 28).

Turner, Sharon. *The sacred history of the world* . . . Vol. 1. London, 1832. (DAR 30.1: 2v.; letter to J. M. Herbert, 2 June 1833).

Ulloa, A. de. See Juan, G. and Ulloa, A. de.

Volney, Constantin François Chasseboeuf. *Voyage en Syrie et Égypt pendant les années 1783, 1784 et 1785.* 2 vols. 2d ed. Paris, 1787. (*Red notebook*, pp. 8e, 52).

Wallis, Samuel. *An account of a voyage round the world in the years 1766, 1767, and 1768,* in Hawkesworth, John *An account of the voyages* . . . London, 1773. (Robert FitzRoy's letter to the South African Christian Recorder, 28 June 1836, *Collected papers* 1: 26).

Webster, William Henry Bayley. *Narrative of a voyage to the southern Atlantic Ocean, in the years 1828, 29, 30, performed in H.M. Sloop Chanticleer* . . . 2 vols. London, 1834. (DAR 32.1: 7; *Red notebook*, p. 33e).

Weddell, James. *A voyage towards the South Pole performed in the years 1822 — 24.* London, 1825. (DAR 32.2: 109v.).

Werner, G. See Syme, P.

[Whateley, Richard]. *A view of the Scripture revelations concerning a future state* . . . *by a country pastor [R. W.]* . London, 1829. (Letter from Caroline Darwin, 28 October [1833]).

§Whewell, William. Essay towards a first approximation to a map of cotidal lines. *Philosophical Transactions of the Royal Society of London* 123 (1833): 147–236. (Letter to J. S. Henslow, March 1834; *Narrative* 2 (Appendix): 227). 'Philosophical tracts', Darwin Library – CUL.

§Wood, James. *Elements of algebra.* 8th ed. Cambridge, 1825. (Letter to J. S. Henslow, 30 October 1831).

The volumes listed below, though mentioned in the notes or letters, have been omitted from the above list because the evidence that they were on board is not conclusive. The references to some of them may be derived from secondary sources that have not been located.

Administration du Muséum Royal d'Histoire Naturelle. *Instruction pour les voyageurs . . . sur la manière de recueillir de conserver et d'envoyer les objets d'histoire naturelle.* Paris, 1818. 'Philosophical tracts', Darwin Library – CUL †.

Buckland, William. Considerations of the evidences of a recent deluge. *Transactions of the Geological Society of London* 5 (1821): 516–44. (DAR 34.2: 150v.).

Caldcleugh, Alexander. Geology of the Island of Juan Fernández. *Proceedings of the Geological Society of London* 1 (1834): 21–6. (DAR 35.2: 357).

Carne, Joseph. On the relative age of the veins of Cornwall. *Transactions of the Royal Geological Society of Cornwall* 2 (1822): 49–128. (*Red notebook*, p. 20e).

Clarke, A. *Narrative of a journey in the interior of China, and of a voyage to and from that country in the years 1816 and 1817.* London, 1818. (DAR 38.2: 906).

Dalyell, Thomas. *Observations on the Planariae.* Edinburgh, 1814. (*Voyage*, p. 89).

Fox, Robert Were. On the electro-magnetic properties of metalliferous veins in the mines of Cornwall. *Philosophical Transactions of the Royal Society of London* 120 (1830): 399–414. (*Red notebook*, p. 20e).

Hall, Basil. See below, Playfair, John.

Humboldt, Alexander von. *Tableaux de la nature.* Translation by J. B. B. Eyriès of *Ansichten der Natur,* 2d ed., 1826. 2 vols. Paris, 1828. (Letter to Catherine Darwin, 5 July [1832]).

Linnaeus, (Carl von Linné). [?*Systema Vegetabilium*] Ed. 15a. Göttingen, 1797. (Letter from E. A. Darwin, 18 August [1832]). Darwin Library – Down.

Lister, Joseph Jackson. Some observatons on the structure and functions of tubular and cellular Polypi, and of Ascidiæ. *Philosophical Transactions of the Royal Society of London* 126 (1834): 365–88. (*Red notebook*, p. 24; CD refers to the author as 'L. Jackson').

Playfair, John. Account of the structure of the table mountain, and other parts of the Peninsula of the Cape. Drawn up by Professor Playfair, from observations made by Captain Basil Hall, . . . *Transactions of the Royal Society of Edinburgh* 7 (1815): 269–78. (DAR 38.2: 906).

['Proteus']. The Bahama Islands. *United Service Journal and Naval and Military Magazine* 3 (1834): 215–26. (*Red notebook*, p. 27).

Sarmiento de Gambóa, Pedro. *Viage al estrecho de Magallanes*. Madrid, 1768. ('Vide (Sarmiento Viage) Spanish edition' in Syms Covington, MS 'Journal 1831—6', p. 30; Mitchell Library, Sydney, N. S. W.). Darwin Library—Down.

Southey, Robert. *History of Brazil*. London, 1810—19. (Robert FitzRoy's letter to the *South African Christian Recorder*, 28 June 1836, *Collected papers* 1: 22—3).

Taxidermy.(Letter to Susan Darwin, [6 September 1831] 'A little book, if I have got it in bedroom, Taxidermy'.). Possibly Swainson, William. *The naturalist's guide for collecting and preserving subjects of natural history and botany*. London, 1822.

Vancouver,George.*A voyage of discovery to the North Pacific Ocean* . . . 3 vols. London, 1798. (DAR 38.1: 858).

The number of titles in the two lists would come to at least 245 volumes, ranging in size from pamphlets to bulky volumes such as those in the *Encyclopædia Britannica*, and, depending on how many volumes of the larger sets such as Griffith's *Animal kingdom* were on board, the total might be as large as 275. If one allows, conservatively, two inches per volume, approximately 46 linear feet of shelf space would be needed to accommodate them.

[1] Printed with the permission of the Trustees of the National Maritime Museum, Greenwich.

[2] The editors are grateful to David Stanbury for this information and for informing them of the existence of the *Beagle* library regulations.

APPENDIX V
Darwin's early notes on coral reef formation

In the *Autobiography* (pp. 98—9) CD describes how he arrived at his coral reef theory as follows:

> No other work of mine was begun in so deductive a spirit as this; for the whole theory was thought out on the west coast of S. America before I had seen a true coral reef. I had therefore only to verify and extend my views by a careful examination of living reefs. But it should be observed that I had during the two previous years been incessantly attending to the effects on the shores of S. America of the intermittent elevation of the land, together with denudation and the deposition of sediment. This necessarily led me to reflect much on the effects of subsidence, and it was easy to replace in imagination the continued deposition of sediment by the upward growth of coral. To do this was to form my theory of the formation of barrier-reefs and atolls.

No statement of the theory that could be described as 'thought out' has been found in the extensive notes on geological observations that survive from the time CD spent on the west coast of South America. There are, however, several references in the field notebooks and letters of the South American period, which, though fragmentary and indirect, give evidence that CD had the main points of the theory clearly in mind before he left that continent, and that he looked forward to verifying it when he could observe the Pacific islands.

The central idea of his theory, as it was later formulated, was that the islands were formed by the upward growth of coral as the Pacific Ocean floor gradually subsided. A letter from Robert Edward Alison, who had assisted CD in his observations on the elevation of the Chilean coast, makes clear that by April 1835, when he was still at Valparaiso, CD had expressed to his friend his expectation that the Pacific islands would furnish evidence of general subsidence:

'I wish much' Alison wrote from Valparaiso, 'to hear of your report respecting the islands in the Pacific, and it will be curious if you find a sinking of the land there, & a rising here' (letter from R. E. Alison, 25 June 1835).

Possibly at about the same time, but certainly before leaving for the Galápagos early in September, CD jotted down some notes about coral formations in the field notebook labelled 'Santiago Book' (Down House no. 1.18). The medium is ink, which indicates that the notes were made when CD was in residence ashore or on board the *Beagle* and not in the field. His spelling of 'Pacific' suggests that he was writing before the late summer of 1835, for about that time he adopted 'Pacifick' as his normal spelling. An early date is also suggested by the context, which is concerned with subjects of primary interest to CD during his time in Chile: crustal movements, elevation, and subsidence. The passages deal with general geological speculations involving both European and Chilean formations as well as the

Pacific coral reefs. Coral formations are treated as evidence for subsidence, rather than as formations whose origin and structure have to be explained. The tone is hypothetical and speculative:

> As in Pacific a Corall bed. forming as land sunk. would abound with. those genera which live near the surface. (mixed with those of deep water) & what would more easily be told the Lamelliform: Corall forming, Coralls.— I should conceive in Pacific. wear & tear of Reefs must form strata of mixed. broken sorts & perfect deep-water shells (& Milleporæ).—
>
> Parts of reefs themselves would remain amidst these deposits, & filled up with infiltrated calcareous matter.— Does such appearance correspond to any of the great Calcareous formations of Europe.—
>
> Is there a **large** proportion of those Coralls which only live near surface.— If so we may suppose the land sinking: I believe much conglomerate on the other hand is an index of bottom [*above del* 'land'] coming near the surface. If so Red Sandstone Epoch of England. will point out this: Mountain limestone the epoch of depression.— Do the [*over* 'these'] numerous alternations of these two grand classes of rock point out a corresponding opposite & repeated motion of the surface of that part of the Globe.
>
> [Notebook no. 1.18: 6–8.]

The discussion of conglomerate formations continues for several pages, then the following entry occurs:

> May we not imagine each band of conglomerates marks an epoch when that part of the ocean's bottom was near to a continent or shoal water; & that having again being depressed. calcareous fine sediments were deposited. (if under circumstances to allow of corall reefs, such would be very abundant).—
>
> [Notebook no. 1.18: 12.]

Then, a few pages later, comes the statement:

> The Test of depression *in strata [*added*]. is where great thickness has. shallow. coralls growing in situ: this could only happen. when bottom of ocean was subsiding:
>
> [Notebook no. 1.18: 15.]

The emphasis in these notes on global crustal changes strongly suggests that CD's theory of coral reef formation originated as a consequence and corollary of his chief preoccupation at the time: the elevation of the South American continent. CD had by that time become a 'zealous disciple' of Lyell (letter to W. D. Fox, July 1835) and, having found ample direct evidence that elevation of the continent had occurred, he invoked Lyell's principle of compensatory change in the earth's crust and hypothesised a corresponding subsidence in the Pacific. The coral islands would thus furnish proof both of that subsidence and, indirectly, of the elevation of South America.

Paradoxically, CD's adoption of the principle of compensatory crustal changes led him to depart from Lyell's own view of the geology of the Pacific. In his chapter on coral reefs in the second volume of the *Principles of geology*, Lyell had adopted the prevailing view of the time that the Pacific was a region of general volcanic uplift and that the reefs had been formed by corals building on volcanic mountains that had subsided as a result of local earthquake action. The circular atolls were 'nothing more than the crests of submarine volcanoes, having the rims and bottoms of their craters overgrown by corals' (*Principles of geology* 2: 290). Lyell was, of course, unaware of the evidence that CD had accumulated of the extent of elevation of South America, and the observations of the south Pacific islands that were then available seemed to support the view that widespread uplift had occurred there.[1] It was, moreover, considered impossible that tiny coral organisms could build reefs that were hundreds of fathoms deep. It is interesting that CD seems never to have had any doubt that such construction could take place.

CD was certainly familiar with Lyell's chapter and with the observations of Pacific voyagers, notably Beechey 1831, and Quoy and Gaimard, who had shown in their *Zoologie* of Freycinet's *Voyage* (1824), that reef-building took place only in shallow water. As he wrote to his sister Catherine (31 May [1835]), CD had begun reading about 'the South Sea' in May 1835. The *Beagle* library was well stocked with works about earlier voyages and the Admiralty orders had specifically recommended surveys and geological descriptions of the channels and lagoons of the islands (*Narrative* 2: 38–9), so it may be presumed that he was well informed on what was known about them. But of his reading, only one note occurs, in Notebook no. 1.17, written shortly after a visit to Lima in August 1835: 'Corall rapidly growing in Low islands'.

From his reading CD must also have been aware that his view of the relationship of subsidence and coral building, if verified, would be an important contribution to geology. Apparently the first coral reef he saw convinced him that his prediction had been sound, and that he had an explanation of the origin of both annular and barrier reefs. This was in November 1835 at Tahiti and the neighbouring reef of Eimeo (Moorea). It was probably shortly after that visit that he wrote out a draft of his theory in a memorandum headed 'Coral islands' (DAR 41: 1–22, published in Stoddart 1962). It is the first account of the theory of which there is any record. It begins with the statement: 'Although I have personally scarcely seen anything of the ['Corall' *del*] Islands in the Pacifick ocean, I am tempted to make a few observations respecting them.—'

After showing that the observable data were best explained by subsidence he continues:

> Before finally concluding this subject, I may remark that the general horizontal uplifting which I have proved has & *is now* ['activ' *del*] raising upwards the greater part of S. America & as it would appear likewise of N. America, would of necessity be compensated by an equal subsidence in some other part of the world.— Does not the great extent of the Northern & Southern Pacifick include this corresponding Area?—
>
> [Stoddart 1962, DAR 41: 22a.]

The impact of CD's first sighting of a coral island is confirmed by a letter to his sister Caroline dated 29 April 1836, in which he states that 'The subject of Coral formation has for the last half year, been a point of particular interest to me', implying that his interest dates from the time of the visit of the *Beagle* to Tahiti. The letter of 29 April was written shortly after a visit to the Cocos (Keeling) Islands of the Indian Ocean. What CD saw there strengthened his conviction that he had a sound theory and one that was worth publishing. The letter continues: 'I hope to be able to put some of the facts in a more simple & connected point of view, than that in which they have hitherto been considered. The idea of a lagoon Island, 30 miles in diameter being based on a submarine crater of equal dimensions has alway appeared to me a monstrous hypothesis.'

While at Cocos (Keeling) CD wrote an entry in his diary that combines an eloquent description of his response to what he saw with a succinct statement of his theory:

12th. [April 1836] In the morning we stood out of the Lagoon. I am glad we have visited these Islands; such formations surely rank high amongst the wonderful objects of this world. It is not a wonder which at first strikes the eye of the body, but rather after reflection, the eye of reason. We feel surprised when travellers relate accounts of the vast piles & extent of some ancient ruins; but how insignificant are the greatest of these, when compared to the matter here accumulated by various small animals. Throughout the whole group of Islands, every single atom, even from the most minute particle to large fragments of rocks, bear the stamp of once having been subjected to the power of organic arrangement. Capt. FitzRoy at the distance of but little more than a mile from the shore sounded with a line 7200 feet long & found no bottom. Hence we must consider this Isd as the summit of a lofty mountain; to how great a depth or thickness the work of the Coral animal extends is quite uncertain. If the opinion that the rock-making Polypi continue to build upwards as the foundation of the Isd from volcanic agency, after intervals, gradually subsides, is granted to be true; then probably the Coral limestone must be of great thickness. We see certain Isds in the Pacifick, such as Tahiti & Eimeo, mentioned in this journal, which are encircled by a Coral reef separated from the shore by channels & basins of still water. Various causes tend to check the growth of the most efficient kinds of Corals in these situations. Hence if we imagine such an Island, after long successive intervals to subside a few feet, in a manner similar, but with a movement opposite to the continent of S. America; the coral would be continued upwards, rising from the foundation of the encircling reef. In time the central land would sink beneath the level of the sea & disappear, but the coral would have completed its circular wall. Should we not then have a Lagoon Island?— Under this view, we must look at a Lagoon Isd as a monument raised by myriads of tiny architects, to mark the spot where a former land lies buried in the depths of the ocean.

['*Beagle*' diary, pp. 399–400.]

Shortly after returning to England, CD told Lyell of his theory. Lyell was immediately convinced by it. 'I must give up my volcanic crater theory for ever

. . .', he wrote to John Herschel on 24 May 1837, 'the whole theory is knocked on the head & the annular shape & central lagoon have nothing to do with volcanoes nor even with a crateriform bottom . . . Let any mountain be submerged gradually & coral grow in the sea in which it is sinking & there will be a ring of coral & finally only a lagoon in the centre.' (Wilson 1972, p. 449).

CD did not publish a fully developed statement of the theory until 1842, after intensive reading and correspondence with observers of coral reefs in other parts of the world. Though adopted by many geologists, it was a subject of controversy until a century later, when test borings in the Pacific islands finally confirmed CD's predictions of their thickness and structure.

[1] For a summary of the geological observations by voyagers before CD see Stoddart 1976.

MANUSCRIPT ALTERATIONS AND COMMENTS

The alteration notes and comments are keyed to the letter texts by paragraph and line number. To the left of the bracket is that section of the final text to which the note applies. Readers should consult the Note on Editorial Policy in the front matter for details of editorial practice and intent. The following terms are used in the notes as here defined:

del   deleted

illeg   illegible word or part of word

interl   interlined, i.e., inserted between existing text lines

over   overwrites and replaces (as opposed to 'above')

To Robert Waring Darwin [23 October 1825]
1.10 Hawley] 'Hanley' *in copy, see n. 3*
1.12 how] *before* '(to)' *inserted by copyist*
2.1 would] *before* '(never)' *inserted by copyist*
3.6 Hotel in] 'in' *interl*
3.8 [his degree?]] *not in copy*; *ML* 1: 7 *reads* 'his [degree?]'
4.3 free] *after del* 'finished'

To Susan Darwin 29 January [1826]
1.5 intended] *after del* 'had'
1.8 I believe] *interl*
1.9 about] *before del illeg*
1.10 write] *above del* 'send'
1.13 is rapidly] 'is' *above del illeg*
1.14 idle,] *before del* 'lately.'
1.17 talking about] *interl*
1.27 would] *interl*
point omitted after: 1.1 'often'; 1.22 'heard'; 1.24 'his'; 1.24 'stating'; 1.30 'Lord'; 1.33 'ashamed'; 1.35 'charges'; 3.2 'that he'

To Caroline Darwin 8 April [1826]
3.3 never should] *the copy reads* 'should never', *with superscript numbers* '2' *and* '1' *to indicate that the words should be reversed in order. Since it is not clear whether the indications are CD's or the copyist's, they have not been retained in the text.*

To William Darwin Fox 12 [June 1828]
0.1 Friday] *below del* 'Thursday'
1.6 interested:] *comma under colon omitted*
1.9 I took] 'I' *over* 'it'
1.10 I . . . prize] *added*; 'prize' *after blotted* 'prize'
3.2 tell] *above del* 'ask'
3.2 particulars] *after del* 'questions'

3.3 know] *inserted before del* 'hear.'
3.3 something] *final* 's' *del*
3.6 over] *over* 'out'
3.6 My] *over* 'I'
4.2 marks] *above del* 'spots'
4.3 it,] *two points above comma omitted*
4.7 seen] *before del* 'the'
4.10 can] *interl*
5.2 letter:] *comma below colon omitted*
5.9 Miss] *over illeg*
7.2 Shrewsbury] 'bu' *blotted*
point omitted after: 1.3 'what'; 1.7 'sister'; 1.8 'as'; 3.4 'constantly'; 3.5 'pay'; 4.9 'Antennæ'; 5.3 'thing'; 5.5 'how was'; 5.5 'summer'; 5.6 'letter'; 5.7 'same'; 5.8 'Poste'; 5.10 'tell me'; 6.2 'some'; 8.1 'Blaps'; 8.2 'it'

To W. D. Fox [30 June 1828]
1.4 most] *after del* 'a'
1.4 to think] 'to' *inserted in margin before* 'think'
3.1 makes] *after del* 'tempts'
3.2 much] *interl*
6.1 Taken . . . name.—] *added*
7.1 the same . . . specimen] *interl*
7.4 of all . . . can.] *inserted*
8.1 few] *after del* 'sin'
9.5 very pleasant people] *interl*
9.6 when . . . been] *above del* 'after being'
9.17 one with] 'with' *interl*
9.21 them] *after del* 'him'
10.4 Genus.—] *2 points above dash omitted*
point omitted after: 1.2 'you over'; 2.3 'aware'; 2.5 'wanted'; 5.1 'Cychrus'; 7.4 'you'; 7.4 'insects'; 9.1 'day'; 9.10 'science'; 9.15 'Samouelle'; 9.16 'Curtis'; 9.18 'taken'; 9.18 'vide'; 9.19 'band'; 9.20 'seen'; 10.3 'because'; 10.4 'least'; 10.4 'pang'; 10.5 'jealousy'; 11.4 'insects'

To W. D. Fox [29 July 1828]
1.4 stick] 'i' *over* 'u'
2.2 hoping] *after del* 'wishing'
2.6 entomology] *above del* 'insectology'
3.9 small] *interl*
4.3 of this length] *above sketch in MS*
4.6 thorax] *after del* 'thorax'
6.1 Fox] 'ox' *over* 'r'
7.1 (n, . . . relationship)] *set off in large parentheses to left of signature*
point omitted after: 1.4 'God'; 3.10 'Zigena'; 4.2 'odd one'; 4.5 'black'; 4.6 'thorax'; 5.1 'changed'; 5.2 'him'

To W. D. Fox [19 August 1828]
2.2 all] *after del* 'your'
2.3 last] *over* 'least'
3.1 day] *interl*
3.6 not Buprestis] *interl*
5.3 (Littorale] *before del illeg*
6.2 at Osmaston] *interl*
6.3 much] *interl*
6.3 for] *interl*
6.3 to be] 'to' *above del* 'will'
point omitted after: 1.4 'say'; 3.8 'thorax'; 3.9 'abdomen'; 4.2 'direct'; 4.3 'answer'; 5.1 'few'; 5.2 'Cryptocephalus'; 5.3 'allso'; 6.6 'lost'; 6.6 'good'; 6.6 'old'; 9.1 'you'

To John Maurice Herbert [13 September 1828]
1.5 their] 'ir' *over* '[se]'
2.2 black] *interl*
2.3 very] *above del* 'something'
2.5 be so . . . as to] *interl*
2.7 a great many] 'a' *altered from* 'as'; 'great' *above del* 'many'
2.7 of these] *interl*
2.9 over the ferry] *interl*
2.9 near the sea] *interl*
2.10 yellowish] *interl*
2.12 I] *before del illeg*
2.14 me most] 'me' *inserted*
3.1 I] *over* 'in'
3.2 Black] 'Glack' *in MS*
4.2 will] *inserted*
point omitted after: 2.12 'that'; 2.14 *2 points omitted after* 'trouble'

To W. D. Fox [October 1828]
1.1 stuffed] *after del* '&'
2.15 original] *after del illeg*

To W. D. Fox [29 October 1828]

5.3 at] *inserted*
point omitted after: 1.3 'Osmaston'; 1.4 'punctures'; 2.3 'where'; 2.11 '& although'; 3.4 'news'

To J. M. Herbert [3 October 1828]
1.7 When I arrived] *after del* 'After I left'
1.12 which] *after del* 'one'
point omitted after: 1.5 'Drus'; 1.13 'suppose'; 1.20 'pursuits'; 4.2 'stones'

To W. D. Fox [29 October 1828]
1.9 me] *interl*
1.11 mention] *after del* 'tell'
1.12 .'. yours . . . richer] *interl*
1.13 carabi] *above del* 'caabi'
3.1 sends] *interl*
4.1 at X^st Coll:] *interl*
point omitted after: 1.4 'affection'; 1.15 'not'; 1.16 'both new'; 1.18 'day'; 1.20 'find many'; 1.21 'great many'

To E. A. Darwin 21 December 1828
1.1 (Preceded . . . S.E.D.)] *square brackets in MS*

To W. D. Fox [24 December 1828]
1.3 He] 'H' *over* 'h'
1.4 £] *interl*
2.1 cannot] *before omitted comma*
2.2 not] *interl*
3.2 entry] *after del* 'entry'
4.1 recollect] 'reollect' *in MS*
6.4 hope] *after del* 'hope'
6.5 doings] *above del* 'doing'
point omitted after: 1.2 'I'; 1.2 'Dash'; 1.7 'Dromius'; 2.1 'Father'; 2.6 'Darwin'; 3.1 'which'; 4.3 'mine'; 4.6 'know'; 4.7 'your'

To W. D. Fox [7 January 1829]
1.1 to] *over illeg*
6.2 think] 'k' *over* 'g'
7.3 A] *over* 'an'
8.1 I . . . Fox] *inserted*
point omitted after: 3.3 'distinct'; 4.1 'expired'; 4.2 'Father'; 6.4 'difference'; 7.2 'it'

To W. D. Fox [25–9 January 1829]
1.3 have] *interl*
1.5 all your] *after del* 'your'
3.3 have] *interl*
4.1 Marten] *over illeg*
4.3 animals] '¹'a' *over* 't'
5.6 Studies] *over illeg*

5.8 have] *interl*
5.15 next] *interl*
point omitted after: 2.4 'doubt'; 3.1 'My'; 4.1 'probable'; 5.3 'next'; 5.3 'tell'

To W. D. Fox [26 February 1829]
1.3 getting the] 'the' *over* 'of'
2.2 day] *interl*
2.9 good] *interl*
3.8 things] *over illeg*
4.4 I have] 'have' *over* 'no'
5.4 only] *interl after interl del illeg*
5.4 Hope] *before del* 'only'
point omitted after: 1.2 'porter'; 2.2 'Times'; 4.9 'Mycetophagos'; 5.1 'Dyticus'; 5.1 'ordered'

To W. D. Fox [15 March 1829]
1.2 Your] *over illeg*
1.2 however] *interl*
1.4 untimely] *blot over* 'un'
2.1 it is] *interl*
2.1 will] 'w' *over* 'i'
3.4 not] *inserted*
5.6 situated] *interl*
point omitted after: 3.6 'there'; 4.1 'more I see'; 4.6 'till'; 4.6 'thought'; 5.1 'it'; 5.4 'understand'; 7.1 'that'; 7.1 'Volente'

To W. D. Fox 1 April [1829]
1.6 was not] 'not' *interl*
2.6 & as] '&' *after del illeg*
4.4 sky,] *comma over colon*
4.6 sight] 'si' *over* 'sy'
point omitted after: 1.6 'sure'; 2.8 'case.—'; 4.2 'first'; 5.1 'blackguards'

To W. D. Fox [10 April 1829]
1.3 are] *interl*
1.4 it:] 'it' *interl*
1.4 that] 'th' *over* 'wi'
1.5 as] *interl after del* 'have.'
1.5 have] *inserted*
point omitted after: 1.5 'nothing'; 4.12 'you'

To W. D. Fox [12 April 1829]
1.6 always] [1]'a' *over* 'al'
3.1 wrote to] *above del* 'received'
3.1 you] *over* 'your'

To W. D. Fox [23 April 1829]
point omitted after: 1.1 'thought'; 3.1 'that'; 3.2 'well as'; 4.1 'afraid'

To W. D. Fox [18 May 1829]
1.3 before] *interl*
2.3 female] *interl*
2.3 price 5[s]] *interl*
3.1 being] *interl*
3.1 upon] *before del interl* 'being'
4.1 go] *over* 'goes'
4.3 vast] *after del illeg*
4.5 Standish] 'd' *over illeg*
5.3 the] *interl*
8.7 so] *interl*
point omitted after: 2.2 'also'; 2.3 'female'; 7.5 'pleasure'; 7.11 'pretty'; 8.1 'that'; 8.8 'arrangements'

To W. D. Fox 7 June [1829]
1.4 portmanteau] *after del* 'trunk'
1.9 have] *over* 'had'
1.9 (] *over* ')'
1.11 tea] *before omitted comma*
2.6 Brett 10 '' 6] *inserted*
point omitted after: 1.10 'thought'; 1.10 'time'; 2.1 'which'; 2.4 'Clayton'; 4.3 'it'; 4.3 'where'

To W. D. Fox [3 July 1829]
1.3 this] *after del* '[till]'
1.3 &] *before del* 'Woo'
1.4 at last] *interl*
2.7 latterale] [1]'a' *over* 'i', [1]'e' *over* 'o'
2.10 rare] *interl*
2.12 the] *interl*
3.3 Catherine] 'C' *over* 'I'
3.4 to] *above del* 'from'
5.2 already] *above del* 'already'
point omitted after: 1.2 'conclusion'; 2.3 'pretty'; 2.7 'Calathus'; 2.10 'genera'

To W. D. Fox [15 July 1829]
2.3 glad of it] 'it' *after blot*
2.3 Jenyns] 'ns' *over* 'nyns'
2.6 thing] 'think' *in MS*
2.7 may] 'ay' *over* 'y'
5.2 lead] *above del* 'live'
5.2 a perfectly] 'a' *over* 'as'
5.6 in] *interl*
5.9 that from] 'that' *interl*
6.2 many] 'ny' *over* 'y'
point omitted after: 1.2 'now is'; 1.2 'not'; 1.3 'you, that'; 1.10 'sorry'; 2.1 'idle'; 2.7 'whenever'; 2.8 'forget'; 2.9 'little'; 2.10 'him.—'; 3.2 'wonders'; 4.1 'Recollect'; 4.1 'money'; 5.11 'Erasmus'; 6.1 'Be sure'; 6.2 'useful'

To W. D. Fox 29 [July 1829]
2.4 sometime] 's' *over* 'I'
2.6 between] *after del* 'from'
2.9 going] *before del* 'to s'
2.10 if you] 'you' *interl*
4.3 but] *altered from* '&'
point omitted after: 3.1 'hope'; 3.3 'surprized'

To W. D. Fox 26 [August 1829]
2.2 under] *after del* 'St'
3.4 your'] *after del* 'th'
5.3 do] *over* 'th'
6.1 also] *interl*
point omitted after: 1.1 '&'; 1.3 'And'; 2.1 'whether';
2.2 'Newcastle'; 4.4 'is'

To W. D. Fox 4 September [1829]
point omitted after: 1.3 'visit'

To W. D. Fox [22 September 1829]
1.5 it] *after del* 'his'
1.7 & take . . . London] *added*
4.3 you pretty] 'you' *interl*
point omitted after: 1.2 'it'; 1.3 'plans'; 3.2 'mind';
4.2 'insects'; 4.2 'trust'; 4.3 'Moor'; 8.1 'your'; 8.2
'Charles'

To W. D. Fox [15 October 1829]
2.1 was] *after del* '('
2.4 any] *before del* 'thing'
3.4 who] *over* 'she'
point omitted after: 2.1 'Meeting'; 3.1 'de'; 3.1 'Il';
4.1 'very'; 4.2 'it'; 8.1 'Charles'

To W. D. Fox [3 November 1829]
2.4 a great] 'gr' *over* 'de'
point omitted after: 1.1 'Aiken'; 1.3 'Imprimis,'; 2.2
'now I'; 2.6 'attendants'; 2.8 'Wednesday'; 4.1
'idle'

To W. D. Fox [3 January 1830]
1.4 here] *over* 'there'
3.4 wrote] *above del* 'have written'
5.3 lungs] *above del* 'langs'
5.4 for] *before del* 'the'
point omitted after: 2.2 'doubt will be very'; 2.4
'more'; 2.11 'Waterhouse'; 3.6 'skin:'; 4.3
'Senior'; 4.4 'plans'; 6.3 'it'

To W. D. Fox [13 January 1830]
2.1 books] 'b' *over* '[w]'
2.4 that] 'th' *over illeg*
3.3 Demetrias] 'etrias' *interl*

6.3 an] 'n' *inserted*
6.5 one] *interl*
8.2 good] 'g' *over* 'op'
point omitted after: 2.2 'spirit'; 2.7 'supposition'; 6.2
'civil'; 6.8 'hope'; 7.3 'spirits'; 8.4 'himself'; 8.6
'men'

To W. D. Fox [25 March 1830]
1.4 with] *interl*
1.6 joyful] 'joy' *over* 'si'
3.2 whom] *interl*
4.1 Coach] 'C' *over* 'c'
4.4 post.—] *before del illeg*
point omitted after: 1.1 'Go,!!! I am'; 2.3 'our'; 2.5
'beetles'

To W. D. Fox [1 April 1830]
1.2 taking] 'ta' *over del* 'ma'
3.2 many] 'ny' *over* 'y'
point omitted after: 3.5 'of'; 6.2 'reason'

To W. D. Fox [9 May 1830]
2.6 difficulties] *after del* 'dif'
2.10 night] *interl*
point omitted after: 1.2 'with'; 1.4 'benefit'; 2.11
'sight.—'; 3.1 'undecided.—'; 3.2 'else'; 4.2
'Norfolk.—'

To W. D. Fox [31 May 1830]
1.1 write] *over illeg*
1.5 write] *after del* 'wite'
2.1 paid] *after del* 'bee'
2.5 Cannot] *after del illeg*
4.2 have] *above del* 'them'
point omitted after: 4.2 'for you'

To W. D. Fox 25 [August 1830]
4.2 On] 'O' *over* 'a'
4.2 Monday we] 'we' *interl*
4.3 we took] 'we' *interl*
point omitted after: 1.6 'so that'; 1.14 'now'; 2.5
'know'; 4.6 'Monday'; 4.14 'taken some'; 5.2
'believe'

To W. D. Fox [8 September 1830]
1.2 write to me] 'to me' *interl*
1.4 on] 'o' *over* 'f'
4.2 me] *inserted*
point omitted after: 1.2 'say'; 4.3 'hear.'

To W. D. Fox [8 October 1830]
1.3 not] *interl*
2.3 myself] 'my' *over illeg*
2.4 that] *over* 'on'

3.4 absence] 'abs' *over illeg*
3.5 think] 'k' *over* 'g'
4.2 enough] *before del* 'of'
4.6 depends on] 'o' *over* 'f'
point omitted after: 4.4 'suppose'

To W. D. Fox 5 November [1830]
2.6 think he] 'he' *inserted*
point omitted after: 2.4 'Henslowe'; 9.3 'Jenyns'

To W. D. Fox [27? November 1830]
1.3 kept] 'p' *over* 'ep'
2.6 £3] *before del* '4'
point omitted after: 1.2 'articles'; 1.3 'kept'

To W. D. Fox [23 January 1831]
2.1 the] *over* 'a'
point omitted after: 1.3 'passing'; 2.3 'now'; 2.5
'pleasures'; 3.5 'think'; 3.5 'term'

To W. D. Fox [9 February 1831]
1.12 humbug] 'b' *over* 'g'
1.13 in] *over* 'it'
2.2 I possibly] 'I' *interl*
2.4 place] *after del* 'ple'
4.1 to] *interl*
point omitted after: 1.1 'answered'; 1.5 'present'; 2.1
'duty'; 2.2 '*if*'; 3.4 'is'; 5.1 'is'

To W. D. Fox [15 February 1831]
0.2 Tuesday] *after del* 'Tuesda'; *above del*
'Monday'
2.1 told] *after del illeg*
2.5 The Swan] 'S' *over* 's'
point omitted after: 2.7 '12.'

To W. D. Fox [7 April 1831]
2.1 botanizing] 'o' *over* 'a'
point omitted after: 1.4 'possession'; 1.8 'before'

To Caroline Darwin [28 April 1831]
1.10 quietly] *above del* 'smoothly', *with marginal
note* 'copied in error'

To W. D. Fox [11 May 1831]
1.8 practice] *before del* 'I'
2.4 given] *over* 'gave'
4.4 other] 'r' *over* 'rs'

To W. D. Fox [9 July 1831]
2.2 them] *over* 'it'

point omitted after: 1.4 'reasons'; 2.4 'former'; 3.3
'imagination'; 3.3 'fancy'; 3.7 'whom'

To J. S. Henslow [11 July 1831]
1.6 plumb] *after del* 'index'
1.11 suppose,] *before del* '[that]'
1.12 will not] 'not' *interl*
1.13 And] 'd' *over* 'for'
1.15 But] 't' *over* 'p'
1.22 has] 's' *over* 've'
1.30 you] *interl*
1.30 directed] *before del* 'it'
point omitted after: 1.12 'will not'; 1.14 'information'; 1.22 'manages'; 1.26 'A. Ways'; 1.27
'recollect'

To W. D. Fox 1 August [1831]
2.2 think] *interl*
point omitted after: 3.2 'Newcastle,'; 4.1 'am'

To J. S. Henslow 30 [August 1831]
1.3 have] *interl*
1.4 refuse] 'r' *over illeg*
1.7 &] *before del* 'his'
1.8 my] *after del* 'the'
2.5 (in Durham?)] *interl*
point omitted after: 1.6 'are'; 2.4 'Peacock'; 2.8
'Sedgwick'; 2.11 'although'; 4.4 'stock'

To R. W. Darwin 31 August [1831]
2.5 if] *after del* 'I'
3.1 cannot] *after del* 'd'
3.3 send] *after del* 'b'
4.1 kindness,] *before del* 'if I wanted anything'
point omitted after: 2.10 'expence'; 2.10 'think'; 6.1
'Frank'

To Susan Darwin [4 September 1831]
0.2 Morning] *interl*
1.6 he says] *interl*
1.8 that] *interl*
1.10 old;] *semi-colon altered from colon*
2.1 Henslow] *after del* 'To morrow I go Cap
Beaufort, & then shall call on Col: Fitzroy, to
whom Wood gives me letter of introduction to
find out where his brother is.— I believe he is
with the vessel at Plymouth, if so I shall go
there directly.—'
2.1 will] *above del* 'has'
3.3 to the great] *above del* 'now is'
3.6 Henslow] *interl*

4.2 trouble.—] *before del* 'Look, in my bedroom, what is the last number that I have got of Griffiths animal kingdom.—'
4.2 in bedroom] *interl*
4.3 or some such title,] *interl*
4.4 Metereology:] *before del* 'one by Leslie on'
4.5 bills:] *before del* 'to order 3 jointed hoops for catching beettles, like my former ones, only rather stronger, all to fit into one'
5.1 I] *after del* 'Will the Ba: '
5.1 if . . . convenient] *added*
5.2 ditto . . . bank?] *encircled*; 'what bank?' *added*
6.2 against] *interl*
6.2 young] *interl*
6.3 tells me] *interl*
point omitted after: 0.2 'Sunday'; 2.1 'town'; 3.4 'himself'; 4.3 'see'; 8.1 'you'

To Susan Darwin [5 September 1831]
1.4 say] *interl*
1.9 to do] *interl*
1.10 I am certain of] *interl*
1.21 I live] *before del* 'I'
1.21 him] *interl*
2.4 says] *interl*
3.2 to England, . . . way] *interl*
4.1 The] *altered from* 'Their'
4.1 manner] *interl*
5.10 cities] *above del* 'cities'
5.10 S.] *interl*
6.6 up] *over* 'in'
point omitted after: 1.1 'time'; 1.2 'excuse'; 1.2 'altering'; 2.4 'including'; 3.2 'home'; 4.1 'remain'; 5.1 'Beaufort'; 5.1 'start'

To J. S. Henslow [5 September 1831]
1.4 only] *after del* 'on'
1.5 care] *interl*
2.2 as] *over* 'in'
4.2 Foxes] *interl*
point omitted after: 1.2 'prosperous'; 1.3 'praise'; 1.7 'Beaufort'; 2.1 'Fitzroy'; 2.1 'is'; 2.7 'with'; 2.9 'probably'; 3.1 'anything'; 3.2 'sure'; 6.1 'letter'

To Susan Darwin [6 September 1831]
1.5 slippers] 'stippers' *in MS*
1.6 new] *interl*
1.6 (about . . . deep)] *interl*
2.5 to] *interl*
2.12 since] *after del* 'the'

3.2 not] *interl*
3.3 at ye Admiralty] *interl*
point omitted after: 1.7 'compass'; 1.8 'it'; 1.8 'thinks'; 1.9 'would be'; 2.8 'anywhere'; 2.8 'loaded'; 3.3 'writing'; 3.4 'like:'; 3.4 'daresay'; 3.7 'amusement'

To W. D. Fox 6 [September 1831]
1.12 wish] *interl*
2.3 your] *after del* 'an'
2.9 sending *immediately* the] *interl*
2.10 would] *above del* 'were'
2.13 an] *over interl* 'add'
4.1 I have] *after del illeg*
point omitted after: 1.5 'This'; 1.9 'I again'; 2.5 'did'; 2.6 'long'; 2.13 'minutes'; 2.13 'Cambridge'

To Susan Darwin [9 September 1831]
1.2 obliged] *interl*
1.4 letter] *after del* 'second'
2.3 he] *over illeg*
2.8 I] *over* ')'
2.9 even] *after del* 'to'
4.3 of gold could] *interl*
5.3 they really] *interl*
5.15 crowns] *after del* 'a'
7.4 convinced] 'v' *over illeg*
7.15 except nipples] 'nipples' *interl*
point omitted after: 1.4 'letter'; 2.1 'Beaufort'; 3.1 'beginning'; 3.5 'although'; 3.6 'Fitz'; 6.2 'proof'; 6.5 'bad'; 7.11 'bag'; 7.13 'Edward'; 8.3 'bore'

To J. S. Henslow 9 [September 1831]
1.7 pistols] 'pis' *over* 're'
1.10 in books] *before del* 'a g'
1.12 to advice] *interl*
1.13 me] *after del* 'to'
1.17 his] *interl after del* 'of'
1.17 impression] *after del* 'his'
2.9 am to] *interl*
2.15 I am] *interl*
point omitted after: 2.17 'were'

To Charles Whitley [9 September 1831]
2.2 the priviledge] 'the' *over* 'an'
2.10 capital] 'capilat' *in MS*
4.4 3] *over* 'th'
5.4 them] *interl*
7.1 me] *interl*
8.2 Caernarvon] *before del* 'Prof'

To Susan Darwin [14 September 1831]
1.2 of a] 'a' *interl*
1.3 scarcely] 'cely' *over illeg*
1.3 three] 'ree' *over* 'ere'
1.8 anything] *interl*
1.10 feel] *interl before del* 'I think'
2.2 sort] *before del* 'of its class'
2.6 there] 'r' *over* 'y'
4.2 voyage] 'o' *over* 'y'
point omitted after: 1.2 'London.'; 3.5 'next'; 4.4
'hopes'

To Susan Darwin 17 [September 1831]
1.3 go on] 'on' *over* 'to'
1.7 the] *altered from* 'they'
1.7 cabins] *after del* 'ship'
1.13 2] *interl*
2.4 whom] *above del* 'that'
3.1 have been] *above del* 'am'
3.2 ever] *after del* 'wh'
4.3 islands] 'istands' *in MS*
7.1 Shrewsbury] *after del* 'London'
point omitted after: 1.1 'daresay'; 1.15 'good'; 3.5
'mention'; 3.6 'shore'; 5.2 'marked'

To J. S. Henslow 17 [September 1831]
1.2 writing] *after del* 'not'
point omitted after: 1.15 'comfortable'; 1.16 'collection'; 1.16 'morning'; 1.18 'future.'

To W. D. Fox 19 [September 1831]
2.12 overcome it] 'it' *inserted*
3.1 tonight] 'night' *over* 'morr'
3.2 after . . . affairs] *interl*
3.3 (most . . . before)] *interl*
3.7 have] *interl*
3.11 up] *interl*
4.3 live] *interl after del* 'life'
point omitted after: 0.1 'I'; 1.2 'letter'; 2.8 'metereology'; 2.16 'judges;'; 3.5 'you'; 3.8 'spoilt'; 3.10
'forget'; 3.13 'accordingly'; 4.3 'years'; 4.5 'back'

To J. S. Henslow 28 [September 1831]
1.1 &] *after del illeg*
3.1 send] *after del* 'write'
point omitted after: 1.4 'ones'

To J. S. Henslow [4 or 11 October 1831]
1.3 consignment] 'g' *over* 'ge'
1.4 resourse] [1]'s' *over* 'c'
1.6 see] *after del* 't'

2.3 direction] 'ion' *inserted*
2.3 it was] 'it' *interl*
2.3 it off] 'it' *inserted*
3.2 he] *interl*
4.1 before] *before del* 'leaving'
point omitted after: 1.2 'me'; 1.5 'greatest'; 1.5
'Cambridge'; 2.6 'Cambridge'; 2.8 'suppose'; 2.9
'Brown'; 3.1 'Brother'

To J. S. Henslow 30 [October 1831]
1.3 should] *before del* 'have'
1.3 yet] *interl*
1.3 think] *over* 'thought'
1.20 little] *after del* 'of'
1.22 it] *interl*
1.29 2[d] & 3[d]] *interl*
point omitted after: 1.1 'so'; 1.2 'stupid'; 1.6 'such';
1.7 'know'; 1.11 'Father'; 1.20 'painted'; 1.21
'officers'; 1.23 'luxuriously'; 1.29 'parts'; 2.6
'compared'; 2.7 'better'

To J. S. Henslow 15 [November 1831]
1.13 W] *above del* 'E'
1.40 Perhaps] *after del* 'though'
point omitted after: 1.6 'Everything'; 1.6 'made so';
1.16 'So'; 1.17 'hope'; 1.20 'it'; 1.27 'may'; 1.31
'know'; 1.33 'whom'; 1.34 'heard'; 1.35 'sailor';
1.37 'last.'; 1.37 'When'; 1.40 'money'; 2.1
'perhaps'

To Charles Whitley 15 November [1831]
1.4 most] *after del* 'U'
1.5 Phys:[s]] 'y' *altered from* 'i'
1.6 at] *over* 'in'
3.3 things.—] *before del* 'I must do one of two
things'
point omitted after: 2.6 'dinner'; 2.7 'lies'; 2.9 'men';
7.1 'God'; 7.1 'Yours'

To W. D. Fox 17 [November 1831]
1.4 so] *interl*
1.7 it] *interl*
3.12 did] *interl*
3.13 what man] 'man' *interl*
3.15 that time] 'time' *interl*
point omitted after: 1.1 'suppose'; 2.2 'better'; 2.10
'longer'; 2.10 'Wales,'; 2.11 'Islands'; 3.1
'voyages'; 3.4 'prosperous'; 3.5 'serious'; 3.8
'enjoy'; 3.10 'all'; 3.11 'myself, that'; 3.12 'truth';
3.13 'think'; 3.14 'long'; 3.18 'into'

To J. S. Henslow 3 December [1831]
1.9 to know] *interl*
2.3 when] 'w' *over* 'in'
point omitted after: 1.9 'one'; 3.1 'this'; 3.3 'utility'

To R. W. Darwin 8 February — 1 March 1832
1.1 past] *above del* 'from'
2.8 a right] 'a' *interl before del* 'no'
2.23 cried] *interl*
2.28 islands] '*istands*' *in MS*
3.1 a net] 'a' *over illeg*
3.2 animals,] *before del* 'wh'
4.5 voyage] 'v' *over* 'g' *or* 'j'
6.2 away from it] *above del* 'at any place'
6.11 nearly] *after del* 'one'
7.16 there] *interl*
8.3 could] *altered from* 'can'
8.3 you] *before del* 'can'
8.4 tree] *interl*
8.4 of] *over* 'in'
8.4 exactly] *after del* 'the'
8.5 or tamarinds] *interl*
8.6 but of] 'of' *interl*
10.5 & situated] *after del* 'The'
10.10 (copied . . . journal)] *inserted*
10.19 be] *before del* 'th'
11.1 on the way in] *interl*
11.5 date] *interl*
11.8 all] *interl*
12.3 it] *interl*
15.1 I find . . . sisters] *inserted at head of letter,*
   encircled
point omitted after: 1.2 'meeting'; 2.1 'sailed'; 2.3
'proved'; 2.8 'Nobody'; 2.26 'drawback'; 3.7
'pleasanter'; 3.9 'leaving'; 3.10 'find'; 4.10 'log';
6.6 'more'; 11.2 'study'; 11.6 'officer'; 11.16 'Log'

To R. W. Darwin 10 February 1832
1.21 fresh] *interl*
4.3 having] *after del* 'a person'
4.3 a struggle] 'a' *interl*
point omitted after: 1.14 'S! Jago'

To Caroline Darwin 2 — 6 April 1832
1.15 this] *after del* 'it', *over* 'is'
1.18 a sentence] 'a' *over* 'or'
2.3 consists] *over* 'is that'
3.7 living] *over* 'sai'
3.7 on blue] 'on' *over* 'in'
5.6 Wickham] *before omitted* ' ' '
5.7 at them] *before del* ' ' '

6.2 this] *before del* 'have'
7.1 6] *over* '5'
7.10 I shall] 'I' *over* 'w'
7.16 impossible] *after del* 'improbable'
7.16 some] *after del* 'more', 's' *over* 'm'
7.17 ever] *interl*
7.20 (the (enjoyment . . . immense)] *interl*
8.1 had] *interl*
point omitted after: 1.8 'here'; 1.14 'interest'; 3.9
'bargain'; 5.1 'style'; 5.1 'alongside'; 5.13 'soon';
5.14 'those'; 6.12 'bears'; 9.1 'who'

To Caroline Darwin 25 — 6 April [1832]
1.3 returning] *after del* 'goi'
2.6 better] 'b' *over* 'm'
2.10 in] *interl*
2.13 French] *after del* 'en'
2.18 voyage] *after del illeg*
3.14 a man] *interl*
3.17 would] *after del* 'is to me'
5.2 one] *after del* 'I'
6.2 on Monday] *interl*
7.1 from you] 'from' *inserted*
point omitted after: 3.25 'from'; 4.1 'mention'; 6.8
'has'

To W. D. Fox May 1832
1.7 Teneriffe] *before del* 'first'
2.2 send] *above del* 'bring'
2.10 be] *interl*
2.17 delight] *after del* 'dl'
2.17 who] *interl*
3.4 many] 'n' *over* 'y'
3.6 our] *interl*
3.7 we] *over* 'one'
4.8 a mile] 'a' *interl*
point omitted after: 3.4 'life'; 3.4 'been in'; 3.8
'amusements'

To Catherine Darwin May — June [1832]
1.2 latter] *interl*
1.3 May] *after del illeg*
1.10 directed . . . club] *interl*
1.13 when I return] *interl*
1.15 marvelling] *after del* 'wonder'
1.23 who] *above del* 'he'
3.1 Herbert] *interl*
3.7 brother . . . Hunt] *interl*
3.12 Friday] *after del* '3ᵈ Tue'
3.14 should] *above del* 'are'
3.15 Till . . . Video] *interl*

3.16 than] *over* 'these'
3.16 weeks] *interl*
3.23 more] *interl*
4.3 course] 'o' *over* '⸤r⸥'
4.5 far as] 'as' *interl*
point omitted after: 1.8 'story;'; 1.24 'precarious';
2.1 'from'; 3.2 'always'; 3.4 'ship?'; 3.6 'Taffrill'

To J. S. Henslow 18 May — 16 June 1832
1.4 my's] 'y' *over* 'e'
1.9 voyage] 'v' *over* 'g'
1.17 from] *interl*
1.18 arrived] *over* 'came'
1.22 that] *over* '& I'
1.26 &] *inserted*
2.2 some] *above del* 'many'
2.2 Caryophyllea] ²'y' *over* 'i'
2.2 were] *over* 'are'
2.5 equalling] *after del* 'fa'
2.9 will] *over illeg*
2.11 Is] 's' *over* 't'
2.16 out] *interl*
2.23 Abrolhos] 'Abrothos' *in MS*
2.27 a month] 'a' *above del* 'some'
2.31 Rio] *interl*
2.35 preemenence] ³'e' *over* 'a'
3.9 send] *before del* 'home'
4.1 The] *after del* 'Cap'
4.3 not] *interl*
5.3 I] *after del* '&'
7.2 are] *after del* 'will'
7.4 rainy] *before del* 'days'
8.6 interval] 'rva' *over illeg*
point omitted after: 1.21 'coast'; 1.22 'from'; 1.23
'ignorance'; 2.7 '&'; 2.7 'up'; 3.7 'those'; 6.1
'box'; 6.2 'up'; 8.8 'Sedgwick'; 8.8 'know'

To J. M. Herbert [1–6] June 1832
1.2 interval] 'rval' *over* '⸤veral⸥'
1.3 how to] 't' *over* 'f'
1.3 forget one] 'one' *over illeg*
1.10 its open] 'its' *interl*
1.13 the happiness] 'the' *interl*
2.3 St Jago] *after del* 'the'
2.8 blue] *before del* 'of'
2.12 (1ˢᵗ Lieut)] *interl*
2.15 proh] 'r' *over* 'o'
4.4 tailor] 'i' *over* 'y'
4.11 should] *after del* 'you'
4.12 which] 'i' *over* 'o'
4.17 can] *over* 'even'

point omitted after: 2.6 'hammock'; 2.6 'forget the';
2.8 'weather'; 3.1 'boat'; 3.2 'boat'; 3.4 'medley';
3.5 'find'; 3.5 'water'; 4.16 'Lowe'

To Catherine Darwin 5 July [1832]
1.13 brought] 'b' *over* 'g'
1.15 tableaux] 'bl' *over* 'be'
2.12 on] *over* 'in'
2.12 coasts] 'c' *over* 'of'
2.14 dates] *after del* 'al'
point omitted after: 2.6 'Naturalists'; 2.13 'me.'

To Susan Darwin 14 July — 7 August [1832]
1.13 unpropitious] *before del* 'thing'
2.3 Back to M Video] *interl*
2.14 there] 're' *over* 'n'
5.9 to cover] 't' *over* 'f'
5.24 satisfaction] *interl*
5.25 branches] 'br' *over* 'ne'
5.25 previously] *interl*
5.26 several] 'severals' *in MS*
8.20 What] *before del* 'are'
point omitted after: 1.9 'inaccurate'; 2.6 'has'; 2.7
'cloth'; 5.8 'as it'; 8.12 'me'

To J. S. Henslow [23 July —] 15 August [1832]
2.8 accuse] *after del* 'of'
2.9 I] *after del* 'All'
2.14 mihi] *after del* 'mi'
2.14 objects] 'o' *over* 'an'
3.9 ¾ of] *above del* 'half'
3.15 water in] 'in' *above del* 'under'
3.17 one] *above del* '⸤four?⸥'
3.19 me] *interl*
3.23 send] *interl*
4.4 against] 'a' *over* 'f'
4.8 plains] 'ins' *over* 'nes'
5.11 shot] *interl*
5.11 a cwt] 'a' *altered from* 'an'
5.12 the wet] 'the' *interl*
point omitted after: 2.14 'is'; 3.16 'examined'; 3.21
'time'; 4.4 'you'; 5.12 'work'

To Caroline Darwin
    24 October — 24 November [1832]
1.3 any] *before del* 'th'
1.9 is] *interl*
1.10 a fort] 'a' *interl*
1.10 On] *before del illeg*
1.14 now] *after del* 'are'
2.8 me] *before del* 'from'

2.14 animals] *interl*
2.18 are attached] 'are' *interl*
4.8 me] *before del* 'saying'
4.11 or double] *interl*
5.8 HM] *before del* 'S'
6.3 is] *interl*
6.17 involuntarily] *altered from* 'involuntary'
7.3 15th.] '5' *over* '[6]'
7.7 from] *over* 'by'
7.10 known] *above del* 'existing'
7.19 even] *after del* 'where'
8.6 advance] *after del* 'advan'
point omitted after: 1.2 'afforded'; 2.10 'dimensions:'; 2.10 'sure'; 4.4 'food'; 4.11 'time'; 6.13
'mentioned'; 7.11 'plum cake'; 8.6 'here'

To J. S. Henslow
  [*c.* 26 October —] 24 November [1832]
1.3 surveying] *before del* '[in]'
1.4 Capt:] *after del* caret?
1.6 to] *over* 'in'
1.9 as this last] *interl*
1.12 by] *above del* 'bounded'
2.4 to] *over* 'of'
2.5 is] *over* 'in'
2.7 talk to] 'to' *interl*
2.11 1st] *before del* 'I have'
2.14 belonged] *above del* 'was'
2.14 or] *over* '&'
2.17 them] *interl*
2.18 genus] 'gen' *over* 'are'
2.20 large] *after del* 'very'
2.21 an] *interl*
2.27 have] *over* 'has'
3.3 (N^r. 710)] *interl*
3.4 bird] *over* 'ornitho'
3.17 be] *interl*
3.17 any] *interl*
3.18 animal] *interl*
4.1 Bahia Blanca] *interl*
4.2 the specimens] 'the' *over* 'my'
4.3 this letter] *interl*
4.4 the Lecture] 'the' *before del* 'the'
5.8 Novr. 24th.—] *interl*
point omitted after: 1.3 'time'; 1.5 'month'; 1.5
'this'; 1.11 'is:'; 2.3 'government'; 3.18 'Perhaps';
5.6 'frequency'; 5.19 'containing'; 5.21
'exceptions'

To W. D. Fox [12 — 13] November 1832
3.6 ye Tropics] 'ye' *interl*
3.12 Baron] 'B' *over* 'b'

4.1 are] *interl*
4.1 foul] 'l' *over* 'ld'
point omitted after: 1.1 'sooner'; 2.1 'that'; 2.8
'hear'; 3.5 'me'; 4.3 'Europæan'; 5.1 'yourself'

To Caroline Darwin 30 March — 12 April 1833
1.10 the most] 'the' *interl*
2.8 Ansons] *after del* 'Lord'
2.15 it did] 'it' *interl*
2.16 much] 'm' *over* 'a'; *before del* 'goo'
3.7 almost] *interl*
3.20 the voices] 'the' *altered from* 'their'
3.29 views] *interl*
3.31 nearly] 'y' *over* 'ed'
4.2 calm] *after del* 'cl'
5.2 organic] *after del* 'fossils'
5.4 equally] *interl*
6.1 fancy] 'c' *over* 'y'
6.2 we are] 'are' *interl*
8.2 such] *after del* 'to'
point omitted after: 2.8 'climate'; 5.6 'them'

To J. S. Henslow 11 April 1833
0.1 April] 'A' *over* 'M'
1.8 stove] *altered from* 'staved'
1.12 with] *over illeg*
1.20 forget,] *before del* 'the'
1.24 kingdom] 'k' *over* 'w'
6.1 curious to] *before del* 'some great'
point omitted after: 1.4 'time'; 1.6 'days'; 1.16
'houses'; 7.2 'formation'

To Catherine Darwin 22 May — 14 July 1833
2.13 that] *interl*
4.10 go] *over* 'gone'
5.3 be with] 'be' *over* 'go'
5.9 from time] *interl*
6.7 letter.] *before del* 'to you'
7.12 Portuguese] 'P' *over* 'p'
7.13 &] *before del* 'the'
7.16 (I believe)] *interl*
8.5 (who . . . Schooner)] *interl*
8.7 not] *interl*
8.11 Conception] 'C' *over* 'th'
8.19 an] *interl*
9.16 quadrupeds] *above del* 'animals'
9.16 which] *before del illeg*
9.17 quadrupeds] *above del* 'animals'
9.17 Tell] *after del* 'in the'
10.11 very] *before del* '[dl]'
12.6 Hutton] *before del* ' ' '
12.15 must] *before del* 'not'

12.20 &] *before del* 'la'
point omitted after: 4.4 'South-seas'; 4.6 'geology';
5.1 'own'; 5.5 'anyone'; 5.12 'mentioned'; 9.4
'direct'; 10.9 'ones:'; 12.16 'motions.'; 12.17
'daresay'

To W. D. Fox 23 May 1833
4.10 as savage] 'as' *inserted*
5.3 you] *inserted*
5.4 cordially] 'cor' *over illeg*
point omitted after: 1.6 'Beagle'; 4.9 '!'; 6.4 '19'

To J. M. Herbert 2 June 1833
1.12 it] *over illeg*
1.18 glacial] *over illeg*
2.9 care] *interl*
8.3 were] *above del* 'will be'
8.4 case] *after del* 'the'
point omitted after: 1.2 'house'; 1.8 'saw'; 1.11
'place'; 1.12 'at'; 1.14 'country'; 3.10 'stories'; 5.2
'Read'; 5.2 'Heads'; 8.1 'just'; 9.4 'Reflect'

To J. S. Henslow 18 July 1833
1.3 the winter] 'the' *over* 'our'
1.6 hearing] *before del* 'all'
1.15 be] *interl*
2.1–4 come . . . them,] *added where CD must have
left space for insertion of details of the cargo*
2.4 smallest] *after del* 'two'; 'est' *added*
2.4 & flat] *interl*
2.4 contains] 's' *added*
2.5 it is] *above del* 'they are'
2.8 the] *over* 'them'
2.10 then] *interl*
3.2 could] *after del* 'f'
3.9 have] *above del* 'have a very'
3.13 her] 'r' *over illeg*
4.3 the fine] 'the' *over* 'all'
4.6 made . . . voyage] *interl*
point omitted after: 2.2 'contains'; 2.5 'spirits';

To Caroline Darwin 20 September [1833]
2.2 except] *after del* 'b'
2.3 army is] 'is' *above del* 'are'
2.4 progress] *above del* 'journey'
2.15 the Rio] 'the' *above del* 'near'
2.23 a] *before del* 'n'
3.3 the Pampas] 'the' *over* 'this'
4.3 say] *after del* 'remembe'
6.1 My] *over illeg*
point omitted after: 1.5 'having'; 2.7 'be'; 2.9 'companion.'; 5.4 'leaving'

To Caroline Darwin 23 [October 1833]
1.3 Northern] 'N' *over* 'n'
1.10 great] *after del* 'im'
1.10 to] *after del* 'on'
1.11 we] *after del* 'I'
1.11 with] *after del* 'in'
1.14 amongst] *above del* 'with'
1.19 whose] *above del* 'house'
3.2 from . . . misfortunes] *interl*
5.2 whom they] 'they' *interl*
5.3 Indian] *final* 's' *del*
5.3 years old] *interl*
5.4 ugly] 'u' *over illeg*
5.6 English] *interl*
5.8 whether] *before del* 'he will'
5.10 or boat] *interl*
6.3 good] 'd' *over* 'g'
7.1 dear Caroline. Yours] *interl*
9.1 Capt:] *after del illeg*
9.1 some] *after del illeg*
point omitted after: 1.6 'Pampas.'; 1.17 'happened'

To W. D. Fox 25 October 1833
0.2 October] *above del* 'September'
1.11 these] *before del* 'orders'
2.1 who] *interl*
3.3 to] *after del* 'the'
3.9 have] *after del* 'm'
point omitted after: 1.13 'the open'; 3.6 'America is';
4.1 'reestablished'

To Frederick William Hope 1 November 1833
1.3 from] *interl*
1.6 been] 'b' *over* 'c'
1.10 of the] *interl*
1.10 Heteromeri] 'i' *over* 'ous'
1.11 one] *four points omitted after*
1.15 of the] *interl*
1.23 de Janeiro] *interl*
1.27 Netley] 'N' *over illeg*
1.28 amongst] *over* 'in'
1.29 Judging] 'Ju' *over* 'Th'
2.1 I] *over* 'C'
point omitted after: 1.21 'net'; 1.21 'carry'

To J. S. Henslow 12 November 1833
0.2 November] *below del* 'Sept' *and del* 'Oct'
1.8 evaporation] *after del* 'ep'
1.9 require] *after del* 'be'
1.22 4] *over* 'Am'
2.6 one] *over* 'an'
3.1 I shall] *before del* '[mo]'

point omitted after: 3.7 '&c.'; 5.2 'Werner'

To Caroline Darwin 13 November 1833
1.7 now] *interl*
1.13 nature] 'n' *over* 'c'
1.17 I take] 'I' *over* 'of'
1.17 so] *before del* 'ut'
2.4 on the globe] *interl*
3.7 little] *above del* 'few'
point omitted after: 1.8 'formation'; 1.18 'made';
1.22 'mainly'; 4.1 'all &'

To J. S. Henslow March 1834
2.2 worth] *interl*
2.3 printed] *interl*
2.4 will] *before del* 'to'
2.9 depend on] 'on' *over* 'of'
2.10 dependent] *after del* 'in'
2.10 on] *over* 'f'
3.1 far . . . country] *interl*
3.4 are] *over* '&'
4.1 the] *over* 'my'
4.1 assure you] 'you' *interl*
4.4 collected] *after del illeg*
4.10 had] *interl*
4.14 This] *over* 'The'
4.14 bed] *after del* 'Patag'
4.15 covered by] 'by' *above del* 'in'
4.16 more] *over* '7'
4.17 of South . . . America] *interl*
4.18 their] *altered from* 'they'
4.20 hind] 'h' *over illeg*
4.20 are] *above del* 'were'
4.22 Pampas] *before del* 'plains'
4.25 600] '6' *over illeg*
5.1 Ammonite] *above del* 'Nautilus'
5.3 which] *before del* 'I'
5.5 leaves] *after del* 'presen'
5.6 On] *over illeg*
5.6 are] *over* 'is'
5.9 know] *interl*
5.11 to cry] *interl*
5.11 big] 'g' *over* 'd'
5.15 tell] *altered from* 'tells'
5.17 they are] 'are' *interl*
5.17 as] 'a' *over* 'f'
5.20 each] *before del* 'to'
6.3 Corallines in] 'in' *over* 'f'
6.8 another] *above del* 'another'
6.9 their] 't' *over* 'o'
6.11 & scales] '&' *over comma*

7.1 me] *interl*
8.6 who] *interl*
9.2 as] *over* 'by'
10.3 to Rio] *interl*
15.2 just] 'j' *over* '&'
15.5 by water] 'by' *over* '[for]'
point omitted after: 2.1 'astonished'; 3.5 'find'; 4.10
'had'; 4.18 'period'; 6.5 'cells'; 6.12 'geographical'; 7.6 'last'; 8.5 'effect'

To Edward Lumb 30 March 1834
2.3 few] *interl*
2.15 scenes] *after del* 'a'
3.4 M^r] *over* 'Chas'
point omitted after: 1.7 'Keen'; 1.8 'Negro'; 2.3
'besides'; 2.8 'narrower'; 2.9 'Cape'; 3.1 'hear';
3.5 'me'; 3.7 'Charles'; 4.1 'must'; 4.2 'suppose';
4.3 'about'; 4.3 'Chupat'

To Catherine Darwin 6 April 1834
1.2 another] 'other' *added*
1.15 some] *after del* 'the'
2.3 tell] *interl*
3.4 wife] *before omitted* ' ' '
3.12 by] *interl*
4.5 a little time since] *above del* 'the other day'
4.8 an] *interl*
5.11 close] *after del* 'near'
5.12 !!Darwin!!] *exclamation points above and below*
'Darwin' *omitted*
5.13 the Ship] *interl*
6.2 with] *before del illeg*
8.3 cross] *above del* 'enter'
12.1 her] *interl*
12.2 Solomon] 'S' *over* 's'
point omitted after: 1.14 'find'; 2.3 'former'; 2.8
'proportion'; 2.19 'own country'; 2.22 'skins'; 3.4
'wife'; 3.7 'Englishmen'; 4.1 'island'; 4.1 'place';
5.1 'Rio'; 10.3 'Christian'

To Catherine Darwin 20−9 July 1834
2.7 chiefly] *above del* 'principally'
3.3 y^e] *inserted*
3.5 follow] *after del* '[dir]'
3.7 y^e] *inserted*
3.8 view] *after del* 'the'; *before omitted comma*
3.13 Sarmiento] *after del* 'M.'
3.16 exit] *above del* 'passage'
5.2 ever] 'r' *over* 'n'
5.10 more] *above del illeg*
5.14 with it] *interl*
5.15 even] *after del* 'de'

5.15 descriptions] *before del* 'of them'
6.7 lives] 'v' *over* 'f'
7.6 the] *interl*
7.6 others] *before del* 'they'
8.2 most] 'm' *over* 'a'
8.7 setting] *interl*
9.3 he tells me] *before omitted comma*
9.8 me in] *interl*
10.7 the second] *above del* 'one'
10.9 shall] *after del* 'care'
10.14 ?] *above omitted comma*
point omitted after: 1.2 'whole'; 2.5 'collect'; 7.1
'July'; 8.5 'del'; 8.8 'Cader-Idris.— The'; 10.8
'two'; 11.2 'Charles'

To Charles Whitley 23 July 1834
1.5 are doing] 'are' *interl*
1.7 last] *interl*
1.17 before] *above del* 'formerly'
1.17 can] *interl*
2.4 are] *over* 'have'
2.8 hereafter] *after del* 'now'
3.1 more] *above del* 'so'
point omitted after: 4.9 'friends'

To J. S. Henslow 24 July — 7 November 1834
1.1 are] *over* 'were'
1.6 I suppose] *interl*
1.7 Not] *after del* 'Not'
1.9 is] 'in' *in MS*
2.2 the] *over* 'its'
2.4 at] *interl*
2.4 which] *over* 'when'
2.4 times] *interl*
3.3 describe] *after del* 'notes'
3.9 (new to me)] *interl*
3.15 which] *before del* 'we were doubt'
4.7 may] *interl*
4.7 enjoy] *before del* 'to'
5.8 field] *above del* 'mass'
5.9 the great] 'the' *over* 'this'
5.16 wish] 'w' *over illeg*
5.16 it] *over* 'mo'
6.3 like that] *interl before del* 'of'
7.3 2] *interl*
7.5 all] *interl*
8.1 can] *interl*
9.1 in my books] *interl*
9.2 alcyonium] *interl*
9.7 Head] *before del* '& an Escara.'

9.11 contained] 'ed' *over* 'ing'
9.19 me] *interl*
9.23 suppose] *above del* 'believe'
10.5 me;] *before del* 'if'
11.6 at] *interl*
12.8 of greenstone] *interl*
12.21 possesses] 'poss' *over illeg*
14.4 October] *above del* 'September'
14.8 terrestrial] *interl*
14.14 Baths] *above del* 'Springs'
15.5 steer] *above del* 'sail'
point omitted after: 7.4 'different'; 9.17 'coast'; 10.4
'from'; 12.24 'believe'; 14.3 'received it'; 14.20
'doubts'

To Caroline Darwin 9—12 August 1834
1.4 am] *above del* 'was very'
1.9 or markers] *interl*
1.19 shall] *above del* 'should'
2.4 But . . . bulky.] *interl*
3.3 in the] 'in' *interl*
4.1 Geological] 'og' *over* 'g'
4.5 whilst] '1st' *over* 'ch'
5.8 Yours] 's' *added*
point omitted after: 1.13 'Another point'; 1.14
'Surgeons'; 1.15 'all'

To J. S. Henslow 4 October 1834
0.2 1834] '4' *over* '5'

To Caroline Darwin 13 October 1834
0.2 October] *above del* 'August'
1.15 slept] *after del* 'staid'
1.30 expense] 's' *over* 'c'
3.9 voyage] *over illeg*
point omitted after: 2.16 'New'

To Catherine Darwin 8 November 1834
1.3 has] *interl*
1.4 Ship] *before del* 'for'
1.12 spirits, &] *before del* 'as Byn'
1.21 much of] 'of' *over* 'as'
2.5 much] *above del* 'any'
2.10 England] *interl*
2.10 returning to] *above del* 'seeing'
3.13 & in] *above del* 'for the next'
3.16 completion] *after del* 'survey of'
3.16 of S.W coast] *interl*
4.12 however in] *before del* 'there being'
point omitted after: 3.14 'endured'; 4.15 'hind most'

To J. S. Henslow 8 November 1834
2.2 your] *interl*
2.2 to] *after del* 'f'
point omitted after: 1.5 'prospects'

To W. D. Fox [7–11] March 1835
2.11 & Anaspis . . . Osmaston)] *added*
3.4 Valdivia] 'V' *over* 'B'
point omitted after: 1.1 'What'

To Caroline Darwin 10–13 March 1835
1.6 know] *before del* 'cet'
2.1 after receiving this] *interl*
2.2 then] *before del* 'one letter to yᵉ Isle of France'
3.5 tiles] 't' *over* '&'
3.15 one] *after del* 'perhaps'
3.21 what] *after del* 'how'
3.24 within] *after del* '⌈beh⌉'
4.4 it to] 'it' *over* 'the'
4.7 to pick me up,] *interl*
4.8 will] *over* 'is n'
4.13 shall] *after del* 'am'
4.15 my] *interl*
4.15 money] *after del* 'th'
4.16 anyhow] *interl*
5.3 And] *before del* 'after'
5.8 man] *after del* 'of'
6.9 & Commodores] '&' *before del* 'the'
6.17 is] *above del* 'was'
point omitted after: 2.8 'long;'

To J. S. Henslow [10]–13 March 1835
1.3 September] *after del* 'December'
1.12 now] 'n' *over* 'm'
2.11 has been] *above del* 'is'
2.13 recent] *interl*
3.3 Carabi] 'i' *over* 's'
3.6 case] *after del illeg*
3.7 nearly] *after del illeg*
3.9 live] *after del* 'a'
3.18 in the mean time] *interl*

To J. S. Henslow 18 April 1835
1.3 the truth of] *interl*
1.4 will appear] 'I fear' *interl in pencil, probably by JSH after* 'will'
1.9 Eastern] *after del illeg*
1.21 has taken place by] *interl*
3.8 amongst Strata] *interl*
3.8 ejected] *interl*
3.9 show] *above del illeg*

3.11 (here)] *interl*
4.1 Limestones] *interl*
4.4 substance in] 'in' *over* 'of'
5.6 guess] *over* 'give'
5.8 & few] *interl*
6.2 mass] *interl*
6.2 on] *above del illeg*
6.7 first] *above del* 'above'
6.7 pebbles of] *interl*
6.8 (& deposition . . . part)] *interl*
6.10 pass] *interl*
7.9 upper] *interl*
7.12 ought] *over* 'might'
7.16 (particularly characteristic)] *interl*
8.7 30–40] *interl*
8.9 Like Lot's wife] *interl*
9.5 consequence] *before del* 'the'
9.9 or] *above del* '&'
10.3 for] 'f' *over* 'o'
10.4 that] *interl*
11.9 Paris] *before del* 'to'
12.5 chiefly] *interl*
12.10 many] *interl*
12. or berrys] *interl*
12.13 ever] *after del* 'have'
15.3 with] *interl*
point omitted after: 3.11 'rocks'; 4.5 'stupendous';
6.3 'Micaceous'; 6.4 'red'; 6.10 'Uspallata'; 6.10
'also'; 9.1 'proof'; 9.2 'beds'; 9.5 'consequence';
10.6 'formed'; 12.12 'grow'; 15.1 'reach'

To Susan Darwin 23 April 1835
1.11 none] *before del* 'so'
1.15 afford] *before del illeg*
1.20 me] *interl*
1.20 this years] *interl*
2.8 a quantity] 'a' *altered from* 'an'; 'quantity' *after del* 'extent'
2.14 about] *interl*
4.2 without] *before del* 'without'
5.3 so soon] *interl*
5.5 his] *above del* 'their'
6.3 unequal] 'un' *interl*
6.4 trembling] 'g' *over* 'gs'
6.9 to delay] 'to' *interl*
10.1 with] *after del illeg*
point omitted after: 1.19 'up'; 6.1 'Fitz Roy'; 7.12
'South'; 7.15 'going to'; 9.1 'a letter'

To Catherine Darwin 31 May [1835]
2.7 road] *interl*
2.11 or] *over* '&'

4.7 more] 'm' *over* 'e'
point omitted after: 2.7 'it'; 2.11 'buy'; 2.13 'morn-ing'; 3.7 'got'

To Caroline Darwin
  [19] July — [12 August] 1835
1.4 since] *before del* 'Coqu'
1.8 of Chili] 'of' *interl*
2.6 Shortly after] *above del* 'when'
2.8 a great . . . Fitz Roy] *interl*
2.12 nearly] *interl*
2.13 Court-Martial] *before interl del illeg*
3.7 have] *after del* 'now'
4.3 ones] *interl*
5.1 July 12th] *encircled in MS*
5.4 on . . . Deck] *interl*
5.8 have compelled] 'have' *interl*
5.8 have left] 'have' *above del illeg*; 'ft' *over* 'ave'
5.11 y^e] *interl*
5.17 till] *before del* 'the'
point omitted after: 5.2 'to'; 5.9 'left'

To W. D. Fox [9 — 12 August] 1835
1.1 other] *interl*
1.2 was] 'w' *over illeg*
2.11 have] *interl*
3.3 are] *interl*
4.7 Mr Lyells] *after del* 'Ch'
5.4 Europe] *before del* 'ver'
6.5 soon] *interl*
point omitted after: 1.7 'friends'; 2.6 'Beetle',; 3.4 'shorter'; 4.3 'interest'; 4.9 'parts'

To J. S. Henslow 12 [August] 1835
1.2 it] *over illeg*
1.6 Conway] *after del illeg*
1.10 to . . . letter] *interl*
1.16 strata] *interl*
1.16 thick] *interl*
1.21 produced] 'pr' *over* 'de'
2.4 the] *altered from* 'their'
2.14 varying] *interl*
point omitted after: 1.15 'Volcanoes'

To H. S. Fox 15 August 1835
point omitted after: 2.10 'there'; 2.17 'ever'

To [Alexander Burns Usborne]
  [c. 1 — 5 September 1835]
1.4 or fragments] *interl*
1.12 such . . . &c] *interl*

1.17 on Paper] *interl*
1.19 (& reasons for)] *interl*
1.29 bed.] *before del* 'or'
3.2 label] 'tabel' *in MS*
3.3 but] *before del* 'not'
4.3 imbedding] *after del* 'stone'
point omitted after: 1.1 'ocean'; 1.9 'Beach'; 1.13 'manifest'; 1.14 'shells'; 3.1 '*prevailing*'; 4.2 'stratified'

To Susan Darwin 3 [September] 1835
1.4 the extra] *above del* 'more'
2.7 here] *interl*
2.14 sing] *interl*
3.3 will] *interl*
3.6 there] *over* 'they'
point omitted after: 2.15 'ever boy'

To Caroline Darwin 27 December 1835
1.8 the] *over* 'my'
1.9 reckoned] *above del* 'counted'
1.10 but] 'b' *over* '&'
1.12 eight] 'e' *over* 's'
2.3 against] *above del* 'under'
2.7 for] *over* '[five]'
3.3 but] *interl*
4.1 how] 'h' *over* 'I'
5.6 ever] *interl*
5.6 thing] 't' *over* 'f'
5.10 Queen] *after del* 'Island'; 'Island' *above del* 'mountains'
6.2 shall] *after del* 'now are &'
6.9 I] *after del illeg*
6.10 know] *interl*
6.10 that they] *interl*
7.5 so on] 'on' *over* 'as'
8.4 can say,] *interl*
9.2 I received] *interl*
point omitted after: 1.7 'growl'; 2.8 'days.'; 5.5 'seen'; 7.1 'much'; 7.6 'letter'; 8.3 'slumbers'; 8.4 'Quarter'; 8.6 'd—d'

To Philip Parker King [21 January 1836]
point omitted after: 1.1 'evening,'; 1.13 'Monday'; 2.1 'Charles'

To Susan Darwin 28 January 1836
0.2 6] *over* '5'
1.3 drawn] 'w' *over illeg*
1.4 or rather] *interl*
1.13 home,] *before del* 'what'

2.7 an] *interl*
2.9 prophecyed] 'cy' *over* 'z'
2.16 even] *interl*
5.3 foretold by] *interl*
5.4 oracular] 'oracularar' *in MS*; ²'r' *over* 'l'
point omitted after: 1.7 'New'; 3.1 'night'; 3.4 'such';
3.7 'King'; 4.1 'Sound'

To J. S. Henslow [28–9] January 1836
2.9 acre] *before del* 'fetched by Auction'
4.10 layers] *above del* 'streams'
point omitted after: 1.8 'again'; 2.14 'North'

To Catherine Darwin 14 February 1836
1.7 here] *interl*
3.12 If] 'f' *over* 't'
3.13 country,] *before omitted semi-colon*
3.14 home] *after del* 'with'
4.1 letter] *above del* 'year'
5.1 to me] *before del* 'to'
6.2 an] *inserted*
6.2 fact] *interl*
point omitted after: 0.1 'Van'; 1.6 'board'; 1.7
'than'; 1.7 'Sydney—the'; 2.1 'particular'; 2.4
'he'; 2.5 '&'; 2.6 'not'; 2.10 'know'; 2.10 'pleasant'; 3.2 'know'; 3.2 'English'; 3.15 'considering'; 3.15 'important'; 6.2 'unquestioned'

To W. D. Fox 15 February 1836
1.4 otherwise] *interl*
2.1 that] *over* 'which'
4.4 of the] 'of' *interl*
4.8 at] *over* 'of'
4.9 Man] 'Ma' *over* 'Me'
4.12 imperfectly] *after del* 'anything'
4.15 hundreds,] *before del* 'of'
4.22 Emancipists] *above del* 'convicts'
point omitted after: 3.3 'existence'; 3.11 'me'; 3.21
'Knighthood.'; 3.23 'wish'; 4.1 'now'

To Caroline Darwin 29 April 1836
3.8 object] 'c' *interl*
4.4 different] 't' *over* 'ces'
4.4 now] *interl*
4.7 kingdoms] 'g' *over* 'd'
4.10 am just] *after del* 'find'
5.5 or any other] *interl*
5.5 club] 'l' *over* 'u'
5.7 whom] 'm' *added*
6.3 rare] *after del* 'rap'
6.3 (in ships)] *interl*

6.5 you] *interl*
6.11 would have sufficed] *interl*
7.7 entirely] 'en' *over* 'i'
point omitted after: 3.4 'country'; 3.6 'difference';
3.10 'continues'; 4.3 'Helena'; 4.13 'clearness';
6.5 'time'; 7.1 '6'; 7.2 'Good'; 7.3 'converted'

To Catherine Darwin 3 June [1836]
1.8 dozen] 'n' *over* 'y'
5.4 myself] *after del* 'much'
5.7 young] 'g' *over* 'd'
5.7 down] *interl*
6.4 of a] 'a' *inserted*
8.3 for] *interl*
9.3 not know] 'not' *inserted*
point omitted after: 1.11 'knew'; 2.7 'directed to';
3.3 'wall'; 5.6 'fields'; 5.7 'trees'; 6.4 'on'; 8.1
'suspicion'; 8.2 'time'

To J. S. Henslow 9 July 1836
point omitted after: 1.3 'time'; 1.4 'Professor'; 1.4
'Sedgwick'; 1.8 'displayed'; 1.9 'exclaim'; 1.10
'domum' '; 4.4 'me'; 5.3 'exceedingly'; 5.5
'alone'; 5.5 'country,'

To Caroline Darwin 18 July 1836
1.4 at Ascencion] 't' *over* 's'
1.12 cross] *after del* 'reach'
1.15 of England] 'o' *over* 'i'
1.21 Who] *altered from* '[What]'; *before del* 'eve'
1.21 lady's] 'y' *over* 'ie'
2.6 nearly] *interl*
4.2 Club] 'l' *over* 'u'
point omitted after: 1.1 'moment'; 1.4 'chance'; 1.7
'Ship'; 1.12 'Atlantic'; 1.19 'them'; 1.22 'turns';
2.3 'clouds'; 2.4 'part'; 2.5 'ocean'; 3.3 'works';
4.3 'wandering'; 4.7 'time'

To Susan Darwin 4 August [1836]
1.8 November; the] 't' *over* 'I'
1.9 all] 'a' *over* 's'
2.3 be] *after del* 'You'
2.3 know] *over* 'see'
2.10 stately] *after del* 'splen'
2.12 vivid] *after del* 'p'
3.7 from . . . scribble] *interl*
3.11 very] 'y' *over* 'd'
point omitted after: 1.9 'sorts'; 2.3 'surprised'; 2.4
'pleasure'; 2.4 'place'; 2.7 'find'; 2.12 'able to';
2.14 'us'; 3.3 'least'; 3.6 'gives'; 3.6 'hopes'

To Caroline Darwin 24 October [1836]
2.3 is part] *the copy reads* 'in parts'

To Charles Whitley 24 October [1836]
2.3 fulfilling] ¹'l' *over* 'll'
2.4 As] *over* 'I', *before del* 'left'
2.12 a most] 'a' *over* 'as'
2.17 arrangement] ¹'a' *over illeg*
4.1 you] 'ou' *over* 'our'
point omitted after: 1.3 'line'; 2.1 'most'; 2.2
'caterpillars'; 2.3 'did'; 2.5 'letter'; 2.6 'idea'; 2.7
'arrives'; 2.12 'friends'; 2.16 'shall'; 2.17 'present'; 2.18 'I know'; 3.1 'a'; 3.5 'over'

To J. S. Henslow [30 – 1 October 1836]
1.6 card] *before del* 'to'
1.13 I] *before del* 'do'
1.24 people] *after del* 'any one'
1.27 several] *above del* 'some'
1.29 you] *interl*
1.30 see] *after del illeg*
1.31 For] *above del* 'But'
1.36 societies] 'ies' *over* 'y'
2.1 think] *interl*
2.2 which] *interl*
2.2 himself] *interl*
2.16 plan] *interl*
2.18 clearly] *interl*
3.4 announce] *above del* 'send'
3.6 I . . . &] *interl*
3.8 far] *interl*
3.12 the two] 'the' *interl*
3.12 had] *after del* 'worn.'
3.16 that of] *interl*
4.15 was] *altered from* 'all'
4.18 some] *after del* 'those'
point omitted after: 1.12 'animals'; 1.13 'met'; 1.19
'Museum'; 1.23 'Cambridge'; 1.40 'with'; 1.41

'whether'; 2.9 'lodgings'; 2.12 'College'; 2.12
'large'; 3.3 'Perhaps'; 3.5 'leave the heavy'; 3.11
'Botanists'; 7.3 'boxes'

To J. S. Henslow [1 November 1836]
1.3 boxes] *after del illeg*
1.3 Marsh's] *after del* 'the'
3.1 attending] *before del* 'pap'
point omitted after: 3.9 'surprised'

To W. D. Fox 6 November [1836]
4.3 the] *interl*
5.2 capital] 'capilat' *in MS*
5.4 your] *after del* 'the'
point omitted after: 2.1 'know'; 5.4 'world'; 6.7
'proposed'; 8.2 'be'; 9.1 'me'

To Charles Wilkes [7 November 1836]
point omitted after: 1.6 'case'

To W. D. Fox 15 December [1836]
2.2 you] 'y' *over* 'g'
2.4 do] 'd' *over* 'w'
2.6 William] ²'i' *over* 'a'
3.8 him] *interl*
3.12 a little] *interl*
point omitted after: 3.8 'father'; 4.7 'horse!'; 6.1
'Madam'

To Richard Owen 19 December [1836]
2.2 the models] *above del* 'it'
3.4 And] *over* 'It'
3.5 thing] 'think' *in MS*
point omitted after: 1.1 'Sir'; 1.3 'because'; 1.4
'country'; 1.5 'at'; 1.7 'hesitated'; 1.7 'you'; 1.8
'returns'; 1.12 'knight'; 2.2 'because'; 2.5 'think';
2.7 'back'; 2.8 'them'; 3.2 'bone'; 3.5 'chance'

BIBLIOGRAPHY

The following bibliography includes all books and papers referred to in this volume by author—date reference or by short title. Short titles are used for some standard reference works (e.g., *DNB*, *OED*), and for CD's books and papers and editions of his letters and manuscripts (e.g., *Descent, Red notebook*). Other references are given in author—date form.

ADB: *Allgemeine deutsche Biographie*. Under the auspices of the Historical Commission of the Royal Academy of Sciences. 56 vols. Leipzig: Duncker and Humblot. 1875—1912.

Allen, Peter. 1978. *The Cambridge Apostles*. Cambridge: Cambridge University Press.

Alum. Cantab.: *Alumni Cantabrigienses. A biographical list of all known students, graduates and holders of office at the University of Cambridge, from the earliest times to 1900.* Compiled by J. A. Venn. Part II. From 1752 to 1900. 6 vols. Cambridge: Cambridge University Press. 1940—54.

Alum. Oxon.: *Alumni Oxonienses: the members of the University of Oxford, 1715—1886.* By Joseph Foster. 4 vols. Oxford: 1888.

Andrews, Alexander. 1859. *The history of British journalism, from the foundation of the newspaper press in England, to the repeal of the stamp act in 1855.* 2 vols. London.

Apperley, Charles James [Nimrod]. 1837. *Memoirs of the life of the late John Mytton, esq., of Halston, Shropshire . . . with notices of his hunting, shooting, driving, racing, eccentric and extravagant exploits.* 2d ed. London.

Archer, William. 1890. *William Charles Macready*. London.

Ashworth, J. H. 1935. Charles Darwin as a student in Edinburgh, 1825—1827. *Proceedings of the Royal Society of Edinburgh* 55: 97—113.

Aubuisson de Voisins, Jean François d'. 1814. *An account of the basalts of Saxony, with observations on the origin of basalt in general.* Translated, with notes, by P. Neill. Edinburgh.

———. 1819. *Traité de géognosie*. 2 vols. Strasbourg.

Audubon, John James. 1827—38. *The birds of America, from original drawings.* 4 vols. London.

Aust. Dict. Biog.: *Australian dictionary of biography: 1788—1850; 1851—1890.* Edited by Douglas Pike and Bede Nairn. 6 vols. Melbourne: Melbourne University Press. 1966—76.

Autobiography: *The autobiography of Charles Darwin 1809—1882. With the original omissions restored.* Edited by Nora Barlow. London: Collins. 1958.

Babbage, Charles. 1832. *On the economy of machinery and manufactures.* London.

Babington, Charles Cardale. 1841 – 3. Dytiscidæ Darwinianæ. *Transactions of the Royal Entomological Society of London* 3: 1 – 17.

Bagshaw, Samuel. 1851. *History, gazetteer, and directory of Shropshire.* Sheffield.

Bakewell, Robert. 1813. *An introduction to geology, illustrative of the general structure of the earth.* London.

Banks, M. R. 1971. A Darwin manuscript on Hobart Town. *Papers and Proceedings of the Royal Society of Tasmania* 105: 5 – 19.

Barrett, Paul H. 1974. The Sedgwick – Darwin geologic tour of North Wales. *Proceedings of the American Philosophical Society* 118: 146 – 64.

Barrow, John. 1831. *The eventful history of the mutiny and piratical seizure of H.M.S. Bounty: its causes and consequences.* London. Reprint, edited and introduced by Stephen W. Roskill, 1976. London: The Folio Society.

Basalla, G. 1963. The voyage of the *Beagle* without Darwin. *Mariner's Mirror* 49: 42 – 8.

BDR: British diplomatic representatives 1789 – 1852. Edited by S. T. Bindoff, E. F. Malcolm Smith, and C. K. Webster. London: Offices of the Royal Historical Society. 1934.

'Beagle' diary: Charles Darwin's diary of the voyage of H.M.S. Beagle. Edited by Nora Barlow. Cambridge: Cambridge University Press. 1933.

Beaglehole, John Cawte, ed. 1962. *The Endeavour journal of Joseph Banks, 1768 – 1771.* 2 vols. Sydney: Trustees of the Public Library of New South Wales in association with Angus and Robertson.

Beechey, Frederick William. 1831. *Narrative of a voyage to the Pacific and Beering's Strait, to co-operate with the Polar expeditions: performed in His Majesty's Ship Blossom . . . in the years 1825, 26, 27, 28.* 2 pts. London.

Bell, Charles. 1806. *Essays on the anatomy of expression in painting.* London.

——. 1833. *The hand. Its mechanism and vital endowments as evincing design.* London.

——. 1844. *The anatomy and philosophy of expression as connected with the fine arts.* Preface by George Bell, and an appendix on the nervous system by Alexander Shaw. 3d ed. London.

Bénézit, Emmanual. 1976. *Dictionnaire critique et documentaire des peintres, sculpteurs, dessinateurs et graveurs de tous les temps et de tous les pays, par un groupe d'écrivains spécialistes français et étrangers.* Revised edition. 10 vols. Paris: Librairie Gründ.

Berkenhout, John. 1764. *Clavis Anglica linguæ botanicæ; or, a botanical lexicon; in which the terms of botany . . . are applied, derived, explained, contrasted and exemplified.* London.

Besterman, Theodore. 1969. *Voltaire.* London and Harlow: Longmans.

Bewick, Thomas. 1790. *A general history of quadrupeds. The figures engraved on wood by T. Bewick.* [The text by R. Beilby.] Newcastle-upon-Tyne.

——. 1797 – 1804. *History of British birds.* 2 vols. Newcastle-upon-Tyne.

BHGW: *Biographisch-literarisches Handwörterbuch zur Geschichte der exacten Wissen-schaften*. By Johann Christian Poggendorff. Vols. 1 – 4, Leipzig: Verlag von Johann Ambrosius Barth. 1863 – 1904. Vol. 5, Leipzig and Berlin: Verlag Chemie g.m.b.h. 1926.

Bigelow, Jacob. 1831. *Elements of technology*. 2d ed. Boston.

Birds: Part III of *The zoology of the voyage of H.M.S. Beagle*. By John Gould. Edited and superintended by Charles Darwin. London: 1838 – 41.

Blainville, Henri Marie Ducrotay de. 1834. Rapport sur les résultats scientifiques du voyage de M. Alcide d'Orbigny . . . par MM. De Blainville, Brongniart, Savary, Cordier. *Nouvelles annales du Muséum National d'Histoire Naturelle* 3: 84 – 115.

Borthwick, D. E. 1965. Outfitting the United States exploring expedition: Lieutenant Charles Wilkes' European assignment, August – November, 1836. *Proceedings of the American Philosophical Society* 109: 159 – 72.

Bory de Saint-Vincent, Jean Baptiste Georges Marie, ed. 1822 – 31. *Dictionnaire classique d'histoire naturelle*. 17 vols. Paris.

Boswell, James. 1791. *The life of Samuel Johnson*. 2 vols. London.

Bradley, Richard. 1726. *A general treatise on husbandry and gardening*. 2 vols. London.

Brande, William Thomas. 1819. *A manual of chemistry*. London.

Brewster, David. 1823. Observations on the mean temperature of the globe. [Read 7 February 1820.] *Transactions of the Royal Society of Edinburgh* 9: 201 – 25; also *Edinburgh Journal of Science* n.s. 4 (1831): 300 – 20.

——. 1826. Results of the thermometrical observations made at Leith Fort, every hour of the day and night, during the whole of the years 1824 and 1825. [Read 23 January 1826.] *Transactions of the Royal Society of Edinburgh* 10: 362 – 88; *Edinburgh Journal of Science* n.s. 5: 18 – 32.

Bridges, Esteban Lucas. 1948. *Uttermost part of the earth*. London: Hodder and Stoughton.

Brookes, Samuel. 1815. *An introduction to the study of conchology*. London.

Buch, Leopold von. 1813. *Travels through Norway and Lapland . . . Translated . . . by John Black. With notes . . . by Robert Jameson*. London.

Buckland, William. 1832. On the fossil remains of the Megatherium recently imported into England from South America. *Report* of the British Association meeting in Oxford (1832), pp. 104 – 7.

——. 1837. *Geology and mineralogy considered with reference to natural theology: with supplementary notes*. Bridgewater treatises, no. 6. 2d ed. 2 vols. London.

Burchell, William John. 1822 – 4. *Travels in the interior of Southern Africa*. 2 vols. London.

Burke's Landed Gentry: *Burke's genealogical and heraldic history of the landed gentry*. By John Burke. 1 – 17 editions. London: Burke's Peerage Ltd. 1833 – 1952.

Burke's Peerage: *Burke's peerage and baronetage*. By John Burke. 1 – 105 editions. London: Burke's Peerage Ltd. 1826 – 1980.

Burney, James. 1803 – 17. *A chronological history of the discoveries in the South Sea or Pacific Ocean.* 5 vols. London.

Burstyn, H. L. 1975. If Darwin wasn't the *Beagle*'s naturalist, why was he on board? *British Journal for the History of Science* 8: 62 – 9.

Caldcleugh, Alexander. 1825. *Travels in South America, during the years 1819 . . . 21.* 2 vols. London.

Cambridge Philosophical Society 1835. See [Henslow, John Stevens, ed.] 1835.

CDEL: A critical dictionary of English literature, and British and American authors, living and deceased. By S. Austin Allibone. 3 vols. Philadelphia and London: 1859 – 71. Supplement by John Foster Kirk, 2 vols. Philadelphia and London: 1891.

Clark, John Willis and Hughes, Thomas McKenny. 1890. *The life and letters of the Reverend Adam Sedgwick.* 2 vols. Cambridge.

Clift, William. 1835. Some account of the remains of the Megatherium sent to England from Buenos Ayres by Woodbine Parish, Jun. [Read 13 June 1832.] *Transactions of the Geological Society of London* 2d ser. 3: 437 – 50.

Coddington, Henry. 1830. On the improvement of the microscope. [Read 22 March 1830.] *Transactions of the Cambridge Philosophical Society* 3: 421 – 8.

Coldstream, John. 1826. Account of some of the rarer atmospherical phenomena observed at Leith in 1825. *Edinburgh Journal of Science* 5: 85 – 92.

Coleridge, Samuel Taylor. 1825. *Aids to reflection in the formation of a manly character on the several grounds of prudence, morality and religion.* London.

Collected papers: The collected papers of Charles Darwin. Edited by Paul H. Barrett. Chicago and London: University of Chicago Press. 1977.

Colp, Ralph, Jr. 1977. *To be an invalid: the illness of Charles Darwin.* Chicago: University of Chicago Press.

Complete Peerage: The complete peerage of England Scotland Ireland Great Britain and the United Kingdom extant extinct or dormant. By G. E. Cokayne. New edition, revised and much enlarged. Edited by the Hon. Vicary Gibbs and others. 12 vols. London: The St Catherine Press. 1910 – 59.

Conybeare, William Daniel and Phillips, William. 1822. *Outlines of the geology of England and Wales.* Pt 1 (no more published). London.

Coral reefs: The structure and distribution of coral reefs. Being the first part of the geology of the voyage of the Beagle, under the commmand of Capt Fitzroy, R.N. during the years 1832 to 1836. By Charles Darwin. London: 1842.

Cullen, William. 1776 – 84. *First lines of the practice of physic.* 4 vols. Edinburgh.

——. 1814. *Synopsis nosologiae methodicae. Auctore G. Cullen. To which is added an appendix containing a synopsis of the systems of Sauvages, Linnaeus, Vogel . . . and a translation of Cullen's Nosology . . . by John Thomson.* Edinburgh.

Curtis, William. 1771. *Instructions for collecting and preserving insects, particularly moths and butterflies.* London.

——. 1772. *Fundamenta entomologiae . . . being a translation of the Fundamenta entomologiae of Linnaeus.* London.

Cuvier, Georges. 1812. *Recherches sur les ossemens fossiles de quadrupèdes, ou l'on rétablit les caractères de plusieurs espèces d'animaux que les révolutions du globe paroissent avoir détruites.* 4 vols. Paris.

——. 1817. *Mémoires pour servir à l'histoire et à l'anatomie des mollusques.* Paris.

——.1827—35. *The animal kingdom arranged in conformity with its organization . . . with additional descriptions of all the species hitherto named, and of many not before noticed.* By Edward Griffith and others. 16 vols. London.

DAB: Dictionary of American biography. Under the auspices of the American Council of Learned Societies. 20 vols., index, and 7 supplements. New York: Charles Scribner's Sons; London: Oxford University Press. 1928—81.

Dalyell, John Graham. 1814. *Observations on some interesting phenomena in animal physiology, exhibited by several species of Planariae. Illustrated by coloured figures of living animals.* Edinburgh.

Dalzel, Andrew. 1789—97. Ἀναλεκτα Ἑλληνικα. *Sive collectanea Græca.* 2 vols. Edinburgh.

Darling, L. 1978. H.M.S. *Beagle*: further research, or twenty years a-Beagling. *Mariner's Mirror* 64: 315—25.

Darwin, Erasmus. 1791. *The economy of vegetation.* Pt 1 of *The botanic garden.* London.

Darwin, Francis. 1912. FitzRoy and Darwin, 1831—36. *Nature* 88: 547—8.

Darwin and Henslow: *Darwin and Henslow: the growth of an idea.* Edited by Nora Barlow. London: John Murray. 1967.

Darwin Pedigree: *Pedigree of the family of Darwin.* By H. Farnham Burke. Privately printed. 1888.

Daubeny, Charles Giles Bridle. 1826. *A description of active and extinct volcanoes; with remarks on their origin, their chemical phaenomena, and the character of their products.* London and Oxford.

Davy, Humphry. 1800. *Researches, chemical and philosophical; chiefly concerning nitrous oxide . . . and its respiration.* London. Reprint 1972. London: Butterworth.

——. 1830. *Consolations in travel, or the last days of a philosopher.* Edited by John Davy. London.

DBF: *Dictionnaire de biographie français.* Under the direction of J. Balteau, M. Barroux, M. Prevost, and others. 15 vols., (A—Gilbert). Paris: Librairie Letouzey et Ané. 1933—82.

DCF: *Dictionnaire des comédiens français.* By Henry Lyonnet. 2 vols. Paris: E. Jorel. [1911—12.]

De Beer, Gavin, ed. 1959. Darwin's Journal. *Bulletin of the British Museum (Natural History) Historical series* 2: 3—21.

Descent: The descent of man, and selection in relation to sex. By Charles Darwin. 2 vols. London: 1871.

Desmond, Ray. 1977. *Dictionary of British and Irish botanists and horticulturists, including plant collectors and botanical artists.* 3d ed. London: Taylor and Francis.

A diary of the wreck of His Majesty's ship Challenger, on the western coast of South America, in May, 1835. With an account of the subsequent encampment of the officers and crew, during a period of seven weeks, on the south coast of Chili. 1836. London.

Dictionnaire classique d'histoire naturelle. See Bory de Saint-Vincent, J. B. G. M.

DNB: Dictionary of national biography. Edited by Sir Leslie Stephen and Sir Sidney Lee. 63 vols. and 2 supplements (6 vols.). London: Smith, Elder and Co. 1885–1912. *Dictionary of national biography 1912–70.* Edited by H. W. C. Davis, J. R. H. Weaver, and others. 6 vols. London: Oxford University Press. 1927–81.

DNZB: A dictionary of New Zealand biography. Edited by G. H. Scholefield. 2 vols. Wellington: Department of Internal Affairs New Zealand. 1940.

Dod, Charles Roger. 1972. *Electoral facts from 1832 to 1853 impartially stated . . . Edited with an introduction and bibliographical guide to electoral sources, 1832–1885., by H. J. Hanham.* Brighton: Harvester Press.

Donn, James. 1823. *Hortus Cantabrigiensis; or an accented catalogue of indigenous and exotic plants cultivated in the Cambridge Botanic Garden.* 10th ed., with numerous additions and corrections by J. Lindley. London.

DSB: Dictionary of scientific biography. Edited by Charles Coulston Gillispie. 14 vols., supplement I, and index. New York: Charles Scribner's Sons. 1970–80.

Earle, Augustus. 1832. *A narrative of a nine months' residence in New Zealand, in 1827; together with a journal of a residence in Tristan d'Acunha.* London.

EB: Encyclopaedia Britannica. 11th ed. 29 vols. Cambridge: Cambridge University Press. 1910–11.

Ehrenberg, Christian Gottfried. 1834. *Über die Natur und Bildung der Coralleninseln und Corallenbänke im rothen Meere.* Berlin.

Eiseley, L. 1959. Charles Darwin, Edward Blyth, and the theory of natural selection. *Proceedings of the American Philosophical Society* 103: 94–158.

Emma Darwin: Emma Darwin: a century of family letters 1792–1896. Edited by Henrietta Litchfield. Revised edition. 2 vols. London: John Murray. 1915.

Emma Darwin (1904): *Emma Darwin: a century of family letters 1792–1896.* Edited by Henrietta Litchfield. 2 vols. Privately printed: Cambridge University Press.

Eton School Lists: The Eton School lists, from 1791 to 1850. By H. E. C. Stapylton. 2d ed. London: 1864.

Eyton, Thomas Campbell. 1836. *A history of the rarer British birds.* 2 pts. London.

——. 1838. *A monograph on the Anatidae, or duck tribe.* London.

Falkner, Thomas. 1774. *A description of Patagonia, and the adjoining parts of South America.* Hereford. Facsimile reprint with introduction and notes by Arthur E. S. Neumann, 1935. Chicago: Armann and Armann.

FitzRoy, Robert. 1836. Sketch of the surveying voyages of His Majesty's ships Adventure and Beagle, 1825–1836. *Journal of the Royal Geographical Society of London* 6: 311–42.

FitzRoy, Robert. 1837. Extracts from the diary of an attempt to ascend the River Santa Cruz, in Patagonia, with the boats of H.M.S. Beagle. *Journal of the Royal Geographical Society of London* 7: 114–26.

Fleming, John. 1822. *The philosophy of zoology; or, a general view of the structure, functions, and classification of animals.* 2 vols. Edinburgh.

——. 1828. *A history of British animals.* Edinburgh.

Flügel, Johann Gottfried, ed. 1838. *A complete dictionary of the English and German and German and English languages.* 2d ed., improved and augmented. 2 vols. Leipzig.

Foggo, John. 1826. Results of a meteorological journal kept at Seringapatam during the years 1814 and 1816. *Edinburgh Journal of Science* 5: 249–58.

——. 1827. On the dew-point hygrometer formerly described in this journal, vol. IV, p. 127. *Edinburgh Journal of Science* 7: 36–44.

Fossil Mammalia: Part I of *The zoology of the voyage of H.M.S. Beagle.* By Richard Owen. Edited and superintended by Charles Darwin. London: 1838–40.

Freeman, Richard Broke. 1977. *The works of Charles Darwin: an annotated bibliographical handlist.* 2d ed., revised and enlarged. Folkestone: W. Dawson and Sons; Hamden, Conn.: Archon Books, Shoe String Press.

——. 1978. *Charles Darwin: a companion.* Folkestone: W. Dawson and Sons; Hamden, Conn.: Archon Books, Shoe String Press.

——. 1978–9. Darwin's Negro bird-stuffer. *Notes and Records of the Royal Society of London* 33: 83–6.

Freycinet, Louis Claude Desaulses de. 1824–44. *Voyage autour du monde, entrepris par ordre du Roi.* 9 vols. Plates, 4 vols. (1824–6). Paris.

Galton, Francis. 1874. *English men of science: their nature and nurture.* London.

Gay, Claude. 1833. Aperçu sur les recherches d'histoire naturelle faites dans l'Amérique du Sud, et principalement dans le Chili, pendant les années 1830 et 1831. *Annales des sciences naturelles* 28: 369–393.

——. 1836. Extrait d'une lettre à M. de Blainville, datée Valdivia, le 5 juillet 1835, concernant les habitudes des sangsues au Chili, et la tendance que montrent les reptiles dans le même pays, à devenir vivipares. *Comptes rendus hebdomadaires des séances de l'Académie des sciences Paris* 2: 322.

Genest, John. 1832. *Some account of the English stage, from the Restoration in 1660 to 1830.* 10 vols. Bath and London.

Geological Society, London. 1808. *Geological enquiries.* London.

Geology of 'Beagle': *Geological observations on coral reefs, volcanic islands, and on South America* etc. A combined edition of *Coral reefs, Volcanic islands,* and *South America.* London: 1851.

Gilbert, Pamela. 1977. *A compendium of the biographical literature on deceased entomologists.* London: British Museum (Natural History).

Goebel, Julius. 1927. *The struggle for the Falkland Islands: a study in legal and diplomatic history.* New Haven: Yale University Press. Reprint, edited and introduced by J. C. J. Metford, 1982. Port Washington, N.Y.: Kennikat.

Good, John Mason. 1822. *The study of medicine.* 4 vols. London.

Gould, John. 1832 — 7. *The birds of Europe.* 5 vols. London.

——. 1834. *A monograph of the Ramphastidae, or family of toucans.* London.

Grant, Robert Edmond. 1827. Notice regarding the ova of the Pontobdella muricata, Lam. *Edinburgh Journal of Science* 7: 160 — 1.

Grove, George. 1980. *The new Grove dictionary of music and musicians.* Edited by Stanley Sadie. 20 vols. London: Macmillan Publishers.

Gruber, Howard Ernest and Barrett, Paul H. 1974. *Darwin on man. A psychological study of scientific creativity . . . Together with Darwin's early and unpublished notebooks.* New York: E. P. Dutton and Co.

Gruber, J. W. 1969. Who was the *Beagle*'s naturalist? *British Journal for the History of Science* 4: 266 — 82.

Guilding, Lansdown. 1825. Description of a new species of *Onchidium.* [Read 4 November 1823.] *Transactions of the Linnean Society of London* 14: 322 — 4 and Tab. IX.

H[aken, Johann Christian Ludwig]. 1805. *Robinsons des Jüngern, Beschreibung seiner Reise nach Otahite und den Südsee-Inseln. Ein nützliches Lesebuch für die Jugend.* Altona.

Haller, Albrecht von. 1757 — 66. *Elementa physiologiae corporis humani.* 8 vols. Lausanne.

Hanham 1972. See Dod 1972.

Harbord, Edward, Baron Suffield. 1825. *Considerations on the game laws.* Norwich and London.

Harker, A. 1907. Notes on the rocks of the *Beagle* collection. *Geological Magazine* 5th ser. 4: 100 — 6.

Harris, James. 1751. *Hermes: or a philosophical inquiry concerning language and universal grammar.* London.

HDA: Historical dictionary of Argentina. By Lone S. Wright and Lisa M. Nekhom. Latin American Historical Dictionaries, No. 17. Metuchen, N.J. and London: The Scarecrow Press. 1978.

Head, Francis Bond. 1826. *Rough notes taken during some rapid journeys across the Pampas and among the Andes.* London.

Heber, Reginald. 1828. *Narrative of a journey through the upper provinces of India from Calcutta to Bombay 1824 — 25. (With notes upon Ceylon.) An account of a journey to Madras and the Southern provinces, 1826, and letters written in India.* (Edited by Amelia Heber.) 2 vols. London.

Henry, William. 1810. *The elements of experimental chemistry.* 6th ed. 2 vols. London.

——. 1823. 9th ed. 2 vols. London.

[Henslow, John Stevens, ed.] 1835. *Extracts from letters addressed to Professor Henslow by C. Darwin, esq. read at a meeting of the Cambridge Philosphical Society 16 November, 1835.* Privately printed for the Cambridge Philosophical Society. Reprinted 1960 and in *Collected papers* 1: 3 — 16.

Henslow, John Stevens. 1837. Description of two new species of *Opuntia*; with remarks on the structure of the fruit of Rhipsalis. *Magazine of Zoology and Botany* 1: 466 – 9.

———. 1838. Florula Keelingensis: an account of the native plants of the Keeling Islands. *Annals and Magazine of Natural History* 1: 337 – 47.

Herschel, John Frederick William. 1831. *A preliminary discourse on the study of natural philosophy*. Published in D. Lardner's *Cabinet cyclopædia*. London.

Hewitson, William Chapman. [1833 – 42]. *British oology; being illustrations of the eggs of British birds, with figures of each species, as far as practicable, drawn and coloured from nature, accompanied by descriptions of the materials and situation of their nests, number of eggs, etc.* 2 vols. and supplement. Newcastle-upon-Tyne.

Hobbs, John L. 1960. The Haycocks changed the face of Shrewsbury. *The Shropshire Magazine* (February 1960): 17 – 18.

Hooker, Joseph Dalton. 1844 – 7. *Flora Antarctica*. Pt. 1 of *The botany of the Antarctic voyage of H.M. Discovery-Ships Erebus and Terror . . . 1839 – 43, under the command of . . . Sir James Clark Ross.* 2 vols. London.

———. 1846a. An enumeration of the plants of the Galapagos Archipelago; with descriptions of those which are new. [Read 4 March, 6 May, and 16 December 1846.] *Transactions of the Linnean Society of London* 20 (1851): 163 – 233.

———. 1846b. On the vegetation of the Galapagos Archipelago, as compared with that of some other tropical islands and of the continent of America. [Read 1 and 15 December 1846.] *Proceedings of the Linnean Society of London* 1 (1849): 314. *Transactions of the Linnean Society of London* 20 (1851): 235 – 62.

Hooker, W. J. and Arnott, G. A. W. 1836, 1841. Contributions towards a flora of South America and the islands of the Pacific. *Companion to the Botanical Magazine* 2: 41 – 52, 250 – 4; *Journal of Botany* 3: 19 – 47, 310 – 48.

Howell, Thomas J. 1816. *The stranger in Shrewsbury; or an historical and descriptive view of Shrewsbury and its environs.* Shrewsbury.

Humboldt, Alexander von. 1805 – 34. *Voyages aux régions équinoxiales du Nouveau Continent, fait en 1799 . . . 1804 par A. de Humboldt et A. Bonpland.* 34 vols. Paris.

———. 1814 – 29. *Personal narrative of travels to the equinoctial regions of the New Continent during the years 1799 – 1804, by A. de Humboldt, and A. Bonpland; with maps, plans, . . . written in French by A. de H., and translated into English by H. M. Williams.* 7 vols. London.

———. 1817. Des lignes isothermes et de la distribution de la chaleur sur le globe. *Mémoires de physique et de chimie de la Société d'Arcueil* 3: 462 – 602. Also *Edinburgh Philosophical Journal* 3 (1820): 1 – 20, 256 – 74; 4 (1821): 23 – 37, 262 – 81; 5 (1821): 28 – 39.

———. 1828. *Tableaux de la nature*. Translated from the German by J. B. B. Eyriès. 2d ed. 2 vols. Paris.

———. 1831. *Fragmens de géologie et de climatologie asiatiques.* 2 vols. Paris.

Humboldt, Alexander von and Bonpland, A. 1811 — 33. *Recueil d'observations de zoologie et d'anatomie comparée.* Pt 2 of *Voyage aux régions équinoxiales.* 2 vols. Paris.

Hunter, John. 1778. *The natural history of the human teeth.* 2d ed. London.

Hutton, James. 1795. *Theory of the earth, with proofs and illustrations.* 2 vols. Edinburgh and London.

Jenyns, Leonard. 1835. *A manual of British vertebrate animals.* Cambridge and London.

——. 1862. *Memoir of the Reverend John Stevens Henslow.* London.

Jespersen, P. H. 1948 — 9. Charles Darwin and Dr. Grant. *Lychnos,* pp. 159 — 67.

Journal and remarks: Vol. III of *Narrative of the surveying voyages of His Majesty's Ships Adventure and Beagle.* By Charles Darwin. Edited by Robert FitzRoy. London: 1839.

Journal of researches: *Journal of researches into the geology and natural history of the various countries visited by H.M.S. Beagle . . . from 1832 to 1836.* By Charles Darwin. London: 1839.

Judd, John Wesley. 1909. Darwin and geology. In Seward, Albert Charles, ed., *Darwin and modern science.* Cambridge: Cambridge University Press.

——. 1911. Charles Darwin's earliest doubts concerning the immutability of species. *Nature* 88: 8 — 12.

Keevil, John Joyce. 1943. Robert McCormick, R.N., the stormy petrel of naval medicine. *Journal of the Royal Naval Medical Service* 29: 36 — 62.

——. 1957 — 63. *Medicine and the Navy, 1220 — 1900.* 4 vols. (Vols. 3 and 4 by C. Lloyd and J. L. S. Coulter.) Edinburgh and London: E. and S. Livingstone.

Kerr, Robert. 1792. *The animal kingdom, or, zoological system of . . . C. Linnaeus. Class I. Mammalia . . . being a translation of that part of the Systema Naturæ . . . with numerous additions from more recent zoological writers.* London.

Keynes, Richard Darwin, ed. 1979. *The 'Beagle' record: selections from the original pictorial records and written accounts of the voyage of H.M.S. Beagle.* Cambridge: Cambridge University Press.

Kirby, William and Spence, William. 1815 — 26. *An introduction to entomology; or, elements of the natural history of insects.* 4 vols. London.

——. 1828. 5th ed. 4 vols. London.

Kotzebue, Otto von. 1821. *Entdeckungs-Reise in die Süd-See und nach der Berings Strasse zur Erforschung einer nordöstlichen Durchfahrt. Unternommen in den Jahren 1815 . . . 1818.* 3 vols. Weimar.

——. 1830. *Neue Reise um die Welt, in den Jahren 1823 . . . 26.* 2 vols. Weimar.

Lacordaire, Jean Théodore. 1830. Mémoire sur les habitudes des Coléoptères de l'Amérique méridionale. *Annales des Sciences Naturelles* 20: 185 — 291; 21: 149 — 94.

Lamarck, Jean Baptiste Pierre Antoine de Monet de. 1815 — 22. *Histoire naturelle des animaux sans vertèbres.* 7 vols. Paris.

——. 1835 — 45. 2d ed. 11 vols. Paris.

Lamouroux, Jean Vincent Félix. 1821. *Exposition méthodique des genres de l'ordre des Polypiers.* Paris.

Larousse XX: Larousse du XXe siècle. Under the direction of Paul Augé. 6 vols. Paris: Librairie Larousse. 1928 – 33.

Latham, John. 1781 – 1802. *A general synopsis of birds.* 3 vols. and 2 supplements. London.

Levaillant, François. 1790. *Voyage dans l'intérieur de l'Afrique par le Cap de Bonne Espérance, dans les années 1780 – 85.* 2 vols. Paris.

———.1790. *Travels from the Cape of Good Hope into the interior parts of Africa.* Translated from the French. 2 vols. London.

Levy, Paul. 1979. *G. E. Moore and the Cambridge Apostles.* London: Weidenfeld and Nicolson.

Lingwood, P. 1984. The duties of natural history. *Biology Curators Group Newsletter.*

Linnaeus, (Carl von Linné). 1772. *Fundamenta entomologiae.* See Curtis 1772.

———. 1783a. *A system of vegetables . . . with their character and differences, translated from the thirteenth edition . . . of the Systema vegetabilium . . . and from the Supplementum plantarum . . . by a botanical society at Lichfield.* 2 vols. Lichfield and London.

———. 1783b. *Philosophia botanica.* 2d ed. Vienna.

———. 1789 – 96. *Caroli Linnæi . . . Systema naturæ, sive regna tria naturæ systematice proposita per classes, ordines, genera, et species.* 13th ed. by J. F. Gmelin. (Bound in 10 vols.) Lyons.

———. 1797. *Systema vegetabilium secundum classes, ordines, genera, species, cum characteribus et differentiis.* 15th ed. Göttingen.

Linnean Society. 1908. *The Darwin – Wallace celebration held on Thursday, 1st July, 1908 by the Linnean Society of London.* London: printed for the Linnean Society.

Lister, Thomas Henry. 1826. *Granby.* 3 vols. London.

Living Cirripedia: A monograph on the sub-class Cirripedia with figures of all the species. (Vol. I) *The Lepadidæ; or pedunculated cirripedes.* By Charles Darwin. London: 1851. (Vol. II) *The Balanidæ (or sessile cirripedes); the Verrucidæ,* etc. By Charles Darwin. London: 1854.

Lizars, John. 1822. *A system of anatomical plates of the human body, accompanied with descriptions, and physiological, pathological, and surgical observations.* 3 vols. Edinburgh.

LL: The life and letters of Charles Darwin, including an autobiographical chapter. Edited by Francis Darwin. 3 vols. London: 1887.

Loss of His Majesty's frigate *Challenger.* 1835. *Nautical Magazine* 4: 789 – 96.

The loss of the Kent Indiaman, by fire in the Bay of Biscay. 1826. London.

Lothrop, Samuel Kirkland. 1928. *The Indians of Tierra del Fuego.* Contributions from the Museum of the American Indian, Heye Foundation, vol. 10. New York: Museum of the American Indian, Heye Foundation.

Lowe, Richard Thomas. 1833. Primitiae faunae et florae Maderae et Portus Sancti. [Read 15 November 1830.] *Transactions of the Cambridge Philosophical Society* 4: 1 – 70.

Lowth, Robert. 1762. *A short introduction to English grammar; with critical notes.* London.

Lubbock, John William. 1830—2. Researches in physical astronomy. [Read 29 April and 9 December 1830; 19 May, 9 June, and 17 November 1831; 9 February, 7 June, and 21 June 1832.] *Philosophical Transactions of the Royal Society of London* 120: 327—57; 121: 17—66, 231—82, 283—98; 122: 1—49, 229—36, 361—81, 601—7.

Lyell, Charles. 1830—3. *Principles of geology, being an attempt to explain the former changes of the earth's surface, by reference to causes now in operation.* 3 vols. London.

Lyell, Katherine Murray, ed. 1881. *Life, letters and journals of Sir Charles Lyell, Bart.* 2 vols. London.

McCormick, Robert. 1884. *Voyages of discovery in the Arctic and Antarctic seas, and round the world.* 2 vols. London.

Mackintosh, James. 1830—2. First 3 vols. of *The history of England* in the *Cabinet cyclopaedia.* London. Vols. 4—10 (1835—40) continued from the late James Mackintosh by William Wallace and Robert Bell.

Macleay, William Sharp. 1819—21. *Horæ entomologicæ; or, essays on the annulose animals, etc.* Vol. 1, pts 1, 2 (no more published). London.

Manzoni, Alessandro F. T. A. 1827. *I promessi sposi. Storia milanese del secolo XVII.* 3 vols. Milan.

Marsham, Thomas. 1802. *Coleoptera britannica, sistens insecta Coleoptera britanniae indigena.* Vol. 1. London.

Martin, Arthur Patchett. 1893. *Life and letters of the Right Honourable Robert Lowe, Viscount Sherbrooke.* 2 vols. London.

Martineau, Harriet. 1833. *Cinnamon and pearls.* No. 20 in vol. 5 of *Illustrations of political economy,* 25 nos. in 6 vols., 1832—4. London.

——. 1833—4. *Poor laws and paupers illustrated.* 4 pts. London.

——. 1837. *Society in America.* 3 vols. London.

Mellersh, Harold Edward Leslie. 1968. *FitzRoy of the Beagle.* London: Rupert Hart-Davis.

Miers, John. 1826. *Travels in Chile and La Plata, including accounts respecting the geography, geology, statistics, government, finances, agriculture, manners and customs, and the mining operations in Chile.* 2 vols. London.

ML: More letters of Charles Darwin: a record of his work in a series of hitherto unpublished letters. By Francis Darwin and Albert Charles Seward. 2 vols. London: John Murray. 1903.

Modern English Biography: Modern English biography containing many thousand concise memoirs of persons who have died since the year 1850. By Frederic Boase. 3 vols. and supplement (3 vols.). Truro: Netherton and Worth. 1892—1921.

Moore, Thomas. 1825. *Memoirs of the life of the Right Honourable Richard Brinsley Sheridan.* London.

Morris, John and Sharpe, Daniel. 1846. Description of eight species of brachiopodous shells from the palaeozoic rocks of the Falkland Islands. *Quarterly Journal of the Geological Society of London* 2: 274−8.

Munsche, P. B. 1981. *Gentlemen and poachers: the English game laws 1671−1831.* Cambridge: Cambridge University Press.

Murchison, Roderick Impey. 1839. *The Silurian system, founded on geological researches in the counties of Salop, Hereford, Radnor, Montgomery, Caermarthen, Brecon, Pembroke, Monmouth, Gloucester, Worcester, and Stafford; with descriptions of the coalfields and overlying formations.* 2 pts. London.

Murray, Lindley. 1824. *English grammar.* 5th ed. 2 vols. York.

Narbrough, John. 1694. *An account of several late voyages and discoveries to the south and north towards the Streights of Magellan, the South Seas, . . . also towards Nova Zembla, Greenland or Spitsberg.* 2 pts. London.

Narrative: Narrative of the surveying voyages of His Majesty's Ships Adventure and Beagle between the years 1826 and 1836. Edited by Robert FitzRoy. 3 vols. and appendix. London: 1839.

NBU: Nouvelle biographie universelle. Under the direction of Jean Chrétien Ferdinand Hoefer. 46 vols. Paris: 1852−66.

NDB: Neue deutsche Biographie. Edited by the Historical Commission of the Bavarian Academy of Sciences. 12 vols., (A−Kre). Berlin: Duncker and Humblot. 1953−80.

Newton, Isaac. 1704. *Opticks; or, a treatise of the reflexions, refractions, inflexions and colours of light.* London.

Noehden, Georg Heinrich. 1829. *Rabenhorst's pocket dictionary of the German and English languages.* Revised by H. E. Lloyd. 3d ed. 2 pts. London.

O'Byrne, William R. 1849. *A naval biographical dictionary: comprising the life and services of every living officer in Her Majesty's Navy, from the rank of admiral of the fleet to that of lieutenant, inclusive.* London.

OED: The Oxford English dictionary. Being a corrected re-issue with an introduction, supplement, and bibliography of *A new English dictionary.* Edited by James A. H. Murray, Henry Bradley, W. A. Craigie, C. T. Onions. 12 vols. and supplement. Oxford: Clarendon Press. 1970. Supplement. Edited by R. W. Burchfield. 3 vols., (A−Scz). Oxford: Clarendon Press. 1972−82.

Orbigny, Alcide Dessalines d'. 1835−47. *Voyage dans l'Amérique méridionale — le Brésil, la République orientale de l'Uruguay, la République argentine, la Patagonie, la République du Chili, la République de Bolivia, la République du Pérou — executé pendant les années 1826 . . . 1833.* 9 vols. Paris.

'Ornithological notes': Darwin's ornithological notes. Edited by Nora Barlow. *Bulletin of the British Museum (Natural History) Historical Series* 2 (1963): 201−78.

Owen, Richard. 1853. *Descriptive catalogue of the osteological series contained in the Museum of the Royal College of Surgeons in England.* 2 vols. London.

Owen, Richard Startin. 1894. *The life of Richard Owen. With the scientific portions revised by C. Davies Sherborn, also an essay on Owen's position in anatomical science by the Right Hon. T. H. Huxley.* 2 vols. London.

Parish, Woodbine. 1834. An account of the discovery of portions of three skeletons of the Megatherium in the province of Buenos Ayres in South America. Followed by a description of the bones by William Clift. [Read 13 June 1832.] *Proceedings of the Geological Society of London* 1: 403 – 4.

Parish, Woodbine. 1838. *Buenos Ayres, and the provinces of the Rio de la Plata: their present state, trade, and debt; with some account from original documents of the progress of geographical discovery in those parts of South America during the last sixty years.* London.

Parkes, Samuel. 1806. *Chemical catechism for the use of young people: with copious notes for the assistance of the teacher, to which are added a vocabulary of chemical terms, useful tables, and a chapter of amusing experiments.* London.

Partridge, Eric Honeywood. 1973. *The Routledge dictionary of historical slang.* Abridged by J. Simpson. London: Routledge and Kegan Paul.

Pemberton, Christopher Robert. 1806. *A practical treatise on various diseases of the abdominal viscera.* London.

Pennant, Thomas. 1771. *Synopsis of quadrupeds.* Chester.

——. 1781. *History of quadrupeds.* 2 vols. London.

——. 1793. 3d ed. 2 vols. London.

Phillips, William. 1816. *Elementary introduction to the knowledge of mineralogy: including some account of mineral elements and constituents; explanations of terms in common use; brief accounts of minerals, and of the places and circumstances in which they are found. Designed for the use of the student.* London.

——. 1823. 3d ed. London.

——. 1837. 4th ed. Augmented by R. Allan. London.

Plarr, Victor Gustave. 1930. *Plarr's lives of the fellows of the Royal College of Surgeons of England.* Revised by Sir D'Arcy Power. 2 vols. London: Simpkin Marshall.

Playfair, John. 1802. *Illustrations of the Huttonian theory of the earth.* Edinburgh and London.

Poole, John. 1825. *Paul Pry, a comedy, in three acts.* In *Duncombe's edition of the British Theatre,* vol. 1. London.

Porter, D. M. 1980a. Charles Darwin's plant collections from the voyage of the *Beagle. Journal of the Society for the Bibliography of Natural History* 9: 515 – 25.

——. 1980b. The vascular plants of Joseph Dalton Hooker's *An enumeration of the plants of the Galapagos Archipelago; with descriptions of those which are new. Botanical Journal of the Linnean Society* 81: 79 – 134.

——. 1981. Darwin's missing notebooks come to light. *Nature* 291: 13.

——. 1982. Charles Darwin's notes on plants of the *Beagle* voyage. *Taxon* 31: 503 – 6.

——. 1983. More Darwin *Beagle* notes resurface. *Archives of Natural History* 11: 315 – 16.

Porter, D. M. (in press). The *Beagle* collector and his collections. In Kohn, David, ed., *The Darwinian heritage*. Princeton: Princeton University Press.

Proctor, Robert. 1825. *Narrative of a journey across the Cordillera of the Andes, and of a residence in Lima, and other parts of Peru, in the years 1823 and 1824.* London.

Provincial Medical Directory: The provincial medical directory. London: 1847.

Rabenhorst's pocket dictionary of the German and English languages. See Noehden 1829.

Radcliffe, Ann. 1794. *The mysteries of Udolpho, a romance interspersed with some pieces of poetry.* 4 vols. London.

Ramsay, A. C., Bristow, H. W., Geikie, A. and Bauerman, H. 1862. *A descriptive catalogue of the rock specimens in the Museum of Practical Geology.* 3d ed. London.

Ray, Gordon Norton, ed. 1945−6. *The letters and private papers of William Makepeace Thackeray.* 4 vols. London: Oxford University Press.

Réaumur, René-Antoine Ferchault de. 1734−42. *Mémoires pour servir à l'histoire des insectes.* 6 vols. Paris.

Red notebook: The red notebook of Charles Darwin. Edited by Sandra Herbert. *Bulletin of the British Museum (Natural History) Historical Series* 7 (1980): 1−164. Published as a separate volume, Ithaca, N.Y.: Cornell University Press. 1980.

Rennie, James. 1832. *Alphabet of insects.* London.

Reptiles: Part V of *The zoology of the voyage of H.M.S. Beagle.* By Thomas Bell. Edited and superintended by Charles Darwin. London: 1842−3.

Richardson, Samuel. 1747−8. *Clarissa; or, the history of a young lady.* 7 vols. London.

Romilly, Joseph. 1967. *Romilly's Cambridge diary 1832−42. Selected passages from the diary of the Reverend Joseph Romilly, fellow of Trinity College and registrary of the University of Cambridge.* Chosen, introduced and annotated by J. P. T. Bury. Cambridge: Cambridge University Press.

Samouelle, George. 1819. *The entomologist's useful compendium; or, an introduction to the knowledge of British insects.* London.

Sarjeant, William A. S. 1980. *Geologists and the history of geology. An international bibliography from the origins to 1978.* 5 vols. London: Macmillan.

Schweigger, August Friedrich. 1826. Observations on the anatomy of the Corallina opuntia and some other Corallines. (Transl.) *Edinburgh New Philosophical Journal* 1: 220−4.

Scoresby, William. 1820. *An account of the Arctic regions, with a history and description of the northern whale-fishery.* 2 vols. Edinburgh.

Scrope, George Julius Poulett. 1825. *Considerations on volcanoes, the probable cause of their phenomena, the laws which determine their march, the disposition of their products, and their connection with the present state and past history of the globe; leading to the establishment of a new theory of the earth.* London.

Sedgwick, Adam. 1833a. *Discourse on the studies of the University of Cambridge.* Cambridge and London.

——. 1833b. [Abstract of an account of the geology of North Wales.] *Philosophical Magazine and Journal of Science* 2: 381.

Selby, Prideaux John. [1818—]34. *Illustrations of British ornithology; or, figures of British birds . . . in their full natural size.* Edinburgh.

Shepperson, G. 1961. The intellectual background of Charles Darwin's student years at Edinburgh. In Banton, Michael, ed., *Darwinism and the study of society: a centenary symposium.* London: Tavistock Publications; Chicago: Quadrangle Books.

Sherer, Moyle. 1823. *Recollections of the Peninsula.* London.

———.1825. *The story of a life.* London.

Shrewsbury School Register: Shrewsbury School register 1734—1908. Edited by J. E. Auden. Oswestry: Woodall, Minshall, Thomas and Co., Caxton Press. 1909.

Smith, Andrew. 1838—49. *Illustrations of the zoology of South Africa, consisting chiefly of figures and descriptions of the objects of natural history collected during an expedition into the interior of South Africa in the years 1834—1836; fitted out by the Cape of Good Hope Association for exploring Central Africa.* 5 vols. London.

———. 1939—40. *The diary of Dr. Andrew Smith, director of the 'Expedition for exploring Central Africa', 1834—1836.* Edited by Percival R. Kirby. 2 vols. Cape Town: Van Riebeeck Society.

Smith, James Edward. 1790—1814. *English botany; or, coloured figures of British plants, with their essential characters, synonyms, and places of growth.* 2,592 plates by James Sowerby. 36 vols. London.

Smith, Sydney. 1968. The Darwin collection at Cambridge with one example of its use: Charles Darwin and Cirripedes. *Actes du XIe Congrès International d'Histoire des Sciences* (1965), pp. 96—100.

Somerville, Martha. 1873. *Personal recollections from early life to old age, of Mary Somerville, with selections from her correspondence. By her daughter Martha Somerville.* London.

South America: Geological observations on South America. Being the third part of the geology of the voyage of the Beagle, under the command of Capt. Fitzroy, R.N. during the years 1832 to 1836. By Charles Darwin. London: 1846.

Southey, Robert. 1823. *History of the Peninsular War.* Vol. 1. London.

Stanbury, David. 1977. *A narrative of the voyage of H.M.S. Beagle: being passages from the 'Narrative' written by Captain Robert FitzRoy . . . selected and edited by David Stanbury.* London: The Folio Society.

———. 1979. Notes—H.M.S. *Beagle. Mariner's Mirror* 65: 355—7.

Stark, John. 1828. *Elements of natural history.* 2 vols. Edinburgh and London.

Stenton, Michael. 1976. *Who's who of British members of Parliament, 1832—1885.* Brighton: Harvester Press.

Stephens, James Francis. 1827—46. *Illustrations of British entomology; or, a synopsis of indigenous insects, containing their generic and specific distinctions; with an account of their metamorphoses, times of appearance, localities, food, and economy, as far as practicable embellished with coloured figures of the rarer and more interesting species.* 11 vols. and supplement. London.

Stephens, James Francis. 1829. *A systematic catalogue of British insects: being an attempt to arrange all the hitherto discovered indigenous insects in accordance with their natural affinities.* 2 pts in 2 vols. London.

——. 1833. Description of *Chiasognathus grantii*, a new lucanideous insect forming the type of an undescribed genus, together with some brief remarks upon its structure and affinities. [Read 16 May 1831.] *Transactions of the Cambridge Philosophical Society* 4: 209 – 17.

Stoddart, D. R., ed. 1962. Coral islands, by Charles Darwin, with introduction, map and remarks. *Atoll Research Bulletin* No. 88, Pacific Science Board, National Academy of Sciences – National Research Council, Washington, D.C.

——. 1976. Darwin, Lyell and the geological significance of coral reefs. *British Journal for the History of Science* 9: 199 – 218.

Sulivan, Henry Norton. 1896. *Life and letters of the late Admiral Sir Bartholomew James Sulivan, . . . 1810 – 1890.* London.

Sulivan, J. A. 1979. Notes – H.M.S. *Beagle. Mariner's Mirror* 65: 76.

Sulloway, F. J. 1982a. Darwin and his finches: the evolution of a legend. *Journal of the History of Biology* 15: 1 – 53

——. 1982b. Darwin's conversion: the *Beagle* voyage and its aftermath. *Journal of the History of Biology* 15: 325 – 96.

——. 1982c. The *Beagle* collections of Darwin's finches (Geospizinae). *Bulletin of the British Museum (Natural History) Zoology series* 43: 49 – 94.

——. 1983. Further remarks on Darwin's spelling habits and the dating of *Beagle* voyage manuscripts. *Journal of the History of Biology* 16: 361 – 90

Swainson, William. 1822. *The naturalist's guide for collecting and preserving subjects of natural history and botany . . . particularly shells.* London.

Syme, Patrick. 1814. *Werner's nomenclature of colours, with additions, arranged so as to render it highly useful to the arts and sciences, particularly zoology, botany, chemistry, mineralogy, and morbid anatomy. Annexed to which are examples selected from well-known objects in the animal, vegetable, and mineral kingdoms.* Edinburgh.

——. 1821. 2d ed. Edinburgh.

Tapparelli, Massimo (Marchese d'Azeglio). 1833. *Ettore Fieramosca, o la disfida di Barletta.* Florence.

Tasch, P. 1950. Darwin and the forgotten Mr Lonsdale. *Geological Magazine* 87: 292 – 6.

Tennyson, Alfred. 1830. *Poems, chiefly lyrical.* London.

Thackray, A. 1974. Natural knowledge in cultural context: the Manchester model. *American Historical Review* 79: 672 – 709.

Thomson, K. S. 1975. H.M.S. *Beagle, 1820 – 1870. American Scientist* 63: 664 – 72.

Todd, Robert Bentley. 1835 – 59. *The cyclopaedia of anatomy and physiology.* 5 vols. London.

Trevelyan, George Macaulay. 1942. *English social history: a survey of six centuries, Chaucer to Queen Victoria.* London: Longmans and Co.

Turner, Sharon. 1832 — 7. *The sacred history of the world, as displayed in the creation and subsequent events to the deluge. Attempted to be philosophically considered in a series of letters to a son.* 3 vols. London.

Ure, Andrew. 1821. *A dictionary of chemistry.* London.

——. 1823. 2d ed., with additions. London.

Vigors, Nicholas Aylward. 1825. Observations on the natural affinities that connect the orders and families of birds. [Read 3 December 1823.] *Transactions of the Linnean Society of London* 14: 395 — 517.

Volcanic islands: *Geological observations on the volcanic islands visited during the voyage of H.M.S. Beagle, together with some brief notices of the geology of Australia and the Cape of Good Hope. Being the second part of the geology of the voyage of the Beagle, under the command of Capt. Fitzroy, R.N. during the years 1832 to 1836.* London: 1844.

Voyage: *Charles Darwin and the voyage of the Beagle.* Edited by Nora Barlow. London: Pilot Press. 1945.

Walford, Edward. 1860. *The county families of the United Kingdom; or, Royal manual of the titled and untitled aristocracy of Great Britain and Ireland.* London.

Walker, Francis. 1839. *Monographia Chalciditum.* 2 vols. (Vol. 2. Species collected by C. Darwin.) London.

Ward, Thomas Humphry. 1926. *History of the Athenaeum, 1824 — 1925.* London: printed for the Athenaeum Club.

Waterhouse, Charles Owen. 1875. On some new genera and species of hetero-merous Coleoptera (Helopidæ) from Tierra del Fuego. *Transactions of the Royal Entomological Society of London*, pp. 331 — 7.

Waterhouse, George Robert. 1837 — 40. Description of some new species of exotic insects. [Read 5 December 1836.] *Transactions of the Royal Entomological Society of London* 2: 188 — 96.

Waterton, Charles. 1825. *Wanderings in South America, the North-West of the United States, and the Antilles, in the years 1812, 1816, 1820 and 1824. With original instructions for the perfect preservation of birds, etc., for cabinets of natural history.* London.

Wedgwood, Barbara and Hensleigh. 1980. *The Wedgwood circle, 1730 — 1897: four generations of a family and their friends.* London: Studio Vista.

Wellesley Index: *The Wellesley index to Victorian periodicals 1824 — 1900.* Edited by Walter E. Houghton. 3 vols. Toronto: University of Toronto Press; London: Routledge and Kegan Paul. 1966 — 79.

Whately, Richard. 1829. *A view of the Scripture revelations concerning a future state; laid before his parishioners, by a country pastor.* London.

Whewell, William. 1833. Essay towards a first approximation to a map of cotidal lines. [Read 2 May 1833.] *Philosophical Transactions of the Royal Society of London* 123: 147 — 236.

——. 1835. *Architectural notes on German churches. A new edition. To which is now added, notes written during an architectural tour in Picardy and Normandy.* Cambridge.

White, Gilbert. 1789. *The natural history and antiquities of Selborne, in the county of Southampton.* 2 vols. London.

Williamson, George Charles, ed. 1903 – 5. *Bryan's dictionary of painters and engravers.* New ed., revised and enlarged. 5 vols. London: George Bell and Sons.

Wilson, Leonard G. 1972. *Charles Lyell, the years to 1841: the revolution in geology.* New Haven and London: Yale University Press.

Winslow, J. H. 1975. Mr Lumb and Masters Megatherium: an unpublished letter by Charles Darwin from the Falklands. *Journal of Historical Geography* 1: 347 – 60.

Winstanley, Denys Arthur. 1940. *Early Victorian Cambridge.* Cambridge: Cambridge University Press.

Withering, William. 1796. *An arrangement of British plants; according to the latest improvements of the Linnaean system.* 3d ed. 4 vols. Birmingham.

Wood, William. 1821. *Illustrations of the Linnaean genera of insects.* 2 vols. London.

Young, Thomas. 1807. *A course of lectures on natural philosophy and the mechanical arts.* 2 vols. London.

Zoology: The zoology of the voyage of H.M.S. Beagle, under the command of Captain Fitzroy, during the years 1832 to 1836. Published with the approval of the Lords Commissioners of Her Majesty's Treasury. Edited and superintended by Charles Darwin. 5 parts. London: 1838 – 43.

BIOGRAPHICAL REGISTER
AND INDEX TO CORRESPONDENTS

The dates of letters to and from any correspondent are listed under the correspondent in chronological order. Dates of letters from the correspondent are printed in italic type; those of letters to the correspondent are in roman.

Abercromby, James (1776 – 1858). Politician. M.P. for Edinburgh, 1832 – 9. Speaker of the House of Commons, 1835 – 9. Created Baron Dunfermline in 1839. (*DNB*.)

Airy, George Biddell (1801 – 92). Plumian Professor of Astronomy and Director of the Observatory, Cambridge University, 1828 – 35. Astronomer Royal, 1835 – 81. FRS 1836. (*DNB, DSB*.)

Alderson, Edward Hall (1787 – 1857). Judge of the Court of Common Pleas, 1830 – 4. Baron of the Exchequer, 1834 – 57. (*DNB*.)

Alderson, Georgina. *See* Drewe, Georgina.

Alexander, Mr. Monsieur Alexandre, mimic and ventriloquist. (*DCF*.)

Alexander, William John (1797 – 1873). B.A., Trinity College, Cambridge, 1825. Admitted Lincoln's Inn, 1822. Called to the Bar, 1825. On the Oxford circuit. (*Alum. Cantab., Modern English Biography*.)

Alison, Robert Edward. English resident of Valparaiso who wrote on South American affairs. Helped CD with geological observations.
   [June? 1834]; 25 June 1835

Alison, William Pulteney (1790 – 1859). Professor of 'institutes of medicine' at Edinburgh University, 1822 – 42. (*DNB*.)

Allen, Baugh. *See* Allen, Lancelot Baugh.

Allen, Caroline (1768 – 1835). Daughter of John Bartlett Allen. Married Edward Drewe, rector of Willand, Devon, in 1793. (*Emma Darwin*.)

Allen, Catherine (1765 – 1830). Daughter of John Bartlett Allen. Married James Mackintosh in 1798. (*Emma Darwin*.)

Allen, Frances (Fanny) (1781 – 1875). Daughter of John Bartlett Allen. (*Emma Darwin*.)

Allen, Jessie (1777 – 1853). Daughter of John Bartlett Allen. Married Jean Charles Léonard Simonde de Sismondi in 1819. (*Emma Darwin*.)

Allen, John Bartlett (1733 – 1803). Of Cresselly, Pembrokeshire. (*Burke's Landed Gentry 1952*.)

Allen, Lancelot Baugh (1774 – 1845). Warden of Dulwich College, 1805 – 11; Master, 1811 – 20. Police magistrate, 1819 – 25. A Clerk in the Chancery, 1825 – 42. Son of John Bartlett Allen. (*Alum. Cantab*.)

Allen, Louisa Jane. *See* Wedgwood, Louisa Jane.

Althorp, Lord. *See* Spencer, John Charles.

Anson, Frances. Married Sir George Anson (1769 – 1849) in 1800.

Anson, Francis Harcourt (d. 1831). Naval officer. Son of Sir George Anson (1769 – 1849).

Anson, George (1697 – 1762). Circumnavigated the globe, 1740 – 4. First Lord of the Admiralty, 1751 – 6, 1757 – 62. Admiral of the Fleet, 1761. (*DNB*.)

Anson, George (1769 – 1849). Army officer. M.P. for Lichfield, 1806 – 41. Friend of the Owens of Woodhouse. (Stenton 1976.)

Argand, Aimé (1750 – 1803). Swiss physicist and chemist. Inventor of the first scientifically constructed oil burner, known as the Argand lamp. (*BHGW*, *DBF*.)

Armstrong, John (1784 – 1829). Physician to the London Fever Institution, 1819 – 24. Lectured on medicine at private medical schools in London from 1821. (*DNB*.)

Ash, Edward John (d. 1851). B.A., Christ's College, Cambridge, 1819; Fellow, 1819; Proctor, 1829. Steward of Christ's College in 1831. Rector of Brisley, Norfolk, 1838 – 51. (*Alum. Cantab.*)

Askew, Henry William (1808 – 90). B.A., Emmanuel College, Cambridge, 1832. Justice of the Peace for Cumberland, Lancashire, and Argyll. (*Alum. Cantab.*)

Aspull, George (1813 – 32). Pianist and composer. (Grove 1980.)

Aubuisson de Voisins, Jean François d' (1769 – 1841). French geologist, mineralogist, and mining engineer. (*DBF*, *DSB*.)

Audubon, John James (1785 – 1851). American ornithologist and illustrator. (*DAB*, *DSB*.)

Austen, Jane (1775 – 1817). Novelist. (*DNB*.)

Austria, Emperor of. *See* Francis II.

Azeglio, Massimo Taparelli, Marchese d' (1798 – 1866). Italian statesman and author. (*EB*.)

Babbage, Charles (1792 – 1871). Mathematician and pioneer in the design of mechanical computers. FRS 1816. (*DNB*, *DSB*.)

Babington, Charles Cardale (1808 – 95). Botanist and archaeologist. B.A., St John's College, Cambridge, 1830. Professor of botany, Cambridge, 1861 – 95. An expert on taxonomy. FRS 1851. (*DNB*, *DSB*.)

Bakewell, Robert (1768 – 1843). Geologist and mineralogist. (*DNB*, *DSB*.)

Balcarce, Juan Ramón González (1773 – 1836). Argentinian military officer. Minister of War and Navy under Juan Manuel de Rosas. Elected governor of Buenos Aires in 1832 but ousted by Rosas' supporters. (*HDA*.)

Banks, Joseph (1743 – 1820). Naturalist, patron of science, and President of the Royal Society, 1778 – 1820. Accompanied Cook, as naturalist, on circumnavigation of the globe, 1768 – 71. FRS 1766. (*DNB*, *DSB*.)

Barker, Thomas Alfred (1808–91). M.D., Edinburgh, 1829. Matriculated Downing College, Cambridge, 1830; M.B., 1835; M.D., 1840. Physician and lecturer on medicine at St Thomas's Hospital, 1840–68. (*Alum. Cantab.*, *Modern English Biography*.)

Barrow, John (1764–1848). Geographer, traveller, and naturalist. Founder of the Royal Geographical Society. Secretary of the Admiralty, 1804–45. (*DNB*, Sarjeant 1980.)

Basket, Fuegia (1821–83?). A Fuegian girl brought to England in 1830 by Robert FitzRoy. In 1833 she returned to Tierra del Fuego in the *Beagle*. Married York Minster. (Freeman 1978.)

Bates, Henry Walter (1825–92). Entomologist. Travelled in South America, 1848–59. Developed the concept of mimicry. Assistant Secretary, Royal Geographical Society, 1846–92. FRS 1881. (*DNB, DSB*.)

Batty, Robert (1789?–1848). Army officer, topographical writer, and artist. Exhibited at the Royal Academy, 1825–32. FRS 1822. (Bénézit 1976, *DNB*.)

Bayley, Major. Shrewsbury friend of CD.

Beadon, Richard a'Court (1809–90). B.A., St John's College, Cambridge, 1832. Clergyman. (*Alum. Cantab.*)

Beauclerk, Amelius (1771–1846). Naval officer; Admiral, 1830. Commander-in-chief at Plymouth, 1836–9. (*DNB*.)

Beaufort, Francis (1774–1857). Naval officer; retired as Rear-Admiral in 1846. Hydrographer to the Admiralty, 1832–55. One of the founders of the Royal Astronomical Society and of the Royal Geographical Society. FRS 1814. (*DNB*.)

1 September [1831]; 1 September [1831]

Beaumont, Thomas Wentworth (1792–1848). M.P. for Northumberland, 1818–26 and 1830–7. A founder of the *Westminster Review*, 1824. (*DNB*.)

Beechey, Frederick William (1796–1856). Naval officer and geographer. Participated in exploration and surveying voyages to the Arctic, Africa, South America, and Ireland. President, Royal Geographical Society, 1855. (*DNB*.)

Begnis, Giuseppe de (1793–1849). Italian opera singer. (Grove 1980.)

Bell, Charles (1774–1842). Anatomist who investigated the human nervous system and expression of emotions. Co-owner and principal lecturer at the Great Windmill Street School of Anatomy, 1812–25. Surgeon at the Middlesex Hospital, 1812. Professor of surgery at Edinburgh University, 1836. FRS 1826. (*DNB, DSB*.)

Bell, Thomas (1792–1880). Dental surgeon at Guy's Hospital, 1817–61. Professor of zoology at King's College, London, 1836. President, Linnean Society, 1853–61. Described the reptiles from the *Beagle* voyage. FRS 1828. (Desmond 1977, *DNB*.)

Bériot, Charles Auguste de (1802–70). Belgian violinist and composer. Second husband of Maria Malibran. (Grove 1980.)

Berkeley, Miles Joseph (1803–89). Clergyman and botanist. B.A., Christ's College, Cambridge, 1825. Expert on British fungi. FRS 1879. (*DNB, DSB.*)

Berkenhout, John (1730?–91). Physician, naturalist, and miscellaneous writer. (*DNB.*)

Beverley, Robert Mackenzie (d. 1868). Matriculated at Trinity College, Cambridge, 1816; LL.B., 1821. Achieved notoriety by the publication in 1833 of a pamphlet on the corrupt state of the University. Author of other controversial works. (*Alum. Cantab., Modern English Biography.*)

Bewick, Thomas (1753–1828). Wood engraver and naturalist who produced and illustrated many natural history books. (*DNB.*)

Biddulph, Charlotte Myddelton. Mother of Robert Myddelton Biddulph. (*Burke's Landed Gentry 1879.*)

Biddulph, Charlotte Elizabeth Myddelton (d. 1871). Sister of Robert Myddelton Biddulph. (*Burke's Landed Gentry 1879.*)

Biddulph, Fanny Myddelton. *See* Owen, Fanny Mostyn.

Biddulph, Fanny Charlotte Myddelton (1833–1900). Eldest child of Fanny and Robert Myddelton Biddulph. (*Burke's Landed Gentry 1952.*)

Biddulph, Robert Myddelton (1805–72). M.P. for Denbighshire, 1832–5 and 1852–68. Colonel of the Denbighshire Militia, 1840–72. Lord Lieutenant of Denbighshire, 1841–72. Aide-de-camp to the Queen, 1869–72. Married Fanny Owen in 1832. (*Modern English Biography.*)

Biddulph, Thomas Myddelton (1809–78). Army officer. Master of Queen Victoria's household, 1851, and keeper of the Queen's privy purse, 1867. General, 1877. Brother of Robert Myddelton Biddulph. (*DNB.*)

Bigelow, Jacob (1786–1879). American physician and botanist. Professor of materia medica, Harvard Medical School, 1815–55. Held the chair of application of science to the useful arts, Harvard, 1816–27. (*DAB.*)

Biscoe, John (1794–1843). Sea captain with the London whaling and sealing firm of Enderby. Circumnavigated the Antarctic continent, 1830–2. (*Aust. Dict. Biog.*)

Blackie, John Stuart (1809–95). Scottish man of letters. Advocate, 1834. Professor of Greek at Edinburgh University, 1852–82. (*DNB.*)

Blainville, Henri Marie Ducrotay de (1777–1850). French anatomist and zoologist. Professor of comparative anatomy, Muséum d'Histoire Naturelle, 1832. (*DBF, DSB.*)

Blakeway, John Brickdale (1765–1826). Barrister and clergyman. Author of topographical and antiquarian works on Shrewsbury. (*DNB.*)

Blane, Robert (1809–71). B.A., Trinity College, Cambridge, 1831. Army officer. (*Alum. Cantab., Modern English Biography.*)

Blasis, Virginia de (1804–38). Italian soprano.

Blumenbach, Johann Friedrich (1752–1840). German natural historian, anthropologist, and comparative anatomist. Professor of medicine at Göttingen University from 1778. (*DSB, NDB.*)

Blunt, Thomas. Shrewsbury chemist.

Blyth, Edward (1810−73). Naturalist. Curator of the Museum of the Royal Asiatic Society of Bengal, Calcutta, 1841−62. (*DNB, DSB.*)

Boat Memory (d. 1830). A Fuegian man brought to England in 1830 by Robert FitzRoy. Died of smallpox in Plymouth Naval Hospital soon after arriving. (Freeman 1978.)

Bohn, Henry George (1796−1884). London bookseller and publisher. (*DNB.*)

Boothby, Brooke (1784−1829). Prebendary of Southwell and rector of Kirkby-in-Ashfield, Nottinghamshire. Married Louisa Henrietta Vernon in 1816. (*Alum. Oxon., Burke's Peerage 1980.*)

Boringdon, Lord. *See* Parker, Edmund.

Bory de Saint-Vincent, Jean Baptiste Georges Marie (1778−1846). French army officer and naturalist. Editor of the *Dictionnaire classique d'histoire naturelle*, 1822−31. Led several botanical collecting expeditions and contributed to the knowledge of island faunas, the zoogeography of the seas, and the classification of man. (*DBF, DSB.*)

Boswell, James (1740−95). Scottish barrister. Biographer of Samuel Johnson. (*DNB.*)

Boughey, Anastasia Elizabeth (d. 1893). Sister of Thomas Fletcher Fenton Boughey and Anne Henrietta Boughey. Married Edward Joseph Smythe Jr in 1840. (*Burke's Peerage 1980.*)

Boughey, Anne Henrietta (d. 1879). Sister of Thomas Fletcher Fenton Boughey. Married Everard Robert Bruce Feilding in 1832. (*Burke's Peerage 1980.*)

Boughey, Thomas Fletcher Fenton, 3d Baronet (1809−80). High sheriff of Staffordshire, 1832. (*Alum. Oxon., Modern English Biography.*)

Bowring, John (1792−1872). Linguist, writer, and traveller. Editor of the *Westminster Review*, 1824. FRS 1856. (*DNB.*)

Bradley, Richard (d. 1732). Professor of botany at Cambridge University, 1724. Published horticultural works. FRS 1712. (*DNB, DSB.*)

Brande, William Thomas (1788−1866). Professor of chemistry at the Royal Institution, 1813−52. Author of several textbooks based on his courses of lectures. FRS 1809. (*DNB, DSB.*)

Brereton, Thomas (1782−1832). Army officer. Court-martialled for negligence and inaction during the Reform riots in Bristol in 1831. Committed suicide before the trial was concluded. (*DNB.*)

Brewster, David (1781−1868). Scottish physicist who specialised in optics. Invented the kaleidoscope, 1816. Assisted in organising the British Association for the Advancement of Science, 1831. FRS 1815. (*DNB, DSB.*)

Briggs, Mark. Darwin family coachman at The Mount, Shrewsbury.

Brisbane, Matthew (d. 1833). First British resident on the Falkland Islands. Murdered 23 August 1833 in an uprising of imported South American labour at Port Louis. (Freeman 1978.)

Briscoe, Richard. Shrewsbury chemist.

Bristowe, Anna Maria (1826—62). Daughter of Mary Ann and Samuel Ellis Bristowe. William Darwin Fox's niece. (*Darwin Pedigree.*)

Bristowe, Mary Ann (1800—29). Sister of William Darwin Fox. Married Samuel Ellis Bristowe in 1821. (*Darwin Pedigree.*)

Bristowe, Samuel Ellis (1800—55). Of Beesthorpe, Nottinghamshire. Justice of the Peace. (*Darwin Pedigree.*)

Broderip, William John (1789—1859). Magistrate and naturalist. A founder of the Zoological Society, 1826. FRS 1828. (*DNB*, Sarjeant 1980.)

Brodie, Benjamin Collins (1783—1862). Surgeon and physiologist. Published important papers on animal heat. Lectured on surgery at the Great Windmill Street School of Anatomy, 1808—30. FRS 1810. (*DNB, DSB.*)

Brongniart, Alexandre (1770—1847). French geologist and zoologist. Director of the Sèvres porcelain factory, 1800—47. Professor of mineralogy at the Muséum d'Histoire Naturelle from 1822. (*DBF, DSB.*)

Brookes, Samuel. Conchologist. (*CDEL.*)

Brougham, Henry Peter, Baron Brougham and Vaux, (1778—1868). Reformer. A founder of the *Edinburgh Review*, 1802. Lord Chancellor, 1830—4. FRS 1803. (*DNB.*)

Broughton, Clement Francis. Vicar of Uttoxeter, Staffordshire, 1830—54. (*Alum. Cantab.*)

Brown, Robert (1773—1858). Botanist. Librarian to Joseph Banks, 1810—20. Keeper of the botanical collections, British Museum, 1827—58. FRS 1810. (*DNB, DSB.*)

Browne, William Alexander Francis (1805—85). M.D., Edinburgh, 1826. President of the Plinian Society, Edinburgh, when CD was a member. Lunatic asylum superintendent who adopted a new method of treating the insane. Commissioner in lunacy for Scotland, 1857—70. (*Modern English Biography.*)

Bruce, Robert de, VIII (1274—1329). King and liberator of Scotland. (*DNB.*)

Brudenell, Mrs. Actress.

Bruegel, Pieter (the elder) (1525?—69). Flemish painter and draughtsman. (Bénézit 1976.)

Buch, Christian Leopold von (1774—1853). Widely travelled German geologist. (*DSB, NDB.*)

Buckland, William (1784—1856). Reader in geology at Oxford University, 1818—49. President of the Geological Society, 1824—5 and 1840—1. Dean of Westminster from 1845. Author of the Bridgewater treatise on geology (1836). (*DNB, DSB.*)

Buffon, Georges-Louis Leclerc, Comte de (1707—88). French naturalist and scientific administrator. (*DBF, DSB.*)

Bulkeley Owen, Thomas Bulkeley. *See* Owen, Thomas Bulkeley Bulkeley.

Bull, Ole Borneman (1810—80). Norwegian violinist and composer. (Grove 1980.)

Bulwer-Lytton, Edward George Earle Lytton, 1st Baron Lytton, (1803–73). Popular novelist and political figure. (*DNB.*)

Burchell, William John (1781–1863). Explorer and naturalist. Collected plants in St Helena, South Africa, and Brazil. (Desmond 1977, *DNB.*)

Burnett, James, Lord Monboddo, (1714–99). Scottish judge and man of letters. (*DNB.*)

Burnett, William (1779–1861). Physician-general of the navy, 1824?–41. (*DNB.*)

Burney, James (1750–1821). Naval officer. Sailed with Cook on his second (1772–5) and third (1776–80) voyages. Retired as Rear-Admiral. Wrote accounts of his voyages of discovery. (*DNB.*)

Burton, Eliza (Elizabeth Blanche) (d. 1877). Daughter of Mary and Henry Burton. (*Burke's Landed Gentry 1952.*)

Burton, Henry (1755–1831). Clergyman. Vicar of Atcham, Shropshire, 1799–1831. (*Alum. Cantab.*)

Burton, Mary ('Ma'am'). Wife of Henry Burton. (*Burke's Landed Gentry 1952*).

Butler, Fanny. *See* Kemble, Frances Anne.

Butler, Pierce (d. 1867). Of Germantown, Pennsylvania. Inherited a large plantation in Georgia. His marriage, 1834, to Fanny Kemble ended in divorce, 1849.

Butler, Richard, 2d Earl of Glengall, (1794–1858). Irish politician and author of comic plays. (*Complete Peerage.*)

Butler, Samuel (1774–1839). Educator and clergyman. Headmaster of Shrewsbury School, 1798–1836. Bishop of Lichfield and Coventry, 1836–9. (*DNB.*)

Butler, Thomas (1806–86). Clergyman. Son of Samuel Butler. Attended Shrewsbury School, 1815–25; Assistant master, 1829–34. Rector of Langar with Barnston, Notts., 1834–76. (*Alum. Cantab., Shrewsbury School Register.*)

Butterton, George Ash (1805–91). B.A., St John's College, Cambridge, 1827; Fellow, 1828–37. Clergyman and headmaster. (*Alum. Cantab.*)

Button, Jemmy. One of the Fuegians Robert FitzRoy took to England in 1830 and returned to Tierra del Fuego in 1833. (Freeman 1978.)

Buxton, Thomas Fowell (1786–1845). Philanthropist and politician. Advocated prison reform and the abolition of slavery. (*DNB.*)

Bynoe, Benjamin (1804–65). Naval surgeon, 1825–63. Assistant surgeon in the *Beagle*, 1832–7; surgeon, 1837–43. FRCS 1844. (Desmond 1977, Plarr 1930.)

Byron, George Gordon, 6th Lord (1788–1824). Poet. (*DNB.*)

Caldcleugh, Alexander (d. 1858). Business man and plant collector in South America. FRS 1831. (Desmond 1977.)
     [28 August – 5 September 1834]
Caldwell, Margaret Emma (Emma) (d. 1830). Married Henry Holland in 1822. (*Burke's Landed Gentry 1846.*)

Cameron, Jonathan Henry Lovett (1807—88). Shrewsbury school-friend of CD. B.A., Trinity College, Cambridge, 1831. Rector of Shoreham, Kent, 1860—88. (*Alum. Cantab.*)

Candolle, Augustin-Pyramus de (1778—1841). Swiss botanist. Professor of natural history, Academy of Geneva, 1816—35. (*DSB.*)

Caradori-Allan, Maria Caterina Rosalbina (1800—65). Alsatian soprano. (*DNB*, Grove 1980.)

Carlisle, Anthony (1768—1840). Surgeon to the Westminster Hospital, 1793—1840. Professor of anatomy at the Royal Academy, 1808—24. Vice-president of the Royal College of Surgeons four times, and president in 1828 and 1837. FRS 1800. (*DNB, DSB.*)
    [19 December 1836]; *23 December 1836*

Carlos, Don (1788—1855). Second surviving son of Charles IV and claimant of the throne of Spain. Escaped to England with Dom Miguel in 1834 following the civil war in Portugal. (*EB.*)

Carlyle, Jane Baillie Welsh (1801—66). Married Thomas Carlyle in 1826. (*DNB.*)

Carlyle, Thomas (1795—1881). Essayist and historian. (*DNB.*)

Caroline Amelia Elizabeth, of Brunswick-Wolfenbüttel, (1768—1821). Queen of George IV. (*DNB.*)

Carr, John (1785—1833). B.A., Trinity College, Cambridge, 1807. Headmaster of Durham School, 1812—33. Professor of mathematics at Durham University, 1833. Vicar of Brantingham, Yorkshire, 1818—33. (*Alum. Cantab.*)

Cavendish, George Henry (1810—80). Matriculated at Trinity College, Cambridge, 1829. M.P. for North Derbyshire, 1834—80. (*Alum. Cantab., Modern English Biography.*)

Cavendish, Thomas (1560—92). Circumnavigated the globe, 1586—8. (*DNB.*)

Cecil, Brownlow, 2d Marquis of Exeter, (1795—1867). Succeeded his father as Marquis in 1804. Matriculated St John's College, Cambridge, 1812; M.A., 1814. Lord Lieutenant of Northamptonshire, 1842—67. (*Alum. Cantab., Complete Peerage.*)

Chaffers, Edward Main. Master of the *Beagle*.

Chafy, William (1779—1843). Master of Sidney Sussex College, Cambridge, 1813—43. (*DNB.*)

Chambers, William Frederick (1786—1855). Physician to St George's Hospital, London, 1816—39. FRS 1828. (*DNB.*)

Chantrey, Francis Legatt (1781—1841). Sculptor and painter. Leading portrait sculptor from 1811. FRS 1818. (Bénézit 1976, *DNB.*)

Chapman, Benjamin (1798—1871). B.A., Christ's College, Cambridge, 1829. Rector of Westley Waterless, Cambridgeshire, 1829—36; vicar of Leatherhead, Surrey, 1837—71. (*Alum. Cantab.*)

Chester, Harry (1806—68). Clerk in the Privy Council Office, 1826—58. Son of Sir Robert Chester. (*Alum. Cantab., Modern English Biography.*)

Chester, Robert (1768 – 1848). Magistrate and deputy lieutenant for Hertford-shire. (*Burke's Landed Gentry 1846*.)

Children, John George (1777 – 1852). Mineralogist. Keeper of the zoological collections, British Museum, 1823 – 40. President of the Entomological Society 1834 – 5. FRS 1807. (*DNB*, Sarjeant 1980.)

Clark, William (1788 – 1869). Anatomist and clergyman. Professor of anatomy at Cambridge University, 1817 – 66. FRS 1836. (*DNB*.)

Clarke, Edward Daniel (1769 – 1822). Geologist, traveller, and antiquary. Professor of mineralogy at Cambridge University, 1808 – 22. (*DNB*, *DSB*.)

Claude Lorrain (1600 – 82). French landscape painter. (Bénézit 1976.)

Clayton, Robert. Cambridge fishmonger and wild-fowl dealer.

Clegg, Anne. *See* Hill, Anne.

Clift, William (1775 – 1849). Naturalist. Curator of the Hunterian Museum at the Royal College of Surgeons, 1793 – 1844. FRS 1823. (*DNB*, *DSB*.)

Clive, Edward, 1st Earl of Powis, (1754 – 1839). Eldest son of Baron Robert Clive. M.P. for Ludlow, 1774 – 94. Governor of Madras, 1798 – 1803. Lord Lieu-tenant of Shropshire, 1804 – 39. (*DNB*.)

Clive, Edward (1799 – 1877). Fourth son of William Clive of Styche. Captain in the army. Cousin of Edward Clive, 1st Earl of Powis. (*Eton School Lists*.)

Clive, George Arthur (d. 1881). Sixth son of William Clive of Styche. Clergyman. Educated at Eton and St John's College, Cambridge. Rector of Montford and Shrawardine, Shropshire, 1840 – 81. Cousin of Edward Clive, 1st Earl of Powis. (*Alum. Cantab.*)

Clive, Henry Bayley (1800 – 70). Fifth son of William Clive of Styche. Educated at Eton and St John's College, Cambridge. A magistrate for Shropshire. M.P. for Ludlow, 1847 – 52. Cousin of Edward Clive, 1st Earl of Powis. (*Alum. Cantab.*)

Clive, Marianne. *See* Tollet, Marianne.

Clive, Richard. Eldest son of William Clive of Styche. Served in the Indian Civil Service. Chief Secretary to the Government at Madras in 1831. Died in India. Cousin of Edward Clive, 1st Earl of Powis. (*Eton School Lists*.)

Clive, Robert, Baron (1725 – 74). Military leader and administrator in India, 1747 – 60. Governor of Bengal, 1757 – 60, 1765 – 6. M.P. for Shrewsbury, 1760 – 74. Father of Edward Clive, 1st Earl of Powis. (*DNB*.)

Clive, Robert Henry (1789 – 1854). Second son of Edward Clive, 1st Earl of Powis. M.P. for Ludlow, 1818 – 32; South Shropshire, 1832 – 54. (*Alum. Cantab.*)

Clive, Robert Herbert (1796 – 1867). Third son of William Clive of Styche. Served in the Bengal Civil Service. Military Secretary at Madras in 1831. Cousin of Edward Clive, 1st Earl of Powis. (*Eton School Lists*.)

Clive, William (1745 – 1825). Of Styche, Shropshire. Brother of Baron Robert Clive.

Clive, William (1795 – 1883). Second son of William Clive of Styche. Vicar of Welshpool, Montgomeryshire, 1819 – 65; vicar of Montford, Shropshire, 1831 – 5. Cousin of Edward Clive, 1st Earl of Powis. Married Marianne Tollet in 1829. (*Alum. Cantab.*)

Clutton, Ralph (d. 1886). Clergyman. B.A., Emmanuel College, Cambridge, 1826; Fellow, 1828 – 44. (*Alum. Cantab.*)

Cobbett, William (1763 – 1835). Essayist, politician, and agriculturist. (*DAB, DNB.*)

Coddington, Henry (d. 1845). Mathematician and clergyman. Tutor at Trinity College, Cambridge, 1822 – 33. Wrote on optics. FRS 1829. (*DNB.*)

Colbeck, William Royde (d. 1885). B.A., Emmanuel College, Cambridge, 1827; Fellow, 1829. Clergyman. (*Alum. Cantab.*)

Colburn, Henry (d. 1855). London publisher. (*DNB.*)

Coldstream, John (1806 – 63). Physician. M.D., Edinburgh, 1827. Practitioner in Leith, 1829 – 47. Friend of CD at Edinburgh University. (*DNB.*)
      28 February 1829; 13 September 1831

Coleridge, Samuel Taylor (1772 – 1834). Poet and philosopher. (*DNB.*)

Combe, George (1788 – 1858). Scottish phrenologist. Founded the Phrenological Society, 1820, and the *Phrenological Journal*, 1823. (*DNB.*)

Compson, James Edward (1793 – 1834). Vicar of St Chad's, Shrewsbury, 1826 – 34.

Congreve, Mary (1745 – 1823). Shrewsbury friend of the Darwin family.
27 October [1821]

Congreve, Sarah Adelaide (1737 – 1836). Elder sister of Mary Congreve.

Congreve, William, 1st Baronet (1743 – 1814). Comptroller of the Royal Laboratory, Woolwich, and superintendent of military machines. (*Burke's Peerage 1849.*)

Congreve, William, 2d Baronet (1772 – 1828). Inventor of the Congreve rocket, 1808. M.P. for Plymouth, 1820 – 8. (*DNB.*)

Constable, Archibald (1774 – 1827). Edinburgh publisher. (*DNB.*)

Conybeare, William Daniel (1787 – 1857). Geologist and clergyman. Greatly enlarged and improved William Phillips' compilation of English stratigraphy. FRS 1832. (*DNB, DSB.*)

Cook, James (1728 – 79). Commander of several voyages of discovery. Circumnavigated the world, 1772 – 5. FRS 1776. (*DNB, DSB.*)

Cookesley, Henry Parker (d. 1887). B.A., Trinity College, Cambridge, 1831. Clergyman. (*Alum. Cantab.*)

Cooper, James Fenimore (1789 – 1851). American novelist. (*DAB.*)

Corbet, Dryden Robert (1805 – 59). Of Sundorne Castle, Shropshire. (*Burke's Landed Gentry 1879.*)

Corbett, Joseph (d. 1838). Archdeacon of Shropshire, 1792 – 1838. (*Alum. Cantab.*)

Corfield, Richard (1781—1865). Rector of Pitchford, Shropshire, 1812—65. (*Alum. Cantab.*)

Corfield, Richard Henry (1804—97). Son of Richard Corfield. Attended Shrewsbury School, 1816—19. CD stayed at his house in Valparaiso in 1834 and 1835. (*Shrewsbury School Register.*)
    26—7 June 1835; 14—18 July 1835

Corfield, William (1811—87). Son of Richard Corfield. Matriculated at Christ's College, Cambridge, 1830; B.A., 1834. Clergyman. (*Alum. Cantab.*)

Corregio, Antonio Allegri (1489?—1534). Italian painter. (Bénézit 1976.)

Cotes, John (1799—1874). M.P. for North Shropshire, 1832—4. (*Modern English Biography.*)

Cotton, Eloisa (d. 1872). Sister of William Mostyn Owen Sr. Married Henry Calveley Cotton in 1815. Fanny and Sarah Owen's aunt. (*Burke's Peerage 1980.*)

Cotton, Francis Vere (1799—1884). Naval officer. Captain 1841; Admiral 1875. Brother of Henry Calveley Cotton. (*Burke's Peerage 1879, Modern English Biography.*)

Cotton, Henry Calveley (1789—1850). Vicar of Great Ness, Shropshire. (*Alum. Oxon., Burke's Peerage 1980.*)

Cotton, Matilda Eloisa (Mattie) (d. 1892). Daughter of Eloisa and Henry Calveley Cotton. Fanny and Sarah Owen's cousin. (*Burke's Peerage 1980.*)

Cotton, Richard (d. 1839). Shrewsbury naturalist. FGS. FLS.

Covington, Syms (1816?—61). Became CD's servant in the *Beagle* in 1833 and remained with him as assistant, secretary, and servant until 1839, when he emigrated to Australia. (Freeman 1978.)

Cramer, François (1772—1848). Violinist. One of the first professors at the Royal Academy of Music. (*DNB*, Grove 1980.)

Craven, William, 2d Earl of Craven, (1809—66). Lord Lieutenant of Warwickshire, 1854—6. (*Modern English Biography.*)

Crewe, Frances (d. 1845). Married Robert Wedgwood in 1835. (*Emma Darwin.*)

Crewe, Willoughby (1797—1850). Rector of Mucklestone, Shropshire and Astbury, Cheshire. (*Alum. Oxon., Burke's Peerage 1879.*)

Cullen, William (1710—90). Physician. Lectured on chemistry and medicine at Glasgow University and later at Edinburgh University. (*DNB, DSB.*)

Cumberland, Duke of. *See* Ernest Augustus.

Cuming, Hugh (1791—1865). Naturalist and traveller. Collected shells and living orchids in the Pacific, on the coast of Chile, and in the Philippine Islands. Returned to England in 1839. (Desmond 1977, *DNB.*)

Cumming, James (1777—1861). Professor of chemistry at Cambridge University, 1815—60. Noted as an excellent teacher. FRS 1816. (*DNB, DSB.*)

Curtis, John (1791—1862). Entomologist and artist. Executed engravings for many eminent naturalists. President of the Entomological Society, 1855. (*DNB.*)

Curtis, William (1746—99). Botanist and entomologist. (Desmond 1977, *DNB*.)

Cuvier, Georges (1769—1832). French systematist, comparative anatomist, palaeontologist, and administrator. (*DBF, DSB*.)

Dallas, Charles (d. 1855). Governor of St Helena, 1828—36. (*Modern English Biography*.)

Dallas, Davidona Eleanor. Daughter of General Charles Dallas. Married Francis Harding in 1833.

Dalton, John (1766—1844). Chemist and natural philosopher. Originator of the chemical atomic theory. (*DNB, DSB*.)

Dalyell, John Graham (1775—1851). Antiquary and naturalist. (*DNB*.)

Dalzel, Andrew (1742—1806). Classical scholar. Professor of Greek at Edinburgh University, 1779—1805. (*DNB*.)

Daniell, John Frederic (1790—1845). Meteorologist, chemist, and physicist. Professor of chemistry, King's College, London, 1831—45. FRS 1813. (*DNB, DSB*.)

Darby, Mary. Only daughter and heiress of Samuel Darby of Colebrookdale. Married Edward Pryce Owen in 1825. (*Burke's Landed Gentry 1846*.)

Darlington, Earl of. *See* Vane, Henry.

Darnell, Daniel (d. 1897). B.A., Trinity College, Cambridge, 1834. Clergyman. (*Alum. Cantab.*)

Darwin, Caroline Sarah (1800—88). CD's sister. Married Josiah Wedgwood III in 1837. (*Darwin Pedigree*.)
> 2 *[January 1826]*; 6 January 1826; *[27 February 1826]*; *[22 March 1826]*; 8 April [1826]; *11 April [1826]*; [28 April 1831]; [31?] October [1831]; 12 November [1831]; *20—31 December [1831]*; *12[—29] March [1832]*; 2—6 April 1832; 25—6 April [1832]; *12—28 June [1832]*; *12[—18] September 1832*; 24 October — 24 November [1832]; *13 January 1833*; *7 March [1833]*; 30 March — 12 April 1833; *1—4 May 1833*; *1 September 1833*; 20 September [1833]; 23 [October 1833]; *28 October [1833]*; 13 November 1833; *30 December [1833]*; *9—28 March [1834]*; 9—12 August 1834; *30 September 1834*; 13 October 1834; *29 December [1834]*; *28 January [1835]*; 10—13 March 1835; *30 March 1835*; [19] July — [12 August] 1835; 27 December 1835; *29 December [1835]*; *28 March 1836*; 29 April 1836; 24 October [1836]; 18 July 1836; [9 November 1836]; [7 December 1836]

Darwin, Catherine. *See* Darwin, Emily Catherine.

Darwin, Catty. *See* Darwin, Emily Catherine.

Darwin, Charlotte Maria Cooper (1827—85). Daughter of William Brown Darwin. Cousin of William Darwin Fox. (*Darwin Pedigree*.)

Darwin, Edward (1782—1829). Army officer. Eldest son of Erasmus and Elizabeth Collier Darwin. (*Darwin Pedigree*.)

Darwin, Elizabeth Collier (1747—1832). Widow of Colonel Edward Sacheverel Chandos-Pole. Became Erasmus Darwin's second wife in 1781. (*Darwin Pedigree*.)

Darwin, Emily Catherine (1810–66). CD's sister. Became Charles Langton's second wife in 1863. (*Darwin Pedigree.*)

> [*c. June 1823*]; [*26 October 1825*]; 4 December [*1825*]; 15 January [*1826*]; 11 April [*1826*]; 20–31 December [*1831*]; 8 January – 4 February 1832; 26–27 April [*1832*]; [May–June 1832]; 5 July [1832]; 25 July [– 3 August] 1832; 14 October [*1832*]; 22 May – 14 July 1833; 29 May 1833; 27 September 1833; 29 October 1833; 27 November 1833; 27–30 January 1834; 6 April 1834; 20–9 July 1834; 29 October 1834; 8 November 1834; 28 January [*1835*]; 31 May [1835]; 30 October 1835; 29 January 1836; 14 February 1836; 3 June [1836]; 27 [December 1836]

Darwin, Emma. *See* Wedgwood, Emma.

Darwin, Erasmus (1731–1802). Physician, botanist, and poet. CD's grandfather. Advanced an evolutionary theory similar to that subsequently expounded by Lamarck. FRS 1761. (*DNB, DSB.*)

Darwin, Erasmus Alvey (1804–81). CD's brother. Attended Shrewsbury School, 1815–22. Matriculated Christ's College, Cambridge, 1822; M.B., 1828. At Edinburgh University, 1825–6. Qualified but never practised as a physician. Lived in London from 1829 to his death. A close friend of Hensleigh Wedgwood and Thomas Carlyle. (*Alum. Cantab.*)

> 25 [*October 1822*]; 14 November 1822; 8 December 1822; 5 [*March 1823*]; 18 May 1823; [*10 January 1825*]; [*17 January 1825*]; [*24 January 1825*]; [*1 February 1825*]; [*24 February 1825*]; [*8 March 1825*]; [*June 1825*]; [*9 March 1826*]; [*9 March 1826*]; [*29 September 1826*]; 10 October [*1826*]; 21 December 1828; 18 August [*1832*]; 23 January [*1833*]

Darwin, Frances Anne Violetta. *See* Galton, Frances Anne Violetta.

Darwin, Francis (1848–1925). CD's son. B.A., Trinity College, Cambridge, 1870. Collaborated with CD on several botanical projects, 1875–82. Lecturer in botany, Cambridge University, 1884; reader, 1888–1904. Edited CD's letters. FRS 1882. (*DNB, DSB.*)

Darwin, Francis Sacheverel (1786–1859). Son of Erasmus and Elizabeth Collier Darwin. Justice of the Peace and deputy lieutenant of Derbyshire. Knighted, 1820. (*Alum. Cantab.*)

Darwin, George Howard (1845–1912). CD's son. B.A., Trinity College, Cambridge, 1868. Plumian Professor of Astronomy and Experimental Philosophy, Cambridge University, 1883–1912. FRS 1879. (*DNB, DSB.*)

Darwin, Henrietta Emma (1843–1927). CD's daughter. (*Burke's Landed Gentry 1952.*)

Darwin, Jane (1746–1835). Married William Alvey Darwin in 1772. Mother of Ann Fox. (*Darwin Pedigree.*)

Darwin, Jane Harriett (1794–1866). Married Francis Sacheverel Darwin in 1815. (*Darwin Pedigree.*)

Darwin, Lady. *See* Darwin, Jane Harriett.

Darwin, Marianne. *See* Parker, Marianne.

Darwin, Mrs. *See* Darwin, Elizabeth Collier.

Darwin, Robert Waring (1766–1848). Physician; M.D., Leiden, 1785. Had a large practice in Shrewsbury, and resided at The Mount which he built *c.* 1796–8. Third son of Erasmus Darwin by his first wife, Mary Howard. Married Susannah, daughter of Josiah Wedgwood I, in 1796. CD's father. FRS 1788. (Freeman 1978.)

> [23 October 1825]; *30 – 1 August 1831*; 31 August 1831; 31 August [1831]; *1 September 1831*; 8 February – 1 March 1832; 10 February 1832; *1 February 1833*; *7 March 1833*; *28 December 1835*

Darwin, Susan Elizabeth (1803–66). CD's sister. Lived at The Mount, Shrewsbury, until her death. (*Darwin Pedigree.*)

> *4 December [1825]*; *2 [January 1826]*; 29 January [1826]; *[27 March 1826]*; [4 September 1831]; [5 September 1831]; [6 September 1831]; [9 September 1831]; [14 September 1831]; 17 [September 1831]; *20 – 31 December [1831]*; *12 February [– 3 March] 1832*; *12 May [– 2 June] 1832*; 14 July – 7 August [1832]; *15[– 18] August 1832*; *12 – 18 November 1832*; *3 – 6 March 1833*; *22 – 31 July 1833*; *15 October 1833*; 3 December [1833]; *12[– 28] February 1834*; *[23] May 1834*; *[24] November 1834*; *16 February 1835*; 23 April 1835; 3 [September] 1835; *22 November 1835*; 28 January 1836; *12 February 1836*; 4 August [1836]

Darwin, Susannah (1765–1817). Daughter of Josiah Wedgwood I. Married Robert Waring Darwin in 1796. CD's mother. (*Darwin Pedigree.*)

Darwin, Violetta. *See* Galton, Frances Anne Violetta.

Darwin, William Alvey (1726–83). Brother of Erasmus Darwin. Father of Ann Fox. (*Darwin Pedigree.*)

Darwin, William Brown (1774–1841). Barrister. Son of William Alvey Darwin and brother of Ann Fox. Married Elizabeth de St Croix (1790–1868). (*Alum. Cantab.*)

Daubeny, Charles Giles Bridle (1795–1867). Professor of chemistry at Oxford University, 1822–55; professor of botany, 1834, of rural economy, 1840. FRS 1822. (*DNB, DSB.*)

Davy, Humphry (1778–1829). Professor of chemistry at the Royal Institution, 1802–13. President of the Royal Society, 1820–7. FRS 1803. (*DNB, DSB.*)

Dawes, Richard (1793–1867). B.A., Trinity College, Cambridge, 1817. Mathematical tutor and bursar of Downing College, 1818. Vicar of Tadlow, Cambs., 1820–40. Dean of Hereford, 1850–67. (*Alum. Cantab., DNB.*)

Denman, Thomas, 1st Baron (1779–1854). Lawyer and politician. Drafted Reform Bill, 1831. Lord Chief Justice, 1832–50. (*DNB.*)

Derbishire, Alexander. Mate of the *Beagle*; left the ship at Rio de Janeiro in 1832.

Deshayes, Gérard Paul (1797–1875). French palaeontologist and conchologist. Collected and studied living and fossil molluscs. (*DBF, DSB.*)

Despard, John (1745–1829). Army officer. Fought in the Seven Years' War and the American War of Independence. General, 1814. (*DNB.*)

Dixon, Manley (d. 1837). Admiral. Commander-in-chief at Plymouth, 1830–3.

Don, David (1800–41). Professor of botany, King's College, London, 1836–41. Linnean Society librarian, 1822–41. (Desmond 1977, *DNB.*)

Donn, James (1758–1813). Botanist. Curator of the Cambridge botanic garden, 1790–1813. (Desmond 1977, *DNB.*)

Douglas, Charles D. Surveyor. Resident of Chiloé.
        5 January 1836; 24 February 1835

Downes, John (d. 1890). B.A., Christ's College, Cambridge, 1833. Clergyman. (*Alum. Cantab.*)

Downes, William James. Left Shrewsbury School, 1823. Entered Worcester College, Oxford, 1823. (*Shrewsbury School Register.*)

Drewe, Caroline. *See* Allen, Caroline.

Drewe, Georgina. Daughter of Caroline Allen and Edward Drewe. Married Edward Hall Alderson in 1823. (*Emma Darwin.*)

Drewe, Marianne (179?–1822). Daughter of Caroline Allen and Edward Drewe. Married Algernon Langton in 1820. (*Emma Darwin.*)

Dugard, Thomas (1777–1840). Physician to the Shropshire Infirmary, 1811–40. Honorary member of the Geological Society of London, 1807.

Duncan, Adam (1812–67). Matriculated Trinity College, Cambridge, 1829; M.A., 1834. M.P. for Southampton, 1837–41; Bath, 1841–52; Forfarshire, 1854–9. A Lord of the Treasury, 1855–8. Styled Viscount Duncan from 1831 when his father was created Earl of Camperdown. Succeeded as Earl in 1859. (*Alum. Cantab.*)

Duncan, Andrew, the elder (1744–1828). Professor of physiology at Edinburgh University, 1790–1821. (*DNB.*)

Duncan, Andrew, the younger (1773–1832). Professor of materia medica at Edinburgh University, 1821–32. (*DNB.*)

Earle, Augustus (1793–1838). Artist and traveller. Artist in the *Beagle*, 1831–2. (*Aust. Dict. Biog.*, Bénézit 1976.)

Eck, Frederick W.
        [September 1834]

Edmonston, John. Negro slave who had travelled with Charles Waterton. After being freed he lived at 37 Lothian Street, Edinburgh. He made a living as a taxidermist and taught CD how to stuff birds. (Freeman 1978.)

Edward. Servant of the Darwin family at The Mount, Shrewsbury.

Ehrenberg, Christian Gottfried (1795–1876). German naturalist, microscopist, and traveller. Studied the development of coral reefs. (*DSB*, *NDB.*)

Elliston, Robert William (1774–1831). Actor. Manager of the Drury Lane Theatre, 1819–26. (*DNB.*)

Empson, William (1791–1852). Barrister. Professor of general polity and the laws of England at the East India Company College, Haileybury, 1824–52. Editor of the *Edinburgh Review*, 1847–52. (*DNB.*)

Ernest Augustus, Duke of Cumberland and King of Hanover, (1771 – 1851). Fifth son of George III. Chancellor of Trinity College, Dublin, 1805. Field marshal in the British army, 1813. (*DNB*.)

Evans, George. Of Portrane. M.P. for Dublin County, 1832 – 41. (Stenton 1976.)

Evans, George De Lacy (1787 – 1870). Army officer and politician. M.P. for Rye, 1830 – 1; Westminster, 1833 – 41 and 1846 – 65. (*DNB*.)

Exeter, Marquis of. *See* Cecil, Brownlow.

Eyre, Samuel Parr (d. 1826). Admitted to St John's College, Cambridge, 1822. Died 1 February, 1826, aged 21. (*Alum. Cantab., Shrewsbury School Register*.)

Eyton, Charles James (1812 – 54). Admitted pensioner at St John's College, Cambridge, 1829. Brother of Thomas Campbell Eyton. (*Alum. Cantab*.)

Eyton, Thomas (1777 – 1855). Barrister. Recorder of Wenlock. High sheriff of Shropshire, 1840. Father of Thomas Campbell Eyton. (*Alum. Cantab*.)

Eyton, Thomas Campbell (1809 – 80). Shropshire naturalist and collector of bird skins and skeletons. Friend and Cambridge contemporary of CD. (*DNB*.)
12 November 1833

Eyton, William Archibald (1813 – 69). Admitted pensioner at St John's College, Cambridge, 1829. Army captain. Brother of Thomas Campbell Eyton. (*Alum. Cantab*.)

Falkner, Thomas (1707 – 84). Jesuit missionary in South America, 1740 – 68. (*DNB*.)

Feilding, Everard Robert Bruce (1799 – 1854). Rector of Stapleton, Shropshire, 1824 – 54. (*Alum. Oxon*.)

Fielding, Antony Vandyke Copley (1787 – 1855). Water-colour painter and fashionable drawing-master. (Bénézit 1976, *DNB*.)

Fielding, E. R. B. *See* Feilding, E. R. B.

Fielding, Henry (1707 – 54). Novelist, journalist, and barrister. (*DNB*.)

Fitzgerald, Edward (1809 – 83). B.A., Trinity College, Cambridge, 1830. Poet and translator. (*DNB*.)

FitzRoy, Frances. *See* Trevor, Frances.

FitzRoy, George Henry, 4th Duke of Grafton, (1760 – 1844). M.P. for Cambridge University, 1784 – 1811. Lord Lieutenant of Suffolk, 1790. Uncle of Robert FitzRoy. (*DNB*.)

Fitzroy, Mary Henrietta. *See* O'Brien, Mary Henrietta.

FitzRoy, Robert (1805 – 65). Naval officer, hydrographer, and meteorologist. Commander of the *Beagle*, 1828 – 36. Author of a narrative of the surveying voyages of the *Adventure* and *Beagle*, 1839. M.P. for Durham, 1841 – 3. Governor of New Zealand, 1843 – 5. Chief of the meteorological department of the board of trade, 1854. Vice-admiral, 1863. FRS 1851. (*DNB, DSB*.)
1 September [1831]; [19 September 1831]; *23 September 1831*; [4 or 11 October 1831]; [10 October 1831]; *[1833?]*; *24 [August 1833]*; *4 October 1833*; [28 August 1834]; 6 October [1836]; *[19 –]20 October [1836]*; *30 December 1836*

Fitzwilliam, Charles William Wentworth, 3d Earl (1786–1857). M.P. for Yorkshire, 1807–30; Northamptonshire, 1831–3. FRS 1811. (*DNB*.)

Fitzwilliam, Richard, 7th Viscount (1745–1816). Bequeathed to Cambridge University his collection of pictures, engravings, and books which became the basis of the present Fitzwilliam Museum begun in 1837. (*DNB*.)

Fleming, John (1785–1857). Zoologist and geologist. Professor of natural philosophy at King's College, Aberdeen, 1834. Professor of natural science at New College (Free Church), Edinburgh, 1845–57. (*DNB, DSB*.)

Fletcher, Harriet (1799–1842). Daughter of Richard Fletcher. Married William Darwin Fox in 1834. (*Darwin Pedigree*.)

Fletcher, Richard (1768–1813). Served in the Royal Engineers during the Peninsular War. Created Baronet in 1811. (*DNB*.)

Flourens, Marie-Jean-Pierre (1794–1867). French physiologist and historian of science. Permanent secretary of the Academy of Sciences, 1833. (*DBF, DSB*.)

Flügel, Johann Gottfried (1788–1855). German lexicographer. (*EB*.)

Foggo, John. Author of papers on entomology and meteorology.

Forbes, Edward (1815–54). Zoologist, botanist, and invertebrate palaeontologist. Professor of botany, King's College, London, 1842. Palaeontologist with the Geological Survey, 1844–54. Professor of natural history, Edinburgh University, 1854. FRS 1845. (*DNB, DSB*.)

Forester, Isabella Elizabeth Annabella (d. 1858). Sister of John George Weld Forester. Married Major-General George Anson, nephew of George Anson (1769–1849), in 1830. (*Burke's Peerage 1980*.)

Forester, John George Weld, 2d Baron (1801–74). Of Willey Park, Shropshire. M.P. for Wenlock, 1826–8. Succeeded to the barony in 1828. (*Modern English Biography*.)

Forster, Edward (1765–1849). Botanist and banker. Vice-president of the Linnean Society, 1828. FRS 1821. (Desmond 1977, *DNB*.)

Forsyth, Charles. Midshipman in the *Beagle*. (Freeman 1978).

Fox, Ann (1777–1859). Second wife of Samuel Fox of Thurlston Grange near Derby. Mother of William Darwin Fox. (*Darwin Pedigree*.)

Fox, Eliza (1801–86). Sister of William Darwin Fox. (*Darwin Pedigree*.)

Fox, Eliza Ann (1836–74). Eldest child of Harriet and William Darwin Fox. (*Darwin Pedigree*.)

Fox, Emma (1803–85). Sister of William Darwin Fox. (*Darwin Pedigree*.)

Fox, Frances Jane (b. 1806). Sister of William Darwin Fox. (*Darwin Pedigree*.)

Fox, Harriet. *See* Fletcher, Harriet.

Fox, Henry Stephen (1791–1846). British diplomat. Minister plenipotentiary, Buenos Aires, 1831–2; Rio de Janeiro, 1833-6; Washington, D.C., 1836–44. (*DNB*.)

31 October 1833; 25 July 1834; 15 August 1835

Fox, Julia (b. 1809). Sister of William Darwin Fox. (*Darwin Pedigree*.)

Fox, Samuel (1765 — 1851). Of Thurlston Grange near Derby. Justice of the Peace. Father of William Darwin Fox. (*Darwin Pedigree.*)

Fox, William Darwin (1805 — 80). Clergyman. Matriculated Christ's College, Cambridge, 1824; B.A., 1829. CD's second cousin. A close friend at Cambridge who shared CD's enthusiasm for entomology. Maintained an active interest in natural history throughout his life and provided CD with much information. Rector of Delamere, Cheshire, 1838 — 73. Spent the last years of his life at Sandown, Isle of Wight. (*Alum. Cantab.*)

> 12 [June 1828]; [30 June 1828]; [29 July 1828]; [19 August 1828]; [October 1828]; [29 October 1828]; [24 December 1828]; [7 January 1829]; [25 — 9 January 1829]; [26 February 1829]; [15 March 1829]; 1 April [1829]; [10 April 1829]; [12 April 1829]; [23 April 1829]; [18 May 1829]; 7 June [1829]; [3 July 1829]; [15 July 1829]; 29 [July 1829]; 26 [August 1829]; 4 September [1829]; [22 September 1829]; [15 October 1829]; [3 November 1829]; [3 January 1830]; [13 January 1830]; [25 March 1830]; [1 April 1830]; [9 May 1830]; [31 May 1830]; 25 [August 1830]; [8 September 1830]; [8 October 1830]; 5 November [1830]; [27? November 1830]; [23 January 1831]; [9 February 1831]; [15 February 1831]; [7 April 1831]; [11 May 1831]; [9 July 1831]; 1 August [1831]; 6 [September 1831]; 19 [September 1831]; 17 [November 1831]; May [1832]; *30 June 1832*; *29 August — 28 September 1832*; [12 — 13] November 1832; *23 January 1833*; 23 May 1833; 25 October 1833; *1 November 1834*; [7 — 11] March 1835; [9 — 12 August] 1835; 15 February 1836; 6 November [1836]; 15 December [1836]

Fox, William Johnson (1786 — 1864). Preacher, politician, and author. (*DNB.*)

Francis II (1768 — 1835). The last Roman Emperor, and, as Francis I, first Emperor of Austria. (*EB.*)

Fremantle, Charles Howe (1800 — 69). Naval officer. Captain, 1826; Admiral, 1864. (*Modern English Biography.*)

Frere, Mary (1781 — 1864). Wife of William Frere. Noted for her beautiful voice. Established a salon in the Master's Lodge, Downing College, Cambridge, and arranged private theatricals in the College hall. (*Modern English Biography.*)

Frere, William (1775 — 1836). Barrister. Master of Downing College, Cambridge, 1812 — 36. Vice-Chancellor of Cambridge University, 1819. (*DNB.*)

Freycinet, Louis Claude Desaulses de (1779 — 1842). French navigator. (*DBF, EB.*)

Gaimard, Joseph Paul (1796 — 1858). French naval surgeon and naturalist. Participated in many voyages of scientific exploration. A colleague of Jean Quoy. (*DBF, DSB.*)

Gall, Franz Joseph (1758 — 1828). Physician, neuroanatomist, and psychologist whose theories were applied in the form of phrenology. (*DBF, DSB.*)

Galton, Bessy. *See* Galton, Elizabeth Ann.

Galton, Elizabeth Ann (b. 1808). Daughter of Samuel Tertius and Violetta Galton. CD's cousin. (*Darwin Pedigree.*)

Galton, Erasmus (b. 1815). Son of Samuel Tertius and Violetta Galton. CD's cousin. (*Darwin Pedigree.*)

Galton, Frances Anne Violetta (Violetta) (1783–1874). Daughter of Erasmus and Elizabeth Collier Darwin. Married Samuel Tertius Galton in 1807. (*Darwin Pedigree.*)

Galton, Francis (1822–1911). Scientist and founder of the eugenics movement. Son of Samuel Tertius and Violetta Galton. CD's cousin. FRS 1860. (*DNB, DSB.*)

Galton, Lucy Harriot (1809–48). Daughter of Samuel Tertius and Violetta Galton. Married James Moilliet of Choney Court, Hereford, in 1832. (*Darwin Pedigree.*)

Galton, Samuel John (1753–1832). Of Duddeston House, Warwickshire. Father of Samuel Tertius Galton. (*Burke's Landed Gentry 1879.*)

Galton, Samuel Tertius (1783–1844). Deputy lieutenant of Warwickshire. Married Frances Anne Violetta Darwin in 1807. (*Alum. Cantab.*, *Darwin Pedigree.*)

Galton, Violetta. *See* Galton, Frances Anne Violetta.

García, Manuel José (1784–1848). Argentinian jurisconsult, administrator, and diplomat. Minister of treasury under Rosas. (*HDA.*)

Garland, Lewes (or Lewis) (d. 1844). B.A., Trinity College, Cambridge, 1827. (*Alum. Cantab.*)

Gay, Claude (1800–73). French naturalist and traveller who surveyed the flora and fauna of Chile. Professor of physics and chemistry at Santiago College, 1828–42. (*DBF.*)

George IV (1762–1830). King of Great Britain, 1820–30. (*DNB.*)

Gibbon, Edward (1737–94). Historian. (*DNB.*)

Giffard, Caroline Mallet (1802–41). Became John Mytton's second wife in 1821. (*Burke's Landed Gentry 1952.*)

Giffard, Louisa Paulina Charlotte (1807–79). Sister of Caroline Mallet Giffard. Married Thomas Fletcher Fenton Boughey in 1832. (*Burke's Landed Gentry 1952.*)

Gifford, Harriet Maria (d. 1857). Daughter of Caroline and Edward Drewe. Married Robert, 1st Baron Gifford, in 1816. (*Burke's Peerage 1980.*)

Gifford, Robert, 1st Baron (1779–1826). Barrister and politician. Attorney-General, 1819–24. Lord Chief Justice of Common Pleas, 1824. (*DNB.*)

Glasspoole, Frederick Bream. M.D., Edinburgh, 1827. Physician in Hereford. (*Provincial Medical Directory.*)

Gloucester, Duke of. *See* William Frederick.

Gmelin, Johann Georg (1709–55). Naturalist and explorer. Professor of chemistry and natural history at the Academy of Sciences, St Petersburg, 1731–47. Professor of medicine, botany, and chemistry at Tübingen University, 1749. (*DSB, NDB.*)

Goldie, George (1784—1853). M.D., Edinburgh, 1808. Practised in York, 1815—49. Physician to the York county hospital, 1822—33. In charge of the cholera hospital, York, during the epidemic of 1831. (*Modern English Biography*.)

Goldsmith, Oliver (1728—74). Dramatist, poet, and novelist. (*DNB*.)

Gooch, William (1769--1851). Colonel. Father-in-law of William Venables Vernon Harcourt. (\Burke's Peerage 1980\.)

Good, John Mason (1764—1827). Physician and author. FRS 1808. (*DNB*.)

Gordon, Henry Percy (1806—76). Barrister. B.A., Peterhouse, Cambridge, 1827. Deputy lieutenant for the Isle of Wight. FRS 1830. (*Alum. Cantab.*, *Modern English Biography*.)

Gore, Philip Yorke. Chargé d'affaires at Buenos Aires, 1832—4. (*BDR*.)

Gore, William Ormsby. *See* Ormsby-Gore, William.

Goulburn, Henry (1784—1856). M.P. for Cambridge University, 1831—56. Chancellor of the Exchequer, 1828—34, 1841—6. Home Secretary under Peel, 1834—5. (*DNB*.)

Gould, John (1804—81). Self-taught ornithologist and artist. Taxidermist to the Zoological Society of London, 1826—81. Described the birds collected on the *Beagle* and *Sulphur* expeditions. FRS 1843. (*DNB*, *DSB*.)

Grafton, Duke of. *See* FitzRoy, George Henry.

Graham, John (1794—1865). CD's tutor at Christ's College, Cambridge. Master of the College, 1830—48. Vice-Chancellor of the University, 1831 and 1840. Bishop of Chester, 1848—65. (*DNB*.)

Graham, Robert (1786—1845). Physician and botanist. Regius Professor of Botany at Edinburgh University, 1820—45. Physician to the Edinburgh Infirmary. (*DNB*.)

Grant, Robert Edmond (1793—1874). Scottish physician and zoologist. Befriended CD in Edinburgh. Professor of comparative anatomy and zoology, University College, London, 1827—74. FRS 1836. (*DNB*, Sarjeant 1980.)

Gray, George Robert (1808—72). Zoologist expert on insects and birds. Assistant in the zoological department of the British Museum, 1831—72. FRS 1865. (*DNB*, Gilbert 1977.)

Greig, Samuel (d. 1807). Commissioner of the Russian navy and Russian consul general in Great Britain. First husband of Mary Fairfax Greig Somerville.

Grenville, William Wyndham, Baron (1759—1834). Statesman. Chancellor of Oxford University, 1809—34. (*DNB*.)

Greville, Robert Kaye (1794—1866). Botanist. Made botanical tours in the Scottish highlands. M.P. for Edinburgh, 1856. (Desmond 1977, *DNB*.)

Grey, Charles, 2d Earl (1764—1845). Statesman. Prime Minister of the Whig administration, 1831—4. (*DNB*.)

Grey, William Scurfield (1808—76). B.A., St John's College, Cambridge, 1831. Barrister. High sheriff, Durham, 1867. (*Alum. Cantab.*)

Griffith, Charles. British consul at Buenos Aires, 1834. (*BDR*.)

Griffith, Edward (1790—1858). Naturalist and solicitor. (*DNB*.)

Guilding, Lansdown (1797 – 1831). Botanist and zoologist. Colonial chaplain, St Vincent. (Desmond 1977.)

Haddington, Lord. *See* Hamilton, Thomas.

Hall, Jeffry Brock (1807 – 86). B.A., Christ's College, Cambridge, 1830. (*Alum. Cantab.*)

Haller, (Victor) Albrecht von (1708 – 77). Anatomist, physiologist, and botanist. Professor of anatomy, surgery, and medicine at Göttingen University, 1736 – 53. (*DSB, NDB.*)

Hamilton, Edward William Terrick (1809 – 98). B.A., Trinity College, Cambridge, 1832; Fellow, 1834 – 42. Resident of New South Wales, 1840 – 55, and governor of the Australian Agricultural Company, 1857 – 98. M.P. for Salisbury, 1865 – 9. (*Alum. Cantab., Aust. Dict. Biog.*)

Hamilton, Hamilton Charles James (1779 – 1856). Minister plenipotentiary at Buenos Aires, 1834 – 6; Rio de Janeiro, 1836 – 46. (*BDR, Modern English Biography.*)

Hamilton, Thomas, 9th Earl of Haddington, (1780 – 1858). Politician. Lord Lieutenant of Ireland, 1834 – 5. First Lord of the Admiralty, 1841 – 6, and Lord Privy Seal, 1846. (*DNB.*)

Hamilton, William (1788 – 1856). Scottish metaphysician and logician. Professor of logic and metaphysics at Edinburgh University, 1836 – 56. (*DNB, DSB.*)

Hamilton, William Rowan (1805 – 65). Mathematician. Astronomer Royal of Ireland and Andrews Professor of Astronomy at Trinity College, Dublin, 1827. (*DNB, DSB.*)

Hammond, Josiah (1803 – 75). Studied at St Bartholomew's Hospital and practised at Market Street, Cambridge. Surgeon to Addenbrooke's Hospital and to the Gaol. FRCS 1843. (Plarr 1930.)

Hamond, Robert Nicholas (1809 – 83). Naval officer; Lieutenant, 1827. Married Sophia Caroline Musters, sister of Charles Musters, in 1836. His elder brother, Anthony, had married Mary Ann Musters, sister of Charles Musters, in 1828. (*Burke's Landed Gentry 1952*, O'Byrne 1849.)

Handel, George Frideric (Georg Friedrich) (1685 – 1759). English (naturalised) composer of German birth. (*DNB*, Grove 1980.)

Harbord, Edward, 3d Baron Suffield, (1781 – 1835). Philanthropist and politician. Carried reforms concerning prison discipline and game-laws. (*DNB.*)

Harcourt, Edward Vernon (1757 – 1847). Archbishop of York from 1808. (*DNB.*)

Harcourt, William Venables Vernon (1789 – 1871). Son of Edward Harcourt, Archbishop of York. Rector of Wheldrake, Yorkshire, 1824 – 37. General Secretary to the first meeting of the British Association for the Advancement of Science in York, 1831. Married Matilda Mary Gooch in 1824. FRS 1824. (*DNB.*)

Harding, Francis (1799 – 1875). Entered navy 1812. Served in H.M.S. *War-spite* on the South American station until 1833. Captain, 1841. Retired as vice-admiral, 1867. (*Modern English Biography*, O'Byrne 1849.)

Harding, John (d. 1866). Incumbent of St George's, Shrewsbury, 1832 – 66. (*Alum. Oxon.*)

Harris, James (1709 – 80). Philosopher and philologist. (*DNB*.)

Harris, James. British trader at the Rio Negro, Patagonia, from whom Robert FitzRoy hired two boats for surveying parts of the coast too shallow for the *Beagle*.

Harris, William Snow (1791 – 1867). Electrician and inventor. Made improvements to lightning-conductors. FRS 1831. (*DNB*.)

Harrowby, Lord. *See* Ryder, Dudley.

Haviland, John (1785 – 1851). Regius Professor of Medicine at Cambridge University, 1817 – 51. (*DNB*.)

Hawkins, Caesar Henry (1798 – 1884). Surgeon. Taught anatomy at the Great Windmill Street School of Anatomy. Surgeon to St George's Hospital, 1829 – 61. FRS 1856. (*DNB*, Plarr 1930.)

Hawley, Richard Maddock. M.D., Edinburgh, 1807. Fellow, Royal College of Physicians, Edinburgh, 1827.

Haycock, Edward (1791 – 1870). Noted Shropshire architect. County Surveyor, 1834 – 66. Friend of Robert Waring Darwin.

Haycock, Mrs John Hiram (née Elizabeth Trevitt). Mother of Edward Haycock.

Haydn, (Franz) Joseph (1732 – 1809). Austrian composer. (Grove 1980.)

Head, Francis Bond (1793 – 1875). Colonial governor and author. Travelled in South America as manager of the Rio Plata Mining Association, 1825 – 6. (*DNB*.)

Heaviside, James William Lucas (1808 – 97). B.A., Sidney Sussex College, Cambridge, 1830; Fellow, 1832; Tutor, 1833 – 8. Professor of mathematics at the East India Company College, Haileybury, 1838 – 57. Canon of Norwich, 1860 – 97. (*Alum. Cantab.*, *Modern English Biography*.)

Heawood, Edward (d. 1847). B.A., Christ Church, Oxford, 1815. Died rector of Halstead, Kent. (*Alum. Oxon.*)

Heber, Reginald (1783 – 1826). Bishop of Calcutta, 1822 – 6. Wrote of his extensive travels in India. (*DNB*.)

Henry VIII (1491 – 1547). King of England, 1509 – 47. (*DNB*.)

Henry, William (1774 – 1836). Manchester chemist. FRS 1808. (*DNB*, *DSB*.)

Henry, William Charles. Son of William Henry. Studied medicine at Edinburgh University and chemistry under Justus von Liebig in Germany. Known mainly for his biography of John Dalton.

Henslow, George (1835 – 1925). Clergyman and teacher. Lecturer in botany at St Bartholomew's Medical School, 1886 – 90. Younger son of John Stevens Henslow. (*Alum. Cantab.*, Desmond 1977.)

Henslow, Harriet. *See* Jenyns, Harriet.

Henslow, John Prentis. Solicitor. Father of John Stevens Henslow.

Henslow, John Stevens (1796—1861). Clergyman, botanist, and mineralogist.
Professor of mineralogy at Cambridge University, 1822—7; Professor of
botany. Extended and remodelled the Cambridge Botanic Garden. Curate of
Little St Mary's Church, Cambridge, 1824—32; vicar of Cholsey-cum-
Moulsford, Berkshire, 1832—7; rector of Hitcham, Suffolk, 1837—61. CD's
teacher and friend. (*DNB, DSB.*)

   [11 July 1831]; [6 or 13 August 1831]; *24 August 1831*; 30 [August 1831]; [2
   September 1831]; [5 September 1831]; 9 [September 1831]; 17 [September
   1831]; 28 [September 1831]; [4 or 11 October 1831]; *25 October 1831*; 30
   [October 1831]; 15 [November 1831]; *20 November 1831*; 3 December [1831];
   6 February 1832; 18 May — 16 June 1832; [23 July —] 15 August [1832]; [c.
   26 October —] 24 November 1832; *15—21 January 1833*; 23 January [1833];
   1 February 1833; 11 April 1833; 18 July 1833; *31 August 1833*; [20—7]
   September 1833; 12 November 1833; 1 March 1834; 2 May 1834; *22 July
   1834*; 24 July — 7 November 1834; 4 October 1834; 8 November 1834;
   [10]—13 March 1835; 18 April 1835; 12 [August] 1835; [28—9] January
   1835; 28 December 1836; [30—1 October 1836]; [1 November 1836]; 9 July
   1836; 6 October [1836]

Henslow, Leonard Ramsay (1831—1915). Clergyman. Elder son of John Stevens
Henslow. (*Alum. Cantab.*)

Henslow, Mrs. *See* Jenyns, Harriet.

Herapath, John (1790—1868). Mathematician and journalist. (*DNB, DSB.*)

Herbert, Edward, 1st Baron of Cherbury, (1583—1648). Philosopher, historian,
and diplomatist. (*DNB.*)

Herbert, John Maurice (1808—82). B.A., St John's College, Cambridge, 1830;
Fellow, 1832—40. Barrister, 1835. County Court Judge, South Wales, 1847—
82. (*Alum. Cantab., Modern English Biography.*)

   [13 September 1828]; [3 October 1828]; *[early May 1831]*; *15—17 April 1832*;
   [1—6] June 1832; *1[—4] December 1832*; 2 June 1833; *[28 March] 1834*; *[19
   November 1836]*

Herschel, John Frederick William (1792—1871). Astronomer, mathematician,
chemist, and philosopher. Member of many learned societies. Carried out
astronomical observations at the Cape of Good Hope, 1834—8. Master of the
Mint, 1850—5. FRS 1813. (*DNB, DSB.*)

Hewitson, William Chapman (1806—78). Entomologist and oologist. (*DNB*,
Gilbert 1977.)

Hey, Samuel (d. 1852). B.A., Queens' College, Cambridge, 1809. Vicar of
Ockbrook, Derbyshire, 1816—52. (*Alum. Cantab.*)

Hey, William (1811—82). B.A., St John's College, Cambridge, 1834. Clergyman,
teacher, and entomologist. Son of Samuel Hey. (*Alum. Cantab.*)

Hildyard, Frederick (1803–91). Attended Shrewsbury School, 1814–21. B.A., Trinity College, Cambridge, 1825. Clergyman. Assistant master at Shrewsbury School, 1825–6. (*Alum. Cantab.*)

Hill, Anne (1815–91). Married Rowland Hill in 1831. (*Complete Peerage.*)

Hill, John (1802–91). B.A., Oriel College, Oxford, 1824. Rector of Great Bolas, Shropshire, 1831 – 77. Brother of Rowland Hill. Married Charlotte Kenyon in 1833. (*Alum. Oxon., Burke's Peerage 1980.*)

Hill, Rowland, 2d Viscount (1800–75). M.P. for Shropshire, 1821–32; North Shropshire, 1832–42. Lord Lieutenant of Shropshire, 1845–75. (*Alum. Oxon., Complete Peerage.*)

Hobhouse, John Cam, Baron Broughton de Gyfford, (1786–1869). Statesman and author. M.P. for Westminster, 1820–33; Nottingham, 1834–47; Harwich, 1848–51. FRS 1814. (*DNB.*)

Hodgson, Nathaniel Thomas Lumley (1808–86). B.A., Christ's College, Cambridge, 1832. (*Alum. Cantab.*)

Holden, James Richard (1807–76). B.A., Christ's College, Cambridge, 1830. Rector of Pleasley, Derbyshire, 1831–56. (*Alum. Cantab.*)

Holland, Bessy. Sister of Henry Holland.

Holland, Edward (1806–75). CD's second cousin. B.A., Trinity College, Cambridge, 1829. M.P. for East Worcestershire, 1835–7; Evesham, 1855–68. President of the Royal Agricultural Society. (*Alum. Cantab.*)

Holland, Emma. *See* Caldwell, Margaret Emma.

Holland, Henry (1788–1873). Physician. Distant cousin of the Darwins and Wedgwoods. Physician in ordinary to Queen Victoria, 1852. President of the Royal Institution. FRS 1816. (*DNB.*)

Holland, Mrs Edward. *See* Isaac, Sophia.

Holland, Saba (1802–66). Daughter of Sydney Smith. Henry Holland's second wife, 1834. Published a memoir of her father, 1855. (*DNB.*)

Hood, Thomas Samuel. British Consul general at Montevideo.

Hook, Theodore Edward (1788–1841). Novelist and wit. Editor of *John Bull*, 1820. Editor of the *New Monthly Magazine*, 1836–41. (*DNB.*)

Hooker, Joseph Dalton (1817–1911). Botanist. Assistant director, Royal Botanic Gardens, Kew, 1855–65; Director, 1865–85. Worked chiefly in taxonomy and plant geography. Son of William Jackson Hooker. Friend and confidant of CD. FRS 1847. (*DNB, DSB.*)

Hooker, Mr. Landowner in Uruguay.

Hooker, William Jackson (1785–1865). Botanist. Regius Professor of Botany, University of Glasgow, 1820. Established the Royal Botanic Gardens at Kew, 1841, and served as first director. Father of Joseph Dalton Hooker. FRS 1812. (*DNB, DSB.*)

Hope, Frederick William (1797–1862). Entomologist and clergyman. Gave his collection of insects to Oxford University and founded a professorship of zoology, 1849. (*DNB*, Gilbert 1977.)

1 November 1833; *15 January 1834*

Hope, Henry. *See* Hope, Thomas Henry.

Hope, Louisa. *See* Leighton, Louisa.

Hope, Thomas Charles (1766 – 1844). Professor of chemistry at Edinburgh University, 1799 – 1843. (*DNB, DSB.*)

Hope, Thomas Henry (1794 – 1871). High sheriff of Gloucestershire, 1837. Brother of Frederick William Hope. Married Louisa Leighton in 1833. (*Alum. Cantab.*)

Hore, William Strong (1807 – 82). B.A., Queens' College, Cambridge, 1830. Clergyman and botanist. (*Alum. Cantab.*, Desmond 1977.)

Horner, Leonard (1785 – 1864). Geologist and educationist. A founder of the *Edinburgh Review*, 1802. Warden of University College, London, 1827 – 31. President of the Geological Society, 1845 – 7, 1860 – 2. Father-in-law of Charles Lyell. FRS 1813. (*DNB, DSB.*)

Horsfield, Thomas (1773 – 1859). Naturalist. Keeper of the East India Company's museum, London, 1820 – 59. (Desmond 1977, *DNB.*)

Horton, Robert John Wilmot (1784 – 1841). Politician. Governor of Ceylon, 1831 – 7. (*DNB.*)

Hubbersty, Nathan (1803 – 81). B.A., St John's College, Cambridge, 1826. Assistant master at Shrewsbury School, *c.* 1826 – 8. Headmaster of Wirksworth Grammar School, 1832 – 51. (*Alum. Cantab.*)

Hughes, Charles. Attended Shrewsbury School, 1818 – 19. Resident in Buenos Aires, 1832 – 3. (*Shrewsbury School Register.*)
    2 November [1832]

Humboldt, Friedrich Wilhelm Heinrich Alexander von (1769 – 1859). Eminent German naturalist and traveller. Explored South America, 1799 – 1804. (*DSB, NDB.*)

Hume, Joseph (1777 – 1855). Radical politician. M.P. for Aberdeen, 1818 – 30; Middlesex, 1830 – 7. (*DNB.*)

Hummel, Johan Nepomuka (1778 – 1837). Hungarian pianist, conductor, and composer. (Grove 1980.)

Hunt, Thomas (d. 1859). Rector of Wentnor and West Felton, Shropshire, 1817 – 59. (*Alum. Cantab.*)

Hunter, John (1728 – 93). Surgeon and anatomist. His collection of zoological specimens formed the basis of the Hunterian Museum of the Royal College of Surgeons. FRS 1767. (*DNB, DSB.*)

Hunter, William (1718 – 83). Anatomist. Brother of John Hunter. First professor of anatomy, Royal Academy, 1768. FRS 1767. (*DNB, DSB.*)

Hustler, William (1787 – 1832). B.A., Jesus College, Cambridge, 1811; Fellow, 1811 – 32. University Registrary, 1816 – 32. (*Alum. Cantab.*)

Hutton, James (1726 – 97). Scottish natural philosopher and geologist. Propounded a uniformitarian view of geological history in his *Theory of the earth* (1795). (*DNB, DSB.*)

Huxley, Thomas Henry (1825 – 95). Zoologist, comparative anatomist, essayist, and educator. Lecturer, Royal School of Mines, from 1854. Fullerian Professor at the Royal Institution, 1863 – 7. Hunterian Professor at the Royal College of Surgeons, 1863 – 9. Powerful and polemical supporter of Darwinism. FRS 1851. (*DNB, DSB*.)

Iliff, Frederick (1799 – 1869). Clergyman and teacher. Assistant master at Shrewsbury School. (*Alum. Cantab., Modern English Biography*.)

Impey. CD's college servant at Cambridge.

Isaac, John. Banker. Brother of Sophia Isaac.

Isaac, Sophia (d. 1851). Married Edward Holland in 1832.

Jameson, Robert (1774 – 1854). Geologist and mineralogist. Regius Professor of Natural History and keeper of the museum at Edinburgh University, 1804 – 54. Editor of the *Edinburgh Philosophical Journal*, 1824 – 54. (*DNB, DSB*.)

Jenyns, George Leonard (1763 – 1848). Vicar of Swaffham Prior, Cambridgeshire, 1787 – 1848; Prebendary of Ely, 1802 – 48. Inherited Bottisham Hall, Cambridgeshire, from his second cousin in 1787. (*Alum. Cantab*.)

Jenyns, Harriet (1797 – 1857). Daughter of George Leonard Jenyns and sister of Leonard Jenyns. Married John Stevens Henslow in 1823. (*Burke's Landed Gentry 1879*.)

Jenyns, Leonard (1800 – 93). Naturalist and clergyman. Son of George Leonard Jenyns. Vicar of Swaffham Bulbeck, Cambridgeshire, 1828 – 49. Member of many scientific societies. Described the *Beagle* fish specimens. Adopted the name Blomefield in 1871. A friend of CD at Cambridge. (Desmond 1977, *DNB*.)

Johnson, Henry. Physician. Attended Shrewsbury School. M.D., Edinburgh, 1829. Senior physician, Shropshire Infirmary. (*Provincial Medical Directory*.)

Johnson, Samuel (1709 – 84). Writer and lexicographer. (*DNB*.)

Jones, John Edward (1806 – 62). Sculptor (*DNB*.)

Keats, John (1795 – 1821). Poet. (*DNB*.)

Keen, Mr. Landowner in Uruguay.

Kemble, Frances Anne (Fanny) (1809 – 93). Actress and author. Married Pierce Butler in 1834. (*DAB, DNB*.)

Kennedy, Benjamin Hall (1804 – 89). Clergyman and teacher. B.A., St John's College, Cambridge, 1827; Fellow, 1828 – 32. Headmaster of Shrewsbury School, 1836 – 66. Regius Professor of Greek at Cambridge University, 1867 – 89. (*Alum. Cantab., DNB*.)

Kenyon, Charlotte (1813 – 84). Daughter of Thomas Kenyon of Pradoe. Married John Hill in 1833. (*Burke's Peerage 1980*.)

Kenyon, George, 2d Baron (1776 – 1855). Barrister. (*Modern English Biography*.)

Kenyon, Lloyd (1804 – 36). Captain in the Royal Horse Guards. Son of Thomas Kenyon of Pradoe. (*Burke's Peerage 1879*.)

Kenyon, Lloyd, 3d Baron (1805 – 69). M.P. for St Michael's, Cornwall, 1830 – 2. (*Complete Peerage, Modern English Biography.*)

Kenyon, Louisa Charlotte. Wife of Thomas Kenyon of Pradoe. (*Burke's Peerage 1980.*)

Kenyon, Thomas (1780 – 1851). Of Pradoe, Shropshire. Brother of George Kenyon. (*Burke's Peerage 1980.*)

Kenyon, Thomas (1805 – 59). Army officer. Son of Thomas Kenyon of Pradoe. (*Burke's Peerage 1879.*)

Kerr, Robert (1755 – 1813). Scientific writer and translator. (*DNB.*)

King, Philip Gidley (1817 – 1904). Eldest son of Philip Parker King. Midshipman in the *Beagle*, 1831 – 6. Lived in Australia from 1836. (*Aust. Dict. Biog.*)

King, Philip Parker (1793 – 1856). Naval officer and hydrographer. Commander of the *Adventure* and *Beagle* on the first surveying expedition to South America 1826 – 30. Settled in Australia. Rear-admiral, 1855. FRS 1824. (Desmond 1977, *DNB.*)
[21 January 1836]

Kinnaird, George William Fox, 9th Baron (1807 – 78). Succeeded to the Scottish peerage in 1826. Privy councillor, 1840. Lord Lieutenant of Perthshire, 1866. (*DNB.*)

Kirby, Henry (d. 1858). B.A., Clare College, Cambridge, 1817; Fellow, 1817; Senior Proctor, 1829 – 30. Rector of Great Waldingfield, 1842 – 58. (*Alum. Cantab.*)

Kirby, William (1759 – 1850). Clergyman and entomologist. One of the first Fellows of the Linnean Society, 1788. FRS 1818. (*Alum. Cantab., DNB.*)

Knowles, James Sheridan (1784 – 1862). Irish-born actor and dramatist. (*DNB.*)

Kotzebue, Otto von (1787 – 1846). Russian explorer and navigator. Published accounts of his voyages. (*EB.*)

Kynaston, John Roger (1797 – 1866). Captain, North Shropshire Yeomanry Cavalry, 1831 – 50. Succeeded as 3d Baronet in 1839. Brother of Letitia Kynaston. (*Burke's Peerage 1849, Modern English Biography.*)

Kynaston, Letitia (d. 1834). Of Hardwick, Shropshire. (*Burke's Peerage 1849.*)

Lacordaire, Jean Théodore (1801 – 70). French traveller and naturalist. Professor of zoology, University of Liége, Belgium, 1835. (Gilbert 1977, *NBU.*)

Laennec, Théophile-René-Hyacinthe (1781 – 1826). Leading French physician and anatomist. Appointed to the chair of medicine in Paris and a lectureship at the Collège de France, 1822. (*DSB.*)

Laffer, John Athanasius Herring (d. 1861). Attended Shrewsbury School. B.A., Christ's College, Cambridge, 1833. Vicar of St Gennys, Cornwall, 1834 – 61. (*Alum. Cantab.*)

Lamarck, Jean Baptiste Pierre Antoine de Monet de (1744 – 1829). French naturalist. Held various botanical positions at the Jardin du Roi, 1788 – 93. Professor of zoology, Muséum d'Histoire Naturelle, 1793. Believed in spontaneous generation and the progressive development of animal types and propounded a theory of transmutation. (*DSB.*)

Lamb, Henry William, 2d Viscount Melbourne, (1779−1848). Statesman. Home Secretary under Lord Grey, 1830−4. Prime Minister, 1835−41. (*DNB*.)

Lambert, Mr.
    [*c. July 1835*]

Lamouroux, Jean Vincent Félix (1776−1825). French naturalist. Professor of natural history, Caen, 1810. Contributed many articles to the *Dictionnaire classique d'histoire naturelle*. (*DSB*.)

Langton, Algernon (1781−1829). Army officer; Major, 1817. B.A., Downing College, Cambridge, 1828. Ordained priest, 1824. Uncle of Charles Langton. Married Marianne Drewe in 1820. (*Alum. Cantab*.)

Langton, Bennet (b. 1822). Son of Marianne Drewe and Algernon Langton. (*Emma Darwin*.)

Langton, Charles (1801−86). B.A., Trinity College, Oxford, 1824. Rector of Onibury, Shropshire, 1832−40. Married Charlotte Wedgwood in 1832. After her death, he married CD's sister Catherine in 1863. (*Alum. Oxon*.)

Langton, Charlotte. *See* Wedgwood, Charlotte.

Laplace, Pierre-Simon, Marquis de (1749−1827). French mathematician, physicist, and cosmologist. (*DSB*.)

Lardner, Dionysius (1793−1859). Professor of natural philosophy and astronomy, London University, 1827. Edited the *Cabinet cyclopaedia*, 1829−49. FRS 1828. (*DNB*.)

Latham, Anne. Laundry maid at The Mount, Shrewsbury. Married Mark Briggs in 1832.

Latham, John (1740−1837). Physician and ornithologist. FRS 1775. (*DNB*.)

Lawson, John (d. 1890). Shrewsbury school-friend of Erasmus Alvey Darwin and CD. Vicar of Seaton Carew, Durham, 1835−90. (*Alum. Oxon., Shrewsbury School Register*.)

Leighton, Baldwin, 7th Baronet (1805−71). M.P. for South Shropshire, 1859−65. Married Mary Parker of Sweeney Hall, Shropshire, in 1832. (*Burke's Peerage 1980, Modern English Biography*.)

Leighton, Clare. Daughter of Louisa Anne and Francis Knyvett Leighton Sr. (*Burke's Landed Gentry 1846*.)

Leighton, Colonel. *See* Leighton, Francis Knyvett, Sr.

Leighton, Francis (1801−70). Matriculated Trinity College, Cambridge, 1821; B.A., 1827. Ordained deacon, 1826. Rector of Cardeston, Shropshire, 1828−70. (*Alum. Cantab*.)

Leighton, Francis Knyvett (1807−81). B.A., Oxford, 1828; Fellow of All Souls' College, 1829−43. Vice-Chancellor of Oxford University, 1866−70. Canon of Westminster, 1868−81. (*Alum. Oxon., Modern English Biography*.)
    [*3 June 1829*]

Leighton, Francis Knyvett, Sr (1772−1834). Army officer. Father of Francis Knyvett, Clare, and Louisa Leighton. (*Burke's Landed Gentry 1846*.)

Leighton, Louisa. Sister of Francis Knyvett Leighton. Married Henry Hope in 1833. (*Burke's Landed Gentry 1846.*)

Leighton, Louisa Anne. Daughter of St Leger Aldworth, 1st Viscount Doneraile. Married Francis Knyvett Leighton Sr in 1805.

Leighton, Mary (1799–1864). Of Sweeney Hall, Shropshire. Married Sir Baldwin Leighton in 1832. (*Burke's Landed Gentry 1879.*)

Lennon, Patrick. Merchant in Rio de Janeiro.

Lennox, Charles Gordon-, 5th Duke of Richmond, (1791–1860). Army officer and politician. Postmaster-general, 1830–4. President, Royal Agricultural Society, 1845–60. (*DNB.*)

Leslie, John (1766–1832). Mathematician and natural philosopher. Professor of natural philosophy at Edinburgh University, 1819. (*DNB, DSB.*)

Levaillant (Le Vaillant) François (1753–1824). French explorer and naturalist. (*DSB.*)

Lindley, John (1799–1865). Botanist and horticulturist. Professor of botany at London University, 1828–60. Horticultural editor of *Gardeners' Chronicle* from 1841. FRS 1828. (*DNB, DSB.*)

Linnaeus (Carl von Linné) (1707–78). Swedish botanist and zoologist. Enunciated principles for defining species and genera and proposed a system for the classification of the natural world that formed the basis for the scientific study of living organisms. (*DSB.*)

Lister, Thomas Henry (1800–42). Novelist and dramatist. (*DNB.*)

Liston, John (1776?–1846). Comic actor. (*DNB.*)

Litchfield, Henrietta Emma. *See* Darwin, Henrietta Emma.

Lizars, John (1787?–1860). Surgeon and teacher of anatomy. Professor of surgery, Royal College of Surgeons, Edinburgh, 1831. (*DNB.*)

Lloyd, Hannibal Evans (1771–1847). Philologist and translator. (*DNB.*)

Lloyd, Henry James (1794–1853). Fifth son of Francis Lloyd of Leaton Knolls, Shropshire. Clergyman. Married Elizabeth Miles of Leigh Court, Somerset, in 1832. (*Alum. Cantab.*)

Locke, John (1632–1704). Philosopher. (*DNB, DSB.*)

Londonderry, Lord. *See* Stewart, Charles William.

Lonsdale, William (1794–1871). Geologist. Served the Geological Society from 1829 to 1842, first as Curator and Librarian, and after 1838 as Assistant Secretary and Librarian. (*DNB, DSB.*)

López, Estanislao (1786–1838). Caudillo of Santa Fé province, Argentina, 1818–38. (*HDA.*)

Lowe, Henry Porter (1810–87). B.A., Trinity Hall, Cambridge, 1833. High sheriff of Nottinghamshire, 1859. Adopted the name Sherbrooke in 1847. (*Alum. Cantab.*)

Lowe, Richard Thomas (1802–74). Clergyman and botanist who lived on Madeira, 1832–52. (Desmond 1977, *DNB.*)

Lowe, Robert, 1st Viscount Sherbrooke, (1811–92). Younger brother of Henry Porter Lowe. B.A., University College, Oxford, 1833. Politician, lawyer, and colonialist. Lived in Australia, 1842–50. Leader writer with *The Times*, 1850–68. M.P., 1850–80. FRS 1871. (*DNB*.)

Lowth, Robert (1710–87). Clergyman. Professor of poetry at Oxford University, 1741–50. Bishop of London, 1777. (*DNB*.)

Lubbock, John William, 3d Baronet (1803–65). Astronomer, mathematician, and banker. First Vice-Chancellor of London University, 1837–42. CD's neighbour at Down. FRS 1829. (*DNB, DSB*.)

Lumb, Edward. British merchant in Buenos Aires. (Freeman 1978.)
13 November 1833; 30 March 1834; *2 May 1834*; *8 May 1834*

Lyell, Charles (1797–1875). Uniformitarian geologist. Professor of geology, King's College, London, 1831–3. President of the Geological Society, 1834–6 and 1849–50. Scientific mentor and friend of CD. FRS 1826. (*DNB, DSB*.)
26 December 1836

Lyell, Katherine Murray (1817–1915). Daughter of Leonard Horner. Married Henry Lyell, brother of Charles Lyell, 1848. Collected plants in India. Edited *Life, letters and journals of Sir Charles Lyell* (1881). (Desmond 1977.)

Macaulay, Thomas Babington, 1st Baron (1800–59). Historian and politician. Regular contributor to the *Edinburgh Review*. (*DNB*.)

McCormick, Robert (1800–90). Naval surgeon, explorer, and naturalist. Wrote accounts of his voyages. Surgeon in the *Beagle*, 1831–2. (*DNB*, Plarr 1930.)

Macgillivray, William (1796–1852). Assistant and secretary to Robert Jameson, 1823. Conservator of the Royal College of Surgeons' Museum, Edinburgh, 1831–41. Professor of natural history, Aberdeen, 1841. (*DNB*.)

Mackay, Mrs. CD's landlady at 11 Lothian Street, Edinburgh.

Mackenzie, Henry (1745–1831). Scottish novelist and miscellaneous writer. Comptroller of taxes for Scotland, 1804–31. (*DNB*.)

Mackintosh, Catherine. *See* Allen, Catherine.

Mackintosh, Frances (Fanny). *See* Wedgwood, Frances Mackintosh.

Mackintosh, James (1765–1832). Philosopher and historian. Professor of law and general politics at the East India Company College, Haileybury, 1818–24. Married Catherine Allen in 1798. (*DNB*.)

Mackintosh, Robert (1806–64). Son of James and Catherine Allen Mackintosh. Brother of Frances Mackintosh Wedgwood. (*Emma Darwin*.)

Maclear, Thomas (1794–1879). Astronomer and physician. Royal astronomer at the Cape of Good Hope, 1834–70. FRS 1831. (*DNB, DSB*.)

Macleay, William Sharp (1792–1865). Zoologist and diplomat. Emigrated to New South Wales in 1839. Originator of the quinary system of taxonomy. (*Aust. Dict. Biog., DNB*.)

Macready, William Charles (1793–1873). Actor. (*DNB*.)

Magellan, Ferdinand (1480–1521?). The first circumnavigator of the globe. (*EB*.)

Mainwaring, Charles Kynaston (1803–61). Of Oteley Park, Shropshire. High sheriff of Shropshire, 1829. Married Frances Salusbury of Galltfaenan, Denbighshire, in 1832. (*Burke's Landed Gentry 1952.*)

Mainwaring, Julia (d. 1851). Her elder sister, Sarah, succeeded her uncle to Whitmore Hall, Staffordshire, in 1825. (*Burke's Landed Gentry 1879.*)

Malibran, Maria Felicia Garcia (1808–36). Spanish mezzo-soprano. (Grove 1980.)

Maling, Harriot (1790–1825). Youngest daughter of Erasmus and Elizabeth Collier Darwin. Married Thomas James Maling in 1811. Died in Valparaiso, South America. (*Darwin Pedigree.*)

Maling, Thomas James (1778–1849). Naval officer. Served on the South American station, 1823–7. Vice-admiral, 1841. (*Darwin Pedigree*, O'Byrne 1849.)

Malthus, Thomas Robert (1766–1834). Clergyman and political economist. First professor of history and political economy at the East India Company College, Haileybury, 1805–34. Quantified the relationship between growth in population and food supplies in his *Essay on population* (1798). FRS 1819. (*DNB, DSB.*)

Manners-Sutton, Charles, 1st Viscount Canterbury, (1780–1845). Barrister and politician. Speaker of the House of Commons, 1817–35. Elected M.P. for Cambridge University in December 1832. Created Baron Bottesford and Viscount Canterbury in 1835. (*DNB.*)

Manzoni, Alessandro Francesco Tommaso Antonio (1785–1873). Italian poet and novelist. (*EB.*)

Marindin, Samuel (1807–52). Attended Shrewsbury School, 1821–5. B.A., Trinity College, Cambridge, 1829. Army officer and clergyman. Married Isabella Colvile of Ochiltree and Craigflower, Ayrshire, in 1834. (*Alum. Cantab., Burke's Landed Gentry 1952.*)

Marryat, Frederick (1792–1848). Naval officer, novelist, and editor. (*DNB.*)

Marsh, Anne. *See* Marsh-Caldwell, Anne.

Marsh-Caldwell, Anne (1791–1874). Novelist. Married Arthur Cuthbert Marsh (d. 1849) in 1817. Sister of Emma Holland. Added Caldwell to her surname in 1858. (*DNB.*)

Marsham, Thomas (d. 1819). Entomologist. (*DNB*, Gilbert 1977.)

Martens, Conrad (1801–78). Landscape painter. Replaced Augustus Earle as artist in the *Beagle*, 1833–4. Settled in Australia in 1835. (*Aust. Dict. Biog.*, Bénézit 1976.)

Martineau, Harriet (1802–76). Author, reformer, and traveller. (*DNB.*)

Massie, Edward (1805–93). Shrewsbury school-friend of Erasmus Alvey Darwin and CD. Fellow and tutor of University College, Durham, 1841–5. Curate of Gawsworth, Cheshire. (*Modern English Biography, Shrewsbury School Register.*)

Mathews, Andrew (d. 1841). Gardener for the Horticultural Society of London at Chiswick. Collected plants in Peru and Chile, 1830–41. (Desmond 1977.)

Matthew, Henry (1807–61). B.A., Sidney Sussex College, Cambridge, 1832. President of the Union, 1830. Clergyman from 1837. (*Alum. Cantab.*)
    [2 February 1831]; [14 February 1831]; [March or April 1831]

Matthew, John (d. 1837). Rector of Kilve with Stringston, Somerset, 1797– 1837. Father of Henry Matthew. (*Alum. Oxon.*)

Matthews, Richard (1811–93). Missionary on board the *Beagle*, sent to establish a mission at Tierra del Fuego. Abandoned this aim in 1833 and rejoined the *Beagle*. Missionary in New Zealand from 1835. (*DNZB.*)

Mayer, Mrs. Proprietor of a girls' school at Greville House, Paddington Green, London.

Mayo, Herbert (1796–1852). Neurologist. Lecturer on anatomy at the Great Windmill Street School of Anatomy. Surgeon at the Middlesex Hospital, 1827–42. Professor of anatomy, King's College, London, 1830–6. FRS 1828. (*DNB, DSB.*)

Melbourne, Viscount. *See* Lamb, Henry William.

Miers, John (1789–1879). Botanist and engineer. Travelled and worked in South America, 1819–38. FRS 1843. (Desmond 1977, *DNB.*)

Miguel, Dom Maria Evarist (1802–66). King of Portugal overthrown in 1834 when he escaped to England with Don Carlos of Spain. (*EB.*)

Mill, John Stuart (1806–73). Philosopher and political economist. His *Logic* (1843) was an influential work in the philosophy of science. (*DNB, DSB.*)

Miller, William Hallowes (1801–80). Mineralogist and crystallographer. Professor of mineralogy at Cambridge University, 1832–80. FRS 1838. (*DNB, DSB.*)

Milton, John (1608–74). Poet. (*DNB.*)

Mogg, Charles William Cumberland (1804–92). Physician. M.B., Caius College, Cambridge, 1833. Practised in London and Derbyshire. Officer of health, North London, 1849. (*Alum. Cantab.*)

Monboddo, Lord. *See* Burnett, James.

Moncrieff, William Thomas (1794–1857). Dramatist. (*DNB.*)

Monro, Alexander, tertius, (1773–1859). Anatomist. Professor of medicine, surgery, and anatomy at Edinburgh University, 1817–46. (*DNB.*)

Montagu, John William, 7th Earl of Sandwich (1811–84). Admitted Trinity College, Cambridge, 1827. Captain, Huntingdon Militia, 1831. Ensign, Grenadier Guards, 1832–5. (*Alum. Cantab., Modern English Biography.*)

Moore, Thomas (1779–1852). Irish poet and songwriter, known for his patriotic verse. Biographer of Sheridan. (*DNB.*)

Moreno, Manuel (1790–1857). Argentinian physician, politician, and diplomat. First chemistry professor, University of Buenos Aires, 1822–8. Argentina's minister in England, 1828–35. (*HDA.*)

Morley, Lord. *See* Parker, John.

Morris, John (1810 – 86). Geologist. Originally a pharmaceutical chemist in Kensington. Professor of geology, University College, London, 1854 – 77. (*DNB*, Sarjeant 1980.)

Mosley, Frances. *See* Wedgwood, Frances Mosley.

Mosley, John Peploe (1766 – 1834). Rector of Rolleston, Staffordshire, 1799 – 1834. Father of Frances and Peploe Paget Mosley. (*Alum. Cantab., Burke's Peerage 1980.*)

Mosley, Peploe Paget (1793 – 1868). Brother of Frances Mosley Wedgwood. Succeeded his father, John Peploe Mosley, as rector of Rolleston, Staffordshire, 1834 – 68. (*Alum. Cantab., Burke's Peerage 1980.*)

Mulgrave, Lord. *See* Phipps, Constantine John.

Münchausen, Hieronymus von (1720 – 97). Soldier known for narrating extravagant sporting adventures.

Murchison, Charlotte (d. 1869). Married Roderick Impey Murchison in 1815.

Murchison, Roderick Impey (1792 – 1871). Geologist noted for his work on the Silurian system. A leading figure in the Geological Society, British Association for the Advancement of Science, and Royal Geographical Society. FRS 1826. (*DNB, DSB.*)

Murillo, Bartolomé Esteban (1618 – 82). Spanish painter. (Bénézit 1976.)

Murray, George Augustus Frederick John (1814 – 64). Admitted as nobleman at Trinity College, Cambridge, 1832. Expelled for gambling, 1834. Army officer. Succeeded his father as Lord Glenlyon, 1837, and his uncle as 6th Duke of Atholl, 1846. (*Alum. Cantab.*)

Murray, John (1808 – 92). Publisher and author of guide-books. CD's publisher after 1845. (*DNB.*)

Murray, Lindley (1745 – 1826). Grammarian. (*DAB, DNB.*)

Musters, Charles (d. 1832). Fourth son of John Musters of Colwick Hall, Nottinghamshire. Volunteer 1st Class in the *Beagle*. Died of fever at Rio de Janeiro. (*Burke's Landed Gentry 1879.*)

Musters, Mary Anne (d. 1832). Mother of Charles Musters. (*Burke's Landed Gentry 1879.*)

Mytton, Caroline Mallet. *See* Giffard, Caroline Mallet.

Mytton, John (Jack) (1796 – 1834). Sportsman and eccentric. Served in the army, 1816 – 17. M.P. for Shrewsbury, 1818 – 20. High sheriff of Shropshire and Merionethshire. Dissipated a fortune and died of delirium tremens in the King's Bench prison in 1834. Cousin of Fanny and Sarah Owen. (*DNB.*)

Mytton, Mrs. *See* Giffard, Caroline Mallet.

Nancy. CD's childhood nurse who was kept on as a servant of the Darwin family at The Mount, Shrewsbury.

Narbrough, John (1640 – 88). Admiral. Commissioner of the navy, 1680 – 7. (*DNB.*)

Nelson, Horatio (1758 – 1805). Vice-admiral killed at the battle of Trafalgar. (*DNB.*)

Newman, Edward (1801−76). Naturalist. A founder of the Entomological Society, 1826. Natural history editor of the *Field*, 1858−76. (Desmond 1977, *DNB*.)

Newton, Isaac (1642−1727). Natural philosopher. (*DNB, DSB*.)

Noehden, Georg Heinrich (1770−1826). German grammarian. (*NBU*.)

Normanby, Lord. *See* Phipps, Constantine Henry.

Northumberland, Duke of. *See* Percy, Hugh.

Oakeley, William (1806−51). Of Oakeley, Shropshire. (*Burke's Landed Gentry 1879*.)

O'Brien, Mary Henrietta. Married Robert FitzRoy in 1836.

O'Connell, Daniel (1775−1847). Irish lawyer and politician. Leader of the movement for Catholic emancipation. (*DNB*.)

Orbigny, Alcide Charles Victor Dessalines d' (1802−57). French palaeontologist who travelled widely in South America, 1826−34. Professor of palaeontology, Muséum d'Histoire Naturelle, 1853. (*DSB*.)

Ormsby-Gore, William (1779−1860). Conservative M.P. for North Shropshire, 1835−57. Contested North Shropshire unsuccessfully in 1832. (*Modern English Biography*.)

Owen, Arthur Mostyn (1813−96). Son of William Mostyn Owen Sr of Woodhouse. Attended Shrewsbury School, 1827−8, and the East India Company College, Haileybury, 1829−31. Served in the Indian Civil Service, 1832−48. High sheriff of Shropshire, 1876. (*Modern English Biography, Shrewsbury School Register*.)

Owen, Caroline Mostyn (d. 1897). Third daughter of William Mostyn Owen Sr of Woodhouse. (*Burke's Landed Gentry 1952*.)

Owen, Charles Mostyn (1818−94). Son of William Mostyn Owen Sr of Woodhouse. B.A., Trinity College, Oxford, 1842. Army officer. Served in South Africa during the Kaffir War of 1845−7. Chief constable of Oxfordshire. (*Modern English Biography*.)

Owen, Edward Pryce (1788−1863). Artist, etcher, and clergyman. Vicar of Wellington and rector of Eyton-upon-the-Wildmoors, Shropshire, 1823−40. Son of Archdeacon Hugh Owen. (Bénézit 1976, *DNB*.)

Owen, Emma Mostyn (d. 1888). Fourth daughter of William Mostyn Owen Sr of Woodhouse. (*Burke's Landed Gentry 1952*.)

Owen, Fanny (Frances) Mostyn. Second daughter of William Mostyn Owen Sr of Woodhouse. Married Robert Myddelton Biddulph in 1832. A close friend and neighbour of CD before the *Beagle* voyage. (*Burke's Landed Gentry 1952*.)
  [January 1828]; [late January 1828]; [9 March 1828]; [July? 1828]; [26 October 1828]; 27 January [1830]; 27 January [1830]; [8 September 1830]; 4 October [1830]; [8 April 1831]; [22 September − 2 October 1831]; [26 September 1831]; [6 October 1831]; 2 [December 1831]; 1 March 1832; [c. 21 October 1833]

Owen, Francis Mostyn (b. 1815). Son of William Mostyn Owen Sr of Wood-house. Attended Shrewsbury School, 1829 – 31. Army officer; Captain, 1845. (*Burke's Landed Gentry 1952, Shrewsbury School Register.*)

Owen, Harriet Elizabeth Mostyn. Wife of William Mostyn Owen Sr. (*Burke's Landed Gentry 1952.*)

Owen, Henry Mostyn (1820 – 43). Son of William Mostyn Owen Sr of Wood-house. Army officer. Died in India. (*Burke's Landed Gentry 1952.*)

Owen, Hugh (1760 – 1827). Archdeacon of Shropshire, 1821. Author of works on Shrewsbury. (*DNB.*)

Owen, Laura Mostyn. Youngest sister of William Mostyn Owen Sr. (*Burke's Landed Gentry 1952.*)

Owen, Marianne Bulkeley (d. 1865). Married Thomas Bulkeley Bulkeley Owen in 1824. (*Burke's Landed Gentry 1879.*)

Owen, Richard (1804 – 92). Comparative anatomist. Assistant-conservator of the Hunterian Museum, Royal College of Surgeons, 1827; Hunterian professor, 1836 – 56. Superintendent of the Natural History departments, British Museum, 1856 – 84. Described the *Beagle* fossil mammal specimens. FRS 1834. (*DNB, DSB.*)
    19 December [1836]

Owen, Sarah Harriet Mostyn. Eldest daughter of William Mostyn Owen Sr of Woodhouse. Married firstly Edward Hosier Williams (d. 1844) in 1831 and secondly Thomas Chandler Haliburton in 1856. A close friend and neighbour of CD before the *Beagle* voyage. (*Burke's Landed Gentry 1952.*)
    31 [December 1827]; 18 February [1828]; 30 March [1828]; [27 – 30 September 1831]; 26[– 31] August 1832; 21 October 1833

Owen, Sobieski Mostyn (d. 1890). Youngest daughter of William Mostyn Owen Sr of Woodhouse. (*Burke's Landed Gentry 1952.*)

Owen, Thomas Bulkeley Bulkeley (1790 – 1867). Of Tedsmore Hall, Shropshire. (*Burke's Landed Gentry 1879.*)

Owen, William Mostyn (1806 – 68). Eldest son of William Mostyn Owen Sr of Woodhouse. Major, Royal Dragoons. (*Burke's Landed Gentry 1952.*)

Owen, William Mostyn, Sr. Lieutenant, Royal Dragoons. Squire of Woodhouse, Shropshire. (*Burke's Landed Gentry 1952.*)
    1 March 1832; 10 April – 1 May 1834; 5 October [1836]; 19 December 1836

Paganini, Niccolò (1782 – 1840). Italian violinist and composer. Gave a series of concerts in Britain from May 1831 to September 1832. (Grove 1980.)

Paley, William (1743 – 1805). Anglican clergyman and philosopher who pro-pounded an elegant and popular system of natural theology. (*DNB, DSB.*)

Palmerston, Lord. *See* Temple, Henry John.

Panting, Thomas (d. 1836). Solicitor, Shrewsbury. Friend of the Darwin family. (*Shrewsbury School Register.*)

Parish, Woodbine (1796 – 1882). Diplomat. Chargé d'affaires at Buenos Aires, 1825 – 32. FRS 1824. (*DNB.*)

Parker, Charles (b. 1831). Fourth son of Henry and Marianne Parker. CD's nephew. (*Darwin Pedigree.*)

Parker, Edmund, Viscount Boringdon, (1810–64). Son of John Parker, 1st Earl of Morley. Lord-in-waiting to Queen Victoria, 1846. (*DNB.*)

Parker, Francis (1829–71). Third son of Henry and Marianne Parker. CD's nephew. Solicitor in Chester. (*Darwin Pedigree.*)

Parker, Henry (1788–1856). M.D., Edinburgh, 1814. Physician to the Shropshire Infirmary. Married Marianne Darwin in 1824. (*Darwin Pedigree, Provincial Medical Directory.*)

Parker, Henry, Jr (1827–92). Second son of Henry and Marianne Parker. CD's nephew. Fellow of Oriel College, Oxford, 1851–85. (*Darwin Pedigree*, Freeman 1978.)

Parker, John, 1st Earl of Morley, (1772–1840). Succeeded to the peerage in 1788. Frequent speaker in the House of Lords and after 1827 an active supporter of parliamentary reform. FRS 1795. (*DNB.*)

Parker, Marianne (1798–1858). CD's eldest sister. Married Henry Parker in 1824. (*Darwin Pedigree.*)
[23 February 1826]; [c. 13 March 1826]; 12[–29] March [1832]

Parker, Mary Susan (1836–93). Daughter of Henry and Marianne Parker. CD's niece. (*Darwin Pedigree.*)

Parker, Miss. *See* Leighton, Mary.

Parker, Robert (b. 1825). Eldest son of Henry and Marianne Parker. CD's nephew. (*Darwin Pedigree.*)

Parkes, Samuel (1761–1825). Manufacturing chemist in London. Published manuals of chemistry between 1806 and 1815. (*DNB.*)

Peacock, George (1791–1858). Tutor in mathematics at Trinity College, Cambridge, 1823–39. Lowndean Professor of Geometry and Astronomy at Cambridge University, 1837. Dean of Ely, 1839–58. FRS 1818. (*DNB, DSB.*)
[6 or 13 August 1831]; [c. 26 August 1831]

Pearson, Henry (1807–56). Shrewsbury school-friend of Erasmus Alvey Darwin and CD. B.A., Trinity College, Cambridge, 1830. Barrister and author of mathematical and legal works. (*Alum. Cantab., Modern English Biography.*)

Peel, Robert, 2d Baronet (1788–1850). Prime Minister, 1834–5 and 1841–6. (*DNB.*)

Peile, Thomas Williamson (1806–82). Clergyman and teacher. Attended Shrewsbury School, 1821–4. B.A., Trinity College, Cambridge, 1828. Assistant master at Shrewsbury School, 1828–9. Tutor at Durham University, 1834–41. (*DNB.*)

Pelham, John Cressett (d. 1838). M.P. for Shropshire, 1822–32; Shrewsbury, 1835–7. Contested Shrewsbury unsuccessfully in 1832. (Stenton 1976.)

Pemberton, Christopher Robert (1765–1822). Physician at St George's Hospital, 1800–8. FRS 1797. (*DNB.*)

Pennant, Thomas (1726–98). Traveller and naturalist. FRS 1767. (*DNB, DSB.*)

Percy, Hugh, 3d Duke of Northumberland, (1785 – 1847). Politician and diplomat. High Steward of Cambridge University, 1834 – 40; Chancellor, 1840 –
7. (*DNB*.)

Phillips, Thomas (1770 – 1845). Portrait painter. Professor of painting, Royal
Academy, 1825 – 32. (Bénézit 1976, *DNB*.)

Phillips, William (1775 – 1828). Mineralogist and geologist. London printer and
bookseller. FRS 1827. (*DNB, DSB*.)

Philpott, Henry (1807 – 92). B.A., St Catharine's College, Cambridge, 1829;
Fellow, 1829 – 45; Master, 1845 – 60. Bishop of Worcester, 1860 – 90. (*Alum.
Cantab., DNB*.)

Phipps, Constantine Henry, 1st Marquis of Normanby, (1797 – 1863). Statesman. A popular novelist in his youth. (*DNB*.)

Phipps, Constantine John, 2d Baron Mulgrave, (1744 – 92). Naval officer and
politician. A Lord of the Admiralty, 1777. (*DNB*.)

Place, Francis (1771 – 1854). Radical reformer. (*DNB*.)

Playfair, John (1748 – 1819). Mathematician and geologist. Professor of mathematics at Edinburgh University, 1785 – 1805; Professor of natural philosophy, 1805. Expounded the geological theories of James Hutton in a short and
clear form, 1802. FRS 1807. (*DNB, DSB*.)

Poole, John (1786? – 1872). Dramatist and miscellaneous author. (*DNB*.)

Powell, John Allan (d. 1859). Solicitor at Lincoln's Inn from 1806. Attorney for
George IV at the trial of Queen Caroline in 1820. (*Modern English Biography*.)

Powis, Earl of. *See* Clive, Edward.

Price, James. Of Pwllycrochan, Denbighshire. Clergyman. Father of John Price.

Price, John (1803 – 87). Welsh scholar, naturalist, and schoolmaster. Attended
Shrewsbury School. B.A., St John's College, Cambridge, 1826. Assistant
master at Shrewsbury School, 1826 – 7. Private tutor in Chester. (*Alum.
Cantab., Modern English Biography*.)
    [*July 1826*]

Proctor, George (d. 1858). B.A., Christ's College, Cambridge, 1831. Clergyman.
(*Alum. Cantab*.)

Proctor, Robert. Traveller in South America. Wrote an account of his journeys in
Peru. Uncle of George Proctor. (*CDEL*.)

Pulleine, Robert (1806 – 68). B.A., Emmanuel College, Cambridge, 1829. Curate
of Spennithorne, Yorkshire, 1830 – 45; rector of Kirkby-Wiske, 1845 – 68.
(*Alum. Cantab*.)

Quiroga, Juan Facundo (1788 – 1835). Controller of the northern provinces of
Argentina. Moved to Buenos Aires in 1832 with the aim of making Argentina a
federal republic. Assassinated in 1835. (*HDA*.)

Quoy, Jean René Constant (1790 – 1869). French naval surgeon and naturalist.
Participated in many voyages of scientific exploration. (*DSB*.)

Radcliffe, Ann (1764 – 1823). Novelist. (*DNB*.)

Ram Mohan Roy (1774—1833). Indian religious and social reformer. In England as the agent of the Emperor of Delhi, 1830—3. (*EB.*)

Ramsay, Marmaduke (d. 1831). B.A., Jesus College, Cambridge, 1818; Fellow and tutor, 1819—31. (*Alum. Cantab.*)

Ramsay, William (1793—1871). Naval officer. Commander of the *Black Joke* tender; captured the *Marinerito* Spanish slave brig in 1831. Returned to England in 1832. Commander of the *Dee* Steamer in the West Indies, 1834—7. Brother of Marmaduke Ramsay. (*Modern English Biography.*)

Raphael (Raffaello Sanzio) (1483—1520). Italian painter. (Bénézit 1976.)

Raspe, Rudolf Erich (1737—94). Author of geological and literary works. (*DNB, DSB.*)

Réaumur, René-Antoine Ferchault de (1683—1757). French mathematician, experimental physicist, and naturalist. (*DSB.*)

Rennie, James (1787—1867). Professor of natural history at King's College, London, 1830—4. Emigrated to Australia, 1840. (Desmond 1977, *DNB.*)

Richardson, Samuel (1689—1761). Novelist. (*DNB.*)

Richmond, Duke of. *See* Lennox, Charles Gordon-.

Ritchie, David. A member of the Plinian Society at the University of Edinburgh (1829).

Rivadavia, Bernadino (1780—1845). Argentinian statesman. First national president of Argentina, 1826—7. Resigned because of opposition to his policy of a unified Argentina. Lived in Spain from 1829. Attempted to return to Argentina in 1834 but was not allowed to disembark. (*HDA.*)

Rivington, John (1720—92). Publisher. (*DNB.*)

Rivington, John (1779—1841). Publisher. Grandson of John Rivington (1720—92). (*DNB.*)

Robinson, George Alington (d. 1888). B.A., Christ's College, Cambridge, 1830. Clergyman. (*Alum. Cantab.*)

Rodwell, John Medows (1808—1900). Matriculated Caius College, Cambridge, 1826; B.A., 1830. Clergyman and orientalist. (*DNB.*)

Romilly, Joseph (1791—1864). Fellow of Trinity College, Cambridge, 1815. Registrary of the University, 1832—61. Kept a diary of University life from 1820 to 1863. (*DNB.*)

Roper, Henry W. (1806—83). B.A., St John's College, Cambridge, 1829. Army officer. (*Alum. Cantab.*)

Rosas, Juan Manuel de (1793—1877). Governor of Buenos Aires, 1829—32 and 1835—52, who ruled as dictator of Argentina. Led a campaign against the Indians in order to gain more territory, 1833—5. (*HDA.*)

Ross, John (1777—1856). Naval officer and Arctic navigator. Went in search of a North-West Passage, 1818 and 1829—33. (*DNB.*)

Rossini, Gioacchino (Antonio) (1792—1868). Italian composer. (Grove 1980.)

Roussin, Albin Reine, Baron (1781—1854). French naval officer and diplomat. Surveyed the coast of Brazil, 1819. Admiral, 1840. (*Larousse XX.*)

Rowlett, George (d. 1834). Purser on board the *Beagle*. (Freeman 1978.)

Royston, Richard (1599 – 1686). Founder of a publishing firm of the same name. Bookseller to Charles I, Charles II, and James II. (*DNB*.)

Russell, John, 1st Earl (1792 – 1878). Statesman. Introduced Reform Bill twice in 1831 but it was not passed until 1832. Prime Minister, 1846 – 52 and 1865 – 6. (*DNB*.)

Ryder, Dudley, 1st Earl of Harrowby, (1762 – 1847). Statesman. Supported parliamentary reform. (*DNB*.)

Ryle, Jane Harriett. *See* Darwin, Jane Harriett.

Sacchini, Antonio (Maria Gasparo Gioacchino) (1730 – 86). Italian composer and leading figure in serious opera. (Grove 1980.)

Saint-Hilaire, Augustin François César Prouvençal (Auguste de) (1779 – 1853). French naturalist. Surveyed the flora and fauna of Brazil, 1816 – 22. (*DSB*.)

Saint-Pierre, Jacques Henri Bernardin de (1737 – 1814). French man of letters. (*EB*.)

Salt, Thomas. Partner in the Shrewsbury law firm of Dukes and Salt.

Salwey, Charlotte Margaretta (d. 1858). Married Richard Betton of Overton House, Shropshire, in October 1831. (*Burke's Landed Gentry 1879*.)

Samouelle, George (d. 1846). Entomologist. (*CDEL*, Gilbert 1977.)

Sandwich, Lord. *See* Montagu, John William.

Sarmiento, Pedro (1532 – 1608?). Spanish navigator who surveyed the Straits of Magellan and set up the unsuccessful colony later known as Port Famine. (*Larousse XX*.)

Savigny, Marie-Jules-César Lelorgne de (1777 – 1851). French zoologist. (*DSB*.)

Schweigger, August Friedrich (1783 – 1821). German botanist. (*ADB*.)

Scoresby, William (1789 – 1857). Master-mariner, author, and clergyman. FRS 1824. (Desmond 1977, *DNB*.)

Scott, Walter (1771 – 1832). Scottish novelist and poet. (*DNB*.)

Scrope, George Julius Poulett (1797 – 1876). Geologist and political economist. M.P. for Stroud, Gloucestershire, 1833 – 68. Carried out pioneering work in vulcanology. His ideas helped to shape Lyell's uniformitarian theories. FRS 1826. (*DNB, DSB*.)

Secker, Isaac Onslow (1799 – 1861). Barrister. Metropolitan police magistrate at Southwark, 1846 – 9; Greenwich and Woolwich courts, 1849 – 60; Marylebone, 1860 – 1. (*Alum. Cantab*.)

Sedgwick, Adam (1785 – 1873). Geologist and clergyman. Woodwardian Professor of Geology at Cambridge University, 1818 – 73. Canon of Norwich, 1834 – 73. FRS 1821. (*DNB, DSB*.)
   4 September 1831; 18 September 1831

Selby, Prideaux John (1788 – 1867). Naturalist. High sheriff of Northumberland, 1823. (Desmond 1977, *DNB*.)

Sellow, Friedrich (1789 – 1831). Prussian naturalist who worked and travelled in Brazil and Uruguay, 1814 – 31.

Selwyn, George Augustus (1809–78). B.A., St John's College, Cambridge, 1831. Bishop of New Zealand, 1841–67. Bishop of Lichfield, 1867–78. Selwyn College, Cambridge, was erected in his memory. (*DNB.*)

Selwyn, William (1806–75). B.A., St John's College, Cambridge, 1828. Clergyman. Lady Margaret Professor of Divinity at Cambridge University, 1855–75. Brother of George Augustus Selwyn. FRS 1866. (*DNB.*)

Sergeant, Frederick Thomas (d. 1863). B.A., Corpus Christi College, Cambridge, 1827. Called to the bar, 1830. (*Alum. Cantab.*)

Seward, Albert Charles (1863–1941). Botanist and geologist. Lecturer of botany at Cambridge University, 1890; Professor, 1906–36. Joint editor (with Francis Darwin) of *More letters of Charles Darwin* (1903). FRS 1898. (*DNB, DSB.*)

Seymour, Michael (1802–87). Naval officer on the South American station, 1827–9 and 1833–5. Captain of the *Challenger* wrecked on the coast of Chile in 1835. Admiral, 1864. (*DNB.*)

Sharpe, Daniel (1806–56). Geologist, business man, and antiquary. President of the Geological Society, 1856. FRS 1850. (*DNB,* Sarjeant 1980.)

Sharpe, Edmund (1809–77). B.A., St John's College, Cambridge, 1833. Architect and railway engineer. (*DNB.*)

Sheepshanks, Thomas (1796–1875). Clergyman and teacher. Assistant master at Shrewsbury School, 1820–5. (*Alum. Cantab., Modern English Biography.*)

Shelley, Percy Bysshe (1792–1822). Poet. (*DNB.*)

Sherer, Moyle (1789–1869). Army officer, travel writer, and novelist. (*DNB.*)

Sheridan, Richard Brinsley (1751–1816). Dramatist and parliamentary orator. (*DNB.*)

Simpson, George (d. 1888). B.A., Christ's College, Cambridge, 1830. Clergyman. (*Alum. Cantab.*)

[26] *January [1831]*

Sismondi, Jean Charles Léonard Simonde de (1773–1842). Swiss historian. Married Jessie Allen in 1819. (*EB.*)

Sismondi, Jessie. *See* Allen, Jessie.

Slaney, Elizabeth Frances (d. 1870). Eldest daughter and co-heiress of Robert Aglionby Slaney. Married Thomas Campbell Eyton in 1835. (*Burke's Landed Gentry 1879.*)

Slaney, Robert Aglionby (1791–1862). An advocate for rural and economic reform. M.P. for Shrewsbury, 1826–34, 1837–41, 1847–62. High sheriff of Shropshire, 1854. (*DNB.*)

Smith, Adam (1723–90). Political economist. (*DNB.*)

Smith, Andrew (1797–1872). Army surgeon stationed in South Africa, 1821–37. An authority on South African zoology. Director-general, army medical department, 1853–8. (*DNB.*)

Smith, Charles Hamilton (1776–1859). Army officer and naturalist. (*DNB.*)

Smith, James Edward (1759 – 1828). Botanist. Founded the Linnean Society in 1788 and held the office of president until his death. FRS 1785. (*DNB, DSB.*)

Smith, Saba. *See* Holland, Saba.

Smith, Sydney (1771 – 1845). Essayist and wit. A founder of the *Edinburgh Review*, 1802. Canon of St Paul's, London, 1831. (*DNB.*)

Smith, Thomas. Admitted sizar at Emmanuel College, 1821; B.D., 1831. Vicar of Chipping Sodbury, Gloucestershire, 1822 – 58. (*Alum. Cantab.*)

Smyth, William (1765 – 1849). Private tutor to Richard Brinsley Sheridan's eldest son, 1793 – 1806. Regius Professor of Modern History at Cambridge University, 1807 – 49. (*DNB.*)

Smythe, Edward Joseph (1787 – 1856). High sheriff of Shropshire, 1831. (*Burke's Peerage 1880.*)

Smythe, Edward Joseph, Jr (1813 – 41). Eldest son of Edward Joseph Smythe (1787 – 1856). Married Anastasia Boughey in 1840. (*Burke's Peerage 1880.*)

Snyders, Frans (1579 – 1657). Flemish painter. (Bénézit 1976.)

Solander, Daniel Carl (1733 – 82). Swedish botanist. Assistant librarian, British Museum, 1763. Travelled with Joseph Banks to the South Seas in the *Endeavour*, 1768 – 71, and to Iceland, 1772. Keeper of the natural history department, British Museum, 1773. (*DNB, DSB.*)

Somerville, Mary Fairfax Greig (1780 – 1872). Writer on science. (*DNB, DSB.*)

Soult, Nicolas Jean de Dieu, Duke of Dalmatia, (1769 – 1851). Marshal of France. Minister of war under Louis Philippe, 1830 – 4, 1840 – 4. (*EB.*)

South, James (1785 – 1867). Physician and astronomer. FRS 1821. (*DNB, DSB.*)

Southey, Robert (1774 – 1843). Poet, historian, and miscellaneous writer. (*DNB.*)

Sowerby, George Brettingham (1788 – 1854). Conchologist and artist. Produced catalogues of shells and molluscs. (*DNB*, Sarjeant 1980.)

Sowerby, George Brettingham, Jr (1812 – 84). Eldest son of George Brettingham Sowerby. Conchologist and scientific illustrator. (*DNB*, Sarjeant 1980.)

Spence, William (1783 – 1860). Entomologist. A founder of the Entomological Society, 1833; President, 1847. FRS 1834. (*DNB*, Gilbert 1977.)

Spencer, John Charles, Viscount Althorp and 3d Earl Spencer, (1782 – 1845). Politician. Chancellor of the Exchequer, 1830 – 4. Leader of the Whigs in the House of Commons, 1830 – 4. Succeeded to the earldom in 1834. (*DNB.*)

Stanley, Edward George Geoffrey Smith, 14th Earl of Derby, (1799 – 1869). Statesman. M.P. for North Lancashire, 1832 – 44. Colonial Secretary, 1833 – 4 and 1841 – 4. Prime Minister, 1852, 1858 – 9, 1866 – 8. (*DNB.*)

Stark, John. Natural history writer. (*CDEL.*)

Stebbing, George James. Son of a mathematical instrument maker at Plymouth. Supernumerary, at Robert FitzRoy's expense, in the *Beagle*.

Stephens, Catherine (1794 – 1882). English soprano and actress. (Grove 1980.)

Stephens, James Francis (1792 — 1852). Entomologist and zoologist. Employed in the Admiralty office, Somerset House, 1807 — 45. Assisted in arranging the insect collection at the British Museum. (*DNB*, Gilbert 1977.)

Sterne, Laurence (1713 — 68). Clergyman and humorist. Published sermons and novels. (*DNB*.)

Stewart, Charles William, 3d Marquis of Londonderry, (1778 — 1854). Army officer, politician, diplomat, and author. (*DNB*.)

Stokes, John Lort (1812 — 85). Naval officer. Served in the *Beagle* as midshipman, 1826 — 31; mate and assistant surveyor, 1831 — 7; lieutenant, 1837 — 41; commander, 1841 — 3. Admiral, 1877. (Desmond 1977, *DNB*.)

Stokes, Pringle (d. 1828). Commander of the *Beagle*, 1826 — 8. Committed suicide during the *Beagle*'s first voyage to South America.

Stuart-Wortley-Mackenzie, James Archibald, 1st Baron Wharncliffe, (1776 — 1845). Statesman. Opposed parliamentary reform but supported the Reform Bill on its second reading believing resistance to be hopeless. (*DNB*.)

Suffield, Baron. *See* Harbord, Edward.

Sulivan, Bartholomew James (1810 — 90). Naval officer and hydrographer. Lieutenant in the *Beagle*, 1831 — 6. Surveyed the Falkland Islands, 1838 — 46. Admiral, 1877. (*DNB*.)

   [*17 January — 7 February 1832*]

Sutcliffe, Thomas (1790? — 1849). Adventurer in South America who held various military and administrative posts in Chile, 1822 — 38. Appointed governor on the island of Juan Fernandez in 1834 but forced to return to England in 1838. (*DNB*.)

Swainson, William (1789 — 1855). Naturalist and author. Collected extensively in Sicily and Brazil. Used the quinary system of William Sharp Macleay. Emigrated to New Zealand in 1840. FRS 1820. (*DNB*, *DSB*.)

Syme, Patrick (1774 — 1845). Flower painter. Translated Werner's *Nomenclature of colours* (1814). (Bénézit 1976, *DNB*.)

Tapparelli, Massimo. *See* Azeglio, Massimo Taparelli, Marchese d'.

Temple, Henry John, 3d Viscount Palmerston, (1784 — 1865). Statesman. M.P. for Cambridge University, 1811 — 31. Foreign Secretary, 1830 — 41 and 1846 — 51. Home Secretary, 1852 — 5. Prime Minister, 1855 — 8 and 1859 — 65. (*DNB*.)

Tennyson, Alfred, 1st Baron (1809 — 92). Matriculated Trinity College, Cambridge, 1828, but left in 1831 without taking a degree. Poet Laureate, 1850. (*DNB*.)

Thackeray, George (1777 — 1850). Provost of King's College, Cambridge, 1814 — 50. (*DNB*.)

Thackeray, William Makepeace (1811 — 63). Novelist. (*DNB*.)

Theakston, Joseph (1772 — 1842). Sculptor. Assisted Francis Legatt Chantrey, 1818 — 42. (Bénézit 1976, *DNB*.)

Thompson, Harry Stephen (later Meysey Thompson) (1809–74). B.A., Trinity College, Cambridge, 1832. Prominent agriculturist and a founder of the Royal Agricultural Society, 1838. (*Alum. Cantab., DNB.*)

Thompson, Leonard (1806–78). B.A., Trinity College, Cambridge, 1829. Married Lady Mary Wentworth Fitzwilliam in 1831. (*Alum. Cantab.*)

Thompson, Thomas Charles (1811–85). B.A., Trinity College, Cambridge, 1834. Clergyman. Brother of Harry Stephen Thompson. (*Alum. Cantab.*)

Thomson, Thomas (1773–1852). Regius Professor of Chemistry at Glasgow University, 1818–52. Edited his own journal, *Annals of Philosophy*, 1813–20. FRS 1811. (*DNB, DSB.*)

Titian (Tiziano Vecellio) (1487?–1576). Italian painter. (Bénézit 1976.)

Todd, Robert Bentley (1809–60). Physician, physiologist, and medical editor. Professor of physiology at King's College, London, 1836–53. FRS 1838. (*DNB.*)

Tollet, Eliza (Elizabeth). Daughter of George Tollet. Older sister of Georgina and Ellen Tollet. (*Burke's Landed Gentry 1846.*)

Tollet, Ellen Harriet (d. 1890). Daughter of George Tollet and younger sister of Georgina Tollet. A close friend of the Wedgwood family. (*Burke's Landed Gentry 1846.*)

Tollet, George (1767–1855). Of Betley Hall, Staffordshire. Agricultural reformer. Close friend of Josiah Wedgwood II. (Freeman 1978.)

Tollet, Georgina. Daughter of George Tollet. A close friend of the Wedgwood family. (*Burke's Landed Gentry 1846.*)

Tollet, Marianne. Daughter of George Tollet. Married William Clive in 1829. (*Burke's Landed Gentry 1846.*)

Townley, Richard Greaves (1786–1855). M.P. for Cambridgeshire, 1831–41 and 1847–52. (*Alum. Cantab.*)

Townsend, George (1788–1857). Clergyman and author. Prebendary of Durham, 1825–57. (*DNB.*)

Trevor, Frances (1803–78). Sister of Robert FitzRoy. Married George Rice Trevor in 1824. (*Complete Peerage.*)

Trevor, George Rice (1795–1869). M.P. for Carmarthenshire, 1820–31 and 1832–52. Married Robert FitzRoy's sister, Frances, in 1824. Became 4th Baron Dynevor in 1852. (*Alum. Cantab.*, Stenton 1976.)

Turner, James Farley (d. 1841). Shrewsbury school-friend of CD. B.A., Christ's College, Cambridge, 1831. Clergyman. (*Alum. Cantab.*)
   [*28 July 1826*]

Turner, Sharon (1768–1847). Historian. (*DNB.*)

Ure, Andrew (1778–1857). Professor of chemistry and natural philosophy at the Andersonian Institution, Glasgow, 1804–30. Chemical consultant in London, 1830–57. Editor of the *Dictionary of chemistry*. FRS 1822. (*DNB, DSB.*)

Usborne, Alexander Burns. Master's Assistant in the *Beagle*. Surveyed the coast of Peru after the *Beagle* had left for the Galápagos Islands, 1835 – 6. (Freeman 1978.)

[c. 1 – 5 September 1835]

Vane, Henry, Earl of Darlington, (1788 – 1864). Army officer and politician. M.P. for South Shropshire, 1832 – 42. (*Alum. Oxon.*)

Vaughan, William (1799 – 1866). Attended Shrewsbury School. B.A., St John's College, Cambridge, 1822. Vicar of Astley, Shropshire, 1827 – 61 and rector of Pontesbury, 1829 – 66. (*Alum. Cantab.*)

Venables, Richard Lister (1809 – 94). B.A., Emmanuel College, Cambridge, 1831. Rector of Whitney, Herefordshire, 1834 – 43. Married Mary Augusta Dalrymple, daughter of General A. M. Poltoratzky of Russia, in 1834. (*Alum. Cantab.*)

Vernon, Henry, 3d Baron (1747 – 1829). Of Sudbury Hall, Derbyshire. (*Burke's Peerage 1980.*)

Vernon, Louisa Henrietta (d. 1861). Daughter of Henry Vernon. Married Brooke Boothby in 1816. (*Burke's Peerage 1980.*)

Vidal, Pedro Pablo (1777 – 1848). Argentinian priest. Served as legislator in Buenos Aires, 1830 – 4. (*HDA.*)

Vigors, Nicholas Aylward (1785 – 1840). Irish zoologist and politician. FRS 1826. (*DNB.*)

Voltaire, François Marie Arouet de (1694 – 1778). French man of letters. (*DSB, EB.*)

Wakefield, John Mort (d. 1853). Clergyman and teacher. Assistant master at Shrewsbury School, 1826 – 52. (*Alum. Cantab., Shrewsbury School Register.*)

Wale, Alexander Malcolm (1797 – 1884). B.A., St John's College, Cambridge, 1819; Fellow, 1821 – 31; Senior Proctor, 1828 – 9. Clergyman. (*Alum. Cantab.*)

Walker, Francis (1809 – 74). Entomologist. (Gilbert 1977.)

Wallace, Alfred Russel (1823 – 1913). Independently discovered natural selection. Author of pioneering work on protective coloration, mimicry, and zoogeography. FRS 1893. (*DNB, DSB.*)

Walpole, John. Consul general and plenipotentiary at Santiago, Chile, 1837 – 41; Chargé d'affaires and consul general, 1841 – 7. (*BDR.*)

Warter, Henry Degory (1770 – 1853). Deputy lieutenant of Shropshire. (*Burke's Landed Gentry 1952.*)

Warter, Henry Diggory. *See* Warter, Henry Degory.

Warter, John Wood (1806 – 78). Clergyman and antiquary. Son of Henry Degory Warter and Shrewsbury school-friend of CD. (*DNB.*)

23 December 1824

Waterhouse, George Robert (1810 – 88). Naturalist. A founder of the Entomological Society, 1833. Curator, London Zoological Society, 1836 – 43. On staff of the British Museum (Natural History), 1843 – 80. Described CD's entomological specimens from the *Beagle* voyage. (*DNB*, Gilbert 1977.)

Waterton, Charles (1782 – 1865). Naturalist and traveller. (*DNB, DSB.*)
Watkins, Frederick (1808 – 88). Clergyman. Attended Shrewsbury School, 1823 – 6. Admitted Christ's College, Cambridge, 1825; matriculated, 1826; kept 2 terms: admitted Emmanuel College, 1827; B.A., 1830. Archdeacon of York, 1874 – 88. (*Alum. Cantab.*)
   [*18 September 1831*]; 18 August 1832
Wavel, William (d. 1829). Devonshire physician. (*BHGW.*)
Way, Albert (1805 – 74). Antiquary and traveller. B.A., Trinity College, Cambridge, 1829. Fellow, Society of Antiquaries, 1839; Director, 1842 – 6. A founder of the Archaeological Institute, 1845. (*Alum. Cantab., DNB.*)
Weber, Carl Maria Friedrich Ernst von (1786 – 1826). German composer, conductor, pianist, and critic. (Grove 1980.)
Wedgwood, Allen. *See* Wedgwood, John Allen.
Wedgwood, Bessy. *See* Wedgwood, Elizabeth (Bessy).
Wedgwood, Catherine (Kitty) (1774 – 1823). Daughter of Sarah and Josiah Wedgwood I. CD's aunt. (*Emma Darwin.*)
Wedgwood, Charlotte (1797 – 1862). Daughter of Bessy and Josiah Wedgwood II. Married Charles Langton in 1832. (*Burke's Peerage 1980.*)
   22 September [1831]; 12 January – 1 February 1832; 27 [September] 1832
Wedgwood, Eliza. *See* Wedgwood, Sarah Elizabeth (Eliza).
Wedgwood, Elizabeth. *See* Wedgwood, Sarah Elizabeth (Elizabeth).
Wedgwood, Elizabeth (Bessy) (1764 – 1846). Eldest daughter of John Bartlett Allen. Married Josiah Wedgwood II in 1792. (*Emma Darwin.*)
Wedgwood, Emma (1808 – 96). Youngest daughter of Bessy and Josiah Wedgwood II. Married CD, her cousin, in 1839. (*Burke's Peerage 1980.*)
   [*24 October 1836*]; [*28 October 1836*]; [*21 November 1836*]; *17 December 1836*
Wedgwood, Frances (Fanny) (1806 – 32). Daughter of Bessy and Josiah Wedgwood II. (*Burke's Peerage 1980.*)
Wedgwood, Frances Crewe. *See* Crewe, Frances.
Wedgwood, Frances Julia (Snow) (1833 – 1913). Writer. Daughter of Hensleigh and Frances Mackintosh Wedgwood. (*Burke's Peerage 1980.*)
Wedgwood, Frances Mackintosh (Fanny) (1800 – 89). Daughter of James and Catherine Mackintosh. Married Hensleigh Wedgwood in 1832. (*Burke's Peerage 1980*, Freeman 1978.)
   [24 October 1836]; [28 October 1836]; [21 November 1836]; 17 December 1836
Wedgwood, Frances Mosley (d. 1874). Married Francis Wedgwood in 1832. (*Burke's Peerage 1980.*)
Wedgwood, Francis (Frank) (1800 – 88). Master-potter and partner in the works at Etruria until 1876. Son of Bessy and Josiah Wedgwood II. Married Frances Mosley in 1832. (*Alum. Cantab.*)
Wedgwood, Godfrey (1833 – 1905). Son of Francis and Frances Mosley Wedgwood. (*Emma Darwin.*)

Wedgwood, Harry. *See* Wedgwood, Henry Allen.

Wedgwood, Henry Allen (Harry) (1799 – 1885). B.A., Jesus College, Cambridge, 1821. Barrister. Son of Bessy and Josiah Wedgwood II. Married Jessie Wedgwood in 1830. (*Alum. Cantab.*)

Wedgwood, Hensleigh (1803 – 91). B.A., Christ's College, Cambridge, 1824; Fellow, 1829 – 30. Philologist and barrister. Metropolitan police magistrate at Lambeth, 1832 – 7. Registrar of metropolitan carriages, 1838 – 49. Son of Bessy and Josiah Wedgwood II. Married Frances Mackintosh in 1832. (*DNB.*)
    [16] November [1836]; *[20 December 1836]*

Wedgwood, James Mackintosh (Bro) (1834 – 64). Son of Hensleigh and Frances Mackintosh Wedgwood. (*Emma Darwin.*)

Wedgwood, Jane. *See* Wedgwood, Louisa Jane.

Wedgwood, Jessie (1804 – 72). Daughter of John and Louisa Jane Wedgwood. Married Henry Allen Wedgwood in 1830. (*Emma Darwin.*)

Wedgwood, John (1766 – 1844). Banker and horticulturist. A founder of the Horticultural Society of London, 1804. Son of Sarah and Josiah Wedgwood I. Married Louisa Jane Allen in 1794. (*Burke's Peerage 1980*, Desmond 1977.)

Wedgwood, John Allen (Allen) (1796 – 1882). Son of John and Louisa Jane Wedgwood. Vicar of Maer, Staffordshire, 1825 – 63. (*Alum. Cantab.*)

Wedgwood, Josiah, I (1730 – 95). Master-potter. Founded the Wedgwood pottery works at Etruria, Staffordshire. CD's grandfather. FRS 1783. (*DNB, DSB.*)

Wedgwood, Josiah, II (1769 – 1843). Of Maer Hall, Staffordshire. Master-potter of Etruria. First M.P. for Stoke-on-Trent, 1832 – 4. Married Elizabeth (Bessy) Allen in 1792. (*Burke's Peerage 1980, Emma Darwin.*)
    30 – 1 August 1831; *31 August 1831*; 1 September 1831; [5 October 1836]

Wedgwood, Josiah, III (1795 – 1880). Of Leith Hill Place, Surrey. Son of Bessy and Josiah Wedgwood II. Married Caroline Sarah Darwin in 1837. (*Burke's Peerage 1980.*)

Wedgwood, Kitty. *See* Wedgwood, Catherine.

Wedgwood, Louisa Frances (1834 – 1903). Eldest daughter of Henry Allen and Jessie Wedgwood. (*Emma Darwin.*)

Wedgwood, Louisa Jane (Jane) (1771 – 1836). Daughter of John Bartlett Allen. Married John Wedgwood in 1794. (*Emma Darwin.*)

Wedgwood, Mrs Josiah, I. *See* Wedgwood, Sarah.

Wedgwood, Robert (1806 – 81). Clergyman. Son of John and Louisa Jane Wedgwood. Married Frances Crewe in 1835. (*Alum. Cantab.*)

Wedgwood, Sarah (1734 – 1815). Married Josiah Wedgwood I in 1764. (*Emma Darwin.*)

Wedgwood, Sarah Elizabeth (Eliza) (1795 – 1857). Eldest daughter of John and Louisa Jane Wedgwood. (*Emma Darwin.*)

Wedgwood, Sarah Elizabeth (Elizabeth) (1793 — 1880). Eldest daughter of Bessy and Josiah Wedgwood II. (*Burke's Peerage 1980.*)
  [16] November [1836]
Wedgwood, Sarah Elizabeth (Sarah) (1778 — 1856). Youngest daughter of Sarah and Josiah Wedgwood I. CD's aunt. (*Emma Darwin.*)
  23 December [1836]
Wedgwood, Snow. *See* Wedgwood, Frances Julia.
Wedgwood, Thomas (1771 — 1805). Son of Josiah Wedgwood I. Published researches on heat and light, 1791 — 2. Developed the concept of photography. Patron of Samuel Taylor Coleridge and John Leslie. (*DNB.*)
Wedgwood, Thomas Josiah (Tom) (1797 — 1862). Son of John and Louisa Jane Wedgwood. Colonel in the Guards. (*Emma Darwin.*)
Wellesley, Arthur, 1st Duke of Wellington, (1769 — 1852). Army officer and statesman. Field marshal, 1813. Chancellor of Oxford University, 1834 — 52. (*DNB.*)
Wellington, Duke of. *See* Wellesley, Arthur.
Werner, Abraham Gottlob (1749 — 1817). German mineralogist and geologist. Taught at the Mining Academy, Freiberg, 1775 — 1815. (*DSB.*)
Westwood, John Obadiah (1805 — 93). Entomologist and palaeographer. Hope Professor of Zoology at Oxford University, 1861 — 93. Entomological referee for the *Gardeners' Chronicle.* (*DNB*, Gilbert 1977.)
Wharncliffe, Lord. *See* Stuart-Wortley-Mackenzie, James Archibald.
Whately, Richard (1787 — 1863). Clergyman, logician, and political economist. Drummond Professor of Political Economy at Oxford University, 1829 — 31. Archbishop of Dublin, 1831 — 63. (*DNB, DSB.*)
Whewell, William (1794 — 1866). Mathematician and historian and philosopher of science. Tutor at Trinity College, Cambridge, 1823 — 38; Master, 1841 — 66. Professor of mineralogy at Cambridge University, 1828 — 32. FRS 1820. (*DNB, DSB.*)
White, Adam (1817 — 79). Naturalist. Employed in the zoological department of the British Museum, 1835 — 63. (*DNB*, Gilbert 1977.)
White, Gilbert (1720 — 93). Naturalist and clergyman. Author of *The natural history and antiquities of Selborne* (1789). (*DNB, DSB.*)
Whitley, Charles Thomas (1808 — 95). Attended Shrewsbury School, 1821 — 6. B.A., St John's College, Cambridge, 1830. Reader in natural philosophy and mathematics at Durham University, 1833 — 55. Vicar of Bedlington, Northumberland, 1854 — 95. (*Alum. Cantab., Modern English Biography.*)
  [9 September 1831]; *13 September 1831*; 23 [September 1831]; 15 November [1831]; 23 July 1834; *5 February 1835*; 24 October [1836]
Whitley, John (d. 1830). Of Liverpool. Father of Charles Thomas Whitley.
Whitmore, Ainslie Henry (1801 — 43). B.A., Christ's College, Cambridge, 1830. Clergyman. (*Alum. Cantab.*)

Whitmore, Thomas (1782 — 1846). M.P. for Bridgnorth, Shropshire, 1806 — 31. (*Alum. Oxon.*)

Wickham, John Clements (1798 — 1864). Naval officer and magistrate. First-lieutenant in the *Beagle*, 1831 — 6. Commander of the *Beagle*, 1837 — 41, surveying the Australian coast. Emigrated to Australia in 1842. Police magistrate in New South Wales, 1843 — 57; Government Resident, 1857. (*Aust. Dict. Biog.*, Desmond 1977.)

Wilberforce, William (1759 — 1833). Philanthropist and politician. Parliamentary leader of movement for abolition of slavery, 1787, and a founder of the Anti-Slavery Society, 1823. (*DNB.*)

Wilkes, Charles (1798 — 1877). American naval officer and explorer. Head of the Depot of Charts and Instruments, Washington, 1833 — 6. Travelled to Europe in 1836 to procure scientific instruments for the Pacific exploring expedition, 1838 — 42, under his command. (*DAB.*)
    [7 November 1836]

William IV (1765 — 1837). King of Great Britain and Ireland, 1830 — 7. (*DNB.*)

William Frederick, 2d Duke of Gloucester, (1776 — 1834). Army officer. Field marshall, 1816. Chancellor of Cambridge University, 1811 — 34. (*DNB.*)

Williams, Edward Hosier (d. 1844). Of Eaton Mascott, near Shrewsbury. Solicitor in partnership with J. A. Powell and others at Lincoln's Inn. Married Sarah Owen in 1831.

Williams, Richard. Brother of Edward Hosier Williams.

Williams, Sarah. *See* Owen, Sarah Harriet Mostyn.

Willis, Robert (1800 — 75). Engineer and archaeologist. Fellow of Caius College, Cambridge, 1826. Jacksonian Professor of Natural and Experimental Philosophy at Cambridge University, 1837 — 75. (*DNB, DSB.*)

Wilmer, Bradford (1808 — 86). B.A., Christ's College, Cambridge, 1831. (*Alum. Cantab.*)

Wilmot, Robert, 2d Baronet (d. 1834). Of Osmaston, Derbyshire. Father of Robert John Wilmot Horton. (*Burke's Peerage 1927.*)

Wilmot Horton, Robert John. *See* Horton, Robert John Wilmot.

Wilson, Belford Hinton (1804 — 58). Consul general, Lima, Peru, 1832 — 7; chargé d'affaires and consul general, 1837 — 41. Consul general in Venezuela, 1842 — 52. (*Modern English Biography.*)

Wingfield, Edward John (1802 — 30). Attended Shrewsbury School. B.A., Christ Church College, Oxford, 1825. (*Alum. Oxon.*)

Wingfield, John (1769 — 1862). Of Onslow, Shropshire. Army officer. High Sheriff of Shropshire, 1814. Mayor of Shrewsbury, 1833. (*Burke's Landed Gentry 1952.*)

Withering, William (1741 — 99). Physician and botanist. Chief physician to the Birmingham general hospital. Active member of the Lunar Society, Birmingham. FRS 1785. (*DNB, DSB.*)

Wood, Alexander Charles (b. 1810). Matriculated Trinity College, Cambridge, 1831. A colonial land and emigration commissioner. Robert FitzRoy's cousin. (*Alum. Cantab.*)

Wood, William (1774 – 1857). Zoologist, surgeon, and bookseller. FRS 1812. (*DNB.*)

Woodyeare, John Fountain (1809 – 80). B.A., Christ's College, Cambridge, 1831. Clergyman. (*Alum. Cantab.*)

Wordsworth, William (1770 – 1850). Poet. (*DNB.*)

Worsley, Thomas (1797 – 1885). Clerical fellow of Downing College, Cambridge, 1824 – 36; Master, 1836 – 85. Rector of Scawton, Yorkshire, 1826 – 81. (*Alum. Cantab., Modern English Biography.*)

Wynn, Watkin Williams, 5th Baronet (1772 – 1840). M.P. for Denbighshire, 1796 – 1840. Lord Lieutenant of Denbighshire and Merionethshire. (*Alum. Cantab.*)

Wynne, Maurice (d. 1835). Rector of Bangor Is-Coed, Flintshire, and vicar of Much Wenlock, Shropshire. (*Burke's Landed Gentry 1846.*)

Wynne, Rice (1777 – 1846). Apothecary and surgeon. Mayor of Shrewsbury, 1822. (*Provincial Medical Directory.*)

Yarrell, William (1784 – 1856). Zoologist. Engaged in business as newspaper agent and bookseller in London. Wrote standard works on British birds and fishes. (*DNB.*)

Yate, Charles (1804 – 60). B.A., St John's College, Cambridge, 1827; Fellow, 1829 – 42. Clergyman. (*Alum. Cantab.*)

York Minster. A Fuegian man brought to England in 1830 by Robert FitzRoy. In 1833 he returned to Tierra del Fuego in the *Beagle*. (Freeman 1978.)

Yorke, Charles Philip (1799 – 1873). Naval officer and politician. M.P. for Cambridgeshire, 1832 – 4. Became 4th Earl of Hardwicke on the death of his uncle, 1834. Admiral, 1863. (*DNB.*)

Yorke, Simon (1771 – 1834). Of Erthig, Denbighshire. High Sheriff of Denbighshire, 1807. M.P. for Grantham, 1793 – 1802. (*Burke's Landed Gentry 1952.*)

Young, Charles Mayne (1777 – 1856). Actor. (*DNB.*)

Young, Thomas (1773 – 1829). Physician, physicist, and Egyptologist. Professor of natural philosophy at the Royal Institution, 1801 – 3. Physician to St George's Hospital, London, 1811 – 29. (*DNB, DSB.*)

Bibliography of biographical sources

ADB: *Allgemeine deutsche Biographie.* Under the auspices of the Historical Commission of the Royal Academy of Sciences. 56 vols. Leipzig: Duncker and Humblot. 1875 – 1912.

Alum. Cantab.: *Alumni Cantabrigienses. A biographical list of all known students, graduates and holders of office at the University of Cambridge, from the earliest times to 1900.* Compiled by J. A. Venn. Part II. From 1752 to 1900. 6 vols. Cambridge: Cambridge University Press. 1940 – 54.

Alum. Oxon.: *Alumni Oxonienses: the members of the University of Oxford, 1715 – 1886.* By Joseph Foster. 4 vols. Oxford: 1888.

Aust. Dict. Biog.: *Australian dictionary of biography: 1788 – 1850; 1851 – 1890.* Edited by Douglas Pike and Bede Nairn. 6 vols. Melbourne: Melbourne University Press. 1966 – 76.

BDR: *British diplomatic representatives 1789 – 1852.* Edited by S. T. Bindoff, E. F. Malcolm Smith, and C. K. Webster. London: Offices of the Royal Historical Society. 1934.

Bénézit, Emmanual. 1976. *Dictionnaire critique et documentaire des peintres, sculpteurs, dessinateurs et graveurs de tous les temps et de tous les pays, par un groupe d'écrivains spécialistes français et étrangers.* Revised edition. 10 vols. Paris: Librairie Gründ.

BHGW: *Biographisch-literarisches Handwörterbuch zur Geschichte der exacten Wissenschaften.* By Johann Christian Poggendorff. Vols. 1 – 4, Leipzig: Verlag von Johann Ambrosius Barth. 1863 – 1904. Vol. 5, Leipzig and Berlin: Verlag Chemie g.m.b.h. 1926.

Burke's Landed Gentry: *Burke's genealogical and heraldic history of the landed gentry.* By John Burke. 1 – 17 editions. London: Burke's Peerage Ltd. 1833 – 1952.

Burke's Peerage: *Burke's peerage and baronetage.* By John Burke. 1 – 105 editions. London: Burke's Peerage Ltd. 1826 – 1980.

CDEL: *A critical dictionary of English literature, and British and American authors, living and deceased.* By S. Austin Allibone. 3 vols. Philadelphia and London: 1859 – 71. Supplement by John Foster Kirk, 2 vols. Philadelphia and London: 1891.

Complete Peerage: *The complete peerage of England Scotland Ireland Great Britain and the United Kingdom extant extinct or dormant.* By G. E. Cokayne. New edition, revised and much enlarged. Edited by the Hon. Vicary Gibbs and others. 12 vols. London: The St Catherine Press. 1910 – 59.

DAB: *Dictionary of American biography.* Under the auspices of the American Council of Learned Societies. 20 vols., index, and 7 supplements. New York: Charles Scribner's Sons; London: Oxford University Press. 1928 – 81.

Darwin Pedigree: *Pedigree of the family of Darwin.* By H. Farnham Burke. Privately printed. 1888.

DBF: *Dictionnaire de biographie français.* Under the direction of J. Balteau, M. Barroux, M. Prevost, and others. 15 vols., (A – Gilbert). Paris: Librairie Letouzey et Ané. 1933 – 82.

DCF: *Dictionnaire des comédiens français.* By Henry Lyonnet. 2 vols. Paris: E. Jorel. [1911 – 12.]

Desmond, Ray. 1977. *Dictionary of British and Irish botanists and horticulturists, including plant collectors and botanical artists.* 3d ed. London: Taylor and Francis Ltd.

DNB: *Dictionary of national biography.* Edited by Leslie Stephen and Sidney Lee. 63 vols. and 2 supplements (6 vols.). London: Smith, Elder and Co. 1885 – 1912. *Dictionary of national biography 1912 – 70.* Edited by H. W. C. Davis, J. R. H. Weaver, and others. 6 vols. London: Oxford University Press. 1927 – 81.

DNZB: *A dictionary of New Zealand biography*. Edited by G. H. Scholefield. 2 vols. Wellington: Department of Internal Affairs New Zealand. 1940.

DSB: *Dictionary of scientific biography*. Edited by Charles Coulston Gillispie. 14 vols., supplement I, and index. New York: Charles Scribner's Sons. 1970—80.

EB: *Encyclopaedia Britannica*. 11th ed. 29 vols. Cambridge: Cambridge University Press. 1910—11.

Emma Darwin: *Emma Darwin: a century of family letters 1792—1896*. Edited by Henrietta Litchfield. 2 vols. London: John Murray. 1915.

Eton School Lists: *The Eton School lists, from 1791 to 1850*. By H. E. C. Stapylton. 2d ed. London: 1864.

Freeman, Richard Broke. 1978. *Charles Darwin: a companion*. Folkestone: W. Dawson and Sons Ltd; Hamden Conn.: Archon Books, Shoe String Press.

Gilbert, Pamela. 1977. *A compendium of the biographical literature on deceased entomologists*. London: British Museum (Natural History).

Grove, George. 1980. *The new Grove dictionary of music and musicians*. Edited by Stanley Sadie. 20 vols. London: Macmillan Publishers Ltd.

HDA: *Historical dictionary of Argentina*. By Lone S. Wright and Lisa M. Nekhom. Latin American Historical Dictionaries, No. 17. Metuchen, N.J. and London: The Scarecrow Press, Inc. 1978.

Larousse XX: *Larousse du XXᵉ siècle*. Under the direction of Paul Augé. 6 vols. Paris: Librairie Larousse. 1928—33.

Modern English Biography: *Modern English biography containing many thousand concise memoirs of persons who have died since the year 1850*. By Frederic Boase. 3 vols. and supplement (3 vols.). Truro: Netherton and Worth. 1892—1921.

NBU: *Nouvelle biographie universelle*. Under the direction of Jean Chrétien Ferdinand Hoefer. 46 vols. Paris: 1852—66.

NDB: *Neue deutsche Biographie*. Edited by the Historical Commission of the Bavarian Academy of Sciences. 12 vols., (A—Kre). Berlin: Duncker and Humblot. 1953—80.

O'Byrne, William R. 1849. *A naval biographical dictionary: comprising the life and services of every living officer in Her Majesty's Navy, from the rank of admiral of the fleet to that of lieutenant, inclusive*. London.

Plarr, Victor Gustave. 1930. *Plarr's lives of the fellows of the Royal College of Surgeons of England*. Revised by Sir D'Arcy Power. 2 vols. London: Simpkin Marshall Ltd.

Provincial Medical Directory: *The provincial medical directory*. London: 1847.

Sarjeant, William A. S. 1980. *Geologists and the history of geology. An international bibliography from the origins to 1978*. 5 vols. London: The Macmillan Press Ltd.

Shrewsbury School Register: *Shrewsbury School register 1734—1908*. Edited by J. E. Auden. Oswestry: Woodall, Minshall, Thomas and Co., Caxton Press. 1909.

Stenton, Michael. 1976. *Who's who of British members of Parliament, 1832—1885*. Brighton: Harvester Press.

INDEX

The dates of letters to and from the correspondents are listed in the Biographical register and index to correspondents and are not repeated here.

Robert Darwin = Elizabeth Hill
1682-1754   1702-97

William Alvey Darwin = Jane Brown   Elizabeth Collier $\overset{2}{=}$ Erasmus Darwin $\overset{1}{=}$ Mary Howa
1726-83   1746-1835   Chandos-Pole   1731-1802   1740-
                           1747-1832

Samuel Fox = Anne   Jane Harriett = Francis   Robert Wa
1765-1851   1777-1859   Ryle   Sacheverel   1766-18
                    1794-1866   1786-1859

Samuel Tertius = Frances Anne
Galton   Violetta
1783-1844   1783-1874

                             Elizabeth Ann   Henry Parker = Marian
                             (Bessy) b.1808   1788-1856   1798-1858

Mary Ann = Samuel   Lucy Harriot
1800-29   Ellis Bristowe   1809-48
           1800-55

Eliza   Milicent Adèle   Susan
1801-86   1810-83   Elizabe
                                                1803-

Emma   Emma Sophia
1803-85   b.1811

William = Harriet Fletcher   Darwin
Darwin   1799-1842   b.1814
1805-80

Frances Jane   Erasmus   Erasmus Alv
b.1806   b.1815   1804-8

Julia   Francis   Emily Catheri
b.1809   1822-1911   1810-6